Fundamental and Biomedical Aspects of Redox Processes

Gheorghe Duca
Institute of Chemistry, Moldova State University, Moldova

Ashok Vaseashta
International Clean Water Institute, USA

IGI Global
PUBLISHER of TIMELY KNOWLEDGE

A volume in the Advances in Chemical and Materials Engineering (ACME) Book Series

Published in the United States of America by
IGI Global
Engineering Science Reference (an imprint of IGI Global)
701 E. Chocolate Avenue
Hershey PA, USA 17033
Tel: 717-533-8845
Fax: 717-533-8661
E-mail: cust@igi-global.com
Web site: http://www.igi-global.com

Library of Congress Cataloging-in-Publication Data

Names: Duca, Gheorghe, 1952- editor. | Vaseashta, Ashok, editor.
Title: Fundamental and biomedical aspects of redox processes / Gheorghe Duca and Ashok K. Vaseashta, editor.
Description: Hershey, PA : Engineering Science Reference, [2023] | Includes bibliographical references and index. | Summary: "Redox processes represent a major advancement in water treatment technology representing a new and unique way of water medium treatment by its natural process of electrochemical oxidation/reduction and adsorption action which reduces and/or removes many unwanted contaminants from water. Such processes control the chemical speciation, bioavailability, toxicity, mobility, and adsorption of water pollutants in the environment. At a cellular level, redox signaling occurs when specific molecules like free radicals, reactive oxygen species, and other compounds influence the system and serve as indicators that are involved with a wide range of biological processes. Redox molecules are made by mitochondria responsible for intracellular communication. Cell signaling is vital for health because it ensures things like immune function, tissue repair, cell development, hormone operations, and essentially every single bodily process. These signals allow cells to detect damage, protect themselves, and repair or replace injured cells. Understanding such transformations require complex approaches and extensive research of electronic structure, thermodynamics, kinetics, and catalysis of the redox reactions. Although many studies have been conducted in this area so far, the problem is far from being sufficiently unraveled"-- Provided by publisher.
Identifiers: LCCN 2022039862 (print) | LCCN 2022039863 (ebook) | ISBN 9781668471982 (h/c) | ISBN 9781668471999 (s/c) | ISBN 9781668472002 (eISBN)
Subjects: LCSH: Environmental chemistry. | Oxidation-reduction reaction.
Classification: LCC TD193 .H36 2023 (print) | LCC TD193 (ebook) | DDC 660/.28443--dc23/eng/20221107
LC record available at https://lccn.loc.gov/2022039862
LC ebook record available at https://lccn.loc.gov/2022039863

This book is published in the IGI Global book series Advances in Chemical and Materials Engineering (ACME) (ISSN: 2327-5448; eISSN: 2327-5456)

British Cataloguing in Publication Data
A Cataloguing in Publication record for this book is available from the British Library.

For electronic access to this publication, please contact: eresources@igi-global.com.

Advances in Chemical and Materials Engineering (ACME) Book Series

J. Paulo Davim
University of Aveiro, Portugal

ISSN:2327-5448
EISSN:2327-5456

Mission

The cross disciplinary approach of chemical and materials engineering is rapidly growing as it applies to the study of educational, scientific and industrial research activities by solving complex chemical problems using computational techniques and statistical methods.

The **Advances in Chemical and Materials Engineering (ACME) Book Series** provides research on the recent advances throughout computational and statistical methods of analysis and modeling. This series brings together collaboration between chemists, engineers, statisticians, and computer scientists and offers a wealth of knowledge and useful tools to academics, practitioners, and professionals through high quality publications.

Coverage

- Ceramics and glasses
- Multifuncional and Smart Materials
- Wear of Materials
- Biomaterials
- Fracture Mechanics
- Industrial Chemistry
- Thermo-Chemical Treatments
- Polymers Engineering
- Heat Treatments
- Artificial Intelligence Methods

IGI Global is currently accepting manuscripts for publication within this series. To submit a proposal for a volume in this series, please contact our Acquisition Editors at Acquisitions@igi-global.com or visit: http://www.igi-global.com/publish/.

The Advances in Chemical and Materials Engineering (ACME) Book Series (ISSN 2327-5448) is published by IGI Global, 701 E. Chocolate Avenue, Hershey, PA 17033-1240, USA, www.igi-global.com. This series is composed of titles available for purchase individually; each title is edited to be contextually exclusive from any other title within the series. For pricing and ordering information please visit http://www.igi-global.com/book-series/advances-chemical-materials-engineering/73687. Postmaster: Send all address changes to above address.

Titles in this Series

For a list of additional titles in this series, please visit: http://www.igi-global.com/book-series/advances-chemical-materials-engineering/73687

Handbook of Research on Corrosion Sciences and Engineering
Younes El Kacimi (University Ibn Tofail, Morocco) and Lei Guo (Tongren University, China)
Engineering Science Reference • © 2023 • 673pp • H/C (ISBN: 9781668476895) • US $325.00

Handbook of Research on Advanced Functional Materials for Orthopedic Applications
R. Ranjith (SNS College of Technology, India) and J. Paulo Davim (University of Aveiro, Portugal)
Engineering Science Reference • © 2023 • 400pp • H/C (ISBN: 9781668474129) • US $325.00

Diversity and Applications of New Age Nanoparticles
Suneev Anil Bansal (Department of Mechanical Engineering, Bharat Institute of Engineering and Technology, India) Virat Khanna (Department of Mechanical Engineering, MAIT, Maharaja Agrasen University, India) Nilanthy Balakrishnan (School of Chemical and Physical Sciences, Keele University, UK) and Pallav Gupta (Department of Mechanical Engineering, A.S.E.T., Amity University, India)
Engineering Science Reference • © 2023 • 419pp • H/C (ISBN: 9781668473580) • US $270.00

Polymer-Inorganic Nanostructured Composites Based on Amorphous Silica, Layered Silicates, and Polyionenes
Kostyantyn M. Sukhyy (Ukrainian State University of Chemical Engineering, Ukraine) and Elena A. Belyanovskaya (Ukrainian State University of Chemical Engineering, Ukraine)
Engineering Science Reference • © 2023 • 304pp • H/C (ISBN: 9781799897286) • US $260.00

Development, Properties, and Industrial Applications of 3D Printed Polymer Composites
R. Keshavamurthy (Dayananda Sagar College of Engineering, India) Vijay Tambrallimath (Manipal Institute of Technology, Manipal Academy of Higher Education, Manipal, India) and J. Paulo Davim (University of Aveiro, Portugal)
Engineering Science Reference • © 2023 • 319pp • H/C (ISBN: 9781668460092) • US $270.00

The Classes of Higher Dimensional Polytopes in Chemical, Physical, and Biological Systems
Gennadiy Vladimirovich Zhizhin (Russian Academy of Natural Sciences, Russia)
Engineering Science Reference • © 2022 • 366pp • H/C (ISBN: 9781799883746) • US $215.00

Advances in the Modelling of Thermodynamic Systems
Elhoucine Essefi (University of Gabes, Tunisia) and Ibtissem Jendoubi (Faculty of Sciences of Bizerte, Tunisia)
Engineering Science Reference • © 2022 • 309pp • H/C (ISBN: 9781799888017) • US $270.00

IGI Global
PUBLISHER of TIMELY KNOWLEDGE

701 East Chocolate Avenue, Hershey, PA 17033, USA
Tel: 717-533-8845 x100 • Fax: 717-533-8661
E-Mail: cust@igi-global.com • www.igi-global.com

Editorial Advisory Board

Igor Povar, *Institute of Chemistry, Chisinau, Moldova*
Viorica Railean-Plugaru, *Nicolaus Copernicus University, Poland*
Lidia Romanciuc, *Institute of Chemistry, Chisinau, Moldova*
Rodica Sturza, *Technical University, Chisinau, Moldova*
Boris Tsukerblat, *Ben-Gurion University of the Negev, Israel*
Vasilie Vasylechko, *Physico-Chemical Institute of Environmental and Human Protection, Odessa, Ukraine*
Tat'yana Rakitskaya, *Physico-Chemical Institute of Environmental and Human Protection, Odessa, Ukarine*
Inga Zincovscaia, *Joint Institute for Nuclear Research, Dubna, Russia*

Table of Contents

Detailed Table of Contents

Since the middle of the last century, special attention has been devoted to the study of phenomenological and theoretical aspects of the elementary act of electron and/or proton transfer. The initially proposed model, based on the Frank-Condon principle, suffered from a violation of the law of energy conservation, until R.A. Marcus assumed that energy levels can be adjusted via fluctuations in the polar environment. This model had hard limitations on multielectron transfers. A little later H. Taube drew attention that the model of "solid balls" refers only to a small class of reactions. An important class of compounds with unusual reactivity and the possibility of synchronous multi-electron transfers turned out to be complexes with partial charge transfer. Now one can see the notable interest and, accordingly, a significant number of publications relate to multielectronic (up to 8 electrons!) synchronous transfers in organized nanostructured systems used in electrochemical energy storage devices with high energy density.

Molecular oxygen is the greenest of all possible oxidizing agents, but its specific electronic configuration eliminates the possibility of its direct reaction with organic substances. Single-electron transfer between triplet O2 and singlet organics is prohibited according to the Wigner spin selection rule. Moreover, attaching the first electron is disadvantageous even from a thermodynamic point of view. The authors will not consider the high-temperature processes, however, and will focus on the mechanisms for generating active oxygen species under normal temperature conditions. All methods of "soft" activation of oxygen can be conditionally divided into three main groups: photo-excitation; direct or sensitized initiation of a radical-chain oxidative process; and catalysis by transition metal ions with the inclusion of oxygen in the coordination sphere. New perspectives on molecular oxygen activation pathways and their application in some oxidative reactions including degradation of organic pollutants, organic synthesis, and environmentally significant reactions are discussed.

The paper discusses the decomposition mechanisms of hydrogen peroxide (HP) under the action of light, with the involvement of metal ions and their coordination compounds, as well as the participation of hydrogen peroxide in the peroxidase oxidation of various chemical substances. It is shown that these reactions are accompanied by the formation of partial charge transfer complexes and other short-living species as intermediates. Water self-purification capacity following the chemical way is discussed. New indicators for estimating the quality of natural water bodies have been proposed. HP plays an essential role in strengthening the immune system of living organisms, serving in the fight against various diseases, including COVID-19. In this regard, a kinetic model is proposed to evaluate and predict viral infections. We have experimentally demonstrated that HP can find broad applications in agriculture, specifically, in seed treatment, plant growth acceleration, the food industry, etc.

The authors show that the instability of the transition (activated) states of chemical reactions are due to the vibronic coupling of the ground electronic state with the appropriate low-lying excited states, resulting in the pseudo Jahn-Teller effect (PJTE). In application to reactions relevant to environmental problems, the PJTE origin of their activation barrier is illustrated by several examples, including the proton transfer in deprotonated H3O2- and protonated H5O2+(H2O)4 water clusters and in a number of proton-bound dimers. The results can serve as a general parametrized analytical model of the potential energy profile in simulating proton transfer processes in such systems. It also includes modeling the polarizing effect of the solvent, serving as an additional stabilizing factor in intramolecular H-bonds. Similarly, it is shown that the transition states of photochemical reactions are formed by the PJTE interaction of the relevant excited state with a higher in energy electronic state. Examples of electron transfer and charge transfer by coordination are also shown.

The electronic spin has a crucial importance in life chemistry, although relatively rarely encountered in several processes, while the vast majority of organic molecules (including the biological systems) are based on closed-shell spinless configurations. To advocate for the incidence of the spin it is suggestive to point out that the oxygen molecule, O2, the fuel in the quasi-totality of organisms, has a spin-triplet ground state and all the processes triggered by respiration are including, then, spin interactions. In the investigation of subtle spin-related molecular and supramolecular interactions, a wise technique consists in introducing stable open-shell organic radicals as “spies” of a given environment, under electronic paramagnetic resonance measurements. The nitroxide-based spin carriers are frequently used as spin probes, the TEMPO group, (2,2,6,6-Tetramethylpiperidin-1-yl)-oxyl, being the most widely used in this view. Here the authors illustrate a case study on a series of nitroxide-based biradicals, corroborating the EPR measurements with state-of-the-art modeling.

The relationships between electronic structure particularities and antioxidant activity of some chemical compounds are considered on the basis of the density functional theory (DFT) calculations of reactions between some natural phenolic antioxidants (gallic, caffeic, p-coumaric, and other acids), as well as dihydroxyfumaric acid, and the stable radicals ABTS•+ and DPPH• in water, ethanol, and methanol. Particular attention is paid to the processes of the primary proton transfer between neutral molecules of the above antioxidants and molecules of these solvents. The important factors are also considered: the formation of the charge transfer complexes (CTC) between the anionic forms of these antioxidants and the radicals, and possible tautomeric effects essentially changing the pre-reactional structures of the investigated antioxidant molecules and their anions. The quantitative relationships between the electronic structure parameters and the antioxidant activity of some short chain n-alkanes with radicals OH• and NO3• are also discussed.

It is shown that the isomorphic substitution of one of the paramagnetic ions of di-homonuclear cluster with antiferromagnetic exchange interaction by an ion, whose wave functions are reversed in time with respect to the wave functions of the substituted ion, leads to the formation of a di-heteronuclear cluster with ferromagnetic exchange interaction. This method of spin rearming can be used for the synthesis of new di-heteronuclear clusters with ferromagnetic exchange interaction. The redox-induced processes can affect the time-reversal symmetry. Particularly in the reduction-half reaction + 2 Cu the time-reversal symmetry for the ion, existing before the start of the reaction, disappears after the reaction due to the transformation of the ion with spin S = 1/2 into a neutral Cu atom without spin. An analogy is found between six incomplete time-reversal operators for systems with spins S = 1/2, 3/2, 2, 5/2.

The chapter presents a thermodynamic approach for the complex chemical and redox equilibria investigation of mono- and two-phase systems containing metal ions of slightly soluble solid phases in different oxidation states. For heterogeneous systems, this approach utilizes thermodynamic relationships combined with original mass balance constraints. The forms of occurrence of metal ions are determined by the major chemical reactions in the aquatic environment such as hydrolysis, oxidation, reduction, and precipitation. On the basis of the introduced generalized equation of the overall redox reaction the physical meaning of the overall electrode potential under conditions other than standard ones, was revealed. The diagrams of complex chemical equilibria have been used for the graphical representation of complex equilibria in aqueous solutions. The developed approach can be applied to calculating potential–pH (Pourbaix) diagrams, based on the thermodynamic analysis of chemical equilibria in homogeneous and heterogeneous multicomponent systems.

This chapter presents a review of the research that was conducted on a natural antioxidant - dihydroxyfumaric acid in the frames of the laboratory of physical and quantum chemistry. The structure and chemistry of dihydroxyfumaric acid are presented in view of its importance in biological cycles, as well as in prebiotic chemistry models. Technological applications are briefly reviewed for wine, food, and tobacco industries. Organic syntheses of new derivatives are presented, and their calculated antioxidant/antiradical activities towards the DPPH and ABTS radicals are discussed. Computational chemistry techniques are employed to gain insights into the isomerism and tautomerism of dihydroxyfumaric acid, and indices of global and local reactivity are calculated for the acid and its new derivatives, in order to theoretically explain the experimental results.

A detailed discussion of the synergistic, additive, and antagonistic antioxidant interactions is presented as the main subject of the chapter. Relevant and recent data is considered in order to emphasize the importance and actuality of the subject. Also, the prooxidant character of antioxidants is shown as a problem that can be eliminated by utilizing low concentrations of antioxidants and by combining them. The reaction mechanisms involved in the antioxidant interactions produced after a combination of various compounds are stated and exemplified. Thus, the regeneration mechanism, the complexes and adducts formation, and complementary effects are responsible for a specific type of antioxidant interaction. On the basis of the beneficial effects of the combination of natural compounds, the general idea of the advantageous consumption of antioxidants from foods is supported.

Herein, the authors summarize the performance of the innovative semiconducting materials produced by the electrospinning-calcination method and applied in photocatalytic systems for advanced oxidation. Specific details regarding the synthesis, characterization, application, and mechanism of action of these semiconducting catalysts are highlighted. A special emphasis was placed on analyzing the kinetics of the processes related to advanced oxidation. In this regard, the application of the pseudo-first-order kinetic (PFO) model and the Langmuir-Hinshelwood (LH) model was presented and discussed. Likewise, the advantages of adopting the design of experiments methodology are highlighted to build the empirical models which can be used for process optimization.

Redox balance is important for maintaining physiological activities in organisms. Once the level of redox-active species is abnormal, it may destroy the balance of the internal environment or further induce biological pathological changes. Therefore, it is of great scientific significance to pay attention to the changes of redox-active species in vivo. Coordination compounds are very broad and important molecular systems that can undergo a wide variety of electronic transitions, with the charge transfer from metal to ligands and their associated excited states being particularly significant. These characteristics give coordination compounds the possibility of regulating redox reactions in vivo. After cisplatin, ruthenium complexes have been extensively studied in order to be alternative molecular systems with less toxicity to normal cells. This chapter will mainly introduce in vivo redox systems, redox process detection as well as ruthenium anti-tumor drugs for redox regulation.

A model of the kinetics of the body's immune response to a viral load is proposed. The model takes into account both cellular and humoral components of the immune response. The interaction of T- and B-type lymphocytes, as well as the effect of delaying the formation of antibodies, were also taken into account. Numerical analysis of the model system with some values of the model parameters showed the possibility of implementing three main modes of regular behavior in the system: 1) a mode of unlimited growth in the amount of antigen and a critical decrease in the number of T-type lymphocytes (unfavorable prognosis of the disease); 2) the mode of dynamic balance in the system "antigen - T-lymphocytes" (satisfactory prognosis of the disease); 3) a mode of rapid reduction in the amount of antigen, adequate growth, and reaching stationary values of the number of immune cells (favorable prognosis of the disease).

The ability of antioxidants to scavenge free radicals is well known, but their possible therapeutic action is still unclear because so far most studies are in vitro on cultured cells (where the antioxidant concentrations are typically much larger than those that can be reached in the human body) or in vivo on animals. This chapter is focused on ascorbic acid (vitamin C). Vitamin C exhibits dual reactivity with pro-oxidant activity registered at concentrations above 1 mM while at lower concentrations it has an antioxidant effect. Vitamin C appears to be effective in preventing respiratory diseases, shortening recovery time after surgeries, increasing the efficiency of the immune response to infections, improving glycemic control in type 2 diabetes, decreasing the need for fluids in the treatment of burns, and protecting the skin. Although countless beneficial effects of antioxidants have been proposed, far more studies are required before most of these claims should be taken seriously, possibly to discover new antioxidants that are more effective against the disease.

Oxidoreductases are a large class of enzymes involved in the catalysis of oxidation/reduction reactions. These oxygen-utilizing biomolecules firmly control the reactions involving molecular oxygen mostly by regulating their cofactors. Flavoenzymes, a large group of oxidoreductases, possess either a flavin mononucleotide or a flavin adenine dinucleotide cofactor non-covalently bound to the protein substrate. Flavin cofactors are versatile due to reactivity and many oxidizing states allowing them to undergo electron transfer reactions. As a result, flavoenzymes can participate in a wide range of biochemical and photochemical processes including electron transport, C-H oxidations, oxidative protein folding, and even repair of cross-linked DNA. Structural studies have shown that the large variety of functions that flavoenzymes possess are limited and dependent on structure and protein folding to ensure the specific electrostatic environment. The present work concerns the structural aspects of specific flavin-dependent oxidoreductases influencing electron transfer reactions.

Oxidoreductases are a special class of enzymes that use the redox mechanism for the efficient transformation of organic substrates. Most oxidoreductases contain metals in the active site and, for optimal functioning, require the participation of a small co-substrate with the ability to donate electrons. From the multitude of enzymes with economic and applicable potential, the authors focused their attention on three particular classes: catalases, peroxidases, and laccases. Catalases and peroxidases contain heme iron in their active sites and most often require electron donors such as oxygen or hydrogen peroxide while laccase contains copper and demands special co-substrates such as 2,2'-azino-bis(3-ethylbenzothiazoline-6-sulfonic acid) or syringaldehyde. The theoretical aspects regarding the mechanism in which the electron transfer of the three enzymes is involved as well as the practical applications of the selected enzymes in the field of environmental remediation will be the subject of this chapter.

New, original, and ecologically pure methods for the synthesis of compounds with hybrid terpene and heteroatomic functional groups or fragments have been developed, starting from accessible (+)-sclareolide, obtained from (-)-sclareol, well-known labdane diterpenoid diol extracted from the waste of Clary sage (Salvia sclarea L.) remaining after extracting the essential oil. The natural origin of terpene compounds supposes biocompatibility, selective biological activity, and low toxicity. The compounds with combined skeleton were obtained by coupling some terpene derivatives (acids, chloroanhydrides, bromides) with azaheterocyclic compounds or heterocyclization of some intermediates such as hydrazides, hydrazincarbothioamides, or thiosemicarbazones. A series of over 120 newly obtained substances were subjected to biological testing, of which 10 showed pronounced antifungal and antibacterial activity, two amides showed pronounced antioxidant activity, and two derivatives with guanidine fragments showed high antitumor activity.

Foreword

The problems related to the environment and the search for solutions to undertake them are crucial in our society. For a better quality of life and protection of our surrounding environment, we must be aware of all kinds of risks against them, including anthropogenic, to prevent, mitigate, and restore damages produced by various human activities. This is possible only by a deep understanding of the chemical processes that take place for and against the protection of the environment. In this context, the book edited by Professors Duca and Vaseashta is welcome as it offers an up-to-date overview of the redox processes involved in ecological chemistry.

There are 18 chapters authored by scientists and researchers who have led significant contributions to this emerging field. The context is divided into two books. The first part of the first book (consisting of 11 chapters) is devoted to several fundamental aspects of redox reactions, covering both theoretical and experimental problems: electron and proton transfer mechanisms, activation of the oxygen molecule, quantum chemical studies on the structure-antioxidant activity, reaction mechanisms, and thermodynamics of redox reactions. Special attention is provided to hydrogen peroxide which plays an important role in environmental chemistry. The second part (consisting of seven chapters) focuses on the biomedical relevance of redox processes involving coordination compounds, enzymes, and vitamins. One of the chapters is devoted to nitrogen and sulfur-containing terpenoid and their applications.

The book is interesting and well-written and is addressed to a broad readership. Students will discover research topics that are exciting and timely, and scientists will find a valuable tool for developing research projects. The book is also useful for teaching, providing teachers and students with essential and recent achievements in ecological chemistry. I congratulate the editors and authors for this excellent work!

Marius Andruh
University of Bucharest, Romania

Foreword

The study of redox processes has become increasingly important as the world continues to face complex environmental, biomedical, and technological challenges. Redox reactions involve the transfer of electrons from one species to another, resulting in changes in the oxidation states of the reactants. The species that loses electrons is said to be oxidized, while the species that gains electrons is said to be reduced. These processes involve the transfer of electrons from one molecule to another and play a critical role in a wide range of chemical reactions in nature and the laboratory. The study of redox reactions has led to the development of many technologies for treatment and environmental remediation. These technologies rely on the use of chemical, physical, and biological processes to reduce or eliminate contaminants from the environment.

The *Handbook of Fundamental and Biomedical Aspects of Redox Processes* by editors Gheorghe Duca and Ashok Vaseashta provides a comprehensive overview of the latest research in this field. This two-book set covers a wide range of topics related to redox processes, including the environmental impacts of redox reactions, the role of redox processes in biological systems, and the use of redox chemistry in technological applications.

The book consists of two parts written by many very well-known scientists and professionals. The first part consists of a theoretical approach to electron and proton transfer processes using quantum and computational chemistry. The second part examines redox processes in living organisms. Almost all contributors are eminent scientists in their respective fields who offer insights into the fundamental principles of redox processes, as well as their practical applications. The book also includes discussions of recent advances in supramolecular chemistry, which explores the interactions between molecules and the properties that emerge from such interactions.

Overall, the *Handbook of Fundamental and Biomedical Aspects of Redox Processes* is an invaluable resource of knowledge for researchers, students, and professionals who are interested in understanding the fundamental principles and practical applications of redox processes. It provides a comprehensive overview of this rapidly evolving field and highlights several ways redox chemistry shapes our surrounding world. Furthermore, redox reactions play a crucial role in the functioning of many biological processes such as respiration, photosynthesis, and digestion. These reactions also play an important role in industrial processes such as metallurgy, electroplating, and battery technology. Some common examples of redox technologies include, for example, water treatment which also includes electrocoagulation, electrochemical oxidation, photocatalysis, and biological treatment. These technologies are effective in treating a wide range of contaminants, including organic pollutants, heavy metals, and pathogens. We foresee many applications to be added to what is already provided in the book. Since the field is rapidly developing, this handbook will serve as a reference for many young researchers.

In summary, redox reactions play a critical role in both biological and industrial processes and have led to the development of many technologies for treatment and environmental remediation. These technologies provide a sustainable and effective way to treat and protect the contaminated environment. We congratulate the editors and chapter contributors for their excellent contributions and we hope that this handbook is successful in its aim to further the field of redox process in the environment, biomedical, and contamination remediation technologies.

Janusz Lipkowski
Cardinal Stefan Wyszynski University, Poland

Foreword

Redox reactions, or oxidation-reduction reactions are one of the most fundamental types of chemical reactions. They play crucial roles both in biology, where they underlie fundamental processes like respiration, photosynthesis, and metabolism, and in technological applications, where they are used in such various processes as wastewater treatment, corrosion prevention, and as the fundamental basis for energy storage in batteries. The rich and diverse landscape of these reactions has prompted generations of chemists, biochemists, and engineers to explore their secrets and harness their power for the elucidation of fundamental processes and exploration of their technological potential. The book *Handbook of Research on Fundamental and Biomedical Aspects of Redox Processes* aims to provide a comprehensive overview of current progress and applications of redox chemistry in diverse fields. Its 18 chapters, written by experts in their respective fields, are organized into two sections: Fundamental Aspects of Redox Processes and Biomedical Aspects of Redox Processes.

The first section covers a wide range of topics related to the basic principles of redox chemistry. As is well-known, the earliest successes in the systematic study of reaction rates of redox reactions were obtained for the treatment of the transfer of individual electrons between chemical species without accompanying transfer of nuclei, which are nowadays successfully described using Marcus' theory, which has enabled the (previously unsuspected) discovery of apparently counter-intuitive decreases in reaction rate accompanying extremely exergonic electron-transfers. More complex redox reactions, which involve electron transfer accompanied by the movement of atoms from one species to the other with a concomitant change in redox states, cannot however be easily described using a unified model yet, due to the considerable perturbation of acceptor (and donor) electronic structure entailed by the transfer of nuclei. The chapters in this section explore the intricacies of these fundamental processes and provide a detailed description of electron and proton transfer mechanisms, their fundamental physical determinants, activation mechanisms of molecular oxygen, and the electronic mechanisms underlying the action of several antioxidants. The second section focuses on the biomedical applications of redox chemistry. The chapters in this section provide in-depth analyses of the role of metal-containing anti-tumoral agents in the regulation of redox reactions in the biomedical environment, as well as of the action of antioxidants in clinical treatments and of the mechanisms of electron transfer and specificity in flavin-containing oxidoreductase enzymes, heme-containing oxidoreductases, and Cu-containing laccases.

The quality and diversity of the contributions present in this handbook on the fundamental and biomedical aspects of redox processes provide a comprehensive overview of the fundamental principles of redox chemistry and their applications in wide-ranging fields. We, therefore, expect this book to become an essential resource for researchers, academics, and students interested in the fascinating diversity of redox processes in the fields of chemistry, biochemistry, environmental science, and materials science.

Pedro Silva
Universidade Fernando Pessoa, Portugal

Preface

Chemical reactions of electron transfer between oxidants and reducing agents – called Redox processes are the basis of the most important processes in nature, for example, living organisms and the environment, and are and are involved in several natural phenomena including chemical, geological, and biomedical processes. The editors have solicited contributions from experts in the field and compiled them into two books. In the first book titled *Handbook of Fundamental and Biomedical Aspects of Redox Processes*, several chapters provide fundamentally compelling arguments that such transformations necessitate complex approaches and extensive research of the electronic structure, thermodynamics, kinetics, and catalysis of redox processes. Although, so far, several studies have been carried out in this area, the problem is far from solved. These reactions are ubiquitous in environmental and biomedical contexts and play an important role in various biological systems. Listed below are some examples of redox processes in environmental contexts, for example:

1. **Biogeochemical Cycles:** Redox reactions play a vital role in biogeochemical cycles, such as the carbon, nitrogen, and sulfur cycles. For example, the process of photosynthesis involves the oxidation of water to produce oxygen, while respiration involves the reduction of oxygen to produce water. Nitrogen fixation and denitrification are also redox reactions that are essential for the cycling of nitrogen in ecosystems.
2. **Oxidative Stress:** Environmental factors such as pollution, radiation, and chemicals can cause oxidative stress in living organisms. This occurs when there is an imbalance between the production of reactive oxygen species (ROS) and the body's ability to detoxify them. Redox reactions are involved in both the generation and detoxification of ROS.

Selected examples of redox processes in environmental contexts are, for example:

1. **Cellular Respiration:** Redox reactions are central to cellular respiration, which is the process by which cells convert glucose into ATP, the main energy currency of the cell. During cellular respiration, glucose is oxidized to carbon dioxide, and oxygen is reduced to water.
2. **Antioxidants:** Antioxidants are molecules that can neutralize ROS and prevent oxidative damage to cells. Many antioxidants function by undergoing redox reactions with ROS. For example, vitamin C can donate electrons to neutralize ROS and become oxidized.

In particular, the role of vibronic coupling in the formation of electron transfer barriers and its possible manipulation is not explored. A comprehensive analysis of materials with reducing and transforming properties has not been carried out comprehensively, especially those in the classes of enodioli, thiols, polyphenols, and vitamins. Also, the problem of preventing negative effects caused by oxidants, such as damage to living cells, DNA, proteins, and lipids, remains largely unresolved. This book bridges existing knowledge gaps and discusses such topics and other related topics from several leading experts, covering fundamental and interdisciplinary studies based on a wide range of chemical and physical methods, including studies from different fields of chemistry, in particular quantum chemistry, computational chemistry, chemical kinetics and thermodynamics, and related fields.

The book describes the following points that are not covered in detail elsewhere.

- Mechanisms of electron and proton transfer.
- Pseudo Jahn-Teller effect in states of transition instability in redox-processes.
- Hydrogen peroxide in ecological chemistry and biomedicine.
- Kinetic models of membrane-cell processes in the presence of immune and vitamin complexes.
- Mechanisms of electron transfer in oxidoreductase enzymes.
- Isomerism and tautomerism of dihydroxy fumaric acid.
- Synergism of antioxidants action.
- Thermodynamics of Redox processes in homogeneous and heterogeneous multicomponent systems.

The expected result is the accumulation of new knowledge on the essence of oxidation-reduction processes, the theoretical disclosure and justification of key mechanisms, and the discovery of new aspects of methods of control of redox reactions, all of which have a major interest in life and health, the quality of food products, as well as the state of the environment, especially for water, since clean water resources continue to deplete at a rapid pace.

The target audience of this book will be composed of professionals and researchers working in the fields of Biomedical Science, Biochemistry, Enzymology, Molecular Biology, Ecological Chemistry, Quantum Chemistry, Physical Chemistry, Thermodynamics, Computational Chemistry, Water Professionals, and various other related disciplines, for example, administrative sciences and management, Education, Sociology, and topics related to public policy and decision support. Moreover, the book will provide insights and support to principals concerned with managing expertise, knowledge, information, and policy development in different types of scientific and educational communities. In addition, the book will be very useful for many researchers, undergraduate /postgraduate students, scientists, policy-makers, industry professionals, practitioners, and some chapters can be adopted as text supplements for independent study. We hope that the information provided by the authors of this book guides them in their research, new discoveries, and many directions, yet undiscovered.

Handbook of Fundamental and Biomedical Aspects of Redox Processes provides a comprehensive overview of contemporary progress and potential applications of redox chemistry in various fields. The book consists of 18 chapters, written by experts in their respective fields, and are organized into two sections: Fundamental aspects of Redox processes, and Biomedical aspects of Redox processes, The first part of the book includes chapters of a theoretical nature, which provide a computational or deductive analysis of some basic redox systems and processes. The book opens with a chapter in which Travin and Duca report on the evolution of the postulates of Markus' theory regarding the elementary act of electron/proton transfer. The new vision refers to synchronous multi-electron transfers in complexes with partial load transfer and nanostructured systems, which offer a much wider range of their applications. Next, the chapter written by Vaseashta, Duca, and Travin, focuses on discussing the wide variety of chains and reaction mechanisms based on the molecular activation of oxygen, emphasizing their specifics and significance for ecological processes. In one chapter, Duca and Travin discuss how hydrogen peroxide has versatile characteristics and can be treated as a wonder chemical. The role of unstable transition states of chemical reactions, with the transfer of protons and electrons through co-ordination and the appearance of the pseudo-Jahn-Teller effect is elucidated in the work of a team of authors Gorinchoy, Balan, Gorbachev, Arsene, Polinger, Duca and Bersuker in one chapter. Another theoretical work proposed by Cimpoesu, Buta, Toader and Ionita are related to the critical importance of electronic spin in biological systems and organic molecules. A quantum chemical study on the links between electronic structure and antioxidant activity is presented in a chapter by Gorbachev, Gorinchoy, Balan, and Arsene. A rather intriguing approach to the reversal of time is proposed by Geru. Povar and Spinu reported the thermodynamics of redox processes in multicomponent aqueous systems. The following series of chapters discuss the results of experimental studies based on original approaches and research hypotheses. Thus, various aspects of the research on dihydroxyfumaric acid, an important substance for biological and food systems, are presented in a chapter elaborated by Bolocan and Duca. Synergetic, antagonistic, and other effects of antioxidant interactions are discussed by Vicol and Duca in their chapter. Pascariu and Cojocaru present new semiconductor materials that can be applied for advanced oxidation. The role of coordination compounds in the regulation of redox reactions in vivo is revealed in a chapter by Benniston and Zeng. An original kinetic pattern was developed by Yaltychenko, Gorinchoy, and Duca in a chapter describing the body's immune response to the viral load and revealing the three possible ways of the system's regular behavior. A case study on the application of vitamin C for clinical treatments is presented by Silagh-Dumitrescu, Oana, and Lehene. Anghel and Erhan discuss the biochemical and photochemical processes of electron transfer with the involvement of biomolecules - flavin-based oxidoreductase. This research is completed by the other paper, presented by Leonties, Aricov, and Raducan in a chapter on the mechanisms of reaction by electron transfer in other oxidationductase enzymes – catalyzes, peroxidases, and laccases. The advanced method of oxidation was proposed for the removal of pharmaceutical contaminants from the water, as described by Gonta, Mocanu, Duca. Several new biologically active terpenoid substances with various possible applications, due to their antifungal and antibacterial activity, have been synthesized and studied, as shown by Aricu, Ciocarlan, Lungu, Cucicova, Blaja, Secara, and Ungur.

In short, redox processes play a key role in a wide range of systems. Understanding these reactions is essential for promoting human health. Specifically, the disclosure of the mechanisms and specifics of redox processes opens the ways of carrying out chemical processes in soft, most favorable conditions, focusing on chemical equilibria as well as protecting the human body from aggressive actions.

Finally, we would like to recognize many contributors for their excellent contributions, as they have been working with passion in this field for a long time. We also thank our reviewers who have tirelessly helped the book with their expert and objective comments. Such observations are essential to ensure that any technical gaps are closed, and chapters appear on an ongoing basis. Both editors recognize the IGI publishers for their constant guidance and help with the book. We certainly hope that many researchers, students, and scientists benefit from these chapters for their ongoing investigations to seek answers from these chapters. We also look forward to the opportunity for many readers to become contributors to our later editions. Of course, efforts to prepare books invariably consume a lot of time away from our families and pets. One of the editors (AV) owes a lot of missed playtimes to his very cute companion dog Isabella, who is an avid frisbee and soccer ball player, and another editor (GD) owes a lot of time to his nephew George and his beautiful garden of flowers, vegetables, and berries - features of his country. We would like to thank some authors from the Republic of Moldova who carried out the research within the national scientific program no.20.80009.5007.27.

Disclaimers: The opinions, recommendations, and propositions presented in various chapters represent the findings, views, and opinions of their authors and may not coincide with the position(s) of the editors. The editors do not promote, endorse, or express their dissent against any product and/or company mentioned in the chapters. Although due precautions are exercised before publication, the editors, however, do not take any responsibility for any infringement of any copyrighted content by the authors of either their own or from elsewhere.

Gheorghe Duca
Institute of Chemistry, Moldova State University, Moldova

Ashok Vaseashta
International Clean Water Institute, USA

Section 1

Fundamental Aspects of Redox Processes

Chapter 1
Electron and Proton Transfer Mechanisms From Marcus to Supramolecular Constructions

Sergey Olegovich Travin
N.N. Semenov Federal Research Center for Chemical Physics, Russian Academy of Sciences, Russia

Gheorghe Duca
Institute of Chemistry, Moldova State University, Moldova

ABSTRACT

Since the middle of the last century, special attention has been devoted to the study of phenomenological and theoretical aspects of the elementary act of electron and/or proton transfer. The initially proposed model, based on the Frank-Condon principle, suffered from a violation of the law of energy conservation, until R.A. Marcus assumed that energy levels can be adjusted via fluctuations in the polar environment. This model had hard limitations on multielectron transfers. A little later H. Taube drew attention that the model of "solid balls" refers only to a small class of reactions. An important class of compounds with unusual reactivity and the possibility of synchronous multi-electron transfers turned out to be complexes with partial charge transfer. Now one can see the notable interest and, accordingly, a significant number of publications relate to multielectronic (up to 8 electrons!) synchronous transfers in organized nanostructured systems used in electrochemical energy storage devices with high energy density.

INTRODUCTION

Since the middle of the last century, special attention has been devoted to the study of phenomenological and theoretical aspects of the elementary act of electron transfer. First significant breakthrough was Libby`s model (1952), according to which the products are in the wrong environment of solvent molecules when an electron is transferred from one reacting ion or molecule to another. Thus, spectroscopy into chemical kinetics was tacitly introduced the Frank-Condon principle. At the same time, the initially proposed model suffered from a clear violation of the law of energy conservation. The "way out" was

DOI: 10.4018/978-1-6684-7198-2.ch001

found by R. A. Marcus, who assumed that energy levels can be adjusted via fluctuations in the environment surrounding the reagents.

With sufficiently large fluctuations corresponding to the intersection of the initial reagents' energy term with the reaction products' term, a transition with zero energy change is possible at the point of their intersection, and it could formally be considered as vertical. A little later, another scientist, in the future also a Nobel laureate, Henry Taube, joined the discussion about the mechanisms of reactions. He noted that the "simple" electron transfer is realized only in systems such as Ne/Ne^{+}.

Taube noted that the charge transfer model between "solid balls" refers only to a relatively small class of redox reactions involving transition metal complexes with an inert and rigid coordination sphere. On the contrary, in the case of labile metal complexes, in which ligand substitution occurs rapidly, and especially in reactions involving non-metallic, molecular oxidants and reducing agents, we deal with the almost instantaneous penetration of one of the reagents into the coordination sphere of the other. In the transition state of inner sphere reactions, one or more ligands are part of the internal coordination sphere of both - oxidized or reduced species. Accordingly, the intermediate is a strong ion pair. Therefore, a single solvate shell is formed around the transient complex, which is weakly sensitive to how the electron density is distributed inside it.

During the discussion of Libby's concept, a third type of mechanism was also proposed, "involving the transfer of a hydrogen atom" from a reducing agent to an oxidizer. In fact, this is a synchronous transfer of an electron and a proton. An important consequence of the Frank-Condon principle is the quadratic dependence of the energy of the reorganization of the medium on the number of transferred electrons, which restricts the amount of simultaneously transferred electrons to one. And, nevertheless, the existence of synchronous two-electron processes, without intermediate formation of free radical products, is proved! Moreover, in enzymatic reactions, even electronic processes are the norm rather than the exception. An important class of compounds with unusual reactivity and the possibility of synchronous multi-electron transfers turned out to be complex with partial charge transfer: they are formed when a redox-active (i.e., capable of oxidation or reduction) ligand enters the coordination sphere of a transition metal ion that has the ability to change the degree of oxidation by more than unity.

Notable interest and, accordingly, a significant number of publications relating to multi-electrons (up to 8 electrons!) synchronous transfers in organized nanostructured systems used in electrochemical energy storage devices with high energy density, including multivalent batteries, since they can potentially provide high capacity. But the most interesting are supramolecular systems, in which rather act the laws of mechanics instead of statistical molecular physics. In this case, electron transfer reactions lead to mechanical work, similar to how it occurs in myoglobin during oxidative phosphorylation. Of course, for supramolecular "machines" models of the Markus type, as well as the application of the Frank-Condon principle, are inapplicable.

We emphasize that both in the Marcus model and in the Taube approach, as well as in all subsequent models, there is an attempt to use the transition state method (another name is the Theory of absolute reaction rates). For "molecular machines" the approach must be modified. The appropriate model includes a special kind of Markov random process. In physical literature, this model got named "ultrametric diffusion". One of the most interesting demonstrations of redox processes in supramolecular systems became an international competition among supramolecular machines, called nanocar races. The first championship in this discipline took place in April 2017, the second one in March 2022, both in Toulouse.

The Electronic Transfer Theory: Why is it Needed?

The ability to anticipate and control the course of a reaction is the eternal dream of all chemists. In this regard, since the middle of the last century, special attention has been devoted to the study of phenomenological and theoretical aspects of the elementary act of electron transfer – an indispensable stage of all redox processes, and often reactions that formally are not attributed as redox.

Apparently, the first significant breakthrough in the theoretical description of the kinetics of electron transfer between two metal ions, which were approximated by rigid charged spheres, was the Marcus theory (Marcus, 1964; Marcus, 1993). He proceeded from Libby`s model finally published in 1952, (Libby, 1952), according to which, when an electron is transferred from one reacting ion or molecule to another, two new molecules or ions formed are in the wrong environment of solvent molecules, since the nuclei do not have time to move during a rapid electron jump. For example, in the reaction of the formation of Fe^{2+} ion, it is in some configurations of a set of nearby dipole solvent molecules that would correspond to the initial Fe^{3+} ion. Similar remarks apply to the newly created reaction counter partner. Spectroscopy into chemical kinetics was tacitly introduced the Frank-Condon principle, according to which:

- Electrons do not exchange energy with nuclei.
- Electrons move much faster than nuclei.
- Electrons always have an equilibrium configuration at any arrangement of nuclei.

Thus, the model implied that the limiting stage of charge transfer is the restructuring of the dipoles of the polar medium, whereas the light particle electron always manages to adjust to the slow movements of the dipoles. At the same time, the initially proposed model suffered from a clear violation of the law of energy conservation: reaction product ions were formed in an "incorrect environment" with high reorganization energy. But the only way how such an event could occur is through the absorption of a quantum of light (a vertical transition that does not save energy). So dark reactions remained unexplained. The Frank-Condon principle is obviously perfectly justified in chemical reactions, as well as in spectroscopy, but something was clearly missing.

In Libby's interpretation (by the way – unreservedly accepted by Marcus), it was assumed that the electronic interaction of reagents causing electron transfer is relatively weak. This point of view is still the one that seems to be the most appropriate for most such reactions until now. It was possible to reconcile the requirements of the law energy conservation of with the Frank-Condon principle, according to which the electron always manages to track the configuration of heavier particles, by introducing fluctuations in the environment surrounding the reagents into the model.

Fluctuations had to occur in various nuclear coordinates, such as the orientation coordinates of individual solvent molecules and, indeed, in any other coordinates, the most likely distribution of products which differs from the distribution of reagents. With sufficiently large fluctuations corresponding to the intersection of the initial reagents' energy term with the reaction products' term, a transition with zero energy change is possible at the point of their intersection, and it could formally be considered vertical. A little later, another scientist, in the future also a Nobel laureate, Henry Taube, joined the discussion about the mechanisms of reactions. He noted that the "simple" electron transfer is realized only in systems such as Ne/Ne^{+}.

Taube noted (Creutz, & Taube, 1969; Taube, 1984; Richardson & Taube, 1984), that all chemical reactions can be divided into two categories: substitution reactions and redox reactions. The latter can always be considered as involving electron transfer, although it is generally assumed that when we consider mechanisms in solution, electron transfer is not as simple as in the case of Ne/Ne +. The rearrangement of atoms is always connected with changes of electron density in each center, and this should be taken into account. Also, it should be borne in mind that quite often substitution reactions are accompanied by redox transformations and vice versa.

Taube drew attention to the fact that the charge transfer model between "solid balls" refers only to a relatively small class of redox reactions involving transition metal complexes with an inert and rigid coordination sphere. On the contrary, in the case of labile metal complexes, in which ligand substitution occurs rapidly, and especially in reactions involving non-metallic, molecular oxidants and reducing agents, we deal with the almost instantaneous penetration of one of the reagents into the coordination sphere of the other. In the transition state of inner sphere reactions, one or more ligands are part of the internal coordination sphere of both - oxidized or reduced species. Accordingly, the intermediate is a strong ion pair. Therefore, a single solvate shell is formed around the transient complex, which is weakly sensitive to how the electron density is distributed inside it.

Thus, the Frank-Condon barrier for electron transfer resulting from the reorganization of the inner sphere is small, and there is an easy electron transfer compared to the mechanism of the outer sphere. During the discussion of Libby's concept, a third type of mechanism was also proposed, "involving the transfer of a hydrogen atom" from a reducing agent to an oxidizer. In fact, this is a synchronous transfer of an electron and a proton, which adds complexity to synchronization, but completely removes the issue of the Frank-Condon barrier, associated with the reorganization of the polar medium.

The quantum chemical theory of proton transfer was created by R.R. Dogonadze, A.M. Kuznetsov, and V.G. Levich in 1967 (Dogonadze et al., 1965; Levich et al., 1970). Within the framework of this concept, the reaction

$$AH + B \rightleftarrows A^{-} + BH^{+} \tag{1}$$

is considered as a transition of the system from the initial to the final surface of potential energy. Each of these surfaces is a function of the proton coordinate. The transition between terms, as in the Marcus model, occurs in accordance with the Frank-Condon principle. The change in the electronic state occurs much faster (in a time of 10^{-15}-10^{-16} s) than the position of the proton changes (time ~ 10^{-14} s) and the position of the solvent molecules surrounding the reagent pair (time 10^{-12}-10^{-13} s). Therefore, the redistribution of the electron density between reacting particles occurs practically at fixed coordinates of the nuclei and is possibly when the energy change in the transition state is close to zero, i.e. at the intersection point of the terms. The transition of the system from the initial to the final state occurs in two fundamentally different ways:

1. adiabatically, i.e., when the splitting of terms is significant, and the process occurs as a smooth movement along the potential energy surface with the overcoming of a potential barrier in a transition state. In this case, the A-H bond is stretched due to the vibrational excitation of the molecule to a critical value r*, at which the former diabatic terms intersected.
2. because of a tunnel jump. To make such a transition, the vibrational levels of the proton in the initial and final states must be equalized. This can be achieved via thermal fluctuations of the orientation of the solvent dipoles. In this case, the A—H bond is changed to the final value by a jump.

We emphasize that both in the Marcus model and in the Taube approach, as well as in all subsequent models, there is an attempt to use the transition state method (another name is the Theory of absolute reaction rates). However, it should be noted that there is more than a good reason to abandon the formal use of the theory of the transition state. Up to now, there are no criteria for the applicability of this theory, which thereby creates the illusion of universality of this method. At the same time, as follows from the most general considerations, such criteria should exist.

It is also important to note the fact that there is a fair amount of terminological confusion in the literature. For this reason, "proton transfer" is often referred to as the reduction addition of a hydrogen atom occurring in a homogeneous medium or on a conducting electrode according to the scheme $AH+B \rightleftarrows A+BH$. In fact, in this variant, there is no charge transfer per se, there is a transfer of a neutral particle. Nevertheless, very often the calculation of the rate constant of either electron transfer is applied to such reactions (assuming that the proton is already "where it should be", somewhere halfway between the donor and the acceptor), or vice versa, the calculation of the rate constant of the proton movement is carried out, assuming that the electron density is always redistributed quickly and "will certainly have time to adjust." From my point of view, both of these extremes are equally groundless. It can only be recalled that the proton transfer itself does not relate to redox, but rather to acid-base equilibria, which, as is known, occur extremely quickly, are completed in nanoseconds and, thus, are of no practical interest for real redox processes.

Marcus's theory in relation to calculating the rates of redox reactions has received universal recognition and is very widespread. Moreover, it is repeatedly retold in a simplified form in hundreds, if not thousands of publications, therefore, I consider it clearly redundant to present it again and make another portion of errors and distortions from myself.

Instead, I propose to focus on some "bottlenecks," which are most often hushed up or interpreted quite loosely. Such places of inconsistencies of impeccable theory with an undereducated experiment of the set, but we will affect only three of them: 1) the vanishingly fine line between external and intra-sphere routes; 2) the inverse Markus region; and 3) the application of the theory to synchronous multi-electron processes

Let's start with the simplest "collection of contradictions," which for this reason we marked with the number 1. Marcus theory formulas work with parabolic energy temperatures that correspond to the collision of two solid balls surrounded by a continuous dielectric continuum. This model can rightfully compete to the extent of its idealization with the household model of the "spherical horse in a vacuum."

Indeed, the formula

$$\lambda_0 = (\Delta e)^2 \left(\frac{1}{2a_1} + \frac{1}{2a_2} - \frac{1}{R} \right) \left(\frac{1}{D_{op}} - \frac{1}{D_s} \right) \tag{2}$$

operates with dimensions of two spherical balls (reagents), which try to approach a distance slightly exceeding the sum of their radii. This result is elementally obtained by integrating the entire volume (with the exception of the energy occupied by spherical reactants) of the electric field, which in turn is directly proportional to the square of its strength.

In practice, it is quite difficult to find a reaction in which both reagents resemble solid conductive balls. Similar properties are, of course, provided for instance by the ferro- and ferricyanide complexes recharging reaction, characterized by an extremely rigid coordination sphere and very low lability constants, i.e., ligand substitution rates. And the point is not only that the fingers on the hands are enough to count most of such cases. The essence is that the absolutely vast majority of reagents of practical interest resemble solid conductive balls no more than a racehorse resembles a sphere in a vacuum. In this sense, it remains more to be surprised why the Markus model as a whole gives satisfactory energy and temperature principal patterns well, and not that in some cases there are quite serious deviations from it.

It is important to emphasize that the energy dictated by the Frank-Condon principle of the reorganization of the polar medium ceases to play not only a key, but also any role in general if both reagents in relation to the medium are "in the house," i.e. form an intermediate complex with charge transfer. Although the Marcus model is very simplified and highly idealized, it provides insights into the main reasons for the manifestation of the linear relation of free energies and deviations from it. It should be noted that almost immediately after the publications of Marcus, the works of the Soviet school appeared, from a group of authors, which included Levich, Dogonadze and Kuznetsov. They gave a significantly stricter version of the justification of already known patterns based on quantum mechanical representations. At the same time, the division of the ongoing processes into two fundamentally different types appeared: adiabatic and non-adiabatic. In the first case, the surface of the potential energy was very different from the terms of the initial reagents, and as a result of their interaction, there was a noticeable disturbance of the terms that greatly distort their parabolic form, which significantly complicated the calculation of the expected activation energy. But the transition of the system from the initial state to the final one occurs with a transmission coefficient close to one. On the contrary, in the non-adiabatic case, the energy of the terms is considered much simpler and more accurate, but the probability of transition (temperature-independent pre-exposure) is determined by the integral of overlapping the wave functions of the donor and acceptor. The criterion for non-adiabaticity is the value of the energy of the donor-acceptor interaction, which should not exceed kT.

The Inverse Marcus Region

Let`s return to the Marcus theory. Rudolph "Rudy" Marcus was born in Canada, Montreal on July 21, 1923. Contrary to his respectful age, Rudy did not retire. For 44 years, out of his 99, he has been teaching at the legendary Caltech (*California Institute of Technology*). Caltech's website even has his contact details!

It should be noted that the above arguments are of high theoretical value, but from the point of view of the "mass consumer," i.e. practical chemical kinetics studying a particular process, do not give the slightest predictive value. We can say that with an accuracy of plus/minus, the order of the velocity constant is determined by the usual linear ratios of free energies (LSE) of the Brønsted type, while attempts at more accurate calculations run into such several complicating factors that "the sheepskin is not worth the dressing." As a result, it has to be stated that the total number of works showing the coincidence of the experimental and design speed constants is very small and their flow has practically dried up to date. Along with it, interest in the theory of absolute reaction rates as a whole faded. This model can rightfully compete to the extent of its idealization with the household model of the "spherical horse in a vacuum."[1]

Figure 1. a and b. Possible presentations of Marcus' inverted region

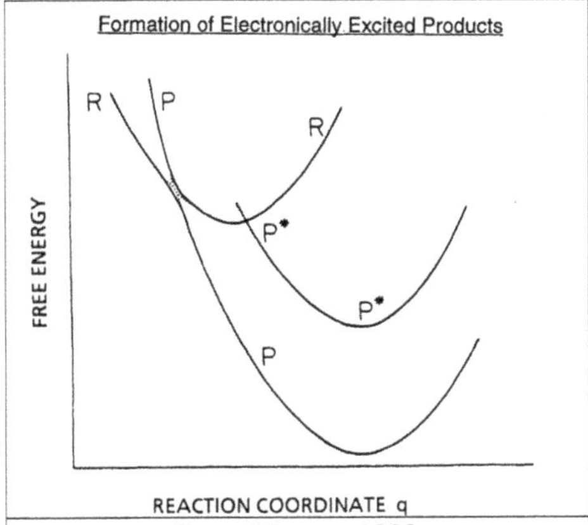

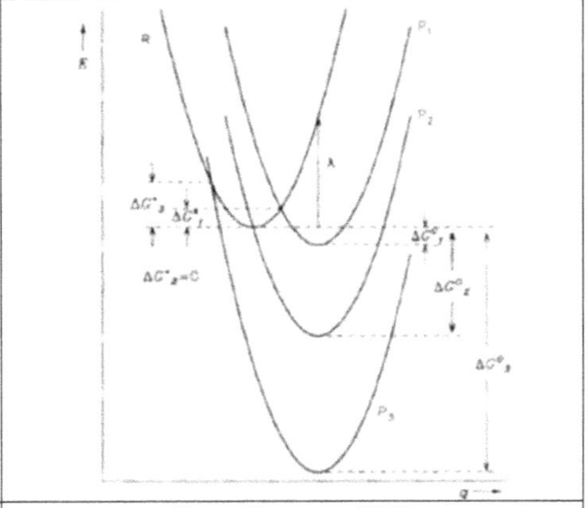

From Marcus, 1993	From Grampp, 1993
A favored formation of an electronically excited state of the products.	The potential energies of the reactants (R) and the products (P1 – P3) of three systems with different free energies of reaction ΔG^0, as a function of the nuclear configuration: 1) Exergonic reaction with $\Delta G^0 > 0$; 2) Exergonic reaction with $\lvert\Delta G^0\rvert = \lambda$; 3) $\lvert\Delta G^0\rvert > \lambda$ - Marcus inverted region

The main "practically valuable" result of Marcus's theory turned out to be a formula for estimating the activation energy of electron transfer reactions, which is algebraically simply obtained by solving the quadratic equation for the point of intersection of terms (such as in Fig. 1):

$$\Delta G = \frac{\left(\Delta G^0 + \lambda\right)^2}{4\lambda}$$

Most clearly, in my taste, the question of the applicability of Marcus's theory to reactions with a large negative thermodynamic effect is posed in work (Goldsmith et al., 2019). It clearly indicates the alternative of switching to either the non-activation or inverse region when crossing the free energy change threshold (Fig. 2).

Figure 2. From Goldsmith et al., 2019. Schematic plot of the logarithm of the rate constant k versus the driving force ΔGo with the free energy barrier given by Eq. (3). The gray vertical line indicates the driving force equal to the reorganization energy λ. Outset are depictions of the normal, activationless, and inverted regions, illustrated by the reactant (blue) and product (red) free energy parabolas as functions of a collective solvent coordinate

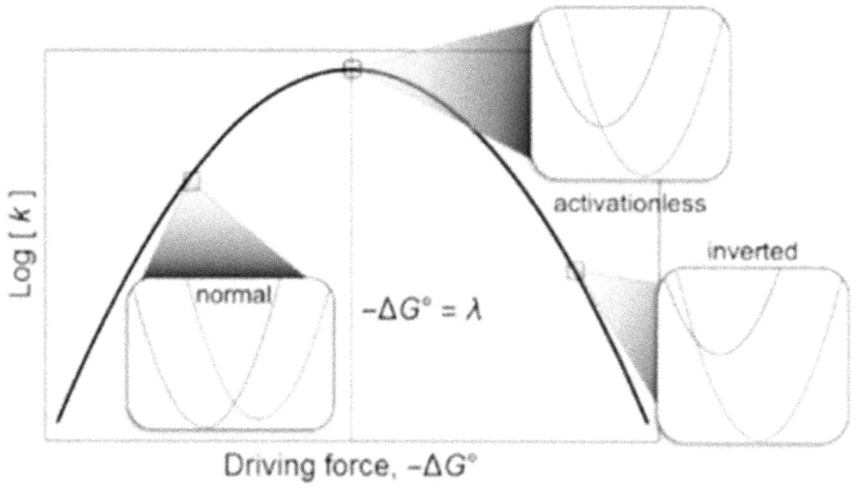

In the normal region, the free energy barrier decreases, and the rate constant correspondingly increases, as the driving force increases. The activationless region occurs when $-\Delta G^{o} = \lambda$ and the barrier is zero. In the inverted region, the free energy barrier increases, and the rate constant correspondingly decreases, as the driving force increases. The signature under the picture means that the driving force is free energy, whereas the reorganization of the polar medium is a barrier, which opposes to driving force and retards the reaction. Anyhow, the authors of work (Goldsmith et al., 2019) had carried out very peculiar theoretical analysis of Proton-Coupled Electron Transfer (PCET). They established conditions, where inverted Marcus behavior can be observed: (1) the overlap integrals of wave functions for the ground state of reactant and the excited product must be negligible; and (2) the reaction free energies contributing significantly to the rate constant must be negative with magnitudes greater than the reorganization energy.

These criteria are typically not satisfied by harmonic or Morse potentials but are satisfied by more realistic asymmetric double well potentials. These insights may help guide the design of more effective energy conversion devices. But in any case, purely theoretical analysis without any experimental measurements can give only necessary, but not obligatory sufficient conditions for inverse Marcus region, so they can't be regarded as a proof of its existence.

Figure 3. a and b. The famous plot from Goldsmith et al., 2019 can be found in hundreds of papers

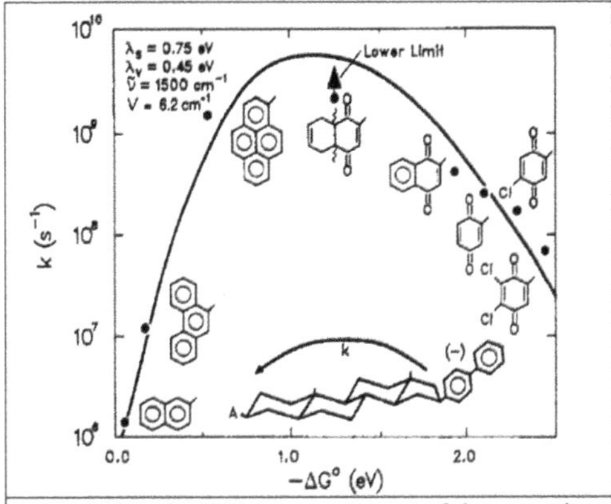

Intramolecular ET rate constants of the negative ions of the compounds listed in Table I as a function of free energy change in MTHF at 296 K. Electrons are transferred from the 4-biphenylyl ion group to the acceptors (A,) having the general structures shown.

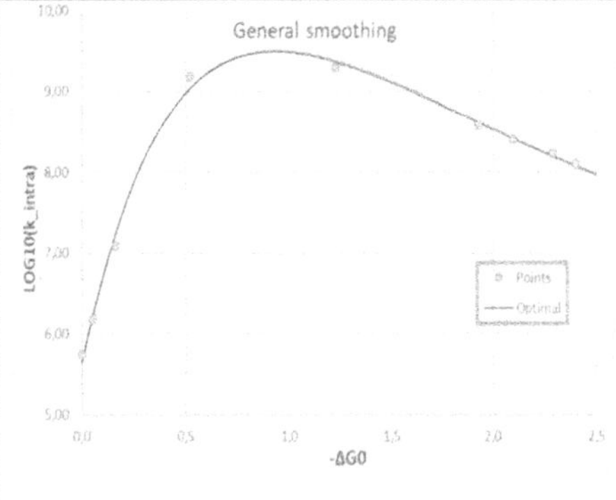

Unbiased smoothing of data, obtained by Miller, Calcaterra and Closs, 1986 demonstrates distinct bias to the left.

An important stage in the study and understanding of the reactions taking place in the inverse Markus region was the work of Closs et al. (1986). A series of molecules with a common A1-Space-A2 structure was synthesized. Pulsed radiolysis was used to convert them to negative ions, and then the rate of intramolecular electron transfer was measured. In the first series of studies, the spatial separator Sp was a rigid poly-conjugated androstane skeleton with acceptors attached to the 3- and 16-positions. The series consisted of eight molecules with A2 = 4-biphenyl and eight different A1-groups differing in electron affinity by 2.4 eV (in fact, the series was a set of nine substances, one of which contained biphenyl groups at both ends). The authors argue that electron transfer rates differ by more than 3 orders of magnitude over a series and depend in a very non-linear way on the free reaction energy. This paper contains just a more peculiar analysis of the same data of the same authors briefly published earlier (Miller et al., 1984).

The logarithm profile of the velocity constant versus free energy maximizes in methyltetrahydrofuran with approximately 1-eV exoergicity. At more negative free energies, velocities drop substantially, proving the existence of an "inverted region." In isooctane, maximum velocity occurs with fewer exoergic reactions. A second series of molecules was developed to study the spacer effect, which ranged from steroid to decalin (eight compounds) and cyclohexane (four compounds). There is no comprehensive dependence on distance. Instead, rates are strongly influenced by donor-acceptor attachment geometry, which affects electronic coupling. Such a strong dependence on stereochemistry suggests that the main component of the bond occurs through the bonds. A semi-quantitative relationship of the magnitude of this relationship with the number of bonds between donor and acceptor has been established. For the type of spacers under investigation, it was found that electron transfer rates were slowed by 1 order of magnitude for each 2 bonds for constant reorganization energy. This relationship is shown to persist only if the stereochemistry of the attachment in the entire series whose terms are compared is maintained constant. Stereoelectronic effects are of considerable magnitude and are predicted to be very large in some special structures.

This work actually "staked out" a convincing proof of the presence of an inverse Markus region and even received the approval of "Rudy" Marcus himself (he referred to this data in his Nobel lecture). Yet even after 30 years or more, there is no definitive conviction of the credibility of the evidence. Not wanting to cast even the slightest shadow over these excellent studies (Closs et al., 1986; Miller et al., 1984), some critical considerations still need to be expressed.

Yes, there is no doubt that the authors conducted very witty and elegant studies in which the constants of the speed of electronic transfer are measured. Apparently, the values of the rate constants themselves are determined correctly, at least there is no reason to suspect the error of the measurements. But the excessively free interpretation of the received figures makes you very much doubt that they confirm the inverse region.

We see that in the studied series there is a tangible decline in speed constants while simultaneously increasing the energy gain in the reaction (supposedly!). It is noteworthy that the energy parameters were not measured directly, but were obtained by extrapolation of data from four different groups, moreover obtained in different solvents different from MTHF, which leaves a lot of room for arbitrariness.

But that's not what matters. Firstly, for intermolecular velocity constants (second order), the entire range of variation does not exceed two times, with a change in free energy by two and a half volts. If the Marcus formula worked at least to some extent for this series, then one could expect a proportionality coefficient of 0.5 between the reaction между ΔG^0 and the activation $\Delta G^{\#}$, that is, a change in the rate constant by an order for each "step" in the free energy 0,12V, or almost 20 orders for the 2,5V range. In fact, it can be stated that, within a small amount of spread, the bimolecular electron transfer rate constants simply do NOT depend on free energy, at least within said range.

Secondly, the diagram wandering from work to work has a sharply asymmetric character (see Figure No.3 b), while the Markus formula assumes a clearly parabolic, i.e. symmetrical relative to the vertical axis (passing through the maximum) picture, i.e., the findings support anything but the existence of an inverse branch of the Marcus formula.

Thirdly, and this is perhaps the most important thing, without any exception, all followers of the existence of the inverse region with great pleasure give the same picture for the dependence of constants. This picture shows the dependency obtained in MTHF medium as a solvent, and only it. And everybody is carefully avoiding the demonstration of the graph for isooctane - this data is available only in tabular form, since the graph is least suitable for a visual demonstration of the inverse branch, rather it even refutes it.

In the paper (Grampp, 1993) Grampp notes, that that the rate constants of strongly exergonic intermolecular electron transfer reactions in solution approach a limiting value k_{diff}, which is the value for diffusion-controlled reactions, which is very expectable and reasonable, but is in the contradiction with the idea of inverse region. This immense discrepancy between Marcus theory and experiment justifiably caused some skeptical comments in the beginning. Grampp insists, that the breakthrough for the Marcus theory with regard to the inverted region was made about twenty years after its publication by Miller, Closs, and Calcaterra in (Closs et al., 1986; Miller et al., 1984).

All examples used to support "inverse region" have in common the intramolecular electron transfer step. In contrast, up to now Rehm-Weller behavior (this is the common name for diffusion-limited pattern) has been found for all charge separation reactions described by the second-order bimolecular rate constant.

In 1989 Russian authors studied the electron back transfer for radical ion pairs in the triplet state and likewise found an inverted effect (but possibly – no! *S.T.*). Through the use of a solvent with a low dielectric constant, the diffusion of the components of the radical ion pair away from each other was hindered. Relatively few publications deal with both the electron transfer reactions discussed above, namely, the charge separation and the electron back transfer.

Several models have been proposed only to be rejected again. For instance, models that take into account quantum mechanical tunneling for the Marcus inverted region were developed, but they could not satisfactorily explain the quantitative differences between Rehm-Weller and Marcus behavior. Nevertheless, Grampp makes a conclusion, that "it can be categorically stated that all he predictions of the Marcus theory on the kinetics of electron transfer have been fully verified". But it is no more than his strong belief. I can only refute his point of view with quoting his paper: "All these models are extremely difficult to test by experiment. So the occurrence or absence of the Marcus inverted region is still a field for wide-ranging theoretical deliberation". It is still so now, almost 30 years later.

Many other authors (Gould et al., 1987; Makita & Hastings, 2017; Parada et al., 2019; Wang et al., 2021) had attacked the fortress of "inverse region", but with the same success. Perhaps, the most "balanced" position regarding the existence of the Markus inverse zone was expressed by Suppan (1992). He had paid attention to the fact, that the existence of the M.I.R. (Marcus Inverse Range) was at first received with skepticism. As an example, one can read the review by Sutin (1986). Suppan explains the negative attitude to the new physical model by the inevitable position for a generation of chemists schooled in the concept of linear free energy relationships, and the fact that for a long time, there was no experimental evidence to confirm the theoretical prediction.

This became so disturbing that a lot of papers appeared from the Marcus group itself in order to explain the absence of the M.I.R. The latter show what has become to be known as "Rehm-Weller behavior" (Rehm & Weller, 1970), as opposed to the predicted "Marcus behavior". That is, the plot of the rate constant versus free energy rises rapidly in the "normal" region up to $\Delta G > 0$, then reaches the diffusional limit (in the case of intermolecular reactions) and stays there no matter how exergonic the process becomes. Figure 4 gives an illustration of this Rehm-Weller behavior, which still seems to be the rule for the vast majority of "honest" electron transfer processes.

Figure 4. The plot of log κ vs. ΔG for the forward (solid symbols) and the back (open symbols) electron-transfer reactions). Circles, squares, and triangles refer to M^{n+} = $Cr(bpy)_3^{3+}$, $Rh(phen)_3^{3+}$, and $Ir(Me_2phen)_2Cl_2^{+}$, respectively. The curves are calculated according to standard kinetic equations (full line) and the Marcus free energy relationship (Eq. 3, broken line)
(Suppan, 1992).

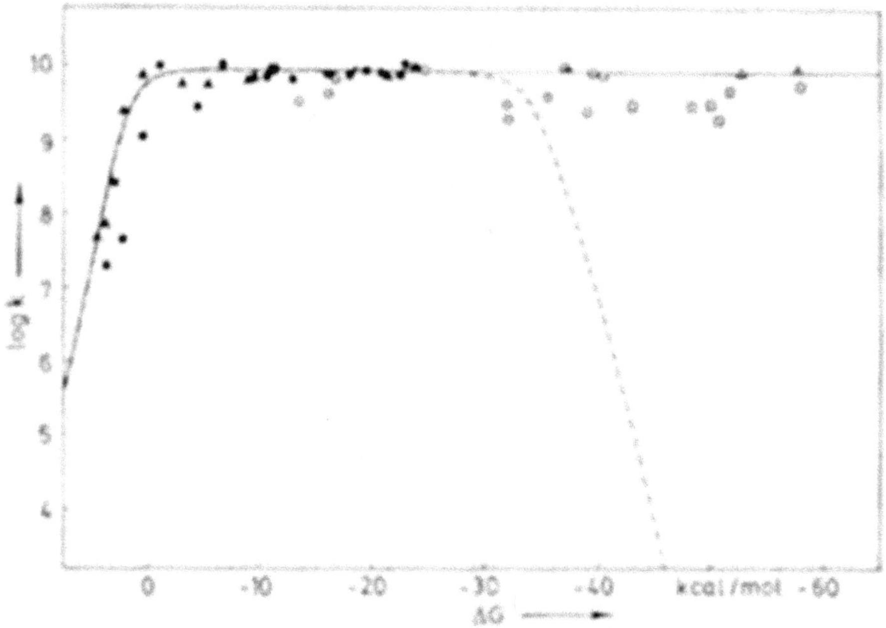

Summing up the above mini discussion on the existence of the reverse Markus range, I consider it necessary to emphasize two features of this topic. The first: "Whoever is looking will always find." It is meant that experimentally proven examples of the drop in velocity constant, despite the increase in thermodynamic gain, there are plenty of such examples. Another question is how much these effects are described (more commonly not described) by Marcus' formula. And the second: model systems used for evidence are so complex, and computational constructions are so florid that they can be used to justify anything - both compliance with the Marcus formula and its violation. In particular, the slowdown of electronic transfer reactions can be associated both with an increase in the non-adiabaticity of the process as the distance grows (Hush, 1985), and due to purely energy reasons (Liang et al., 1990). Although it is most likely in each specific case, the speed is influenced by both.

Another very interesting series of questions arise when discussing processes in which electron transfer in one way or another is associated with proton transfer. I must stress, that this concept is highly ambiguous and uses the same words for absolutely different cases. In the work of Darcy (2018) the authors express the same point of view. Proton-coupled electron transfer (PCET) covers a wide range of reactions involving the transfer(s) of electrons and protons. The best-known PCET reaction, hydrogen atom transfer (HAT), has been studied in detail for more than a century. HAT is generally described as the concerted transfer of a hydrogen atom ($H\bullet \equiv H^+ + e^-$) from one group to another, $Y + H{-}X \rightarrow Y{-}H + X$, but a strict definition of HAT has been difficult to establish. Distinctions are more challenging when the transfer of "H•" involves e− and H+ that transfer to/from spatially distinct sites or even completely separate reagents (multiple-site concerted proton−electron transfer, MS-CPET).

When the PCET thermochemistry is known, HAT and PCET rate constants can be understood and often predicted using linear free energy relationships (the Brønsted catalysis law) and Marcus theory type approaches, what is very surprising for neutral, not charged particle. One must be reminded, that the central idea of Marcus theory is the nonequilibrium polarization of medium, giving reorganization energy. And this is correct only for charged particles transfer. How it works for neutral pieces remains a mystery.

Some authors (Darcy, 2018; Yoneda et al., 2021) inform us, that although photoinduced proton-coupled electron transfer (PCET) plays an essential role in photosynthesis, a full understanding of the mechanism is still lacking due to the complex nonequilibrium dynamics arising from the strongly coupled electronic and nuclear degrees of freedom. Very fine details of the ultrafast photoinduced electron-transfer reactions were obtained in (Biasin et al., 2021) using a photoinduced technique with femtosecond and sub-angstrom resolution. The highest attention to such processes is explained by their importance in natural and artificial energy conversion systems, and they are therefore the subject of intense experimental and theoretical studies across several fields of chemistry. Such research elucidates the key steps governing energy transformations. A thorough understanding of electron transfer in the condensed phase requires detailing the complex couplings between electronic and atomic degrees of freedom in the solute and the surrounding solvent. Femtosecond electronic and vibrational spectroscopic studies have shown that solute–solvent dynamics critically affect energy consumption and the rates of electron transfer.

A very interesting review of recent advances in the design and application of excited-state intramolecular proton-transfer (ESIPT) is presented in (Sedgwick et al., 2018). The authors had described in detail the development of fluorescence-based sensors and imaging agents (probes), important in biology, physiology, pharmacology, and environmental science for the selective detection of biologically and/or environmentally important species. The development of ESIPT-based fluorescence probes is particularly attractive due to their unique properties, which include environmental sensitivity and the potential for ratio metric sensing.

The polar environment can not only inhibit proton transport, creating energy barriers but also serve as an excellent conductor of proton transmission over significant distances through a hydrogen bonding system. In (Lim et al., 2019). Porous crystalline metal–organic frameworks (MOFs) or porous coordination polymers (PCPs) are emerging as a new class of proton conductors with numerous investigations. Some of the MOFs exhibit an excellent proton-conducting performance (higher than 10^{-2} S • cm^{-1}) originating from the interesting hydrogen (H)-bonding networks with guest molecules, where the conducting medium plays a crucial role. In the overwhelming majority of MOFs, the conducting medium is H_2O because of its degenerate conjugate acid–base system ($H_3O^+ + H_2O \leftrightarrow H_2O + H_3O^+$ or $OH^- + H_2O \leftrightarrow H_2O + OH^-$) and the efficient H-bonding ability through two proton donor and two acceptor sites with tetrahedral geometry.

How Many Electrons Can Be Transferred at a Time?

The question is not idle and is not asked out of purely academic curiosity. The fact is that if you go beyond the class of compounds such as "hard balls," i.e. ions and complexes of transition metals, you will have to admit that all chemical bonds are formed by electronic pairs. Therefore, all stable molecules without exception contain an even number of electrons. Accordingly, the transition between them implies the transfer of exactly an even number of electrons (with or without protons - this is the third question). If the transfer of electrons from the donor to the acceptor (i.e. from the reducing agent to the oxidizing agent) is carried out one at a time, then such a reaction will inevitably result in an odd electronic compound, i.e. a radical (or radical ion). This is if only one of the initial reagents was a stable molecule, and its partner was a metal complex with wide electron reception and recoil capabilities and not necessarily one at a time. In the case of the interaction of the molecular oxidizing agent and the reducing agent, the formation of a radical pair is possible.

The radical pair has a high chance of recombining without leaving the solvent cell. If not, the appearance of highly active particles in the reaction medium allows their further non-selective interaction with all components of the system, both stable and intermediate. For a number of organic synthesis processes, such a decrease in selectivity is not good. Hence such an increased interest in the possibilities of synchronous transfer of several electrons at once, two or four (synchronous transfer of three electrons is as uninteresting in this aspect as one).

On the other hand, Marcus theory is very rigid about increasing the number of electrons (or protons) transferred. In formula (2) there is e2, i.e. with an increase in the number of transferred elementary charges by half, the reorganization energy increases fourfold! This means that instead of the typical water environment value of about 40 kcal/Mol, a two-electron transfer will be prevented by a practically impermeable impenetrable barrier of 160 kcal/mol, which even noticeably exceeds the C-H bond energy. It is clear that with such obstacles, there can be no question of four-electron processes. And yet, they exist!

By the way, taking this opportunity I want to draw attention to the following circumstance. Try to calculate the energy of reorganization of some reagents according to formula (2). What to substitute there? Radii in Angstroms? Electron charge in pendants? Nothing will work out! Theorists generally like a system of units in which the speed of light, the constant bar and the charge of an electron are equal to one, they call such a system natural, although it causes rejection in any experimenter who "walks on the ground," so the result must be led to generally accepted units. From the type of formula (2), and it is replicated by thousands of copies in articles by numerous followers of Marcus's theory, it follows that none of them has ever, ever tried to bring the calculation result using this formula to a number. Otherwise, at least someone remembered that this formula lacks the Avogadro number (in the numerator) and the dielectric constant of the vacuum (in the denominator). That's when it all comes together.

Returning to the topic of confrontation between two energy factors (reorganization energy and thermodynamic reaction effect), it can be noted that for single-electron transfers, the reorganization contribution gives preferable opportunities, but not every system can "use" them due to the fact that free radicals are quite expensive products from an energy point of view. Conversely, the terrifyingly high Frank-Condon barrier, which provides reorganization energy, can be overcome if the redox potential of two-electron transfer is much more profitable than the indicators for a single-electron route. For the first time, this simple, but very fruitful thought was expressed namely in our laboratory (where I work to this day) in the mid-70s of the last century (Nadezhdin et al., 1975). This idea was further developed, however, without reference to Russian authors, it received in a number of works, for example, (Koper, 2011). The influence of the Coulomb barrier for sequential two-electron transfer in a nanoengineered photocatalyst I, as discussed in (Wang et al., 2020).

The paper (Bender et al., 2016) provides a very interesting discussion of the considerations that come into play when electrochemical reactions involve the transfer of more than one electron, as is the case in many electrocatalytic reactions. The author justifies that when trying to catalyze the transfer of 1 and 2, or more than two electrons, or more precisely when there is no radical intermediate product of single electron transfer; at least one catalytic intermediate complex arises with necessity. And when it comes to the transfer of four electrons, for example, during photosynthesis, when an oxygen molecule is reduced to water, there must be at least two intermediate complexes.

Bender et al. (2016) presented a mini-review on the ambiguity of recharging products in lithium-oxygen sodium-oxygen cells, which are complex concepts of next-generation batteries. While in the case of non-aqueous Li/O_2 batteries, the cell reaction occurring was unambiguously determined (formation of Li_2O_2), the situation is much less clear in the case of non-aqueous Na/O_2 cells. Charge and discharge hysteresis on voltamperograms is discussed, which characteristics contain both thermodynamic and kinetic information. Potential plateaus correspond to the specific equilibrium of the reaction between different phases, even if their absolute potential value includes overvoltage for from different processes. All recently published data were digitized and normalized to maximum capacity (set by the authors) to improve imaging. However, this normalization "hides" information about discharge depth (DOD) and absolute capacity. It is well known that a longer cycle life can be achieved if only a portion of the available charge ("fine cycling") is cyclical. Thus, quantitative differences in hysteresis may be due to different discharge depths.

The authors divided the available data into three groups based on the main products of the discharge: 1) sodium superoxide (NaO_2) formed during single-electron transfer; 2) sodium peroxide (Na_2O_2) formed by two-electron transfer and 3) hydrated sodium peroxide and other hydrated oxides such as NaOH. Two discharge products, with almost equal free enthalpies of formation, but different numbers of transferred electrons and completely different kinetics, appear to compete, namely NaO_2 and Na_2O_2. Cells forming either superoxide or peroxide have been reported, but it is unclear how the reaction in the cell can be influenced for the selective transfer of one or two electrons. At the same time, the "bottleneck" occurs at completely different stages. These conclusions derive from simple thermodynamic considerations and, through calculations of functional density theory by first principles, lead to fairly simple predictions about the most optimal catalyst for these different classes of reaction.

These concepts can be applied to two-electron hydrogen oxidation and hydrogen release reactions, four-electron oxygen reduction and oxygen release reactions, and multi-electron carbon dioxide reduction, three electrocatalytic reactions potentially important for future energy cycle. Despite the picture, which is elegant from a thermodynamic point of view, the author completely ignores the fact that Marcus's approach with the fundamental influence of the energy of the reorganization of the environment has a very distant relation to the inner-sphere processes that are implied in the formation of intermediate complexes.

Initially, it was believed that all redox reactions are carried out in the form of a sequence of single-electron steps. This was dictated both by the requirements of Marcus's theory and by the rapid development of the Electronic Paramagnetic Resonance (EPR) technique, which allows the registration of single-electron radical products. In cases where such products were not found, it was believed that their search was not carried out thoroughly enough. Skepticism was similarly expressed regarding reactions leading to the appearance of metal ions and complexes in "unusual" oxidation states, such as Cu^{3+} or Fe^{4+}.

Gradually, there is an "addiction" to synchronous multielectronic processes, while ignoring the requirements of Marcus theory. At the same time, the mention of Marcus's formulas is almost always present, but there is no discussion or objection to multi-electronic prohibitions in the works of the last decade. Type: as many electrons are transferred as required - it is necessary; the model doesn't mind.

Initially, it was noted that in many kinetically non-equilibrium systems, the formal redox potential given by the Nernst equation does not describe the real relationship between the initial reagents and the products of electronic transfer (Ardo, 2011). Gradually, they stopped paying attention to non-equilibrium.

At the same time, it is completely meaningless to deny the need to cooperate two steps and, accordingly, synchronous transfer of two electrons in cases where it is necessary to use the total energy of two absorbed photons (see, for example, Chan et al., 2012). It is shown, that singlet fission, the creation of two triplet excitons from one singlet exciton, is being explored to increase the efficiency of solar cells and photodetectors based on organic semiconductors. The efficiency of multielectron transfer is expected to depend critically on other dynamic processes available to the singlet (S1) and the multiexciton, but little is known about these competing channels. The authors had provided preliminary evidence for the extraction of two electrons from the multiexciton.

In opposite way, an interesting example of the production of two triplet excitons from the absorption of a single photon is given in Tritsch et al. (2013). To implement this it is necessary to harvest the two triplets from the singlet fission material. There are kinetic data, that in the tetracene/copper phthalocyanine (CuPc) triplets are produced from singlet fission on the timescale of 45±5 ps. However, the efficiency of triplet energy transfer is limited by a much more effective loss channel due to faster formation (400±100 fs) recombination (2.6±0.5 ps) of charge transfer excitons.

The very important for environmental chemistry from a practical point of view process is the production of hydrogen peroxide from molecular oxygen. In case, if this reduction reaction has a pathway via simultaneous two-electron transfer, and the energy loss is much lower than in the case of step-by-step one-electron changes with the formation of free radicals. In (Sun et al., 2021) a new strategy to achieve decentralized preparation of hydrogen peroxide (H_2O_2) in a green manner is discussed.

As a promising new type of catalytic material the authors propose carbon-based single-atom catalysts, which permits achieve wide-range adjustments of the electronic structure of the active metal centers, so it will be possible to use the reaction of electrochemical production of H2O2 from the selective two-electron transfer oxygen reduction reaction (ORR). Herein, starting from the review of characterizing methods and reaction mechanisms of ORR via two-electron and four-electron transfer pathways, the vital role of binding strength between OOH intermediate and active sites is discussed. Nonetheless, the four-electron transfer pathways look not sufficiently justified. Currently reported carbon-based single-atom catalysts for H_2O_2 production are systematically summarized and critically reviewed. Some predictions for future H_2O_2 production via electrochemical routes are offered.

In general, recently, the largest number of synchronous multielectronic transfer reports have been received specifically for electrodes rather than homogeneous electrochemical reactions. This seems somewhat strange, especially since Evans (2008) drew attention to the fact that for most electrode reactions of organic and organometallic species, only a single oxidation or reduction is encountered within the accessible potential range of the solvent/electrolyte medium that is chosen.

The author of review (Sun et al., 2021) pays special attention to the two principal questions that arise in discussion of single-electron reactions. The first of these is thermodynamics in nature: What are the factors that govern the standard potential for oxidation or reduction? And how do they correlate with Molecular Properties? The second is a kinetic question: What are the factors that determine the rates of heterogeneous or homogeneous electron-transfer reactions? The well-known linear relationships between one-electron redox potentials and logarithms of rate constants are demonstrated, and they obviously obey Marcus' theory.

More interesting picture one can see in the case of two-electron reactions. As mentioned earlier, some molecules can accept not only one but two electrons depending on the molecular structure and the functional groups that are present. For example, there can be two different functional groups that will give rise to reduction steps occurring at two resolvable potentials as in α,p-dinitrocumene, which contains two different kinds of electroactive nitro groups.

The first amusing detail refers to the fact that even if a molecule has exactly the same active groups, their successive redox-potentials will be different.

The second peculiarity concerns that the plot of ΔG vs $Log(k)$ is much more scattered in the case of two-electron reactions compared with that can be seen in the one-electron case. No significant correlation is seen in this case.

One of the most tempting elements from the point of view of multi-electron transfer is manganese, which has multiple oxidation states, some of which are even realized in stable compounds. (Luther, 2005; Luther, 2010). Two-electron transfer thermodynamic calculations of NO_3^- reduction with Fe^{2+} and Mn^{2+} for the first two transfer steps are presented. These reactions have similar pH dependence as found for O_2 reduction to water via two-electron steps. Also cobalt complexes are a point of interest (Marcotte, 2004).

Ferrate(VI) ($Fe^{VI}O_4^-$, Fe(VI)) and ferrate(V) ($Fe^{V}O_4^{3-}$, Fe(V)) have high oxidizing power and upon decomposition form a nontoxic byproduct, Fe(III), which makes them environmentally friendly oxidants in water and wastewater treatment. The kinetics of the reaction between Fe(VI) and I^- was determined using a stopped-flow technique. The second-order rate constants for the oxidation of I- and other inorganic compounds by protonated ferrate(VI) ($HFe^{VI}O_4^-$) and ferrate(V) ($Fe^{V}O_4^{3-}$) ions were correlated with thermodynamic reduction potentials to understand the reaction mechanisms. A linear relationship was found between log(k) and 1-e-reduction potentials for iodide, cyanides, and superoxide while oxy-compounds of nitrogen, sulfur, selenium, and arsenic demonstrated a linear relationship for 2-e- reduction potentials (Sharma, 2010).

Electrochemical reduction of CO_2 (CO_2ER) has received significant attention due to its potential to sustainably produce valuable fuels and chemicals. In (Deng et al., 2022) the two-electron transfer for the adsorption of CO_2 onto the surface of electrocatalysts was elucidated. It is expected, this finding to provide guidance for improving activity through the enhancement of the CO_2 adsorption processes.

Even for persulfate activation by biochar from shrimp shells, the authors (Yu et al., 2020) declare a two-electron transfer path. And of course, two-electron pathway is preferred in (Morales et al., 2000) for Ferredoxin: $NADP^+$:reductase (FNR), a flavine-adeninedinucleotide (FAD) flavoenzyme, uses high energy electrons photoproduced by photosystem I (PSI) and protons to reduce nicotinamide-adenine-dinucleotide phosphate ($NADP^+$) to NADPH following the reaction

$$2Fd\_red + NADP^+ + 2H+ <=> 2Fd\_ox + NADPH + H^+ \quad (4)$$

The reader can derive a lot of useful information about two- and multi-electronic transfers from the references starting from (Kotelnikov, 2011) and up to the end of the list. We want to pay special attention to the publication (Chen et al., 2020), in which the authors insist on an eight-electronic transfer. No more no less! Such a reaction is beyond understanding based on Marcus' theory. That is why we left this selection of links without a detailed discussion.

From Spheric Horses to Molecular Machines

An important consequence of the Frank-Condon principle is the quadratic dependence of the energy of the reorganization of the medium on the number of transferred electrons. It is obvious that all odd-electron reactions lead to the formation of free radicals, which is extremely costly from the point of view of the free energy of the process (the value can reach the level of the energy of breaking a chemical bond). An estimate of the reorganization energy for the external transfer of an electron between not-too-large reagents gives a value about 1 ev. Such a cost leaves single-electron reactions energetically competitive and even preferable, compared to synchronous two-electron transfer. Indeed, when the charge is doubled, the reorganization energy increases fourfold, which even theoretically excludes the possibility of overcoming the barrier when transferring an electron pair.

And, nevertheless, the existence of synchronous two-electron processes, without intermediate formation of free radical products, is proved! Moreover, in enzymatic reactions, even electronic processes are the norm rather than the exception. For example, in photosystem-2, the dimer of super–oxidized chlorophyll leads to the oxidation of two water molecules to molecular oxygen without the intermediate formation of reactive oxygen species – hydroxyl and superoxide radicals and even a relatively stable molecular product - hydrogen peroxide. I.e., there is a synchronous four-electron process.

At the same time, it has been experimentally measured that in redox reactions involving proteins, the reorganization energy, despite their huge size (and in fact, thanks to them), is noticeably less than even in the case of intra-atmospheric reactions and amounts to only tenths of an electronvolt. An important class of compounds with unusual reactivity and the possibility of synchronous multi-electron transfers turned out to be complex with partial charge transfer: they are formed when a redox-active (i.e., capable of oxidation or reduction) ligand enters the coordination sphere of a transition metal ion that has the ability to change the degree of oxidation by more than unity.

One of the relatively new experimentally confirmed results was the detection of ultrafast electron transfer reactions in the case of photoexcitation of one of the reagents. As follows from the Frank-Condon principle, the absorption of a quantum of light transfers the reagent to a configuration-excited state (most likely by vibrational degrees of freedom), from which the transition to the term of finite products occurs without activation or (in the opposite direction) barrier-free. A similar effect can be achieved not only by the transmission of light energy but also by overvoltage on the electrode. In both cases, the speed of the fast process is limited by diffusion, and the observed rate constant is close to the diffusion-controlled one.

Notable interest and, accordingly, a significant number of publications relating to multielectronic (up to 8 electrons!) synchronous transfers in organized nanostructured systems used in electrochemical energy storage devices with high energy density, including multivalent batteries, since they can potentially provide high capacity.

But the most interesting are supramolecular systems, in which rather act the laws of mechanics instead of statistical molecular physics. In this case, electron transfer reactions lead to mechanical work, similar to how it occurs in myoglobin during oxidative phosphorylation with the participation of ATP. Of course, for supramolecular "machines" models of the Markus type, as well as the application of the Frank-Condon principle, are inapplicable.

Formally, to describe the mobility of a supramolecular structure, it is necessary, first of all, to specify a set of its states, called configuration states, and the potential energy of the structure as a function of its states. A macromolecular structure consisting of $\boldsymbol{N}$ elements (atoms, molecular links, or fragments of a polymer chain, depending on the representation of the macromolecule), each of which has $\boldsymbol{m}$ degrees of freedom. To describe the configuration states, it is necessary to introduce a Euclidean space of dimension $\boldsymbol{M} = \boldsymbol{N}*\boldsymbol{m}$ and to map an $\boldsymbol{m}$-dimensional vector $\boldsymbol{R}$ for each state.

The potential energy $\boldsymbol{F}(\boldsymbol{R})$ in the state $\boldsymbol{R}$, generally speaking, it can be determined from the interatomic potentials. The function $\boldsymbol{F}(\boldsymbol{R})$ will be a hypersurface over a multidimensional space of configuration states. It is commonly called the energy landscape. By setting suitable equations of motion on the energy landscape, it would seem that one can try to describe the dynamics of the structure.

In general (and especially for proteins), not only the large-scale structure of the energy landscape is important, but also its fine-scale structure. And the problem here is that the energy landscapes of proteins turn out to be "rugged terrain" when examined in detail.

The number 10^{100} gives quite a suitable idea of the number of local minima on the energy landscape of the protein. It is clear that when modeling the mobility and function of a protein, a complete description of such energy surfaces is impossible even numerically.

Note that all known examples of computer modeling of the dynamics of polyatomic structures, comparable in complexity to a protein, give either a detailed picture of the behavior of the molecular structure, but in a small area of the configuration space, or a rough description of the structure (due to a drastic reduction in the degrees of freedom taken into account), a sketch picture of the behavior as a whole. Modeling in such cases needs, of course, simplification, which, generally speaking, does not have to be universal but should be relevant to the structures under consideration.

Surprisingly, for protein molecules, in recent years it has been possible to build such a simplification. The first technique consists in the approach that not "configurations in general" are considered, but only the states of local minima into which the system inevitably slides when moving through the energy landscape and where it spends almost 100% of the time. But this technique by itself does not lead to the desired result, since the number of even local minima remains astronomically large.

The second simplification is that the entire landscape is divided hierarchically into subsets nested within each other - pools of minima. Technically, such a division is carried out like the topography of natural landscapes using equipotential sections. Using this technique, many states are divided into large pools separated by large energy barriers on the landscape; then large pools are divided into smaller pools embedded in them, separated by lower energy barriers, etc., up to local minima proper. As a result, a tree-like graph is mapped to a multidimensional energy landscape, the "leaves" (nodes on the border of the tree) of which correspond to the local minima of the landscape, and the branches correspond to the points of the "pass" through which transitions between local minima are carried out.

If the vertices of a tree-like graph of the energy landscape are associated with activation barriers, then the condition of transition through the vertex of the minimal subtree is equivalent to the condition of the maximum barrier between states. For the latter reason, a random walk on the boundary of a tree can be understood as a walk on an ultrametric lattice. In the physical literature a Markov random process of this type is called ultrametric diffusion.

Up until the early 2000s, it remained unclear whether the approximation of inter-basin kinetics was adequate or not. Nevertheless, the increase in computing power and the success of the model's predictions make us believe in its applicability and usefulness for describing molecular machines.

There are even international competitions among supramolecular machines. The first race of ***nanocar*** molecules was held in Toulouse on April 28 and 29, 2017 on a 100-nanometer track. The gold-plated race was won by the Swiss team, which was the first to cross the finish line, covering a distance of 133 nanometers in a day, as specified in the article "24 Hours of Toulouse" (2022).

On the silver surface, the nanocar of the Austrian-American team from Rice University and University of Graz set the first speed record of 95 nanometers per hour. The second international Nanocar Race (Nanocar Race II) was held from 24/3/2022, 11:00 am to 25/03/2022, 11:00 am 24 hours a day without stops also in Toulouse at Dupouy Campus. The first place was shared by the NANOHISPA and NIMS-MANA teams, making about 54 laps and covering 678 nm and 1054 nm, respectively. The former demonstrated a lane change for a detour, while the latter crossed the trench and returned. StrasNanocar took third place, covering 476 nm and completing 28 laps.

CONCLUSION

In the present review, we have considered the main theoretical developments and subsequent application to real chemical systems of charge transfer models in the polar environment. We did not consciously address the questions of new continual solvation models suitable for adequately describing the equilibrium and non-equilibrium solvation effects that may occur in real redox reactions. It is also important to emphasize that electron transfer reactions and proton transfer reactions can have significantly different mechanisms in nature. Therefore, their modeling and theoretical description is naturally divided into two more or less independent parts. It is not possible to bring them together with a modern level of understanding of this section of science. Moreover, this applies to the reactions of synchronous transfer of an electron and a proton (the so-called proton assistance), which do not formally lead to the resulting movement of the charge, but are associated with its significant redistribution. A description of such reactions at the quantum mechanical level still awaits new Marcus.

ACKNOWLEDGMENT

This work was performed under the Moldovan National Research Project Nr. 20.80009.5007.27, “Physico-chemical mechanisms of redox processes with electron transfer involved in vital, technological and environmental systems”.

REFERENCES

Andrade, M. F. C., Ko, H. Y., Zhang, L., Car, R., & Selloni, A. (2020). Free energy of proton transfer at the water–TiO 2 interface from ab initio deep potential molecular dynamics. *Chemical Science (Cambridge)*, *11*(9), 2335–2341. doi:10.1039/C9SC05116C PMID:34084393

Ardo, S., Achey, D., Morris, A. J., Abrahamsson, M., & Meyer, G. J. (2011). Non-Nernstian two-electron transfer photocatalysis at metalloporphyrin–TiO2 interfaces. *Journal of the American Chemical Society*, *133*(41), 16572–16580. doi:10.1021/ja206139n PMID:21888402

Bender, C. L., Schröder, D., Pinedo, R., Adelhelm, P., & Janek, J. (2016). One-or Two-Electron Transfer? The Ambiguous Nature of the Discharge Products in Sodium–Oxygen Batteries. *Angewandte Chemie International Edition*, *55*(15), 4640–4649. doi:10.1002/anie.201510856 PMID:26970321

Biasin, E., Fox, Z. W., Andersen, A., Ledbetter, K., Kjær, K. S., Alonso-Mori, R., & Khalil, M. (2021). Direct observation of coherent femtosecond solvent reorganization coupled to intramolecular electron transfer. *Nature Chemistry*, *13*(4), 343–349. doi:10.103841557-020-00629-3 PMID:33589787

Chan, W. L., Tritsch, J. R., & Zhu, X. Y. (2012). Harvesting singlet fission for solar energy conversion: One-versus two-electron transfer from the quantum mechanical superposition. *Journal of the American Chemical Society*, *134*(44), 18295–18302. doi:10.1021/ja306271y PMID:23066740

Chatterjee, A., & König, B. (2019). Birch-Type Photoreduction of Arenes and Heteroarenes by Sensitized Electron Transfer. *Angewandte Chemie*, *131*(40), 14427–14432. doi:10.1002/ange.201905485 PMID:31379035

Chen, G. F., Yuan, Y., Jiang, H., Ren, S. Y., Ding, L. X., Ma, L., & Wang, H. (2020). Electrochemical reduction of nitrate to ammonia via direct eight-electron transfer using a copper–molecular solid catalyst. *Nature Energy*, *5*(8), 605–613. doi:10.103841560-020-0654-1

Closs, G. L., Calcaterra, L. T., Green, N. J., Penfield, K. W., & Miller, J. R. (1986). Distance, stereoelectronic effects, and the Marcus inverted region in intramolecular electron transfer in organic radical anions. *Journal of Physical Chemistry*, *90*(16), 3673–3683. doi:10.1021/j100407a039

Creutz, C., & Taube, H. (1969). Direct approach to measuring the Franck-Condon barrier to electron transfer between metal ions. *Journal of the American Chemical Society*, *91*(14), 3988–3989. doi:10.1021/ja01042a072

Darcy, J. W., Koronkiewicz, B., Parada, G. A., & Mayer, J. M. (2018). A continuum of proton-coupled electron transfer reactivity. *Accounts of Chemical Research*, *51*(10), 2391–2399. doi:10.1021/acs.accounts.8b00319 PMID:30234963

Deng, W., Zhang, P., Seger, B., & Gong, J. (2022). Unraveling the rate-limiting step of two-electron transfer electrochemical reduction of carbon dioxide. *Nature Communications*, *13*(1), 1–9. doi:10.103841467-022-28436-z PMID:35145084

Dogonadze, R. R., Kuznetsov, A. M., & Chernenko, A. A. (1965). Theory of homogeneous and heterogeneous electronic processes in liquids. *Russian Chemical Reviews*, *34*(10), 759–775. doi:10.1070/RC1965v034n10ABEH001561

Eberson, L. (1982). Electron-transfer reactions in organic chemistry. Advances in Physical Organic Chemistry. Academic Press.

Evans, D. H. (2008). One-electron and two-electron transfers in electrochemistry and homogeneous solution reactions. *Chemical Reviews*, *108*(7), 2113–2144. doi:10.1021/cr068066l PMID:18620364

Fourmond, V., Wiedner, E. S., Shaw, W. J., & Léger, C. (2019). Understanding and design of bidirectional and reversible catalysts of multielectron, multistep reactions. *Journal of the American Chemical Society*, *141*(28), 11269–11285. doi:10.1021/jacs.9b04854 PMID:31283209

Goldsmith, Z. K., Soudackov, A. V., & Hammes-Schiffer, S. (2019). Theoretical analysis of the inverted region in photoinduced proton-coupled electron transfer. *Faraday Discussions*, *216*, 363–378. doi:10.1039/C8FD00240A PMID:31017599

Gould, I. R., Ege, D., Mattes, S. L., & Farid, S. (1987). Return electron transfer within geminate radical ion pairs. Observation of the Marcus inverted region. *Journal of the American Chemical Society*, *109*(12), 3794–3796. doi:10.1021/ja00246a055

Grampp, G. (1993). The Marcus inverted region from theory to experiment. *Angewandte Chemie International Edition in English, 32*(5), 691–693. doi:10.1002/anie.199306911

Hours of Toulouse. (2022). 24 Hours of Toulouse. *Nature Nanotechnology, 17*(5), 433. doi:10.103841565-022-01143-y PMID:35581424

Hush, N. S. (1985). Distance dependence of electron transfer rates. *Coordination Chemistry Reviews, 64*, 135–157. doi:10.1016/0010-8545(85)80047-3

Jin, Y., Zhang, Q., Zhang, Y., & Duan, C. (2020). Electron transfer in the confined environments of metal–organic coordination supramolecular systems. *Chemical Society Reviews, 49*(15), 5561–5600. doi:10.1039/C9CS00917E PMID:32643720

Kim, H., Kim, J. S., Heo, J. M., Pei, M., Park, I. H., Liu, Z., & Lee, T. W. (2020). Proton-transfer-induced 3D/2D hybrid perovskites suppress ion migration and reduce luminance overshoot. *Nature Communications, 11*(1), 1–13. doi:10.103841467-020-17072-0 PMID:32632144

Koper, M. T. M. (2011). Thermodynamic theory of multi-electron transfer reactions: Implications for electrocatalysis. *Journal of Electroanalytical Chemistry (Lausanne, Switzerland), 660*(2), 254–260. doi:10.1016/j.jelechem.2010.10.004

Kotelnikov, A. I., Medvedev, E. S., & Goreachev, N. S. (2011). Structural and Dtnamic Aspects of Electron Transfer in Proteins – Highly Organized Natural Nanostructures (in Rus.). *Proc.AS. Ser.Chim, 7*, 1295.

Kresge, A. J. (1975). What makes proton transfer fast. *Accounts of Chemical Research, 8*(10), 354–360. doi:10.1021/ar50094a006

Launay, J. P. (2001). Long-distance intervalence electron transfer. *Chemical Society Reviews, 30*(6), 386–397. doi:10.1039/b101377g

Levich, V. G., Dogonadze, R. R., German, E. D., Kuznetsov, A. M., & Kharkats, Y. I. (1970). Theory of homogeneous reactions involving proton transfer. *Electrochimica Acta, 15*(2), 353–367. doi:10.1016/0013-4686(70)80027-5

Li, X. B., Tung, C. H., & Wu, L. Z. (2019). Quantum dot assembly for light-driven multielectron redox reactions, such as hydrogen evolution and CO_2 reduction. *Angewandte Chemie, 131*(32), 10918–10925. doi:10.1002/ange.201901267 PMID:30821022

Li, Z., Vinayan, B. P., Jankowski, P., Njel, C., Roy, A., Vegge, T., & Zhao-Karger, Z. (2020). Multi-Electron Reactions Enabled by Anion-Based Redox Chemistry for High-Energy Multivalent Rechargeable Batteries. *Angewandte Chemie, 132*(28), 11580–11587. doi:10.1002/ange.202002560 PMID:32220137

Liang, N., Miller, J. R., & Closs, G. L. (1990). Temperature-independent long-range electron transfer reactions in the Marcus inverted region. *Journal of the American Chemical Society, 112*(13), 5353–5354. doi:10.1021/ja00169a054

Libby, W. F. (1952). Theory of Electron Exchange Reactions in Aqueous Solution. *Journal of Physical Chemistry, 56*(7), 863–868. doi:10.1021/j150499a010

Lim, D. W., Sadakiyo, M., & Kitagawa, H. (2019). Proton transfer in hydrogen-bonded degenerate systems of water and ammonia in metal–organic frameworks. *Chemical Science (Cambridge)*, *10*(1), 16–33. doi:10.1039/C8SC04475A PMID:30746070

Liu, C., Ren, L., Yan, B., Luo, L., Zhang, J., & Awasthi, M. K. (2021). Electron transfer and mechanism of energy production among syntrophic bacteria during acidogenic fermentation: A review. *Bioresource Technology*, *323*, 124637. doi:10.1016/j.biortech.2020.124637 PMID:33421831

Liu, R., Liang, Z., Gong, Z., & Yang, Y. (2019). Research Progress in Multielectron Reactions in Polyanionic Materials for Sodium-Ion Batteries. *Small Methods*, *3*(4), 1800221. doi:10.1002mtd.201800221

Luther, G. W. III. (2005). Manganese (II) oxidation and Mn (IV) reduction in the environment—Two one-electron transfer steps versus a single two-electron step. *Geomicrobiology Journal*, *22*(3-4), 195–203. doi:10.1080/01490450590946022

Luther, G. W. III. (2010). The role of one-and two-electron transfer reactions in forming thermodynamically unstable intermediates as barriers in multi-electron redox reactions. *Aquatic Geochemistry*, *16*(3), 395–420. doi:10.100710498-009-9082-3

Makita, H., & Hastings, G. (2017). Inverted-region electron transfer as a mechanism for enhancing photosynthetic solar energy conversion efficiency. *Proceedings of the National Academy of Sciences of the United States of America*, *114*(35), 9267–9272. doi:10.1073/pnas.1704855114 PMID:28814630

Marcotte, S., Villers, D., Guillet, N., Roue, L., & Dodelet, J. P. (2004). Electroreduction of oxygen on Co-based catalysts: Determination of the parameters affecting the two-electron transfer reaction in an acid medium. *Electrochimica Acta*, *50*(1), 179–188. doi:10.1016/j.electacta.2004.07.029

Marcus, R. A. (1964). Chemical and electrochemical electron-transfer theory. *Annual Review of Physical Chemistry*, *15*(1), 155–196. doi:10.1146/annurev.pc.15.100164.001103

Marcus, R. A. (1993). Electron Transfer Reactions in Chemistry: Theory and Experiment (Nobel Lecture). *Angewandte Chemie International Edition in English*, *32*(8), 1111–1121. doi:10.1002/anie.199311113

Miller, J. R., Calcaterra, L. T., & Closs, G. L. (1984). Intramolecular long-distance electron transfer in radical anions. The effects of free energy and solvent on the reaction rates. *Journal of the American Chemical Society*, *106*(10), 3047–3049. doi:10.1021/ja00322a058

Morales, R., Charon, M. H., Kachalova, G., Serre, L., Medina, M., Gómez-Moreno, C., & Frey, M. (2000). A redox-dependent interaction between two electron-transfer partners involved in photosynthesis. *EMBO Reports*, *1*(3), 271–276. doi:10.1093/embo-reports/kvd057 PMID:11256611

Moser, C. C., Keske, J. M., Warncke, K., Farid, R. S., & Dutton, P. L. (1992). Nature of biological electron transfer. *Nature*, *355*(6363), 796–802. doi:10.1038/355796a0 PMID:1311417

Nadezhdin, A. D., Kozlov, Y. N., & Purmal, A. P. (1975). Reactions of one-and two-electron donors with TNM. *Zh. Fiz. Khim*, *49*(9), 2263–2269.

Ovchenkova, A. E., Safonova, E. A., Martynov, A. G., & Senchykhin, I. N., Gorbunova, G., Tsyvadze, A. (2021). New Crown-Substituted Phtalo- Oxanthrocyanate of Ittrium: Synthesis and Study of Supramolecular Aggregation (in Rus.). *Conf.Proc.*

Parada, G. A., Goldsmith, Z. K., Kolmar, S., Pettersson Rimgard, B., Mercado, B. Q., Hammarström, L., & Mayer, J. M. (2019). Concerted proton-electron transfer reactions in the Marcus inverted region. *Science*, *364*(6439), 471–475. doi:10.1126cience.aaw4675 PMID:30975771

Parida, S. K., Mandal, T., Das, S., Hota, S. K., De Sarkar, S., & Murarka, S. (2021). Single electron transfer-induced redox processes involving N-(acyloxy) phthalimides. *ACS Catalysis*, *11*(3), 1640–1683. doi:10.1021/acscatal.0c04756

Rehm, D., & Weller, A. (1970). Kinetics of Fluorescence Quenching by Electron and H-Atom Transfer. *Israel Journal of Chemistry*, *8*(2), 259–271. doi:10.1002/ijch.197000029

Ren, W., Xiong, L., Nie, G., Zhang, H., Duan, X., & Wang, S. (2019). Insights into the electron-transfer regime of peroxydisulfate activation on carbon nanotubes: The role of oxygen functional groups. *Environmental Science & Technology*, *54*(2), 1267–1275. doi:10.1021/acs.est.9b06208 PMID:31846314

Richardson, D. E., & Taube, H. (1984). Mixed-valence molecules: Electronic delocalization and stabilization. *Coordination Chemistry Reviews*, *60*, 107–129. doi:10.1016/0010-8545(84)85063-8

Rutledge, H. L., & Tezcan, F. A. (2020). Electron transfer in nitrogenase. *Chemical Reviews*, *120*(12), 5158–5193. doi:10.1021/acs.chemrev.9b00663 PMID:31999100

Sariciftci, N. S., Smilowitz, L., Heeger, A. J., & Wudl, F. (1992). Photoinduced electron transfer from a conducting polymer to buckminsterfullerene. *Science*, *258*(5087), 1474–1476. doi:10.1126cience.258.5087.1474 PMID:17755110

Sedgwick, A. C., Wu, L., Han, H. H., Bull, S. D., He, X. P., James, T. D., & Yoon, J. (2018). Excited-state intramolecular proton-transfer (ESIPT) based fluorescence sensors and imaging agents. *Chemical Society Reviews*, *47*(23), 8842–8880. doi:10.1039/C8CS00185E PMID:30361725

Sharma, V. K. (2010). Oxidation of inorganic compounds by ferrate (VI) and ferrate (V): One-electron and two-electron transfer steps. *Environmental Science & Technology*, *44*(13), 5148–5152. doi:10.1021/es1005187 PMID:20527775

Sun, K., Xu, W., Lin, X., Tian, S., Lin, W. F., Zhou, D., & Sun, X. (2021). Electrochemical oxygen reduction to hydrogen peroxide via a two-electron transfer pathway on carbon-based single-atom catalysts. *Advanced Materials Interfaces*, *8*(8), 2001360. doi:10.1002/admi.202001360

Suppan P. (1992). The Marcus inverted region. *Photoinduced Electron Transfer IV*, 95-130.

Sutin, N. (1986). Marcus' contributions to electron-transfer theory. *Journal of Physical Chemistry*, *90*(16), 3465–3466. doi:10.1021/j100407a002

Taube, H. (1984). Electron Transfer between Metal Complexes— A Retrospective View(Nobel Lecture). *Angewandte Chemie International Edition in English*, *23*(5), 329–339. doi:10.1002/anie.198403293

Tritsch, J. R., Chan, W. L., Wu, X., Monahan, N. R., & Zhu, X. Y. (2013). Harvesting singlet fission for solar energy conversion via triplet energy transfer. *Nature Communications*, *4*(1), 1–7. doi:10.1038/ncomms3679 PMID:24145737

Volkov, A. G., Bookal, A., Hairston, J. S., Roberts, J., Taengwa, G., & Patel, D. (2021). Mechanisms of multielectron reactions at the plasma/water interface: Interfacial catalysis, RONS, nitrogen fixation, and plasma-activated water. *Electrochimica Acta, 385*, 138441. doi:10.1016/j.electacta.2021.138441

Wang, J., Ding, T., Gao, K., Wang, L., Zhou, P., & Wu, K. (2021). Marcus inverted the region of charge transfer from low-dimensional semiconductor materials. *Nature Communications, 12*(1), 1–9. doi:10.103841467-021-26705-x PMID:34732730

Wang, J., Ding, T., & Wu, K. (2020). Coulomb Barrier for Sequential Two-Electron Transfer in a Nanoengineered Photocatalyst. *Journal of the American Chemical Society, 142*(32), 13934–13940. doi:10.1021/jacs.0c06256 PMID:32672949

Wang, W., & Lee, D. J. (2021). Direct interspecies electron transfer mechanism in enhanced methanogenesis: A mini review. *Bioresource Technology, 330*, 124980. doi:10.1016/j.biortech.2021.124980 PMID:33743275

Weinberg, D. R., Gagliardi, C. J., Hull, J. F., Murphy, C. F., Kent, C. A., Westlake, B. C., Paul, A., Ess, D. H., McCafferty, D. G., & Meyer, T. J. (2012). Proton-coupled electron transfer. *Chemical Reviews, 112*(7), 4016–4093. doi:10.1021/cr200177j PMID:22702235

Xu, C., Zi, Y., Wang, A. C., Zou, H., Dai, Y., He, X., Wang, Z. L., Wang, Y.-C., Feng, P., Li, D., & Wang, Z. L. (2018). On the electron-transfer mechanism in the contact-electrification effect. *Advanced Materials, 30*(15), 1706790. doi:10.1002/adma.201706790 PMID:29508454

Yang, J., Feng, L., Pi, S., Cui, D., Ma, F., Zhao, H. P., & Li, A. (2020). A critical review of aerobic denitrification: Insights into the intracellular electron transfer. *The Science of the Total Environment, 731*, 139080. doi:10.1016/j.scitotenv.2020.139080 PMID:32417477

Yoneda, Y., Mora, S. J., Shee, J., Wadsworth, B. L., Arsenault, E. A., Hait, D., & Fleming, G. R. (2021). Electron–nuclear dynamics accompanying proton-coupled electron transfer. *Journal of the American Chemical Society, 143*(8), 3104–3112. doi:10.1021/jacs.0c10626 PMID:33601880

Yu, J., Tang, L., Pang, Y., Zeng, G., Feng, H., Zou, J., & Tan, J. (2020). Hierarchical porous biochar from shrimp shell for persulfate activation: A two-electron transfer path and key impact factors. *Applied Catalysis B: Environmental, 260*, 118160. doi:10.1016/j.apcatb.2019.118160

ENDNOTES

1 For those who never heard the joke about spherical horses – this is probably the most famous mathematical model of real objects. Anyhow, back to that unfortunate equine… The story comes in many guises, involving various domestic birds and beasts from chickens to camels, but the version I remember involves a super-rich oil tycoon with a passion for horse racing. His horses did well, but he wanted them to do better, and he figured that if physicists could work out how to split the atom, they ought to be clever enough to get his horses running that tiny bit faster as well. He advertised, and in due course, a physicist arrived and set to work with computers and hideously complex formulae. Six months later, the tycoon was starting to wonder when he was going to start seeing results from his very

considerable investment. (Even theoretical physicists need funding, don't you know?) He summoned said physicist to his office. "How's it going?" he asked. "Where are my winners? My world record holders? My timeless, immortal fame?" "Oh, it's going really well," replied the physicist. "In fact, I got a solution just yesterday." "Great!" said the tycoon. "Let's go racing....!" "Er... not quite yet, Sir...." "You see, so far, I've only solved the problem for the case of a spherical horse in a vacuum... .."The text is borrowed from https://memenetwork.wordpress.com/2013/11/19/beware-of-the-spherical-horse-in-a-vacuum/

Chapter 2
Molecular Oxygen Activation Mechanisms

Ashok Vaseashta
https://orcid.org/0000-0002-5649-0067
International Clean Water Institute, USA

Gheorghe Duca
Institute of Chemistry, Moldova State University, Moldova

Sergey Travin
N.N. Semenov Federal Research Center for Chemical Physics, Russian Academy of Sciences, Russia

ABSTRACT

Molecular oxygen is the greenest of all possible oxidizing agents, but its specific electronic configuration eliminates the possibility of its direct reaction with organic substances. Single-electron transfer between triplet O^2 and singlet organics is prohibited according to the Wigner spin selection rule. Moreover, attaching the first electron is disadvantageous even from a thermodynamic point of view. The authors will not consider the high-temperature processes, however, and will focus on the mechanisms for generating active oxygen species under normal temperature conditions. All methods of "soft" activation of oxygen can be conditionally divided into three main groups: photo-excitation; direct or sensitized initiation of a radical-chain oxidative process; and catalysis by transition metal ions with the inclusion of oxygen in the coordination sphere. New perspectives on molecular oxygen activation pathways and their application in some oxidative reactions including degradation of organic pollutants, organic synthesis, and environmentally significant reactions are discussed.

INTRODUCTION

Oxygen is the most common element on Earth. Its content in the earth's crust exceeds 46% and is in absolute terms ~ 2.8×10^{19} tons. Slightly less, namely ~ 1.15×10^{18} tons of oxygen is contained in ocean water. However, only a negligible part of it e.g., at the level of 0.004% of the total content is represented in the atmosphere in free form, as gaseous molecular oxygen. Hence, there are approximately 1.18×10^{15}

DOI: 10.4018/978-1-6684-7198-2.ch002

tons of oxygen, which is relatively small compared to its content in water or sand but nevertheless more than enough, to maintain life on Earth[1,2]. Note that one kilogram of biomass (from the simplest phytoplankton as representatives of the genus homo sapiens) accounts for about five hundred kilograms of pure oxygen, which for a medium-sized person would be enough (based on a norm of 700 g per 70 kg of weight) for 137 years or two lives - this is without "recharging" and refueling due to the activities of oxygen-extracting plants.

It is intuitive to draw attention to the widespread misconception that forests (and primarily in the Amazon or Siberia) are the "lungs" of the planet, supplying oxygen to the rest of life on Earth. In fact, that's not the case. Indeed, a third of all trees in the world are concentrated on more than 6 million hectares of Siberian taiga, although this ecosystem has only a minor impact on the level of carbon dioxide in the atmosphere. It is noteworthy that Taiga is fully lit by the sun for no more than one month a year because it takes any tree about 50 years to grow here! Amazon forests - the largest rainforest array on the planet - do not have such a problem. This huge green massif stretched across the territories of eight countries, and it surpasses all the rainforests taken together in its area. According to experts, the juicy greens of the Amazon basin produce a fifth of the Earth's oxygen. At first glance, this seems a lot, however, it is necessary to assess based on the scientific data.

In fact, oxygen is produced not only by those plants that grow in the forest. All plant organisms, including the inhabitants of water bodies, and residents of the steppes, and deserts constantly produce oxygen. Plants, unlike animals, fungi, and other living organisms, can synthesize organic substances, using the energy of light for this. This process is called photosynthesis. As a result of photosynthesis, oxygen is released. It is a byproduct of photosynthesis. Oxygen is released very, very much, in fact, 99% of the oxygen that is present in the Earth's atmosphere is of plant origin. And only 1% comes from the mantle, the underlying layer of the Earth.

On average, approximately 1 hectare of forests annually releases 4 tons of oxygen and consumes 5 tons of carbon dioxide. A person exhales up to 1 kilogram of carbon dioxide per day, 365 kg per year. Consequently, 1 hectare of forest absorbs carbon dioxide, which is exhaled by 13 people. However, forest plants not only release oxygen but also consume it, and it's not just and also not so much about daily biorhythms. Yes, indeed, if during the day a plant releases oxygen under the influence of photosynthesis, converting carbon dioxide into it, then at night the same process (in fact, not exactly this, but its dark chemical component) goes in the opposite direction: oxygen is consumed, and carbon dioxide is released.

In an extremely simplified form, these gross processes can be represented as:

Day - photosynthesis: $x{\bullet}CO_2+y{\bullet}H_2O$ + solar energy $\rightarrow C_x(H_2O)_y+x{\bullet}O_2\uparrow$
Night - metabolism: $C_x(H_2O)_y+x{\bullet}O_2\rightarrow x{\bullet}CO_2\uparrow+y{\bullet}H_2O$

Of course, this daily cycle does not imply "running in place." The yield of products during the daytime process is always higher than their consumption in the dark. Hence, therefore, as a result, the amount of carbon-containing substance in the plant only increases over time. The amount of "poly" carbon given in the equation is cellulose or fiber - the main building material of plants. Since a plant can only grow "up," during life the process of fiber accumulation dominates the process of its nocturnal consumption, and the release of oxygen prevails over the release of carbon dioxide.

After a plant dies, cellulose accumulated during its life gradually turns into carbon dioxide, even if it is destined to protect itself from forest fires. Hence, a process of slow decay begins, during which all, figuratively speaking, of the oxygen released by it during life will be used up and returned to the state of carbon dioxide. Thus, the higher role of plants in the oxygen balance of the planet is almost zero. Forests, although they adorn human life, do not supply us with oxygen, but only lend it to use the same, while the plant is alive.

The exception to this rule is, perhaps, the trunks of trees that were "lucky" to drown in the swamps. They are waiting for a very slow process of transformation first into peat, and then into coal, oil, and natural gas. And everything else that will be preserved in the bowels of the Earth will not require the return of oxygen from a temporary exchange fund. If in terms of "oxygen", humans live "on loan" for a very short period of a couple of decades, then in terms of fossil fuels this is a very long-term loan, amounting to hundreds of millions of years. It can be argued with certainty that all (or almost all) atmospheric oxygen is nothing more than a by-product of a billion years of accumulation of coal and hydrocarbons in the bowels of the Earth. Or vice versa – depending upon the perspectives (see also our article in the same collection *"Hydrogen Peroxide, a Substance that Conquered the World"* (Duca et al, 2022), about the oxygen catastrophe and the sudden change in the reducing state of the atmosphere to oxidative (Vaseashta et al, 2022).

It should be noted that the biomass of oceanic phytoplankton, indeed, forms the so-called "real lungs" of the planet. After living for two to three days, after its death, plankton loses buoyancy and drowns in the depths of the ocean, inaccessible for mixing and therefore free of oxygen. No return of atmospheric oxygen is provided for the oxidation of bottom organic deposits. Therefore, it is these algae that are genuine suppliers of oxygen, which we can spend both on our own breathing and on the breathing of our industrial furnaces and power plants.

Several important comments on the circle of redox equivalents involving molecular oxygen, primarily concerning the time constants of natural cycles (by the way, they are different for different substances).

1. The oxygen content in the atmosphere is approximately 4500-5000 times higher than its annual supply during photosynthesis. Hence, if not only forests but also oceanic plankton disappeared at once, we would still live as long as it took to create Egyptian pyramids.
2. However, the carbon dioxide in the atmosphere is much less. A total of 850 gigatons per ^{12}C or about 3,100 gigatons in terms of potentially released oxygen from atmospheric carbon dioxide, or 0.26% of what is available, which is negligible on the one hand, and not surprising on the other, given that CO_2 in the atmosphere is much less than O_2. In other words, oxygen bound in the form of carbon dioxide, although it is the main source of "fresh" atmospheric arrivals, does not play a prominent role in the overall oxygen balance, hence, is a paradox.
3. What would happen if everyone suddenly gave up greenhouse gas emissions together? And for the sake of climate warming fighters, they would stop emitting CO_2 into the atmosphere. Not suspended, but completely stopped! It would not be bad, but very bad. A real catastrophe would await humanity in a dozen years. The entire supply of atmospheric carbon dioxide is only six times its annual consumption during photosynthesis. Do you want to be left without forests? Fight greenhouse gases, you're on the right track!

Although life on Earth was born in the anaerobic (non-oxygen) era, the transition to using oxygen as a universal oxidizer from a chemical point of view is not just natural, but even inevitable. The fact is that aerobic life is energetically much more efficient than anaerobic life. This is because oxygen is a much stronger oxidizing agent than reactants involved in anaerobic energy release processes (e.g., in glycolysis). And it is well known that in conditions of a limited number of resources (and it is always not infinite!), the species that survive more efficiently, convert energy into biomass (Travin, 2018). In relation to the evolutionary reversal that occurred, it is appropriate to recall a quote from (Prigozhin & Stengers, 1986).

"Both the deterministic nature of kinetic equations, which make it possible to calculate in advance a set of possible states and determine their relative stability, and random fluctuations that" choose *"one of several possible states near the bifurcation point are closely related. This mixture of necessity and chance constitutes the "history" of the system. "*

It should not be overlooked that it was the powerful thermodynamic capabilities of oxygen as an oxidizing agent that led to the fact that for billions of years, it was not at all in free form, and to this day the amount of oxygen bound in the form of oxides of various elements is orders of magnitude superior to its presence in molecular free form in the atmosphere. It can also be added that it would not be free at all if it were not for serious kinetic obstacles to the reception of electrons by this unique oxidizer.

Kinetic Inertia and Mechanisms of Oxygen Activation

Fortunately, in the main energy state, oxygen has a unique set of molecular orbitals for stable molecules, which is why it is a triplet (the total spin is 1), i.e., in fact, it is a stable biradical. Triplet oxygen molecules contain two unpaired electrons, making triplet oxygen an unusual example of a stable and frequently occurring biradical: it is more stable as a triplet than a singlet. According to the theory of molecular orbitals, the electronic configuration of triplet oxygen consists of two electrons occupying two π -molecular orbitals (MOs) of equal energy (that is, degenerate MOs). In accordance with Hund's rules, they remain unpaired and parallel along the back, which explains the paramagnetism of molecular oxygen. These half-filled orbitals are anti-binding in nature, which means they are reluctant to receive electrons.

The mechanism of singlet-triplet transitions and the limitations imposed by them on the reactivity of molecular oxygen are discussed in detail, in many publications, see, for example, (Kresin, V. Z. and William A. Lester. 1987). Thus, it is a thermodynamically excellent oxidizer, due to quantum mechanical limitations on spin pairing and triplet-singlet evolution, is kinetically inaccessible to "simple" participants in redox transformations and in the vast majority of cases requires enzymatic activation or something very similar to it involving complex transition metal ions (Minaev, 2007). Unless we consider fire situations, where at a temperature increased to thousands of degrees, oxygen does not care which substrate to oxidize.

Values of redox potentials of successive one- and two-electron oxygen reduction transitions to hydrogen peroxide (in various electrolytic dissociation states) have been known for half a century and are given in Table 1. These values are of great importance for many mechanisms of inorganic reactions and even more for biological oxidation processes mechanisms. The role of superoxide radicals in many biological processes is well known, and therefore, since the middle of the last century, numerous attempts have been made to measure or at least calculate or estimate these redox potentials. Now, after many years, it can only be noted that real reactions almost always occur in the internal sphere of the metal complex of the transition metal, therefore, "pure," model values of redox potentials have only an indirect relation to the ongoing oxidation processes.

Table 1. Redox potentials of the $O_2/HO_2/H_2O_2$ systems (Ilan et al, 1976)

No.	Reaction					E⁰(V)
1	O_2	+	e^-	→	O_2^-	-0.33
2	$H^+ + O_2$	+	e^-	→	HO_2	-0.037
3	$H^+ + O_2^-$	+	e^-	→	HO_2^-	1.00
4	$H^+ + HO_2$	+	e^-	→	H_2O_2	-1.42
5	$2H^+ + O_2^-$	+	e^-	→		-1.71
6	$H^+ + O_2$	+	$2e^-$	→	HO_2^-	+0.34
7	$2H^+ + O_2$	+	$2e^-$	→	H_2O_2	+0.69

$pK_{1a} = 4.88$, $pK_{2a} = 11.8$

In accordance with the given values of redox potentials, the following oxygen activation routes are implemented with different probabilities (different speed constants), given in the diagram of Fig. 1.

Figure 1. General scheme of O_2 activation by transient metal ions (complexes)

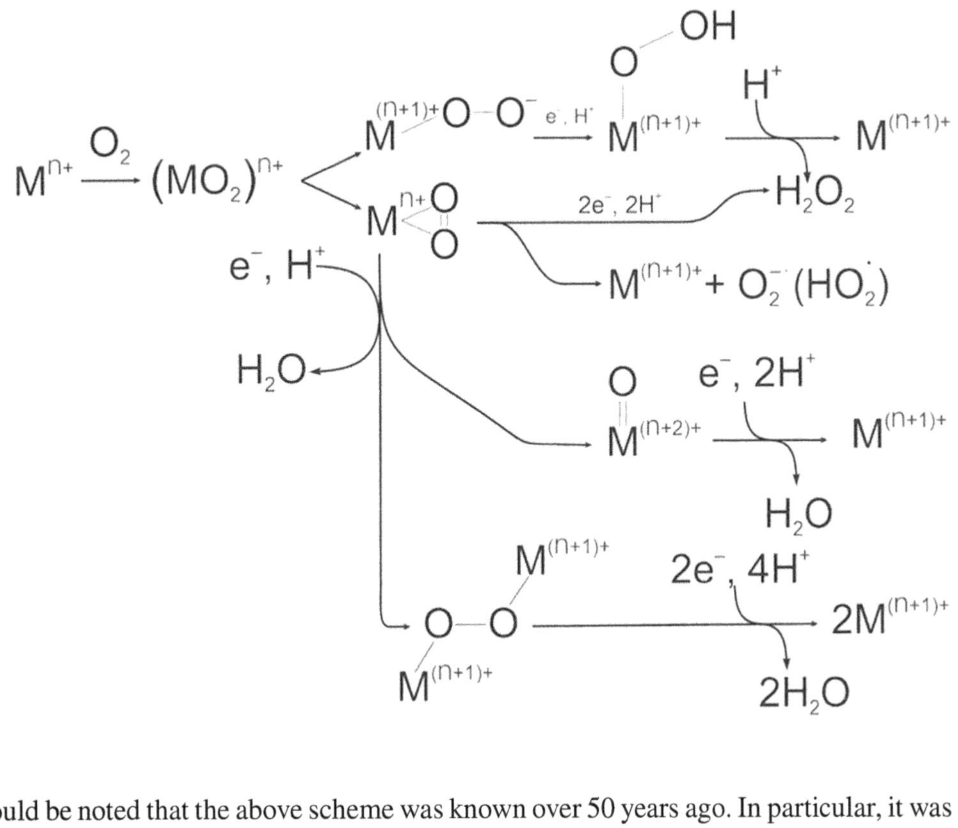

It should be noted that the above scheme was known over 50 years ago. In particular, it was published in several notable references (Veprek-Siska J. 1979, Yee, G.M., Tolman, W.B. (2015). And since then, this scheme has existed almost unchanged as a working model of oxygen activation mechanisms (Sychev et al, 1983). And even regularly published in thematic articles. At the same time, the only important, in our opinion, is an oversimplification. Of the well-known scheme, only one single type of metal-oxygen adduct is considered, although 50 years ago it was known that to acquire an adduct π-type of reactivity of an oxidizing nature, it needs to be converted into a complex with charge transfer of a σ-type (coordinated superoxide radical).

However, the excellent electron-acceptor properties of O_2 have a twofold effect on biological processes. They not only allow biological objects to breathe but also make their presence dangerous for a living cell. Oxidation of almost any cellular compound with molecular oxygen appears to be a thermodynamically favorable process, and if it does not occur, it is only due to high activation barriers for most reactions of this type. On the other hand, these barriers are low for highly reactive electronic respiratory chain carriers such as reduced flavins, non-heme iron proteins, quinols, and especially semiquinones. Unlike the "normal" four-electron reduction of O_2 by cytochrome oxidase (i.e., carried out along a non-radical pathway involving an enzyme), a "parasitic" reaction, for example, reduced flavin and O_2, is a single electron process producing an oxygen anion radical (superoxide, O_2^-). The latter, without having its own serious oxidizing properties, nevertheless initiates reactions that generate H_2O_2 and such a very aggressive oxidizing agent as $OH^·$, which can easily oxidize almost any intracellular substance, including DNA. It is believed that the cell usually protects itself from the harmful side effects of aerobiosis by intercepting and removing precisely the superoxide radical O_2^-, without waiting for the appearance of more dangerous products.

The intermediate position between enzymes containing a metal ion in a protein environment and low-molecular metal complexes is occupied by metalloporphyrins (Skulachev, 1996). The direct environment of the active center is exactly the same as in native enzymes, while the peripheral structure is much simpler and technologically more accessible than in the protein globule. In particular, one of the latest directions is the combination of metalloporphyrins with polyoxomethalates. Polyoxomethallates are cluster anions of an unusual structure. As a rule, they contain a non-metal atom (most often silicon or phosphorus), metal atoms (molybdenum, tungsten, rare earth metals), and oxygen atoms. These anions can have very different bizarre structures.

The work of (Zhu et al, 2016) describes artificially synthesized metalloporphyrins, created according to biomimetic motives, based on active sites of monooxygenase cyctochrome P450. At the same time, it is noted that the set goal to achieve high catalytic activity of artificially created catalysts was not achieved - in fact, the catalytic stability, efficiency, and selectivity of artificial metalloporphyrins is much lower than that of enzymes, because enzymatic catalysis is not only implemented by individual metalloporphyrin centers but also has a serious auxiliary effect of synchronizing environmental centers. Due to this, the systematic modulation of active groups and the microenvironment of active sites necessary to achieve a high catalytic level remains a very difficult topic to follow.

Organic metalloporphyrins and inorganic polyoxometalates (POMs) have been used as active moieties for the synthesis of porous materials to realize highly efficient heterogeneous catalysis (Zhu et al, 2016). The combination of metalloporphyrins and POMs in the same hybrid materials might generate interesting catalytic properties by emerging their unique individual functions. To avoid the random connections between metalloporphyrins, POMs, and lanthanide connecting nodes, a "step-by-step" aggregation strategy was developed. Catalytic experiments demonstrate that the resulting hybrid material exhibits interesting catalytic properties in the heterogeneous epoxidation of olefins, in which the conversion, epoxide selectivity, turnover number, and turnover frequency for the epoxidation of styrene to (1,2-epoxyethyl) benzene are >99%, 94%, 220000, and 22000 h^{-1}, respectively. Although in recent years there has been a tendency to use exotic and/or precious metals in the creation of catalysts that activate molecular oxygen, there remains an interest in traditional transition metals, such as Cu (Zhang et al, 2017).

Commercial zero-valent copper (ZVC) was used as a catalyst for the degradation of acetaminophen by the molecular oxygen O_2 as an oxidant. As expected, it turned out that the involvement of oxygen in the general cycle of transformations occurs exclusively at the stage of its interaction with easily oxidizable ions of monovalent copper. The superoxide radical formed in this stage is not able to oxidize a persistent contaminant, so it does not participate in the reaction with the substrate but is further oxidized to hydrogen peroxide. The "engagement" of the destroyed substrate occurs only under the influence of $OH^{\cdot}$ radicals generated by the interaction of reduced forms of copper with hydrogen peroxide. The H_2O_2, hydroxyl radical $OH^{\cdot}$ and superoxide anion radical $O_2^{\cdot}$ were identified as the main reactive oxygen species (ROSs) generated in the above reaction; however, only $OH^{\cdot}$ caused the decomposition and mineralization of ACT.

In addition, the in-situ generated Cu^+ from ZVC dissolution not only activated O_2 to produce H_2O_2, but also initiated the decomposition of H_2O_2 to generate $OH^{\cdot}$. Meanwhile, the H_2O_2 could also be partly decomposed into $O_2^{\cdot}$, which served as a mediator for copper cycling by reduction of Cu^{2+} to Cu^+ in the ZVC/air system. Therefore, hydroxyl radicals could be continuously generated, and then ACT was effectively degraded.

Conversion from alcohols to the corresponding aldehydes or ketones is one of the most fundamental transformations in organic chemistry. For obvious reasons, many research efforts have been directed at using molecular oxygen as terminal oxidants in recent years. Surprisingly, it turns out that not only the coordination sulfur of transition metal ions can "decelerate" the oxidative abilities of molecular oxygen, but also such an outwardly unremarkable molecule as nitric oxide NO.

Most of the current research is seeking an effective bridge, where transition metals play important roles between molecular oxygen and alcohol substrates. However, Xie and coauthors (Xie et al, 2007) concentrated their interest on molecular nitric oxide (NO) for its ability to activate molecular oxygen, although it widely attracted attention because of its unique biological activity and once was named "the molecule of the year" in 1992 by the editors of the journal *Science*.

It was reported that a triple-component catalytic oxidation system with $NaNO_2$ as an NO equivalent under acidic conditions and O_2 as the terminal oxidant enables the efficient degradation of environmental pollutants, such as 2,4,6- trichlorophenol (TCP). This novel oxidation pathway has been applied to various alcohol substrates and extended to the application for oxidative purification from persistent contaminants. Very recently, the utility of the NO-mediated oxidation mechanism has been applied to other fields including methane oxidation.

An interesting direction in molecular oxygen activation is concerned with neutral (non-ionic) nanoparticles (Woodham et al, 2012 and 2013). Gold nanoparticles and sub-nanoparticles famously act as highly efficient and selective low-temperature oxidation catalysts with molecular oxygen, which look strange in contrast to the nobility of the bulk phase. The origins of this activity and the nature of the active species are still not clear. Gas-phase studies of isolated gold clusters hold promise for disentangling these problems. Here we address the interaction of neutral gold clusters (Au_n: $4 \leq n \leq 21$) with molecular oxygen by probing the highly characteristic O–O vibrational stretch frequencies. This reveals that for selected cluster sizes the oxygen is highly activated with respect to the free moiety. Complementary quantum chemical calculations provide evidence for substantial electron transfer to the O_2 unit and concomitant rearrangement of the parent gold cluster structure upon binding and activation. This gives hint for a model of the interaction between neutral gold clusters and molecular oxygen. Once again, we can express surprise at how the noble metal is ready to part with its electron in favor of an oxidizer such as molecular oxygen, which we know is capable of oxidizing iron or copper, but is considered perfectly safe for gold or platinum.

Compared to the anionic gold clusters, experimentally very little is known about the interactions of neutral gold clusters with molecular oxygen, presumably in part due to the increased experimental difficulty associated with the neutral clusters. Fortunately, anion photoelectron spectroscopy is in part sensitive to the nature of the final state, i.e., the neutral species, albeit in the geometry of the anion. The copper nanoparticles are valuable catalytic precursors for the oxidation with molecular oxygen of a wide range of organic substrates under mild conditions and are largely more efficient than commercial Cu samples (Vitulli et al, 2002).

Within a biomimetic picture, in which 2D metalorganic frameworks are expected to model and reproduce in a tailored way the activity of their biochemical protein counterparts, the fundamental investigation of the adsorption and activation of small ligands at the single-metal atom reaction sites has progressively gained increasing attention. Concerning oxygen, biology offers a variety of tetrapyrrole-based transport and reaction pockets, as e.g., in hemoglobin, myoglobin or cytochrome proteins. Binding and activation of O_2 have been accomplished thanks to complex charge transfer and spin realignment processes, sometimes requiring cooperative mechanisms.

In the very recent paper, we can see the additional demonstration of the ideas, known from the research by Chapsky et al. in the seventies of past century (Ilan, Chapsky and Messel, 1976). Namely, oxygen activation involves the formation of superoxo or peroxo species, undergoing the $2e^-$ or the $4e^-$ reduction processes, depending on the system. Spin inversion is necessary since spin-forbidden transitions contribute to high activation barriers. Most transition metals offer unpaired electrons, both in the ground state or in almost degenerate excited states, allowing reaction with triplet O_2, and spin–orbit coupled configurations, thus providing a quantum mechanical pathway to promote spin inversion. What a happiness that Rudy Marcus did not read this Vesselli`s work (Vesselli, 2021), otherwise there would have been serious objections about $4e^-$ reactions; however, it must be assumed that Vesselli did not read Marcus either.

The development and application of nanomaterials as artificial enzyme mimics have attracted much interest. This is because natural enzymes are usually time-consuming to prepare and purify and is subject to easy denaturation and deactivation. In contrast, nanomaterials are much more stable. They are more robust to high temperatures and other harsh reaction conditions and more suited to industrial applications. It has been found that various nanomaterials have enzyme-like activities. Especially, they can catalyze some redox-type reactions, acting similarly to natural peroxidase, catalase, oxidase, and superoxide dismutation (SOD). According to their chemical compositions, redox-active catalytic nanomaterials are usually categorized into three kinds: carbon-, metal-oxide-, and metal-based nanomaterials.

Compared with the former two kinds, metal nanoparticles (NPs) possess a larger variety of enzyme-mimetic properties. The same metal NPs can mimic multiple enzymes, depending on the reaction conditions. This provides metal NPs with an intriguing environment-responsive ability. As a typical example, Pt NPs can mimic four types of enzymes: peroxidase, catalase, SOD, and oxidase (Shen et al, 2015). Using density functional theory (DFT) calculations, the authors investigate mechanisms of oxidase- and SOD-like properties for metals Au, Ag, Pd, and Pt and alloys $Au_{4-x}M_x$ (x = 1, 2, 3; M = Ag, Pd, Pt). It was found that the simple reaction—dissociation of O_2—supported on metal surfaces can profoundly account for the oxidase-like activities of the metals. The results provide atomistic-level insights into the oxidase- and SOD-like activities of metals and pave the way to the rational design of mimetic enzymes based on metal nanomaterials. Especially, the O_2 dissociative adsorption mechanism will serve as a general way to the activation of molecular oxygen by nano-surfaces and help understand the catalytic role of nanomaterials as pro-oxidants and antioxidants.

Oxygen and Fuel Cells

It is well known that a fuel cell is an electrochemical device, which converts chemical energy into electricity bypassing thermal energy. And the most simple and promising oxidizing agent is definitely molecular oxygen. Fuel cells can be regarded as infinitely rechargeable because they can produce electricity continuously for as long as fuel and oxygen are supplied. Proton-exchange membrane fuel cells (PEMFC) look like the most efficient energy conversion device, and it is expected that in the near future, they could replace traditional internal combustion engines used in automobiles. This technology is clean, and quiet, uses low operating temperatures, and gives energy efficiency and fast startup. For PEMFC operation, oxygen molecules reduced on the cathode play a crucial role in the effective electrochemical processes.

The precious metal Pt has been proven as the most efficient oxygen reduction reaction catalyst with a low reaction activation energy and an enhanced reaction rate of oxygen reduction in an acidic electrolyte. However, there exist obvious drawbacks to Pt catalysts, such as a high cost, low earth abundance, limited stability, and low tolerance toward carbon monoxide and methanol (Liew et al, 2014).

Recently, single-atom catalysts (SACs) have attracted broad research interest as they exhibit high catalytic activity, stability, selectivity, and nearly 100% atom utilization. Fe–N–C electrocatalysts, as a representative of platinum group metal-free (PGM-free) catalysts, exhibit a comparable oxygen reduction reaction activity but insufficient stability to that of commercial Pt/C in proton-exchange membrane fuel cells, due to the unavoidable Fenton's reactions.

(Zhu et al, 2021) reportyв a hard-template approach to synthesize the rare-earth single cerium-atom-doped metal–organic frameworks with a hierarchically macro–meso–microporous structure. Spherical aberration correction electron microscopy confirms the atomic dispersion of Ce sites. Additionally, X-ray absorption spectroscopy (XAS) was employed to further verify the coordination environment of Ce sites, which were stabilized by four-coordinated nitrogen atoms and six-oxygen atoms ($Ce–N_4/O_6$). The Ce sites were embedded in a hierarchically macro– meso–microporous N-doped carbon catalyst, which exhibits a half-wave potential of 0.862 V in oxygen reduction and the highest power density of 0.525 W cm–2 under 2.0 bar H_2/O_2 in the fuel cell test.

The performance of oxygen reduction catalysts (platinum, pyrolyzed iron(II) phthalocyanine (pyr-FePc) and cobalt tetramethoxyphenylporphyrin (pyr-CoTMPP)) is discussed in light of their application in microbial fuel cells. It is demonstrated that the physical and chemical environment in microbial fuel cells severely affects the thermodynamics and the kinetics of electrocatalytic oxygen reduction. The neutral pH in combination with low buffer capacities and low ionic concentrations strongly affects the cathode performance and limits the fuel cell power output. Thus, the limiting current density in Galvanodynamic polarization experiments decreases from 1.5 mA cm-2 to 0.6 mA cm-2 (pH 3.3, Ecathode) 0 V), (Zhao et al, 2006).

CONCLUSION AND IMPLICATIONS FOR FURTHER RESEARCH

In this chapter, we discuss various new perspectives on molecular oxygen activation pathways and their application in some oxidative reactions including degradation of organic pollutants, organic synthesis, and environmentally significant reactions. Applications such as single-atom catalysts and oxygen reduction catalysts are described due to their significance in recent application, such as microbial fuel cells. Furthermore, adaptations here could allow for the very early appearance of oxygenic photosynthesis.

Hence, the hypothesis concerning the occurrence of ROS-utilizing genes/enzymes and at the very least, O_2 tolerance by living organisms cannot be excluded, thus necessitating further studies. This is an ongoing investigation and hence it is likely to lead to many additional investigations.

AUTHOR DISCLOSURE STATEMENT

The authors declare no conflict of interest, financial or otherwise.

REFERENCES

Duca, G. (2012). Homogeneous Catalysis with Metal Complexes: Fundamentals and Applications. Berlin; Heidelberg: Springer.

Duca, G., & Travin, S. (2023). Hydrogen Peroxide, A Substance That Conquered the World. In Duca and Vaseashta (eds.) Handbook of Research on Redox Processes Within Environmental, Technological, and Biomedical Contexts. IGI Publishers.

Ilan Y. A., Czapski G., Meisel D. (1976) The one-electron transfer redox potentials of free radicals. *I. The oxygen/superoxide system //Biochimica et Biophysica Acta (BBA)-Bioenergetics, 430*(2), 209-224.

Kresin, V. Z., & William, A. (1987). On the mechanism of singlet-triplet transitions. *Chemical Physics Letters, 138*(1), 59–64. doi:10.1016/0009-2614(87)80342-1

Liew, K. B., Daud, W. R. W., Ghasemi, M., Leong, J. X., Lim, S. S., & Ismail, M. (2014). Non-Pt catalyst as oxygen reduction reaction in microbial fuel cells: A review. *International Journal of Hydrogen Energy, 39*(10), 4870–4883. doi:10.1016/j.ijhydene.2014.01.062

Minaev, B. F. (2007). Electronic mechanisms of molecular oxygen activation. *Russian Chemical Reviews, 76*(11), 989–1011. doi:10.1070/RC2007v076n11ABEH003720

Prigozhin, I., & Stengers, I. (1986). *Order from Chaos. A new human dialogue with nature*. Progress.

Shen, X., Liu, W., Gao, X., Lu, Z., Wu, X., & Gao, X. (2015). Mechanisms of oxidase and superoxide dismutation-like activities of gold, silver, platinum, and palladium, and their alloys: A general way to the activation of molecular oxygen. *Journal of the American Chemical Society, 137*(50), 15882–15891. doi:10.1021/jacs.5b10346 PMID:26642084

Shen, X., Liu, W., Gao, X., Lu, Z., Wu, X., & Gao, X. (2015). Mechanisms of oxidase and superoxide dismutation-like activities of gold, silver, platinum, and palladium, and their alloys: A general way to the activation of molecular oxygen. *Journal of the American Chemical Society, 137*(50), 15882–15891. doi:10.1021/jacs.5b10346 PMID:26642084

Skulachev, V. P. (1996). Role of uncoupled and non-coupled oxidations in the maintenance of safely low levels of oxygen and its one-electron reductants. *Quarterly Reviews of Biophysics, 29*(2), 169–202. doi:10.1017/S0033583500005795 PMID:8870073

Sychev, A., & Travin, S. Duca Gh, & Skurlatov Yu. (1983). Catalytic reactions and the protection of the environment. Chisinau: Stiinta Publishing.

Travin S. O. (2018) Modern methods of mathematical modeling of chemical and biological processes in the environment. *Chemical safety, 2*(1), 22-49.

Vaseashta, A., Duca, G., & Travin, S. (Eds.). (2022). *Handbook of Research on Water Sciences and Society* (Vols. 1–2). IGI Global. doi:10.4018/978-1-7998-7356-3

Veprek-Siska, J. (1979). The role of metal ions in oxygen activation. *Acta Biologica et Medica Germanica, 38*(2-3), 357–361. PMID:517007

Vesselli, E. (2021). Stabilization and activation of molecular oxygen at biomimetic tetrapyrroles on surfaces: From UHV to near-ambient pressure. *Nanoscale Advances, 3*(5), 1319–1330. doi:10.1039/D0NA00827C PMID:36132852

Vitulli, G., Bernini, M., Bertozzi, S., Pitzalis, E., Salvadori, P., Coluccia, S., & Martra, G. (2002). Nanoscale copper particles derived from solvated Cu atoms in the activation of molecular oxygen. *Chemistry of Materials, 14*(3), 1183–1186. doi:10.1021/cm011199x

Woodham, A. P., Meijer, G., & Fielicke, A. (2012). Activation of molecular oxygen by anionic gold clusters. *Angewandte Chemie International Edition, 51*(18), 4444–4447. doi:10.1002/anie.201108958 PMID:22461317

Woodham, A. P., Meijer, G., & Fielicke, A. (2013). Charge separation promoted activation of molecular oxygen by neutral gold clusters. *Journal of the American Chemical Society, 135*(5), 1727–1730. doi:10.1021/ja312223t PMID:23327337

Xie, Y., Mo, W., Xu, D., Shen, Z., Sun, N., Hu, B., & Hu, X. (2007). Efficient NO equivalent for activation of molecular oxygen and its applications in transition-metal-free catalytic aerobic alcohol oxidation. *The Journal of Organic Chemistry, 72*(11), 4288–4291. doi:10.1021/jo0705824 PMID:17447820

Yee, G. M., & Tolman, W. B. (2015). Transition Metal Complexes and the Activation of Dioxygen. In P. Kroneck & M. Sosa Torres (Eds.), *Sustaining Life on Planet Earth: Metalloenzymes Mastering Dioxygen and Other Chewy Gases. Metal Ions in Life Sciences* (Vol. 15). Springer. doi:10.1007/978-3-319-12415-5_5

Zhang, Y., Fan, J., Yang, B., Huang, W., & Ma, L. (2017). Copper–catalyzed activation of molecular oxygen for oxidative destruction of acetaminophen: The mechanism and superoxide-mediated cycling of copper species. *Chemosphere, 166*, 89–95. doi:10.1016/j.chemosphere.2016.09.066 PMID:27689888

Zhao, F., Harnisch, F., Schröder, U., Scholz, F., Bogdanoff, P., & Herrmann, I. (2006). Challenges and constraints of using oxygen cathodes in microbial fuel cells. *Environmental Science & Technology, 40*(17), 5193–5199. doi:10.1021/es060332p PMID:16999088

Zhu, M., Zhao, C., Liu, X., Wang, X., Zhou, F., Wang, J., Hu, Y., Zhao, Y., Yao, T., Yang, L.-M., & Wu, Y. (2021). Single Atomic Cerium Sites with a High Coordination Number for Efficient Oxygen Reduction in Proton-Exchange Membrane Fuel Cells. *ACS Catalysis, 11*(7), 3923–3929. doi:10.1021/acscatal.0c05503

Zhu, S. L., Xu, X., Ou, S., Zhao, M., He, W. L., & Wu, C. D. (2016). Assembly of a metalloporphyrin–polyoxometalate hybrid material for highly efficient activation of molecular oxygen. *Inorganic Chemistry*, *55*(15), 7295–7300. doi:10.1021/acs.inorgchem.6b00971 PMID:27408952

ENDNOTES

1 https://ugc.berkeley.edu/background-content/oxygen-levels/

2 https://energyeducation.ca/encyclopedia/Atmospheric_oxygen

Chapter 3
Hydrogen Peroxide: A Substance That Conquered the World

Gheorghe Duca
Institute of Chemistry, Moldova State University, Moldova

Serghey Travin
N.N. Semenov Federal Research Center for Chemical Physics, Russian Academy of Sciences, Russia

ABSTRACT

The paper discusses the decomposition mechanisms of hydrogen peroxide (HP) under the action of light, with the involvement of metal ions and their coordination compounds, as well as the participation of hydrogen peroxide in the peroxidase oxidation of various chemical substances. It is shown that these reactions are accompanied by the formation of partial charge transfer complexes and other short-living species as intermediates. Water self-purification capacity following the chemical way is discussed. New indicators for estimating the quality of natural water bodies have been proposed. HP plays an essential role in strengthening the immune system of living organisms, serving in the fight against various diseases, including COVID-19. In this regard, a kinetic model is proposed to evaluate and predict viral infections. We have experimentally demonstrated that HP can find broad applications in agriculture, specifically, in seed treatment, plant growth acceleration, the food industry, etc.

INTRODUCTION

Approximately 2.5 billion years ago, the Earth's atmosphere did not contain free molecular oxygen at all and was purely reductive in nature. But, as proved quite recently, even in the absence of oxygen, which is usually a precursor of hydrogen peroxide, the latter could be formed as a result of tectonic processes of crustal rock fracture during the mechanochemical reactions (Stone et al., 2022). Hydrogen peroxide can be found everywhere, even in outer space in the interstellar medium (Bergman et al., 2011). And it is formed everywhere in the natural aquatic environment (oceans, seas, rivers, lakes, and even swamps) under solar irradiation (Cooper et al., 1988).

DOI: 10.4018/978-1-6684-7198-2.ch003

It was shown experimentally that hydrogen peroxide in the environment is the main oxidizing agent in the processes of chemical self-purification of environmental compartments and the formation of chemical composition appropriate to the biological value of habitation (Duca et al., 1995). Over the past 200 years, oxidative reactions involving H_2O_2 and their mechanisms have been studied using both laboratory and natural systems. Beginning from hydrogen peroxide discovery in 1818 by Jacques Louis Thenard, based on the reaction:

$$BaO_2 + 2HNO_3 = Ba(NO_3)_2 + H_2O_2 \tag{1}$$

Hydrogen peroxide still demonstrates unexpected properties which require additional studies and explanations (Cotton et al., 1994). In this chapter, we attempt to reveal the role of hydrogen peroxide in science, nature, health (biopharmaceuticals), and technology. Comprehensive studies of the important chemical and biological systems involving hydrogen peroxide have been started in "N.N. Semenov" Institute of Chemical Physics, Russian Academy, and Moldova State University.

The history of hydrogen peroxide research as a model for studying redox mechanisms began in the 1950s with the works of Prof. Anatol Purmal, Yurii Skurlatov, Yurii Kozlov, Sergey Travin from the Institute of Chemical Physics of the Russian Academy and Prof. Alexei Sychev, Victor Isac, Jacob Tighineanu, Gheorghe Duca, Maria Gonta, Victor Covaliov, Viorica Gladchi, Lidia Romanciuc from the Moldova State University. These studies are currently being continued in the Institute of Chemistry, research laboratory of Kinetics, Quantum Chemistry, and Physical Research Methods (Duca, 2017, 2007; Duca et al., 2002).

Evolution of the Redox Processes With the Involvement of H_2O_2

When did biologically significant redox processes appear and what is their importance for the evolution of life on Earth? The history of our planet's evolution passed through two main stages: reducing conditions (which are traditionally called anaerobic, although they could be called anoxic - the air at that time consisted of methane, ammonia, and other oxygen-free components) and aerobic oxidizing conditions. During the initial period, the solar energy was stored in the so-called coacervates (or biomimetic protocells - condensed liquid-like droplets formed by liquid–liquid phase separation of molecules through multiple weak associative interactions). They acted as compartments that sequestered and concentrated a wide range of solutes, and their spontaneous formation made coacervates attractive protocell models. These membrane-less components initiated the origin of life on the Earth – this idea was first proposed by Russian biochemist A. Oparin in the 1930ies (Oparin, 1968).

In the second period of life evolution on the Earth, the photosynthetic system appeared, when the anaerobic development mechanisms were replaced with the aerobic ones. Solar energy was preserved in the form of molecular oxygen. Since then, aerobic life on Earth has always dominated anaerobic life.

The reduction of oxygen provides the largest free energy release per electron transfer act, as compared to the reduction of fluorine and chlorine. However, the bonding of O_2 ensures its sufficient stability to be accumulated in a planetary atmosphere, whereas the weakly bonded halogen gases are far more reactive and unable to achieve significant abundance. Consequently, the atmosphere rich in O_2 provides the largest feasible energy source. This universal uniqueness suggests that abundant O_2 is necessary for the high-energy demands of complex life anywhere. On Earth, aerobic metabolism provides about an order of magnitude more energy for a given intake of food than anaerobic metabolism (Catling et al., 2005).

About 5.5 bln years ago the Solar system appeared, and about 4.6 B years – our planet, Terra, or the Earth. In the early periods, acid gases with reducing properties were prevailing in the ocean and atmosphere. These gases appeared as a result of intensive volcanic activity, under the conditions of insufficient free oxygen.

Due to the gases and chemical substances dissolved in water, photochemical reactions of synthesis and decomposition with the formation of organic substances occurred under solar irradiation within the World Ocean. The structure of organic compounds became more and more complicated, thus leading to the development of the first water organisms – anaerobic ones (Holland, 2006). After the appearance of heterotrophic bacteria, the main consumers of organic substances from abiotic syntheses, the evolution further led to the appearance of autotrophic bacteria, which were able to use water as a donor, with oxygen emission into the external environment (Olsson et al., 2017; Van Niel, 1932).

The creation of photosynthetic systems was a new stage in life evolution on Earth when the anaerobic mechanisms of development have been replaced with aerobic ones. Meanwhile, solar energy was preserved in the form of molecular oxygen and hydrocarbons. As a result of photosynthetic microorganisms' activity, the chemical composition of Earth's atmosphere has been restructured. Oxygen gradually oxidized the reducing equivalents in air, leading to the formation of an aerobic atmosphere on the Earth (Catling et al., 2005).

The great oxygen revolution on our planet is thought to have occurred 2.3 billion years ago (Lyons et al., 2014). This very slow-stretched process lasted for hundreds of millions of years. Apparently, an unambiguous evidentiary answer to this question will not be received in the foreseeable future. Hypothetically, it was at that time that hydrogen peroxide appeared in the depths of the planet (Stone et al., 2022). It is considered that within the depths of the planet, there could be a giant tank of oxygen. This discovery may explain the existence of a mysterious geological phenomenon that provokes the water migration into the Earth's depths, due to the activity of tectonic plates under the influence of extreme pressure and temperatures, disintegrated by releasing hydrogen, that leads to the combination of oxygen with iron in the nucleus to create a new ore under the elevated pressure, iron peroxide. Considerable amounts of hydrogen peroxide, which is widely spread on Earth and in the Universe, could appear as well (Slesak et al., 2012). It is also believed that hydrogen peroxide is the basis of water formation on the Earth.

According to Per Bergman, an astronomer at the Onsala Observatory in Sweden, his entire team was delighted to discover the traces of hydrogen peroxide in our galaxy at a distance of 400 light years away from the Earth (Bergman et al., 2011).

On the Earth, hydrogen peroxide was found in the atmosphere (rainwater), hydrosphere (oceans, seas, lakes, rivers, groundwater, etc.), as well as in living organisms. H_2O_2 was detected in the glaciers of Antarctica and is an important component of atmospheric precipitates (Sigg and Neftel, 1988). H_2O_2 content depends on environmental pollution. This discovery is of high importance because, following the reaction between hydrogen peroxide with hydrogen abundant in interstellar space, water can be obtained. Based on this, an explanation could be proposed on how water was formed on the Earth, especially since previously it was believed that it was initially formed in outer space.

STRUCTURE OF H_2O_2

Hydrogen peroxide is a reactive oxygen species and the simplest peroxide, a compound having an oxygen–oxygen single bond. It can decompose slowly when exposed to light and rapidly decomposes in the presence of organic or reactive compounds. The rotation of the two OH hydroxyl groups attached to the O-O bond is only slightly limited by the low potential barrier (1 kcal • mol^{-1}) in trans-configuration, the barrier for the cis-configuration being much higher (Hydrogen Peroxide, 2006), Fig.1.

Figure 1. Structure and dimension of H_2O_2 molecule in the gas and solid phases

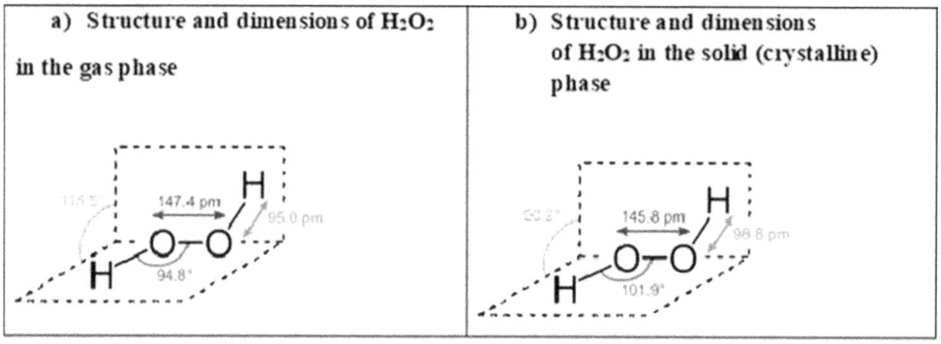

Hydrogen peroxide has a nonplanar molecule with twisted C_2 symmetry that was first shown by Paul-Antoine Giguère in 1950 using infrared spectroscopy (Giguere, 1950). It has the single O–O bond, its molecule has a relatively high rotational barrier of 386 cm−1 (4.62 kJ/mol) for the rotation between the enantiomers via *trans*-configuration, and 2460 cm^{-1} (29.4 kJ/mol) via *cis*-configuration (Hunt et al., 1965). Such "open book" configuration is the simplest example of H_2O_2 structure, demonstrating the ways of the internal rotation or bond torsion (first demonstrated by quantum mechanical calculations by Penney and Sutherland, later confirmed by IR spectroscopy) (Penney and Sutherland, 1934).

The molecular structures of gaseous and crystalline H_2O_2 are significantly different. This difference is attributed to the effects of hydrogen bonding, which is absent in the gaseous state (Dougherty & Anslyn, 2005). Crystals of H_2O_2 are tetragonal with the space group D^4_4 or $P4_12_12$ (Abrahams et al., 1951). In aqueous solutions, hydrogen peroxide differs from pure substances due to the effects of hydrogen bonding between water and hydrogen peroxide molecules. Thus, hydrogen peroxide molecules in liquid form can be arranged in groups, forming clusters (Tachikawa, 2021; Yu & Yang, 2011). The structure and configuration of H_2O_2 molecule were the research subjects for the team from Moldova, led by Dr. Natalia Gorinchoy. She revealed for the first time the reasons behind the low-symmetry linear configuration of this molecule that could be explained based on the pseudo effect of Jahn-Teller (Arsene & Gorinchoy, 2019; Arsene, 2008). Schematic illustration to the covalence origin of the vibronic instability of the linear configuration of H_2O_2 is presented on Fig.2.

Figure 2. Schematic illustration to the covalence origin of the vibronic instability of the linear configuration of H_2O_2

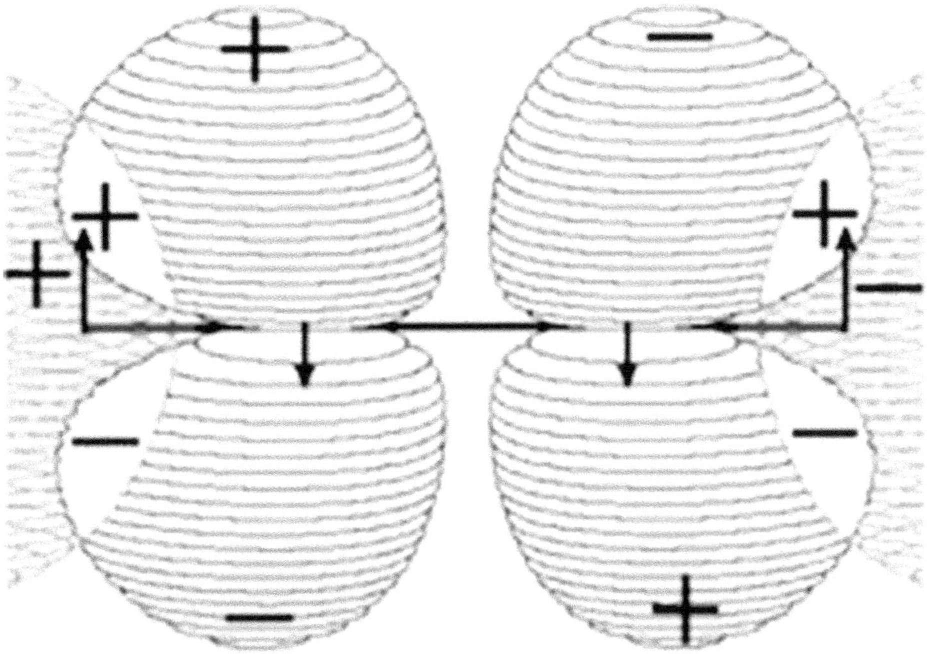

Quantum-chemical calculations have demonstrated that the O-O bond (138 kJ) breakage in hydrogen peroxide molecule is difficult from the thermodynamic viewpoint, and the relevant activation energy is high (Kulkarni et al., 2006). So, to break the hydrogen peroxide molecule, it needs to be activated, just like in the case of molecular oxygen (498 kJ).

$$2H_2O_2 \longrightarrow 2H_2O + O_2 \tag{2}$$

Molecular structure of hydrogen peroxide and its specifics have also been studied in the US, Germany, and other countries (Serra-Maia et al., 2018; Plauck et al., 2016).

HYDROGEN PEROXIDE FORMATION

Schematically, the series of possible electron transfer reactions with the involvement of H_2O_2 (Plauck 2016) could be presented as a four-step scheme, Fig. 3.

Figure 3. A series of electron transfer reactions with participation of H_2O_2. The numbers reflect the free energy gained/lost for electron transfer, eV.

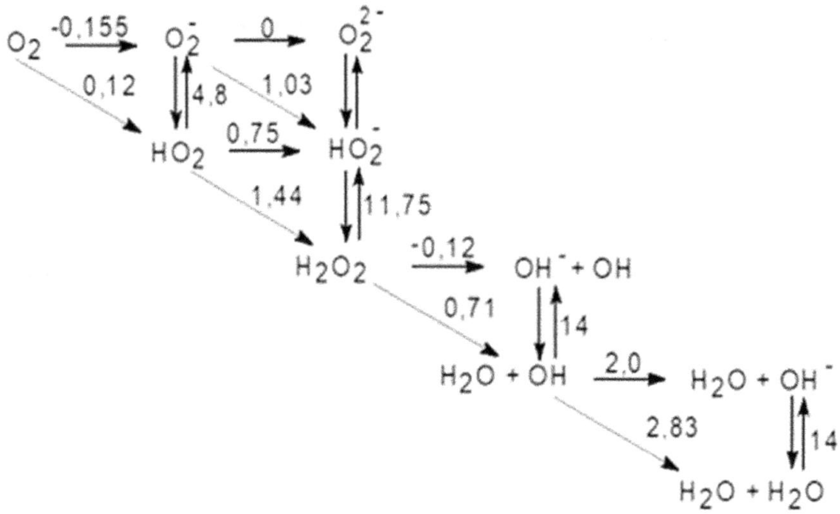

It can be seen that the oxygen molecule, undergoing the series of transformation processes accompanied with the related energy changes, with intermediate generation and decomposition of hydrogen peroxide, makes the basis of water formation (Scholes and Weiss, 1959).

The overall scheme of O_2 and H_2O_2 reduction through two-electron inner-sphere transfer, leading to the formation of water molecule, is presented on Fig.4 (Duca, 2021; Ramaswamy & Mukerjee, 2011):

Figure 4. Two-electron reduction of O_2 and H_2O_2. The numbers reflect the free energy gained/lost for electron transfer, eV.

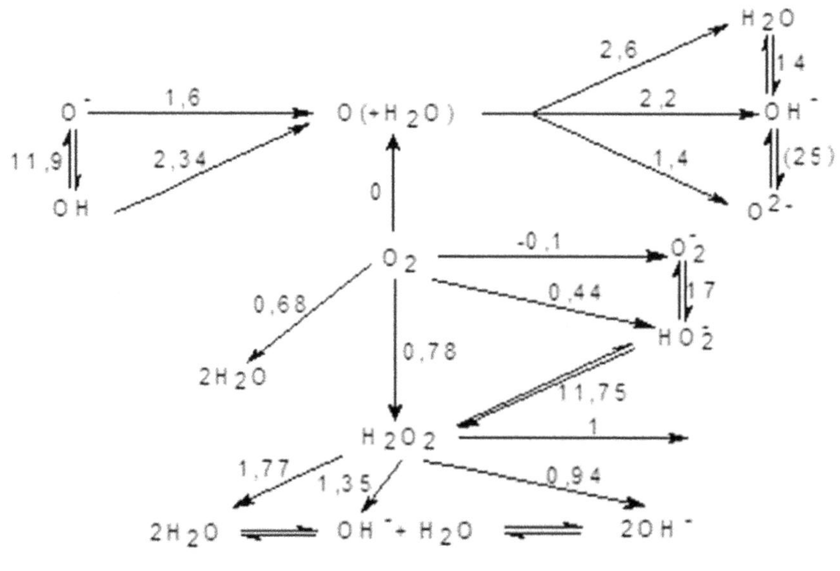

The redox reactions with oxygen and hydrogen peroxide involvement with one or two-electron transfer are typical for the redox-processes, occurring in the natural water environment. These two mechanisms determine the basic principles of hydrogen peroxide formation in the environment as the most profitable substance from the energetic point of view.

H_2O_2 Formation in the Atmosphere

Photochemical formation of hydrogen peroxide in the atmosphere as a result of CH_4 oxidation, with photochemical initiation step, can be presented as a following sequence of reactions (Fujii & Otani, 2017):

$$O_3 + h\nu \longrightarrow O_2 + O^* (\text{la } \lambda < 310 \text{ nm}) \quad (3)$$

$$O^* + H_2O \longrightarrow 2OH^{\cdot} \quad (4)$$

$$O^* + CH_4 \longrightarrow OH^{\cdot} + CH_3^{\cdot} \quad (5)$$

$$OH^{\cdot} + CH_4 \longrightarrow H_2O + CH_3^{\cdot} \quad (6)$$

$$CH_4 + O^* \longrightarrow CH_3O^{\cdot} + H^{\cdot} \quad (7)$$

$$CH_3O^{\cdot} + O_2 \longrightarrow CH_2O + HO_2^{\cdot} \quad (8)$$

$$HO_2^{\cdot} + HO_2^{\cdot} \longrightarrow H_2O_2 + O_2 \quad (9)$$

H_2O_2 in the Hydrosphere

In 1966 the American scientists Van Baalen and Marler had discovered H_2O_2 in the Gulf of Mexico (Baalen & Maarler, 1966). Approximately the same time Russian scientist Sinel`nikov had detected HP in the fresh waters (Sinel`nikov, 1971, 1968). In 1983 hydrogen peroxide was detected in the Volga river by Yu.Skurlatov and others (Shtamm et al., 1991). In 1986 the Moldavian scientists L.Romanciuc, V.Gladchi, Gh.Duca have identified this substance in the rivers Nistru, Prut and lake Beleu in Moldova (Gladchi et al., 2013, Romanciuc et al., 2009).

The main sources of H_2O_2 in the hydrosphere are considered atmospheric precipitation, photochemical processes, and metabolites of aquatic organisms. Measurements performed by US scientists in the Atlantic and Pacific oceans showed that after rainfalls, the H_2O_2 content on the ocean surface increases sharply (10-100 times), but within a few hours it decreases to a stationary level, conditioned by the processes within aquatic basins (Cooper et al., 1988). This level undergoes oscillations, which demonstrates the role of solar radiation in the formation of H_2O_2.

It was shown that the content of hydrogen peroxide in natural waters had seasonal character (Price et al., 1994). It was increased in natural water during the daytime, due to the photochemical processes and more intensive solar irradiation and was decreased in night or when there were higher amounts of the reducers (Miller et al., 2005; Olszyna et al., 1988).

The field studies showed the occurrence of the redox purification processes, based on the content of hydrogen peroxide. In the studied natural waters, the H_2O_2 was absent, or its content was insufficient. Its traces were detected in physiological concentrations available for living and developing organisms. It was confirmed that high pollution degree with chemical substances can interrupt the self-purification chain with OH free radicals, which is lower in the waters of Dniester River than in the lakes (Gladchi et al., 2008).

The ways hydrogen peroxide occurrence and disappearance are summarized on Table 1 (Duca & Vaseashta, 2020; Sychev et al., 1983).

Table 1. Pathways of H_2O_2 formation and decomposition in the natural environment

FORMATION:	**DECOMPOZITION**
➢ *photochemical processes*:	➢ *biotic*
SODC $\longrightarrow O_2^-. \longrightarrow H_2O_2 + O_2$	$H_2O_2 + H_2O_2 \rightarrow O_2 + 2H_2O$
➢ *redox-catalytic processes*:	$H_2O_2 + DH_2 \rightarrow D + 2H_2O$
a) $O_2 + DH_2 \longrightarrow D + H_2O_2$	➢ *abiotic*
b) $2Cu^+ + O_2 + 2H^+ \longrightarrow 2Cu^{2+} + H_2O_2$	$H_2O_2 \longrightarrow 2OH\bullet$
➢ *biologic emissions*:	H2O2 as oxidant:
$R\text{–}CHNH_2COOH + H_2O + O_2 \longrightarrow$	$Me^+ + H_2O_2 \longrightarrow Me(H_2O_2)^+ \longrightarrow MeOH^+ + OH^\bullet$
$\longrightarrow R\text{–}CHOCOO^- + NH_4^+ + H_2O_2$	H2O2 as reducer:
➢ *atmospheric precipitations*	$Me^{2+} + H_2O_2 \longrightarrow Me(HO_2)^+ \longrightarrow Me^+ + HO_2^{-\bullet}$

Based on the experimental data it was concluded that the highest amounts of hydrogen peroxide in natural waters is resulted from photochemical processes, under the solar light action on the dissolved organic substances in the presence of transition metal ions and complexes.

ROLE OF HYDROGEN PEROXIDE IN THE REDOX REACTIONS

According to "Britannica" (Olson, 2022), "redox processes" are defined as reactions accompanied by the changes in oxidation state: an increase in the atom's oxidation number corresponds to oxidation; a decrease, to the reduction. In this generalized theory, there can be three pathways of the changes in oxidation state: oxygen-atom transfer (gain, oxidation; loss, reduction), hydrogen-atom transfer (loss, oxidation; gain, reduction), and electron transfer (loss, oxidation; gain, reduction). Most oxidation-reduction (redox) processes involve the transfer of oxygen atoms, hydrogen atoms, or electrons, all three processes sharing two important characteristics: (1) they are coupled - i.e., in any oxidation reaction a reciprocal reduction occurs, and (2) they involve a characteristic net chemical change - i.e., an atom or electron goes from one unit of matter to another.

Redox-Reaction With Formation of Partial Charge Transfer Complexes

It is already confirmed that the majority of redox processes occur via the intermediate formation of complexes with partial charge transfer (PCTC) (Skurlatov et al., 1989). They can be formed during the interaction between the metal ions and redox reactants, including O_2 and H_2O_2. Therefore, PCTC are intermediate products of inner-sphere electron transfer reaction.

Partial charge transfer complexes are molecular and coordination complexes of the redox partners, with intermediate charge transfer degree, which are transformed in a polar medium into the products of resulting electron transfer under thermal excitation. They represent a distinct class of chemical compounds with the donor-acceptor bond: electron-donor–electron-acceptor complex, characterized by the electronic transition to an excited state, with the partial transfer of electronic charge from donor to acceptor. The degree of charge transfer in complexes increases with the polarity of medium (Orabi, 2021). PCTC occurs due to the asymmetric distribution of electrons in chemical bonds (Karamoschos & Tasis, 2022). A partial charge is more commonly called net atomic charge. It is represented by the Greek lowercase letter δ, namely δ− or δ+.

According to R.J Marcus (Nobel Prize Winner) theory, the rate of charge transfer is linked to the factor *A* which depends on the nature of charge transfer (unimolecular or bimolecular) and Gibbs free activation energy (Marcus, 1964):

$$kCT = Ae^{-\Delta G * kBT} \tag{10}$$

The PCTC formation energy curve is presented on Fig.5.

Figure 5. Energy curve of partial charge transfer complex formation

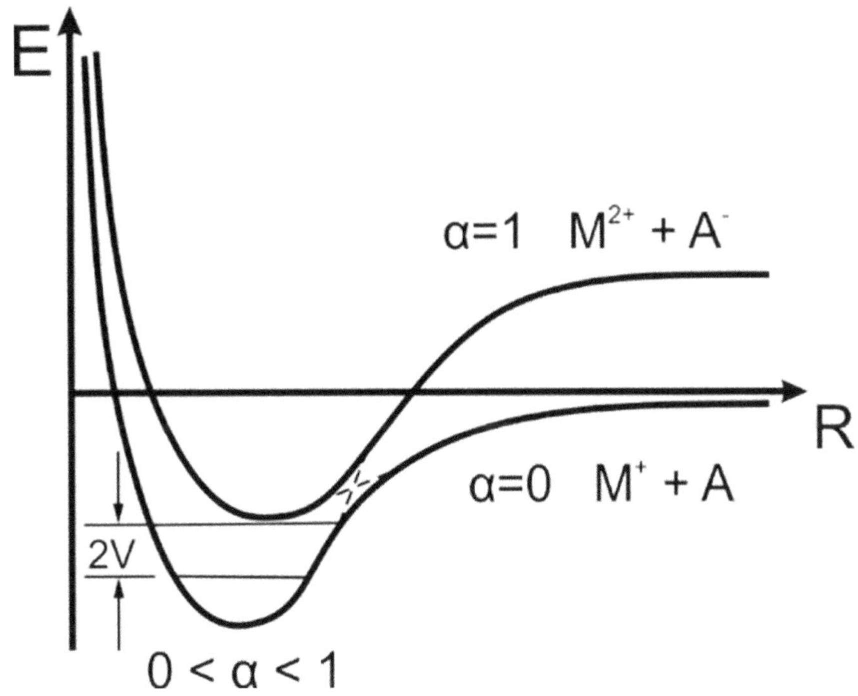

The width of energy gap between the main and excited terms in charge transfer complexes is about 2 Volts. The transition from the initial state to the finite one occurs not in the minimum of the potential pit but in the intersection zone of the terms. Thermodynamic equilibrium is established between the primary complex and complex in the exited state, with partial transfer of one or two electrons from donor to acceptor.

Specific Properties of Partial Charge Transfer Complexes

Based on the theory of partial charge transfer (Skurlatov et al., 1989), thermodynamic equilibrium is established between the primary complex and the complex containing electrons in the excited state, being the result of the partial internal transfer of one or two (unpaired) electrons from donor to acceptor under the room temperature. Due to this phenomenon, PCTC can exhibit the properties of both the initial chemical products and the final charge transfer products, thus demonstrating their specific reactivity. Partial charge transfer was shown to be responsible for activation of participating reactants in the complex. For example, coordination of ascorbic acid with metal improves its reducing capacity (Grinstead, 1960). Similarly, coordination of oxygen with the reducing agent (Cu^{+}) sharply increases its oxidizing capacity (Hongshin et al., 2016).

The formation of partial charge transfer complexes (Bhunia et al., 2020) can proceed by the outer-sphere way as well, which occurs between two species that do not undergo substitution and do not involve the incursion of significant covalent bond formation. It occurs when none of the ligands can function as a bridge. It is faster than the inner sphere pathway because the energetic demands are less. No new bonds are broken or formed. Interaction between two coordination spheres exists but is not as pronounced as for the bridge complex in the inner sphere.

Electron transfer from ligand to metal provokes the ligands' vibrational excitation to bring the molecule in a vibrationally excited state. The most important feature of intermediate complex in the inner-sphere electron transfer reactions is rapidly established thermal equilibrium between the ground and electron-exited states. Mechanisms of exited levels stabilization in charge transfer complexes include the formation of hydrogen or another bridging bond.

Importance and Distribution of Charge Transfer Complexes

Charge transfer processes occur in numerous chemical, biological and material systems. Oxidation and reduction reactions, photoinduced charge transfer in metal complexes and charge injection in solar cells are examples of charge transfer processes. During the formation of partial charge transfer complexes, the system evolves from an initial state "charge on the donor" to a final state "charge on the acceptor," along a reaction coordinate, reaching energetic stabilization. A common feature of partial charge transfer complex formation reactions is activation of substrate or oxidizing agent during coordination with metal ion.

Partial charge transfer complexes are applied for solar energy conversion and photocatalysis. Complexes with partial charge transfer play a significant role in homogeneous catalysis, photochemical processes, solar batteries, wastewater treatment, production of wines, juices, development of chemical-biological processes, etc. Partial charge transfer complexes are formed even in the cells of living organisms (Duca, 2012; Skurlatov et al., 1989).

Redox-Reactions Including Superoxidic or Peroxidic-Type Oxygenated Complexes

Partial mono- or bi-electronic charge transfer in oxygenated complexes of superoxidic or peroxidic type can be represented as superpositions of boundary structures without transfer, or with full charge transfer (Krumova & Cosa, 2016; Skurlatov et al., 1984):

O O^{-} / O^{-} H^{+}

$(MO^{+}_{2})_{I} = (M^{+} \parallel \leftrightarrow M^{3+}(\mid \leftrightarrow M^{3+}) \longrightarrow M^{3+} + HO^{-}_{2}$ (11)

O O^{-} \O

H^{+}

$(MO^{+}_{2})_{II} = (M^{+}...O{=}O \leftrightarrow M^{2+}\text{-}O\text{-}O^{-}) \longrightarrow M^{+} + O^{-}_{2}\ (HO_{2})$ (12)

Binuclear oxygenated complexes can also be represented as superpositions of non-transfer and fully charge-transfer structures:

H^{+}

$(MOOM)^{2+} = (M^{+}...O{=}O...M^{+} \leftrightarrow M^{2+}\text{-}\ O^{-}\text{-}\ O^{-}\text{-}M^{2+}) \longrightarrow 2M^{2+} + HO^{-}_{2}$ (13)

Intrasphere Electron Transfer From M^{+} to H_2O_2

Upon interaction of metal ions with the hydrogen peroxide molecule, coordination of hydrogen peroxide occurs, following the monodentate mechanism, when only one vacancy is available in the coordination sphere of metal, and bidentate mechanism, when two sites are available. Depending on the metal valence, one or two electrons transfer can occur during the complex formation. The change in charge distribution within the complex with partial charge transfer, as well as changes in the electrical structure of the interacting molecules, can provoke significant variations in the electric field in each atom of the complex.

In case of the inter-transfer interaction of H_2O_2 with M^{+} ions that have potentially vacant places in the coordination sphere, it is possible, as in the case of O_2, to coordinate the H_2O_2 molecule in two ways (Enami et al., 2013):

- monodentate, when only one vacancy is present,
- bidentate – when two vacancies are present.

In both cases, depending on the redox characteristics of the metal, it is possible to achieve the mono- and bi-electronic transfer. The most favorable conditions for this are achieved when the degree of charge transfer (+) in the complex is close to 0.5, as in the case of the O_2 covalent bond (Mandal et al., 2020; Duca, 2012). These conditions are met at certain values of the redox potential of the metal ion:

- in case of mono- and binuclear complexes of the peroxide type:

$$E_0(M^{3+}/M^+) \approx E_0(O_2/HO^-_2) = 0{,}44 \text{ eV}; \quad (14)$$

- in case of the superoxide type complexes:

$$E_0(M^{2+}/M^+) \approx E_0(O_2/HO_2) = 0{,}12 \text{ eV}. \quad (15)$$

Intra-Transfer Multielectron Redox Transformations

Peroxide type complexes (MO_2^+) can interact with the particles that show reactivity similar to that of M^{3+} (bi-electronic acceptor) and even atomic oxygen (hydroxyl agent). As a result, bi-electronic oxidation, hydroxylation or epoxidation of the substrates occurs. A typical example is the hydroxylation system:

$$Fe^{3+} - EDTA - DH_2 - O_2 - \text{substrate} \quad (16)$$

Here, the DH_2 donor (e.g., ascorbic acid) reduces the iron ion (coordination number 8). As a result, the interaction of Fe^{2+}EDTA with O_2 results in a bi-electronic charge transfer complex formation inside the sphere where it is possible to break the O-O bond and form (partially) the structure of atomic oxygen (Duca, 2012; Gambardella et al., 2005).

Monoelectronic Inter-Transfer

It was shown that in many environmental and technological systems partial charge transfer complexes can initiate radical processes, affecting the quality of products or the environment. Complexes with partial charge transfer, being intermediate products, are decomposed during the redox processes, causing the formation of free radicals, among which the OH radical is one of the most important. The subsequent reaction pathways are determined by the redox potential of the reaction system.

Thus, in case of the monoelectronic charge transfer, the metal peroxide complex decomposes with the formation of the OH radical (Sridharan et al., 1980):

$$M^+ + H_2O_2 \rightarrow (MH_2O_2)^+ = (M^+...OH\text{-}OH^- \leftrightarrow M^{2+}\text{-}OH^-...OH) \rightarrow M^{2+}OH^- + OH \quad (17)$$

In the absence of interception of intermediate particles, such as MO^+, $M(OH)^{2+}$, MOH^{2+}, $(MH_2O_2)^+$ in a protonic environment, the interaction of M^+ with H_2O_2 is accompanied by the formation of M^{3+} or OH:

$$(2H^+) \rightarrow M^{3+}$$

$$M^+ + H_2O_2 \langle \quad (18)$$

$(H^+) \rightarrow M^{2+} + OH$

The direction of the overall process is determined by the value of the redox potential of the pairs M^{3+}/M^+, M^{2+}/M^+ which is equal, accordingly, to: $E_0(H_2O_2/2H_2O)$+1.77 eV and $E_0(H_2O_2/OH+H_2O)$+0.71 eV (Duca, 2012).

Bielectronic Innersphere Transfer

A complex with partial charge transfer with the metal or atomic oxygen with the highest degree of oxidation can be formed in case of innersphere bi-electron transfer. Another type of intermediate particles formed in such processes are MO^+ which further react with active species in acidic medium or are hydrolyzed in neutral medium (Zhu et al., 2008).

In the case of innersphere b-ielectron transfer, the intermediate complex is formed in which the maximum oxidation state of the metal ion or atomic oxygen is manifested:

OH^-

$$M^+ + H_2O_2 \longrightarrow M(OH)_2^+ = [M^+ \mid \leftrightarrow M^{3+}\langle\ ] \tag{19}$$

OH^-

$$M^+ + H_2O_2 \longrightarrow MO^+ \ldots H_2O = (M^+ \ldots O \leftrightarrow M^{3+}\text{–}O^{2-})\ H_2O \tag{20}$$

In an acidic environment, MO^+ and $M(OH)^{2+}$ particles interact with H+ with the formation of other reactive particles: the complex MOH^{2+} + ($M^{2+} \ldots OH$ + $M^{3+}OH^-$) and the metal ion in the maximum oxidation state M^{3+}.

In a neutral environment, MO+ and $M(OH)^{2+}$ particles are hydrolyzed (because they are in the M^{3+} state) with the formation of hydroxoderivatives. Hydrolysis of these particles can be accompanied by water oxidation with OH radical.

Reactions of MO^+ Type Particles

The other possible reactions with the participation of MO^+ particles involve oxidation, hydroxylation and epoxidation. MO^+ type particles can participate in the bielectronic transfer oxidation of H donors (Duca, 2012; Duca et al., 2002):

$$MO^+ + DH_2 \rightarrow M^+ + D + H_2O \tag{21}$$

Also, these particles can participate in hydroxylation reactions:

$$MO^+ + RH \rightarrow M^+ + ROH \tag{22}$$

and epoxidation:

O

/\

$$MO^+ + >C = C< \rightarrow M^+ + >C - C< \quad (23)$$

HYDROGEN PEROXIDE IN CATALYTIC REACTIONS

Catalytic reactions with participation of metal ions, oxygen, hydrogen peroxide, accompanied with the formation of charge transfer complexes as intermediates, play the key role in vitally important systems. Moldovan researchers have studied the mechanisms of numerous redox reactions in the systems involving the metal ions, substrates, oxygen, and hydrogen peroxide, with intermediate formation of PCTC (Shvydkiy et al., 2022; Duca & Travin, 2020; Duca, 2012) It has been shown that such reactions take place in natural water, wines, juices and other complex systems, determining the formation of their quality (Fig.6).

Figure 6. Series of reactions with the involvement of charge transfer complexes and metal complexes as catalysts
(Duca et al., 1997).

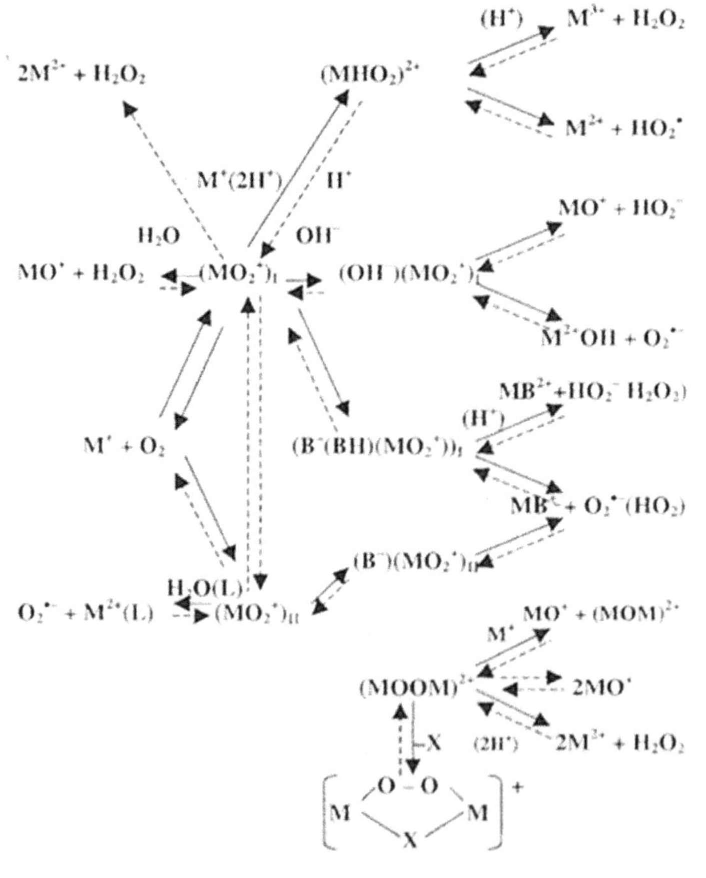

The Rates of H_2O_2 Formation and Decomposition in Natural Waters

The rate of H_2O_2 formation in natural waters is a function of the biotic and abiotic contribution (Gladchi et al., 2008; Shtamm et al., 1991; Sinel`nikov, 1971) and can be represented as follows:

$$W_f = W_{f,A} + K_{fB}B \tag{24}$$

where: $W_{f,A}$ - rate of abiotic H_2O_2 formation reactions; B - microalgae biomass; K_{fB} - effective rate constant of H_2O_2 formation in the biotic process.

H_2O_2 decomposition rate depends on the several factors and primarily on H_2O_2 concentration, presence of microalgae and intensity of solar radiation (Duca, 2021). The rate of H_2O_2 decomposition in a biological reaction is proportionate to algae biomass contents:

$$w_d = (\alpha_{d,1} + \alpha_{d,2}[H_2O_2])B \tag{25}$$

and

$$\alpha_{d,1}B = w_r = \alpha_r B \tag{26}$$

where: w_r – rate of reducing equivalents formation, as consumers of H_2O_2 in dehydrogenation reactions; $\alpha_{d,1}$, $\alpha_{d,2}$, α_r – effective rate constants of appropriate processes.

The Fenton System

The first process studied with the involvement of hydrogen peroxide was the Fenton reaction, with H_2O_2 as an oxidizing agent, and iron (Fe^{2+}) as a catalyst and reducing agent. This system was studied in detail by us (Shvydkiy et al., 2022; Duca, 2021, 2012; Romanciuc et al., 2009).

The Fenton's system is broadly used to destroy many organic substances in nature and technology (Tian et al., 2022; Skurlatov et al., 1989):

$$Fe^{2+} + H_2O_2 \rightarrow Fe^{3+} + OH^{\cdot} + OH^{-} \tag{27}$$

$$Fe^{2+} + OH^{\cdot} \rightarrow Fe^{3+} + OH^{-} \tag{28}$$

$$OH^{\cdot} + H_2O_2 \rightarrow HO^{2\cdot} + H_2O \tag{29}$$

$$Fe^{3+} + H\dot{O}_2 \rightarrow Fe^{2+} + H^{+} + O_2 \tag{30}$$

$$Fe^{3+} + O_2 \rightarrow Fe^{2+} + O_2 \tag{31}$$

$$Fe^{2+} + HO^{2\cdot} \rightarrow Fe^{3+} + HO^{2-} \quad (32)$$

$$HO^{2\cdot}(O^{2\cdot -}){+}H_2O_2{\rightarrow}H_2O{+}O_2 {+}OH \quad (33)$$

$$2Fe^{2+} + H_2O_2 + 2H+ \rightarrow 2Fe^{3+} + 2H_2O \quad (34)$$

Catalytic Redox Process in Natural Water

Catalytic redox processes in natural waters can occur through several mechanisms, such as: cyclic mechanism, activated mechanism or radical-chain mechanism (Shvydkiy et al., 2022; Tsuneda & Taketsugu, 2018; Duca, 2012; Sychev et al., 1983). Appropriate schemes of various mechanisms are given below on the Table 2 (Serra-Maia et al., 2018).

Table 2. Mechanisms of catalytic redox processes with the involvement of hydrogen peroxide

<table>
<tr>
<td>Cyclic mechanism
$– Fe(III) – O – Fe(III) + DH_2 \longrightarrow – Fe(II) \cdots Fe(II) – + D + H_2O$
$– Fe(II) \cdots Fe(II) – + O_2 (H_2O_2) \longrightarrow – Fe(III) – O – Fe(III) – + H_2O_2 (H_2O)$</td>
<td>Activated mechanism
$Cu^+ + O_2 \longrightarrow (CuO_2)^+$
$Cu^+ + H_2O_2 \longrightarrow (CuH_2O_2)^+ \longrightarrow (CuO)^+ + H_2O$
$CuO_2^+ + DH_2 \longrightarrow Cu^+ + D + H_2O_2$
$CuO^+ + DH_2 \longrightarrow Cu^+ + D + H_2O$</td>
</tr>
<tr>
<td>Radical-chain mechanism
a) The reducer doesn't form a complex compound with the metal
Chain initiation
$DH_2 \longrightarrow DH^- + H^+$
$CuOH^+ + DH^- \longrightarrow Cu^+ + D^{-\cdot} (R) + H_2O$
Chain propagation
$Cu^+ + H_2O_2 \longrightarrow CuO^+ + H_2O$
$Cu^+ + O_2 \longrightarrow CuO_2^+$
$CuO^+ + DH^- \longrightarrow Cu^+ + D + OH^-$
$CuO_2^+ + DH^- \longrightarrow Cu^+ + D + HO_2^-$
$R + O_2 \longrightarrow RO_2 \longrightarrow P + O_2^{-\cdot}$
$Cu^{2+} + O_2^{-\cdot} \longrightarrow Cu^+ + O_2$
$Cu^+ + H_2O_2 \longrightarrow CuO^+ + H_2O \longrightarrow CuOH^+ + OH^\bullet$
$DH^- + OH^\bullet \longrightarrow P + H_2O$
Chain termination
$CuO^+ + Cu^+ \longrightarrow 2Cu^{2+}$
$CuO_2 + Cu^+ \longrightarrow 2Cu^{2+}$</td>
<td>Radical-chain mechanism
b) The reducer forms a complex compound with the metal
Chain initiation
$Cu^{2+} + DH^- \longrightarrow CuDH^+$
$CuDH^+ + H_2O_2 \longrightarrow Cu^{2+} + D^{-\cdot} (R) + OH^\bullet$
Chain propagation
$OH^\bullet + DH^- \longrightarrow D^{-\cdot} (R) + H_2O_2$
$R + O_2 \longrightarrow RO_2$
$RO_2 + CuDH^+ \longrightarrow Cu^{2+} + OH^\bullet + D$
Chain termination
$CuO^+ + Cu^+ \longrightarrow 2Cu^{2+}$
$CuO_2 + Cu^+ \longrightarrow 2Cu^{2+}$</td>
</tr>
</table>

Based on the redox processes studied, we concluded that the redox catalytic processes in natural waters can be schematically represented as a chemical reactor. The redox substances penetrate in such micro-reactor from the atmosphere, water, and bottom deposits, which serve the sources of oxidative or reductive species. In dependence on these species, water can be either in the reductive, or in the oxidative state, which can affect the development of living organisms in it. These states can be assessed using the formula (Duca, 2012):

$$rH_2 = \frac{0.058pH + Eh}{0.029} \qquad (35)$$

If the value of the molecular hydrogen pressure $rH_2 > 28$, water is considered to be in the oxidant state; in case of $rH_2 < 28$ – there is a reducing state of water environment.

H_2O_2 and the Kinetic Redox State of Waters

As stated above, the redox processes in water depend on the presence of oxidants, free radicals, solar light and reducing substances, which determine its reduction or oxidation character. In natural waters permanently occur the biotic and abiotic processes of formation and transformation of oxidizing equivalents and reducers, as well as penetration of substances outside the water system (Duca et al., 2020; Duca, 2012; Bahadir & Duca, 2009).

As a result of interactions between these components, the redox state is determined by the ratio between the flow of oxidizing equivalents – hydrogen peroxide (W_{ox}) and flow of reducers (DH_2) which interact with it (W_{red}).

The redox state of natural waters can be considered as normal (oxidant) only when H_2O_2 is present, along with the dissolved oxygen, in physiological concentrations determined as favorable for hydrobionts: $0.3\ \mu g/l < H_2O_2 \leq 100\ \mu g/l$.

For normal regime of oxygen, in dependence on H_2O_2, reducers and strong oxidants concentrations, the following redox dynamic states can be established in natural water:

- *oxidant* - when H_2O_2 is present.
- *reducing* (Red_{ef}) - when H_2O_2 is absent.
- *superoxidant* (Ox_{ef}) - products and semi-products of technological cycles are detected (Cr(VI), $KMnO_4$, Cl_2 etc.).

In dependence on nature of oxidant, the redox processes can be subdivided to:

- Redox process with participation of oxygen, which can assessed using the parameters Eh, rH_2;
- Redox processes with participation of hydrogen peroxide, which are quantified by H_2O_2 contents;
- Redox processes with participation of $OH^\bullet$ radicals, evaluated by inhibition capacity indicator, $\Sigma k_i S_i$;
- Redox processes with participation of solar light ($h\nu$).

Redox Processes With the Participation of $OH^\bullet$ Radicals

It was found that hydrogen peroxide formation and decomposition in water bodies is accompanied with the formation of free radicals (Gladchi et al., 2013, 2008; Duca et al., 2008, 2002, 1994). The most active of them is hydroxyl radical - $OH^\bullet$, which participate in numerous photochemical and redox-catalytical processes.

Photochemical Processes

$$SOD \longrightarrow OH^{\bullet} \quad (36)$$

$$H_2O_2 \longrightarrow 2OH^{\bullet} \quad (37)$$

$$NO_3^- \longrightarrow NO_2 + OH^{\bullet} \quad (38)$$

$$NO_2^- \longrightarrow NO + OH^{\bullet} \quad (39)$$

$$Fe(OH)^{2+} \longrightarrow Fe^{2+} + OH^{\bullet} \quad (40)$$

Redox-Catalytic Processes

$$Cu+ + H_2O_2 \longrightarrow CuOH+ + OH\bullet \quad (41)$$

$$CuDH+ + H_2O_2 \longrightarrow CuOH+ + OH\bullet + D.- \quad (42)$$

$$FeDFH_3+ + H_2O_2 \longrightarrow FeDFH_2+ + 2OH\bullet \quad (43)$$

The $OH^{\bullet}$ radicals play the determinant role in chemical self-purification of natural waters (Bunduchi et al., 2022; Travin et al., 2019; Duca, 2012; Lelieveld et al., 2004). Interaction of $OH^{\bullet}$ radicals with the components of environment:

$$RH + OH^{\bullet} \longrightarrow H_2O + R^{\cdot} \; (k = \sim 10^9 \; M^{-1} \cdot s^{-1}) \quad (44)$$

Concerning the properties of radicals resulted from the interactions with $OH^{\bullet}$ particles, the components of natural waters can be divided in two groups:

- *Substances which radicals interact with O_2, or the transition metal ions, generating $OH^{\bullet}$.*

$$R^{\bullet} + O_2 \longrightarrow RO^{\bullet}_2 \quad (45)$$

$$RO^{\bullet}_2 + RH \longrightarrow ROOH + R^{\bullet} \quad (46)$$

$$ROOH \longrightarrow OH^{\bullet} + RO^{\bullet} \quad (47)$$

$$R^{\bullet} + Me^{2+} \longrightarrow P + Me^{+} \quad (48)$$

$$Me^{+} + H_2O_2 \longrightarrow MeOH^{+} + OH^{\bullet} \quad (49)$$

- *Substances which radicals have lower reactivity or which, as result of transformations, pass in the non-reactive form. They finally can become the traps of $OH^{\bullet}$ H radicals.*

$$R^{\bullet} + R^{\bullet} \longrightarrow R - R \quad (50)$$

SELF-PURIFICATION OF NATURAL WATERS

Natural waters oppose the external intervention through the complex mechanism, including a series of self-purification processes (Duca et al., 2020, 1997; Lelieveld et al., 2004):

- physical processes of mass-transfer – *physical self-purification.*
- processes of metabolism and co-metabolism – *biological self-purification.*
- chemical processes of substances transformation - *chemical self-purification:*
    - *Hydrolysis*, where water is involved in the reaction not only as a solvent, but also as a reagent;
    - *Redox processes*, with participation of O_2, H_2O_2, transition metal ions and $OH^{\bullet}$ radicals;
    - *Direct photolysis, sensibilized and induced,* transformations under the influence of solar light.

Chemical self-purification of water involve series of redox transformations catalyzed with the transition valence metal ions, accompanied with the formation of partial charge transfer complexes as intermediates. As a result of pollutants decomposition in water due to these reactions, the natural ecological balance is equilibrated.

The mechanisms of natural water self-purification processes were studied in detail in rivers (Nistru, Prut, Botna, etc.) and lakes in Moldova, demonstrating the formation of complexes with partial charge transfer, as well as their role in restoring water quality, Fig.7.

Figure 7. Mechanisms of chemical self-purification in natural waters, involving series of redox processes

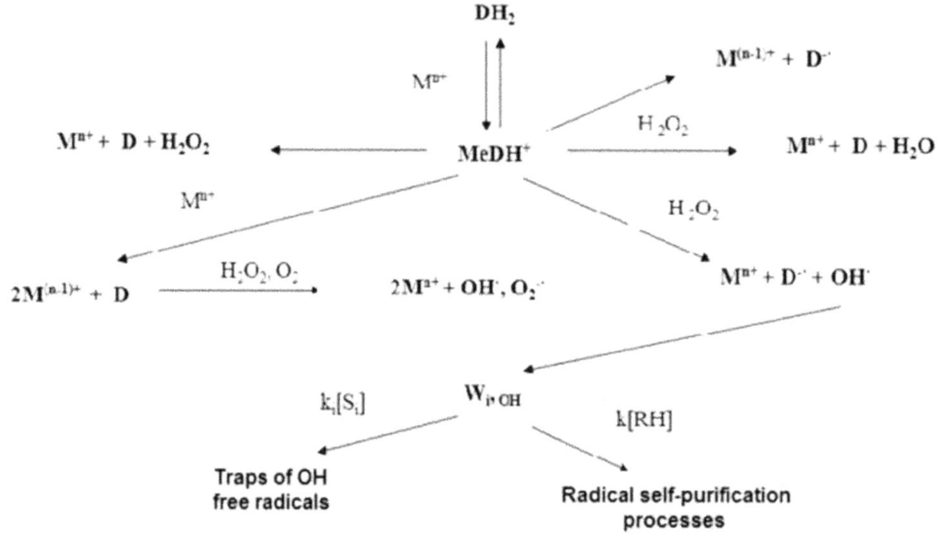

Importance of the Redox Indicators for Natural Water Quality Evaluation

The traditional/classic hydrochemical and hydrobiologic indicators of natural water quality show the presence and amounts of substances, but do not offer information on their transformations and impact on biological value of habitation (Duca & Romanciuc, 2017; Duca et al., 2002, Vaseashta et. al, 2022, Vaseashta et al. 2021).

Limited admissible concentrations (LAC) used as reference values are determined in the distilled water in conditions different from natural waters. In addition, the detected LAC values cannot precisely reflect the condition of natural water bodies and their appropriateness for living organisms. Thus, due to the high capacity to form stable compounds, hydrobionts can face the deficiency in biologically accessible forms even at much higher concentrations of copper ions than LAC. In dependence on qualitative composition of organic substances, expressed in COD (chemical oxygen demand), natural water can be biologically valuable at high values of COD, and inversely, toxic with low COD.

Our studies had shown that the fresh water quality indicators need to be revised and completed with new parameters, to adjust the water monitoring system, specifically: contents in hydrogen peroxide, contents in reducing substances which efficiently react with hydrogen peroxide, concentration of $OH^{\bullet}$ radicals and high reactive particles. In addition to the traditional indicators, to manage and predict the natural water quality, we need to consider: the redox state of water, its inhibiting capacity and ability to self-purification (Duca & Travin, 2020; Duca et al., 1997; Bahadir & Duca, 2009).

The cycle of H_2O_2 in water bodies is strongly linked with transformations of $OH^{\bullet}$ radicals, the most reactive particles of biogeochemical cycle with significant impact on self-purification process. The concentrations of registered particles in the natural waters are presented on Table 3.

Table 3. Detected amounts of oxygen-containing species in natural waters

Particle	O_2	HO_2^-	H_2O_2	$OH^\bullet$
Concentration	$\leq 4{,}4{\cdot}10^{-4}$	10^{-9} - 10^{-8} M	10^{-6} - 10^{-5} M	10^{-17} - 10^{-15} M

The fluxes of oxidizing and reducing equivalents in biosphere are equilibrated in photosynthesis and respiration, and in natural waters, along with the global redox system, also functions a redox quasi-stationery subsystem, hydrogen peroxide and peroxidasic reducing equivalents.

OH Radicals and Inhibiting Capacity in Water Quality Assessing

OH radicals interact with pollutants which could become the radical traps. The kinetics of this process can become an interesting parameter which reveals the inhibiting capacity of polluted water. So, the parameter of inhibiting capacity $\Sigma k_i[S_i]$ can be used to evaluate the water pollution in correlation with inhibitors of the radical processes (Bunduchi et al., 2022; Duca et al., 2019; Travin & Duca, 2019).

Out of all the organic substances presented in water, this indicator characterizes only those involved in the radicalic self-purification of waters. Effective constant of interaction rate of $OH^\bullet$ radicals (k_{ef}) with “radicals’ traps” is:

$$k_{ef} = \Sigma k_i[S_i] \tag{51}$$

where k_i – constant of $OH^\bullet$ radicals interaction with “traps” S_i; $\Sigma k_i[S_i]$ – parameter characterizing the inhibiting capacity of self-purification processes with the involvement of $OH^\bullet$ radicals.

According to the inhibiting capacity criterion, water quality can be evaluated as follows:

- clean water, $\Sigma k_i[S_i] < 3{\cdot}10^4\ s^{-1}$;
- moderately polluted water, $\Sigma k_i[S_i] = 3{\cdot}10^4 - 3{\cdot}10^5\ s^{-1}$;
- polluted water, $\Sigma k_i[S_i] > 3{\cdot}10^5\ s^{-1}$.

H_2O_2 and the Redox-Toxicity Phenomena

The role of hydrogen peroxide in the processes inside the natural water bodies, namely, maintenance of biologic value of habitation, was demonstrated by the large scale toxicologic test (Duca, 2012; Frog et al.; 2012; Skurlatov et al., 1984). The toxic factor identified was *hydrogen peroxide deficiency* in water. It was found that the exogen oxidant (H_2O_2) presence in water is necessary for all the vital processes, including the fats transformation into glycogen in water biota.

Thus, when hydrogen peroxide is contained in water (oxidizing environment), fish fry survival is close to 100%. When the period of the unstable redox state comes (hydrogen peroxide appears during the daytime and is missing in the morning), the yield of fish fry decreases sharply. When water reaches the quasi-reducing state (H_2O_2 is not present at a daytime), fish fry survival tends to zero. It should be stressed that the death of the fish fry occurs at the “normal” hydrochemical values. These tests confirm the necessity to revise and complete the traditional water quality indicators.

HYDROGEN PEROXIDE IN LIVING SYSTEMS

For many years, H_2O_2 was considered a cytotoxic compound of oxygen metabolic cycle, which, along with superoxide anion radical ($O^{2\cdot-}$) and $OH^{\bullet}$ radical, makes part of the group of oxygen-containing active forms with negative impact on human cells (Lismont et al., 2019; Pizzino et al., 2017). Its concentration in cells and tissues is maintained at the level that ensures the redox homeostasis, assisted by the antioxidant systems within the living cell: catalase and glutathione peroxidase (Andréset al., 2022).

The earlier data indicated H_2O_2 concentrations of about 10^{-9}-10^{-7} M in cells and tissues. Currently, there is evidence that the concentration of H_2O_2 in various tissues of the body is higher, making 3.5+10^{-7} M in blood and 10^{-4} M in vitreous body of eye, brain tissues, urine.

In the human body, the main way of H_2O_2 decomposition is catalyzed by the enzyme catalase, resulting in oxygen formation. This reaction is the basis of the multiple uses of H_2O_2:

$$\text{Ferment–Fe-OH} + H_2O_2 \longrightarrow \text{Ferment-Fe-O-OH} + H_2O \quad (52)$$

$$\text{Ferment–Fe-O-OH} + H_2O_2 \longrightarrow \text{Ferment-Fe-OH} + O_2 + H_2O \quad (53)$$

In case of oxidative stress, large amounts of hydrogen peroxide toxic for tissues, are produced in the liver of living organisms (Rost et al., 2007). Two main enzymes – catalase and peroxidase – are involved in decomposition of H_2O_2 to water and oxygen:

$$H_2O_2 + \text{Fe(III)-E} \longrightarrow H_2O + \text{O-Fe(IV)-E} \quad (54)$$

$$H_2O_2 + \text{O-Fe(IV)-E} \longrightarrow H_2O + O_2 + \text{Fe(III)-E} \quad (55)$$

Catalase, present in the peroxisomes of all aerobic cells, but absent in chloroplasts, helps prevent destructive changes at the cellular level, such as the breakdown of the DNA chain, which may eventually lead to different forms of cancer or other chronic and degenerative diseases (Lismont, 2019). Peroxidase catalyzes the oxidation of various substances with the help of hydrogen peroxide in living cells (Halliwell & Gutteridge, 2015). It is involved in the respiration of plants.

These two biologic systems (catalytic and peroxidasic) were used by us as models to study the kinetics and mechanisms of hydrogen peroxide catalytic decomposition (Duca & Travin, 2020; Raducan et al., 2012):

$$H_2O_2 + DH_2 \longrightarrow D + 2H_2O \quad (56)$$

PRODUCTION OF HYDROGEN PEROXIDE

About 4.3 million tons are produced annually of hydrogen peroxide, according to Merchant Research and Consulting Ltd, 2015, and the world leader in this sector is China (The Essential Chemical Industry online), Fig.8.

Figure 8. Annual production of hydrogen peroxide (million tons)

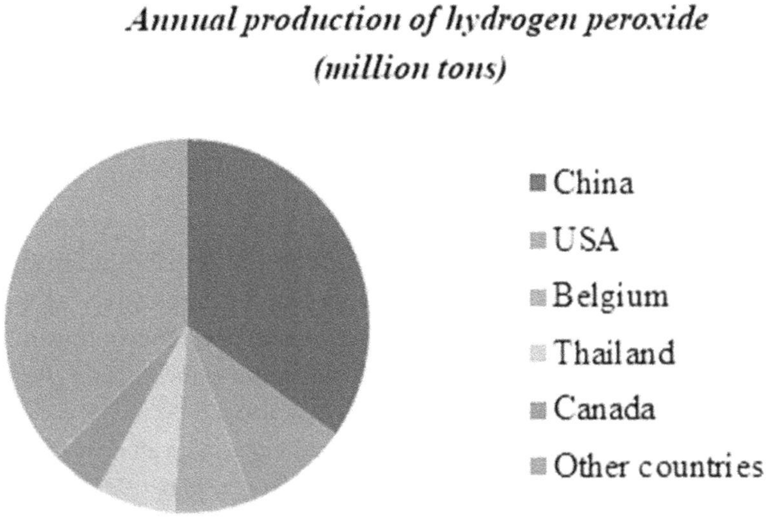

To produce commercial hydrogen peroxide, several methods can be applied:

1. 1. In laboratory conditions, reaction of barium oxide with water in acidic environment (Cotton, 1994):

$$BaO_2 + 8H_2O + H_2SO_4 \longrightarrow BaSO_4 + H_2O_2 + 8H_2O \quad (57)$$

2. The anthraquinone method, developed by BASF Company (Literature Laboratory; Tomonori et al., 2011) including the stages of:
   a) hydrogenation and subsequent oxidation of anthraquinone
   b) extraction of hydrogen peroxide solution
   c) purification and concentration of hydrogen peroxide.
3. 3. The first method of hydrogen peroxide production on the industrial scale was the electrolysis of sulfuric acid (Harper & Daigler, 1968), according to equations below

2HSO4- ↔ H2S2O8 + 2 e (cathode)

H2S2O8 + H2O ↔ H2SO4 + H2SO5

H2SO5 + H2O ↔ H2SO4 + H2O2 2H+ + 2 e ↔ H+ (anode) (58)

H2S2O8 + 2H2O → H2O2 + 2H2SO4

(NH4)2S2O8 + 2H2O → (NH4)2SO4 + H2SO4 + H2O2

4. 4. Catalytic oxidation of isopropyl alcohol (Skinner & Stinle, 1953):

$$(CH_3)_2CHOH + O_2 \longrightarrow (CH_3)_2CO + H_2O_2 \quad (59)$$

APPLICATIONS OF HYDROGEN PEROXIDE

Hydrogen peroxide demonstrates a series of specific effects which makes it broadly used in different areas. Thus, due to the antimicrobial properties without long-term toxicity it is used as antibacterial agent in medicine, pharmaceuticals, veterinary, industry for disinfection, sterilization, disease control, conservation, etc. When combined with other agents, it can improve their action (Abdollahi & Hosseini, 2014; Lobo et al., 2010; Campbell, 1997; Block, 1991). Hydrogen peroxide was proposed for use in agriculture. The research team (Duca et al., 1995, 1988) has elaborated an efficient plant growth stimulator entitled "Cerera", which involves copper ions and hydrogen peroxide. This agent was proposed for the treatment of tomato, lattice, onion, soya, etc. seeds. As a result, the first sprouts appeared 5-7 days earlier; crops were harvested 10-12 days earlier and overall crop productivity was 132% higher as compared to control. The use of H_2O_2 in agriculture was approved in the second half of 20th century (Commission Implementing Regulation, 2017) as an ecological fungicide and bactericide, stimulator for seed germination, source of oxygen for soil.

CONCLUSION

Hydrogen peroxide is one of the most significant substances on our planet, playing role in the appearance and development of the life, in functioning of great variety of vitally important systems, environmental processes that maintain the balance of reduction and oxidative species, natural waters self-purification, waste waters treatment, etc.

Its molecule being rather simple by chemical composition, however, provides possibilities of numerous redox reactions realization in different systems. Revealing the mechanisms of these processes was a subject of study for various research teams, and specifically in the Republic of Moldova. The mechanisms of hydrogen peroxide transformations are laid in the base of explanation and management of the redox balance in natural water environment, adequate to the habitat value, as well as in the other living and technological systems.

The overall environmental impact of hydrogen peroxide uses is beneficial and advantageous as it easily breaks down into non-toxic simple substances - water and oxygen. Due to its oxidizing properties, hydrogen peroxide can be applied to neutralize certain types of toxins, macromolecules and drugs.

The overall action of hydrogen peroxide significantly reduces any risk of developing resistance to the biocide over time, unlike many other types of biocides.

ACKNOWLEDGMENT

This research has been performed under the Moldova National Research Project Nr. 20.80009.5007.27 "Physical-chemical redox processes with electron transfer in the vital, technological and environmental processes".

REFERENCES

Abdollahi, M., & Hosseini, A. (2014). Hydrogen peroxide. Encyclopedia of toxicology. doi:10.1016/B978-0-12-386454-3.00736-3

Abrahams, S. C., Collin, R. L., & Lipscomb, W. N. (1951). The cristal structure of hydrogen peroxide. *Acta Crystallographica*, *4*(1), 15–20. doi:10.1107/S0365110X51000039

Andrés, C. M. C., Pérez de la Lastra, J. M., Juan, C. A., Plou, F. J., & Pérez-Lebeña, E. (2022). Chemistry of hydrogen peroxide formation and elimination in mammalian cells, and its role in various pathologies. *Stresses*, *2*(3), 256–274. doi:10.3390tresses2030019

Arsene, I. (2008). The theoretical study of some reactions with the participation of OH and HO_2 radicals. *Chemistry Journal of Moldova*, *3*(2), 109–113. doi:10.19261/cjm.2008.03(2).18

Arsene, I., & Gorinchoy, N. (2019). DFT study of the entire reaction cycle of H_2O_2 decomposition and O_2 generation catalyzed by Fenton reagent. *Chemistry Journal of Moldova*, *14*(1), 88–97. doi:10.19261/cjm.2018.543

Baalen, C. V., & Marler, J. E. (1966). Occurrence of hydrogen peroxide in seawater. *Nature*, *211*(5052), 951–951. doi:10.1038/211951a0

Bahadir, A. M., & Duca, Gh. (2009). *The role of ecological chemistry in pollution research and sustainable development*. Springer. https://link.springer.com/content/pdf/10.1007/978-90-481-2903-4.pdf doi:10.1007/978-90-481-2903-4

Bergman, P., Parise, B., Liseau, R., Larsson, B., Olofsson, H., Menten, K. M., & Güsten, R. (2011). Detection of interstellar hydrogen peroxide. *Astronomy & Astrophysics*, *531*, L8. doi:10.1051/0004-6361/201117170

Bhunia, S., Rana, A., Ghosh, S., Ivancich, A., & Dey, A. (2020). A designed second-sphere hydrogen-bond interaction that critically influences the O–O bond activation for heterolytic cleavage in ferric iron–porphyrin complexes. *Chemical Science, The Royal Society of Chemistry, 11*(10), 2681-2695. https://hal.archives-ouvertes.fr/hal-02887080/document

Block, S. S. (1991). *Peroxide hydrogen compounds in disinfection, sterilization and preservation* (S. S. Block, Ed.). Vol. 167). Lea & Febinger.

Bunduchi, E., Duca, Gh., & Gladchi, V. (2022). New kinetic parameters for natural water quality assessment. In *Handbook of research on water sciences and society* (Vol. 14). IGI Global. doi:10.4018/978-1-7998-7356-3.ch011

Campbell, D. W., II. (1997). Hydrogen peroxide: medical miracle. Ed.: Hampton Roads Publishing Company.

Catling, D. C., Glein, C. R., Zahnle, K. J., & McKay, C. P. (2005). Why O_2 is required by complex life on habitable planets and the concept of planetary" oxygenation time. *Astrobiology, 5*(3), 415–438. doi:10.1089/ast.2005.5.415 PMID:15941384

Commission Implementing Regulation (EU) (2017). Approving the Basic Substance Hydrogen Peroxide in Accordance With Regulation (EC) No 1107/2009 of the European Parliament and of the Council concerning the placing of plant protection products on the market and amending the Annex to Commission Implementing Regulation (EU) No 540/2011. *Official Journal of EU.09.03.2017*, L63/95. https://eur-lex.europa.eu/legal-content/EN/TXT/PDF/?uri=CELEX:32017R0409&from=LT

Cooper, W. J., Zika, R. G., Petasne, R. G., & Plane, J. M. C. (1988). Photochemical formation of hydrogen peroxide in natural waters exposed to sunlight. *Environmental Science & Technology, 22*(10), 1156–1160. doi:10.1021/es00175a004 PMID:22148607

Cotton, F. A., Wilkinson, G., & Gaus, P. L. (1994). *Basic inorganic chemistry* (3rd ed.). John Wiley and Sons, Inc.

Dougherty, D. A., & Anslyn, E. V. (2005). Modern physical organic chemistry. *Universitas Scientiarum*, 122.

Duca, G. (2007). Ecological chemistry in new independent states of the former Soviet Union. *Clean (Weinheim), 35*(1), 17–25. doi:10.1002/clen.200600007

Duca, Gh. (2012). *Homogeneous catalysis with metal complexes: fundamentals and applications*. Springer. doi:10.1007/978-3-642-24629-6

Duca, G. (2017). Ecological and environmental chemistry. *Chemistry Journal of Moldova, 12*(1), 9–19. doi:10.19261/cjm.2017.427

Duca, G. (2021). Hydrogen peroxide in ecological and environmental chemistry. *Chemistry Journal of Moldova, 16*(2), 28–45. doi:10.19261/cjm.2021.918

Duca, G., Blonschi, V., Gladchi, V., & Travin, S. (2020). Dynamics of different sulfur forms in natural waters and their influence on the redox state. *American Journal of Physical Chemistry, 9*(3), 52. doi:10.11648/j.ajpc.20200903.12

Duca, G., Mereuta, A., Velisco, N., Tanaselia, C., & Mitina, T. (2019). Determination of quality indicators of prut river water. *Chemistry Journal of Moldova, 14*(1), 61–67. doi:10.19261/cjm.2018.541

Duca, Gh., & Romanciuc, L. (2017). An overview of ecological chemistry conferences within 1985-2016. In: *Abstract Book of the 6th International Conference on Ecological & Environmental Chemistry*, 21-28. Chisinau.

Duca, Gh., Romanciuc, L., Goreacea, N., & Borodaev, R. (1997). *Model of self-purification processing in natural water.* In: Metal compounds in environment and life, VII-th Int. Hans Wolfgang Nurnberg Memorial Symp., Modena (Italy), 51.

Duca, G., Secara, N., & Duca, D. (2008). Physical methods of fast reactions investigation. *Chemistry Journal of Moldova*, *3*(1), 22–30. doi:10.19261/cjm.2008.03(1).15

Duca, Gh., Skurlatov, Yu., & Miziti, A. (1994). Introduction in ecological chemistry (Rus.). Moscow: Vysshaia shkola, 400p.

Duca, Gh., Skurlatov, Yu., & Sychev, A. (2002). *Redox catalysis and ecological chemistry*. CEP USM.

Duca, Gh., Skurlatov, Yu., Sychev, A., & Romanciuc, L. (1988). *Method of plants growth stimulation.* Patent Nr.1544334 (SU).

Duca, G., & Travin, S. (2020). Reactions' mechanisms and applications of hydrogen peroxide. *American Journal of Physical Chemistry*, *9*(2), 36–44. doi:10.11648/j.ajpc.20200902.13

Duca, Gh., & Vaseashta, A. (2020). *Handbook of research on emerging developments and environmental impacts of ecological chemistry*. IGI Global., doi:10.4018/978-1-7998-1241-8

Duca, M., Port, A., Romanciuc, L., & Duca, Gh. (1995). The redox state of aquatic medium and plants germination process regulation. Self-purification processes in natural waters, 245-256. Chisinau.

Enami, S., Sakamoto, Y., & Colussi, A. J. (2013). Fenton chemistry at aqueous interfaces. *Proceedings of the National Academy of Sciences of the United States of America*, *111*(2), 623–628. doi:10.1073/pnas.1314885111 PMID:24379389

Frog, B. N., Skurlatov, Yu. I., Shtamm, E. V., & Vichutinskaya, E. V. (2012). The influence of water-soluble sulfur compounds on the toxic properties of natural and waste waters (Rus.). *Vestnik MGSU*, *6*, 105–113. doi:10.22227/1997-0935.2012.6.105-113

Fujii, M., & Otani, E. (2017). Photochemical generation and decay kinetics of superoxide and hydrogen peroxide in the presence of standard humic and fulvic acids. *Water Research*, *123*, 642–654. doi:10.1016/j.watres.2017.07.015 PMID:28710981

Gambardella, F., Ganzeveld, I., Jos, W., & Heeres, H. J. (2005). Kinetics of the reaction of FeII (EDTA) with oxygen in aqueous solutions. *Industrial & Engineering Chemistry Research*, *44*(22), 8190–8198. doi:10.1021/ie050448e

Giguere, P. A. (1950). The infra-red spectrum of hydrogen peroxide. *The Journal of Chemical Physics*, *18*(1), 88–92. doi:10.1063/1.1747464

Gladchi, V., Duca, G., Goreaceva, N., Bunduchi, E., & Lis, A. (2013). Chemical composition from the Dniester River tributaries. *Chemistry Journal of Moldova*, *8*(1), 23–32. doi:10.19261/cjm.2013.08(1).03

Gladchi, V., Goreaceva, N., Duca, G., Bunduchi, E., Romanciuc, L., Mardari, I., & Borodaev, R. (2008). The study of redox conditions in the Dniester River. *Chemistry Journal of Moldova*, *3*(1), 70–76. doi:10.19261/cjm.2008.03(1).14

Grinstead, R. R. (1960). The oxidation of ascorbic acid by hydrogen peroxide. Catalysis by ethylenediaminetetraacetato-Iron(III). *Journal of the American Chemical Society*, *82*(13), 3464–3471. https://pubs.acs.org/doi/10.1021/ja01498a057. doi:10.1021/ja01498a057

Halliwell, B., & Gutteridge, J. M. C. (2015). *Antioxidant defends synthesized in vivo'. Free Radicals in Biology and Medicine* (5th ed.). Academic. doi:10.1093/acprof:oso/9780198717478.003.0003

Harper, W. S., & Daigler, D. W. (1968). *Production of hydrogen peroxide from aqueous acidic solution obtained by hydrolysis.* Patent US 1968-01-23. https://patents.google.com/patent/US3694154A/en

Holland, H. D. (2006). The oxygenation of the atmosphere and oceans. *Philosophical Transactions of the Royal Society of London. Series B, Biological Sciences*, *361*(1470), 903–915. doi:10.1098/rstb.2006.1838 PMID:16754606

Hongshin, L., Hyejin, L., Jiwon, S., Hyungeun, K., Yun, S., Jaehong, K., & Changha, L. (2016). Activation of oxygen and hydrogen peroxide by copper (II) coupled with hydroxylamine for oxidation of organic contaminants. *Environmental Science & Technology*, *50*(15), 8231–8238. doi:10.1021/acs.est.6b02067 PMID:27387011

Hunt, R. H., Leacock, R. A., Peters, C. W., & Hecht, K. T. (1965). Internal-rotation in hydrogen peroxide: The far-infrared spectrum and the determination of the hindering potential. *The Journal of Chemical Physics*, *42*(6), 1931–1946. doi:10.1063/1.1696228

Hydrogen Peroxide. Safety and Handling. (2006). Technical data sheet HH_2323, Revised 0305 Solvay Chemicals, Inc. www.solvaychemicals.us

Karamoschos, N., & Tasis, D. (2022). Photocatalytic evolution of hydrogen peroxide: A minireview. *Energies*, *15*(17), 6202. doi:10.3390/en15176202

Krumova, K., & Cosa, G. (2016). Chapter 1: Overview of reactive oxygen species, In: Singlet oxygen: Applications in biosciences and nanosciences, 1, 1-2. Royal Society of Chemistry. doi:10.1039/9781782622208-00001

Kulkarni, A. D., Pathak, R. K., & Bartolotti, L. J. (2006). Effect of additional hydrogen peroxide to H_2O_2... $(H_2O)n$, n=1 and 2 complexes: Quantum chemical study. *The Journal of Chemical Physics*, *124*(21), 214309. doi:10.1063/1.2202098 PMID:16774409

Lelieveld, J., Dentener, F. J., Peters, W., & Krol, M. C. (2004). On the role of hydroxyl radicals in the self-cleansing capacity of the troposphere. *Atmospheric Chemistry and Physics*, *4*(9/10), 2337–2344. www.atmos-chem-phys.org/acp/4/2337/. doi:10.5194/acp-4-2337-2004

Lismont, C., Revenco, Iu., & Fransen, M. (2019). Peroxisomal hydrogen peroxide metabolism and signaling in health and disease. *International Journal of Molecular Sciences*, *20*(15), 3673. doi:10.3390/ijms20153673 PMID:31357514

Lobo, V., Patil, A., Phatak, A., & Chandra, N. (2010). Free radicals, antioxidants and functional foods: Impact on human health. *Pharmacognosy Reviews*, *4*(8), 118–126. doi:10.4103/0973-7847.70902 PMID:22228951

Lyons, T. W., Reinhard, C. T., & Planavsky, N. J. (2014). The rise of oxygen in Earth's early ocean and atmosphere. *Nature, 506*(7488), 307–315. www.nature.com/doifinder/10.1038/nature13068. doi:10.1038/nature13068 PMID:24553238

Mandal, M., Kawashima, K., Saito, K., & Ishikita, H. (2020). Redox potential of the oxygen-evolving complex in the electron transfer cascade of photosystem II. *The Journal of Physical Chemistry Letters, 11*(1), 249–255. doi:10.1021/acs.jpclett.9b02831 PMID:31729876

Marcus, R. A. (1964). Chemical and electrochemical electron-transfer theory. *Annual Review of Physical Chemistry, 15*(1), 155–196. doi:10.1146/annurev.pc.15.100164.001103

Miller, G. W., Morgan, C. A., Kieber, D. J., King, D. W., Snow, J. A., Heikes, B. G., Mopper, K., & Kiddle, J. J. (2005). Hydrogen peroxide method intercomparision study in seawater. *Marine Chemistry, 97*(1–2), 4–13. doi:10.1016/j.marchem.2005.07.001

Nishimi, T., Kamachi, T., Kato, K., Kato, T., & Yoshizawa, K. (2011). Mechanistic Study on the Production of Hydrogen Peroxide in the Anthraquinone Process. *European Journal of Organic Chemistry, 22*(22), 4113–4120. doi:10.1002/ejoc.201100300

Olson, M. V. (2022). Oxidation-reduction reaction. *Encyclopedia Britannica.* https://www.britannica.com/science/oxidation-reduction-reaction

Olsson, L., Levit, G. S., & Hosfeld, U. (2017). *The "Biogenetic Law' in zoology: from Ernst Haeckel's formulation to current approaches. Theory Biosci.* Springer-Verlag Berlin Heidelberg. doi:10.100712064-017-0243-4

Olszyna, K. J., Meagher, J. F., & Bailey, E. M. (1988). Gas-phase, cloud and rain-water measurements of hydrogen peroxide at a high-elevation site. *Atmospheric Environment, 8*(8), 1699–1706. doi:10.1016/0004-6981(88)90398-8

Oparin, A. I. (1968). Life, its nature, origin and development (2nd). Ed.: M.: Nauka.

Orabi, E. (2021). Molecular dynamics investigation of the structural flexibility of H_2O_2 and H_2S_2 in response to medium polarity. *Journal of Molecular Liquids, 329*, 115469. doi:10.1016/j.molliq.2021.115469

Penney, W. G., & Sutherland, G. B. (1934). A note on the structure of H_2O_2 and H_4N_2 with particular reference to electric moments and free rotation. *Transactions of the Faraday Society, 30*, 898–902. doi:10.1039/tf934300898b

Pizzino, G., Irrera, N., Cucinotta, M., Pallio, G., Mannino, F., Arcoraci, V., Squadrito, F., Altavilla, D., & Bitto, A. (2017). Oxidative stress: Harms and benefits for human health. *Oxidative Medicine and Cellular Longevity, 8416763*, 1–13. doi:10.1155/2017/8416763 PMID:28819546

Plauck, A., Stangland, E. E., Dumesic, J. A., & Mavrikakis, M. (2016). Active sites and mechanisms for H_2O_2 decomposition over Pd catalysts. *Proceedings of the National Academy of Sciences of the United States of America, 113*(14), E1973–E1982. doi:10.1073/pnas.1602172113 PMID:27006504

Price, D., Worsfold, P. J., Fauzi, R., & Mantoura, C. (1994). Determination of hydrogen peroxide in sea water by flow-injection analysis with chemiluminescence detection. *Analytica Chimica Acta, 298*(1), 121–128. doi:10.1016/0003-2670(94)90050-7

Raducan, A., Leonties, A. R., Puiu, M., & Oancea, D. (2012). Kinetics of hydrogen peroxide decomposition by catalase: Hydroxylic solvent effects. *Bioprocess and Biosystems Engineering, 35*(9), 1523–1530. doi:10.100700449-012-0742-0 PMID:22565543

Ramaswamy, N., & Mukerjee, S. (2011). Influence of inner- and outer-sphere electron transfer mechanisms during electrocatalysis of oxygen reduction in alkaline media I. *The Journal of Physical Chemistry C, 115*(36), 18015–18026. doi:10.1021/jp204680p

Romanciuc, L., Gladchi, V., Goreaceva, N., Bunduchi, E., & Duca, Gh. (2009). Water quality monitoring and management in Moldova. In: Bahadir, A.M., Duca, G. (eds) The role of ecological chemistry in pollution research and sustainable development. NATO Science for Peace and Security Series C: Environmental Security. Springer, Dordrecht. doi:10.1007/978-90-481-2903-4_6

Rost, D., Welker, A., Welker, J., Millonig, G., Berger, I., Autschbach, F., Schuppan, D., & Mueller, S. (2007). Liver-homing of purified glucose oxidase: A novel in vivo model of physiological hepatic oxidative stress (H_2O_2). *Journal of Hepatology, 46*(3), 482–491. doi:10.1016/j.jhep.2006.09.025 PMID:17188390

Scholes, G., & Weiss, J. (1959). *Oxygen effects and formation of peroxides in aqueous solutions. Radiation research supplement.* Proceedings of the International Congress of Radiation Research, Burlington, Vermont. https://www.jstor.org/stable/3583638

Serra-Maia, R., Bellier, M., Chastka, S., Tranhuu, K., Subowo, A., Rimstidt, J. D., Usov, P. M., Morris, A. J., & Michel, F. M. (2018). Mechanism and kinetics of hydrogen peroxide decomposition on platinum nanocatalysts. *ACS Applied Materials & Interfaces, 10*(25), 21224–21234. doi:10.1021/acsami.8b02345 PMID:29851338

Shtamm, E., Purmal, A., & Skurlatov, Yu. (1991). The role of hydrogen peroxide in natural water environment (Rus.). *Successes of Chemistry, 60*(11), 2373–2406.

Shvydkiy, V., Dolgov, S., Dubovik, A., Kozlov, M., Povkh, A., Shishkina, L., & Duca, G. (2022). New aspects for the estimation of the state of natural water. *Chemistry Journal of Moldova, 17*(2), 35–42. doi:10.19261/cjm.2022.973

Sigg, A., & Neftel, A. (1988). Seasonal variations in hydrogen peroxide in polar ice cores. *Annals of Glaciology, 10*, 157–162. doi:10.3189/S0260305500004353

Sinel'nikov, V. E. (1968). Luminescent analysis of natural and polluted waters (Rus.). Obninsk: Gidrometcentr SSSR.

Sinel'nikov, V. E. (1971). Hydrogen peroxide level in river water, and methods for detecting it (Rus.). *Hydrobiological Journal, 7*(1), 115–119.

Skinner, J. R., & Stinle, S. E. (1953). *Method for the production of hydrogen peroxide.* Patent US, 1953-12-23. https://patents.google.com/patent/US2871103A/en

Skurlatov, Yu., Duca, Gh., Batir, D., & Travin, S. (1989). Partial charge transfer coordination compounds in the redox processes (Rus.). *Coordination Chemistry*, *15*(3), 291–307.

Skurlatov, Yu., Ernestova, L., Shtamm, E., Shpotova, T., & Kalinin, V. (1984). Redox state and seasonal toxicity of natural water. *Proc. of Acad. Sci. USSR*, *276*(4), 1014–1016.

Slesak, I., Slesak, H., & Kruk, J. (2012). Oxygen and hydrogen peroxide in the early evolution of life on earth: In silico comparative analysis of biochemical pathways. *Astrobiology*, *12*(8), 775–784. doi:10.1089/ast.2011.0704 PMID:22970865

Sridharan, U. C., Reimann, B., & Kaufman, F. (1980). Kinetics of the reaction OH + $H_2O_2 \rightarrow HO_2$ + H_2O. *The Journal of Chemical Physics*, *73*(3), 1286–1293. doi:10.1063/1.440240

Stone, J., Edgar, J. O., Gould, J. A., & Telling, J. (2022). Tectonically-driven oxidant production in the hot biosphere. *Nature Communications*, *13*(1), 1–10. doi:10.103841467-022-32129-y PMID:35941147

Sychev, A., Travin, C., Duca, Gh., & Skurlatov, Iu. (1983). *Catalytic reactions and environmental protection*. Stiinta.

Tachikawa, H. (2021). Formation of hydrogen peroxide from $O^-(H_2O)_n$ clusters. *The Journal of Physical Chemistry A*, *125*(21), 4598–4605. doi:10.1021/acs.jpca.1c02883 PMID:34033487

Tian, X., Luo, T., Nie, Y., Shi, J., Tian, Y., Dionysiou, D. D., & Wang, Y. (2022). New insight into a Fenton-like reaction mechanism over sulfidated β-FeOOH: Key role of sulfidation in efficient iron (III) reduction and sulfate radical generation. *Environmental Science & Technology*, *56*(9), 5542–5551. doi:10.1021/acs.est.2c00132 PMID:35412804

Travin, S., Duca, G., & Gladchi, V. (2019). Self-purification of aquatic media from hexachlorocyclohexane in a radical process. *Chemistry Journal of Moldova*, *14*(1), 47–53. doi:10.19261/cjm.2018.537

Tsuneda, T., & Taketsugu, T. (2018). Theoretical investigations on hydrogen peroxide decomposition in aquo. *Physical Chemistry Chemical Physics*, *20*(38), 24992–24999. doi:10.1039/C8CP04299C PMID:30239545

Van Niel, C. B. (1932). On the morphology and physiology of the purple and green sulphur bacteria. *Archives of Microbiology*, *3*(1), 1–112. doi:10.1007/BF00454965

Vaseashta, A., Duca, G., & Travin, S. (Eds.). (2022). *Handbook of Research on Water Sciences and Society* (Vols. 1–2). IGI Global. doi:10.4018/978-1-7998-7356-3

Vaseashta, A., Gevorgyan, G., Kavaz, D., Ivanov, O., Jawaid, M., & Vasović, D. (2021). Exposome, Biomonitoring, Assessment and Data Analytics to Quantify Universal Water Quality. In A. Vaseashta & C. Maftei (Eds.), *Water Safety, Security and Sustainability. Advanced Sciences and Technologies for Security Applications*. Springer. doi:10.1007/978-3-030-76008-3_4

Yu, C. Y., & Yang, Z. (2011). A systemic investigation of hydrogen peroxide clusters $(H_2O_2)_n$ (n = 1–6) and liquid-state hydrogen peroxide: Based on atom-bond electronegativity. *The Journal of Physical Chemistry A*, *115*(12), 2615–2626. doi:10.1021/jp111284t PMID:21388113

Zhu, Q., Lian, Y., Thyagarajan, S., Rokita, S. E., Karlin, K. D., & Blough, N. V. (2008). Hydrogen peroxide and dioxygen activation by dinuclear copper complexes in aqueous solution: Hydroxyl radical production initiated by internal electron transfer. *Journal of the American Chemical Society*, *130*(20), 6304–6305. doi:10.1021/ja800080z PMID:18433125

Chapter 4
Pseudo Jahn-Teller Effect in Transition States of Redox Processes

Natalia Gorinchoy
https://orcid.org/0000-0003-4529-9061
Institute of Chemistry, Moldova State University, Moldova

Iolanta Balan
https://orcid.org/0000-0002-8704-1344
Institute of Chemistry, Moldova State University, Moldova

Mikhail Gorbachev
https://orcid.org/0009-0005-3058-5497
Institute of Chemistry, Moldova State University, Moldova

Ion Arsene
Ion Creangă Pedagogical State University, Moldova

Victor Polinger
University of Washington, USA

Gheorghe Duca
Institute of Chemistry, Moldova State University, Moldova

Isaac Bersuker
University of Texas at Austin, USA

ABSTRACT

The authors show that the instability of the transition (activated) states of chemical reactions are due to the vibronic coupling of the ground electronic state with the appropriate low-lying excited states, resulting in the pseudo Jahn-Teller effect (PJTE). In application to reactions relevant to environmental problems, the PJTE origin of their activation barrier is illustrated by several examples, including the proton transfer in deprotonated H3O2- and protonated H5O2+(H2O)4 water clusters and in a number of proton-bound dimers. The results can serve as a general parametrized analytical model of the potential energy profile in simulating proton transfer processes in such systems. It also includes modeling the polarizing effect of the solvent, serving as an additional stabilizing factor in intramolecular H-bonds. Similarly, it is shown that the transition states of photochemical reactions are formed by the PJTE interaction of the relevant excited state with a higher in energy electronic state. Examples of electron transfer and charge transfer by coordination are also shown.

DOI: 10.4018/978-1-6684-7198-2.ch004

INTRODUCTION

The role of the chemical reactions' transition (activated) states in environmental processes is crucial for revealing the latter's nature and their possible manipulation. Most of these reactions involve either electron, or proton transfer, or both. On the other hand, the vibronic coupling theory, mostly the Jahn-Teller effect (JTE) and the pseudo JTE (PJTE), were shown to be the only source of spontaneous symmetry breaking of high-symmetry configurations of polyatomic systems, see e.g. (Bersuker & Polinger, 1989); (Bersuker, 2006); (Bersuker, 2013); (Bersuker, 2016); (Bersuker, 2021). In the application of this fundamental principle to chemical-reaction transition states, it was shown that the instability of their ground states at the barrier point is due to the PJTE interaction with an appropriate low-lying excited state (Bersuker & Polinger, 1989); (Bersuker, 1984). This new understanding allowed for further developments of the theory of the PJTE origin of instability of the transition states of chemical reactions. As the only source of instability and spontaneous symmetry breaking of high-symmetry configurations of molecular systems and solids, the JTE and PJTE may be related to a huge variety of physical, chemical, and biological phenomena, but so far, only a limited number of such possibilities were explored.

In this Chapter we demonstrate the role of the JTE and PJTE in forming the transition or activated states of chemical reactions relevant to environmental problems, as well as some examples from spectroscopy and photochemistry. As a whole, this Chapter also contributes significantly to the development of the general theory of the PJTE in chemical reactivity.

In Section 2 we introduce the PJTE in the form applicable to the redox problems under consideration. It includes, in particular, modeling the polarizing effect of the solvent, responsible for an additional stabilizing effect on intramolecular H-bond. In Section 3, a simple case of the reaction, H + H2 ® H2 + H in a linear collision (Bersuker, 1987) is used to show the feasibility of the idea about the PJTE origin of the negative curvature of the potential energy surface at the point of the transition state, and the existence in this nuclear configuration in the stable excited states that cause the instability of the activated complex of the reaction (Bersuker, 1980). It is also shown that in some cases of intermediate states of chemical reactions, it is necessary to take into account the JTE in the degenerate excited state. We also show an example of applying the PJTE approach to the problem of transition states of photochemical reactions (Wang, 2019), which is also important in environmental problems.

In some more recent works, the PJTE was shown to control the proton transfer between two molecular systems, often termed as the Hydrogen (H) bond (Gorinchoy, 2021); (Geru, 2012), (García-Fernández, 2008). This approach, in turn, allowed us to estimate the main parameter of the H-bond, the energy barrier for the proton transfer between the interacting molecular systems. In Section 4, several examples illustrate this methodology by showing how the PJTE influences the energy barrier for the proton transfer. Illustrative examples include the proton transfer in deprotonated $H_3O_2^-$ and protonated $H_5O_2^+(H_2O)_4$ water clusters, which involve a strong low-barrier hydrogen bond - a phenomenon often introduced to explain the surprisingly high rates of some enzyme-catalyzed reactions; several proton-bound dimers $MH^+ + M$ « $M + HM^+$, and other similar systems are also considered. Section 5 of this Chapter analyzes the role of the JTE and PJTE is analyzed in the charge-transfer processes, when common bonding and antibonding molecular orbitals are formed between the interacting reactants, leading to a redistribution of the electronic density in the activated complex.

THE PSEUDO-JAHN-TELLER EFFECT

Basic Formulations Added Covalence Bonding as the Driving Force

Any polyatomic system's equilibrium geometry corresponds to the point of the minimum of the adiabatic potential energy surface (APES), where its first derivatives are zero and the second derivatives are positive. This stable configuration does not necessarily coincide with the highest-symmetry or any high-symmetry configuration. We define the latter as the reference configuration corresponding to the point of the APES, where the first derivatives are zero too, but the second derivative in some directions may be either positive or negative, meaning the reference configuration may be unstable. In application to the transition state of chemical reactions, the reference configuration corresponds to the ground state nuclear configuration at the barrier point $Q = 0$. Here Q is the system's normal coordinate transforming by the irreducible representation Γ_Q of the symmetry group of the system in its reference configuration. The symmetry-adapted nuclear coordinates are well-defined by the group theory (Wilson et al., 1980, Ch. 6). Denote by $E(Q)$ the energy of the system as a function of Q. At the point of minimum of the APES at $Q = 0$, the curvature K of $E(Q)$ in the Q direction,

$$K = (¶^2E \,/\, ¶Q^2)_0 \tag{1}$$

is positive, $K > 0$. Very often, the geometry of the system at this equilibrium point on the APES does not coincide with the highest possible (or even with any high) symmetry expected from general symmetry considerations. The PJTE is the general driving force for all these distortions of high-symmetry configurations in non-degenerate states (Bersuker, 2021); (Bersuker, 2016); (Bersuker, 2013), (Bersuker, 2006); (Bersuker, 1989).

If at the point of high-symmetry configuration on the APES the system is structurally unstable with respect to some nuclear displacements Q, then $K < 0$ in this direction. Let us analyze this possibility. The general formula for the energy is $E = á\Psi_0|H|\Psi_0ñ$, where H is the Hamiltonian of the system, and Ψ_0 is the adiabatic wavefunction of the nondegenerate ground state. Substituting E in Eq. (1), we get:

$$K = \Psi_0|\left(\partial^2 H \,/\, \partial Q^2\right)_0|\Psi_0 + 2\Psi_0|\left(\partial H \,/\, \partial Q\right)_0|\Psi_0\rangle, \tag{2}$$

where $\Psi_0¢= (¶\Psi_0/¶Q)_0$, and all the functions in Eq. (2) are now taken at the point of high-symmetry configuration. Denoting the first term in (2) by K_0,

$$K_0 = \Psi_0|(¶^2H \,/\, ¶Q^2)_0|\Psi_0 \tag{3}$$

and the second one as K_v,

$$K = K_0 + K_v, \tag{4}$$

we notice that $K_v < 0$. Indeed, presenting the Hamiltonian of the system as the power series,

$$H = H_0 + \left(\frac{\partial H}{\partial Q}\right)_0 Q + ... \tag{5}$$

where H_0 is the Hamiltonian of the system with fixed nuclei in the high-symmetry configuration, while the second and following terms stand for the coupling with the nuclear displacements (vibronic coupling), and presenting Ψ_0¢ in terms of the second order perturbation theory with respect to the vibronic coupling, we get directly:

$$K_v = -2\sum_n \frac{\left|\Psi_0 | (\partial H / \partial Q_\alpha)_0 | \Psi_n\right|^2}{E_n - E_0}, \tag{6}$$

where Ψ_n are the wavefunctions of the excited states and E_n are their energies. This K_v term thus contributes to the instability of the system. Therefore, if $K_0 > 0$, the K_v contribution is the only source of instability. Note that, as mentioned above, we define the reference high-symmetry configuration for the displacements Q as the point of the APES at which $(¶E/¶Q)_0 = 0$, meaning zero distortion force in this direction. The matrix elements in Eq. (6) are off-diagonal vibronic coupling constants (Bersuker & Polinger, 1989)

$$F_{0n} = \Psi_0|(¶H / ¶Q_{\alpha})_0|\Psi n \tag{7}$$

that mix the ground and excited states under the nuclear displacements Q; therefore, K_v is termed vibronic contribution. Together with the K_0 value and the energy gap $\Delta_{0n} = E_n - E_0$ between the mixing states, F_{0n} is the main parameter of the PJTE. Note that the vibronic coupling constant F_{0n} in of Eq. (7) is non-zero only if the product of irreducible representations $\Gamma_0 \times \Gamma_n$ of the electron states Ψ_0 and Ψ_n contains the irreducible representation Γ_Q of the displacement Q. This selection rule provides an essential insight into which excited electron states may contribute to the PJTE instability of the APES at $Q = 0$.

In a series of papers, beginning in 1980 [see in (Bersuker & Polinger, 1989); (Bersuker I. B., 2006)], it was proved that for any polyatomic system in the high-symmetry configuration.

$$K_0 > 0, \tag{8}$$

and hence the vibronic contribution K_v is the only source of instability of any polyatomic system in nondegenerate states. This is the theorem of instability of high-symmetry configurations. If $K_0 > 0$ for the high-symmetry configuration of any polyatomic system, then a negative curvature $K = (K_0 + K_v) < 0$ can be achieved only due to the negative vibronic coupling component K_v, and only if $| K_v | > K_0$. It follows that any distortion of the high-symmetry configuration is due to, and only to the mixing of its ground state with excited electronic states by the distortive nuclear displacements, realized via the vibronic coupling terms in Eq. (6). It means that to rationalize the origin of molecular shapes by reasonably starting with some high-symmetry (force-equilibrated) configuration, we notice that there is no way to get this configuration distorted unless we involve its excited electronic states via the vibronic coupling. The latter softens the system with respect to certain nuclear displacements ($K_v < 0$), and if this softening is larger than the original (non-vibronic) stiffness K_0 in this direction, the system becomes unstable with respect to the distortions under consideration, leading to its equilibrium geometry of lower symmetry, or to dissociation.

As nuclear displacements lay in the base of the formation, deformation, and transformation of polyatomic systems, there should be at least one excited state involved (along the ground state) in these processes or, if we consider these properties of the system in the excited state, another excited state or the ground state should be involved. It follows that at least two electronic states of the reference configuration should be involved in the rationalization of structural changes of any polyatomic system. This is the two-electronic-state paradigm in the formation, deformation, and transformation of molecular systems and solids, or compressed, two electronic states in transformations (TEST) paradigm (Bersuker, 2013).

In the simplest case of the PJTE when only one nondegenerate excited state contributes significantly to the instability of the ground state in the reference configuration, we have two interacting electronic states Ψ_{01} and Ψ_{02}, with an energy gap between them E_{02} - E_{01} = Δ. Then, by taking the energy read-off from the ground state level, we get a secular equation for the energy of the two states as follows:

$$\begin{vmatrix} (1/2)K_{01}Q^2-\varepsilon & FQ \\ FQ & (1/2)K_{02}Q^2-\varepsilon+\Delta \end{vmatrix}=0, \tag{9}$$

where K_{01} and K_{02} are the primary force constants in the ground and excited states, respectively. The solution of this 2×2 equation is straightforward:

$$\varepsilon_{1,2}=\frac{1}{4}(K_{01}+K_{02})Q^2+\frac{\Delta}{2}\mp\frac{1}{2}\sqrt{\left[\frac{1}{2}(K_{01}-K_{02})Q^2-\Delta\right]^2+4F^2Q^2} \tag{10}$$

These two functions at $Q \gg 0$ are strongly controlled by the relative values of four parameters: Δ, F, and K_0^Γ, K_{01}, and K_{02}. More precisely, they are determined by the relation between the energy gap Δ and the ratio $2F^2/K_{01}$. First, by evaluating the second derivatives, we get the curvatures of the two functions (10) at $Q = 0$ as follows:

$$K_1=K_{01}-\frac{2F^2}{\Delta},\ K_2=K_{02}+\frac{2F^2}{\Delta} \tag{11}$$

Thus, the vibronic coupling decreases the curvature of the APES of the lower state by $2F^2/\Delta$, while that of the higher state increases by the same amount. If

$$\Delta < 2F^2/K_{01}, \tag{12}$$

the curvature of the lower state becomes negative, meaning instability in the Q direction. This is the PJTE in the case of two interacting electronic states (Bersuker, 2013). It leads to the softening (always) and instability (under the criterion (12)) of the lower state and hardening (always) of the upper one (Figure 1). It is important to emphasize that in the condition of instability (12), there is no limitation on any of the three parameters, taken separately (notably, the most visible energy gap Δ), as their magnitude may be compensated by the combination of the other two of them.

Figure 1. Schematic illustration of the adiabatic potential energy curves in the PJTE in the simplest cases when two electronic states are coupled via one active mode Q. It may be regarded as a relevant cross-section of the APES for a realistic system with several nuclear degrees of freedom. The two states in the PJTE have different symmetry, Q is a low-symmetry (distortive) coordinate, and the energy gap 2Δ at Q = 0 may be relatively very large. E_{pJT} *is the PJTE stabilization energy.*

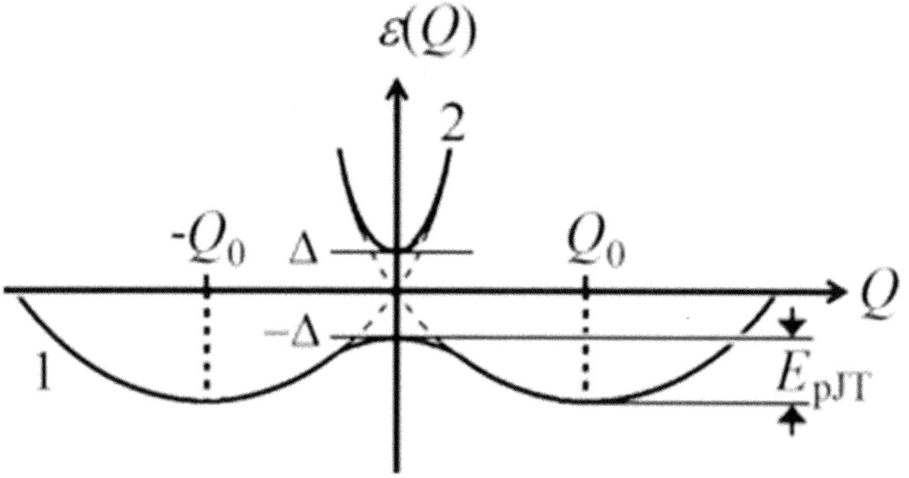

Note, however, that in any polyatomic system there are always many excited states of relevant symmetry, albeit at large energy gaps $\Delta_{1n} = E_n - E_1$, which mix to the ground one by vibronic coupling. Their contribution to the PJTE problem can be taken into account by the second-order perturbation corrections p_1 and p_2, which can be given as follows [see, for example, (Liu, 2010); (Hermoso, 2014); (Ilkhani, 2015)]:

$$-p_1 = -2\sum_n \frac{|F_{1n}|^2}{E_n - E_1} \text{ and } -p_2 = -2\sum_n \frac{|F_{2n}|^2}{E_n - E_2}, \tag{13}$$

where F_{1n} and F_{2n} are the respective vibronic coupling constants between the states |1 and |2 and the high-energy excited states |n. They may be rather small, just lowering the curvature of the mixing states from K_{01} and K_{02} to $K_{01} - p_1$ and $K_{02} - p_2$, and not producing instability of the ground state. With such values of K_{01} and K_{02}, Eq. (10) for fitting the parameters takes the form

$$\varepsilon_{1,2} = \frac{1}{4}(K_{01} - p_1 + K_{02} - p_2)Q^2 + \frac{\text{”}}{2} \mp \frac{1}{2}\sqrt{\left[\frac{1}{2}(K_{01} - p_1 - K_{02} + p_2)Q^2 - \text{”}_{12}\right]^2 + 4F_{12}^2Q^2} \tag{14}$$

At small Q the resulting values of the curvatures of the APES of the two states K_1 and K_2 that are coupled by the PJT interaction in the Q direction, are (cf. Eq. (11)):

$$K_1 = \left(K_{01} - p_1\right) - 2F_{12}^2 / \Delta, K_2 = \left(K_{02} - p_2\right) + 2F_{12}^2 / \Delta. \qquad (15)$$

In many cases, the picture of PJTE instability may be simplified by employing the molecular orbital (MO) presentation instead of (or in addition to) the energy terms. It illustrates the role of individual electronic orbitals in the origin of instability and, in many cases, allows for qualitative estimates of the PJTE with much less computation. In the MO presentation, the electron-nuclear interaction part V in the Hamiltonian H, not shown explicitly in Eq. (5), is taken as a sum of one-electron terms $V = \Sigma_i V_i$, and the two wavefunctions Ψ_0 and Ψ_n in the definition of the vibronic coupling constant F_{0n} = áΨ_0|(¶H/¶Q)$_0$|Ψ_n〉 are taken in the form of single Slater determinants, composed by the one electron MO wavefunctions |φ_iñ. Then, by assuming that the excited states Ψ_n follow from a one-electron excitation α ® m from the occupied MO |φ_α〉to the unoccupied MO, |φ_mñ, we get for the vibronic coupling constant F_{0n} in Eq. (7) the following expression:

$$F_{0n} = \Psi_0 | \left(\partial H / \partial Q\right)_0 | \Psi_n = \sqrt{2} f^{\alpha m}, \qquad (16)$$

if the occupied MO has two electrons, and

$$F_{0n} = \Psi_0 | (\partial H / \partial Q)_0 | \Psi_n = f^{\alpha m} \qquad (17)$$

if the occupied MO is with one electron. Here

$$f^{\alpha m} = \varphi\alpha |_{(\partial} V / \partial Q)0 | \varphi_m \qquad (1_8)$$

is the orbital vibronic coupling constant (Bersuker, 2006). Similar to Fon, ${}_f$αm ^{is} nonzero only if the product of irreducible representations of the two MOs |φαñ ${}_a$nd |φmñ, Γα × Γm, ${}_c$ontains the representation ΓQ o${}_f$ the nuclear coordinate Q. According to Eq. (6), the contribution of the PJTE instability due to the vibronic coupling between these two states is:

$$K_v^{0n} = -4 \frac{\left(f^{\alpha m}\right)^2}{E_n - E_0}, \qquad (19)$$

if |φ_αñ is the double occupied MO (a spin-singlet ground state Ψ_0), and

$$K_v^{0n} = -2 \frac{\left(f^{\alpha m}\right)^2}{E_n - E_0}, \qquad (20)$$

if |φ_αñ is a single occupied MO (doublet ground state Ψ_0). In general,

$$K_v^{0n} = -\left(q_\alpha - q_m\right)\frac{\left(F_{0n}\right)^2}{E_n - E_0}, \tag{21}$$

where q_i are the occupation numbers of the corresponding MO active in the respective PJTE mixing.

Note that only occupied with unoccupied energy levels interact to produce the PJTE. If the two interacting orbitals are both occupied, then, as follows from Eq. (21), $q_\alpha = q_m = 2$, and the PJTE destabilization of the lower state is compensated by the equivalent stabilization of the upper state resulting in a total zero effect. If the electron populations of the lower and upper states are different, $q_\alpha{}^{1}\, q_m$, the PJTE is approximately proportional to their difference $q_\alpha - q_m$. Note also that, as mentioned above, the realization of the PJTE is limited by the condition that the low-symmetry atomic displacements Q produce nonzero vibronic coupling F in Eq. (7), meaning sufficiently strong overlap between the wavefunctions, which are orthogonal in the high-symmetry configuration, and this was shown to improve the covalence bonding between the corresponding atoms (Bersuker & Polinger, 1989); (Bersuker, 2006); (Bersuker, 2016). In other words, the PJTE is the driving force for atomic rearrangements (distortions) that lower the total energy of the system, mainly by improving the covalence bindings with possible additional contributions from charge redistributions (polarizations) (Bersuker & Polinger, 1989); (Bersuker, 2020). All the possible (qualitatively visible) reasons of distortions of high-symmetry configurations in the absence of electronic degeneracy are particular cases of the PJTE.

In other words, the PJTE is the driving force for atomic rearrangements (distortions) that lower the total energy of the system, mainly by improving the covalent bindings with possible additional contributions from charge redistributions (polarizations) (Bersuker & Polinger, 1989); (Bersuker, 2006). All the possible (qualitatively visible) reasons of distortions of high-symmetry configurations in the absence of electronic degeneracy are particular cases of the PJTE.

Added Covalence Bonding as the Driving Force of the PJTE

It was shown (and multiply demonstrated on different occasions) that the driving force of the PJTE is added (improved) covalence bonding (Bersuker et al., 1979); (Bersuker & Stavrov, 1988); (Bersuker, 1993): the PJTE distortion takes place when the change in some interatomic distances by distortion results in better interatomic orbital overlap, and hence in energy gain by the added covalency to the bonding between the atoms. Indeed, in the starting high-symmetry configuration the wavefunctions of the electronic states, ground and excited, are orthogonal by definition. As a result of the distortion, their orthogonality is removed, producing a nonzero overlap between them. If for two near-neighbor atoms the ground state wavefunction pertains (mainly) to one of them, while the excited state wavefunction belongs (mainly) to the other one, their overlap contributes to the covalent bonding between them. Also, if the gain in energy due to this additional covalent bonding is larger than the loss in the deteriorating inner bonds by distortion, the system becomes spontaneously distorted. It was also shown earlier that if the two wavefunctions mixed by the nuclear displacements belong (mainly) to the same atom, then the corresponding term in the PJTE contribution to the instability has the meaning of the electronic polarization of that atom induced by these displacements. Calculations show that the atomic polarization contribution is by orders-of-magnitude smaller than that of added covalence (Bersuker et al., 1984); (Bersuker et al., 1992); (Bersuker, 1993).

Figure 2 illustrates this effect in the above cases of the displacement of the central proton towards one of the electronegative atoms.

Figure 2. Schematic illustration to the PJTE-induced added covalent bonding in reactions of proton transfer: (a) with the H atom in the central position between two electronegative atoms (F, O, N) at a fixed (relatively large) interatomic distance, its 1s orbital overlap with the appropriate linear combination of the p-orbitals of the latter (which produce an excited state MO) is zero (positive and negative overlaps compensate each other), so these orbitals do not contribute to the hydrogen—electronegative-atom bonding; (b) the PJTE displacement of the proton to one of the atoms results in their nonzero overlap producing a covalent contribution to this bonding.

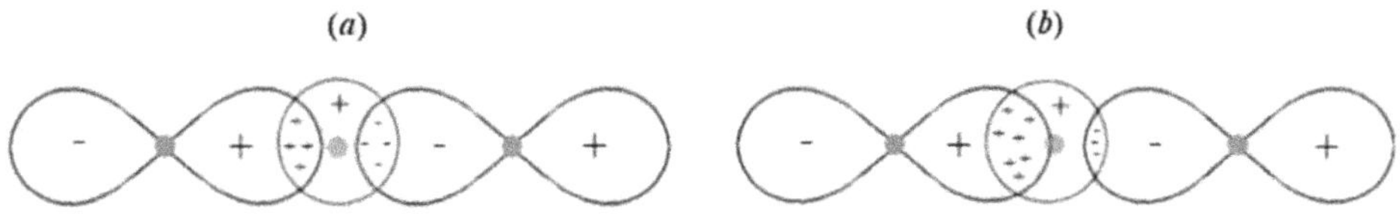

The Effect of Solvation

Usually, the formation of a hydrogen bond includes the emergence of a dipole moment in the system. This circumstance is most noticeable in the case of a symmetric case, A–H–A. In the initial configuration, when the hydrogen atom is at an equal distance from the bounded fragments A, the electron density is equally distributed between the two fragments, and the dipole moment of the system is zero. However, the displacement of the hydrogen atom from the center breaks the symmetry of the charge distribution (Fig. 2), causing a local dipole moment.

This circumstance is significant when a polar-unstable hydrogen bond occurs in a polar medium. The occurrence of a local dipole moment polarizes the medium. In turn, the polarized medium creates a local electric field acting upon the off-center displaced hydrogen atom lowering its electrostatic energy. Since the electric field of the hydrogen dipole decreases with distance from the system, this effect is local. As a result, a shell of polarized molecules forms around the fragment A–H×××A enhancing the local electrostatic field acting on the dipole. The mutual augmentation of the dipole-induced polarization and medium-enhanced dipole instability is called solvation and resembles the polaron effect.

Solvation determines the relative stability of conformers, determines the rate of chemical reactions, and, therefore, plays a significant role in chemistry and biochemistry (see, for example, Dogonadze (1985) and Marcus (2015)). Interpreting the hydrogen bond in terms of the pseudo Jahn-Teller effect allows a simple phenomenological explanation of solvation. In this model, the polar medium effectively changes one of three parameters of the system, the energy gap Δ, the coupling constant F, and the force constant K_0.

Since the gap Δ follows from the energy spectrum of the system at $Q = 0$, when the dipole moment is zero, its magnitude does not depend on the polar property of the medium. The value of the force constant K_0 is determined mainly by the covalent repulsion of the electron shells of approaching atoms. It is effective at a relatively short interatomic distance and does not depend on the induced electric field of the medium. Due to the PJTE, the off-center displacement of hydrogen is $Q_0 = F/K_0$, and the dipole moment of the hydrogen bond is proportional to Q_0. Therefore, the effect of polarization of the medium is in a strengthening of the vibronic coupling. We describe it by a factor ε, replacing the vibronic constant F with εF. In its physical meaning, the dimensionless factor $\varepsilon > 1$ is like the dielectric constant of the polar medium, and its magnitude is proportional to the actual dielectric constant. In specific media, this empirical parameter should be measured experimentally or determined by fitting from quantum mechanical calculations (see Section 4 below).

INSTABILITY OF ACTIVATED STATES OF CHEMICAL REACTIONS AS A PSEUDO JAHN-TELLER EFFECT

From the general picture of the PJTE in Figure 1 for the case of two interacting electronic states, ground and excited, via only one active nuclear displacement coordinate Q, we see that along this coordinate the ground state APES of the system has two minima with an energy barrier between them, similar to the curve (cross-section of the APES) between the reactants and products of a chemical reaction along the (steep-slop) reaction path. Involving the statement that the instability of the configuration at the maximum of the barrier is (uniquely) due to the PJTE (Section 2), and allowing, for simplicity, that there is only one excited state, which is effectively coupled to the ground state to produce the instability in a given direction, we come to eqs.(11) and (12) in Section 2 for the curvatures of the reaction curves near the point of the barrier maximum along the reaction coordinate Q. From these equations, it follows that in such cases the curvature of the active excited state in this direction is positive, meaning that, dependent on the APES in the other cross-sections, there may be stable (observable) excited states of unstable intermediates of chemical reactions (Bersuker, 1980).

Instability of the Transition State of the Reaction H + H_2 ® H_2 + H

The feasibility of the idea about the PJTE origin of the negative curvature of the potential surface at the point of the transition state of a chemical reaction and the existence in this nuclear configuration of stable excited states that cause the instability of the activated complex of the reaction (Bersuker, 1980), can be demonstrated in a simple case of the reaction H + H_2® H_2 + H in a linear collision (Bersuker et al., 1987). In this case the the saddle point on the H_3 potential surface corresponds to a linear symmetric nuclear configuration of $D_{¥h}$ symmetry, with the reaction coordinate corresponding to the low-symmetry σ_u-type nuclear displacement of the $D_{¥h}$ point group (Figure 3). Indeed, the vibrational frequency analysis in the equilibrium $D_{¥h}$ nuclear configuration shows the presence of an imaginary frequency with the value 4884.5 cm^{-1}, which corresponds to the σ_u-type displacement.

Figure 3. Reaction coordinate Q_{σ_u} *= 0.814z1-0.407z2-0.407z3*

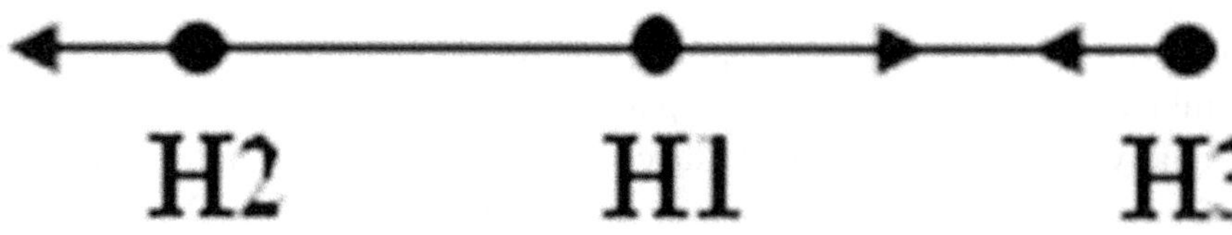

Figure 4 shows the results of *ab initio* calculations (at the CAS(3,6) level with the 6-311+G(p) basis set) of the cross-sections of the APESes along the reaction coordinate Q_{σ_u} .

Figure 4. Cross-sections (APESes) for the ground $^2\Sigma_u$ and the first excites $^2\Sigma_g$ states for the H_3 system along the reaction coordinate $Q(\sigma_u)$

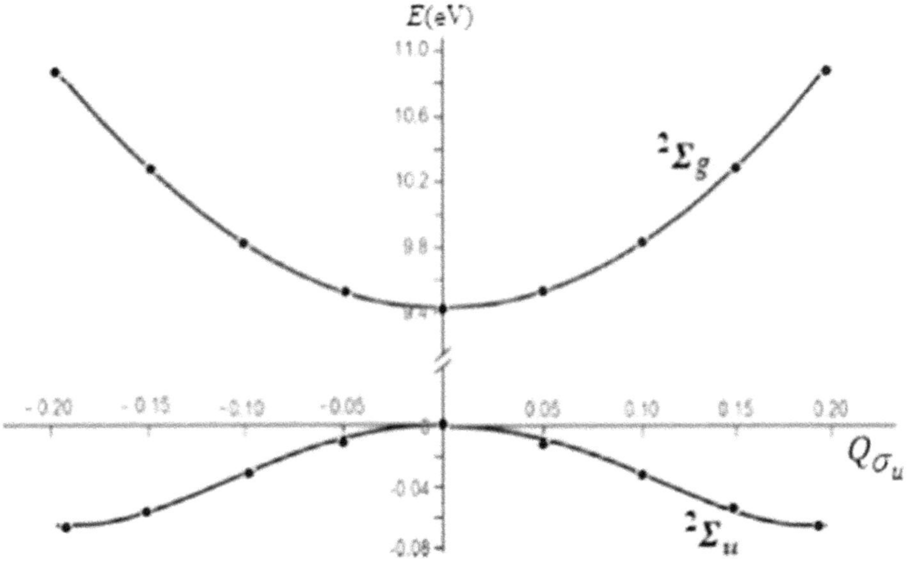

One can see the main trend: the instability of the activated state of the reaction at Q = 0 is accompanied by a stable excited state. The parameters of the PJTE, estimated by fitting the solutions of the secular equation of the type (10) to the *ab initio* calculated energy profiles, are: K_0(gr)=29.30 eV/Å^2, K_1(ex)=58.18 eV/ Å^2, $F^{(01)}$=13.18 eV/Å, where K_0(gr) and K_1(ex) are the non-vibronic contributions to the curvature of the APES in the ground and the excited states, respectively, and $F^{(01)}$ is the vibronic coupling constant, $F^{(01)} = {}^2\Sigma_u|\left(\partial H / \partial Q_{\Sigma_u}\right)_0|^2\Sigma_g$. With these parameters the vibronic contribution to the curvature of the ground state APES, K_v, is $K_v = -2F^2/\Delta = -36.83$eV/Å^2. Then, the resulting values of the curvature of the APES for the ground state K(gr)= –7.53 eV/A^2, and for the excited state K(ex)=95.01 eV/A^2. At the barrier maximum at Q=0 the electronic configuration of the ground $^2\Sigma_u$ state is $(1\sigma_g)^2(1\sigma_u)$, while in the excited state $^2\Sigma_g(1\sigma_g$® $1\sigma_u)$ it becomes $(1\sigma_g)(\sigma_u)^2$, so their primary force constants (and the corresponding curvatures) are significantly different. The obtained numerical results fully confirm the instability of the configuration at the maximum of the energy barrier and the existence of the stable excited state.

Vibronic Origin of the Metastability of the Hydronium Radical and the Energy Barriers of the Dissociation, H_3O·®H_2O+H.

The qualitative features of the electronic structure, the form of the APES of the hydronium radical and possible ways of its dissociation, H_3O•®H_2O+H, can be explained by considering the general three-mode

Jahn-Teller problem (A_1′+A_2″+E′)Ä(a_2′+*e*′+*e*′) in the high-symmetry nuclear configurations of the D_{3h} and C_{3v} symmetry (Ogurtsov et al., 2003); (Ogurtsov et al., 2007). Calculations of the electronic structure of the $H_3O^•$ in the D_{3h} conformation show that in this nuclear configuration the system has the nondegenerate ground state $^2A_1'$ and the low-lying excited double-degenerate $^2E'$and nondegenerate $^2A_2''$states with the energy gaps of 2.71eV, and 8.2eV, respectively. In addition, the vibrational frequencies analysis also shows that the D_{3h} configuration of the $H_3O^•$ system is unstable with respect to a_2''-type nuclear displacement (with the imaginary frequency at 452.71*i*), which transforms its planar configuration into a pyramidal one. In the basis set of four electronic states, the ground state lA_1¢ñ, two degenerate states, lE¢xñ and lE¢yñ of the excited E¢ term, and the excited state lA_2¢¢ñ, and using as the coordinates of the two *e*¢-type active vibronic coupling modes, q_x and q_y (Bersuker & Polinger, 1989); (Polinger et al., 1992), we get the following 4 ´ 4 potential energy Hamiltonian matrix for the $H_3O^•$ system as follows:

$$\hat{H} = \left[\frac{1}{2}K_1\left(q_x^2 - q_y^2\right) + \frac{1}{2}K_2 q_{a_2''}^2\right]\hat{I} + \begin{pmatrix} \Delta & 0 & 0 & V_2 q_{a_2''}^2 \\ 0 & \delta + Vq_x & Vq_y & V_1 q_x \\ 0 & Vq_y & \delta + Vq_x & V_1 q_y \\ V_2 q_{a_2''}^2 & V_1 q_x & V_1 q_y & -2\delta \end{pmatrix} \tag{22}$$

Here *d* and *D* are the parameters determining the energy gaps between the A_1¢and E¢ and between the A_1¢ and A_2¢¢ terms, respectively: 3*d* = *E*(E¢) - *E*(A_1¢), *D* + 2*d* = *E*(A_2¢¢) - *E*(A_1¢), *E*(A_1¢), with *E*(E¢) and *E* (A_2¢¢) as the total energies of the lA_1¢ñ, lE¢ñ and lA_2¢¢ñ states, K_1 and K_2 are the primary force constants of the *e*¢ and a_2¢¢displacements in the ground state, *V* is the linear vibronic coupling constant in the excited E¢ state, F_1 and F_2 are the vibronic constants of the A_1¢- E¢ and A_1¢- A_2¢¢ PJT coupling, respectively, and $\hat{I}$ is the identity matrix.The eigenvalues of this matrix form a four-sheet hypersurface, the APES in the space of three coordinates (q_x, q_y and $q_{a_2''}$), which includes the JTE on the E term and the PJTE coupling between all of them.

Three cross-sections of the APES of this hydronium radical were considered: *a*) along the generalized coordinate $q_x = q_{e_x'}$ associated with the *e*¢-mode in the D_{3h} nuclear configuration ($q_{a_2''}$ =0), which transforms the D_{3h} configuration into the C_{2v} one; it corresponds to one of the dissociation passes, $H_3O^•$®H_2O+H, from the planar structure, b) along the $q_{a_2''}$ coordinate, corresponding to the a_2¢¢ mode ($q_{e_x'}$ =0); and *c*) along the generalized coordinate $q_x = q_{e_x}$ of the *e*-type in the C_{3v} configuration, which also corresponds to dissociation of $H_3O^•$ to H_2O + H from the pyramidal structure.

For the case (*a*) the matrix of the Hamiltonian (22) is reduced to the following 3´3-matrix:

$$\hat{H} = \frac{1}{2}K_1 q_x^2 + \begin{pmatrix} \delta + Vq_x & 0 & V_1 q_x \\ 0 & \delta + Vq_x & 0 \\ V_1 q_x & 0 & -2\delta \end{pmatrix} \tag{23}$$

with the roots

$$\varepsilon_{1,2}\left(q_x\right)=\frac{1}{2}K_1q_x^2+V\frac{q_x}{2}-\frac{\delta}{2}\mp\frac{1}{2}\sqrt{\left(Vq_x-\delta\right)+8\delta^2+8\delta Vq_x+4V_1^2q_x^2}\,, \qquad (24)$$

$$\varepsilon_3\left(q_x\right)=\frac{1}{2}K_1q_x^2+\delta-Vq_x.$$

Along the a_2¢¢ displacements (case b) the APES cross-section curves as functions of the $q_{a_2''}$ coordinate are as follows:

$$\varepsilon_{1,2}\left(q_{a_2''}\right)=\frac{1}{2}K_2q_{a_2''}^2+\frac{\Delta-2\delta}{2}\mp\frac{1}{2}\sqrt{\left(\Delta+2\delta\right)^2+4V_2^2q_{a_2''}^2} \qquad (25)$$

The expressions in the APES cross-sections for the above case (*c*) coincide with those for the case (*a*), except for the different values of the parameters in the Eqs. (24) and (25).

Figure 5 shows the calculated cross sections of the APES of $H_3O^{\bullet}$ along the above considered coordinates in the region of small values of q_{ex}. First, the calculations confirm the presence of the local minimum in both the D_{3h} and C_{3v} configuration (planes (*a*) and (*c*)). Dissociation barriers in both cases are found to be very low: 0.04 kcal/mole and 0.13 kcal/mole for the planar (D_{3h}) and the pyramidal (C_{3v}) structures, respectively. In the plane (*a*) the cross-section of the APES along the generalized *e′* coordinate is shown. At $q_{e_x'} \neq 0$ the D_{3h} symmetry is reduced to the C_{2v}, the E¢ doublet splits, and one of its components (A_1) has a strong vibronic admixture to the ground state, resulting in the instability of the later with respect to dissociation to H_2O + H. The parameters of the vibronic coupling were estimated by fitting *ab initio* calculated data for the APES to the Eqs. (24). The values of parameters of the JTE and the PJTE obtained in this way are presented on the right side of Figure 5.

In the plane (*b*) of Figure. 5, the cross section of the APES along the $q_{a_2''}$ coordinate is shown (as $q_{a_2''}$ value, the distance *d* between the O atom and the H_3 plane is used). In this case, due to the pseudo JTE admixture of the excited $^2A_2$¢¢ term, the ground electronic ^2A¢ state becomes unstable with respect to the out-of-plane displacement of the oxygen atom. At *d*=0.41Å the curve has a minimum, the energy of stabilization being equal to 3.74 kcal/mole. As earlier, the values of the parameters of the PJT coupling, K_2 and V_2, were obtained by fitting the Eq. (25) to the *ab initio* potential curve.

The plane (c) in Figure 5 shows the cross-section of the APES along the q_{e_x} coordinate starting from the C_{3v} configuration. As in the case (a), there is a local minimum at $q_{e_x} = 0$. The calculated height of the barrier for dissociation to H_2O+H is equal to 0.13 kcal/mole. Fitting the *ab initio* data for the APES to the Eqs. (24) yields the shown parameters of the vibronic coupling for this case.

Figure 5. Three cross-sections of the APESes of hydronium radical along: (a), (b), and (c) coordinates

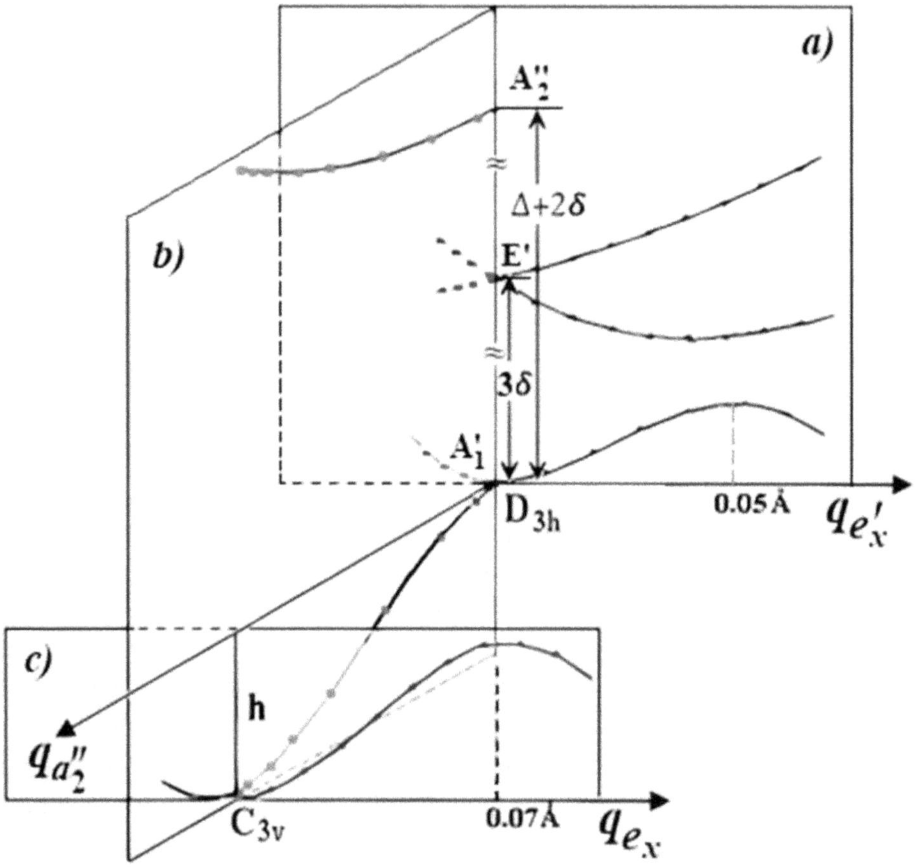

Thus, based on the *ab initio* calculations of the electronic structure and the potential energy surfaces, it was shown that the $H_3O^{\bullet}$ radical is metastable in the nuclear configurations of the D_{3h} and C_{3v} symmetry, and the origin of the metastability of these configurations and the barriers to the dissociation $H_3O^{\bullet}$®H_2O+H are attributed to the common action of the pseudo Jahn-Teller effect in the ground state and the Jahn-Teller effect in the first excited degenerate state. The dissociation from the D_{3h} nuclear configuration can be realized in two ways: directly to H_2O+H with a small barrier (Figure 6) or through the C_{3v} configuration, $H_3O^{\bullet}(D_{3h})$®$H_3O^{\bullet}(C_{3v})$®H_2O+H, with a smaller barrier at the first step ($H_3O^{\bullet}(D_{3h})$® $H_3O^{\bullet}(C_{3v})$), but a higher barrier height at the second step, ($H_3O^{\bullet}(C_{3v})$® H_2O+H).

Figure 6. Potential energy cross-sections along the two possible reaction pathways for the dissociation of hydronium radical, $H_3O^{\cdot}$ ® H_2O + O. The dark-blue curve corresponds to the one-step dissociation from the planar-triangular D_{3h} to the C_{2v} configuration by elongating the radial distance O–H with one of the three hydrogen atoms (a). The green curve and the red curve correspond to the two-step process from the planar-triangular D_{3h} to the metastable triangular pyramid C_{3v} (the green part) and then, shown in red, dissociation to the C_{2v} configuration (b).

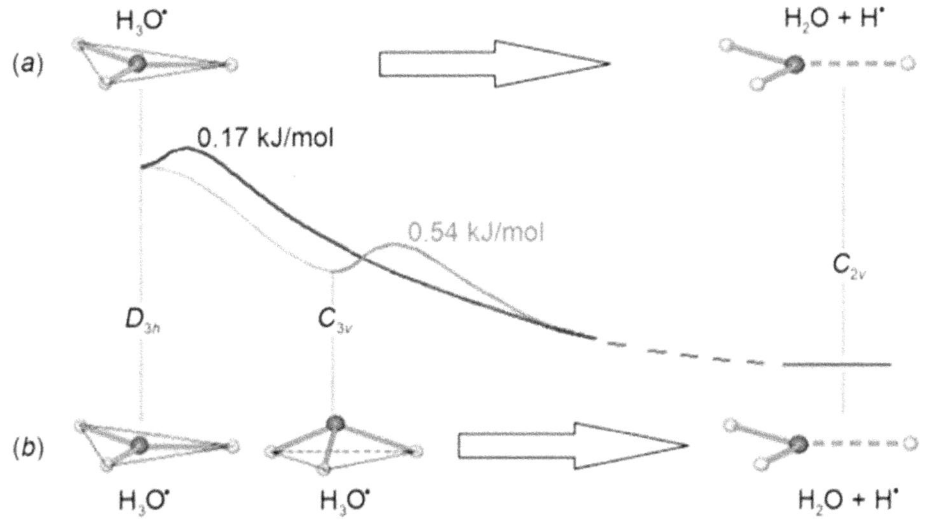

The PJTE in intermediates of photochemical reactions

Many environmental processes are influenced by light, the corresponding chemical reactions taking place in the exited electronic states of the environmentally active molecular systems. Obviously, the excited states of the latter are subject to the PJTE interaction with the higher in energy states, and the same is true for the activated (intermediate) states of the photochemical reactions that take place after the photoexcitation. It follows that the same considerations with regard to the PJTE origin of the energy barrier of chemical reaction, outlined above for environmental species in the ground state, are valid also for photochemical reactions, provided they are applied to the photo-excited state.

The general theoretical aspect of the PJTE origin of the energy barriers of photochemical reaction was considered in detail before (Wang, 2019). Figure 7 shows its main features. Upon excitation from the stable ground state S_0 to the excited state S_1, the latter, due to (and dependent on) the strength of the PJTE coupling to the higher in energy state S_2 becomes either just softened (meaning, lowering its curvature), or unstable (with negative curvature) with respect to the distorting displacement Q. For the sake of induced photo-reactivity, and handling mainly charge-neutral organic molecules, the Q displacements leading to reactivity are expected to be polarizing. Therefore the photo-process of light induced polar structures and related enhanced reactivity was called "sudden polarization" (Wang, 2019). As a result of strong sudden polarization the system acquires well space-separated positive and negative charges in a zwitterion configuration.

Figure 7. The adiabatic potential energy (APE) curves illustrating the pseudo Jahn-Teller effect in the excited state S_1: due to the vibronic coupling to a higher in energy electronic state S_2; dependent on the strength of the coupling, the system becomes either unstable (red curve (a)), or just softened (blue curve (b)) in the direction of the symmetrized nuclear displacements Q. The dashed curves show the APE when the PJTE is ignored; Δ is the energy gap between the mixing states, and S_0 is the stable ground state.

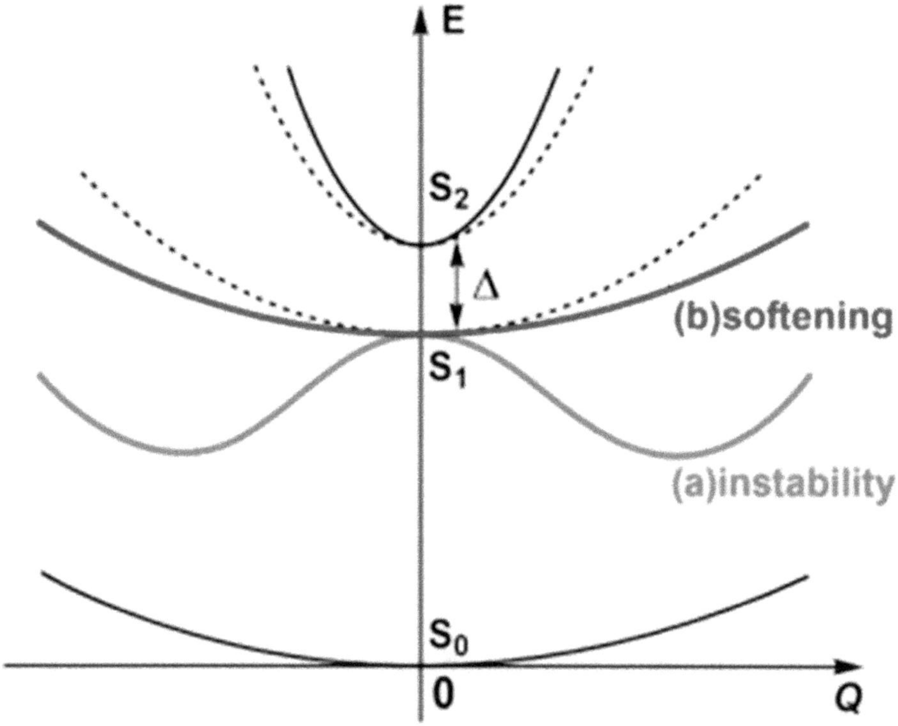

The work (Wang, 2019) shows the many details of the process of sudden polarization and zwitterion formation in organic molecules, involving multimode PJTE and ab initio calculations of the polar-modes contributions to the sudden polarization and zwitterion formation. Figure 8 illustrates the ab initio calculated formation of the zwitterion configuration of fluoro-stilbene upon photo-excitation. It is seen that its potential energy curves follow exactly the predictions of the PJTE in application to the photo-exited state (the zwitterion mode Q_{int} iz an adjusted combination of the polar modes of the molecules (Wang, 2019)).

Figure 8. Potential energy curves of S_0, S_1 and S_2 states of F-stilbene along the zwitterion mode Q_{int}. (Wang, 2019). There are two equivalent minima in the excited state S_1 at Q_{int} = ±0.06 Å with the dipole moments at the minima μ = ±1.88 Debye (cf. Figure 7)

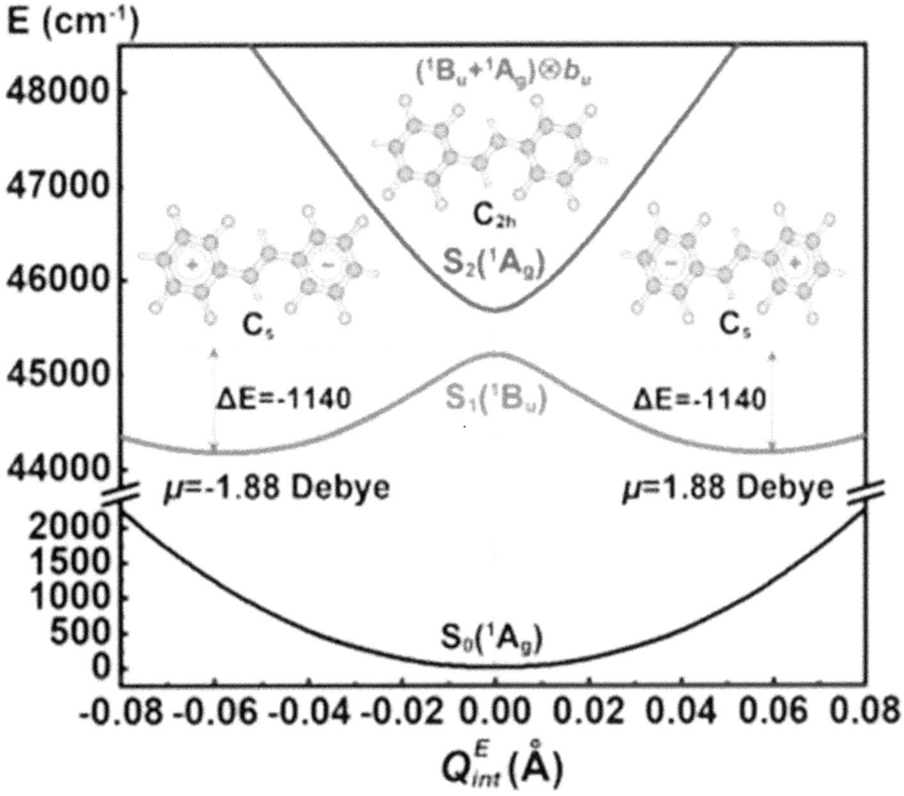

THE H-BOND IN ENVIRONMENTAL REDOX PROCESSES AS A PSEUDO JAHN-TELLER EFFECT

Proton transfers underlies many important chemical transformations, including reduction-oxidation processes, high proton mobility in aqueous solutions, protons diffusion through membrane protein channels, etc.; they play a crucial role in vital phenomena occurring in the atmosphere and in living systems [see, e.g., (Chen et al., 2015); (Elliott & Paddison, 2007), (Jacquemin et al., 2014) and references therein]. A large number of experimental and theoretical works are devoted to the study of proton-active molecular systems, the most widespread of which are protonated aqua-clusters of the type $H^+(H_2O)_n$ or $H_3O^+(H_2O)_n$ [(Elliott et al, 2007), (Jacquemin et al, 2014); (Grabowski, 2015); (Pylaeva et al., 2017)].

Proton-Transfer Energy Barrier in Hydrogen-Bounded $[FHF]^-$

The bifluoride anion plays an important role in the study of the hydrogen bond. Earlier it was established (both experimentally and by the benchmark quantum-chemical calculations) that the $[FHF]^-$ anion is linear and centrosymmetric in the gas phase, whereas in the crystal structures the proton position displacement from the mid-point are observed (Panich, 1995). The relationship between the F-H and F-F distances has been discussed in Ref. (Cornaton, 2016); it was shown that the longer the F-F distance, the greater the displacement of the H atom from the central position to one of the fluorine atoms.

A detailed analysis of the PJTE origin of the proton-transfer energy barrier in the model FHF^- system was carried out in the work (Gorinchoy, 2021). In particular, it was confirmed that the equilibrium nuclear configuration of the $[FHF]^-$ anion is the linear one of the $D_{¥h}$ symmetry with the position of the H-atom exactly in the middle of F•••F distance (no imaginary frequency is observed). Calculated equilibrium F•••F distance is equal to 2.24 Å. At the same time, the vibrational-frequency analysis of $[FHF]^-$ anions at increased F•••F distances indicates the presence of one imaginary frequency (Table 1), which corresponds to a low-symmetry σ_u-type distortion, associated mainly with the displacement of the central proton towards one of the fluorine atoms. This means that the $D_{¥h}$ configuration is unstable with respect to symmetrized σ_u-type displacements of its atoms. It can also be seen from Table 1 that with increasing F-F distance, the height and width of the proton transfer barrier also increased.

Table 1. The values of imaginary frequencies ω, the height h and the width 2Qmin of the proton transfer energy barrier at different F···F distances.

R_{F-F} (Å)	ω (cm^{-1})	h (eV)	$2Q_{min}$ (Å)
2.24	-	-	0.00
2.64	1333.84	0.340	0.70
2.84	1523.78	0.759	1.00

The ground state wavefunction of the system belongs to the $^1\Sigma_g$ representation of the $D_{¥h}$ symmetry point group; the valence electron configuration is $\ldots(1\pi_u)^4(1\pi_g)^4(3\sigma_u)_2(4\sigma_g)^0(5\sigma_g)^0(2\pi_u)^0(6\sigma_g)^0(2\pi_g)^0$. Some of these valence molecular orbitals are shown in Figure 9(*a*). According to the PJTE, excited states that produce the instability of the ground state nuclear configuration should have the $^1\Sigma_u$ symmetry: $\Sigma_g \times \Sigma_u = \Sigma_u$.

It was shown that the main contribution to the instability of the symmetric $D_{¥h}$ structure at increased F•••F distances and the proton displacement to one of the fluorine atoms are due to the pseudo Jahn–Teller mixing of the ground electronic state $^1\Sigma_g$ with the lowest excited state of $^1\Sigma_u$ symmetry, which is formed by one-electron excitation from the HOMO $3\sigma_u$ to the LUMO $4\sigma_g$ (Figure 6(*a*)), the energy gaps Δ_{01} between this state and the ground one varies in the range of 10÷13 eV, depending on the F-F distance (Gorinchoy, 2021). Thus, the PJTE problem is simplified to a two-level one, $(1^1\Sigma_g + 1^1\Sigma_u)Ä\sigma_u$.

Ab initio calculated energy profiles (APES cross-sections) of the systems along the coordinate of instability σ_u for several distances F•••F are shown in Figure 9(*b*-*d*). It can be seen that at the equilibrium distance R_{F-F} equal to 2.24Å, the adiabatic potential is a single-well one, meaning the proton is located in the middle between the fluorine atoms (Figure 9(*b*)). Two other curves in Figures 6(*c*) and 6(*d*) are the cuts of the APES for R_{F-F} = 2.64Å and R_{F-F} = 2.84Å, respectively. It is seen that with increasing F-F distances, the symmetric $D_{¥h}$ nuclear configuration becomes unstable with respect to the displacement of the central proton to one of the fluorine atoms, and the APES of the ground state becomes a double-well one. The height and width of the energy barrier to the asymmetric displacement of the proton also increase. It can also be seen that the APES of the corresponding $^1\Sigma_u$ excited states, admixed to the ground state in the PJT effect, have a minimum at the point Q=0.

Figure 9. (a) The MO energy level scheme for the [FHF]$^-$ system in the ground $^1\Sigma_g$ electronic state with indication of the one-electron excitations to the $^1\Sigma_u$ terms; (b),(c),(d) show cross-sections of the APES of [FHF]$^-$ along the σ_u-type distortion for R(F···F)=2.24Å, 2.64Å, and 2.84Å, respectively; both ab initio and analytical (Eq.(10)) calculations are shown by points and curves, respectively.

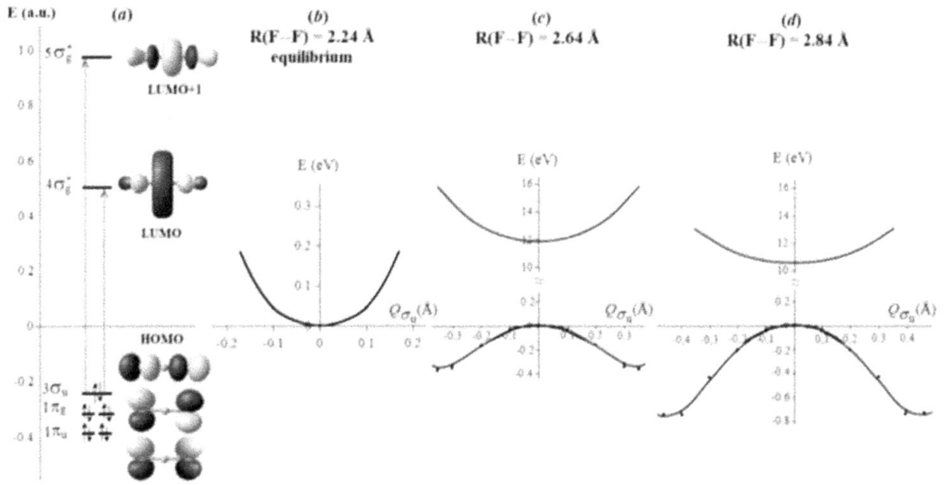

The values of the PJTE parameter are obtained by fitting Eq. (10) to the ab initio calculated energy profiles, see Table 2. The parameters are as follows: Δ is the energy gaps between the ground $^1\Sigma_g$ and the excited $^1\Sigma_u$ terms, with the primary force constants K_{01} and K_{02}, respectively in them, F_{12} is the vibronic coupling constants, and K_{gr} is the resulting value of the curvature of the ground state APES,, which at small Q values can be approximately estimated according Eq. (11). It is seen that the obtained values of K_{01} from Table 2 are positive for all F-F distances, showing that vibronic mixing of the ground state with only the first excited states produce the instability of the ground state D¥h nuclear configuration. The absolute value of the curvature of the adiabatic potential K_{gr} increases with increasing distance F ••• F, which results in an increase in the height and width of the energy barrier (Table 2).

Table 2. PJTE coupling constants F12, energy gaps Δ, primary force constants K01 and K02, and resulting force constants Kgr in the reference D¥h configuration of the [FHF]− anion, obtained by fitting Eq. (10) to the ab initio calculated energy profiles.

$R_{F\text{-}F}$ (Å)	Δ (eV)	F_{12} (eV/Å)	K_{01} (eV/Å²)	K_{02} (eV/Å²)	K_{gr} (eV/Å²)
2.24	13.055	-	-	-	+9.172
2.44	12.796	16.193	36.541	52.259	-4.443
2.64	11.826	15.980	30.847	32.796	-12.339
2.84	10.570	13.956	21.782	21.848	-15.073

Thus, we can conclude that the hydrogen bonds in hydrogen bi-halides $[XHX]^-$ directly follow the predictions of the pseudo Jahn-Teller effect: all calculated potential energy curves along the off-center proton displacement fully coincide with those predicted from the general vibronic coupling theory of the pseudo Jahn-Teller interaction of the ground electronic state with the appropriate low-lying excited state.

PJTE Origin of Proton-Transfer Energy Barriers in Protonated, $(H_5O_2)^+$ and $(H_5O_2)^+(H_2O)_4$, and Deprotonated, $H_3O_2^-$, Water Clusters

The $H_5O_2^+$and $H_3O_2^-$ ions have been the subject of extensive experimental and theoretical studies because of their important role in proton transfer in aqueous solutions. They both display stable configurations with one hydrogen atom located between the two oxygen atoms, as in (O--H--O)$^+$ and (O--H--O)$^-$, respectively. Below we show that the proton transfer in these systems is controlled by the PJTE (see also (Geru, 2012)).

For the dihydronium cation $H_5O_2^+$, the starting symmetry is D_{2h} or D_{2d}. Their geometry optimization (in the 6-311G* basis set) shows that the D_{2d} configuration is lower in energy as compared with the D_{2h} one by 5.51 kJ/mol (Geru, 2012). In this geometry, the out-of-center displacement of the central proton towards one of the oxygen atoms is a b_2-type distortion (Figure 10).

Figure 10. Low-symmetry b_2-type distortion of the $H_5O_2^+$system, which transforms the symmetric D_{2d} nuclear configuration into the C_{2v} one

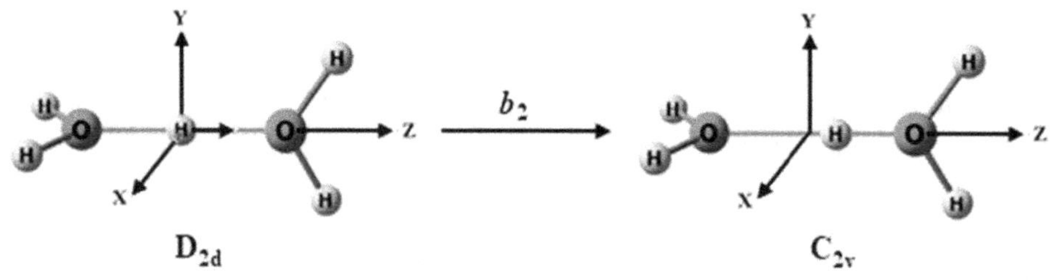

The ground-state wavefunction of $H_5O_2^+$ system belongs to the $^1A_1$ representation of the D_{2d} symmetry point group. Then only excited states of $^1B_2$ symmetry contribute to the instability of the D_{2d} configuration with respect to this distortion: $A_1 \times B_2 = B_2$. Calculations of the total energy of $H_5O_2^+$ as a function of the proton displacement (at the R_{O-O}= 2.34Å) shows a single-well adiabatic potential (Fig. 11). This means that the shared proton is located midway between the oxygen atoms, in accordance with the reported experimental data (James, 2005) and high-level ab initio calculations (Dicken, 2005).

However, with increasing the O–O distance, the APES becomes a double-well one. In Figure 11, the cross-sections of the APES of the dihydronium cation along the b_2 coordinate are shown for several distances R_{O-O}. The excited states of B_2 symmetry that contribute significantly to the instability of the ground state, are formed by one-electron excitation from the HOMO-1 of b_2 symmetry to the LUMO+3 of a_1 symmetry.It is seen that, with increasing the O-O distance, the height and the width of the barrier to proton transfer increases. The parameters of the PJTE are presented in Table 3.

Figure 11. (a) The MO energy level scheme for the $H_5O_2^+$ system in the ground $^1A_1$ electronic state with indication of the one-electron excitations to the $^1B_2$ terms; (b), (c) and (d) show the cross sections of the APES of $[H_5O_2]^+$along the b_2-type distortion for R(O···O) = 2.34Å, 2.36Å, and 2.38Å, respectively; both ab initio and analytical (by Eq.(10)) calculations are shown by points and curves, respectively.

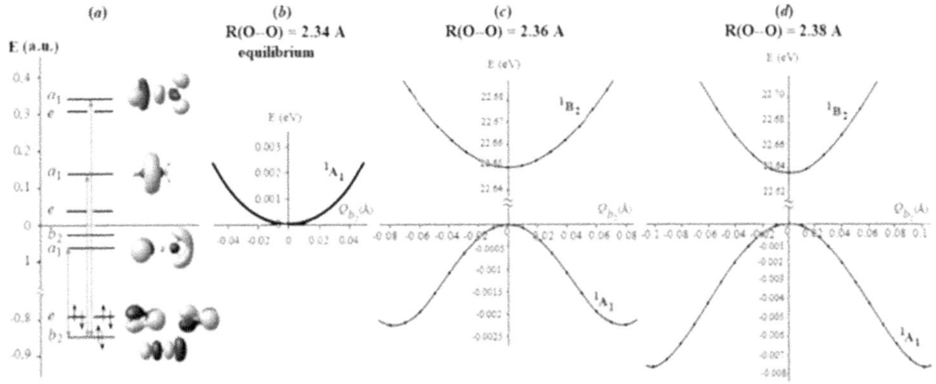

Again, the values of the PJTE parameter obtained by fitting Eq.(10) to the ab initio calculated energy profiles are shown in Table 3. It is seen that the obtained values of K_{01} from Table 3 is positive for all O-O distances, showing that the vibronic mixing of the ground state with only one excited state produce the instability of the ground state D_{2d} nuclear configuration. The absolute value of the curvature of the adiabatic potential K_{gr} increases with increasing distance O•••O, which results in an increase in the height (ΔE_{PJTE}) and width ($2Q_{min}$) of the energy barrier (Table 3).

Table 3. PJTE coupling constants F_{12}(eV/Å), energy gaps Δ(eV), energy gain ΔE_{PJTE} (kJ/mol), barrier width $2Q_{min}$(Å), primary force constants K_{01} and K_{02}, and resulting force constants K_{gr}(eV/Å²) in the reference D_{2d} configuration of $H_5O_2^+$obtained by fitting Eq. (10) to the ab initio calculated energy profiles. Parameters in the last row are given for the $H_5O_2^+(H_2O)_4$ cluster.

R_{O-O} (Å)	Δ	F_{12}	K_{01}	K_{02}	K_{gr}	ΔE_{PJTE}	$2Q_{min}$
2.34	22.685	-	-	-	+ 1.665	-	-
2.36	22.654	24.800	52.755	107.324	-1.540	0.217	0.16
2.38	22.634	24.282	49.207	121.425	-2.380	0.736	0.20
2.382	23. 843	21.358	41.516	106.125	-3.252	0.256	0.24

In the $H_5O_2^+(H_2O)_4$ cluster, which takes into account the first solvation sphere around the Zundel cation, the equilibrium O-O distance in its D_{2d} nuclear configuration is equal to 2.382Å (Figure 12). Calculations show the presence of one imaginary frequency with the value 477 cm^{-1}, which corresponds mainly to the off-center proton displacement leading to the stable C_{2v} nuclear configuration of the cluster.

Figure 12. A general view of the $H_5O_2^+(H_2O)_4$ cluster and the low-symmetry b_2-type distortion which transforms the reference D_{2d} nuclear configuration into the C_{2v} one

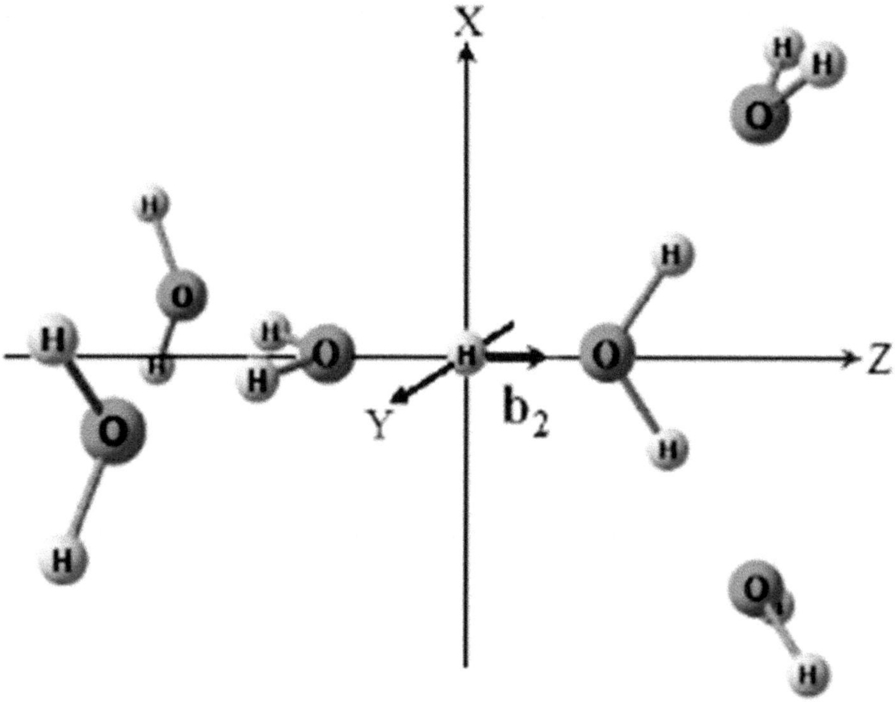

As in the case of the Zundel cation, the vibronic origin of the proton transfer barrier in this system is revealed by solving the PJT $({}^1A_1 + {}^1B_2)$Äb_2 problem. The cross sections of the APESes for the ground $({}^1A_1)$ and corresponding excited $({}^1B_2)$ states of the cluster are presented in Figure 13. Calculated parameters of the PJTE in this case are presented in the last row of the Table 3.

Figure 13. (a) The MO energy level scheme for the $H_5O_2^+(H_2O)_4$ system in its ground $^1A_1$ electronic state with indication of the one-electron excitation to the $^1B_2$ term; (b) Cross-sections of the APESes for the ground $^1A_1$ and corresponding excited $^1B_2$ states of the cluster along the b_2-type distortion

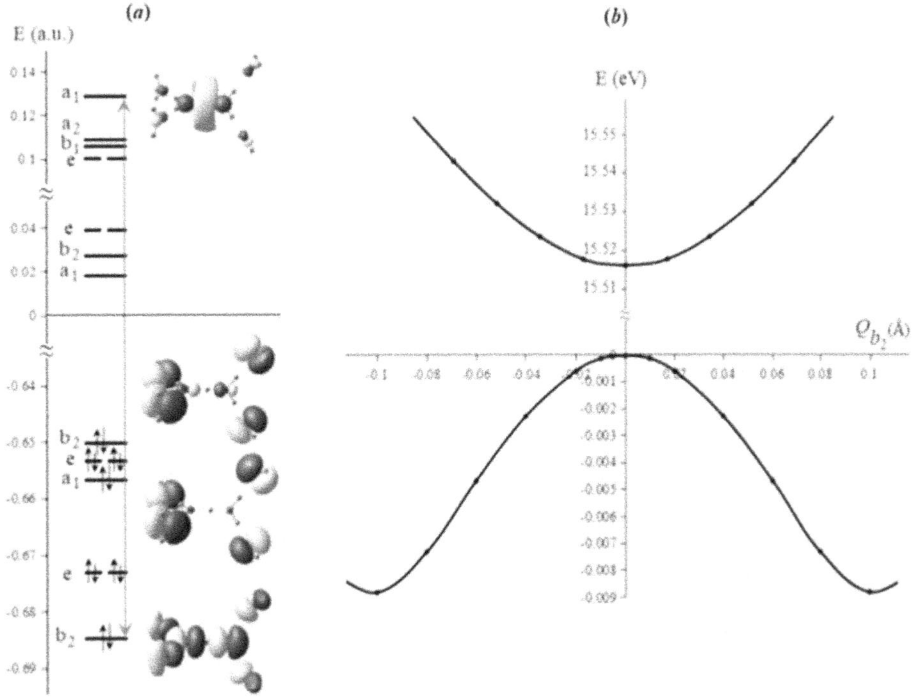

Thus, accounting for only the first solvation sphere around the Zundel cation in the $H_5O_2^+(H_2O)_4$ cluster leads to a double-well adiabatic potential energy surface along the hydrogen-bond coordinate, and to the instability of the on-center position of the shared proton.

To reveal the PJTE origin of proton transfer in the deprotonated water cluster $H_3O_2^-$, both proton-centered highest-symmetry nuclear configurations, cis-configuration of C_{2v} symmetry and trans-configuration of C_{2h} symmetry, can be chosen as the reference ones. These two configurations are transformed into one another by inversion rotation around the O-O axis. We started with the C_{2h} configuration, since (in our calculations) it is lower in energy than the C_{2v} structure by approximately 4 kJ/mol. *Ab initio* calculations of the $H_3O_2^-$ in the planar nuclear configuration of C_{2h} symmetry show that this configuration is unstable with respect to the two nuclear displacements (Figure 14). The distortion of b_u symmetry with the imaginary frequency 748.87cm^{-1}corresponds to proton motion along the O-H-O hydrogen bond and leads to the planar configuration of C_s symmetry. The second one of a_u symmetry (97.36*i*), corresponds to the out-of-plane displacements of the atoms (torsion rotation) resulting in the symmetrical C_2 structure. In their turn, both Cs and C_2 nuclear configurations are also unstable. The first one of Cs symmetry is unstable with respect to the a'' distortion leading to the equilibrium C_1 structure. The nuclear configuration of C_2 symmetry is unstable with respect to displacement of the central proton towards one of the oxygen atoms (b symmetry) also resulting in the equilibrium C_1 configuration (Figure 14).

Figure 14. Sequential symmetry breaking of the reference high-symmetry C_{2h} nuclear configuration of $H_3O_2^-$ anion.

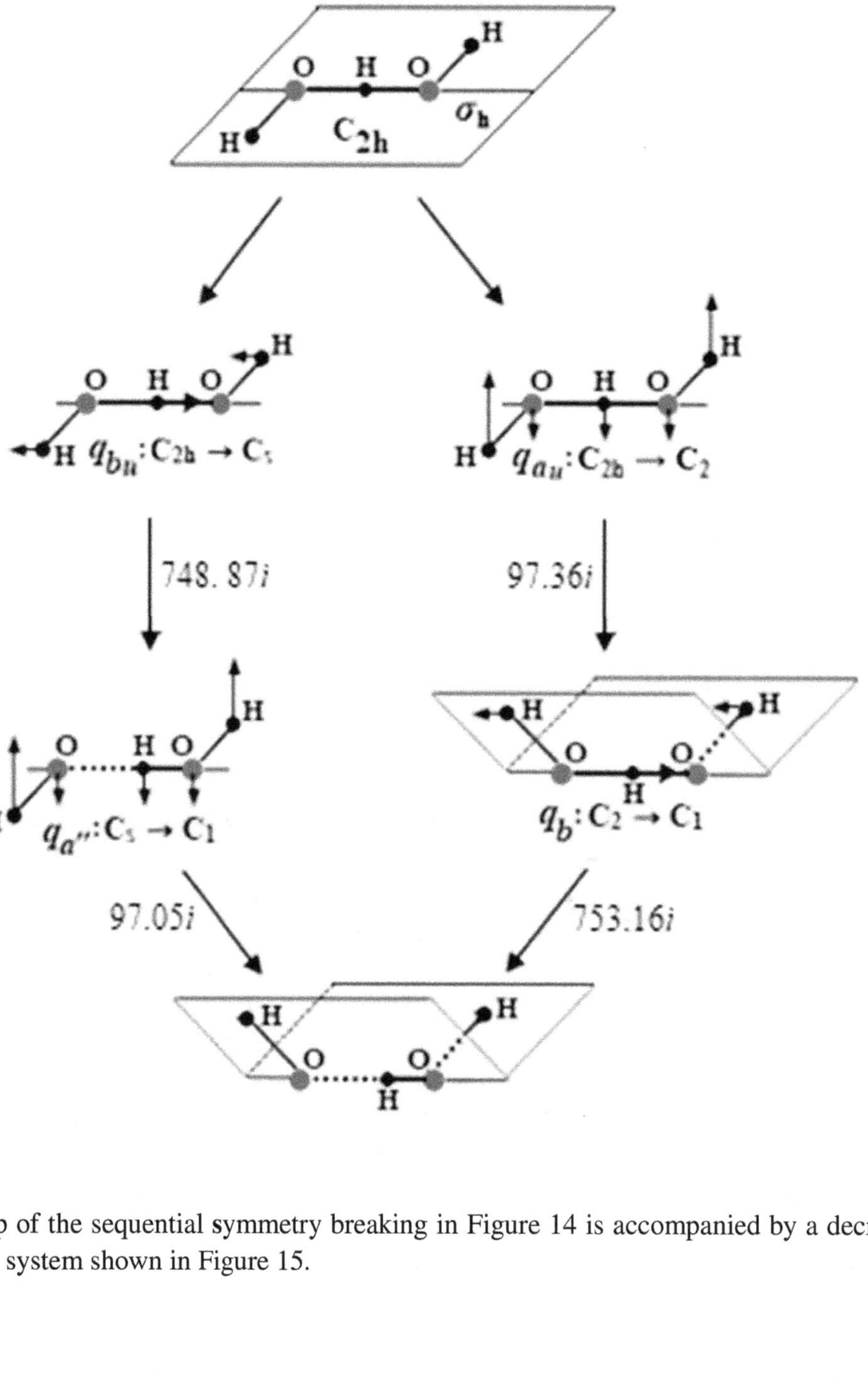

Every step of the sequential symmetry breaking in Figure 14 is accompanied by a decrease in the energy of the system shown in Figure 15.

Figure 15. Decrease in energy at each stage of transformation of high-symmetry C_{2h} nuclear configuration into the equilibrium structure of C_1 symmetry.

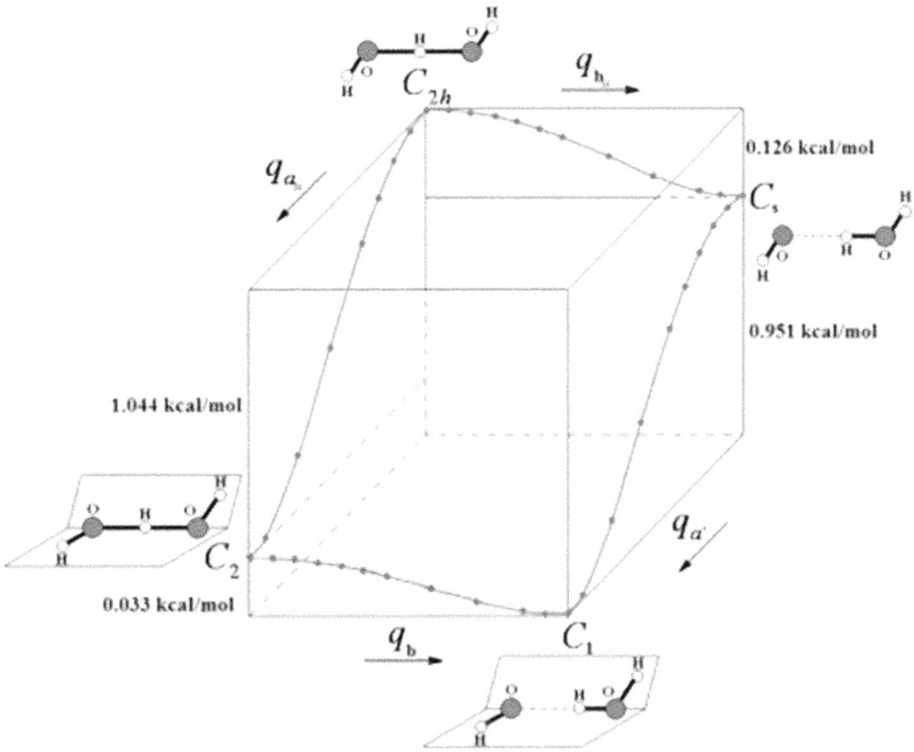

It is seen that the proton transfer in the hydrogen bond O-H-O can take place in both planar C_{2h} nuclear configuration, along the b_u coordinate (C_{2h}®C_s), and in the configuration of C_2 symmetry (C_2®C_1). For these two distortions, the corresponding cross-sections of the APESes for the ground and corresponding excited states were calculated. In the case of the C_{2h}®C_s transition, the main contribution to the instability of the symmetric C_{2h} nuclear configuration is due to the pseudo Jahn–Teller mixing of the ground electronic state ^1A_g with the excited state ^1B_u, formed by one-electron excitation HOMO (b_u) ® LUMO+2 (a_g) (Figure 16).

Figure 16. Molecular orbitals scheme in the C_{2h} geometry and the APESes along the Q_{bu} coordinate for the ground ${}^1A_g$ and excited ${}^1B_u$ states of the $H_3O_2^-$ system.

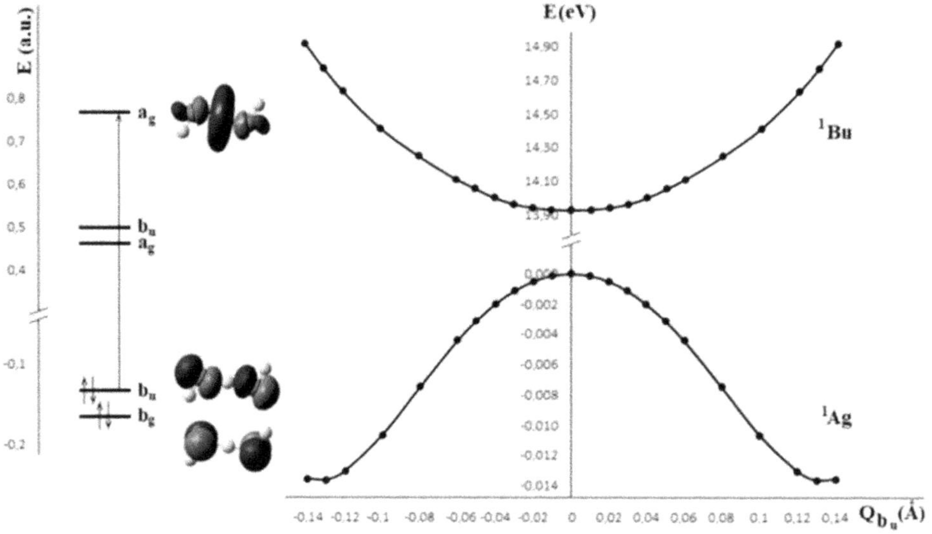

For the C_2®C_1 distortion the instability of the ground state ${}^1A_1$ of the $H_3O_2^-$ system is provided by its PJTE coupling with the excited state of ${}^1B$ symmetry, formed by one-electron excitation HOMO(*b*)®LUMO+2 (*a*) (Figure 17).

Figure 17. Molecular orbitals scheme in the C_2 nuclear configuration and the APES cross-section along the coordinate Q_b for the ground ${}^1A$ and excited ${}^1B$ states of the $H_3O_2^{\cdot-}$ system.

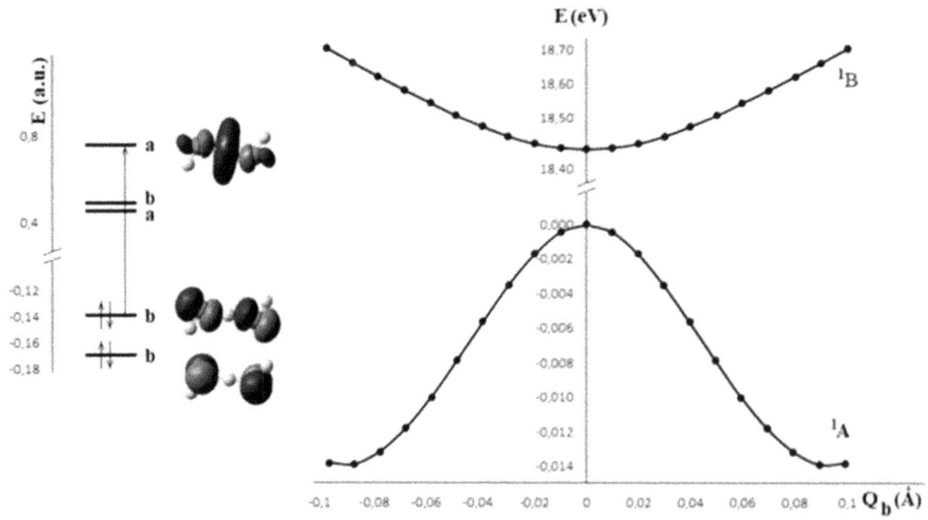

Thus, in both cases we have two-level PJTE problems, $(^1A_g + {}^1B_u)\ddot{A}b_u$ and $(^1A + {}^1B)\ddot{A}b$. Having determined the excited states which are active in the PJTE, and with the calculated cross-sections of the adiabatic potential surfaces (Figures 16, 17), we can estimate the values of the PJTE parameters by fitting Eq. (10) to the ab initio calculated energy profiles (Table 4).

Table 4. PJTE coupling constants F_{12} (eV/Å), energy gaps Δ (eV), energy gain ΔE_{PJTE} (kJ/mol), primary force constants K_0 and resulting force constants K_{gr} (eV/Å^2) in the reference C_{2h} and C_2 configurations of $H_3O_2^-$ obtained by fitting Eq. (10) to the ab initio calculated energy profiles.

q_r	Δ_r	ΔE_{PJTE}	K_0	F_r	$K_r = K_0\text{-}2F^2/\Delta_r$
Q_{bu} (C_{2h}®C_s)	13.93	0.527	12.12	11.01	-3.47
Q_b (C_2®C_1)	18.4	0.137	6.86	10.28	-4.63

Moreover, the calculations show that for each fixed torsion angle **δ**, the intermediate nuclear configurations of C_2 symmetry are also unstable with respect to the off-center proton's displacement (there is one imaginary frequency corresponding to distortion of the *b*-type of the anion $H_3O_2^-$). In Figure 18 the potential energy surface of the ground state of the $H_3O_2^-$ system in the two-dimensional space of the two normal coordinates of twisting (**δ**) and displacement (**Δr**) of the central proton towards one of the oxygen atoms is shown. Thus, the potential barrier to proton transfer during the conversion from the nuclear configuration of C_{2h} symmetry to the equilibrium C_1 structure can be determined by solving the two-mode PJTE problem$(^1A_g + {}^1B_u + {}^1A_u)\ddot{A}(b_u + a_u)$.

Figure 18. Schematic presentation of the APES of the ground state of $H_3O_2^-$ in the two-dimentional space of the two normal coordinates of twisting δ (the dihedral angle between the planes in nuclear configurations of C_2 symmetry) and displacement of the central proton towards one of the oxygen atoms Δr.

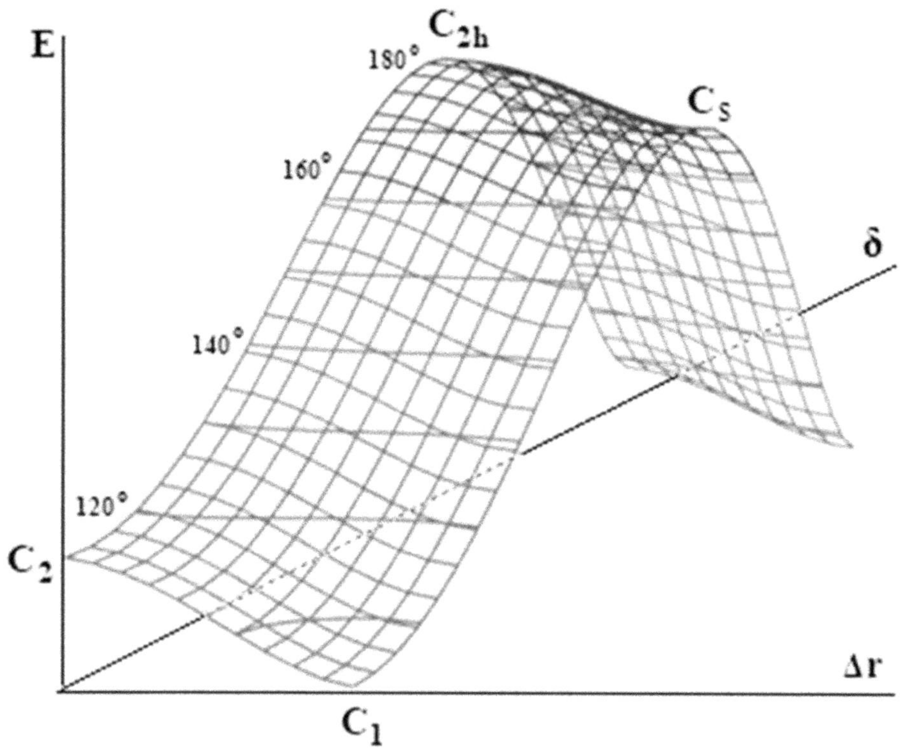

Energy Barriers of Proton Transfer in the Proton-Bound Dimers

In the aqueous solutions of some M molecules (M=pyridine, butanone, etc.), the transfer of the proton from the hydronium cation to the M molecule takes place via the formation of the hydrated MH^+ cation, while in an excess of M, protonated dimers of the MHM^+ species are formed. The latter are not stable, they have a short lifespan. Many researchers have reported that proton-bound dimers are rarely hydrated, or hydration destroys these dimers (Valadbeigi, 2019), (Valadbeigi, 2014). These proton-bound dimers are also observed in ion mobility and mass spectrometry studies. A quantitative analysis of the PJT origin of the proton transfer in the proton-bound ammonia dimer $N_2H_7^+$ was first carried out in the work (García-Fernández, 2008). Below, the PJTE is employed to rationalize the results of ab initio calculations of adiabatic potential energy surfaces for two proton-centered dimers, $(But)_2H^+$ and $(Pyr)_2H^+$ (But = 2-Butanone; Pyr = pyridine) and to demonstrate its efficiency in considering proton transfer in such proton-bound clusters (Balan, 2022).The geometry optimization of these proton-centered $(C_5H_5N)_2H^+$ and $(C_3H_8CO)_2H^+$ dimers in their high-symmetry nuclear configurations (D_{2h} and D_{2d} symmetry for proton-bound pyridine dimer and C_i symmetry in the case of the protonated dimer of butanone), and the analysis of the vibrational frequencies indicate the presence of one imaginary frequency ω in both cases; it corresponds mainly to displacement of the central proton to one of the oxygen or nitrogen atoms (Figures 19, 20, and Table 5).

Figure 19. Low-symmetry b_2-type distortion of the $(C_5H_5N)_2H^+$ system, which transforms the symmetric D_{2d} nuclear configuration into the C_2 one.

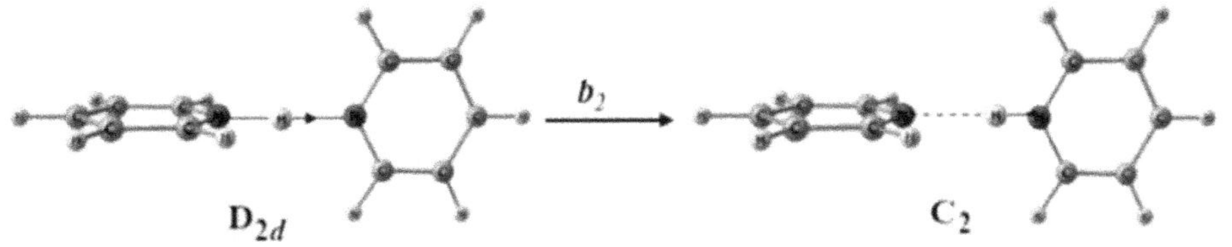

Figure 20. Low-symmetry b_u-type distortion of the $(C_3H_8CO)_2H^+$ system, which transforms the symmetric C_{2h} nuclear configuration into the C_1 one.

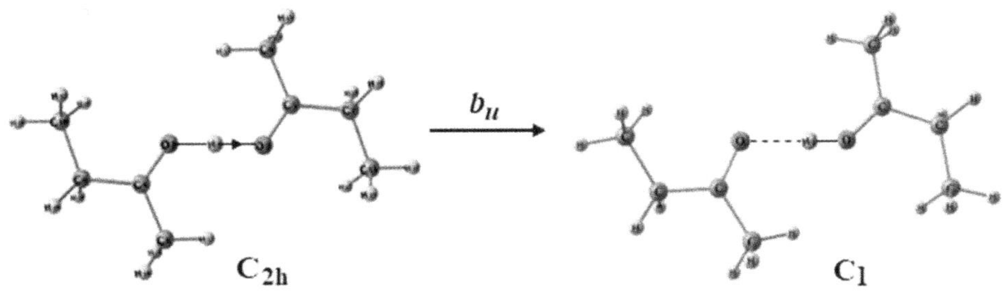

Table 5. Optimized geometric and electronic parameters for the $(C_5H_5N)_2H^+$ and $(C_3H_8CO)_2H^+$ systems calculated by the SCF-HF/6-311G(p) method

	$(C_5H_5N)_2H^+$		$(C_3H_8CO)_2H^+$	
	D_{2h}	D_{2d}	C_{2h}	
$R_{N\text{-}N}$ (Å)	2.565	2.552	$R_{O\text{-}O}$ (Å)	2.365
$R_{N\text{-}H}$ (Å)	1.282	1.276	$R_{O\text{-}H}$ (Å)	1.183
q_H	0.688	0.714	q_H	0.898
q_N	-0.289	-0.335	q_O	-0.523
ω (cm^{-1})	1512.80 (b_{1u})	1451.09 (b_2)	ω (cm^{-1})	1055.43 (b_u)
E (a.u.)	-493.9016	-493.9058	E (a.u.)	-462.4968

In the case of the two proton-bound pyridine molecules $(C_5H_5N)_2H^+$, calculations show that the D_{2d} structure is lower in energy, as compared with the D_{2h} configuration, by 11.026kJ/mol. For this reason, the nuclear configuration of D_{2d} symmetry was chosen as the starting one. In this configuration the ground-state wavefunction of the system belongs to the $^1A_1$ representation of the D_{2d} symmetry point group, corresponding to a closed-shell electronic configuration. The out-of-center displacement of the central proton toward one of the nitrogen atoms is a b_2-type distortion (Figure 19). This means that only the excited states of the $^1B_2$ symmetry contribute to the instability of the D_{2d} configurations with respect to this distortion: $A_1 \times B_2 = B_2$.

Ab initio calculated energy profiles for the ground A_1 and excited B_2states of the $(C_5H_5N)_2H^+$ systems along the coordinate of instability b_2 are shown in Figure 21.

Figure 21. The MO energy level scheme for the $(C_5H_5N)_2H^+$ system in the ground $^1A_1$ state with indication of the one-electron excitation to the $^1B_2$ terms, and cross-sections of the APESes along the b_2-type distortion; both ab initio and analytical calculations are shown by points and curves, respectively (Balan, 2022).

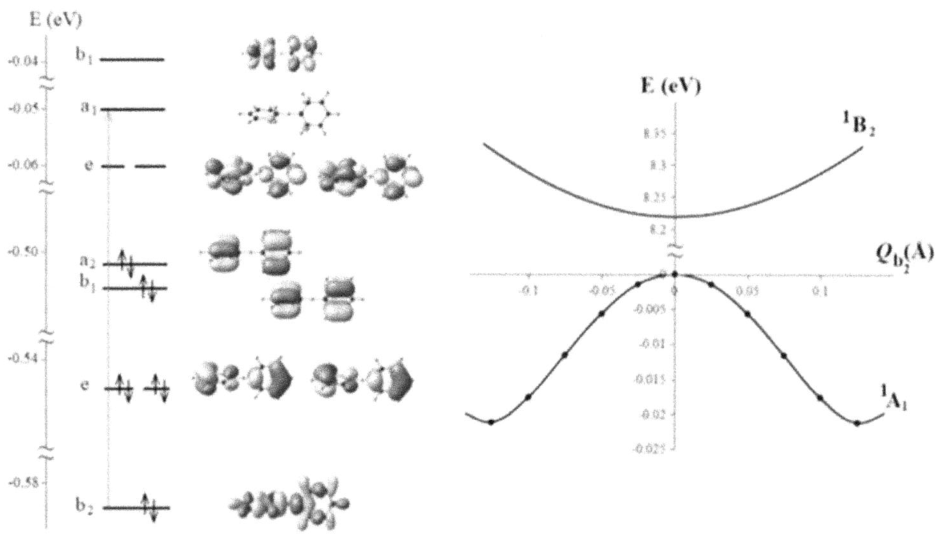

The values of the PJTE parameter are obtained by fitting Eq. (10) to the ab initio calculated energy profiles. They are as follows: the energy gap between the ground$^1A_1$ and the excited $^1B_2$ terms $\Delta = 8.219$ eV, K_{01}= 17.593 eV/Å^2, K_{02} = 131.412 eV/Å^2, the vibronic coupling constants F=9.689 eV/Å, and the resulting force constant at Q=0, $K_{gr} = K_{01} - 2F^2/\Delta = -5.251$ eV/Å^2.

The ground electronic state of the proton-bound dimer $(C_3H_8CO)_2H^+$ in the reference nuclear configuration of C_i symmetry is the singlet ^1A_g state. The out-of-center displacement of the central proton towards one of the oxygen atoms is of the b_u-type (Figure 20). Thus, the vibronic origin of the proton transfer barrier in this system can be revealed by solving the two-level (A_g+B_u)Äb_u PJTE problem. Ab initio calculated energy profiles for the ground ^1A_g and the excited ^1B_u states of the $(C_3H_8CO)_2H^+$ systems along the coordinate of instability b_u are shown in Figure 22.

Figure 22. The MO energy level scheme for the $(C_3H_8CO)_2H^+$ dimer in the ground ^1A_g state with indication of the one-electron excitation to the ^1B_u terms, and cross-sections of the APESes along the b_u-type distortion; both ab initio and analytical (by Eq.(15)) calculations are shown by points and curves, respectively (Balan, 2022).

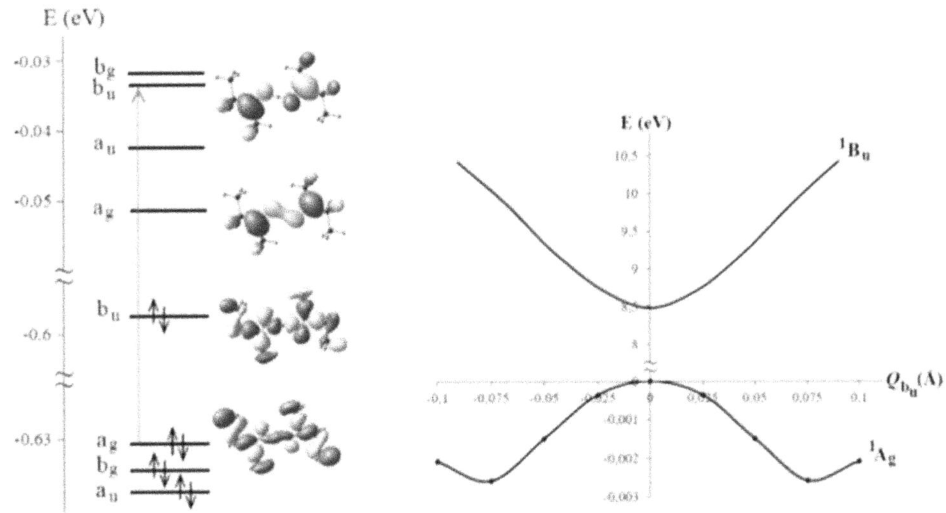

By fitting Eq. (10) to the ab initio calculated energy profiles (Figure 22) one can obtain the parameters of the PJTE in this case. They are: the energy gaps Δ between the ground $^1A_1$ and the excited $^1B_2$ terms Δ=8.478 eV, K_{01} = 18.652 eV/Å², K_{02} = 99.53 eV/Å², the vibronic coupling constant F = 9.253 eV/Å, and the resulting force constant at Q = 0, K_{gr} = -1.546 eV/Å².

THE JANN-TELLER AND PSEUDO JAHN-TELLER EFFECTS IN THE ELECTRON-TRANSFER PROCESSES

In this Section the role of the JTE and PJTE is analyzed in the electron-transfer triggered processes (such as, for example, in the formation of the coordination compounds), when common bonding and corresponding antibonding molecular orbitals are formed between the interacting reactants, that leads to redistribution of the electronic density in the activated complex.

The Jahn-Teller Effect in [(triphos)RhH(η^1:η^2-P_4)], the Intermediate in Hydrogenation of White Phosphorus, [(triphos)RhH_3]+P_4®[(triphos)Rh(η^3-P_3)]+PH_3

As an example of the role of the JTE in intermediate states of chemical reactions, we show here the reaction of the trihydride complex [(triphos)MH_3], M= Rh, Ir, and triphos=1,1,1-tris(diphenylphosphanylmethyl) ethane, MeC(CH_2PPh_2)$_3$, with white phosphorus, resulting in phosphane and the [(triphos)MP_3] complex. Based on the experimental data, it was assumed [(Peruzzini et al., 1998); (Barbaro, 1999); (Barbaro, 2003); (Barbaro, 2008)] that one of the stages of these processes includes the formation of the intermediate complex [(triphos)RhH(η^1:η^2-P_4)], shown in Figure 23 (Gorinchoy, 2011); (Balan & Gorinchoy, 2011).

Figure 23. Structure of the intermediate [(triphos)RhH(η1:η2-P4)] complex in the ab initio optimized geometry

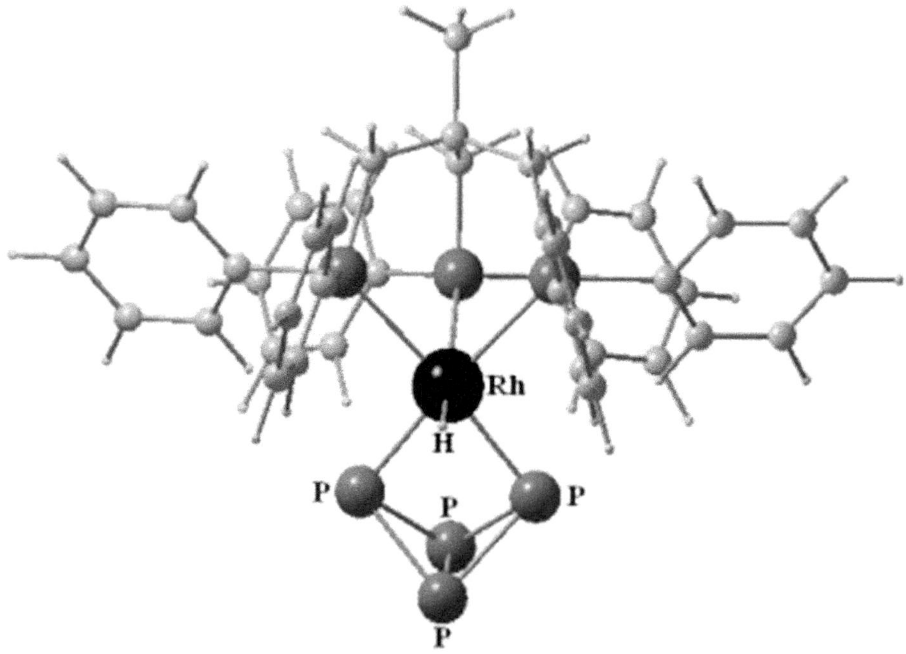

Calculations showed that the geometrical structure of P_4 is significantly changed by its coordination to the complex: one of the P-P bonds (P_1-P_4) is broken, and the tetrahedron P_4 is opened to form the *butterfly* geometry. In Figure 24 the correlation diagram between the molecular orbitals of the complex [(triphos)RhH(η^1:η^2-P_4)] and those of its fragments, P_4 and [(triphos)RhH], is shown.

Figure 24. MO energy-level scheme for bonding between P_4 and [(triphos)RhH]
(Balan & Gorinchoy, 2011)

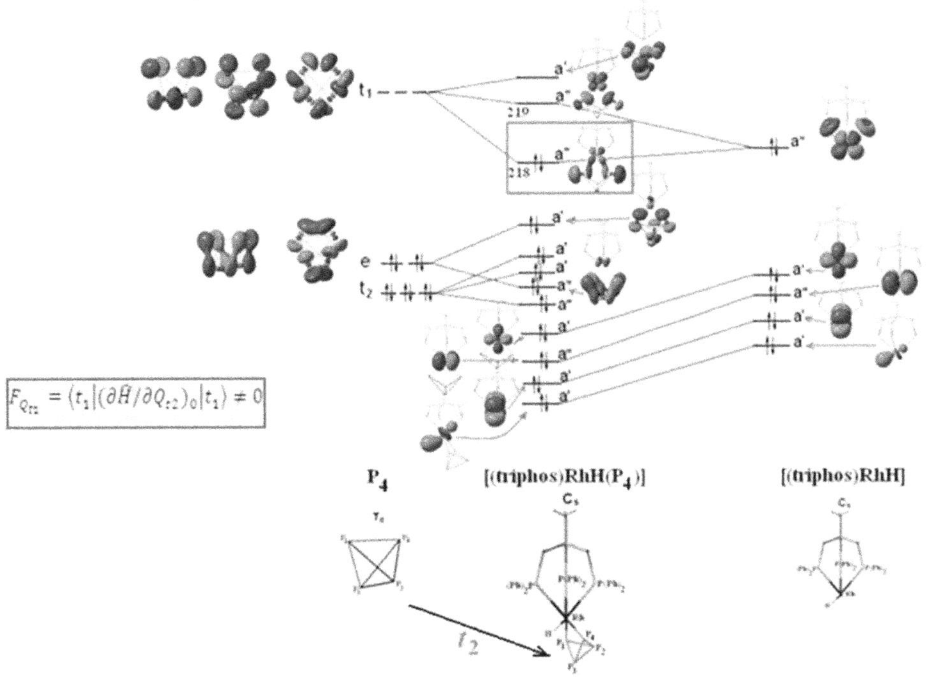

The ground electronic state of a free tetrahedral P_4 molecule is nondegenerate, and it has a closed shell electronic configuration (the left side in Figure 24). When P_4 is coordinated to the complex, only one bonding MO (218 ***a''***) provides for the Rh-P_4 binding. This π-type MO is composed from the filled ***a''*** MO of the complex [(triphos)RhH], which is mainly the $4d_{xz}$ AO of the atom of Rh (the right-hand side in Figure 24), and the appropriate component of the empty t_1 orbital of P_4. Due to forming of this (218 ***a''***) MO, the electron density $\Delta q(t_1) = 0.68\bar{e}$ is transferred from the $4d_{xz}$ AO of the Rh atom to one of the unoccupied t_1 orbitals of the threefold-degenerate T_1 term of P_4 (a π-type back donation). This leads to a nonzero distorting force, $F = \Delta q(t_1) f_{t2}^{t1} \neq 0$, acting on the phosphorus nuclei in the new electron distribution. Here f_{t2}^{t1} is the matrix element of the diagonal vibronic constant $f_{t2}^{t1} = t_1|(\partial H / \partial Q_{t_2})_0|t_1$. This force triggers the JTE t_2-type distortion, resulting in the observed *butterfly* geometry of coordinated P_4 molecule (Gorinchoy, 2011). The distortion of the P_4 molecule is accompanied by a decrease in energy, thereby stabilizing the intermediate complex.

PJTE in Molecular Coordination: Example of Coordinated Formaldehyde

The role of the PJTE in the transition states of chemical reactions can be demonstrated also by the example of formation of the (model) iron complex $Fe(CO)_2(PH_3)_2(\eta^2\text{-}CH_2O)$ (Gorinchoy, 2014). Metal-formaldehyde complexes have been postulated to be the pivotal intermediates in the catalytic activation of carbon monoxide. Some formaldehyde complexes were synthesized and structurally identified (Rosi, 1988); (Buhro, 1983); (Werner, 1998), showing that the η^2-coordinated CH_2O molecule is distorted to a nonplanar (C_S) shape, with a non-zero angle Đα between the CH_2 plane and the elongated C-O bond. A general view of the $Fe(CO)_2(PH_3)_2(\eta^2\text{-}CH_2O)$ complex with the used coordinate system is shown in Figure 25.

Figure 25. Optimized structure of $Fe(CO)_2(PH_3)_2(\eta^2\text{-}CH_2O)$; the calculated value of the C-O distance is 1.33 Å (versus 1.22 Å for the free molecule), and Đα=31.3°
(Gorinchoy, 2014).

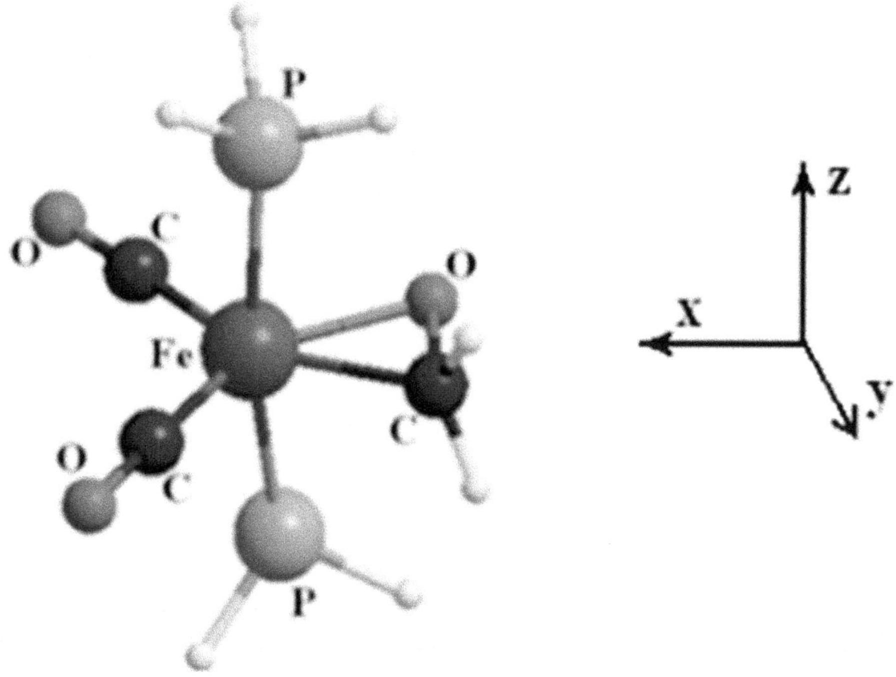

To reveal the main electronic redistribution details accompanying the formation of MOs of the complex under investigation, consider the MO energy-level scheme of the active valence zone of the whole complex $Fe(CO)_2(PH_3)_2(\eta^2\text{-}CH_2O)$ and that of its fragments, $Fe(CO)_2(PH_3)_2$ and CH_2O, shown in Figure 26.

Figure 26. MO energy-level scheme for bonding between H_2CO and $Fe(CO)_2(PH_3)_2$. For the sake of simplicity, only the main correlations are shown
(Gorinchoy, 2014).

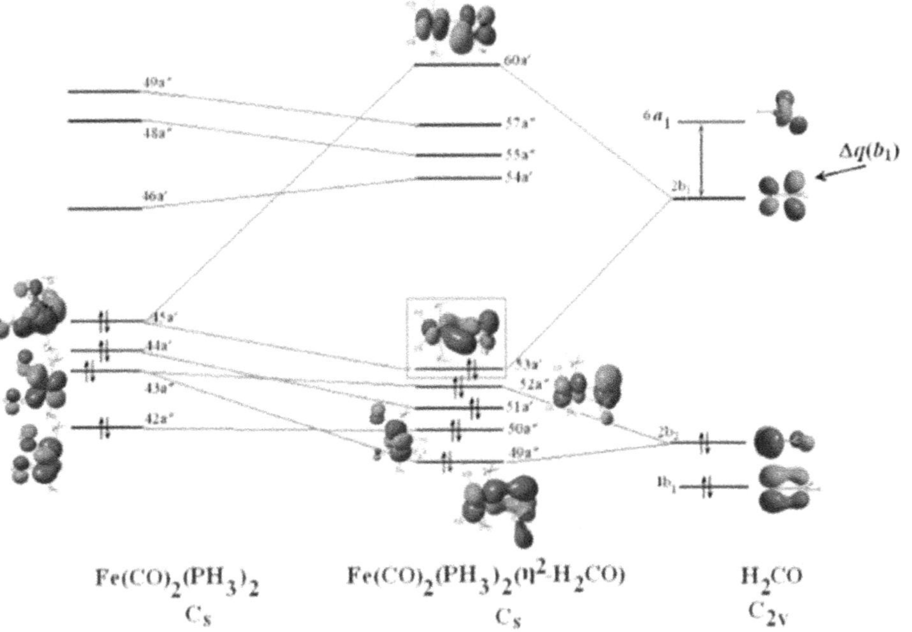

In the transition state, the occupied $2b_2$ MO of CH_2O interacts with the occupied $3d_{xz}$ MO of Fe, giving rise to the two occupied π-type MOs of the $Fe(CO)_2(PH_3)_2(\eta^2\text{-}CH_2O)$ complex: the bonding MO 49a ¢¢ and the antibonding MO 52a ¢¢ (Figure. 26). Both these MOs are double occupied by electrons, and they compensate each other in the bonding, hence their total contribution to the Fe-CH_2O bonding is very small. The results of the DFT calculations yield Δq_π »0. This small value Δq_π characterizes the slightly asymmetrical charge distribution in the mutual compensating bonding (49a¢¢) and antibonding (52a¢¢) MOs. The main contribution to the Fe-CH_2O bonding is provided by forming of the π-type bonding MO 53a¢ from the corresponding MOs of the above fragments, namely, from the filled MO $3d_{xy}$ of the $Fe(CO)_2(PH_3)_2$ and the empty antibonding $2b_1(\pi^*)$-MO of CH_2O. Due to forming of this (53a¢) MO the electron density is transfers from the $3d_{xy}$-AO of the atom of Fe to the π^*-MO of the formaldehyde molecule (a π-type back donation). The orbital charge transfer is quite significant, $\Delta q_{\pi^*}(I) = 0.52\bar{e}$. It is seen from Figure 26 that only this 53a¢ bonding MO of the whole complex is uncompensated by the antibonding orbital (the empty MO 60a¢ in our calculations).

Therefore, one can conclude that Fe-formaldehyde bonding in the studied complex is provided by the π- back donation from the transition metal to the ligand. Thus, as a result of coordination, the lower unoccupied MO $2b_1$ acquires the charge $\Delta q(2b_1)$. Since the distortion mode leading to the non-planar configuration of formaldehyde molecule is also of the b_1-symmetry, the orbital vibronic coupling constant mixing the partly occupied $2b_1$ orbital with the unoccupied MO $6a_1$ becomes nonzero:

$$f = f_{b_1}^{(2b_1,6a_1)} = 2b_1|\left(\partial H / \partial Q_{b_1}\right)_0|6a_1 \neq 0. \qquad (26)$$

Then the curvature of the adiabatic potential of coordinated formaldehyde molecule with respect to the bending mode can be estimated as follows (Gorinchoy, 2011); (Gorinchoy, 2014):

$$K^{coord} = K^{free} - \Delta q \frac{f^2}{\Delta_{2b1,6a1}}. \tag{27}$$

The value of the OVCC $f_{b_1}^{(2b_1,6a_1)}$ can be obtained from consideration of free CH_2O molecule in its reduced or excited states, which have a non-planar pyramidal nuclear configuration (due to the PJTE induced by the occupation of its LUMO $2b_1(\pi^*)$. Its value is calculated to be 3.03eV/Å (Gorinchoy, 2014). Calculated value of the curvature of the APES of free formaldehyde molecule with respect to the bending is equal to $K_{b_1}^{free}$ (CH_2O) = 0.56eV/Å^2, and the value of the energy gap is equal to $\Delta E_{2b1,6a1}$ = 6.18eV. With these numerical data we get the negative resulting values for the curvature of the AP of coordinated CH_2O molecule with respect to the bending mode: K= -0.21 eV/Å^2. It means that the CH_2O molecule is distorted to a pyramidal structure with an energy gain of ΔE_{PJT}, thereby stabilizing the $Fe(CO)_2(PH_3)_2(\eta^2\text{-}CH_2O)$ complex.

CONCLUSION

In this review chapter, we applied the fundamental (based on first principles) approach to one of the central problems of modern chemistry – the origin of chemical reactivity - taking as examples the basic chemical reactions in environmental processes. We show that the properties of the main attribute to chemical transformations, the transition (activated) state of the chemical reactions, are controlled by the vibronic coupling between its ground electronic state with the appropriate low-lying excited state, leading to the pseudo-Jahn-Teller effect (PJTE). In application to the main chemical transformations in environmental processes, including proton transfer, electron transfer, and molecular coordination, we show how, invoking the PJTE theory, we can estimate the dependence of the reaction barriers on the molecular parameters, to rationalize the experimental data and to predict new processes. The functional dependence of the potential energy on the instability coordinate, following the PJTE theory with parameters estimated by means of quantum chemical calculations, can serve as a parametrized analytical model of the adiabatic potential, used to simulate the proton and electron transfer processes in such systems. It provides also for the possibility to manipulate these reactions by means of external influence that targets the PJTE parameters, similar to already realized PJTE-based manipulation processes in other transformations (Bersuker, 2021).

ACKNOWLEDGMENT

This work was performed under the Moldovan National Research Project Nr. 20.80009.5007.27, "Physico-chemical mechanisms of redox processes with electron transfer involved in vital, technological and environmental systems".

REFERENCES

Balan, I., Gorbachev, M., Gorinchoy, N., & Arsene, I. (2022). *Pseudo Jahn-Teller origin of the proton-transfer energy barrier in the proton-bound dimers.* In Proceedings of the International Scientific-Practical Conference Training by research for a prosperous society. Chişinău, Moldova.

Balan, I., & Gorinchoy, N. (2011). Ab initio study of chemical activation and hydrogenation of white phosphorus in reaction with rhodium trihydride complex. *Chemistry Journal of Moldova*, *6*(2), 84–90. doi:10.19261/cjm.2011.06(2).18

Barbaro, P., Caporali, M., Ienco, A., Mealli, C., Peruzzini, M., & Vizza, F. (2008). Dynamic behaviour of the [(Triphos)Rh(η^1:η^2-P_4RR¢)]$^{n+}$ complexes [Triphos = MeC(CH_2PPh_2)$_3$; R = H, Alkyl, Aryl; R¢ = Lone Pair, H, Me; *n* = 0, 1]: NMR and computational studies. *European Journal of Inorganic Chemistry*, *2008*(9), 1392–1399. doi:10.1002/ejic.200701070

Barbaro, P., Ienco, A., Mealli, C., Peruzzini, M., Scherer, O. J., Schmitt, G., Vizza, F., & Wolmershäuser, G. (2003). Activation and functionalization of white phosphorus at rhodium: experimental and computational analysis of the [(triphos)Rh (η^1:η^2-P_4RR¢)]Y complexes (triphos = MeC(CH_2PPh_2)$_3$; R = H, Alkyl, Aryl; R¢= 2 Electrons, H, Me). *Chemistry (Weinheim an der Bergstrasse, Germany)*, *9*(21), 5195–5210. doi:10.1002/chem.200305091 PMID:14613129

Barbaro, P., Peruzzini, M., Ramirez, J. A., & Vizza, F. (1999). Rhodium-mediated functionalization of white phosphorus: A novel formation of C–P bonds. *Organometallics*, *18*(21), 4237–4240. doi:10.1021/om990550+

Bersuker, I. B. (1980). Are activated complexes of chemical reactions experimentally observable ones? *New Journal of Chemistry*, *4*(3), 139–145.

Bersuker, I. B. (1993). The concept of vibronic interactions. *New Journal of Chemistry*, *17*(1-2), 3–6.

Bersuker, I. B. (2006). *The Jahn-Teller Effect*. Cambridge University Press. doi:10.1017/CBO9780511524769

Bersuker, I. B. (2013). Pseudo Jahn–Teller Effect – A two-state paradigm in formation, deformation, and transformation of molecular systems and solids. *Chemical Reviews*, *113*(3), 1351–1390. doi:10.1021/cr300279n PMID:23301718

Bersuker, I. B. (2016). Spontaneous symmetry breaking in matter induced by degeneracies and pseudo-degeneracies. *Advances in Chemical Physics*, *160*, 159–208. doi:10.1002/9781119165156.ch3

Bersuker, I. B. (2021). Jahn–Teller and Pseudo-Jahn–Teller effects: From particular features to general tools in exploring molecular and solid state properties. *Chemical Reviews*, *121*(3), 1463–1512. doi:10.1021/acs.chemrev.0c00718 PMID:33353296

Bersuker, I. B., Gorinchoi, N. N., & Polinger, V. Z. (1984). On the origin of dynamic instability of molecular systems. *Theoretica Chimica Acta*, *66*(3), 161–172. doi:10.1007/BF00549666

Bersuker, I. B., Gorinchoi, N. N., & Polinger, V. Z. (1987). Vibronic origin of basic state instability and stable excited states of transition complex of the H + H_2 ® H_2 + H chemical reaction. *Khimicheskaya Fizika*, *6*(6), 733–736.

Bersuker, I. B., Gorinchoi, N. N., & Polinger, V. Z. (1992). Pseudo Jahn-Teller origin of square-planar configuration instability of main-group-element hydrides. *Journal of Molecular Structure*, *270*, 369–380. doi:10.1016/0022-2860(92)85040-N

Bersuker, I. B., & Polinger, V. Z. (1989). *Vibronic Interactions in Molecules and Crystals*. Springer-Verlag. doi:10.1007/978-3-642-83479-0

Bersuker, I. B., & Stavrov, S. S. (1988). Structure and properties of metalloporphyrins and hemoproteins: The vibronic approach. *Coordination Chemistry Reviews*, *88*, 1–68. doi:10.1016/0010-8545(88)80001-8

Bersuker, I. B., Stavrov, S. S., & Vekhter, B. G. (1979). Pseudo Jahn-Teller origin of the trigger mechanism of conformational changes in the process of hemoglobin oxygenation. *Biofizica*, *24*, 413–418.

Buhro, W. E., Patton, A. T., Strouse, C. E., Gladysz, J. A., McCormick, F. B., & Etter, M. C. (1983). Syntheses, chemical properties, and x-ray crystal structures of rhenium formaldehyde and thioformaldehyde complexes. *Journal of the American Chemical Society*, *105*(4), 1056–1058. doi:10.1021/ja00342a070

Chen, C., Shi, T., Chang, W., & Zhao, J. (2015). Essential roles of proton transfer in photocatalytic redox reactions. *ChemCatChem*, *7*(5), 724–731. doi:10.1002/cctc.201402880

Cornaton, Y., & Marquardt, R. (2016). A global analytical representation of the potential energy surface of the FHF^- anion. *The Journal of Physical Chemistry A*, *120*(30), 5959–5968. doi:10.1021/acs.jpca.6b05325 PMID:27400137

Diken, E. G., Headrick, J. M., Roscioli, J. R., Bopp, J. C., Johnson, M. A., & McCoy, A. B. (2005). Fundamental excitations of the shared proton in the $H_3O_2^-$and $H_5O_2^+$ complexes. *The Journal of Physical Chemistry A*, *109*(8), 1487–1490. doi:10.1021/jp044155v PMID:16833469

Dogonadze, R. R. (Ed.). (1985). *The Chemical Physics of Solvation*. Elsevier.

Elliott, J. A., & Paddison, S. J. (2007). Modelling of morphology and proton transport in PFSA membranes. *Physical Chemistry Chemical Physics*, *9*(21), 2602–2618. doi:10.1039/b701234a PMID:17627306

García-Fernández, P., García-Canales, L., García-Lastra, J. M., Junquera, J., Moreno, M., & Aramburu, J. A. (2008). Pseudo-Jahn-Teller origin of the low barrier hydrogen bond in $N_2H_7^+$. *The Journal of Chemical Physics*, *129*(12), 124313. doi:10.1063/1.2980053 PMID:19045029

Geru, I., Gorinchoy, N., & Balan, I. (2012). Pseudo Jahn–Teller origin of the proton tunneling in Zundel cation containing water clusters. *Ukrainian Journal of Physics*, *57*(11), 1149–1149.

Gorinchoy, N. (2014). Jahn-Teller and pseudo Jahn-Teller origin of structural distortions of coordinated molecules. *Revue Roumaine de Chimie*, *59*(3-4), 273–280.

Gorinchoy, N., Balan, I., Polinger, V., & Bersuker, I. (2021). Pseudo Jahn-Teller origin of the proton-transfer energy barrier in the hydrogen-bonded [FHF]-system. *Chemistry Journal of Moldova*, *16*(1), 115–120. doi:10.19261/cjm.2021.834

Gorinchoy, N. N., Balan, I. I., & Bersuker, I. B. (2011). Jahn–Teller, pseudo Jahn–Teller, and Renner–Teller effects in systems with fractional charges. *Computational & Theoretical Chemistry*, *976*(1-3), 113–119. doi:10.1016/j.comptc.2011.08.013

Grabowski, S. J. (2015). $[FHF]^-$ – The strongest hydrogen bond under the influence of external interactions. *Crystals*, *6*(1), 3. doi:10.3390/cryst6010003

Hermoso, W., Ilkhani, A. R., & Bersuker, I. B. (2014). Pseudo Jahn–Teller origin of instability of planar configurations of hexa-heterocycles $C_4N_2H_4X_2$ (X = H, F, Cl, Br). *Computational & Theoretical Chemistry*, *1049*, 109–114. doi:10.1016/j.comptc.2014.10.007

Huang, X., Braams, B. J., Carter, S., & Bowman, J. M. (2004). Quantum calculations of vibrational energies of $H_3O_2^-$ on an ab initio potential. *Journal of the American Chemical Society*, *126*(16), 5042–5043. doi:10.1021/ja049801i PMID:15099067

Ilkhani, A. R., Hermoso, W., & Bersuker, I. B. (2015). Pseudo Jahn-Teller origin of instability of planar configurations of hexa-heterocycles. Application to compounds with 1,2- and 1,4-C_4X_2 skeletons (X= O, S, Se, Te). *Chemical Physics*, *460*, 75–82. doi:10.1016/j.chemphys.2015.02.017

Jacquemin, D., Zuniga, J., Requena, A., & Céron-Carrasco, J. P. (2014). Assessing the importance of proton transfer reactions in DNA. *Accounts of Chemical Research*, *47*(8), 2467–2474. doi:10.1021/ar500148c PMID:24849375

James, T., & Wales, D. J. (2005). Protonated water clusters described by an empirical valence bond potential. *The Journal of Chemical Physics*, *122*(13), 024516. doi:10.1063/1.1869987 PMID:15847464

Liu, Y., Bersuker, I. B., Zou, W., & Boggs, J. E. (2010). Pseudo Jahn–Teller versus Renner–Teller effects in the instability of linear molecules. *Chemical Physics*, *376*(1-3), 30–35. doi:10.1016/j.chemphys.2010.07.029

Ogurtsov, I., Gorinchoy, N., & Balan, I. (2007). Vibronic origin of the $H_3O^\bullet$ metastability. *Journal of Molecular Structure*, *838*(1-3), 107–111. doi:10.1016/j.molstruc.2007.01.065

Ogurtsov I., Gorinchoy N., Balan I., Munteanu G. (2003) Comparative ab initio study of the pseudo Jahn-Teller instability of planar D_{3h} configuration of $H_3O^\bullet$ radical and H_3O^+ cation. *Scientific annals of the State University of Moldova. Chemical-Biological sciences series,* 366-371. University of Moldova.

Panich, A. M. (1995). NMR study of the F–H···F hydrogen bond. Relation between hydrogen atom position and F–H···F bond length. *Chemical Physics*, *196*(3), 511–519. doi:10.1016/0301-0104(95)00128-B

Peruzzini, M., Ramirez, J. A., & Vizza, F. (1998). Hydrogenation of white phosphorus to phosphane with rhodium and iridium trihydrides. *Angewandte Chemie International Edition*, *37*(16), 2255–2257. doi:10.1002/(SICI)1521-3773(19980904)37:16<2255::AID-ANIE2255>3.0.CO;2-H PMID:29711443

Polinger, V. Z., Gorinchoi, N. N., & Bersuker, I. B. (1992). MO LCAO analysis of the vibronic instability of the $CuCl_5^{3-}$ trigonal bipyramidal configuration. Critical view on the angular overlap model in vibronic problems. *Chemical Physics*, *159*(1), 75–87. doi:10.1016/0301-0104(92)80061-Y

Pylaeva, S. A., Elgabarty, H., Sebastiani, D., & Tolstoy, P. M. (2017). Symmetry and dynamics of FHF– anion in vacuum, in CD_2Cl_2 and in CCl_4. Ab initio MD study of fluctuating solvent–solute hydrogen and halogen bonds. *Physical Chemistry Chemical Physics*, *19*(38), 26107–26120. doi:10.1039/C7CP04493C PMID:28929144

Rosi, M., Sgamellotti, A., Tarantelli, F., & Floriani, C. (1988). Interaction between iron (0) and formaldehyde, thioformaldehyde, and acetone: Ab initio calculations on the model compounds $Fe(CO)_2(Ph_3)_2(\eta^2\text{-}CH_2O)$, $Fe(CO)_2(Ph_3)_2(\eta^2\text{-}CH_2S)$, $Fe(Ph_3)_4(\eta^2\text{-}CH_2O)$, and $Fe(Ph_3)_4(\eta^2\text{-}CMe_2O)$. *Inorganic Chemistry*, *27*(1), 69–73. doi:10.1021/ic00274a016

Valadbeigi, Y., Farrokhpour, H., & Tabrizchi, M. (2014). Effect of hydration on the kinetics of proton-bound dimer formation: Experimental and theoretical study. *The Journal of Physical Chemistry A*, *118*(36), 7663–7671. doi:10.1021/jp506140m PMID:25137125

Valadbeigi, Y., Ilbeigi, V., Michalczuk, B., Sabo, M., & Matejcik, S. (2019). Effect of basicity and structure on the hydration of protonated molecules, proton-bound dimer and cluster formation: An ion mobility-time of flight mass spectrometry and theoretical study. *Journal of the American Society for Mass Spectrometry*, *30*(7), 1242–1253. doi:10.100713361-019-02180-z PMID:31049871

Wang, Y., Liu, Y., & Bersuker, I. B. (2019). Sudden polarization and zwitterion formation as a pseudo-Jahn–Teller effect: A new insight into the photochemistry of alkenes. *Physical Chemistry Chemical Physics*, *21*(20), 10677–10692. doi:10.1039/C9CP01023H PMID:31086864

Werner, H., Steinmetz, M., Peters, K., & von Schnering, H. G. (1998). A novel route to formaldehyde and thioformaldehyde rhodium complexes. *European Journal of Inorganic Chemistry*, *1998*(11), 1605–1617. doi:10.1002/(SICI)1099-0682(199811)1998:11<1605::AID-EJIC1605>3.0.CO;2-H

Wilson, E. B., Decius, J. C., & Cross, P. C. (1980). *Molecular vibrations: the theory of infrared and Raman vibrational spectra*. Dover Publications.

Chapter 5
The Spin Chemistry of Nitroxide-Based Organic Biradicals:
EPR Data and Computational Experiments

Fanica Cimpoesu
Institute of Physical Chemistry Ilie Murgulescu, Romania

Maria-Cristina Buta
Institute of Physical Chemistry Ilie Murgulescu, Romania

Ana Maria Toader
Institute of Physical Chemistry Ilie Murgulescu, Romania

Gabriela Ionita
Institute of Physical Chemistry Ilie Murgulescu, Romania

ABSTRACT

The electronic spin has a crucial importance in life chemistry, although relatively rarely encountered in several processes, while the vast majority of organic molecules (including the biological systems) are based on closed-shell spinless configurations. To advocate for the incidence of the spin it is suggestive to point out that the oxygen molecule, O^2, the fuel in the quasi-totality of organisms, has a spin-triplet ground state and all the processes triggered by respiration are including, then, spin interactions. In the investigation of subtle spin-related molecular and supramolecular interactions, a wise technique consists in introducing stable open-shell organic radicals as "spies" of a given environment, under electronic paramagnetic resonance measurements. The nitroxide-based spin carriers are frequently used as spin probes, the TEMPO group, (2,2,6,6-Tetramethylpiperidin-1-yl)-oxyl, being the most widely used in this view. Here the authors illustrate a case study on a series of nitroxide-based biradicals, corroborating the EPR measurements with state-of-the-art modeling.

DOI: 10.4018/978-1-6684-7198-2.ch005

INTRODUCTION: LINK BETWEEN SPIN COUPLING AND ELECTRON TRANSFER

The electronic spin is a fundamental component of the quantum nature of matter, determining the chemical bond itself and intervening in all the magnetic properties of molecules and materials (Putz et al., 2018). This chapter will not directly address the electron-transfer issue, focusing - in turn - to spin interactions, which are tacitly important in oxidation processes, although this aspect is often overlooked. For instance, the simple oxidation (burning) of organic compounds in air implies a spin coupling problem, since the substrates are usually spin singlets (i.e., having closed-shell electronic structures), while the oxygen molecule has a spin-triplet ground state (Cotton et al., 1999). Then, the direct interaction is spin-forbidden, this being the reason why organic systems (including living organisms) do not undergo spontaneous combustion in the atmosphere, while most of them are thermodynamically unstable with respect to final burning products (say CO_2 and H_2O, also singlet structures), the spin singlet-triplet interdiction determining kinetic stability. Thus, living on earth owes much to spin rules, the issue deserving special attention, a line of structural issues on this paradigm being presented in the body of the chapter. Practically, one may say that the catalysts used in various industrial oxidation processes perform a spin-catalysis (Buchachenko & Berdinsky, 2002; Minaev & Agren, 1996), helping to cross the spin-forbidden gates.

In many other redox processes, it is expected to find at a closer look, hidden spin-coupling mechanisms. For instance, the celebrated case of Fenton systems (Fenton, 1894; Hayyan et al., 2016), may be qualitatively foreseen in this sense. Thus, the process starts with ferrous, Fe(II) based systems, whose aqueous form is expected to be a high-spin $[Fe(H_2O)_6]^{2+}$ complex, with spin quintet (S=2) ground state. It transforms in ferric Fe(III) species, with different spin multiplicity (sextet, S=5/2, in high-spin form, or doublet with S=1/2 in low-spin configuration). The generation of HO· and HOO· radical species (spin doublets) is intimately related to the spin balance determined by the coordination chemistry of iron complexes. To the best of our knowledge, the underlying mechanisms of Fenton processes are not yet been investigated in the key of the determinant role of spin states and spin coupling effects.

Continuing the pledge on the intrinsic connection between electron-transfer and spin coupling mechanisms, one may invoke the case of Cytochrome P450 nitrogen oxide (NO) reductase (Sigel et al., 2007). The active site implies a ferric porphyrin complex (with low-spin doublet state), which binds the NO radical molecule (also spin-doublet), and further interacts with reduced nicotinamide adenine dinucleotide (NADH), a first step being formally interpreted as the reduction of the nitroxide by a transferred hydride anion (leading to HNO as transient molecule), further evolution (coupling with another NO) leading to the reduced state of nitrogen, in the nitrous oxide N_2O (Shaik et al. 2010; Riplinger & Neese, 2011).

Suggesting then that the subtle connection between electron transfer and spin couplings is yet a vastly uncharted area, the body of this chapter offers methodological illustrations of dealing with spin chemistry, hopefully illuminating the way for the advocated undertaking. The work has as a red line the idea of spin interactions, met in various systems and circumstances, illustrating their correlation with the structure of molecular and supramolecular edifices, and exploring the causal mechanisms.

Exchange Spin Coupling

The nominal keyword of spin coupling (SC) refers to a phenomenology generally described by the Heisenberg spin Hamiltonian (Heisenberg, 1928; van Vleck & Sherman, 1935; Anderson, 1959):

$$\hat{H}_{\mathrm{SC}} = -2\sum_{a}\sum_{b<a} J_{ab}\hat{S}_{a}\cdot\hat{S}_{b} \qquad (1)$$

where a and b represent the spin sites, J_{ab} is the so-called effective exchange integral, the spin on each center being taken in the operator form (marked by upper hat). The SC Hamiltonian is used in molecular magnetism (Kahn, 1993; Coronado et al., 1996; Miller & Drillon, 2001; Willett et al., 1985; Ruiz et al., 1997; Desplanches et al., 2002; Triki et al., 2005), the modern face of magnetochemistry. However, during such intensive and extensive use, it was almost forgotten that the SC phenomenology is a much more general paradigm, including the basic mechanisms of the chemical bonding itself. The formula (1) can be considered the semi-empirical facet of the Valence Bond (VB) theory (Heitler & London, 1927; Shaik & Hiberty, 2008; Cooper, 2001; Raimondi & Cooper, 1999; Cooper et al., 1999; Thorsteinsson & Cooper, 1998), the very first quantum model of molecular electronic structure. Must point out that, in spite of its merits, the VB method is rarely used nowadays, although modern versions and computer codes are available (Li et al., 2007; Li & McWeeny, 2002; Song et al., 2012; Song et al., 2005), since the methods based on delocalized molecular orbitals (MOs) took over the business of quantum chemical computations.

A scheme suggesting the subtle interplay between exchange and charge transfer effects is shown in Figure 1. Consider two sites (a and b), with one electron and one orbital, each. Besides, for simplicity, a and b are considered equivalent, with equal orbital (one-electron) energies. Under the such simplified assumption, the eigenvalues emerging from equation (1) are $E(S)$=-$J_{ab}S(S+1)$, where the spin quantum number takes the following values: S=0 (singlet, having relative 0 energy, in model) and S=1 (triplet, having the -2J_{ab} relative energy).

Figure 1. Generic illustration of configuration interaction (CI) between spin-flip ground configuration (GC) and charge transfer (CT) states involved in the exchange coupling mechanisms, exemplified on a two-sites two-electrons idealized model

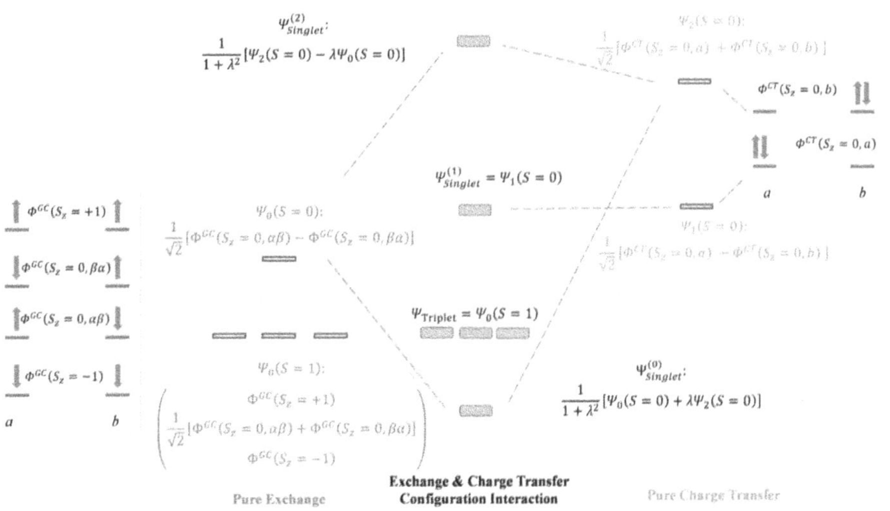

The left part of the diagram shows the ground configurations (GC) conserving one electron per site, running only on the spin-flip combinatorics. An alternative notation of GC configurations is: $\Phi^{GC}(S_z=+1)=|a^\alpha b^\alpha|$, $\Phi^{GC}(S_z=0,\alpha\beta)=|a^\alpha b^\beta|$, $\Phi^{GC}(S_z=0, \beta\alpha)=|a^\beta b^\alpha|$ and $\Phi^{GC}(S_z=-1)=|a^\beta b^\beta|$, where α and β are the respective spin-up and spin-down projections, the vertical bar symbols suggesting that the shown orbital and spin combinations (spin-orbitals) are anti-symmetrized in a Slater determinant (expanded with the given spin-orbital list on lines and running electron labels on columns, or *vice versa*). The right side of the diagram shows charge transfer (CT) configurations, also called ionic forms, for obvious reasons: $\Phi^{CT}(S_z=0, a)=|a^\alpha a^\beta|$, and $\Phi^{CT}(S_z=0, b)=|b^\alpha b^\beta|$. In the ideal case of no spin jumping between sites, the J_{ab} value in the Hamiltonian from equation (1) is positive, leading to the stabilization of the spin triplet as ground state, a situation recognized in magneto-chemistry as ferromagnetic configuration (*F*). In this case $J_{ab} = J_{ab}^F$ is a pure exchange integral, describing an effect without equivalent and intuition in terms of classical electrostatic interactions between two clouds of charges, which, from certain algebraic reasons must be $J_{ab}^F > 0$. Allowing the configuration interaction with CT states, the spin singlet may be stabilized as definite ground state, as shown in the middle of Figure 1. The CI can be qualitatively understood as similar to a mutual "repeal" of configurations with the same spin and symmetry, the lower ones getting deeper in energy, while, in counterpart, the higher levels are further raised. This rough trend is qualitatively shown by the dashed lines connecting the corresponding GC and CT levels with the final lowest and highest singlet states. In the given idealized scheme, one CT singlet combination and the GC triplet set do not benefit from any CI interaction (because of spin and orbital symmetry selection rules). If equation (1) is forced to account for the final situation, one arrives at a negative J_{ab}, in order to make the triplet with $-2J_{ab}$ energy higher than the singlet, fixed in 0.

In molecular magnetism, the singlet is thought to arise from a combination of configurations ascribed as $\Phi^{GC}(S_z=0,\alpha\beta)$ and $\Phi^{GC}(S_z=0,\beta\alpha)$ in Figure 1, i.e., a situation with antiparallel spins on different sites, named antiferromagnetic (AF). As a parenthesis, note that, in a complementary linear combination, the $\Phi^{GC}(S_z=0,\alpha\beta)$ and $\Phi^{GC}(S_z=0,\beta\alpha)$ configurations are appearing also in the $S_z=0$ component of the triplet, aside the $S_z=\pm1$ components, realized by the αα and ββ parallel spin projections.

The pure exchange part, always positive, $J_{ab}^F > 0$, is conceivably contained in all instances in the total J_{ab}, its value becoming negative by the prevalence of the antiferromagnetic component, $J_{ab}^{AF} << 0$, due to CI inserting the CT contributions.

$$J_{ab} = J_{ab}^{F} + J_{ab}^{AF} \tag{2}$$

In certain circumstances, relative rare, implying accidental or symmetry-driven null overlap of the orbitals, i.e., a situation called orbital orthogonality, the J_{ab}^{AF} negative contribution vanishes, letting the whole exchange ferromagnetic (Kollmar & Kahn, 1993; Kahn et al., 1989; Glaser et al., 2006), leading to high-spin ground state (triplet, in the above scheme). Otherwise, in most frequent conjunctures, the singlet is stabilized as the ground state, by the aforementioned CI mechanism. Since the heuristic virtues of spin coupling as a paradigm of the chemical bonding itself were predicated previously, one can understand now the point, in-depth. In the basic chemistry curriculum, is said that the covalent bond is established by spin-pairing. The above sketched CI is the phenomenology behind the metaphor of spin-pairing as an effective bonding engine.

Dipolar Spin Coupling

As pointed out, the exchange effects are pure quantum manifestations, without any equivalent in classical physics. However, there exists a classical facet of spin-spin interaction, *via* magnetic moments (say $\boldsymbol{\mu}_a$ and $\boldsymbol{\mu}_b$ on a and b sites), generated by the carried spins, namely a dipole-dipole term equated as follows:

$$E_{av}^{\mathrm{dip}} \propto \left(\frac{\boldsymbol{\mu}_a \cdot \boldsymbol{\mu}_b}{r_{ab}^3} - 3\frac{\left(\boldsymbol{\mu}_a \cdot \boldsymbol{r}_{ab}\right)\left(\boldsymbol{\mu}_b \cdot \boldsymbol{r}_{ab}\right)}{r_{AB}^5} \right) \tag{3}$$

In the above equation, the bold notation on $\boldsymbol{\mu}_a$, $\boldsymbol{\mu}_b$ and $\boldsymbol{r}_{\mathrm{ab}}$ symbolizes their vectorial nature, the dot corresponding to the scalar product. Aiming qualitative discussion, the proportionality symbol, $\propto$, was used to avoid detailing factors termed with speed of light and vacuum permeability. Skipping the complete derivation (Bencini & Gatteschi, 1990), one notes that the above classical term is converted, for the two-site case, in the following spin operator:

$$\hat{H}_{dip} = D_{ab}\left(\hat{S}_{z_a} \cdot \hat{S}_{z_b} - \frac{1}{3}\hat{S}_a \cdot \hat{S}_b \right), \tag{4}$$

where the D_{ab} coupling parameter is proportional to integrated spin densities, σ, around radical sites:

$$D_{ab} \propto \; -\int\int \sigma_a\left(\boldsymbol{r}_a\right)\frac{1}{r_{ab}^3}\sigma_b\left(\boldsymbol{r}_b\right)d\boldsymbol{r}_a d\boldsymbol{r}_b \tag{5}$$

Note the presence of scalar product $\hat{S}_a \cdot \hat{S}_b$ between total spins, in formula (4), similarly to the case of Heisenberg exchange coupling in equation (1). This means that the summed Hamiltonians contain a D_{ab} component into the effective J_{ab}, but this is quantitatively negligeable relative to the scale in which the exchange spin coupling is measured (i.e., magnetic susceptibility data) or computed.

The dipolar factor, D_{ab} is relevant in the small energy scale (the 10^{-1}- 10^{-2} cm^{-1} range) of electronic paramagnetic resonance (EPR) spectra (Bencini & Gatteschi, 1990; Abragam & Bleaney, 1970). The presence of a part termed with the help of spin projections, $\hat{S}_{z_a} \cdot \hat{S}_{z_b}$, which is expanded simply as the $\mathrm{S}_{z_a} \cdot \mathrm{S}_{z_b}$ numerical product, raises the $2S$+1 degeneracy of a S spin state (provide that $S \geq 1$), the dipolar term inducing a so-called Zero Field Splitting (ZFS) effect. The ZFS can be generically written as follows:

$$\hat{H}_{\mathrm{ZFS}} = D\left(\hat{S}_z^2 - \frac{1}{3}S\left(S+1\right) \right) \tag{6}$$

this being equivalent with the equation (4), putting D=-(1/2)|D_{ab}| and assuming a formal shift of the whole resulting spectrum. On the other hand, the equation (6) accounts more general circumstances, where the ZFS comes from spin-orbit coupling, in transition metal complexes, the *D* parameter being tuned by ligand field factors.

Such interactions are exceptionally interesting, the ZFS determining the magnetic anisotropy features (Ferbinteanu & Cimpoesu, 2014), leading to the special occurrence of Single Molecule Magnets (SMMs) (Christou et al., 2000; Gateschi & Sessoli, 2003) and Single Chain Magnets (SCMs) (Caneschi et al., 2001; Ferbinteanu et al., 2005), i.e., realizing true magnets, miniatured down to molecular and nano scale. However, in spite of the fact that the direction is very challenging, pointing toward futuristic applications claimed as "spintronics" (Troiani & Affronte, 2011; Bogani & Wernsdorfer, 2010), the ZFS effects specific to coordination chemistry, due to the interplay of spin-orbit and ligand field factors, will not be debated here.

Electronic Structure Modeling

Computational simulations are acknowledged nowadays as a bright future of development in all branches of human activity, science being the most legitimate terrain to such prospects. Aiming to understand and even predict certain properties, desired for applications, chemistry has the advantage of a strong edifice affording such goals following correct causal clues, (Putz et al., 2018), by means of electronic structure methods developed in the field of quantum chemistry. Molecular magnetism is particularly tied to the electronic structure approaches, emerging in parallel with the increased availability of modern computers, at the border between past and current millennia (Kahn, 1993; Miller & Drillon, 2001).

There are multiple methods, with various degrees of approximation, adapted to the nature (particularly to the size) of approached systems. Small molecules benefit from very rigorous methods, close to the first principles of quantum mechanics, called generically *ab initio*. The extremely large systems, as proteins or big supramolecular assemblies (e.g., solvent boxes) are inherently confined to very crude approximations, belonging to the molecular mechanics (Allinger & Burkert, 1982; Mackerell, 2004), ignoring the quantum effects. Molecules of moderate size (tens, or even approaching the range of one-two hundreds of atoms) can be tackled at intermediate levels of precision, often satisfactory in reproducing realistic properties, such as heats of formation, vibration, or electronic spectra.

One may hope that all the chemists are acquainted with the idea of atomic orbitals (AOs) and with the generalization of these objects to molecular orbitals (MOs). The most common intuition is that the orbitals, forming a stack of increasing energies, are progressively populated with electrons. In most of molecules, especially the organic compounds, the orbitals are doubly occupied, up to the frontier of the highest occupied molecular orbital (HOMO), below the sequence of empty (virtual) functions, starting with lowest unoccupied molecular orbital (LUMO). A single configuration of this sort is called the closed shell. In relatively fewer cases, the frontier consists in singly occupied molecular orbitals (SOMOs), having the case of open-shell configuration.

Without entering details, one must point out that the technical scaffold behind MO-based methods, which are optimized to account for the electron structure with the help of a single configuration (occupation scheme) belongs to the Hartree-Fock (HF) frame. Nowadays, HF is regarded as obsolete, since it lacks the so-called electron correlation which comes from the physical truth that the electronic structure should be rigorously regarded as being of multi-configurational nature, namely pertaining to a Configuration Interaction (CI) formulation. As support of this point, one may recall the Figure 1, suggesting that the singlet is obtained by the combination of different configurations, at least the $\Phi^{GC}(S_z=0,\alpha\beta)$ and $\Phi^{GC}(S_z=0,\beta\alpha)$, besides the $\Phi^{CT}(S_z=0,a)$ and $\Phi^{CT}(S_z=0,b)$ intervening in a second order manner.

The configurations in Figure 1 are AO-based. One may switch to a MO picture forming the following bonding (σ) and antibonding (σ*) combinations:

$$\sigma = \frac{1}{\sqrt{2\left(1+S_{ab}\right)}}\left(a+b\right) \tag{7a}$$

$$\sigma^{*} = \frac{1}{\sqrt{2\left(1-S_{ab}\right)}}\left(a-b\right) \tag{7b}$$

where S_{ab} is the overlap integral between *a* and *b* AOs.

In the frame named restricted Hartree-Fock (RHF), the singlet ground state is approximated as the $|\sigma^{\alpha}\sigma^{\beta}|$ single configuration. The "restricted" denomination stands for the fact the same orbital can host α and/or β spins. Expanding the RHF configuration into AOs, one may observe that both GC and CT configurations are included (see again Figure 1):

$$\Psi_{RHF} = \left|\sigma^{\alpha}\sigma^{\beta}\right| = \frac{1}{2\left(1+S_{ab}\right)}\left(\left|a^{\alpha}a^{\beta}\right|+\left|a^{\alpha}b^{\beta}\right|+\left|b^{\alpha}a^{\beta}\right|+\left|b^{\alpha}b^{\beta}\right|\right) \tag{8}$$

However, these are not contained in the right proportion, noticing that all GC and CT configurations with a S_z=0 spin projection are contributing, each, with 25% to the ground state. This is an illustration of the alleged lack of electron correlation in HF scheme, the situation being corrected by allowing the configuration interaction (i.e., turning to a multi-configuration approach), which, in a MO basis, implies a partial mixing of the high energy configuration with a doubly occupied antibonding orbital.

A partial alleviation comes from the so-called unrestricted Hartree-Fock (UHF) scheme, allowing different orbitals for spin-up and spin-down spins, e.g., σ and σ', keeping the single determinant idea, $|\sigma^{\alpha}\sigma'^{\beta}|$. The cost of unrestricted version is that the wavefunctions with *S* total spin are no longer eigenfunctions of the spin square operators, with *S*(*S*+1) eigenvalues. This aspect will be detailed a bit later.

Very popular nowadays is the so-called Density Functional Theory (DFT) (Koch & Holthausen, 2001), which in the Kohn-Sham (KS) technical realization (Parr & Yang, 1989) retains much from the HF frame, keeping the single-determinant construction, having also restricted and unrestricted schemes (RKS and UKS). The DFT inserts certain empirical corrections aiming to correct the lack of correlation in HF, while keeping the technical and conceptual simplicity related with the single configuration wave-functions.

Note that DFT is supposed to work only for non-degenerate ground state configurations. There is no place here for details on DFT, pointing only that there is a large variety of proposed empirical corrections, identified by acronyms usually collecting the initial of authors, the most popular being the B3LYP functional (named after a version among those proposed by Becke, Lee, Yang and Parr) (Becke, 1992; Lee et al., 1988).

Once the ideas of unrestricted orbitals and DFT calculations were introduced, it is the place to mention a technique important for the simulation of spin coupling quantities, the so-called Broken-Symmetry (BS) procedure (Noodleman, 1981; Noodleman & Davidson, 1986; Ruiz et al., 1999). As pointed out, the electronic structure of spin-coupled systems is properly accounted in the multiconfigurational methods (which were not much detailed here). If such a method is used, then the spin coupling parameter following the phenomenology of Heisenberg Hamiltonian (eq. 1) is retrieved by the difference of computed singlet and triplet energies, $2J_{ab}= E_{singlet} - E_{triplet}$.

Using the DFT, of single-configurational nature, the singlet is not well accounted if consider a doubly occupied MO configuration, like restricted $|\sigma^{\alpha}\sigma^{\beta}|$ or unrestricted $|\sigma^{\alpha}\sigma'^{\beta}|$. The BS consist in enforcing the calculation of $\Phi^{GC}(S_z=0,\alpha\beta)=|a^{\alpha}b^{\beta}|$ or $\Phi^{GC}(S_z=0,\beta\,\alpha)=|a^{\beta}b^{\alpha}|$ configurations, in unrestricted frame. These are called BS states. Separately, none of these represent a good realization the of singlet, but their DFT-computed energies and spin distribution can offer information about the J_{ab} coupling parameter. The state with maximum spin, generically called High Spin (HS), here the triplet, realized by the UKS calculation of the $|a^{\alpha}b^{\alpha}|$ configuration, is needed to complete the scheme, estimating the exchange parameter by the Yamaguchi-Onishi formula (Mitani et al., 2000; Onishi et al., 2001):

$$J_{ab} = -\frac{E_{HS} - E_{BS}}{S^2_{HS} - S^2_{BS}} \tag{9}$$

All the intervening quantities can be picked-up from the regular DFT calculations, in UKS regime, performed on HS=$|a^{\alpha}b^{\alpha}|$ and BS=$|a^{\alpha}b^{\beta}|$ (or $|a^{\beta}b^{\alpha}|$) configuration. The S^2 quantities are called expectation values of spin-squared operator. For the HS, these are close to the $S(S+1)$ value, i.e., about 2. For the BS configurations, in weakly bonded biradical systems, is expected to have S^2_{BS}~1. Note that, in spite of the fact that in a BS configuration the spin projection is S_z=0, the discussed expectation value is not null, as would be for a true S=0 state.

Completing the briefing on electronic structure methods must point that the so-called basis sets are a necessary ingredient of the procedure, i.e., different ways of defining AOs whose combination will render the optimized MOs. The vast part of existing approaches is based on the so-called Gaussian Type Orbitals (GTOs) each element having a rich variety of possible choices. As consequence of heavy use of GTOs, one of the most popular electronic structure codes is called Gaussian (Frisch et al. 2016). The GTOs are also labeled by specific acronyms, such as 6-31G, 6-311+G*, cc-pVTZ, cc-pV5Z, (Ditchfield et al., 1971; McLean & Chandler, 1980; Kendall et al., 1992). The first case is a rather poor choice, the following ones being moderate, rich and very rich in components, a fact impacting the computer effort (time and memory) in a given calculation.

Nitroxide-Based Radicals

Since, in the following, the nitroxide-based stable radicals will be taken as case studies, the specifics of their electronic structure (Maretti et al., 2007), are presented now, as concretization of the above discussion on electronic structure methods. The spin carrier in such systems is the NO group incorporated in the organic skeleton, as is the case abbreviated as TEMPO (2,2,6,6-tetramethylpiperidin-1-yl)-oxyl. The molecular skeleton of TEMPO is shown in panel (a) of Figure 2.

Idealized, the smallest molecule from this class is H_2NO. The radical is hosted in an antibonding π^*-type orbital on the N-O axis (Fig. 2b,b'). Alternatively, in the spirit of VB theory, the nitroxides can the thought as superposition of two configurations (resonance structures), namely N(:)-O(·) ⇔ N(·)-O(:), with approximately comparable weights (see Fig. 2c,c').

Figure 2. Description of the NO group as a spin carrier: (a) the example of TEMPO radical; MO description, with real orbitals in (b); from ROB3LYP/6-311+G calculation and qualitative scheme in (b'); the VB approach, qualitative scheme in (c') and optimized localized orbitals in (c), from ab initio two-configuration VB calculation with 6-311+G* basis*

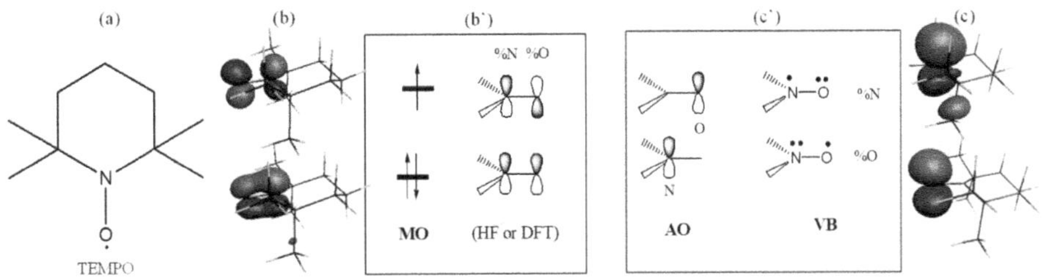

Table 1. Spin populations on N and O atoms of TEMPO molecule, from different calculation modes

Basis\Method	Atom	ROHF	ROB3LYP	UHF	UB3LYP
6-311+G*	O	0.631	0.497	0.656	0.522
	N	0.323	0.409	0.382	0.454
6-31G	O	0.688	0.522	0.679	0.533
	N	0.277	0.393	0.368	0.439

For the sake of completeness, the spin populations on nitrogen and oxygen atoms from TEMPO unit, exemplified by typical calculation methods (and basis sets), are shown in Table 1. The results are mutually similar, but the DFT (with B3LYP functional), pointing to closer values of the N and O spin percentages, can be credited as more realistic. Having supplementary degrees of freedom, the unrestricted modes are accepted as better, in the quantitative sense. The basis set changing does not bring severe mutations of the result, such relative comparisons being retrieved generally, along with many other computation examples.

A partial reason for choosing the nitroxide systems in the following debate is the relative similarity of this group with nitric oxide (NO), also a spin doublet (radical molecule), pointed previously to be important in several redox processes in living organisms. The stable nitroxide radicals, as the above-mentioned TEMPO iconic example, are not themselves investigated in redox reactions, here, serving in first instance only as test of the pure exchange coupling part, supposed to be interplayed with the electron transfer, in the roughly related case of the nitric oxide processes.

Another possible speculation is that the TEMPO-based and similar molecules can be used as "spies" in the nitric-oxide active complex systems, in the spirit of spin-probe techniques (Benelli & Gatteschi, 2002, Ferrando-Soria et al., 2017; Ionita et al., 2010; Ionita et al., 2014; Ionita et al., 2015; Constantin et al., 2019), to get valuable information. At the same time, the TEMPO derivatives are already proven as relevant in biochemical respects, acting as antioxidants (Hinzmann et al., 2020) i.e., having relevance in certain redox balances.

The organic nitroxides are also used in the oxidation of alcohols to aldehydes, ketones or carboxylic acids, obtaining excellent yields (Hinzmann et al., 2020). One of the most important applications of TEMPO is its role in the oxidation of several organic substrates, such as amines, phosphines, phenols, anilines, or primary and secondary alcohols (De Nooy et al., 1996). TEMPO can also be employed in nitroxide-mediated radical polymerization (NMP) (Fischer, 2001; Bertin et al., 2011), a controlled free radical polymerization technique that allows better tuning of the final molecular weight distribution. The TEMPO free radical can be added at the end of a growing polymer, creating a chain that stops expanding. But the linkage between the polymer chain and TEMPO is weak, so it can be broken upon heating, allowing the polymerization to continue, enabling an elegant engineering of the process.

Magneto-Structural Correlations With Nitroxide Spin Carriers

This section proposes causal explanations of molecular structure *vs.* spin coupling in prototypic nitroxide-based dimeric systems. Although there is a large number of nitroxide-based radicals reported with interesting properties, such as ferromagnetism, or magnetic phase transitions (Tamura et al., 1991; Sugawara et al., 1994; Matsushita et al., 1997; Caneschi et al., 1995), to the best of our knowledge these are not paralleled by structure-property correlations, in spite of the fact that, as will be shown in the following, the rationalization is seducingly simple (Buta et al., 2019). In this perspective, organic magnetism with stable radicals (Hicks, 2006; Mansikkamäki & Tuononen, 2018; Toader et al., 2019; Buta et al., 2021) grows as a solid competitor of the "classical", inorganic and coordination magnetochemistry.

As a first step, going to an extreme simplification of molecular models, two molecules of H_2NO, in different configurations are proposed (see Figure 3). The angles between the lobes of the *p* orbitals on N...N and O...O couples are either 0° or 90°, to illustrate the case of overlap (in σ or π patterns), or the orthogonality (symmetry-determined null overlap).

The configuration in upper panel (a) shows a π-type overlap of the distant pairs O…O and N…N, the bottom of (a) showing the quenched overlap. The geometry in Figure 3b assumes a coaxial arrangement of the NO groups, in such a way that only the oxygen atoms are directly interacting, achieving π-type overlap at the coplanar orientation of the molecules (upper panel) and canceling at the mutual angle of 90° (orbital orthogonality achieved simultaneously with the perpendicularity of the molecular planes, shown in bottom part). The third geometry (Figure 3c), presents in upper side a σ-type overlap, at an 0° angle of the O...O and N...N pairs and orbital orthogonality (zero overlap) at 90°, in bottom half. The orthogonality situation in panel (c) is practically the same as that in panel (a).

Figure 3. Selected geometric configurations for H_2NO nitroxide model dimers. (a) – with J given by the combined π-type overlap of O…O and N…N; (b) with J given by the π-type overlap of O…O; (c) with J given by the combined σ-type overlap of O…O and N…N. The crossed linkages are marking the occurrence of orbital orthogonality (null mutual overlap of NO groups)

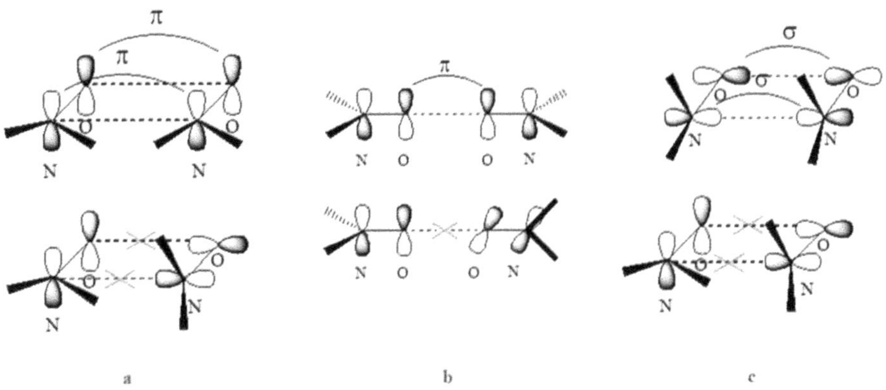

For the geometries suggested in Figure 3, the coupling parameter J_{ab} was calculated by BS-DFT method (L. Noodleman,1981; Noodleman & Davidson, 1986; Ruiz et al. 1999), at different O...O distances. Varying the angles between the pairs of p-p lobes, one may continuously pass from geometries shown in upper vs. bottom parts of the (a)-(c) panels in Figure 2, the results being shown in the figures 4, 5 and 6.

At this point, we will insert the idea that the antiferromagnetic component of the total spin-coupling (see equation 2) is proportional with the square of S_{ab} overlap of the orbitals located on the interacting sites (Hay et al., 1975; Crawford et al. 1976):

$$J_{ab}^{AF} = -k_{ab} \cdot S_{ab}^2 \tag{10}$$

It is easy to accept that the overlap integrals, in various conjunctures depicted in Figure 3, are tuned by the angle between the p-type orbitals.

In all three cases, one may observe the pattern imposed by the dichotomy in a fixed positive contribution (J^F) and a negative term (J^{AF}) proportional to the square of a cosine between p lobes, obeying the phenomenology suggested in equation 10. Thus, the coupling is positive in a domain approaching the orbital orthogonality and becomes maximal antiferromagnetic (negative values, larger absolute value) when the mutual overlap of the p lobes between the two NO groups is optimal.

Figure 4. $J^{(a)}$ coupling parameter at different O…O distances, varying as a function of the angles between the p lobes, for the type (a) geometry of the $(H_2NO)_2$ dimer (see Figure 3), assuming the gradual variation of the degree of π overlap in the O...O and N...N couples. Marked points = calculated by BS-DFT, solid lines = fitted with the phenomenological model $J^{(a)}=J_{(a)}^{F}+J^{0}_{(a)}{}^{AF}\cos^2\tau$

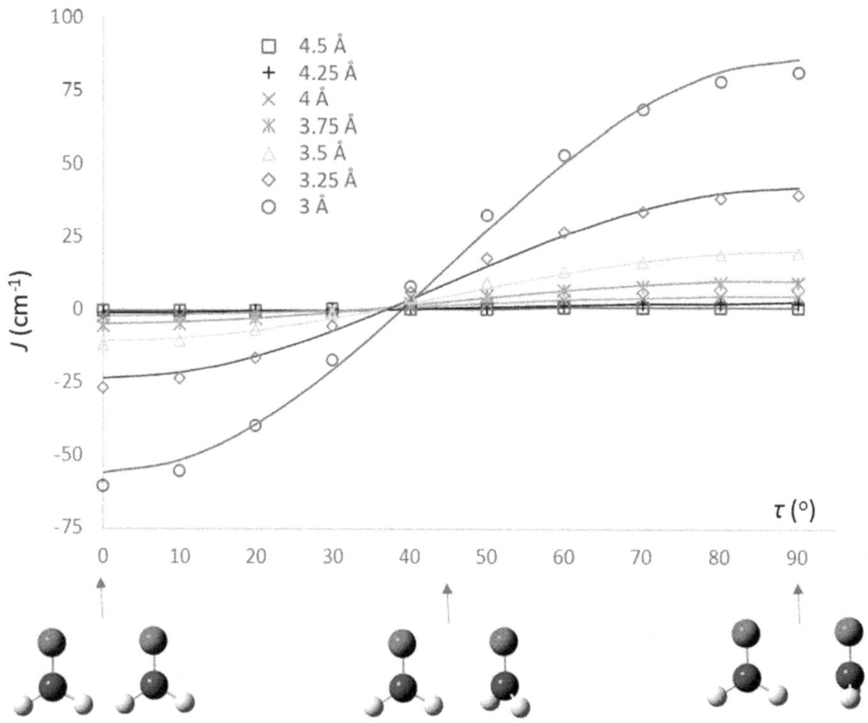

Figure 5. Coupling parameter $J^{(b)}$ at different O...O distances, for type (b) geometry of the $(H_2NO)_2$ dimer (see Fig 3), varying the angles between the p lobes. Marked points = calculated by BS-DFT, solid lines = fit by the formula $J^{(b)} = J_{(b)}^{F} + J^{0}_{(b)}{}^{AF}\cos^2\tau$

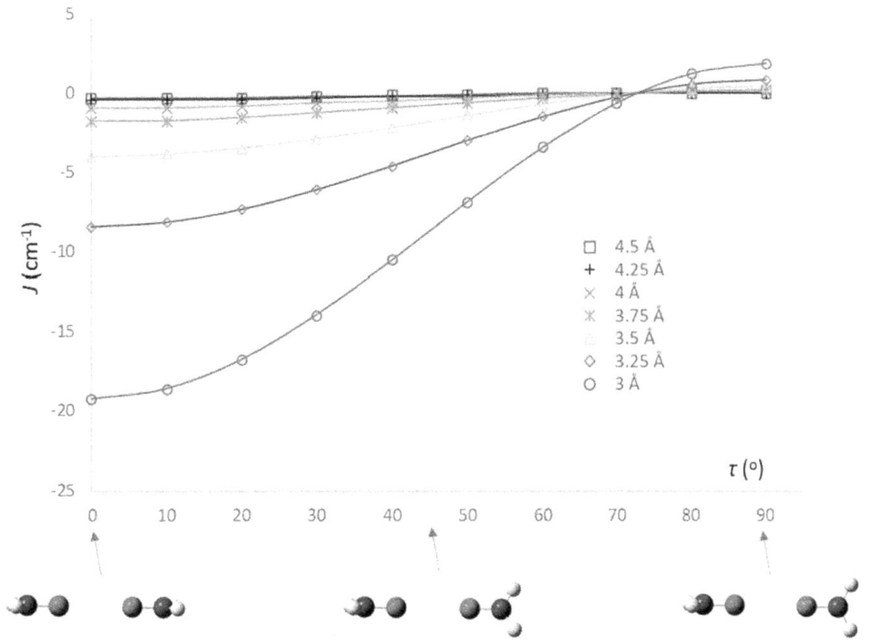

Figure 6. The coupling parameter $J^{(c)}$ at different O...O distances, varying as a function of the angles between the p lobes, for the type (c) geometry of the $(H_2NO)_2$ dimer (see Fig 3), assuming the gradual variation of the degree of σ-type overlap in the couples O...O and N...N. Marked points = calculated, solid lines = fitted with the model $J^{(c)} = J_{(c)}^{F} + J^{0}{}_{(c)}^{AF} cos^2\tau$

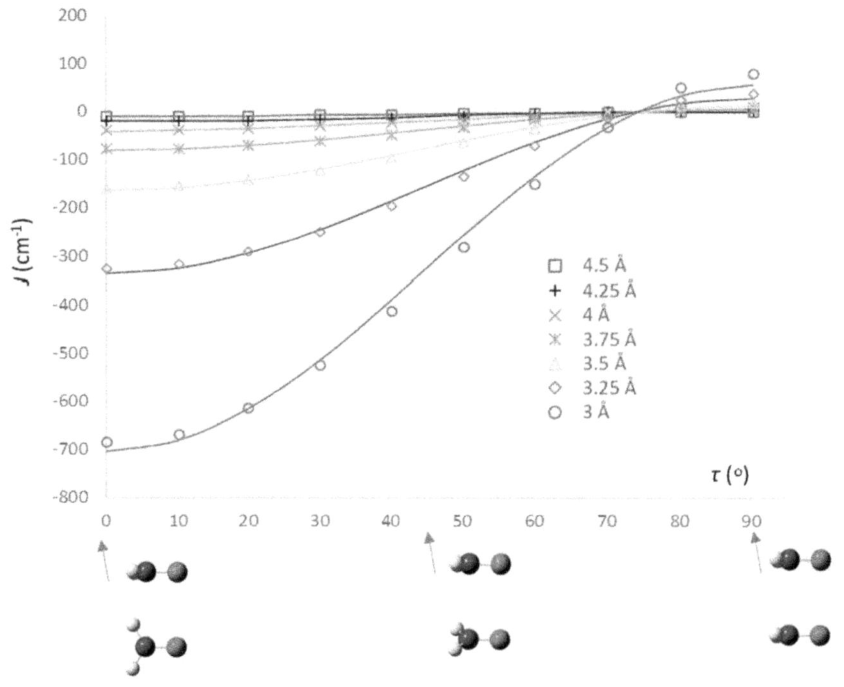

One may observe that all the curves have the same pattern, namely a variation between a minimum negative value, achieved when the molecular fragments are coplanar and a positive maximum, reached when these are perpendicular. This dependence is rationalized elegantly, in the form $J=J_F+J^0{}_{AF}\cdot\cos^2\tau$ for all cases, with ingredients specific to idealized geometries, according to types (a), (b) and (c). In the absence of orbital overlap (at $\tau=90°$), acts only a positive component, J_F, the degree of overlap (increased with the approach towards $\tau=0°$) bringing with it a negative term, J_{AF}. The amplitude of the variation depends on the type of overlap involved, being smaller in case (b), where a π-type overlap between the oxygen atoms acts, compared to geometry (a), where the π-type overlap between the nitrogen atoms also participates. The maximum variation is achieved in case (c) because the overlap of type σ is, in general, stronger than the π- type one.

In this series of simplified models, the $J=J^F+J^{AF}\cdot\cos^2\tau$ pattern is repeating, also, for every defined intermolecular distance. Considering the variation with distance (inter-atomic or inter-planar), it can be observed an exponential attenuation of the absolute value of the J parameter, as well as of the J^F or J^{AF} components. This data is collected in Figure 7, as a function of the distance r between the oxygen atoms. From the graphs, one can see that, at distances greater than 4.5 Å, the two radicals no longer interact with each other, the coupling parameter becoming practically zero.

The deliberate simplicity of the idealized system chosen here allowed geometries that cannot be achieved in all circumstances, e.g., with TEMPO fragments, due to steric crowding. Thus, geometries of type (a) and (b) can be attained, with realistic organic radicals, only at large distances. In spite of this disclaimer, the regularities extracted with the help of simplified computational experiments are valuable.

Figure 7. J (r), $J_F(r)$ și $J_{AF}(r)$ dependence on inter-radical distance for the three configurations from Figure 3

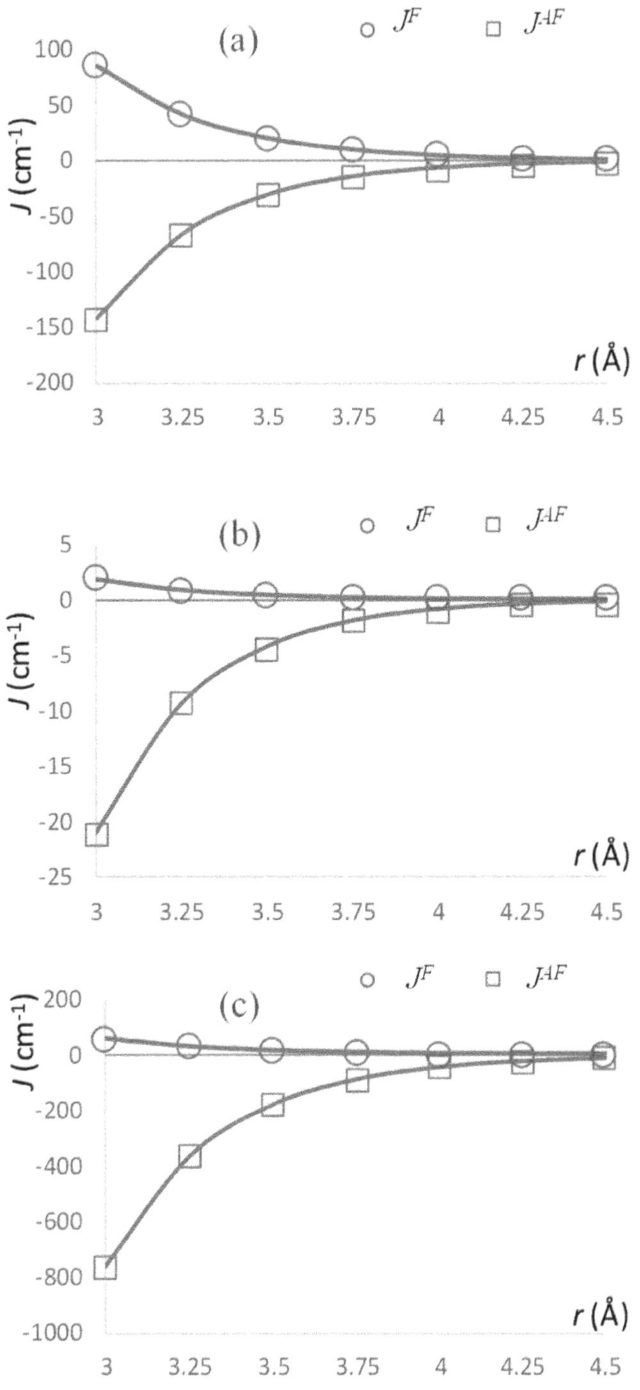

Linked Biradical Systems

Whether in previous instance, idealized systems were illustrated, for the sake of extracting clear correlation between geometries of dimeric molecular ensemble and the underlying spin coupling, the following case studies are advancing toward more realistic systems, considering the biradicals shown in Figure 8 (Caproiu et al., 2008), based on NO spin carriers from TEMPO moieties.

Figure 8. Structure of selected biradical compounds: 1- the meta-$C_6H_4(SO_2NH\text{-}TEMPO)_2$; 2-dibenzo furan-2,8-$(SO_2NH\text{-}TEMPO)_2$

1

2

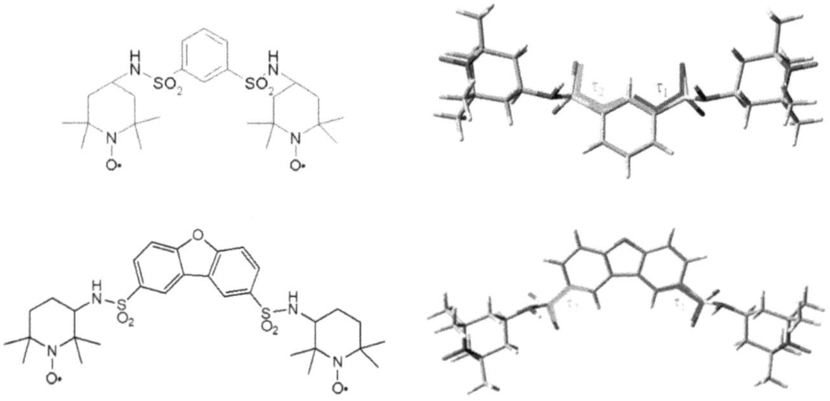

For the sake of completeness, the full spin Hamiltonian describing biradicals is ascribed as follows (Abe, 2013):

$$\hat{H} = g_a\beta\hat{S}_a\cdot\vec{B} + g_b\beta\hat{S}_b\cdot\vec{B} + A_a\hat{S}_a\cdot\hat{I}_a + A_b\hat{S}_b\cdot\hat{I}_b - 2J_{ab}\hat{S}_a\cdot b + D_{ab}\hat{S}_{z_a}\cdot\hat{S}_{z_b} \tag{11}$$

One may recognize in the last two terms the above discussed Heisenberg and dipolar coupling formulas. The middle couple of terms stands for the hyperfine interaction, formally similar to the spin coupling, except that, instead of two electron-based spins, one operator is substituted by the spin of a nucleus, generically labeled *I*. For the nitroxide-based compounds there is the case of $I = 1$ nuclear spin on the nitrogen atom. The first two components in the equation (11) are the Zeeman terms of the two sites, interacting with the external field *B*, modulated by the gyromagnetic Landé factor.

Figure 9. EPR spectra of compounds 1 and 2, in dichloromethane. The units from the abscissa are in milli-Tesla

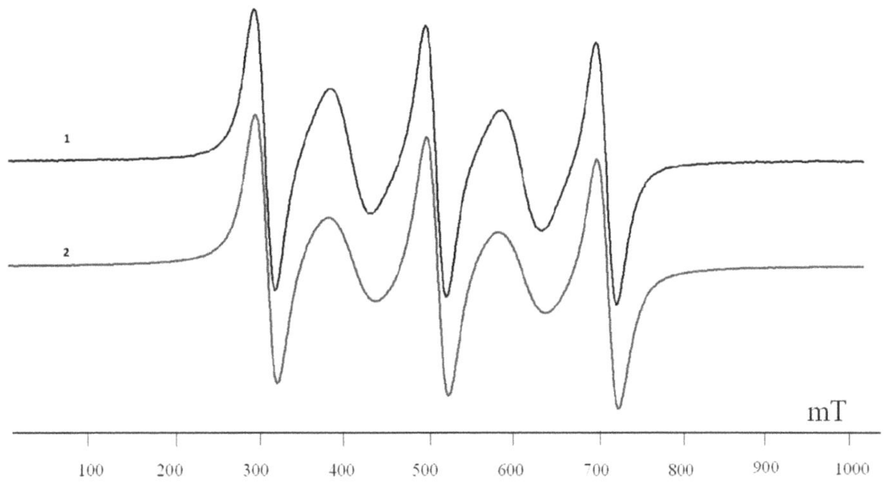

The pattern of the EPR spectra of compounds **1** and **2**, shown in Figure 9, corresponds to an interplay between exchange and hyperfine coupling. More specifically, the signature of this situation is the occurrence of the two broad bands between the central and marginal signals. Otherwise, for uncoupled systems, the typical monomeric TEMPO-alike EPR shape consists in three signals separated by distinct plateaus in between. Considering the idea of magneto-structural correlations, one may suppose the existence of a multitude of configurational isomers, in each of the presented compounds. It is almost impossible to deal with all the configurational species that exist in the system, especially when a solvent is involved. However, with the help of chemical intuition, one may identify the most relevant conformational degrees of freedom. Thus, for both compounds, the dihedral angles marked by τ_1 and τ_2 in Figure 8 can be proposed as key parameters. In the frame of this hypothesis, a certain completeness can be achieved performing the calculation of the potential energy surfaces and related variation of coupling parameters, sweeping the full rotations along the τ_1 and τ_2. The results reported in the following are achieved with the Gaussian 09 program (Frisch et al. 2016), using B3LYP as functional and 6-31G* basis set.

The modeling consists in the constrained optimization of the molecular geometry, for each value of the variables τ_1 and τ_2, taken on a grid between 0° and 360°, with a step of 30°. Thus, the calculation is relatively demanding, involving 169 instances of optimizing a medium-sized molecule. Each optimization implies iterative search of null energy gradient against the variation of all bond lengths, all bond angles and of the dihedrals, except the variables τ_1 and τ_2, which are kept fixed at each point of the computed surface. The 3D energy diagram obtained in this way is called relaxed potential energy surface.

Figure 10 provides an overview of the energy landscape of compound **1**, as a function of the independent rotation of the two SO_2-NH-TEMPO fragments. One may notice that only relatively small variations of energy are involved, with an absolute amplitude of 5.22 kcal/mol between recorded minima and maxima. This means that, under normal conditions (e.g., room temperature), in principle, all these conformations can be visited. In consequence, the data measured in the EPR spectrum should be considered as an average of the parameters attributable to the instantaneous configurations.

After the mapping of optimized geometries, one may estimate the exchange effects using the Broken-Symmetry (BS) methodology. Figure 11 shows that, in a large part of the conformational map, the coupling is practically zero, as is intuitively expected in the case of distant NO groups. But, on the diagonal of the variation after τ_1 and τ_2, in the area where the maxima produced by the mutual steric hindrance of the TEMPO groups are recorded (where the spin carriers come in a relative proximity), one may observe that the coupling parameter is sizeable.

Figure 10. Relaxed potential energy surface (PES) for compound 1, resulted by scanning the dihedral angles C-C-S-O, τ_1 and τ_2, while optimizing the other parameters of the molecular geometry. The conformation noted as minimum and maximum energies are given in kcal/mol, relative to the minimum value recorded (set in zero)

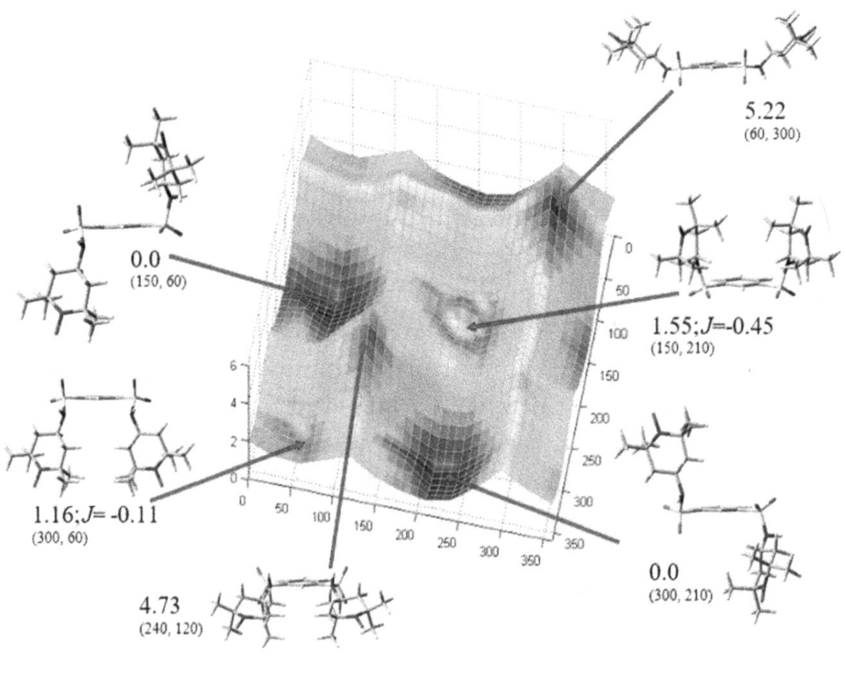

These have low absolute values (tenths of cm^{-1}) but are on a scale detectable in EPR-type experiments. Most situations correspond to the antiferromagnetic coupling ($J < 0$), but there are also areas with small positive values, i.e., weak ferromagnetic situations.

One may illustrate an estimate of the dipolar coupling parameter, generating the effect of zero field splitting (ZFS) on the triplet state of the biradical. In this view, is reasonable to proceed simplistically, taking the distance between the NO moieties (measured between the middle points of each bond) of the two TEMPO groups, using then the estimation from the following equation,

$$|D_{ab}| = 1.299\frac{g^2}{r_{ab}^3} \tag{12}$$

adjusted to yield values in Gauss, when distance is introduced in Ångstroms. The gyromagnetic factor is conventionally taken g=2. This mapping shows relatively important D values, of approximately 30-40 Gauss, for the conformation zones in which a sensitive spin coupling is also encountered, altogether a TEMPO steric hindrance.

Figure 11. The map of exchange coupling (J) and dipolar (D) parameters in compound 1, estimated in the conformations of the geometries constrained by the dihedral angles C-C-S-O and relaxed by all the other geometric variables

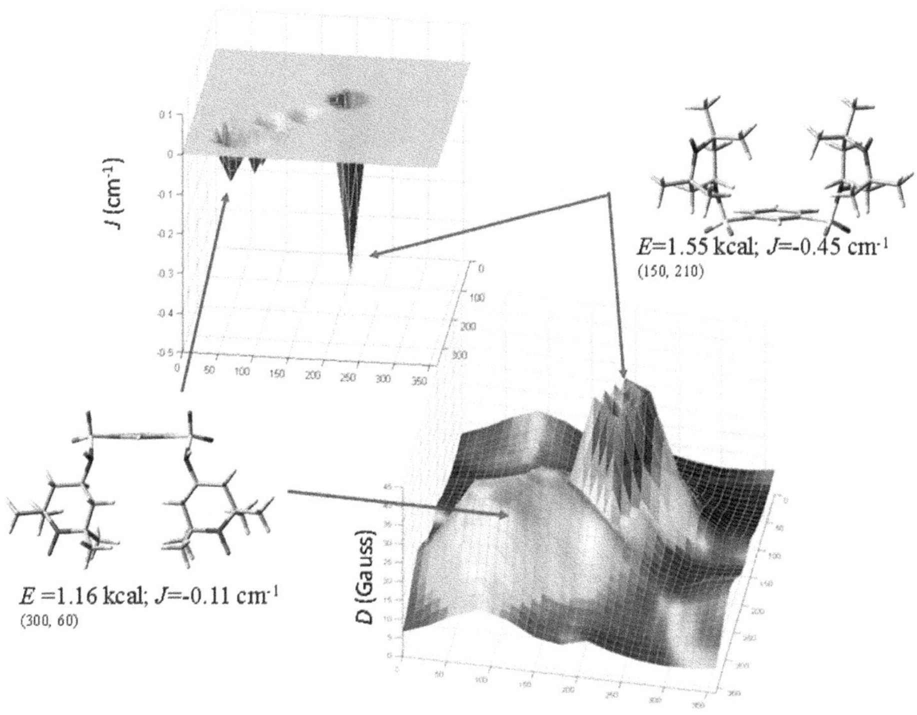

A similar treatment was performed for the system with dibenzo furan-2,8-di-sulfonyl fragment as central structure (see the structure in the bottom part of the Figure 8), taking similarly the dihedrals responsible for the overall rotation of the TEMPO fragments. The potential energy surface from Figure 12 shows that the preferred geometry of this system is the one with the largest distance between the radicals.

Figure 12. Potential energy surface for compound 2

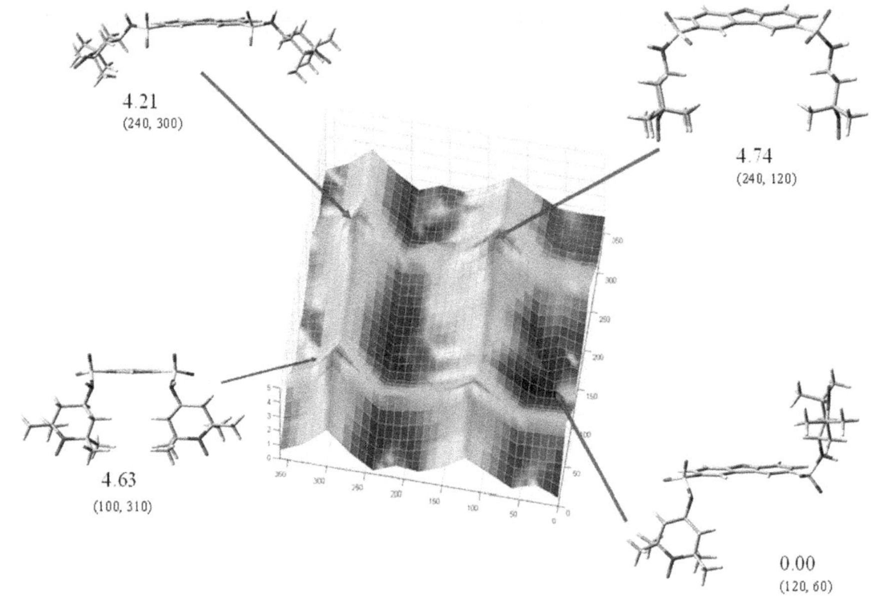

In the case of compound **2**, the fragment that connects the TEMPO radicals is bigger than the one from system **1**, imposing larger distances between the unpaired spins, expecting then to obtain an overall diminished *J* coupling. The BS-DFT estimation of the geometry-dependent exchange coupling parameter for the dibenzo furan-2,8-(SO_2NHTEMPO)$_2$ system, from Figure 13, confirms this fact.

Figure 13. BS-DFT data for the geometry-dependent coupling parameter of compound 2

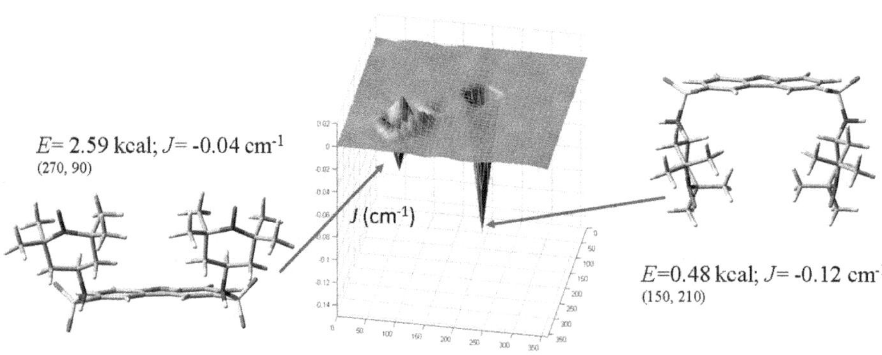

The areas where $J \neq 0$ correspond to the forced strained geometry, where the unpaired electron distributions are closer to each other. The scale of larger module of the peak is about four times smaller than in the previous example. In these conditions, the dipolar coupling becomes negligible, because of the rather abrupt dropping in magnitude, driven by the dependence on the inverse third power of inter-center distance. The presented case study, dealing with modeling in selected flexible biradical systems, illustrates a part of the problem, but it cannot systematically cover all conformational possibilities. A possible way of extending the search, by the compromise of calling simpler modeling methods, is illustrated in the next section.

Facing Complexity With Simpler Methods: Dynamics by Molecular Mechanics

To meet the realistic evolution of conformations, by sampling the randomness induced also by the solvent, must consider molecular dynamic modeling (Rapaport 2004; Allen & Tildesley, 2017). In the frame of previously exposed electronic structure methods, the computing effort would be prohibitive, making then recourse to much simplified procedures from the class of molecular mechanics (Allinger & Burkert, 1982; Mackerell, 2004). molecular mechanics takes empirical parametric expressions for the energy variation with respect of certain bond lengths, bond angles, dihedrals and, besides, for non-covalent dispersion forces between atom pairs. The electronic structure, in itself, is completely ignored, falling then in the domain of classical interactions, forgetting about the expected quantum nature of the behavior at molecular and supramolecular scales. This roughness is the price paid to advance into the intrinsic complexity of large systems.

This analysis is exemplified on the compound **1** from previous section, put in a box of dichloromethane molecules (CH_2Cl_2), according to the solvent used in the laboratory handling. Dynamics by molecular mechanics methods involves choosing a force field that is suitable for the systems at hand. There are many molecular mechanics versions, making the choice generally puzzling. In our case, the option was limited by the condition of finding all the atoms implied by the composition of compound **1** and solvent, altogether. Among the existing repositories, is not easy to find a parametrization including the nitrogen, sulphur and chlorine, concomitantly. Thus, the *oplsaa* force field (Jörgensen et al., 1996) seemed the only one that contains the required characteristics for all the functional groups present in the analyzed system.

A software convenient for this type of calculation is *Tinker* (Rackers et al., 2018), having also a handy graphical interface for controls and representation. Figure 14 shows a screenshot obtained from the film of molecular dynamics on the biradical **1**, put in a box with 248 molecules of CH_2Cl_2 molecules. The size of the cubic box, with a side of 30Å on edges, was chosen to retrieve the density of dichloromethane (correspondingly subtracting the estimated molecular volume of the biradical). The system is assumed in a periodic box regime, i.e., what happens in a unit box, evolves in similar copies in the surrounding cells, extended to infinity. The super-cell extension is needed for the reasonable estimation of intermolecular potentials, particularly at borders. The specific mathematical models accomplishing this goal, like Ewald summation, are not to be detailed here.

In statistical mechanics, the most used environments are the microcanonical ensemble (called also NVE, for conserving the number of particles, *N*, volume, *V*, and energy, *E*) and the canonical ensemble (NVT, keeping a constant temperature, *T*) (Müller-Kirsten, 2013). In the NVE, all microstates have the same probability, $P=1/n$, where n is the number of dynamic configurations. In NVT, the system is in thermal equilibrium with a heated bath, at a certain temperature. Exchanging energy with the thermostat, the microstates are allowed to show variable energies. Then, the states will have different probabilities, $P=e^{(F-E)/kT}$, where E is the total energy of the configuration, k is the Boltzmann constant, and F is the free energy. Because of a simpler realization, the discussed simulation followed the NVE regime.

Figure 14. Screenshot of a cell unit in a molecular dynamic simulation of biradical 1 in dichloromethane as solvent

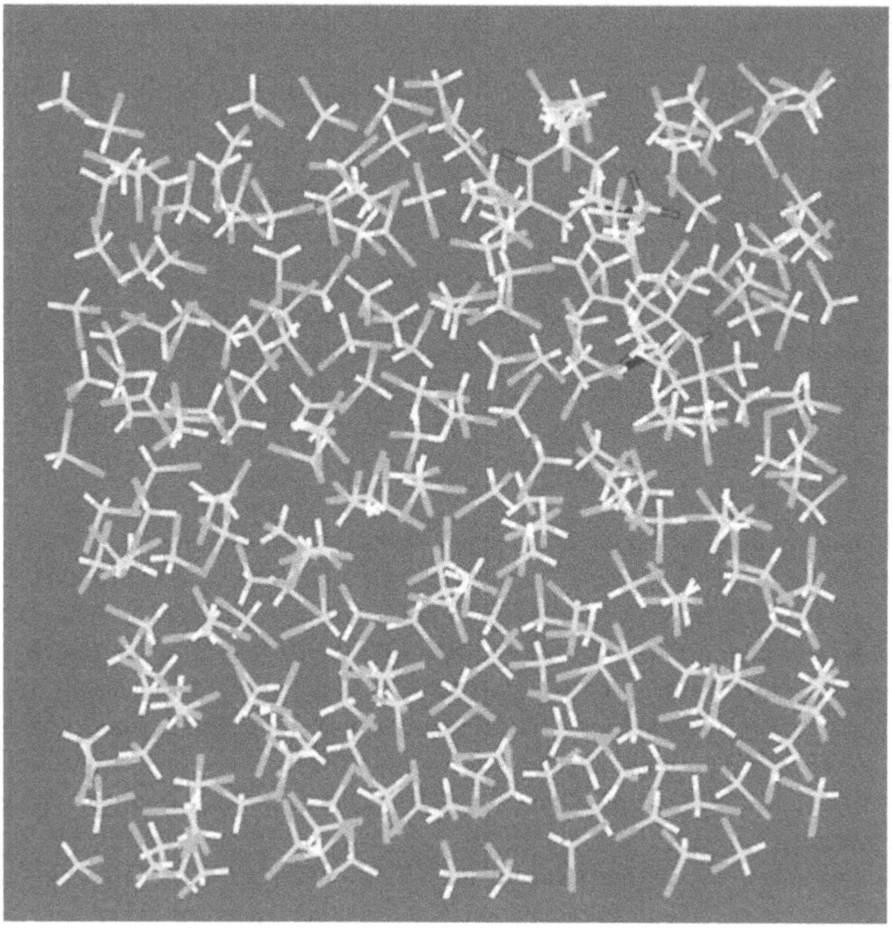

Given the large volume of information contained in the dynamic evolution, further processing is needed to abstract certain conclusions from the movie of the action. Thus, exploring the "black box" of molecular dynamics calculations, a significant parameter is taken as the probability of finding the two radicals at a certain distance. Figure 15 compares the dynamics of biradical in vacuum with its behavior in solvent. In the case without solvent, the two radicals can be found in a larger domain of distribution, including the close encounters that are favoring the relatively larger exchange coupling. It turns out that the radical arms move very little in the solvent, expecting then that in real situations only a small average of coupling is expected.

Figure 15. Distribution of the O...O distance between two spin carriers of the biradical 1, from molecular dynamics simulations. Broader line with two peaks is the biradical in vacuum, the sharper and higher curve representing the biradical in CH_2Cl_2

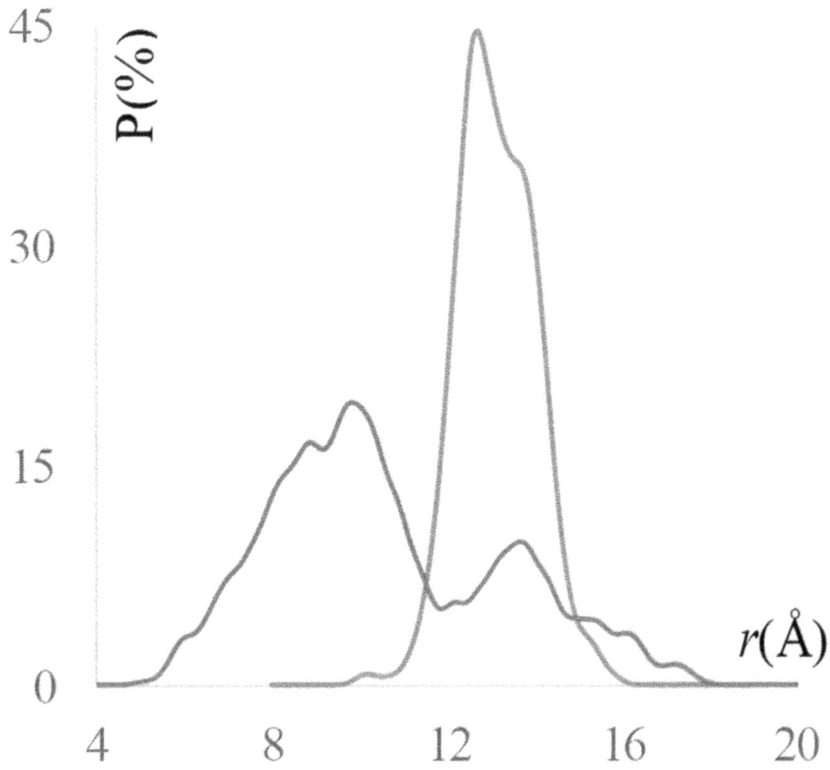

Figure 16 shows an interesting exploration, drawing the iso-surfaces of maximum probability for two features. The central object, hearth-shaped, shows the position of the middle point drawn between the two NO groups. It engulfs the central molecular fragment. From this, one may conclude that the two TEMPO fragments prefer to stay in the *trans* configuration. If a *cis* configuration had been preferred, then two profiles, above and below the central benzene nucleus, would be visible. The other components, consisting if a four-lobe ensemble, draw the probability of finding in space each of the two NO groups. The analysis is made considering the dynamics in the solvent box. Unfortunately, in the molecular mechanics dynamic modeling, the exchange coupling cannot be estimated, but one may set as possible challenge the goal of coming back to the issue, by using models fitted in similar manner to that sketched in the section devoted to the magneto-structural correlations with nitroxides.

Figure 16. The probability density of the placement of the barycenter of the two NO bonds (the four lobe-shaped elements) and the midpoint between the two NO groups (central hearth-shaped surface)), as a result of the molecular dynamics in CH_2Cl_2 solvent

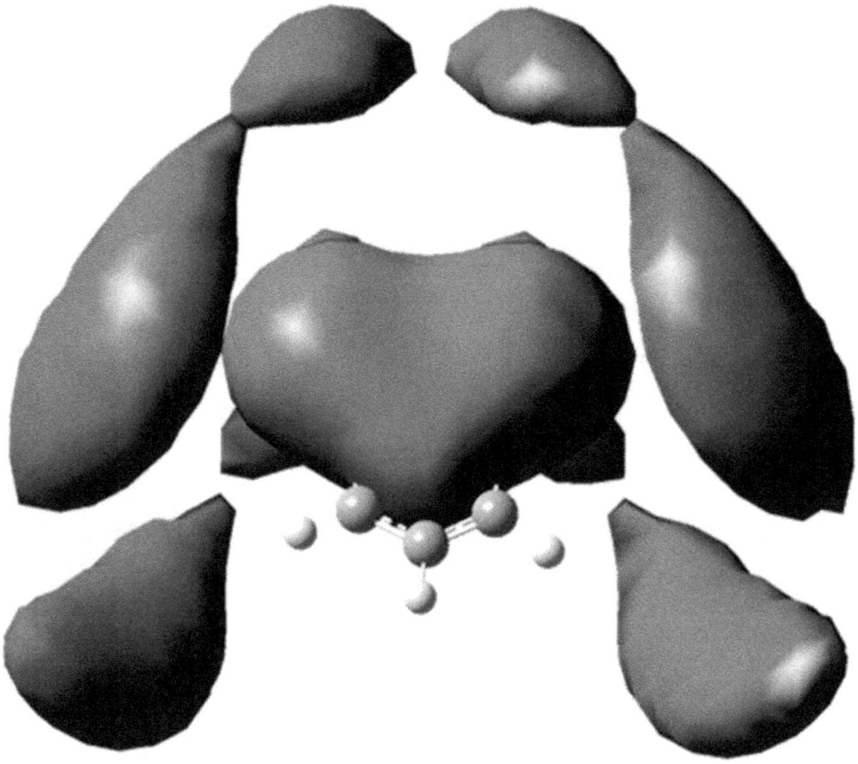

CONCLUSION

The chapter presented in an intentionally pedantic, even didactical, manner several issues related with basic principles and state-of-the-art in modeling the spin coupling and related effects, choosing the nitroxide-based organic radicals as playground. The relative diversity of debated theoretical aspects serves to throw some light on experimental EPR data, illustrated for a selected couple of TEMPO-based biradicals, even though still confined to a qualitative and semiquantitative level of suggestion. As a first step, it was brought in discussion the spin distribution in TEMPO units, as function of computation setting, noticing in all circumstances an almost equal percentage of the N and O atoms. Several illustrative computation experiments rationalized the dependence of the exchange coupling, tuned by mutual position of two nitroxide groups. The investigation was continued mapping the geometry dependence of exchange and dipolar coupling in a couple of realistic biradicals, establishing semiquantitative correlations with experimental EPR spectroscopy data.

To get closer to the complexity encountered in solutions, molecular dynamics calculations were performed, considering a biradical in a box of dichloromethane. Interpreting the collected data, the most probable conformation of the biradical was established. The visited roads evidenced other lanes to be explored, as hints for further research in the domain, drawing extended connections between the paradigms exposed in the different sections of this chapter. The discussion of exchange coupling between nitroxide organic radicals is, somewhat evasively but yet righteously, proposed as relevant collateral in paving the way for the consideration of such effects in biochemical redox mechanism involving nitric oxide, NO, also a radical molecule. The coaxial idea is that the spin coupling is implicit in all the electron transfer processes, an aspect usually ignored, which deserves deepened insight.

ACKNOWLEDGMENT

The realization of this work was done with resources from research grant PN-III-P4-PCE-2021-1881, from Romanian financing office, UEFISCDI.

REFERENCES

Abe, M. (2013). Diradicals. *Chemical Reviews, 113*(9), 7011–7088. doi:10.1021/cr400056a PMID:23883325

Abragam, A., & Bleaney, B. (1970). *Electron Paramagnetic Resonance.* Clarendon Press.

Allen, M. P., & Tildesley, D. J. (2017). *Computer Simulation of Liquids* (2nd ed.). Oxford University Press. doi:10.1093/oso/9780198803195.001.0001

Allinger, N. L., & Burkert, U. (1982). *Molecular Mechanics.* American Chemical Society Publication.

Anderson, P. W. (1959). New Approach to the Theory of Superexchange Interactions. *Physical Review, 115*(1), 2–13. doi:10.1103/PhysRev.115.2

Becke, A. D. (1992). Density-functional thermochemistry. I. The effect of the exchange-only gradient correction. *The Journal of Chemical Physics, 96*(3), 2155–2160. doi:10.1063/1.462066

Bencini, A., & Gatteschi, D. (1990). *EPR of Exchange Coupled Systems.* Springer-Verlag.

Benelli, C., & Gatteschi, D. (2002). Magnetism of lanthanides in molecular materials with transition-metal ions and organic radicals. *Chemical Reviews, 102*(6), 2369–2388. doi:10.1021/cr010303r PMID:12059272

Bertin, D., Gigmes, D., Marque, S. R., & Tordo, P. (2011). Kinetic subtleties of nitroxide mediated polymerization. *Chemical Society Reviews, 40*(5), 2189–2198. doi:10.1039/c0cs00110d PMID:21234490

Bogani, L., & Wernsdorfer, W. (2010). Molecular spintronics using single-molecule magnets. *Nanoscience and technology: a collection of reviews from nature journals*, 194-201.

Bomfleur, B., Sironi, M., Raimondi, M., & Voitländer, J. (2000). Ab initio calculation of the zero-field splitting parameter D of benzene and naphthalene. *The Journal of Chemical Physics, 112*(3), 1066–1069. doi:10.1063/1.480660

Buchachenko, A. L., & Berdinsky, V. L. (2002). Electron Spin Catalysis. *Chemical Reviews, 102*(3), 603–612. doi:10.1021/cr010370l PMID:11890751

Buta, M. C., Frecus, B., Enache, M., Humelnicu, I., Toader, A. M., & Cimpoesu, F. (2021). Intra- and Inter-Molecular Spin Coupling in Phenalenyl Dimeric Systems. *The Journal of Physical Chemistry A*, *125*(32), 6893–6901. doi:10.1021/acs.jpca.1c02705 PMID:34353026

Buta, M. C., Toader, A. M., Frecus, B., Oprea, C. I., Cimpoesu, F., & Ionita, G. (2019). Molecular and Supramolecular Interactions in Systems with Nitroxide-based Radicals. *International Journal of Molecular Sciences*, *20*(19), 4733. doi:10.3390/ijms20194733 PMID:31554219

Calzado, C. J., Cabrero, J., Malrieu, J. P., & Caballol, R. (2002). Analysis of the magnetic coupling in binuclear complexes. I. Physics of the coupling. *The Journal of Chemical Physics*, *116*(7), 2728–2747. doi:10.1063/1.1430740

Caneschi, A., Ferraro, F., Gatteschi, D., le Lirzin, A., Novak, M. A., Rentschler, E., & Sessoli, R. (1995). Ferromagnetic order in the sulfur-containing nitronyl nitroxide radical, 2-(4-thiomethyl) phenyl-4, 4, 5, 5-tetramethylimidazoline-l-oxyl-3-oxide, NIT(SMe)Ph. *Advanced Materials*, *7*(5), 476–478. doi:10.1002/adma.19950070515

Caneschi, A., Gatteschi, D., Lalioti, N., Sangregorio, C., Sessoli, R., Venturi, G., Vindigni, A., Rettori, A., Pini, M. G., & Novak, M. A. (2001). Cobalt (II)-nitronyl nitroxide chains as molecular magnetic nanowires. *Angewandte Chemie International Edition*, *40*(9), 1760–1763. doi:10.1002/1521-3773(20010504)40:9<1760::AID-ANIE17600>3.0.CO;2-U PMID:11353503

Caproiu, M. T., Ionita, G., Draghici, C., & Ionita, P. (2008). Synthesis and characterisation of several di-, tri-, and tetra-radicals linked by flexible or rigid linkers. *ARKIVOC*, *14*(14), 158–165. doi:10.3998/ark.5550190.0009.e16

Chapyshev, S. V., Walton, R., Sanborn, J. A., & Lahti, P. M. (2000). Quintet and Septet State Systems Based on Pyridylnitrenes: Effects of Substitution on Open-Shell High-Spin States. *Journal of the American Chemical Society*, *122*(8), 1580–1588. doi:10.1021/ja993131c

Christou, G., Gatteschi, D., Hendrickson, D. N., & Sessoli, R. (2000). Single-molecule magnets. *MRS Bulletin*, *25*(11), 66–71. doi:10.1557/mrs2000.226

Constantin, M. M., Corbu, C. G., Tanase, C., Codrici, E., Mihai, S., Popescu, I. D., Enciu, A. M., Mocanu, S., Matei, I., & Ionita, G. (2019). Spin probe method of electron paramagnetic resonance spectroscopy–a qualitative test for measuring the evolution of dry eye syndrome under treatment. *Analytical Methods*, *11*(7), 965–972. doi:10.1039/C8AY02783H

Cooper, D. L. (Ed.). (2001). *Valence Bond Theory*. Elsevier Science.

Cooper, D. L., Gerratt, J., & Raimondi, M. (1991). Applications of spin-coupled valence bond theory. *Chemical Reviews*, *91*(5), 929–964. doi:10.1021/cr00005a014

Cooper, D. L., Thorsteinsson, T., & Gerratt, J. (1999). Modern VB representations of CASSCF wave functions and the fully-variational optimization of modern VB wave functions using the CASVB strategy. *Advances in Quantum Chemistry*, *32*, 51–67. doi:10.1016/S0065-3276(08)60406-2

Coronado, E., Delhaès, P., Gatteschi, D., & Miller, J. S. (1996). *Molecular Magnetism: From Molecular Assemblies to the Devices*. Springer. doi:10.1007/978-94-017-2319-0

Cotton, F. A., Wilkinson, G., & Murillo, C. A. (1999). *Manfred Bochmann Advanced Inorganic Chemistry* (6th ed.). Wiley-Interscience.

Crawford, V. H., Richardson, H. W., Wasson, J. R., Hodgson, D. J., & Hatfield, W. E. (1976). Relation between the singlet-triplet splitting and the copper-oxygen-copper bridge angle in hydroxo-bridged copper dimers. *Inorganic Chemistry, 15*(9), 2107–2110. doi:10.1021/ic50163a019

de Nooy, A. E., Besemer, A. C., & van Bekkum, H. (1996). On the use of stable organic nitroxyl radicals for the oxidation of primary and secondary alcohols. *Synthesis, 1996*(10), 1153–1176. doi:10.1055-1996-4369

Desplanches, C., Ruiz, E., Rodriguez-Fortea, A., & Alvarez, S. (2002). Exchange Coupling of Transition-Metal Ions through Hydrogen Bonding: A Theoretical Investigation. *Journal of the American Chemical Society, 124*(18), 5197–5205. doi:10.1021/ja0178160 PMID:11982384

Ditchfield, R., Hehre, W. J., & Pople, J. A. (1971). Self-Consistent Molecular Orbital Methods. 9. Extended Gaussian-type basis for molecular-orbital studies of organic molecules. *The Journal of Chemical Physics, 54*(2), 724–728. doi:10.1063/1.1674902

Fenton, H. J. H. (1894). Oxidation of tartaric acid in presence of iron. *Journal of the Chemical Society Transactions, 65*(65), 899–911. doi:10.1039/CT8946500899

Ferbinteanu, M., & Cimpoesu, F. (2014). Quantum Nanosystems. Structure, Properties, and Interactions. In *Magnetic Anisotropy in Case Studies*. Apple Academics.

Ferbinteanu, M., Miyasaka, H., Wernsdorfer, W., Nakata, K., Sugiura, K. I., Yamashita, M., Coulon, C., & Clérac, R. (2005). Single-chain magnet (NEt4)[Mn2 (5-MeOsalen) 2Fe (CN) 6] made of MnIII– FeIII– MnIII trinuclear single-molecule magnet with an ST= 9/2 spin ground state. *Journal of the American Chemical Society, 127*(9), 3090–3099. doi:10.1021/ja0468123 PMID:15740148

Ferrando-Soria, J., Vallejo, J., Castellano, M., Martinez-Lillo, J., Pardo, E., Cano, J., Castro, I., Lloret, F., Ruiz-Garcia, R., & Julve, M. (2017). Molecular magnetism, quo vadis? A historical perspective from a coordination chemist viewpoint. *Coordination Chemistry Reviews, 339*, 17–103. doi:10.1016/j.ccr.2017.03.004

Fischer, H. (2001). The persistent radical effect: A principle for selective radical reactions and living radical polymerizations. *Chemical Reviews, 101*(12), 3581–3610. doi:10.1021/cr990124y PMID:11740916

Frisch, M. J., Trucks, G. W., Schlegel, H. B., Scuseria, G. E., Robb, M. A., Cheeseman, J. R., Scalmani, G., Barone, V., Mennucci, B., Petersson, G. A., Nakatsuji, H., Caricato, M., Li, X., Hratchian, H. P., Izmaylov, A. F., Bloino, J., Zheng, G., Sonnenberg, J. L., [...] & Fox, D. J. (2009). *Gaussian 09 Revision A.2*. Scientific Research Publishing.

Gateschi, D., & Sessoli, R. (2003). Quantum Tunneling of Magnetization and Related Phenomena in Molecular Materials. *Angewandte Chemie International Edition, 42*(3), 268–297. doi:10.1002/anie.200390099 PMID:12548682

Gherman, O., & Saliu, L. (1976). *Fizică statistică*. Editura Tehnică.

Glaser, T., Theil, H., Liratzis, I., Weyhermüller, T., & Bill, E. (2006). Ferromagnetic Coupling by Orthogonal Magnetic Orbitals in a Heterodinuclear CuIIVIV=O Complex and in a Homodinuclear CuIICuII Complex. *Inorganic Chemistry*, *45*(13), 4889–4891. doi:10.1021/ic0606328 PMID:16780307

Harrison, J. F. (1971). An Ab Initio Study of the Zero Field Splitting Parameters of 3 B 1 Methylene. *The Journal of Chemical Physics*, *54*(12), 5413–5417. doi:10.1063/1.1674842

Hay, P. J., Thibeault, J. C., & Hoffmann, R. (1975). Orbital Interactions in Metal Dimer Complexes. *Journal of the American Chemical Society*, *97*(17), 4884–4899. doi:10.1021/ja00850a018

Hayyan, M., Hashim, M. A., & Al Nashef, I. M. (2016). Superoxide ion: Generation and chemical implications. *Chemical Reviews*, *116*(5), 3029–3085. doi:10.1021/acs.chemrev.5b00407 PMID:26875845

Heisenberg, W. (1928). On the theory of ferromagnetism. *The European Physical Journal A*, *49*(9-10), 619–636. doi:10.1007/BF01328601

Heitler, W., & London, F. (1927). Wechselwirkung neutraler Atome und homöopolare Bindung nach der Quantenmechanik. *The European Physical Journal A*, *44*(6-7), 455–472. doi:10.1007/BF01397394

Her, J.-H., Stephens, P. W., Ribas-Ariño, J., Novoa, J. J., Shum, W. W., & Miller, J. S. (2009). Structure and Magnetic Interactions in the Organic-Based Ferromagnet Decamethylferrocenium Tetracyanoethenide, [FeCp*2]•+[TCNE]•−. *Inorganic Chemistry*, *48*(8), 3296–3307. doi:10.1021/ic801679m PMID:19361235

Hicks, R. G. (2007). What's new in stable radical chemistry? *Organic & Biomolecular Chemistry*, *5*(9), 1321–1338. doi:10.1039/b617142g PMID:17464399

Hicks, R. G. (Ed.). (2010). *Stable Radicals*. Wiley. doi:10.1002/9780470666975

Hinzmann, A., Stricker, M., Busch, J., Glinski, S., Oike, K., & Gröger, H. (2020). Selective TEMPO-Oxidation of Alcohols to Aldehydes in Alternative Organic Solvents. *European Journal of Organic Chemistry*, *2020*(16), 2399–2408. doi:10.1002/ejoc.201901365

Ionita, G., Ariciu, A. M., Smith, D. K., & Chechik, V. (2015). Ion exchange in alginate gels–dynamic behaviour revealed by electron paramagnetic resonance. *Soft Matter*, *11*(46), 8968–8974. doi:10.1039/C5SM02062J PMID:26399427

Ionita, G., Ariciu, A. M., Turcu, I. M., & Chechik, V. (2014). Properties of polyethylene glycol/cyclodextrin hydrogels revealed by spin probes and spin labelling methods. *Soft Matter*, *10*(11), 1778–1783. doi:10.1039/c3sm52004h PMID:24651650

Ionita, G., & Chechik, V. (2010). Exploring polyethylene glycol/cyclodextrin hydrogels with spin probes and EPR spectroscopy. *Chemical Communications*, *46*(43), 8255–8257. doi:10.1039/c0cc02759f PMID:20890495

Jorgensen, W. L., Maxwell, D. S., & Tirado-Rives, J. (1996). Development and testing of the OPLS all-atom force field on conformational energetics and properties of organic liquids. *Journal of the American Chemical Society*, *118*(45), 11225–11236. doi:10.1021/ja9621760

Kahn, O. (1993). *Molecular Magnetism*. Wiley-VCH.

Kahn, O., Pei, Y., & Journaux, Y. (1989). Orthogonality of the Magnetic Orbitals and Ferromagnetic Interaction. *Molecular Crystals and Liquid Crystals Incorporating Nonlinear Optics*, *176*(1), 429–441. doi:10.1080/00268948908037500

Kendall, R. A., Dunning, T. H. Jr, & Harrison, R. J. (1992). Electron affinities of the first-row atoms revisited. Systematic basis sets and wave functions. *The Journal of Chemical Physics*, *96*(9), 6796–6806. doi:10.1063/1.462569

Koch, W., & Holthausen, M. C. (2001). *A Chemist's Guide to Density Functional Theory*. VCH. doi:10.1002/3527600043

Kollmar, C., & Kahn, O. (1993). Ferromagnetic spin alignment in molecular systems: An orbital approach. *Accounts of Chemical Research*, *26*(5), 259–265. doi:10.1021/ar00029a006

Lee, C., Yang, W., & Parr, R. G. (1988). Development of the Colle-Salvetti correlation-energy formula into a functional of the electron density. *Physical Review. B*, *37*(2), 785–789. doi:10.1103/PhysRevB.37.785 PMID:9944570

Li, J., Duke, B., & McWeeny, R. (2007). *VB2000 Version 2.0*. SciNet Technologies.

Li, J., & McWeeny, R. (2002). VB2000: Pushing valence bond theory to new limits. *International Journal of Quantum Chemistry*, *89*(4), 208–216. doi:10.1002/qua.10293

Li, L., Turnbull, M. M., Landee, C. P., Jornet, J., Deumal, M., Novoa, J. J., & Wikaira, J. L. (2007). Synthesis, structure, and magnetic behavior of bis (2-amino-5-fluoropyridinium) tetrachlorocuprate (II). *Inorganic Chemistry*, *46*(26), 11254–11265. doi:10.1021/ic701645y PMID:18020329

Mackerell, A. D. (2004). Empirical force fields for biological macromolecules: Overview and issues. *Journal of Computational Chemistry*, *25*(13), 1584–1604. doi:10.1002/jcc.20082 PMID:15264253

Malrieu, J. P., Caballol, R., Calzado, C. J., De Graaf, C., & Guihery, N. (2014). Magnetic interactions in molecules and highly correlated materials: Physical content, analytical derivation, and rigorous extraction of magnetic Hamiltonians. *Chemical Reviews*, *114*(1), 429–492. doi:10.1021/cr300500z PMID:24102410

Mansikkamäki, A., & Tuononen, H. M. (2018). The role of orbital symmetries in enforcing ferromagnetic ground state in mixed radical dimers. *The Journal of Physical Chemistry Letters*, *9*(13), 3624–3630. doi:10.1021/acs.jpclett.8b01507 PMID:29897759

Maretti, L., Ferbinteanu, M., Cimpoesu, F., Islam, S. S., Ohba, Y., Kajiwara, T., Yamashita, M., & Yamauchi, S. (2007). Spin Coupling in the Supramolecular Structure of a New Tetra (Quinoline– TEMPO) Yttrium (III) Complex. *Inorganic Chemistry*, *46*(3), 660–669. doi:10.1021/ic0610841 PMID:17257008

Matsushita, M. M., Izuoka, A., Sugawara, T., Kobayashi, T., Wada, N., Takeda, N., & Ishikawa, M. (1997). Hydrogen-bonded organic ferromagnet. *Journal of the American Chemical Society*, *119*(19), 4369–4379. doi:10.1021/ja964083o

McLean, A. D., & Chandler, G. S. (1980). Contracted Gaussian-basis sets for molecular calculations. 1. 2nd row atoms, Z=11-18. *The Journal of Chemical Physics*, *72*(10), 5639–5648. doi:10.1063/1.438980

Miller, J. S., & Drillon, M. (Eds.). (2001). *Magnetism: Molecules to Materials*. Wiley-VCH.

Minaev, B., & Agren, H. (1996). Spin-catalysis phenomena. *International Journal of Quantum Chemistry, 57*(3), 519–532. doi:10.1002/(SICI)1097-461X(1996)57:3<519::AID-QUA25>3.0.CO;2-X

Misochko, E. Y., Korchagin, D. V., Bozhenko, K. V., Chapyshev, S. V., & Aldoshin, S. M. (2010). A density functional theory study of the zero-field splitting in high-spin nitrenes. *The Journal of Chemical Physics, 133*(6), 064101. doi:10.1063/1.3474574 PMID:20707555

Mitani, M., Mori, H., Takano, Y., Yamaki, D., Yoshioka, Y., & Yamaguchi, K. (2000). Density functional study of intramolecular ferromagnetic interaction through m-phenylene coupling unit (I): UBLYP, UB3LYP and UHF calculation. *The Journal of Chemical Physics, 113*(10), 4035–4050. doi:10.1063/1.1286418

Nimura, S., & Yabe, A. (1999). In P. M. Lahti (Ed.), *Molecular Magnetism of Organic-Based Materials* (p. 127). Dekker.

Noodleman, L. (1981). Valence bond description of antiferromagnetic coupling in transition metal dimers. *The Journal of Chemical Physics, 74*(10), 5737–5743. doi:10.1063/1.440939

Noodleman, L., & Davidson, E. R. (1986). Ligand spin polarization and antiferromagnetic coupling in transition metal dimers. *Chemical Physics, 109*(1), 131–143. doi:10.1016/0301-0104(86)80192-6

Novoa, J. J., Deumal, M., & Jornet-Somoza, J. (2011). Calculation of microscopic exchange interactions and modeling of macroscopic magnetic properties in molecule-based magnets. *Chemical Society Reviews, 40*(6), 3182–3212. doi:10.1039/c0cs00112k PMID:21321725

Onishi, T., Takano, Y., Kitagawa, Y., Kawakami, T., Yoshioka, Y., & Yamaguchi, K. (2001). Theoretical study of the magnetic interaction for M–O–M type metal oxides. Comparison of broken-symmetry approaches. *Polyhedron, 20*(11-14), 1177–1184. doi:10.1016/S0277-5387(01)00591-5

Parr, R. G., & Yang, W. (1989). *Density functional theory of atoms and molecules*. Oxford University, Press.

Putz, M. V., Cimpoesu, F., & Ferbinteanu, M. (2018). *Structural Chemistry, Principles, Methods, and Case Studies*. Springer.

Rackers, J. A., Laury, M. L., Lu, C., Wang, Z., Lagardère, L., Piquemal, J.-P., Ren, P., & Ponder, J. W. (2018). TINKER 8: A Modular Software Package for Molecular Design and Simulation. *Journal of Chemical Theory and Computation, 14*, 5273–5289. doi:10.1021/acs.jctc.8b00529 PMID:30176213

Raimondi, M., & Cooper, D. L. (1999). Ab initio modern valence bond theory. *Topics in Current Chemistry, 203*, 105–120. doi:10.1007/3-540-48972-X_6

Rapaport, D. C. (2004). *The Art of Molecular Dynamics Simulation* (2nd ed.). Cambridge University Press. doi:10.1017/CBO9780511816581

Riplinger, C., & Neese, F. (2011). The Reaction Mechanism of Cytochrome P450 NO Reductase: A Detailed Quantum Mechanics/Molecular Mechanics Study. *ChemPhysChem, 12*(17), 3192–3203. doi:10.1002/cphc.201100523 PMID:22095732

Ruiz, E., Alemany, P., Alvarez, S., & Cano, J. (1997). Structural Modeling and Magneto-Structural Correlations for Hydroxo-Bridged Copper(II) Binuclear Complexes. *Inorganic Chemistry*, *36*(17), 3683–3688. doi:10.1021/ic970310r PMID:11670060

Ruiz, E., Cano, J., Alvarez, S., & Alemany, P. (1999). Broken symmetry approach to calculation of exchange coupling constants for homobinuclear and heterobinuclear transition metal complexes. *Journal of Computational Chemistry*, *20*(13), 1391–1400. doi:10.1002/(SICI)1096-987X(199910)20:13<1391::AID-JCC6>3.0.CO;2-J

Shaik, S., Cohen, S., Wang, Y., Chen, H., Kumar, D., & Thiel, W. (2010). P450 enzymes: Their structure, reactivity, and selectivity-modeled by QM/MM calculations. *Chemical Reviews*, *110*(2), 949–1017. doi:10.1021/cr900121s PMID:19813749

Shaik, S., & Hiberty, P. C. (2008). *A Chemist's Guide to Valence Bond Theory*. John Wiley & Sons Inc.

Sigel, A., Sigel, H., & Sigel, R. K. O. (2007). *The Ubiquitous Roles of Cytochrome P450 Proteins*. Wiley. doi:10.1002/9780470028155

Song, L., Chen, Z., Ying, F., Song, J., Chen, X., Su, P., Mo, Y., Zhang, Q., & Wu, W. (2012). XMVB 2.0: An ab initio Non-Orthogonal Valence Bond Program. Xiamen University, China.

Song, L., Mo, Y., Zhang, Q., & Wu, W. (2005). XMVB: A program for ab initio nonorthogonal valence bond computations. *Journal of Computational Chemistry*, *26*(5), 514–521. doi:10.1002/jcc.20187 PMID:15704237

Sugawara, T., Matsushita, M. M., Izuoka, A., Wada, N., Takeda, N., & Ishikawa, M. (1994). An organic ferromagnet: A-phase crystal of 2-(2′, 5′-dihydroxyphenyl)-4, 4, 5, 5-tetramethyl-4, 5-dihydro-1 H-imidazolyl-1-oxy-3-oxide (α-HQNN). *Journal of the Chemical Society. Chemical Communications*, (14), 1723–1724. doi:10.1039/C39940001723

Tamura, M., Nakazawa, Y., Shiomi, D., Nozawa, K., Hosokoshi, Y., Ishikawa, M., Takahashi, M., & Kinoshita, M. (1991). Bulk ferromagnetism in the β-phase crystal of the p-nitrophenyl nitronyl nitroxide radical. *Chemical Physics Letters*, *186*(4-5), 401–404. doi:10.1016/0009-2614(91)90198-I

Thorsteinsson, T., & Cooper, D. L. (1998). Modern valence bond descriptions of molecular excited states: An application of CASVB. *International Journal of Quantum Chemistry*, *70*(4-5), 637–650. doi:10.1002/(SICI)1097-461X(1998)70:4/5<637::AID-QUA10>3.0.CO;2-#

Toader, A. M., Buta, C. M., Frecus, B., Mischie, A., & Cimpoesu, F. (2019). Valence Bond Account of Triangular Polyaromatic Hydrocarbons with Spin: Combining Ab Initio and Phenomenological Approaches. *The Journal of Physical Chemistry C*, *123*(11), 6869–6880. doi:10.1021/acs.jpcc.8b12250

Triki, S., Gomez-Garcia, C. J., Ruiz, E., & Sala-Pala, J. (2005). Asymmetric azido-copper(II) bridges: Ferro- or antiferromagnetic? Experimental and theoretical magneto-structural studies. *Inorganic Chemistry*, *44*(15), 5501–5508. doi:10.1021/ic0504543 PMID:16022548

Troiani, F., & Affronte, M. (2011). Molecular spins for quantum information technologies. *Chemical Society Reviews*, *40*(6), 3119–3129. doi:10.1039/c0cs00158a PMID:21336365 van Vleck, J. H., & Sherman, A. (1935). The Quantum Theory of Valence. *Reviews of Modern Physics*, *7*(3), 167–228. doi:10.1103/RevModPhys.7.167

Verdaguer, M., Bleuzen, A., Marvaud, V., Vaissermann, J., Seuleiman, M., Desplanches, C., Scuiller, A., Train, C., Garde, R., Gelly, G., Lomenech, C., Rosenman, I., Veillet, P., Cartier, C., & Villain, F. (1999). Molecules to build solids: High TC molecule-based magnets by design and recent revival of cyano complexes chemistry. *Coordination Chemistry Reviews*, *190–192*, 1023–1047. doi:10.1016/S0010-8545(99)00156-3

Willett, R. D., Gatteschi, D., & Kahn, O. (1985). *Magneto-structural correlations in exchange coupled systems*. Reidel Publishing Co.

Chapter 6
Electronic Structure–Antioxidant Action Relationships for Chemical Compounds:
A Quantum Chemical Study

Mikhail Gorbachev
https://orcid.org/0009-0005-3058-5497
Institute of Chemistry, Moldova State University, Moldova

Iolanta Balan
https://orcid.org/0000-0002-8704-1344
Institute of Chemistry, Moldova State University, Moldova

Natalia Gorinchoy
https://orcid.org/0000-0003-4529-9061
Institute of Chemistry, Moldova State University, Moldova

Ion Arsene
Ion Creangă Pedagogical State University, Moldova

ABSTRACT

The relationships between electronic structure particularities and antioxidant activity of some chemical compounds are considered on the basis of the density functional theory (DFT) calculations of reactions between some natural phenolic antioxidants (gallic, caffeic, p-coumaric, and other acids), as well as dihydroxyfumaric acid, and the stable radicals ABTS•+ and DPPH• in water, ethanol, and methanol. Particular attention is paid to the processes of the primary proton transfer between neutral molecules of the above antioxidants and molecules of these solvents. The important factors are also considered: the formation of the charge transfer complexes (CTC) between the anionic forms of these antioxidants and the radicals, and possible tautomeric effects essentially changing the pre-reactional structures of the investigated antioxidant molecules and their anions. The quantitative relationships between the electronic structure parameters and the antioxidant activity of some short chain n-alkanes with radicals OH• and NO3• are also discussed.

DOI: 10.4018/978-1-6684-7198-2.ch006

INTRODUCTION

Practically in most cases, the experimental investigation of the antioxidant activity of chemical compounds establishes only its relative values in the series of the studied chemical substances supposed to be antioxidants. However, the data obtained in this way do not allow us to make an adequate forecast in the search for new antioxidants. At the same time, it is well known that many properties of any molecular system are determined by the particularities of its electronic structure. Therefore, in order to reveal a possible mechanism of redox reactions for oxidant-antioxidant systems, it is advisable to develop a theoretical model that could describe some correlation between the value of antioxidant ability and the calculated electronic structure parameters, such as, for example, the charges on the atoms (q_i), the energies of the frontier (and the nearest) molecular orbitals (E_i), interatomic distances in its structure, etc. In addition, for this purpose many DFT-based quantum reactivity descriptors as, for instance, chemical hardness (η), softness (σ), electronegativity (χ), Fukui indexes, etc., are also being used. Most often all these reactivity descriptors are calculated by the DFT B3LYP method (Stephens, 1994)which allows one the adequate description of the electronic ground state properties of various organic chemical antioxidants and other biologically active molecules (see the reviews (Spiegel, 2022), (Villaverde, 2020). The following usage of the multiple regression analysis applied to establish the Structure-Activity Relationship (SAR) or Quantitative Structure-Activity Relationship (QSAR) between the antioxidant activity and the above reactivity descriptors leads only to the revealing of some new correlation dependency. However, this way doesn't approach us to the understanding of the sought antioxidant reaction mechanism.

In this connection, we will consider a number of recently published papers that are devoted to establishing the dependence QSAR based on the above-mentioned descriptors. In Ref. (Muliadi, 2021) devoted to the study of myristicin derivatives the following correlation Equation (1) was found that describes the logarithm of their distribution coefficient P in the octanol-water binary system, i.e., the possibility of penetration of an antioxidant into the lipid cell layer:

$$\log_{10} P = -2.600 + 0.006\ IW - 1.558\ q(C8) - 6.532\ EHOMO + 0.014\ PSA + 0.133\ MD, \quad (1)$$

where $n = 6$, and $R^2 = 0.926$.

In Eq.(1) for each of the studied in (Muliadi, 2021)myristicin derivatives, q(C8) and EHOMO are respectively the charges on its carbon atom C8 and the energy of its HOMO obtained by means of the DFT B3LYP calculations. IW, PSA, and MD are respectively its Wiener Index, its polar surface area, and its dipole moment. One can see, that all these one-particle variables of Eq.(1) were determined by means of multiple regression analysis.

In the work (Djeradi, 2014), which is devoted to DFT-based reaction ability combined with QSAR of flavonoid derivatives as antioxidants, it was shown that the so-called condensed-to-atom Fukui functions (Pratim, 2009) play an important role in the found in (Djeradi, 2014) QSAR dependence. Further, in the work (Sarkar, 2012) (which is also devoted to the study of flavonoid compounds) it was found the QSAR equation describing adequately the radical scavenging antioxidant activity of a wide series of flavonoids using DFT based quantum chemical descriptors – the so-called "group frontier electron density" along with parameters of molecular hardness and the parameter of "group philicity" for a nucleophilic attack (Sarkar, 2012). It is to note here that both the above group parameters are related to different cycles of the studied flavonoid molecular systems.

Another very interesting QSAR dependence was found in the work (Filipović, 2015) on the basis of DFT calculations performing by taking into account the solvent - water. This dependence has the following form:

$$VCEAC = 1059 - 2.09\ PA - 1.71\ ETE + 65.32\ nOHvic, \tag{2}$$

with $n = 21$, $R = 0.962$, $s = 29.1$

In Eq.(2) which describes antioxidant potency of selected hydroxybenzoic acids and simple phenolics, VCEAC is vitamin C equivalent antioxidant capacity for each of the studied compounds, PA and ETE are respectively its proton affinity and its electron transfer enthalpy, nOHvic is the number of vicinal phenolic OH groups for any of the investigated molecular systems.

In its turn, in the work (Lafridi, 2020) it was shown the importance of the energy values of the lowest unoccupied molecular orbitals ELUMO along with molecular hardness and one-particle polar surface area for revealing of the QSAR, which connects antiradical properties of phenolic compounds (flavonoids) with the above electronic descriptors obtained on the basis of DFT calculations. Thus, the given work demonstrates the crucial role of the lowest unoccupied molecular orbitals in QSAR. At the same time, in many cases, the influence of the highest occupied molecular orbitals, HOMOs, of studied compounds on their biological activity is also very significant. So, in the work (Mikulski, 2014) it has been shown the main role of the HOMOs of chlorogenic acid, silybin, and all geometric stereoisomers of cynarine on the antioxidant activities of these hepatoprotective compounds. However, it is to be noted here that in a number of cases the antioxidant ability of compounds under study is related to the molecular orbitals, which differ from their frontier molecular orbitals (their HOMOs or their LUMOs). So, for instance, it was shown in the work (Costa, 2018) that the energy levels the HOMO-1 and the LUMO+1 molecular orbitals of the investigated caffeine analogs play also an important role in their antioxidant activity.

Considering all the results on the determination of the antioxidant QSAR in the above works, we can conclude that, despite the fairly good statistical characteristics of the found correlations, they do not bring us any closer to understanding the molecular mechanisms of the corresponding antioxidant action. In most cases, logical conjectures and intuitive motives that led certain authors to establish the dependencies QSAR almost always remain outside the published material. Below, using the examples of some works of the authors of this chapter, we will try to partially fill this gap.

ANTIOXIDANT ACTIVITIES OF SOME ORGANIC ACIDS IN SOLVENTS

Reaction of Dihydroxyfumaric Acid With the Radical DPPH$^{\bullet}$

First of all, it should be noted that the vast majority of redox antiradical reactions occurring both in the external environment and in living organisms are usually carried out in the liquid phase.

The solvent influence on the antioxidant action of dihydroxyfumaric acid (DHFA) in its reaction with the free radical DPPH• is considered in our work (Gorbachev, 2015), where special attention was paid to the following well-known experimental fact: when various acids are being added to purple alcohol solutions of DPPH•, their purple color disappears almost instantly (see the review (Yordanov, 1996)). Bearing in mind the given circumstance we have supposed that the above color change is caused by the protonation of DPPH• by adding of acids. Another experimental fact that confirmed the correctness of this assumption is associated with the results of kinetic studies of the reaction rate of the radical DPPH• with dihydroxyfumaric acid (Secara, 2010). The kinetic stopped-flow data presented in the work (Secara, 2010) indicate that in the first seconds after the start of the reaction between DPPH• and DHFA its rate is proportional to the square root of the initial acid concentration. However, in accordance with the law of mass action, the concentration of protons arising in the initial acid solution due to the dissociation of its neutral molecules is also proportional to the square root of their concentration. These data also confirm the working hypothesis about the primacy of the reaction of protonation of the neutral radical DPPH• by protons from the water-ethanol medium, which appeared in it due to the dissociation of the acid. The initial DFT calculations (for the gas phase) also demonstrate the absence of the direct hydrogen transfer from the oxygen atom of DHFA (belonging to one of its central hydroxyl groups) to the corresponding nitrogen atom of DPPH• because of the significant energy barrier (Arsene, 2021). Therefore, in the DFT B3LYP calculations of the work (Gorbachev, 2015) the protonated radical DPPH• was involved. Figure 1 demonstrates the calculated charge values for some DPPH• heteroatoms. A comparison of their values indicates that the protonation of the DPPH• molecule will most likely occur with the participation of its oxygen atom O4.

Figure 1. The calculated charge values on some heteroatoms of DPPH•

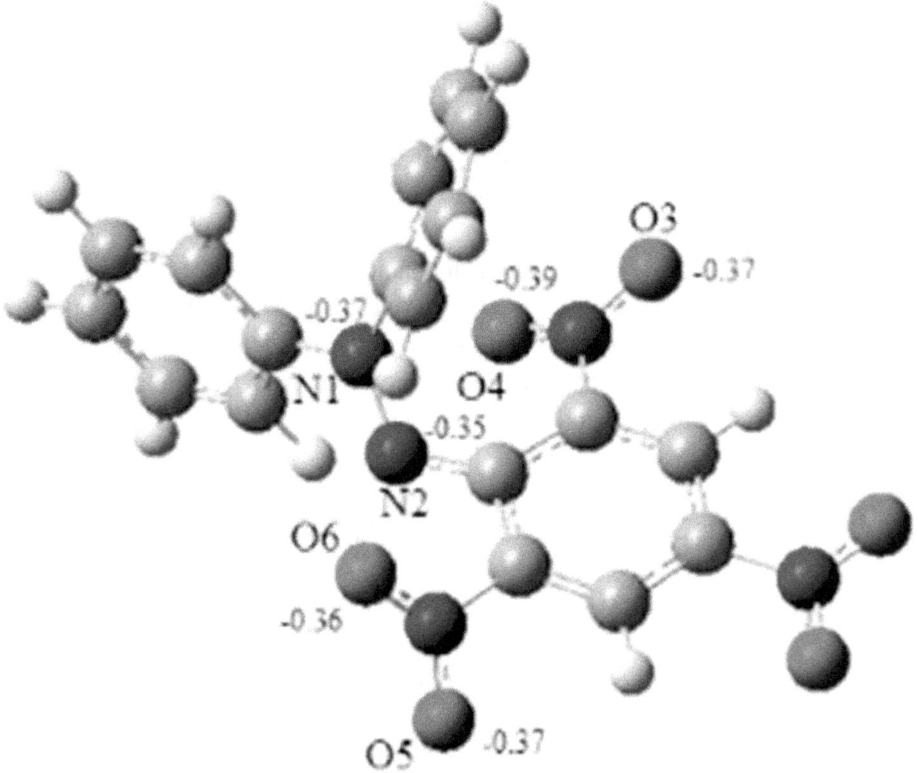

Here it is to be noted, that some particularities of the electronic structure of the protonated via the oxygen atom O4 radical DPPH• (denoted hereafter as [DPPH-H]•+) can quite simply explain the above mentioned color change of DPPH• in acid media. Indeed, in the DPPH• the spin density of its unpaired electron is predominantly localized on the atom N2, while in the considered [DPPH-H]•+ system it is mainly shared between the atom O3 and the nitrogen atom connected with O3 (see Figure 1). It is obvious that the given shift of the spin density must change the absorption spectrum in its visible region. Further DFT calculations have shown that the protonated radical [DPPH-H]•+ can easily react with the anion of DHFA (denoted as [AH]–). In its turn, the united system ([DPPH-H]•+•••[AH]–) is the charge-transfer complex (CTC) shown in Figure 2.

Figure 2. The mutual spatial orientation and the summarized charges in the CTC [DPPH-H]$^{\bullet+}$•••[AH]$^{-}$

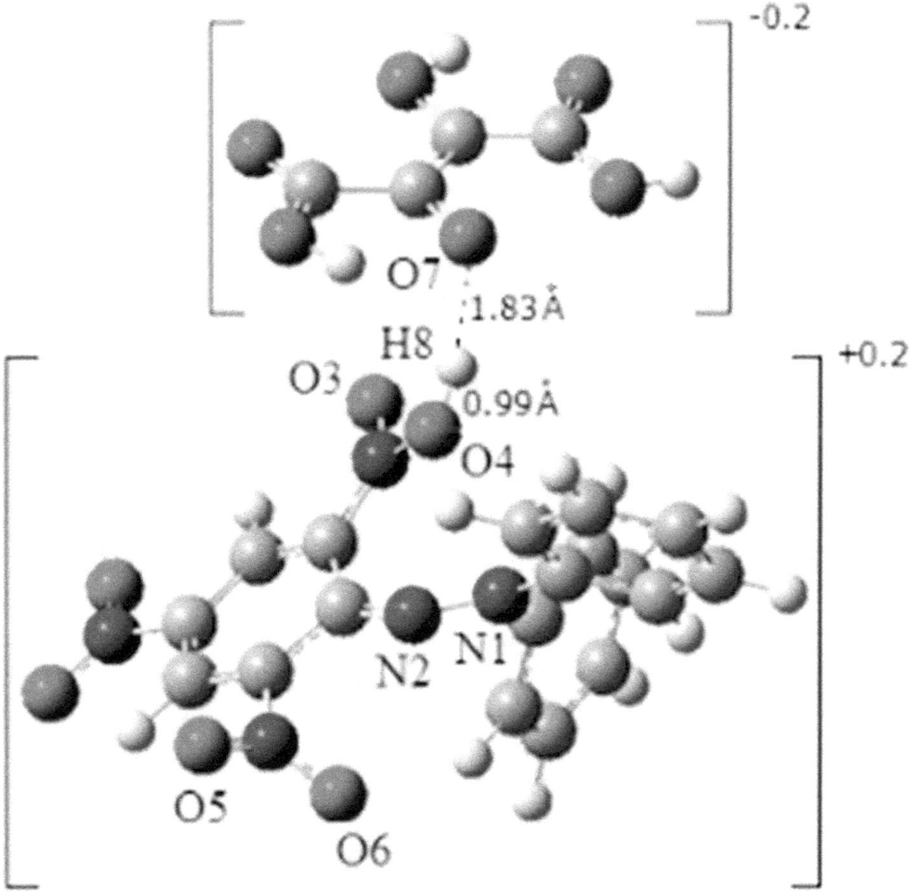

The formation of this CTC is accompanied by the transfer of the negative charge 0.8e from [AH]– to [DPPH-H]•+. The spin density of the CTC is mainly localized on the carbon atoms belonging to the double C=C bond of [AH]–. In its turn, due to the subsequent tautomeric transition of the hydrogen atom from O4 to N2 the considered CTC is easily broken down into the corresponding molecules of 1,1-diphenyl-2-picrylhydrazine and the radical-acid AH•. Further, any two molecules of the radical acid AH• easily form their dimmer (AH)2 which has four carboxyl groups. The electrolytic dissociation of any of these groups in water-alcohol media gives the anion [A2H]–. However, our DFT calculations show that [A2H]– is an unstable particle. It decays into one new anion [AH]– of DHFA and one molecule of dioxosuccinic acid (Secara, 2010).

Thus, based on the DFT calculations, the following seven-stage mechanism (Scheme 1) was proposed (Gorbachev, 2015) for the reaction of dihydroxyfumaric acid (AH2 on Scheme 1) with the free radical DPPH•.

Scheme 1

1) AH2 ↔ [AH]– + H+
2) DPPH• + H+ → [DPPH-H]•+
3) [DPPH-H]•+ + [AH]– → CTC
4) CTC → AH• + 1,1-diphenyl-2-picrylhydrazine
5) 2AH• → (AH)2
6) (AH)2 ↔ [A2H]– + H+
7) [A2H]– → [AH]– + dioxosuccinic acid

Here it is to be noted that the mechanism described by Scheme 1 has an ionic character. All its stages are accompanied by the corresponding energy release (Gorbachev, 2015). Both these reasons make it possible to understand the high reaction rate of between DHFA (AH2) and the first molecule of DPPH• (the middle reaction time is about 7-10 s (Secara, 2010). Moreover, as the volume concentration of H+ formed in stage (1) is proportional to the square root of [AH2] (due to the law of mass action), the rate of the reaction (2) which defines the total rate of the interaction between AH_2 and the first molecule of DPPH• will be also proportional to the product of the concentrations [DPPH•][H+], i.e. to [DPPH•][AH2]1/2. Indeed, this theoretical conclusion is completely confirmed by the experimental data of Ref. (Secara, 2010).

Interaction Between Some Food Acids and the Free Radical ABTS$^{•+}$

The detailed mechanism of the interaction of a number of antioxidant food acids (gallic, ferulic, caffeic, vanillic, cinnamic, syringic, p-coumaric) with the radical cation ABTS•+ is revealed by means of the DFT calculations in (Gorbachev, 2022). In this work we have investigated (by means of the DFT B3LYP method) the interaction of some antioxidant phenolic food acids and their anions (which were considered in their most stable tautomeric forms) with the stable radical ABTS•+ in both the phases: the gas phase and ethanol. The calculations have shown that for all the investigated molecular systems of the types: [an acid & ABTS•+] or [an anion & ABTS•+] the formation of the complexes takes place. In these complexes, the structural fragment of ABTS•+ is connected with a structural fragment of an acid or with a structural fragment of an anion by means of one of the two sulfo-groups of ABTS•+. Figures 3 and 4 show the complexes of ABTS•+ with the neutral molecule of gallic acid and its anion, respectively.

Figure 3. The spatial distribution of the spin density of the unpaired electron in the complex formed by ABTS•+ and a molecule of gallic acid

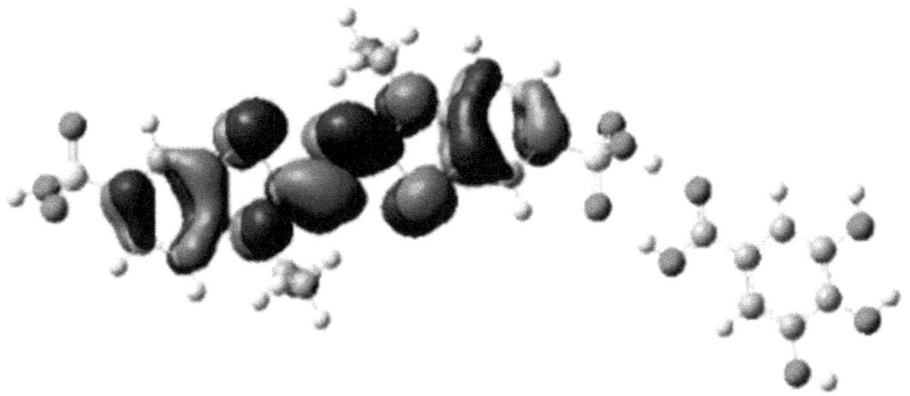

Figure 4. The spatial distribution of the spin density of the unpaired electron in the complex formed by ABTS•+ and an anion of gallic acid

The analysis of the spin density distributions in the given complexes shows that there is a fundamental difference between the complex with neutral molecules of the studied acids and those formed by their anions. In the case of the anions of the studied phenolic food acids, the practically complete charge transfer of electrons from the studied anions to the radical cation ABTS•+ takes place (compare Figs. 3 and 4). The given charge transfer occurs in both the gas phase and in ethanol. Based on the calculated data, a new correlation was found between the residual charges on the structural anionic fragments of the studied complexes [an anion & ABTS•+] and the known experimental data (Singh, 2018) describing the antioxidant activity IC_{50} (mmol/mL) of the studied acids in their reaction with ABTS•+.

In Table 1 the experimental values of IC_{50} for all the phenolic acids along with the corresponding above-mentioned residual charges are presented (Gorbachev, 2022).

Table 1. Residual charges on anions' fragments and charge transfers from the anions of the studied food acids to $ABTS^{\bullet+}$ in the gas phase and in ethanol

Phenolic Acid	IC_{50} (mmol/mL)	Residual charge/Charge transfer (ē)	
		in the gas phase	in ethanol
p-Coumaric	8.21	0.020/0.980	0.047/1.047
Gallic	8.85	0.023/0.977	0.038/1.038
Ferulic	9.47	0.022/0.978	-0.073/0.927
Syringic	13.97	0.017/0.983	-0.078/0.922
Caffeic	18.04	0.021/0.979	-0.103/0.897
Vanillic	132.10	0.020/0.980	-0.113/0.887
Cinnamic	342.46	0.017/0.983	-0.128/0.872

One can see that a clear-cut correlation between the values of IC_{50} and the values of the corresponding residual charges on the anionic structural fragments takes place only in the case of the ethanol medium. In other words, for each of the studied food acids, a greater antioxidant activity IC_{50} corresponds to a greater negative charge transferred from its anion to ABTS•+. However, there is no such correlation in the case of the gas phase.

So, we can draw the following main conclusions about the antiradical activity of the studied phenolic food acids in their test reaction with ABTS•+. Firstly, the antiradical properties of these acids are caused by the reaction between their anions and ABTS•+; and, secondly, the solvent (ethanol) plays the predominant role in the definition of the quantities of their antioxidant activities.

Thus, at the first stage of the mechanism of the antioxidant action of the studied acids (in both DPPH• and ABTS•+ cases), they dissociate with the formation of the corresponding anions. These anions then form charge-transfer complexes with the corresponding radicals, resulting in them in their diamagnetic derivatives. It means that in both the cases under study, the so-called SPLET mechanism takes place.

FREE RADICAL SCAVENGING ACTION OF SOME SHOT-CHAIN N-ALKANES

Another example of establishing the antioxidant QSAR dependence based on the use of appropriate physicochemical models can serve the work (Gorbachev, 2021). It is devoted to revealing the general molecular orbital nature that determines both the quantities: surface tension σ of short-chain n-alkanes at their normal boiling points and the rate of their reaction with hydroxyl radicals in the gas phase (in the atmosphere) at 298 K. In the work (Gorbachev, 2021) we have taken into attention that many physicochemical properties of chemical compounds are being defined by characteristics of certain specific occupied molecular orbitals (SMOs). So, for example, these SMOs can determine such physical quantities of liquid chemical compounds as surface tension (Gorbachev, 2017), (Gorbachev, 2014), aqueous solubility (Abdur Rauf, 2015), etc. It is important to be noted here, that the above-mentioned occupied SMOs are not necessarily the highest occupied MOs (HOMOs) of the compounds under study. So, in the works (Gorbachev, 2017), (Gorbachev, 2014) it was shown that the occupied MOs which are responsible for the studied physicochemical properties can be HOMO-1, HOMO-2, etc. It was shown in (Gorbachev, 2021) that the hydrogen atoms of short-chain n-alkanes enter their SMOs with large values of the coefficients of their atomic orbitals (usually with the same "+" or "-" sign). Thus, namely, these SMOs contribute essentially to the dispersion attraction existing between molecules of these n-alkanes. Further, based on the above concept of SMOs (and using the numerical values of their orbital energies Eorb) it was possible to quantitatively describe (see(Gorbachev, 2021)) the surface tension of short-chain n-alkanes at their normal boiling points.

The spatial form of these specific SMOs for some short-chain n-alkanes is presented in Figure 5. It is easy to see that for any of the investigated n-alkanes the atomic 1s- orbitals of all its hydrogen atoms, entering the LCAO of its SMOs, really have the same sign.

Figure 5. The view of the HOMOs (the upper row) and the SMOs (the lower row) for some of the studied n-alkanes

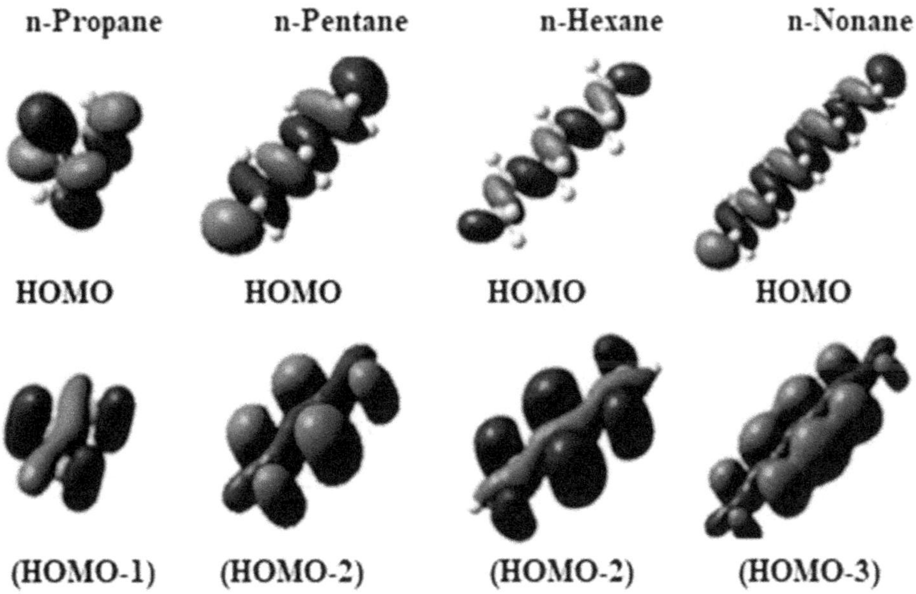

The work (Gorbachev, 2021) deals with the establishment of the antiradical QSAR based on the DFT study of the rate constants k of the reaction of the above-considered short-chain n-alkanes with hydroxyl radicals in the gas phase at the temperature $T = 297 \pm 2$ K. It should be noted here that, since the studied n-alkanes are capable of converting hydroxyl radicals into water molecules, they can be considered as antioxidants; and the aforementioned rate constants can serve as a measure of their antioxidant activity. The equation of their reactions (i) with hydroxyl radicals has the following common form:

R1-- CH_2 –R2 + HO• = R1--C•H –R2 + H2O (i)

Here R1 and R2 are linear alkyl substituents (or one hydrogen atom together with one methyl substituent in the case of ethane). In order to find the desired dependence of QSAR, we used a physico-chemical model that included the following theoretical provisions. First of all, we considered that the studied reaction rate constant k can be described by means of the Eyring equation from the transition state theory (see (Eyring, 1935)). Then, using the assumption that the changes in entropy and volume accompanying reactions (i) do not depend on the radicals R1 and R2, we were able to write the equation for k in the following logarithmic form:

$$\log k = C + D\Delta E, \qquad (3)$$

where C and D are some constant coefficients for each of the studied n-alkanes. ΔE is the activation energy that accompanies each of the reactions (i). This value depends on the nature of the n-alkane under study. Further, since our DFT calculations showed that the absolute values of the atomic charges on the hydrogen atoms of the n-alkanes are near to zero, so the reactions (i) belong to the type of orbital-controlled reactions. Then, using for the estimation of ΔE the expression given by Klopman in his work (Klopman, 1968), Eq. (3) can be presented (see (Gorbachev, 2021)) in the following final form:

$$\log_{10} k = C + Z/(Ei - Eorb). \qquad (4)$$

Here Ei is the energy of the molecular orbital of the hydroxyl radical, on which its unpaired electron is located, Eorb is the energy of the SMO for each investigated alkane. The quantity Z is constant for all the studied n-alkanes (Gorbachev, 2021) .

The multiple regression analysis allows one to determine the above parameters in Eq.(4). The corresponding numerical values of C, Z, and Ei are equal to 8.1254, 496.25 kJ/mol, and 1003.13 kJ/mol, respectively (see (Gorbachev, 2021)). The values of $-\log_{10} k$ calculated by means of Equation (4) and the above values of C, Z, and Ei are presented in the last column of Table 2.

Table 2. The values of $-\log_{10} k$ calculated by means of Eq. (4)

	n-Alkane	$k_{exp} \times 10^{12}$ ($cm^3 \cdot mol^{-1} \cdot s^{-1}$)	E_{orb} ($kJ \cdot mol^{-1}$)	$-\log_{10} k_{exp}$	$-\log_{10} k_{calc.}$
1	Ethane	0.222	893.58	12.65	12.66
2	Propane	1.12	873.56	11.95	11.95
3	Butane	2.10	863.34	11.68	11.68
4	Pentane	5.30	845.16	11.28	11.27
5	Hexane	5.91	841.27	11.22	11.19
6	Heptane	7.26	837.02	11.14	11.11
7	Octane	8.24	835.47	11.08	11.09
8	Nonane	9.36	833.73	11.03	11.05
9	Decane	9.90	833.00	11.00	11.04

One can see that Eq.(4) is the desired QSAR dependence describing the antioxidant activity of the studied short-chain n-alkanes in their reaction with hydroxyl radicals. Below, in Figure 6, it is shown the formation of the three-center C-H-O bonding MO between the SMO of n-pentane and the unpaired electron of hydroxyl radical.

Figure 6. The formation of the three-center C-H-O bonding MO between the SMO of n-pentane and the unpaired electron of hydroxyl radical

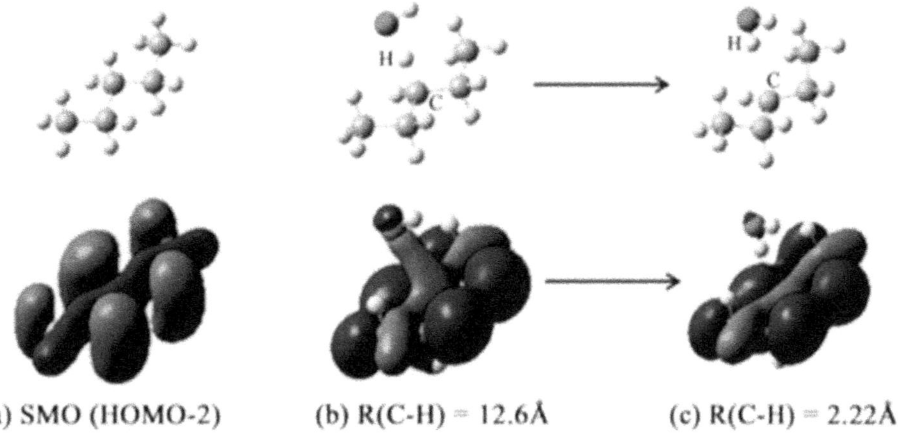

Thus, the occupied Specific Molecular Orbitals (SMOs) of the studied short chain n-alkanes play a decisive role in determining their various properties: the surface tension σ near their normal boiling points and their reaction rate constants k with hydroxyl radicals (at 297 ± 2 °K).

Moreover, the concept of SMOs can be successfully used for establishing of QSARs describing the antioxidant activity of short-chain n-alkanes in their reactions with radicals other than the hydroxyl radical. Such, the QSAR was established between the rate constants K of the reaction of the short-chain C3-C9 n-alkanes with the radical NO_3• in the Earth's atmosphere (Gorbachev, 2022). Here it is to be noted that the nitrate radical NO_3• is one of the main night-time oxidants in the atmosphere. The reactions between NO_3• and hydrocarbons essentially influence air quality and climate. However, in many cases there is an incomplete understanding of the mechanisms of the reactions between NO_3• and volatile organic chemicals. The analysis of the DFT B3LYP data (Gorbachev, 2022) has shown that the following equation holds for the adequate description of the quantity -log10K:

$$-\log_{10}K = A + B/(E_{NO3} - Eorb). \quad (5)$$

Eq. (5) is similar to Eq. (4). However, the numerical values of its constants A, B, and $E_{NO3\bullet}$ (which are equal to 17.5404, 50.9904, and 807.38 kJ/mol, respectively) differ from the constants of Eq. (4). Eq. (5) shows that the specific occupied molecular orbitals, SMOs, of the studied n-alkanes with the same energies Eorb play also the predominant role in their reactions with NO_3•. Table 3 demonstrates the calculation of the quantities $-\log_{10}K$ of the given n-alkanes be means of Eq. (5). The experimental values of $-\log_{10}K$ were taken from the work (Atkinson, 1997).

Table 3. The usage of Eq.(5) for the calculation of $-log_{10}K$ of the n-alkanes under study

No	n-Alkane	E_{orb} (kJ/mol)	$-log_{10}K_{calc.}$	$-log_{10}K_{exp.}$
1	Propane	873.56	16.77	16.77
2	n-Butane	863.34	16.54	16.63
3	n-Pentane	845.16	16.06	16.19
4	n-Hexane	841.27	15.96	16.04
5	n-Heptane	837.02	15.82	15.82
6	n-Octane	835.47	15.72	15.73
7	n-Nonan	833.73	15.64	16.61

Figure 7 below demonstrates the formation of the three-center C-H-O bonding MO between n-pentane and NO_3•

Figure 7. The formation of the three-center C-H-O bonding MO between n-pentane and NO_3•

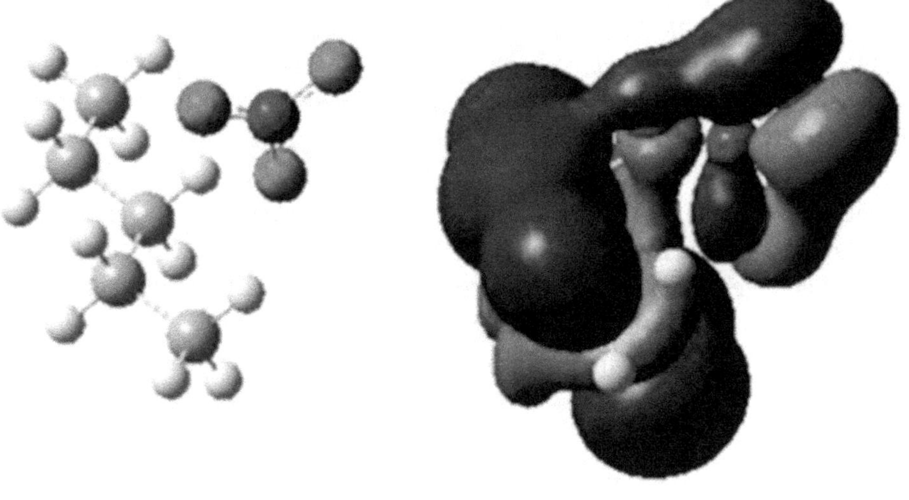

The other work (Gorbachev & Gorinchoy, 2022) (connected with the above considered concept of SMOs) is devoted to the study of the rate constants K of the electrophilic nitration reaction occurring between some short-chain n-alkanes and the nitronium cation NO2+. The given reaction goes in a liquid mixture of concentrated nitric and sulfuric acids at 293 K (Rudakov, 2011). As it was shown in (Gorbachev & Gorinchoy, 2022) the same (described above) SMOs of n-alkanes which are responsible for the abstraction of hydrogen atoms in the case of the reactions of n-alkanes with hydroxyl radicals (Gorbachev, 2021) and NO_3• radicals (Gorbachev, 2022), are also responsible for the transfer of the hydride anion in the reaction of n-alkanes with nitronium cations. The transfer of the corresponding hydride anion H−, whose electrons leave the same SMOs of the investigated n-alkanes, occurs through two the three-center bonds: NO_2^+-H--C, which are shown in Figure 8 for the case of n-pentane.

Figure 8. The interaction of the unoccupied MO of NO_2^+ (a) with the occupied SMO of n-C_5H_{12} (b) leading to the bonding MO of the complex [n-C_5H_{12} & NO_2^+] (c)

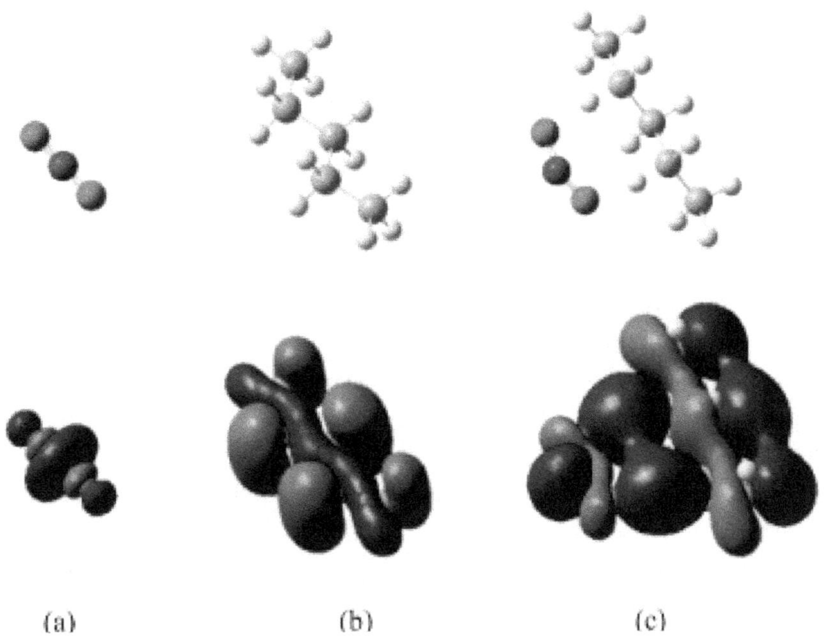

The equation, found in (Gorbachev & Gorinchoy, 2022) and describing the rate constants K, has the following numerical form:

$$-\log_{10}K = 5.5360 + 664.59 \text{ x } (702.88 + \text{Eorb})^{-1} . \quad (6)$$

In Eq. (6) Eorb are the DFT energies (kJ•mol−1) of the above-considered SMOs which can be taken from the work (Gorbachev, 2021). The numerical value 702.88 (kJ•mol^{-1}) is the calculated by the DFT method energy of the virtual MO of NO_2^+ presented in Figure 8. In their turn, the numerical coefficients 5.5360 and 664.59 were defined by means of the multiple regression analysis (Gorbachev & Gorinchoy, 2022). Table 4 contains the experimental values of K (Rudakov, 2011) and $-\log_{10}$K, along with the earlier calculated (and taken from (Gorbachev, 2021) values of Eorb of the studied n-alkanes. It contains also the values of $-\log_{10}$K calculated by means of Eq.(6).

Table 4. The calculation of -$\log_{10}K$ by means of Eq. (6). The data necessary to obtain Equation (6) and the subsequent calculation of -$\log_{10}K$

No	Alkane	K_{exp} (kg·mol^{-1}s^{-1})	E_{orb} (kJ·mol^{-1})	-$\log_{10}K$ Exp.	Calc.
1	Propane	0.0196	873.56	1.71	1.64
2	n-Butane	0.052	863.34	1.28	1.39
3	n-Pentane	0.12	845.16	0.92	0.87
4	n-Hexane	0.18	841.27	0.74	0.73
5	n-Heptane	0.23	837.02	0.64	0.58
6	n-Octane	0.35	835.47	0.45	0.52

Thus, based on all the above-considered theoretical results, one can conclude that for the short-chain n-alkanes their specific occupied molecular orbitals, SMOs, are responsible for the various types of their antioxidant activities.

CONCLUSION

Analyzing all the material presented above, we can draw the following main conclusions.

1. Accounting for the effect of the solvent in quantum chemical calculations plays a decisive role when establishing the desired relationships between the electronic structure and antioxidant activity in the studied series of compounds.
2. For the studied organic acids the so-called SPLET mechanism of their antioxidant action was found to be the most preferable one.
3. The parameters of the electronic structure of antioxidants calculated by the method DFT B3LYP method can be used to successfully establish SAR or QSAR in different series of antioxidant substances.
4. The subsequent application of the method of multiple regression analysis to complete data arrays, including both experimental data on the antioxidant activity of the studied compounds and their molecular electronic parameters (considered either directly or in the form of derived molecular descriptors), allows one to establish correlations that quantitatively describe the antioxidant activity under study.
5. The correlation equations obtained in this way, despite their rather high statistical significance, do not allow us to draw any conclusions regarding the molecular mechanisms of the antioxidant action of the studied compounds.
6. Only SAR or QSAR relationships established on the basis of preliminary physico-chemical models describing possible charge- or orbital-controlled interactions of antioxidant molecules with the studied oxidants make it possible to understand the main details of the mechanisms of the antioxidant reaction under study.

ACKNOWLEDGMENT

This work was performed under the National Research Project Nr. 20.80009.5007.27, "Physico-chemical mechanisms of redox processes with electron transfer involved in vital, technological and environmental systems".

REFERENCES

Abdur Rauf, S. A. (2015). The Effect of n-Methylation of Amino Acids (Ac-X-OMe) on Solubility and Conformation: A DFT Study. *Organic & Biomolecular Chemistry*, *13*, 9993–10006. doi:10.1039/C5OB01565K PMID:26289381

Arsene, I. (2021). DFT Evidence of Ionic Character of Interaction of Dihydroxyfumaric Acid with the Free Radical DPPH*. *Proceedings of the conference Research-based education for a prosperous society,* (*vol. 2*, pp. 9-13). IDSI.

Atkinson, R. (1997). Gas-Phase Tropospheric Chemistry of Volatile Organic Compounds: 1. Alkanes and Alkenes. *Journal of Physical and Chemical Reference Data*, *26*(2), 215–290. doi:10.1063/1.556012

Costa, J. R., Ramos, R., Costa, K., Brasil, D., Silva, C., Ferreira, E., Borges, R., Campos, J., Macêdo, W., & Santos, C. (2018). An in silico study of the antioxidant ability for two caffeine analogs using molecular docking and quantum chemical methods. *Molecules (Basel, Switzerland)*, *23*(11), 1–17. doi:10.3390/molecules23112801 PMID:30380600

Djeradi, H. R., Rahmouni, A., & Cheriti, A. (2014). Antioxidant activity of flavonoids: A QSAR modeling using Fukui indices descriptors. *Journal of Molecular Modeling*, *20*(10), 1–9. doi:10.100700894-014-2476-1 PMID:25311723

Eyring, H. (1935). The Activated Complex in Chemical Reactions. *The Journal of Chemical Physics*, *3*(2), 107–115. doi:10.1063/1.1749604

Filipović, M. M.-M., Marković, Z., Đorović, J., Marković, J. D., Lučić, B., & Amić, D. (2015). QSAR of the free radical scavenging potency of selected hydroxybenzoic acids and simple phenolics. *Comptes Rendus. Chimie*, *18*(5), 492–498. doi:10.1016/j.crci.2014.09.001

Gorbachev, M., Gorinchoy, N. N., & Arsene, I. (2021). Key Role of Some Specific Occupied Molecular Orbitals of Short Chain n-Alkanes in Their Surface Tension and Reaction Rate Constants with Hydroxyl Radicals: DFT Study. *International Journal of Organic Chemistry*, *11*(1), 1–13. doi:10.4236/ijoc.2021.111001

Gorbachev, M. B., Budei, O., Arsene, I., & Gorinchoy, N. N. (2014). Surface Tension of Liquids with π-Conjugated Planar Molecules: DFT Calculation. *Journal of Molecular Liquids*, *190*, 50–53. doi:10.1016/j.molliq.2013.10.017

Gorbachev, M. G. (2015). Antioxidant Properties of Dihydroxyfumaric Acid and Its Dimethyl Ether: a Comparative DFT Study of Their Reactions with the Stable Radical DPPH*. *Chemistry Journal of Moldova, 10*(1).

Gorbachev, M. G. (2022). Molecular orbital nature of solubility of shot chain n-alkanes in water and their reaction rate constants with nitronium cations: A DFT study. *International Journal of Organic Chemistry*, *12*, 92–101. doi:10.4236/ijoc.2022.122008

Gorbachev, M. G. (2022). Molecular Orbital Nature of the Atmospheric Reactions between the NO3• Radical and C3 – C10 n-Alkanes: DFT Study. *The 7th International Conference: Ecological and Environmental Chemistry-2022*. IDSI.

Gorbachev, M. G., Gorinchoy, N., & Balan, I. (2022). Some particularities of the reaction between antioxidant phenolic acids and the free radical ABTS•+: A comparativeDFT study for the gas phase and ethanol. *Chemistry Journal of Moldova*, *17*(1), 24–30. doi:10.19261/cjm.2021.919

Gorbachev, M. G., Gorinchoy, N. N., & Arsene, I. (2017). Surface Tension of Near-Boiling Liquids with Intramolecular πσ*-Hyperconjugation: DFT Calculations. *Journal of Molecular Liquids*, *237*, 108–112. doi:10.1016/j.molliq.2017.04.069

Klopman, G. (1968). Chemical Reactivity and the Concept of Charge- and Frontier-Controlled Reactions. *Journal of the American Chemical Society*, *90*(2), 223–234. doi:10.1021/ja01004a002

Lafridi, H. O. (2020). QSAR modeling of antiradical properties of phenolic compounds using DFT calculations. *Moroccan. J. Chem.*, *8*(4), 841–852.

Mikulski, D. E., Eder, K., & Molski, M. (2014). Quantum-chemical study on relationship between structure and antioxidant properties of hepatoprotective compounds occurring in Cynara scolymus and Silybum marianum. *Journal of Theoretical and Computational Chemistry*, *13*(1), 1–24. doi:10.1142/S0219633614500047

Muliadi, M. B., Basimin, M. Q., & Jayali, A. M. (2021). Density Functional Theory for QSAR Antioxidant Compound Myristicin Derivatives. *Indo.J.Chem.Res.*, *9*(1), 49–56. doi:10.30598//ijcr.2021.9-mul

Pratim, K. (2009). *Chemical Reactivity Theory. A Density Functional View*. CRC Press.

Rudakov, E. V., Volkova, L. K., & Lutsyk, A. I. (2011). Reactivity of C3-C8 alkanes with nitronium ions in sulfuric acid. *Theoretical and Experimental Chemistry*, *47*(4), 244–250. doi:10.100711237-011-9211-0

Sarkar, A. M., Middya, T. R., & Jana, A. D. (2012). A QSAR study of radical scavenging antioxidant activity of a series of flavonoids using DFT based quantum chemical descriptors – the importance of group frontier electron density. *Journal of Molecular Modeling*, *18*(6), 2621–2631. doi:10.100700894-011-1274-2 PMID:22080306

Secara, N. (2010). Stopped-Flow Spectrophotometric Studies of the Kinetics of Interaction of Dihydroxyfumaric Acid with the DPPH Free Radical. *Chemistry Journal of Moldova*, *5*(2), 83–87. doi:10.19261/cjm.2010.05(2).11

Singh, D. V. (2018). Investigation on antioxidant potential of phenolic acids and flavonoids: The common phytochemical ingredients in plants. *Journal of Plant Biochemistry & Physiology*, *6*(3), 1–5. doi:10.4172/2329-9029.1000219

Spiegel, M. (2022). Current Trends in Computational Quantum Chemistry Studies on Antioxidant Radical Scavenging Activity. *Journal of Chemical Information and Modeling*, *62*(11), 2639–2658. doi:10.1021/acs.jcim.2c00104 PMID:35436117

Stephens, P. D., Devlin, F. J., Chabalowski, C. F., & Frisch, M. J. (1994). Ab Initio calculation of vibrational absorption and circular dichroism spectra using density functional force fields. *Journal of Physical Chemistry*, *98*(45), 11623–11627. doi:10.1021/j100096a001

Villaverde, J. J., Sevilla-Morán, B., López-Goti, C., Alonso-Prados, J. L., & Sandín-España, P. (2020). QSAR/QSPR models based on quantum chemistry for risk assessment of pesticides according to current European legislation. *SAR and QSAR in Environmental Research*, *31*(1), 49–72. doi:10.1080/1062936X.2019.1692368 PMID:31766890

Yordanov, N. (1996). Is Our Knowledge about the Chemical and Physical Properties of DPPH Enough to Consider It as a Primary Standard for Quantitative EPR Spectrometry. *Applied Magnetic Resonance*, *10*(1), 339–350. doi:10.1007/BF03163117

Chapter 7
Time-Reversal Symmetry in Spin Systems in the Presence and the Absence of Redox Processes

Ion Geru
Academy of Sciences of Moldova, Moldova

ABSTRACT

It is shown that the isomorphic substitution of one of the paramagnetic ions of di-homonuclear cluster with antiferromagnetic exchange interaction by an ion, whose wave functions are reversed in time with respect to the wave functions of the substituted ion, leads to the formation of a di-heteronuclear cluster with ferromagnetic exchange interaction. This method of spin rearming can be used for the synthesis of new di-heteronuclear clusters with ferromagnetic exchange interaction. The redox-induced processes can affect the time-reversal symmetry. Particularly in the reduction-half reaction + 2 Cu the time-reversal symmetry for the ion, existing before the start of the reaction, disappears after the reaction due to the transformation of the ion with spin S = 1/2 into a neutral Cu atom without spin. An analogy is found between six incomplete time-reversal operators for systems with spins S = 1/2, 3/2, 2, 5/2.

INTRODUCTION

The redox systems are chemical systems, in which reduction and oxidation reactions occur (Amman, 2018). In the decomposition redox reaction AB → A+B the products are formed by the decomposition of the reactants. The reaction

$$2H_2O \rightarrow 2H_2+O_2 \tag{1}$$

is an example of decomposition redox reaction. Particularly, the reaction

DOI: 10.4018/978-1-6684-7198-2.ch007

$$CaCO_3 \rightarrow CaO + CO_2 \tag{2}$$

is a decomposition one, not a redox reaction, since the oxidation state is not changed here (https://biologyonline.com/dictionary/redox-reaction). The combination redox reaction is a process opposite to that in the decomposite reaction. The oxidation state of reactants is increasing at oxidation and decreasing at reduction. The reducing agents are electropositive elements, while the electronegative elements are oxidizing agents. The reaction

$$Zn + CuSO_4 \rightarrow ZnSO_4 + Cu \tag{3}$$

is an example of displacement redox reactions with the oxidation half-reaction

$$\mathrm{Zn} \rightarrow Zn^{2+} + 2e^- \tag{4}$$

and the reduction half-reaction

$$Cu^{2+} + 2e^- \rightarrow Cu. \tag{5}$$

The time-reversal symmetry is one of maim symmetries in physics. Let us consider consider a simple example of displacement redox reactions described by formulas (3) - (5). During the redox reaction the paramagnetic ions Cu^{2+} with the electron configuration $3d^9$ and spin S = 1/2 are transformed into copper atoms with the electron configuration $3d^{10}4s^1$. In this case, the antiunitary time-reversal operator $\boldsymbol{K}$ has the form (Wigner, 1959).

$$\boldsymbol{K} = U^{(1/2)}\boldsymbol{K}_0,\ U^{(1/2)} = i\sigma y \tag{6}$$

where $U^{(1/2)}$ is a unitary 2x2- matrix ($U^{(1/2)} \cdot U^{(1/2)+} = 1$), σy is the imaginary Pauli matrix and K$\boldsymbol{0}$ is the operator of complex conjugation. The anti-unitarity of the operator K is determined by the relation K $(\alpha\psi+\beta\varphi)= \alpha^*\mathrm{K}\psi+\beta^{*\mathrm{K}}\varphi$, where ψ and φ are wave functions, while α and β are complex coefficients.

In this paper, the time-reversal symmetry in spin systems in the presence and absence of the redox processes is studied. In Section 2, a brief discussion of the influence of the redox processes on the time-reversal symmetry is given. In Section 3, the time-reversal symmetry related to Cobalt binuclear compounds with redox-active bridging is studied. In Section 4, the consequences of the decomposition under redox processes of the cluster formed by Mn(II) ions *on the* time-reversal symmetry of the final reactants are considered. In Section 5, it is performed the transformation of the antiferromagnetic exchange interaction dimer cluster Mn(III) – Mn*(ПI)* into *excha*nge interaction of ferromagnetic type in dimer clusters Mn(III) – Co*(ПI)*. Suc*h* a *t*ransformation takes place when one of the ions of a two-nuclear cluster is isomorphically replaced by an ion, whose wave functions are reversed in time with respect to the wave functions of a substituted ion. Section 6 contains main conclusions.

INFLUENCE OF REDOX PROCESSES ON TIME-REVERSAL SYMMETRY

There are redox reactions, for which the time-reversal symmetry exists both before the reaction and after the reaction, although it refers to different ions. An example of such a reaction is the reaction between iron and hydrogen peroxide. Hydrogen peroxide oxidizes ferrous ion $Fe^{2+}(3d^6)$ with the spin S = 2 into the ferric ion $Fe^{3+}(3d^5)$ with the spin S = 5/2 in the presence of an acid (Amman, 2018). In this case, the oxidation half-reaction is $Fe^{2+} \rightarrow Fe^{3+} + e$-. The time-reversal operators for Fe^{2+} ions (S = 2) and for Fe^{3+} ions (S = 5/2) are given by the formulas $\mathbf{K} = U^{(2)}\mathbf{K}_0$ and $\mathbf{K} = U^{(5/2)}\mathbf{K}_0$, respectively, where $U^{(2)}$ and $U^{(5/2)}$ are defined in the Section 3 (see equations (34) and (28)).

Under the action of the operator **K**, all the three spin projection operators $\mathbf{S}_x$, $\mathbf{S}_y$ and $\mathbf{S}_z$ change their signs to the opposite ones (Wigner, 1959). However, there are time-reversal operators under the action of which the signs are reversed not for all three operators $\mathbf{S}_x$, $\mathbf{S}_y$ and $\mathbf{S}_z$. Let us show this using a simple example of a particle with the spin S = 1/2, for which we represent the unitary matrix $U^{(1/2)}$ in the form

$$U^{(1/2)} = i\sigma_y = \sigma_z\sigma_x, \tag{7}$$

where σ_z and σ_x are real Pauli matrices. Taking into account eq. (7), the Wigner time reversal operator can be represented as:

$$\mathbf{K} = \sigma_z\sigma_x\mathbf{K}_0 = \sigma_z\mathbf{K}_0\bullet\sigma_x = \mathbf{K}_{1(x)}\bullet\mathbf{K}_{2(yz)}, \tag{8}$$

where

$$\mathbf{K}_{1(x)} = \sigma_z\mathbf{K}_0,\ \mathbf{K}_{2(yz)} = \sigma_x \tag{9}$$

Under the action of the operator $\mathbf{K}_{1(x)}$, the sign of the operator $\mathbf{S}_x$ is inverted, while this operator does not change the signs of the operators $\mathbf{S}_y$ and $\mathbf{S}_z$:

$$\boldsymbol{K}_{1(x)}\boldsymbol{S}_x\boldsymbol{K}_{1(x)}^{-1} = -\boldsymbol{S}_x,\ \boldsymbol{K}_{1(x)}\boldsymbol{S}_y\boldsymbol{K}_{1(x)}^{-1} = \boldsymbol{S}_y,\ \boldsymbol{K}_{1(x)}\boldsymbol{S}_z\boldsymbol{K}_{1(x)}^{-1} = \boldsymbol{S}_z. \tag{10}$$

Under the action of the operator $\mathbf{K}_{2(yz)}$, the signs of the operators $\mathbf{S}_y$ and $\mathbf{S}_z$ are inverted, while this operator does not change the signs of the operator $\mathbf{S}_x$:

$$\boldsymbol{K}_{2(yz)}\boldsymbol{S}_x\boldsymbol{K}_{2(yz)}^{-1} = \boldsymbol{S}_x,\ \boldsymbol{K}_{2(yz)}\boldsymbol{S}_y\boldsymbol{K}_{2(yz)}^{-1} = -\boldsymbol{S}_y,\ \boldsymbol{K}_{2(yz)}\boldsymbol{S}_z\boldsymbol{K}_{2(yz)}^{-1} = -\boldsymbol{S}_z. \tag{11}$$

It is easy to see that under the action of the operator $\mathbf{K}_{1(x)}\bullet\mathbf{K}_{2(yz)}$ from eq. (8), there takes place the sign inversion of all the three spin projection operators $\mathbf{S}_x$, $\mathbf{S}_y$ and $\mathbf{S}_z$, as it should happen under the action of the Wigner time-reversal operator. Subscripts for operators $\mathbf{K}_{1(x)}$ and $\mathbf{K}_{2(yz)}$ in equations (8) - (11) indicate how many spin projection operators and which ones change signs to the opposite under such time-reversal transformations. In this context, the $\mathbf{K}_{1(x)}$ and $\mathbf{K}_{2(yz)}$ operators are partial (incomplete) time-reversal operators, and **K** is a full time-reversal operator. Thus, the representation of the matrix $U^{(1/2)}$ as a product of the Pauli matrices σ_z and σ_x (see equation (7)) allows us to define two partial time-reversal operators $\mathbf{K}_{1(x)}$ and $\mathbf{K}_{2(yz)}$. The existence of these operators means that the time-reversal symmetry can be reduced, not only violated (broken).

Experimental data indicate that, in some cases, the violation of the time-reversal symmetry takes place (Auber et al., 2001; Lazauskas et al., 2014; Roberts et al., 2015; Titov et al., 2006; Zhu & Ting, 1998). As for the reduction of the Wigner time-reversal symmetry (for example, from **K**-symmetry to $\mathbf{K}_{1(x)}$- or $\mathbf{K}_{2(yz)}$-symmetry), currently there are no such experimental data. Moreover, it follows from general considerations that, along with the operators $\mathbf{K}_{1(x)}$ and $\mathbf{K}_{2(yz)}$, there must be four more incomplete time-reversal operators: $\mathbf{K}_{1(y)}$, $\mathbf{K}_{1(z)}$, $\mathbf{K}_{2(zx)}$, and $\mathbf{K}_{2(xy)}$. However, the existence of these operators cannot be the proofed in such a simple way as it was done in the cases of $\mathbf{K}_{1(x)}$ and $\mathbf{K}_{2(yz)}$ operators. To find $\mathbf{K}_{1(y)}$, $\mathbf{K}_{1(z)}$, $\mathbf{K}_{2(zx)}$, and $\mathbf{K}_{2(xy)}$ operators, let us consider the system of operators **K**, $\mathbf{K}^2$=(U$^{(1/2)}$)2($\mathbf{K}_0$)^2= -e, $\mathbf{K}^3 = \mathbf{K}^2 \bullet \mathbf{K} = -\mathbf{K}$, $\mathbf{K}^4=\mathbf{K}^2 \bullet \mathbf{K}^2$=e, where e is the unit 2x2-matrix defined in the spinor basis. These operators form a fourth-order cyclic symmetry group

$$\mathcal{G}_4^{(1/2)}: \{\mathrm{K}, \mathbf{-e}, \mathbf{-K}, e\}. \tag{12}$$

The group $\mathcal{G}_4^{(1/2)}$ can be extended to a non-Abelian group of order eight by adding the elements $\mathbf{K}_{1(x)}$, $-\mathbf{K}_{1(x)}$, $\mathbf{K}_{2(yz)}$, and $-\mathbf{K}_{2(yz)}$:

$$\mathcal{G}_8^{(1/2)}: \{\mathbf{K}, \text{-e}, \mathbf{-K}, \text{e}, \mathbf{K}_{2(yz)}, -\mathbf{K}_{2(yz)}, \mathbf{K}_{1(x)}, -\mathbf{K}_{1(x)}\}. \tag{13}$$

Based on a simple calculation, one can verify that the group $\mathcal{G}_8^{(1/2)}$ satisfies the postulates that define the symmetry group. However, this group contains only two of the six operators of incomplete time reversal. To find the remaining four operators of partial time reversal, it is necessary to extend the group $\mathcal{G}_8^{(1/2)}$ to a group of higher order, which would contain, along with the operator $\boldsymbol{K}$, all operators of partial time reversal (Geru, 2018). Such a group can be found by constructing the tensor product of the groups $\mathcal{G}_8^{(1/2)}$ and $\mathcal{G}_2$, where the group $\mathcal{G}_2$ is formed by the elements $\boldsymbol{K}_0$ and $\boldsymbol{K}_0^2 = 1$:

$$\mathcal{G}_{16}^{(1/2)} = \mathcal{G}_8^{(1/2)} \otimes \mathcal{G}_2. \tag{14}$$

The extended symmetry group contains the operator $\boldsymbol{K}$ and all six partial time-reversal operators is

$$\mathcal{G}_{16}^{(1/2)}: \{ \boldsymbol{K}, -e, -\boldsymbol{K}, e, \boldsymbol{K}_{2(yz)}, -\boldsymbol{K}_{2(yz)}, \boldsymbol{K}_{1(x)}, -\boldsymbol{K}_{1(x)}, \boldsymbol{K}_{2(zx)}, -\boldsymbol{K}_{1(y)}, -\boldsymbol{K}_{2(zx)}, \boldsymbol{K}_{1(y)}, \boldsymbol{K}_{1(z)}, -\boldsymbol{K}_{1(z)}, \boldsymbol{K}_{2(xy)}, -\boldsymbol{K}_{2(xy)}\}, \tag{15}$$

where the operator $\boldsymbol{K}$ is defined by equation (6). Three anti-unitary partial time-reversal operators $\boldsymbol{K}_{1(x)}$, $\boldsymbol{K}_{1(y)}$, $\boldsymbol{K}_{1(z)}$ and, accordingly, three unitary partial time-reversal operators $\boldsymbol{K}_{2(yz)}$, $\boldsymbol{K}_{2(zx)}$ and $\boldsymbol{K}_{2(xy)}$ have the form:

$$\text{K1(x)} = \sigma\text{zK0, K1(y)} = \text{eK0, K1(z)} = \sigma\text{xK0, K2(yz)}=\sigma\text{x, K2(zx)}=i\sigma\text{y, K2(xy)}= \sigma\text{z}. \tag{16}$$

According to the Cayley table of the group $\mathcal{G}_{16}^{(1/2)}$ (Geru, 2018), the expressions found for six operators of incomplete time reversal (16) are correct. Based on formulas (16), it is easy to show that there are five ways of non-trivial factorization of the Wigner time-reversal operator, that is, the representation of the operator K as a product of two or three partial time-reversal operators:

$$\mathrm{K} = \boldsymbol{K}_{1(x)} \cdot \boldsymbol{K}_{2(yz)} = \boldsymbol{K}_{2(zx)} \cdot \boldsymbol{K}_{1(y)} = \boldsymbol{K}_{2(xy)} \cdot \boldsymbol{K}_{1(z)} = \boldsymbol{K}_{1(x)} \cdot \boldsymbol{K}_{1(y)} \cdot \boldsymbol{K}_{1(z)} = \boldsymbol{K}_{1(x)} \cdot \boldsymbol{K}_{1(z)} \cdot \boldsymbol{K}_{1(y)} \quad (17)$$

TIME-REVERSAL SYMMETRY RELATED TO COBALT BINUCLEAR COMPOUNDS WITH REDOX-ACTIVE BRIDGING

Paramagnetic complexes occur among Co(III) monomers. In particular, (Eiko Toyota & Matsudera, 1980) reported experimental data on 15 paramagnetic cobalt(III) complexes with organic ligands. In (Sinha et al., 2022), a coordination compound containing Co(III) – Co(III) as nonvolative resistive switching device was synthesized and its redox-induced properties were studied. It is known that the presence of redox processes leads to larger coupling constants of exchange interaction in the more oxidized species (McGrady & Stranger, 1997). This is of interest for a purposeful change in the value of the constant *J*.

In crystals the Co^{2+} ion can be in the state with a high spin S = 3/2 (HS) and in the state with a low spin S = 1/2 (LS) (Piwowarska et al., 2019). The HS state corresponds to the case, when the ion is in a "weak" crystal field of octahedral symmetry, has an electronic configuration $(t_{2g})^5(e)^2$ with three unpaired electrons (one in the t_{2g}- state and two in the e- state) and a threefold degenerate ground state T_{1g}. The LS state corresponds to the case when only the Zeeman and hyperfine interactions are taken into account in the spin Hamiltonian. In this case, in a "strong" crystal field of octahedral symmetry, the electronic configuration of the ion Co^{2+} ion has the form $(t_2)^6(e)^1$ with one unpaired electron in the *e*-state. Under these conditions, the ground state of the Co^{2+} ion is an orbital doublet E_g. Note that the electron spin S=1/2, which is attributed to the ion Co^{2+} in the LS state, should be distinguished from the fictitious spin S' =1/2, which is attributed to a specific subset of states of the Co^{2+} ion (McGrady & Stranger, 1997).

In the case where the LS state for monomers Co^{2+} is implemented, the complete time-reversal operator is determined by formula (6). If two LS Co^{2+} ions form a homobinuclear cluster, then the Wigner time-reversal operator is

$$\boldsymbol{K} = U_1^{(1/2)} \boldsymbol{K}_0 \cdot U_2^{(1/2)} \boldsymbol{K}_0 = U_1^{(1/2)} \cdot U_2^{(1/2)}, \quad (18)$$

where it is taken into account that $U_2^{(1/2)}$ is a real 2x2 matrix. In the case of HS state the operator $\boldsymbol{K}$ for monomer Co^{2+} has the form

$$\boldsymbol{K} = U^{(3/2)} \boldsymbol{K}_0 = i \begin{pmatrix} 0 & \sigma_y \\ \sigma_y & 0 \end{pmatrix} \boldsymbol{K}_0. \quad (19)$$

In the supermatrix (19) "0" denotes a 2x2 null matrix. Analogously, in the case of homobinuclear cluster formed by Co^{2+} ions with spins S_1=3/2 and S_2=3/2 we obtain

$$\boldsymbol{K} = U_1^{(3/2)}\boldsymbol{K}_0 \cdot U_2^{(3/2)}\boldsymbol{K}_0 = U_1^{(3/2)} \cdot U_2^{(2/2)} \cdot\cdot, (20),$$

where matrices $U_1^{(3/2)}$ and $U_2^{(3/2)}$ have the same structure as the matrix $U^{(3/2)}$ from eq. (19). To find operators of partial time reversal for spin-3/2 system, it is necessary to generalize relations (7) for this case:

$$U^{(3/2)} = \pounds_z^{(3/2)}\pounds_x^{(3/2)}, \tag{21}$$

where $U^{(3/2)}$ is defined by eq. (19), while matrices $\pounds_z^{(3/2)}$ and $\pounds_x^{(3/2)}$ have the form

$$\pounds_z^{(3/2)} = \begin{pmatrix} \sigma_z & 0 \\ o & \sigma_z \end{pmatrix}, \pounds_x^{(3/2)} = \begin{pmatrix} 0 & \sigma_x \\ \sigma_x & 0 \end{pmatrix}. \tag{22}$$

By analogy with eq. (8), the following relation holds

$$\boldsymbol{K} = \pounds_z^{(3/2)}\pounds_x^{(3/2)}\boldsymbol{K}_0 = \boldsymbol{K}_{1(x)}^{(3/2)} \cdot \boldsymbol{K}_{2(yz)}^{(3/2)}, \tag{23}$$

where operators of the incomplete time reversal $\boldsymbol{K}_{1(x)}^{((3/2)}$ and $\boldsymbol{K}_{2(yz)}^{((3/2)}$ have the form

$$\boldsymbol{K}_{1(x)}^{((3/2)} = \pounds_z^{(3/2)}\boldsymbol{K}_0, \boldsymbol{K}_{2(yz)}^{(3/2)} = \pounds_x^{(3/2)}. \tag{24}$$

Under the action of the operators $\boldsymbol{K}_{1(x)}^{(3/2)}$ and $\boldsymbol{K}_{2(yz)}^{(3/2)}$ the spin projection operators of the spin-3/2 system are transformed according to formulas similar to relations (10) and (11).

The operators $U^{(3/2)}\boldsymbol{K}_0$, $-e^{(3/2)}$, $-U^{(3/2)}\boldsymbol{K}_0$, $e^{(3/2)}$ form a cyclic symmetry group of the fourth order $\mathcal{G}_4^{(3/2)}$ with the identity element $e^{(3/2)}$ given in the spinor basis. By adding the elements

$$\pounds_x^{(3/2)} - \pounds_x^{(3/2)}, \pounds_z^{(3/2)}\boldsymbol{K}_0 - \pounds_z^{(3/2)}\boldsymbol{K}_0,$$

the cyclic group $\mathcal{G}_4^{(3/2)}$ can be extended to non-Abelian symmetry group of the eighth order

Table 1. The Cayley table of the $\mathcal{G}_8^{(3/2)}$ *group*

	$U^{\left(\frac{3}{2}\right)}\mathbf{K}_0$	$-e^{\left(\frac{3}{2}\right)}$	$-U^{\left(\frac{3}{2}\right)}\mathbf{K}_0$	$e^{\left(\frac{3}{2}\right)}$	$£_x^{\left(\frac{3}{2}\right)}$	$-£_x^{\left(\frac{3}{2}\right)}$	$£_z^{\left(\frac{3}{2}\right)}\mathbf{K}_0$	$-£_z^{\left(\frac{3}{2}\right)}\mathbf{K}_0$
$U^{\left(\frac{3}{2}\right)}\mathbf{K}_0$	$-e^{\left(\frac{3}{2}\right)}$	$-U^{\left(\frac{3}{2}\right)}\mathbf{K}_0$	$e^{\left(\frac{3}{2}\right)}$	$U^{\left(\frac{3}{2}\right)}\mathbf{K}_0$	$£_z^{\left(\frac{3}{2}\right)}\mathrm{K}_0$	$-£_z^{\left(\frac{3}{2}\right)}\mathbf{K}_0$	$-£_x^{\left(\frac{3}{2}\right)}$	$£_x^{\left(\frac{3}{2}\right)}$
$-e^{\left(\frac{3}{2}\right)}$	$-U^{\left(\frac{3}{2}\right)}\mathbf{K}_0$	$e^{\left(\frac{3}{2}\right)}$	$U^{\left(\frac{3}{2}\right)}\mathbf{K}_0$	$-e^{\left(\frac{3}{2}\right)}$	$-£_x^{\left(\frac{3}{2}\right)}$	$£_x^{\left(\frac{3}{2}\right)}$	$-£_z^{\left(\frac{3}{2}\right)}\mathbf{K}_0$	$£_z^{\left(\frac{3}{2}\right)}\mathbf{K}_0$
$-U^{\left(\frac{3}{2}\right)}\mathbf{K}_0$	$e^{\left(\frac{3}{2}\right)}$	$U^{\left(\frac{3}{2}\right)}\mathbf{K}_0$	$-e^{\left(\frac{3}{2}\right)}$	$-U^{\left(\frac{3}{2}\right)}\mathbf{K}_0$	$-£_z^{\left(\frac{3}{2}\right)}\mathbf{K}_0$	$£_z^{\left(\frac{3}{2}\right)}\mathbf{K}_0$	$£_x^{\left(\frac{3}{2}\right)}$	$-£_x^{\left(\frac{3}{2}\right)}$
$e^{\left(\frac{3}{2}\right)}$	$U^{\left(\frac{3}{2}\right)}\mathbf{K}_0$	$-e^{\left(\frac{3}{2}\right)}$	$-U^{\left(\frac{3}{2}\right)}\mathbf{K}_0$	$e^{\left(\frac{3}{2}\right)}$	$£_x^{\left(\frac{3}{2}\right)}$	$-£_x^{\left(\frac{3}{2}\right)}$	$£_z^{\left(\frac{3}{2}\right)}\mathbf{K}_0$	$-£_z^{\left(\frac{3}{2}\right)}\mathbf{K}_0$
$£_x^{\left(\frac{3}{2}\right)}$	$-£_z^{\left(\frac{3}{2}\right)}\mathbf{K}_0$	$-£_x^{\left(\frac{3}{2}\right)}$	$£_z^{\left(\frac{3}{2}\right)}K_0$	$£^{\left(\frac{3}{2}\right)}$	$\left(\frac{3}{2}\right)$	$-e^{\left(\frac{3}{2}\right)}$	$-U^{\left(\frac{3}{2}\right)}\mathbf{K}_0$	$U^{\left(\frac{3}{2}\right)}\mathbf{K}_0$
$-£_x^{\left(\frac{3}{2}\right)}$	$£_z^{\left(\frac{3}{2}\right)}\mathbf{K}_0$	$£^{\left(\frac{3}{2}\right)}$	$-£_z^{\left(\frac{3}{2}\right)}\mathbf{K}_0$	$-£_x^{\left(\frac{3}{2}\right)}$	$-e^{\left(\frac{3}{2}\right)}$	$e^{\left(\frac{3}{2}\right)}$	$U^{\left(\frac{3}{2}\right)}\mathbf{K}_0$	$-U^{\left(\frac{3}{2}\right)}\mathbf{K}_0$
$£_z^{\left(\frac{3}{2}\right)}\mathbf{K}_0$	$£_x^{\left(\frac{3}{2}\right)}$	$-£_z^{\left(\frac{3}{2}\right)}\mathbf{K}_0$	$-£_x^{\left(\frac{3}{2}\right)}$	$£_z^{\left(\frac{3}{2}\right)}\mathbf{K}_0$	$U^{\left(\frac{3}{2}\right)}\mathbf{K}_0$	$-U^{\left(\frac{3}{2}\right)}\mathbf{K}_0$	$e^{\left(\frac{3}{2}\right)}$	$-e^{\left(\frac{3}{2}\right)}$
$-£_z^{\left(\frac{3}{2}\right)}\mathbf{K}_0$	$-£_x^{\left(\frac{3}{2}\right)}$	$£_z^{\left(\frac{3}{2}\right)}\mathbf{K}_0$	$£_x^{\left(\frac{3}{2}\right)}$	$-£_z^{\left(\frac{3}{2}\right)}\mathbf{K}_0$	$-U^{\left(\frac{3}{2}\right)}\mathbf{K}_0$	$U^{\left(\frac{3}{2}\right)}\mathbf{K}_0$	$-e^{\left(\frac{3}{2}\right)}$	$e^{\left(\frac{3}{2}\right)}$

$$\mathcal{G}_8^{(3/2)}:\left\{U^{(3/2)}\mathbf{K}_0,\ -e^{(3/2)},\ -U^{(3/2)}\mathbf{K}_0,\ e^{(3/2)},\ £_x^{(3/2)},\ -£_x^{(3/2)},\ £_z^{(3/2)}\mathbf{K}_0,\ -£_z^{(3/2)}\mathbf{K}_0\right\}. \tag{25}$$

The Cayley table of this group is given in Table 1.

According to Table 1, the following relationship holds

$£_z^{(3/2)}\mathbf{K}_0 \cdot £_x^{(3/2)} = U^{(3/2)}\mathbf{K}_0$ (23 a)

which, taking into account the notations (24), coincides with the formula (23) obtained by us in this section, based on other considerations. Thus, based on the Cayley table of the $\mathcal{G}_8^{(3/2)}$ group, it is confirmed that the expression (24) found earlier from other considerations is valid.

The explicit form of the remaining four operators of incomplete time reversal for the spin-3/2 system can be found on the base of the Cayley table of the extended non-Abelian symmetry group of the 16th order $\mathcal{G}_{16}^{(3/2)}$ of the following composition:

$$\mathcal{G}_{16}^{(3/2)}:\{U^{(3/2)}\boldsymbol{K}_0,\ -e^{(3/2)},\ -U^{(3/2)}\boldsymbol{K}_0,\ e^{(3/2)},\ \pounds_x^{(3/2)},\ -\pounds_x^{(3/2)},\ \pounds_z^{(3/2)}\boldsymbol{K}_0,\ -\pounds_z^{(3/2)}\boldsymbol{K}_0,\ U^{(3/2)},\\ -e^{(3/2)}\boldsymbol{K}_0,\ -U^{(3/2)},\ e^{(3/2)}\boldsymbol{K}_0,\ \pounds_x^{(3/2)}\boldsymbol{K}_0,\ -\pounds_x^{(3/2)}\boldsymbol{K}_0,\ \pounds_z^{(3/2)},\ -\pounds_x^{(3/2)}\}\ . \tag{26}$$

Proceeding in the same way as in the case of the group $\mathcal{G}_8^{(1/2)}$, which is a subgroup of the group $\mathcal{G}_{16}^{(1/2)}$ (see eqs. (13)-(16)), based on the structure of the Cayley table of the $\mathcal{G}_{16}^{(3/2)}$ group, we find the following expressions for the four missing partial time reversal operators:

$$\boldsymbol{K}_{1(y)}^{((3/2)} = e^{(3/2)}\boldsymbol{K}_0, \boldsymbol{K}_{1(z)}^{(3/2)} = \pounds_x^{(3/2)}\boldsymbol{K}_0, \boldsymbol{K}_{2(zx)}^{(3/2)} = U^{(3/2)}, \boldsymbol{K}_{2(xy)}^{(3/2)} = \pounds_z^{(3/2)}. \tag{27}$$

These operators have the same structure as the corresponding for spin-1/2 particle operators (see eq. (16)). In order to not to clutter up this chapter of the book, it has been omitted the Cayley table of the $\mathcal{G}_{16}^{(3/2)}$ group, which is similar in structure to the Cayley table of the $\mathcal{G}_{16}^{(1/2)}$ group (Geru, 2018). The four operators (27), together with the previously found operators $\boldsymbol{K}_{1(x)}^{((3/2)} = \pounds_z^{(3/2)}\boldsymbol{K}_0$ and $\boldsymbol{K}_{2(yz)}^{(3/2)} = \pounds_x^{(3/2)}$ from eq. (24), form a complete system of incomplete time-reversal operators for spin-3/2 system.

DECOMPOSITION OF CLUSTERS FORMED BY *MN(II)* IONS WITH CONSEQUENCES ON TIME-REVERSAL SYMMETRY

In the work of Geru (2018) the structure, magnetic and the redox properties of some binuclear *Mn(III)* and *Fe(III)* complexes with antiferromagnetic exchange interaction between components of dimers have been investigated. Same properties were studied also for dinuclear Mn(II) complexes afforded by Schiff base ligands with complex structure (Theil et al., 1997). In the last case, using the EPR method, it was shown that the oxidation by oxygen molecules leads to the release of the Mn(II) binuclear units with formation of the mononuclear Mn(II) and Mn(III) species. In this section, we will find operators of incomplete time reversal related to these mononuclear species.

THE MONOMER *MN(II)*

The monomer *Mn(II)* has an electronic configuration of $3d^5$ and an electron spin of S = 5/2. In this case, the full time-reversal operator $\boldsymbol{K}$ has the form

$$\boldsymbol{K} = U^{(5/2)}\boldsymbol{K}_0. \tag{28}$$

In eq. (28), the unitary 6x6 matrix $U^{(5/2)}$ has all matrix elements equal to zero, except for those located on the secondary diagonal and equal, in turn, to 1 and -1

$$\left(U_{5/2,-5/2}^{(5/2)} = U_{1/2,-1/2}^{(5/2)} = U_{-3/2,3/2}^{(5/2)} = 1,\ U_{3/2,-3/2}^{(5/2)} = U_{-1/2,1/2}^{(5/2)} = U_{-5/2,5/2}^{(5/2)} - 1\right).$$

For a system with the spin S = 5/2, the 8th order discrete non-Abelian group $\mathcal{G}_8^{(5/2)}$ consists of the following elements:

$$\mathcal{G}_8^{(5/2)}:\left\{U^{(5/2)}\boldsymbol{K}_0,\ -e^{(5/2)},\ -U^{(5/2)}\boldsymbol{K}_0,\ e^{(5/2)},\ \pounds_x^{(5/2)}-\pounds_x^{(5/2)},\ \pounds_z^{(5/2)}\boldsymbol{K}_0,\ -\pounds_z^{(5/2)}\boldsymbol{K}_0\right\}, \tag{29}$$

where $\pounds_x^{(5/2)}$ is a 6x6 unitary matrix with non-zero matrix elements on the secondary diagonal, each of which is equal to 1. The matrix $\pounds_z^{(5/2)}$ is a diagonal matrix with matrix elements equal in turn to 1 and -1, starting from the upper left corner of the matrix.

The Cayley table of the of the $\mathcal{G}_8^{(5/2)}$ group (29) coincides with the Cayley table of the group $\mathcal{G}_8^{(3/2)}$ (see Table 1), if in the latter we replace the top index 3/2 of the elements of the group with 5/2. With this remark in mind, we can assume that the following equalities hold:

$$\boldsymbol{K}=\pounds_z^{5/2)}\boldsymbol{K}_0\cdot\pounds_x^{(5/2)}=\boldsymbol{K}_{1(x)}^{(5/2)}\cdot\boldsymbol{K}_{2(yz)}^{(5/2)}, \tag{30}$$

where

$$\boldsymbol{K}_{1(x)}^{(5/2)}=\pounds_z^{(5/2)}\boldsymbol{K}_0,\ \boldsymbol{K}_{2(yz)}^{(5/2)}=\pounds_x^{(5/2)}. \tag{31}$$

The operators $\boldsymbol{K}_{1(x)}^{(5/2)}$ and $\boldsymbol{K}_{2(yz)}^{(5/2)}$ are the partial time-reversal operators for a system with spin S = 5/2. To find the rest of four operators of the incomplete time reversal for a system with spin S = 5/2, it is necessary to extend the group $\mathcal{G}_8^{(5/2)}$ to the non-Abelean group of 16th order $\mathcal{G}_{16}^{(5/2)}$:

$$\mathcal{G}_{16}^{(5/2)}=\mathcal{G}_8^{5/2)}\otimes\mathcal{G}_2. \tag{32}$$

Based on the structure of the Cayley table of the extended symmetry group (32) (the explicit form of this table is not given due to its bulkiness), we have defined four partial time- reversal operators for a system with spin S =5/2:

$$\boldsymbol{K}_{1(y)}^{(3/2)}=e^{(3/2)}\boldsymbol{K}_0,\ \boldsymbol{K}_{1(z)}^{(3/2)}=\pounds_x^{(3/2)}\boldsymbol{K}_0,\ \boldsymbol{K}_{2(zx)}^{(3/2)}=U^{(3/2)},\ \boldsymbol{K}_{2(xy)}^{(3/2)}=\pounds_z^{(3/2)}. \tag{33}$$

These operators, together with the previously found operators $\boldsymbol{K}_{1(x)}^{((3/2)}=\pounds_z^{(3/2)}\boldsymbol{K}_0$ and $\boldsymbol{K}_{2(yz)}^{(3/2)}=\pounds_x^{(3/2)}$, constitute a complete system of six partial time-reversal operators for a system with spin S = 5/2.

THE MONOMER *MN(III)*

The monomer *Mn(III)* has an electronic configuration of $3d^4$ and spin S = 2. The total time-reversal operator for such a system has the form

$$\boldsymbol{K} = U^{(2)}\boldsymbol{K}_0, \tag{34}$$

where the $U^{(2)}$ matrix has the same structure as the $U^{(5/2)}$ matrix, with the difference that there are five, rather than six matrix elements on the secondary diagonal. In this case, the symmetry group has the following composition:

$$\mathcal{G}_8^{(2)}:\left\{U^{(2)}\boldsymbol{K}_0,\ -e^{(2)},\ -U^{(2)}\boldsymbol{K}_0,\ e^{(2)},\ £_x^{(2)},\ -£_x^{(2)},\ £_z^{(2)}\boldsymbol{K}_0,\ -£_z^{(2)}\boldsymbol{K}_0\right\}. \tag{35}$$

The structure of the matrices $U^{(2)}$, $e^{(2)}$, $£_x^{(2)}$, and $£_z^{(2)}$ from eq. (35) is similar to the structure of the matrices $U^{(3/2)}$, $e^{(3/2)}$, $£_x^{(3/2)}$, and $£_z^{(3/2)}$ from eq. (25), with the difference that in the first case the dimension of the matrices is five, not four as in the second case. Besides, unlike the non-Abelian $\mathcal{G}_{16}^{(3/2)}$ and $\mathcal{G}_8^{(5/2)}$ groups, the $\mathcal{G}_8^{(2)}$ group is an Abelian symmetry group because at $S = 2$, according to eq. (34), $\boldsymbol{K}^2 = e^{(2)}$. This corresponds to the absence of Kramers degeneracy of the energy levels of spin systems with integer spin.

After replacing the superscript 2 by 3/2 for all elements of the group $\mathcal{G}_8^{(2)}$, the Cayley table of the group $\mathcal{G}_8^{(2)}$ formally coincides with the Cayley table of the group $\mathcal{G}_8^{(3/2)}$ (see Table 1). Therefore, it follows from the equality

$$\boldsymbol{K} = U^{(2)}\boldsymbol{K}_0 = £_z^{(2)}\boldsymbol{K}_0 \cdot £_x^{(2)} \tag{36}$$

that in the case of systems with spin S = 2, the incomplete time reversal operators $\boldsymbol{K}_{1(x)}^{(2)}$ and $\boldsymbol{K}_{2(yz)}^{(2)}$ have the forms

$$\boldsymbol{K}_{1(x)}^{(2)} = £_z^{(2)}\boldsymbol{K}_0,\ K_{2(yz)}^{(2)} = £_x^{(2)}. \tag{37}$$

The other four partial time reversal operators for the monomer Mn(III), which are determined by extending the group $\mathcal{G}_8^{(2)}$ to an Abelian group $\mathcal{G}_{16}^{(2)}$ of the 16th order $\mathcal{G}_{16}^{(2)} = \mathcal{G}_8^{5/2)} \otimes \mathcal{G}_2$, have the forms

$$\boldsymbol{K}_{1(y)}^{(2)} = e^{(2)}\boldsymbol{K}_0,\ \boldsymbol{K}_{1(z)}^{(2)} = £_x^{(2)}\boldsymbol{K}_0,\ \boldsymbol{K}_{2(zx)}^{(2)} = U^{(2)},\ \boldsymbol{K}_{2(xy)}^{(2)} = £_z^{(2)}. \tag{38}$$

The results obtained in this section for Mn(III) monomer also apply to the Fe(II) monomer with the $3d^6$ electron configuration, in which only two of the six electrons have compensated spins, as a result of which the spin of the Fe(II) monomer has the spin $S = 2$.

TUNABLE TRANSFORMATION OF ANTIFERROMAGNETIC EXCHANGE INTERACTION INTO FERROMAGNETIC ONE IN HOMOBINUCLEAR CLUSTERS

In the exchange-coupled two-nuclear clusters, the exchange interaction occurs between the components of dimers of both antiferromagnetic and ferromagnetic types. In this case, antiferromagnetic exchange occurs much more often than the ferromagnetic one. Apparently, this is due to the fact, that the antiparallel orientation of spins is more favorable than their parallel orientation for achieving a stable state of cluster.

The transition from the antiferromagnetic exchange interaction to the ferromagnetic type exchange interaction can be carried out by using the combined time-reversal operator. The action of this operator is as follows. At the first stage, only the sign of the operator $\boldsymbol{S}_1$ is inverted (the spin operator $\boldsymbol{S}_2$ of the second component of the dimer remains unchanged). Such a transformation is not a symmetry operation. At the second stage, the second ion (with a spin S_2) is isomorphically replaced by an ion, whose wave functions are reversed in time as compared to the wave functions of the replaced ion. The Hamiltonian of a magnetic dimer is invariant under such a two-stage transformation, in which the sign of one of the spin operators $\boldsymbol{S}_1$ or $\boldsymbol{S}_2$ changes and the sign of the exchange interaction constant changes. As a result of this transformation, a homonuclear dimer with an antiferromagnetic exchange interaction will be converted into a heteronuclear dimer with a ferromagnetic exchange interaction between its components. Let us to consider the action of the combined time-reversal operator on the Hamiltonian $\boldsymbol{H}_{exch}$ of the exchange interaction of a two-nuclear *Mn(III) – Mn(III)* cluster, where the electron configuration and spin of the each *Mn(III)* ion are $3d^4$ and $S = 2$, respectively. The magnetic and redox properties of *Mn(III)* and *Fe(III)* dinuclear complexes with shortened salen-type Schiff base ligands have been studied in (Piwowarska et al., 2019). For simplicity, we assume that the main magnetic interaction between ions of the dimer is the isotropic exchange interaction

$$\boldsymbol{H}_{exch} = -J\boldsymbol{S}_1 \bullet \boldsymbol{S}_2. \tag{39}$$

The full time reversal operator for this bihomonuclear cluster has the form

$$\boldsymbol{K} = \boldsymbol{K}_1^{(2)} \cdot \boldsymbol{K}_2^{(2)} = U_1^{(2)} \boldsymbol{K}_0 \cdot U_2^{(2)} \boldsymbol{K}_0, \tag{40}$$

where the subscripts 1 and 2 refer to the two components of the dimer. Under the action of the operator $\boldsymbol{K}_1^{(2)}$ on the Hamiltonian $\boldsymbol{H}_{exch}$ from eq. (39), the sign of the operator $\boldsymbol{H}_{exch}$ is reversed, that is, the Hamiltonian is not invariant under such a transformation:

$$\begin{matrix} \boldsymbol{K}_1^{(2)} \cdot \boldsymbol{H}_{exch} \cdot (\boldsymbol{K}_1^{(2)})^{-1} = -J\boldsymbol{K}_1^{(2)} \cdot \boldsymbol{S}_1 \boldsymbol{S}_2 \cdot (\boldsymbol{K}_1^{(2)})^{-1} = \\ -J\{\boldsymbol{K}_1^{(2)} \cdot S_{1x} (\boldsymbol{K}_1^{(2)})^{-1} S_{2x} + (\boldsymbol{K}_1^{(2)} \cdot S_{1y} (\boldsymbol{K}_1^{(2)})^{-1} S_{2y} + (\boldsymbol{K}_1^{(2)} \cdot S_{1z} (\boldsymbol{K}_1^{(2)})^{-1} S_{2z}\} \end{matrix} = J\boldsymbol{S}_1 \bullet \boldsymbol{S}_2 = -\boldsymbol{H}_{exch}, \tag{41}$$

This result follows from simple direct calculations, given that the matrices $U_1^{(2)}$, S_{1x}, and S_{1z} are real 5x5 matrices, and the matrix S_{1y} is an imaginary 5x5 matrix. A similar result is obtained, when the operator 22 acts on the Hamiltonian $\boldsymbol{H}_{exch}$. This implies the trivial conclusion that the Hamiltonian $\boldsymbol{H}_{exch}$ is invariant with respect to the operator $\boldsymbol{K}$ from eq. (40). However, we can proceed differently to restore the original sign of the Hamiltonian $\boldsymbol{H}_{exch}$ Hamiltonian. To do this, we need to introduce the operator $\boldsymbol{I}_J$, which changes the sign of the exchange interaction constant J to the opposite:

$$\boldsymbol{I}_J J = - J. \tag{42}$$

In order for this operation to have a physical meaning, and not to remain a simple mathematical replacement of J by - J, it should be taken into account that when J changes to - J, the order of the spin levels changes. In the simplest case, when the dimer is formed by ions with spins $S_1 = S_2 = 1/2$, at $J < 0$, the ground state of the homonuclear dimer is a spin singlet ($S = 0$), and the next (excited) level is a spin triplet ($S = 1$). The singlet-triplet splitting is determined by the value of $|J|$. At $J > 0$, the ground state of a heteronuclear two-component cluster is a spin triplet and the first excited state is a spin singlet. In order to transform the antiferromagnetic ordering (J < 0) to ferromagnetic one ($J > 0$) it is necessary to carry out an isomorphic replacement of one of the ions of the homobinuclear cluster by an ion, whose wave functions are reversed in time compared to the wave functions of the replaced ion. In this case, the Hamiltonian $\boldsymbol{H}_{exch}$ is invariant under action of the operator $\boldsymbol{K}_1^{(2)}\boldsymbol{I}_J$ and $\boldsymbol{K}_2^{(2)}\boldsymbol{I}_J$:

$$\boldsymbol{K}_2^{(2)}\boldsymbol{I}_J \cdot \boldsymbol{H}_{exch}\boldsymbol{I}_J^{-1}(\boldsymbol{K}_2^{(2)})^{-1} = \boldsymbol{H}_{exch}, \tag{43}$$

where $\boldsymbol{K}_1^{(2)}\boldsymbol{I}_J$ and $\boldsymbol{K}_2^{(2)}\boldsymbol{I}_J$ are operators of the combined time reversal. Based on the developed approach, it is easy to show how it takes place the transformation of a dimer with an antiferromagnetic exchange interaction into a dimer with a ferromagnetic exchange interaction. In particular, upon isomorphic substitution of one of the *Mn(III)* ions in the *Mn(III)–Mn(III)* cluster with antiferromagnetic exchange interaction for a *Co(III)* ion, the initial cluster transforms into a *Mn(III)–Co(III)* cluster with a ferromagnetic exchange interaction. Note that the electronic configuration of the *Co(III)* ion is $3d^6$, while the electronic configuration of the *Mn(III)* ion is $3d^4$, that is, the number of 3d- electrons of the *Co(III)* ion is equal to the number of 3d- holes in the electron shell of the *Mn(III)* ion.

CONCLUSION

Experimental data (McGrady & Stranger, 1997) indicate that in *Mn(II)* complexes, afforded by Schiff base ligands, an antiferromagnetic exchange between *Mn(II)* centers exists. These data also show that oxidation of dinuclear *Mn(II)* complexes afforded by Schiff base ligands of complex structure by oxygen molecules leads to the release of the constitutive *Mn(II)* binuclear units with formation of mononuclear *Mn (II)* and *Mn (III)* centers. The redox-induced changes in the Heisenberg exchange coupling constants for *Mn (III) – Mn (III)* and *Mn (IV) –Mn (IV)* clusters are caused by changes in the geometry of Mn_2O core. The longer Mn - Mn distance and the more acute O - *Mn - O* angle in *Mn(IV) - Mn(IV)* cluster lead to larger exchange interaction constants in the more oxidized species (McGrady & Stranger, 1997).

An analogy was established between discrete symmetry groups of the eighth order $\mathcal{G}_8^{(1/2)}$, $\mathcal{G}_8^{(3/2)}$, $\mathcal{G}_8^{(2)}$, and $\mathcal{G}_8^{(5/2)}$ containing full and partial time reversal operators. Later, an analogy was established between the groups $\mathcal{G}_8^{(1/2)}$, $\mathcal{G}_8^{(3/2)}$, $\mathcal{G}_8^{(2)}$, and $\mathcal{G}_8^{(5/2)}$, which are extensions of the previous four symmetry groups. Based on this analogy, all six time reversal operators have been found for the Mn(II), Mn(III) and Fe(II) monomers.

The incomplete time reversal operators determine the behavior of the spin projection operators under reduction, but not in the complete absence of complete time reversal symmetry. There are six ways to reduce the Wigner symmetry of time reversal and for its restoring it is necessary to introduce six types of meta-particles (Geru, 2018). Theoretically, it means the possibility of the existence of six forms of meta-matter. The concept of meta-matter was introduced by Wigner (1970), but the meta-particles have not been yet discovered experimentally.

As for the possibility of the transformation of the homonuclear dimer Mn(III)-Mn(III) for wich $J < 0$ into heteronuclear dimer with $J > 0$, such a process is possible because in various compounds there is a dimer Mn(III)-Mn(III) (Sinha et al., 2022) and there is a fairly large number of paramagnetic Co(III) monomers (Eiko Toyota & Matsudera, 1980).

Finally, we note that the change in the sign of the exchange interaction constant of a homonuclear dimer, given by formula (42), is equivalent to the isomorphic substitution of one of the ions of the dimer by an ion whose wave functions are inverted in time with respect to the wave functions of the substituted ion (Geru, 2018). Namely this real isomorphic substitution of one ion of a dimer, rather than the formal replacement of J by $-J$, leads to a change in the type of exchange interaction (from antiferromagnetic in homonuclear dimers to ferromagnetic in heteronuclear dimers). Under the action of each of the operators $\boldsymbol{K}_1^{(2)}\boldsymbol{I}_J$ and $\boldsymbol{K}_2^{(2)}\boldsymbol{I}_J$, the exchange interaction Hamiltonian $\boldsymbol{H}_{exch}$ (39) does not change, that is, it remains invariant (formulas from (43) illustrate this for the case $J = 2$). Such transformations under which the Hamiltonian of the system is invariant are symmetry transformations. Under ordinary symmetry transformations, the geometric, atomic and electron structure of the physical system does not change. In contrast, in the case of the action of operators $\boldsymbol{K}_1^{(2)}\boldsymbol{I}_J$ and $\boldsymbol{K}_2^{(2)}\boldsymbol{I}_J$ (more precisely, the $\boldsymbol{I}_J$ component of these operators), the electron configuration of the substituted ion is replaced by the electronic configuration of the substituting ion.

We also note that the concept of combined time-reversal transformation discussed in this Chapter has experimental confirmation: for the complex $Cu_2(fsa)_2enCH_3COOH$ $J = -650$ cm^{-1} (Galy et al., 1979), while for the complex $CuVO(fsa)_2enCH_3COOH$ J = 118 cm^{-1} (Kahl & Charlot, 1980). Using this approach will increase the number of synthesized complexes containing heteronuclear dimers, which is currently much less than the number of complexes with homonuclear dimers.

REFERENCES

Amman, M. (2018). Principles of Redox Reactions: Basic concepts with Resolved Questions and Problems. *Create Space Publishing Platform,* 26.

Auber, B. (2001). Observation of CP Violation in the B^0 Meson System. *Physical Review Letters, 87*(9), 091801. doi:10.1103/PhysRevLett.87.091801 PMID:11531560

Eiko Toyota, E., & Matsudera, N. (1980). Paramagnetic Co(III) Complexes with Organic Ligands. V. The Isolation of Properties of Salts of the Paramagnetis Bis(2,2-bipyridine)(5nitro-salycilato)cobalt(III) and Paramagnetic "1 and 2"- Isomers of the 5-Nitrosalycitato(triethilenetetramine)-cobalt(III) and Their Derivatives. *Bulletin of the Chemical Society of Japan, 53,* 3517–3521. doi:10.1246/bcsj.53.3517

Galy, J., Jaud, J., Kahn, O., & Tola, P. (1979). Crystal structure and magnetic properties of Cu2(fsa)2en, CH3OH, with H4(fsa)2en = N,N¢-bis(2-hydroxy, 3-carboxybenzilidene)-1,2-diaminoethane. *Inorganica Chimica Acta*, *36*, 229–236. doi:10.1016/S0020-1693(00)89398-6

Geru, I. I. (2018). Time-Reversal Symmetry: Seven Time-Reversal Operators for Spin Containing Systems. Springer Nature. doi:10.1007/978-3-030-01210-6

Kahn, O., & Charlot, M. F. (1980). Overlap density in binuclear complexes: A topological approach of the exchange interaction. *Nuov. J. Chim.*, *4*, 567–576. doi:10.1007/978-94-010-9716-1_11

Lazauskas, R., Song, Y. H., & Gudkov, V. (2014). Parity and time-reversal violation in A = 2 – 4 nuclei. *EPJ Web of Conferences*, *66*, 05014. doi:10.1051/epjconf/20146605014

McGrady, G. E., & Stranger, R. (1997). Redox-Induced Changes in the Geometry and Electronic Structure of Di-. *Journal of the American Chemical Society*, *119*(36), 8512–8522. doi:10.1021/ja964360r

Piwowarska, D., Gnulek, P., & Rudowicz, C. (2019). Origin of the Ground Kramers Doublets for $C0^{2+}d^7$. *Applied Magnetic Resonance*, *50*(6), 797–808. doi:10.100700723-018-1080-4

Rigamonti, P., Zardi, S., Carlino, S., Demartin, F., Castellano, C., Pigani, L., Ponti, A., Ferretti, A. M., & Pasini, A. (2020). Selective formation, reactivity, redox and magnetic properties of. *International Journal of Molecular Sciences*, *21*(21), 7882. doi:10.3390/ijms21217882 PMID:33114262

Roberts, B. M., Dzuba, V. A., & Flambaum, V. V. (2015). Parity and Time-Reversal Violation in Atomic Systems. *Annual Review of Nuclear and Particle Science*, *65*(1), 63–86. doi:10.1146/annurev-nucl-102014-022331

Sinha, S., Sahad, M. E., & Mondal, R. (2022). A Singlet-Diradical Co(III)-Dimer as a Nonvolatile Resistive Switching Device: Synthesis, Redox-Induced Interconversion, and Current–Voltage Characteristics. *Journal of the American Chemical Society*, *144*(44), 20442–20451. doi:10.1021/jacs.2c08941 PMID:36282681

Theil, S., Yerande, R., Chikate, R., Dahan, F., Bousseksou, A., Padhye, S., & Tuchagues, J.-P. (1997). Synthesis, structure, and magnetic and redox properties of linear bis-dinuclear complexes afforded by schiff base ligants containing catecholate and pyridine or imidazole groups. *Inorganic Chemistry*, *36*(27), 6279–6286. doi:10.1021/ic9703538

Titov, V., Mosyagin, N. S., Petrov, A. N., Isaev, T. A., & DeMille, D. (2006). P,T-Parity Violation Effects in Polar Heavy-Atom Molecules. In J.-P. Julien (Eds.), *Recent Advances in the Theory of Chemical and Physical Systems. Progress in Theoretical Chemistry and Physics* (Vol. 15, pp. 253–283). Springer.

Wigner, E. (1959). Theory Groups and Its Applications to the Quantum Mechanics of Atomic Spectra. Cambridge University Press.

Wigner, E. P. (1970). *Symmetries and Reflections*. Indiana University Press.

Zhu, J. X., & Ting, C. S. (1998). Tunneling spectrum into a *d*-wave superconductor with a time-reversal-symmetry-broken surface state. *Physical Review B: Condensed Matter*, *57*(5), 3038–3044. doi:10.1103/PhysRevB.57.3038

Chapter 8
Thermodynamics of Redox Processes in Homogeneous and Heterogeneous Multicomponent Systems

Igor Povar
Institute of Chemistry, Moldova State University, Moldova

Oxana Spinu
https://orcid.org/0000-0002-3166-1931
Institute of Chemistry, Moldova State University, Moldova

ABSTRACT

The chapter presents a thermodynamic approach for the complex chemical and redox equilibria investigation of mono- and two-phase systems containing metal ions of slightly soluble solid phases in different oxidation states. For heterogeneous systems, this approach utilizes thermodynamic relationships combined with original mass balance constraints. The forms of occurrence of metal ions are determined by the major chemical reactions in the aquatic environment such as hydrolysis, oxidation, reduction, and precipitation. On the basis of the introduced generalized equation of the overall redox reaction the physical meaning of the overall electrode potential under conditions other than standard ones, was revealed. The diagrams of complex chemical equilibria have been used for the graphical representation of complex equilibria in aqueous solutions. The developed approach can be applied to calculating potential–pH (Pourbaix) diagrams, based on the thermodynamic analysis of chemical equilibria in homogeneous and heterogeneous multicomponent systems.

DOI: 10.4018/978-1-6684-7198-2.ch008

INTRODUCTION

Although most chemical reactions occurring in natural environments are considered acid–base processes (weathering of aluminosilicates, the solubility of carbonates) involving the transfer of protons, many other reactions involve the transfer of electrons and are known as oxidation–reduction processes. The forms of occurrence of metal ions are determined by the significant chemical reactions in the aquatic environment such as hydrolysis, oxidation, reduction, and precipitation. Depending on pH, redox potential, and total concentration of inorganic and organic ligands, metal ions may undergo various transformations to produce a whole range of chemical species in aqueous systems. Furthermore, depending on the redox conditions, organisms utilize different oxidized constituents as electron acceptors.

Complexing agents change the redox potential of a redox system involving metal ions. Frequently, in the presence of a complexing agent, oxidized species can form more stable complexes than reduced species. Therefore, the redox potential is modified in the presence of a suitable ligand (Bashir et al., 2018). This phenomenon has the potential to develop new various redox reactions in the presence of an appropriate complexing agent (Bashir et al., 2018; Rizvi et al., 2011a; Serjeant, 1984). This process has been applied to potentiometric titrations of some metal ions (Rizvi et al., 2011b; Safavi et al., 2002; Zhang & Zhou, 2019). For example, to determine the suitable pH range for the titration of vanadium (V) with iron (II) in the presence of a ligand, the formal potentials of equimolar mixtures of iron (II) and iron (III), as well as vanadium (IV) and vanadium (V), were measured at various pH, respectively (Umetsu, 1991). By using this aspect, Japanese researchers developed new methods for the potentiometric determination of chromium (VI) with iron (II), copper (II) with iron (II) (Itabashi et al, 1990), vanadium (V) with iron (II) (Umetsu, 1991) and sequential titration of chromium (VI) and iron (III) with cobalt (II) (Katsumata et al., 1997) in the presence of suitable complexing agents. The flow injection analysis methods for the determination of iron (II) (Itabashi et al., 1992) and vanadium (IV) (Itabashi et al., 1991) were developed with photometric detection and elaborated as well for the determination of complexing agents such as EDTA, DTPA, NTA, diphosphate, and others, using ligand effects (Itabashi et al., 1992; Teshima et al., 1993). Based on this ligand effect, Teshima et al. (1992) elaborated a new instantaneous flow injection analysis method for the determination of vanadium (IV) and vanadium (V) using redox reactions of vanadium (IV) with iron (III) and vanadium (V) with iron(II).

Redox processes in the presence of complexing agents control the chemical speciation, along with bioavailability, toxicity, mobility, and adsorption of water pollutants in the environment. Coordination of the redox action of transition metals can be an effective approach for the development of new redox systems for specific applications. *A priori* assessment of the complexation effect of selected ligands on the redox potential is significant for future trace-level applications of potentiometric polymeric membrane electrodes, which are useful analytical tools for heavy metal ion determinations in drinking water at nanomolar total concentrations (Ceresa et al., 2001; Lisak, 2021; Maksymiuk et al., 2020).

In natural aqueous systems redox reactions do not occur in isolation but are coupled through complexation and precipitation reactions to other species. As environmental conditions are determined by a wide range of the chemical composition, ionic strength, different various inorganic and organic impurities, the development of potentiometric sensors for the routine determination of trace heavy metals in environmental samples without strong theoretical knowledge of the complex chemical and redox equilibria is unreasonable. At the same time, ion-selective electrodes (*ISEs*) respond to the activity of ion in the solution, consequently, only ionized species can be detected (Cuartero & Crespo, 2018; Shao et al., 2020). To be ionized, heavy metals usually require acidic conditions. Even in the absence of strongly complexing agents, heavy metals can form hydroxo-complexes in an aquatic environment that are not taken into account by the ISEs. Additionally, the presence of interfering ions and impurities in the sample solution as well as the pH of the sample solution can cause side effects and hence hamper the detection limit of *ISEs*.

Based on the above, it is necessary to mention that the effects of the complexing agents have not been evaluated *quantitatively in full detail* based on *equilibrium considerations*. So far, in numerous works, the optimal conditions have been determined empirically, without resorting to simple relative calculations (Veerabhadraswamy et al., 2018).

1. BACKGROUND

Reactions in which electrons are transferred from one chemical form to another are called redox reactions. Any of these reactions can be considered as the sum of two electrode processes (half-reactions):

$$Ox_1 + z_1\bar{e} = Red_1 z_2,\ E_1^0,\ \Delta G_1^0 \tag{1}$$

$$Ox_2 + z_2\bar{e} = Red_2 - z_1, E_2^0, \Delta G_2^0 \tag{2}$$

$$z_2 Ox_1 + z_1 Red_2 = z_2 Red_1 + z_1 Ox_2,\ E^0,\ \Delta G^0, \tag{3}$$

where *Ox* denotes an oxidizing agent and *Red* denotes a reducing agent, E_i^0 and ΔG_i^0 have their usual meanings. It is possible to draw an analogy between redox reactions and other chemical processes, in particular acid-base reactions. The division of the overall redox reaction (3) into two separate half-reactions (1) and (2) is based on the concept of oxidation. The values *is* and ΔG^0 of the overall reaction (3) is summed from the corresponding values of the half-reactions (1) and (2) as follows:

$$E^0 = E_1^0 - E_2^0 \tag{4}$$

$$\Delta G^0 = z_2 \Delta G_1^0 - z_1 \Delta G_2^0 \tag{5}$$

To determine the change in other thermodynamic functions of reaction (3), specially, the change in enthalpy and entropy, it is necessary to know the corresponding values of the standard thermodynamic characteristics of the formation of reactants and reaction products:

$$\Delta H^0 = \sum_i \vartheta_i \Delta H_f^0(i)$$

$$\Delta S^0 = \sum_i \vartheta_i S_f^0(i),$$

where the stoichiometric coefficients ϑ_i take positive values for the reaction products and negative values for the reactants.

Relations (4) and (5) are of great importance for solving a number of practical problems. Specifically, the magnitude and sign of E^0 and ΔG^0 can be used to judge the direction and extent of the redox reaction under standard conditions. If $E^0 > 0$ ($\Delta G^0 < 0$), then under standard conditions the reaction proceeds from left to right and vice versa, at $E^0 < 0$ ($\Delta G^0 < 0$) under standard conditions reaction (3) is thermodynamically impossible (or the reverse reaction is thermodynamically possible). Under conditions of large absolute values of E^0 and ΔG^0, this conclusion is also valid under conditions other than the standard ones. At the same time, the real conditions for the occurrence of redox processes can differ significantly from the standard ones. First of all, the total initial concentrations (activities) of reactants and reaction products under real conditions are rarely equal to one. In addition, when the reaction conditions change, the nature of both the reactants and the reaction products often changes, since they undergo various side (competing) processes, such as protonation, hydrolysis, complex formation, and the formation of solid phases. Hereinafter these processes will be analyzed separately.

The influence of these factors is taken into account by means of the reaction isotherm equation or the Nernst equation. For the half reaction

$$Ox + z\bar{e} = Red$$

these equations take the form:

$$\Delta G = \Delta G^0 + RTln\frac{C_{Red}}{C_{Ox}} \tag{6}$$

$$E = E^0_{Ox/Red} - \frac{RT}{zF}ln\frac{C_{Red}}{C_{Ox}} \tag{7}$$

Here, C_{Ox} and C_{Red} denote the total concentrations of the oxidized and reduced forms, respectively. The second term in these equations is often commensurate in magnitude with ΔG^0 or $E^0_{Ox/Red}$, and sometimes it may exceed them. The pH of the solution especially strongly affects the value of ΔG^0 and $E^0_{Ox/Red}$, in many cases being predominant.

2. MAIN FOCUS OF THE CHAPTER

2.1. Issues, Controversies, Problems

Forecasting redox reactions is not an easy task. It involves a thorough examination of the reaction's experimental conditions. In particular, one must examine the influence of such factors as the medium acidity, the complexation and the precipitation of the redox species. For instance, depending on its value, the pH can induce the formation of new redox couples. These new couples involve species that do not exist in other domains of acidity. In such a respect, the pH may simply change the redox couple's nature by modifying the acido-basic status of the *Ox* or *Red* form, or both. The superimposition of precipitation and/or complex formation reactions to the redox equilibrium may lead to phenomena, which are of scientific and practical interest.

This chapter is devoted to thermodynamic prediction of the formal electrode potential, under the conditions of the simultaneous occurrence of a series of secondary/side reactions: ligand protonation, complex formation reactions, including polynuclear complex formation, and solid phase formation reactions. This chapter, therefore, is oriented toward understanding the requirement of taking into account, the basic redox reaction of simultaneous chemical processes, depending on the chemical composition and pH of the system.

It should be noted that, though the thermodynamic prediction shows which reactions are possible, it does not indicate anything about the kinetics of the reactions. The developed approach requires that the modeled redox reactions proceed quickly, i.e., that the time to reach thermodynamic equilibrium is short compared to the time-based tenacity of the model, and that the reactions are entirely reversible. Therefore, the usefulness of the derived expressions should be considered by the extent to which the open natural heterogeneous systems have been defined in terms of solid and solution phase composition and the availability of reliable thermodynamic data.

2.2. Thermodynamics of Redox Processes in Homogeneous Multicomponent Systems: General Concepts

Protonation of Oxidized and Reduced Forms

Many oxidized and reduced forms are Brønsted bases and, therefore, undergo protonation reactions when the pH of the solution changes. As a result, the electrode potential of such redox systems is a complex function of solution pH. This dependence is mainly due to two reasons. First, when pH changes, the nature of the oxidized and reduced forms changes as a result of protonation reactions. Second, the concentration of hydrogen ions, which are direct participants in such redox reactions, changes.

Methods for calculating the dependence of the electrode potential (and other thermodynamic functions) on the pH of a solution of systems, in which both the oxidized and reduced forms are Brønsted bases, will be discussed below. Equilibria in such systems are described by the following scheme:

$$Ox + z\bar{e} = Red \qquad (8)$$

$$Ox+iH=H_iOx \quad i=1,2,...,n \tag{9}$$

$$Red+jH=H_jRed \quad j=1,2,...,m \tag{10}$$

Indexes i and j will conventionally designate all quantities related to species H_iOx and H_jRed or to reactions (9) and (10), respectively. In particular, the equilibrium constants of reactions (9) and (10), i.e. overall protonation constants of Ox and Red will be denoted by βi and βj. Under the conditions of reactions (9) and (10), the Nernst equation has the form:

$$E = E^0_{Ox/Red} - \frac{RT}{zF} ln \frac{[Red]}{[Ox]}, \tag{11}$$

where $[Ox]$ and $[Red]$ are the equilibrium concentrations of the oxidized and reduced forms.

The sum of the equations (9) and (10) gives:

$$H_iOx + ze = H_jRed + (i - j)H \tag{12}$$

To determine the latter, the mass balance equations are used (Povar & Spinu, 2016a):

$$C^0_{Ox} = \sum_{i=0}^{n} [H_iOx] = [Ox]\alpha_{Ox} \tag{13}$$

$$C^0_{Red} = \sum_{j=0}^{m} [H_jRed] = [Red]\alpha_{Red}, \tag{14}$$

where

$$\alpha_{Ox} = 1 + \sum_{i=1}^{n} \beta_i [H]^i \tag{15}$$

$$\alpha_{Red} = 1 + \sum_{j=1}^{m} \beta_j [H]^j \tag{16}$$

Here C^0_{Ox} and C^0_{Red} symbolize the total concentrations of oxidized and reduced forms. The equations (13) and (14) are mass balance ones. According to the IUPAC, the alpha coefficient (15) and (16) are called *parasitic* or *side reaction constants* or *coefficients* (McNaught, 1997). Under the analyzed conditions the calculation of the side coefficients is immediate since the stability constants intervening in its expression are known. The significant point to notice is that the α coefficients may change noticeably with experimental conditions, especially with pH changes. Side reactions such as those of the complex formation and those of the formation of hydroxo-complexes may occur simultaneously. In this case, a total side reaction coefficient takes both phenomena into account.

Expressing from equations (13) and (14) $[Ox]$ and $[Red]$ and substituting them into (11), one gets the equation for formal (conditional) potential:

$$E = E^0_{Ox/Red} + \frac{RT}{zF} ln \frac{\alpha_{Red}}{\alpha_{Ox}} - \frac{RT}{zF} ln \frac{C^0_{Red}}{C^0_{Ox}} \tag{17}$$

According to equation (11), the dependence of E on pH is due to a change in the concentration of non-protonated forms $[Ox]$ and $[Red]$. A rigorous thermodynamic derivation of the equation (17) is considered below. Initially, the partial mole fractions of the oxidized and reduced forms and the corresponding Bjerrum functions are introduced as follows:

$$f_i = \frac{[H_iOx]}{\sum_{i=0}^{n}[H_iOx]} = \frac{\beta_i [H]^i}{\alpha_{Ox}}, \quad \bar{n} = \sum_{i=1}^{n} if_i \tag{18}$$

$$f_j = \frac{[H_jRed]}{\sum_{j=0}^{m}[H_jRed]} = \frac{\beta_j [H]^j}{\alpha_{Red}}, \quad \bar{m} = \sum_{j=1}^{m} jf_j, \tag{19}$$

where αO_x and $\alpha Re_{d\,a}$re calculated by the expressions (15) and (16).

Accounting for the protonation of species of oxidized and reduced forms is carried out on the basis of the generalized equation of the overall electrode reaction:

$$\sum_{i=0}^{n} f_iH_iOx + ze = \sum_{j=0}^{m} f_jH_jRed + (\bar{n} - \bar{m})H \tag{20}$$

This equation is a generalization of equation (12) under the conditions when several protonated forms simultaneously participate in the electrode reaction. Obviously, for integer values f_i and f_j equation (20) takes the form (12).

Based on equation (20), the change in the Gibbs energy is calculated under conditions other than standard. For this, the reaction equation (20) is represented as the sum of three components:

1. 1. $Ox+ze=Red$

2. $\sum_{i=0}^{n} f_iH_iOx = Ox + \left(\sum_{i=1}^{n} if_i\right)H$ (21)

3. $Red+\left(\sum_{j=1}^{m} jf_j\right)H=\sum_{j=0}^{m} f_j H_j Red$

In a number of papers (Povar & Spinu, 2016a; Povar & Spinu, 2016c; Povar et al., 2019a; Povar et al., 2022) it was proved that the change in the Gibbs energy of these processes is equal, respectively:

$$\Delta G(1)=\Delta G^0(1)+RTln\frac{C^0_{Red}}{C^0_{Ox}}$$

$$\Delta G(2)=RTln\left(1+\sum_{i=1}^{n}\beta_i[H]^i\right)=RTln\alpha_{Ox}$$

$$\Delta G(3)=-RTln\left(1+\sum_{j=1}^{m}\beta_j[H]^j\right)=-RTln\alpha_{Red}$$

The total change in the Gibbs energy for the generalized reaction of the overall electrode reaction (20) is

$$\Delta G=\Delta G(1)+\Delta G(2)+\Delta G(3)=\Delta G^0(1)+RTln\frac{\alpha_{Ox}}{\alpha_{Red}}+RTln\frac{C^0_{Red}}{C^0_{Ox}} \tag{22}$$

Taking into account that

$$\Delta G=-zFE,\quad E^0_{\frac{Ox}{Red}}=-\frac{\Delta G^0(1)}{zF},$$

the expression (17) for calculating E is obtained. Therefore, the electrode potential for this system is the sum of three components (21), while the dependence on pH is due to the last two.

Further, the calculation of other thermodynamic functions will be considered. Differentiating expression (22) with respect to temperature, one gets for ΔS:

$$\Delta S=-\frac{d\Delta G}{dT}=-\frac{d\Delta G^0(1)}{dT}-Rln\frac{\alpha_{Ox}}{\alpha_{Red}}+RT\left[\frac{dln\alpha_{Ox}}{dT}-\frac{dln\alpha_{Red}}{dT}\right]+Rln\frac{C^0_{Red}}{C^0_{Ox}}.$$

Hence, for the change in enthalpy one gets:

$$\Delta H=\Delta G+T\Delta S=\Delta G^0(1)-T\frac{d\Delta G^0(1)}{dT}+RT^2\left[\frac{dln\alpha_{Red}}{dT}-\frac{dln\alpha_{Ox}}{dT}\right]$$

Taking into consideration that

$$\Delta G^0(1) - T\frac{d\Delta G^0(1)}{dT} = \Delta H^0(1),$$

$$\frac{dln\alpha_{Ox}}{dT} = \sum_{i=1}^{n} f_i \frac{dln\beta_i}{dT} = \left(RT^2\right)^{-1} \sum_{i=1}^{n} f_i \Delta H_i^0$$

$$\frac{dln\alpha_{Red}}{dT} = \sum_{j=1}^{m} f_j \frac{dln\beta_j}{dT} = \left(RT^2\right)^{-1} \sum_{j=1}^{m} f_j \Delta H_j^0,$$

where $\Delta H^0(1)$, ΔH_i^0 and ΔH_j^0 denote standard changes in the enthalpy of reactions (8)-(10), respectively, the expression for ΔH takes the following form:

$$\Delta H = \Delta H^0(1) + \sum_{i=1}^{n} f_i \Delta H_i^0 - \sum_{j=1}^{m} f_j \Delta H_j^0 \tag{23}$$

Differentiating ΔH with respect to temperature, one can obtain an expression for ΔC_p. In this case, it should be taken into account that the partial molar fractions f_i and f_j depend on temperature. The derived formal electrode potential may even reverse the equilibrium, making possible reactions that are impossible under standard conditions, or vice versa.

2.3. Complex Formation of Oxidized and Reduced Forms

Often, both the oxidized and reduced forms form strong complexes with the same ligand *L*, which is usually in large excess compared to the oxidizing and reduced forms. Obviously, the electrode potential in such cases changes significantly. In this section, the formation of mononuclear complexes will be analyzed. The basic properties of the ligand *L*, which is usually a Brønsted base, will be taken into account. Therefore, the following reactions take place in the system:

$$Ox+ze=Red \tag{24}$$

$$Ox+iL=OxL_i,\ \beta i \tag{25}$$

$$Red+jL=RedL_j,\ \beta j \tag{26}$$

$$L+pH=H_pL.\ \beta p \tag{27}$$

Here βi, βj and βp are the equilibrium constants for the corresponding reactions.

The influence of the reactions (25)-(27) on the electrode potential of the half-reaction (24) can be taken into account in the following way. The Nernst equation of the half-reaction (24) under the reaction conditions (25)-(27) takes the form:

$$E = E^0_{Ox/Red} - \frac{RT}{zF} ln \frac{[Red]}{[Ox]} \text{ (28)},$$

where the equilibrium concentrations $[Ox]$ and $[Red]$ are determined from the mass balance conditions:

$$C^0_{Ox} = \sum\nolimits_{i=0}^{n} [OxL_i] = [Ox]\left(1 + \sum\nolimits_{i=1}^{n} \beta_i [L]^i\right) = [Ox]\alpha_{Ox} \tag{29}$$

$$C^0_{Red} = \sum\nolimits_{j=0}^{m} \left[RedL_j\right] = [Red]\left(1 + \sum\nolimits_{j=1}^{m} \beta_j [L]^j\right) = [Red]\alpha_{Red}, \tag{30}$$

where $[L]$ is the equilibrium concentration of ligand, which, under the conditions of ligand protonation in accordance with reaction (27) is determined as follows:

$$C^0_L = \sum\nolimits_{p=0}^{k} \left[H_pL\right] = [L]\left(1 + \sum\nolimits_{p=1}^{k} \beta_p [H]^p\right) = [L]\alpha_L \tag{31}$$

Here C^0_L is the total concentration of ligand. Usually $C^0_L \gg C^0_{Ox}$ and $C^0_L \gg C^0_{Red}$. Determining from equations (29) and (30) $[Ox]$ and $[Red]$ and substituting them into equation (28), one gets the equation:

$$E = E^0_{Ox/Red} + \frac{RT}{zF} ln \frac{\alpha_{Red}}{\alpha_{Ox}} - \frac{RT}{zF} ln \frac{C^0_{Red}}{C^0_{Ox}} \tag{32}$$

This equation has a great practical importance. Provided that βi, βj and βp are known, this equation makes it possible to calculate the dependence of the formal electrode potential on any of four variables C^0_{Ox}, C^0_{Red}, C^0_L and pH. The inverse problem arises in the potentiometric determination of the complex formation constants for OxL_i and $RedL_j$. For this purpose, the dependence of the formal electrode potential on the above variables is experimentally studied.

Equation (28) indicates that the change in the electrode potential is due to a decrease in the concentration of uncomplexed particles of the oxidized and reduced forms. However, such an assumption is not necessary since equation (32) can also be interpreted differently. Indeed, the influence of reactions (25)-(27) on the formal electrode potential can be considered on the basis of the generalized half-reaction equation, which in this case has the form:

$$\sum\nolimits_{i=0}^{n} f_i OxL_i + \bar{k}\left(\bar{n} - \bar{m}\right)H + ze = \sum\nolimits_{j=0}^{m} f_j RedL_j + \left(\bar{n} - \bar{m}\right)\sum\nolimits_{p=0}^{k} f_p H_p L, \tag{33}$$

where f_i, f_j and f_p are partial molar fractions of corresponding species, while $\bar{n}$, $\bar{m}$, and $\bar{k}$ are the Bjerrum functions:

$$f_i = \frac{[OxL_i]}{\sum_{i=0}^{n}[OxL_i]} = \frac{\beta_i[L]^i}{\alpha_{Ox}}, \quad \bar{n} = \sum_{i=1}^{n} if_i$$

$$f_j = \frac{[RedL_j]}{\sum_{j=0}^{m}[RedL_j]} = \frac{\beta_j[L]^j}{\alpha_{Red}}, \quad \bar{m} = \sum_{j=1}^{m} jf_j$$

$$f_p = \frac{[H_pL]}{\sum_{p=0}^{k}[H_pL]} = \frac{\beta_p[H]^p}{\alpha_L}, \quad \bar{k} = \sum_{p=1}^{k} pf_p$$

Equation (33) can be decomposed into three, simpler equations:

I. I. $Ox+Ze=Red$ (1)

II. $Ox + \bar{n}\sum_{p=0}^{k} f_p H_p L = \sum_{i=0}^{n} f_i OxL_i + \bar{k}\bar{n}H$ (-1) (34)

III. $Red + \bar{m}\sum_{p=0}^{k} f_p H_p L = \sum_{j=0}^{m} f_j RedL_j + \bar{k}\bar{m}H$ (1)

Next to each of these equations is a factor by which they must be multiplied to obtain equation (33) when added. The changes in the Gibbs energy of reactions (I) – (III), under conditions other than standard, as shown in the previous section, are equal to:

$$\Delta G(I) = \Delta G^0(I) + RTln\frac{C_{Red}^0}{C_{Ox}^0}$$

$$\Delta G(II) = -RTln\left(1 + \sum_{i=1}^{n}\beta_i[L]^i\right) = -RTln\alpha_{Ox}$$

$$\Delta G(III) = -RTln\left(1 + \sum_{j=1}^{m}\beta_j[L]^j\right) = -RTln\alpha_{Red},$$

where $[L]$ is the equilibrium concentration of ligand and is determined from the equation (31). Then, for the overall change in Gibbs energy is valid the expression:

$$\Delta G = \Delta G(I) + \Delta G(II) + \Delta G(III) = \Delta G^0(I) + RTln\frac{\alpha_{Ox}}{\alpha_{Red}} + RTln\frac{C^0_{Red}}{C^0_{Ox}} \tag{35}$$

Taking into account that the change in Gibbs energy ΔG is related to formal electrode potential E by the relation $\Delta G= -zFE$, it is easy to see that equation (35) is equivalent to equation (32).

Polynuclear Complex Formation

The formation of polynuclear particles of oxidized or reduced forms introduces some features in the behavior of the redox system. This feature lies in the fact that the change in the Gibbs energy or the electrode potential depends on the equilibrium concentration of the oxidized or reduced forms. First, the following electrode reaction scheme is considered.

$$Ox+ze=Red \tag{36}$$

$$iOx+jL=Ox_iL_j,\ \beta i_j \tag{37}$$

$$mRed+nL=Red_mL_n,\ \beta m_n, \tag{38}$$

where L is the complexing agent, hydrogen ions or hydroxyl ions; βi_j nd βmn_a re the equilibrium constants of reactions (37) and (38), respectively. The indices ij *and* mn *de*note all the characteristics related to the complexes Ox*iL*$_{j\ a}$nd Red*mL*$_{n\ o}$r reactions (37) and (38).

It can be shown that the change in the Gibbs energy and the electrode potential of the half-reaction (36), taking into account (37) and (38), are described by the expressions.

$$\Delta G = \Delta G^0 + RTln\frac{[Red]}{[Ox]} \tag{39}$$

$$E = E^0_{Ox/Red} - \frac{RT}{zF}ln\frac{[Red]}{[Ox]}, \tag{40}$$

where ΔG^0 and E are the standard change in the Gibbs energy and the standard electrode potential of the half-reaction (36), $[Ox]$ and $[Red]$ the equilibrium concentrations of the oxidized and reduced forms. The latter are determined from the conditions of the material balance:

$$C_{Ox}^{0}=\sum_{i=1}\sum_{j=0}i[Ox_iL_j]=[Ox]\left(1+\sum_{i=1}\sum_{j=1}i\beta_{ij}[Ox]^{i-1}[L]^{j}\right)=[Ox]\alpha_{Ox} \quad (41)$$

$$C_{Red}^{0}=\sum_{m=1}\sum_{n=0}m[Red_mL_n]=[Red]\left(1+\sum_{m=1}\sum_{n=1}m\beta_{mn}[Red]^{m-1}[L]^{n}\right)=[Red]\alpha_{Red} \quad (42)$$

Here C_{Ox}^{0} and C_{Red}^{0} are the total concentration of oxidized and reduced forms. As can be seen, it is difficult to obtain the functions $\Delta G([L])$ or $E([L])$ explicitly, since to determine $[Ox]$ and $[Red]$ it is necessary to solve the material balance equations (41) and (42), the order of which in the case of the formation of polynuclear complexes is greater than one.

Based on relation (39), the changes in the remaining thermodynamic functions are calculated. For the change in entropy, the following expression is true:

$$\Delta S=-\frac{d\Delta G}{dT}=-\frac{d\Delta G^{0}}{dT}+Rln\frac{[Ox]}{[Red]}+RT\frac{dln[Ox]}{dT}+RT\frac{dln[Red]}{dT} \quad (43)$$

The last two derivatives are determined from the equations (41) and (42):

$$\frac{dln[Ox]}{dT}=-\frac{dln\alpha_{Ox}}{dT}$$

$$\frac{dln[Red]}{dT}=-\frac{dln\alpha_{Red}}{dT}$$

On the other hand, for derivatives of functions αO_x and αRe_{dw} ith respect to temperature, one gets:

$$\frac{dln\alpha_{Ox}}{dT}=\frac{1}{\sum_{i=1}\sum_{j=0}if_{ij}}\sum_{i=1}\sum_{j=1}f_{ij}\frac{dln\beta_{ij}}{dT}$$

$$\frac{dln\alpha_{Red}}{dT}=\frac{1}{\sum_{m=1}\sum_{n=0}mf_{mn}}\sum_{m=1}\sum_{n=1}f_{mn}\frac{dln\beta_{mn}}{dT}$$

Therefore, for the change in entropy, the final expression is:

$$\Delta S=\Delta S^{0}+Rln\frac{[Ox]}{[Red]}-\frac{RT}{\sum_{i=1}\sum_{j=0}if_{ij}}\sum_{i=1}\sum_{j=1}f_{ij}\frac{dln\beta_{ij}}{dT}+\frac{RT}{\sum_{m=1}\sum_{n=0}mf_{mn}}\sum_{m=1}\sum_{n=1}f_{mn}\frac{dln\beta_{mn}}{dT}, \quad (44)$$

Where,

$$\Delta S^0 = -\frac{d}{dT}\left(\Delta G^0\right) = -\frac{d}{dT}\left[\Delta G_f^0\left(Red\right) - \Delta G_f^0\left(Ox\right)\right] = S_f^0\left(Red\right) - S_f^0\left(Ox\right)$$

Substituting into the Gibbs-Helmholtz equation

$\Delta H = \Delta G + T\Delta S,$

expressions for $=\Delta G$ and ΔS, one finds:

$$\Delta H = \Delta H^0 + \frac{RT^2}{\sum_{m=1}\sum_{n=0} mf_{mn}} \sum_{m=1}\sum_{n=1} f_{mn} \frac{dln\beta_{mn}}{dT} - \frac{RT^2}{\sum_{i=1}\sum_{j=0} if_{ij}} \sum_{i=1}\sum_{j=1} f_{ij} \frac{dln\beta_{ij}}{dT}$$

Here, ΔH^0 is the change in the standard enthalpy of the half-reaction. Given that

$$\frac{dln\beta_{ij}}{dT} = \frac{\Delta H_{ij}^0}{RT^2} \frac{dln\beta_{mn}}{dT} = \frac{\Delta H_{mn}^0}{RT^2},$$

The expression for ΔH can be represented as:

$$\Delta H = \Delta H^0 + \$_{Red} \sum_{m=1}\sum_{n=1} f_{mn} \Delta H_{mn}^0 + \$_{Ox} \sum_{i=1}\sum_{j=1} f_{ij} \Delta H_{ij}^0, \tag{45}$$

where

$$\Phi_{Ox} = \frac{1}{\sum_{i=1}\sum_{j=0} if_{ij}}, \quad \Phi_{Red} = \frac{1}{\sum_{m=1}\sum_{n=0} mf_{mn}}. \tag{46}$$

The appearance of functions Φ_{Ox} and Φ_{Red}(46) in equation (45) is due to the fact that the degrees of complex formation αO_x and $\alpha Re_{d\,d}$epend on the equilibrium concentrations of the oxidized and reduced forms, which in turn depend on temperature. In addition, functions (46) ensure the normalization of the total thermal effect per 1 mole of Ox *and* Red*,* respectively.

The scheme of electrode reactions (36-38) can be generalized, taking into account, in particular, the formation of mixed complexes and the main properties of the ligand:

$Ox+ze=Red$

$iOx+jL+nH=Ox_iH_nL_j,\ \beta i_{nj}$

$pRed+rL+qH=Red_pH_qL_t,\ \beta p_{qr}$

$L+kH=H_kL,\ \beta k$

In this case, the change in the Gibbs energy and the electrode potential are described by the same equations (39) and (40), in which $[Ox]$ and $[Red]$ are determined from the conditions of the material balance:

$$C_{Ox}^{0}=\sum_{i=1}\sum_{n=0}\sum_{j=0}i[Ox_iH_nL_i]=[Ox]\left(1+\sum_{i=1}\sum_{n=0}\sum_{j=0}i\beta_{inj}[Ox]^{i-1}[H]^n[L]^j\right)=[Ox]\alpha_{Ox}$$

$$C_{Red}^{0}=\sum_{p=1}\sum_{q=0}\sum_{r=0}p\left[Red_pH_qL_r\right]=[Red]\left(1+\sum_{p=1}\sum_{q=0}\sum_{r=0}p\beta_{pqr}[Red]^{p-1}[H]^q[L]^r\right)=[Red]\alpha_{Red}$$

$$C_L^0=\sum_{k=0}\left[H_pL\right]=[L]\left(1+\sum_{k=1}\beta_k[H]^k\right)=[L]\alpha_L$$

When differentiating ΔG with respect to temperature, in this case, one should consider that not only $[Ox]$ and $[Red]$ are functions of temperature, but also the equilibrium concentration of the ligand $[L]$. Because of their cumbersomeness, expressions for other thermodynamic functions are not given here. From a mathematical viewpoint, the system of the conditions of the material balance is resolvable, but because in one of Nernst equations the numerator of the argument of one of the logarithms exhibits an exponent different from that of the denominator, it is more complicated to solve.

2.4. Thermodynamics of Redox Processes in Heterogeneous Multicomponent Systems

The formation of a precipitate of oxidized or reduced forms changes the Gibbs energy of the electrode reaction by an amount corresponding to the change in the Gibbs energy of formation of the solid phase. In this regard, there is a complete analogy with the effect of complex formation, with the only difference being that instead of the complex formation constant, the solubility product of a slightly soluble compound must be used.

Let the oxidized form forms a solid phase $OxA_{(S)}$ with the anion A. Then the electrode process can be represented as the sum of two reactions:

$$Ox+ze=Red \quad \Delta G_1^0=\Delta G_f^0(Red)-\Delta G_f^0(Ox) \tag{47}$$

$$OxA_{(S)}=Ox+A \quad \Delta G_2^0=-RTlnK_S\left(OxA_{(S)}\right) \tag{48}$$

$$OxA_{(S)}+ze=Red+A \tag{49}$$

The change in the Gibbs energy of the overall process (49) under standard conditions is:

$$E = E^0_{Ox/Red} + \frac{RT}{zF} lnK_S\left(OxA_{(S)}\right)$$

Accordingly, the expression is valid for the standard electrode potential:

$Ox{+}A{+}ze{=}RedA_{(S)}$

Similarly, if the reduced form *Red* forms a solid phase $RedA_{(S)}$, then:

$$\Delta G^0 = \Delta G_f^0\left(Red\right) - \Delta G_f^0\left(Ox\right) + RTlnK_S\left(RedA_{(S)}\right)$$

$$E = E^0_{\frac{Ox}{Red}} - \frac{RT}{zF} lnK_S\left(RedA_{(S)}\right)$$

Under conditions other than standard, a term containing the initial concentrations of the reactants and reaction products is added to these equations. If, in addition to the formation of a solid phase, the reactants or reaction products undergo side reactions, then the latter are taken into account by introducing the corresponding functions α. So, for the reaction scheme:

$Ox{+}ze{=}Red \quad \Delta G(I)$

$OxA_{(S)}{=}Ox{+}A \quad \Delta G(II)$

$Red{+}iL{=}RedL_i \quad \Delta G(III)$

$A{+}jH{=}H_jA \quad \Delta G(IV)$

The changes in the Gibbs energy of these processes are respectively equal to:

$$\Delta G\left(I\right) = \Delta G^0\left(I\right) + RTln\frac{C_{Red}}{C_{Ox}}$$

$$\Delta G\left(II\right) = \Delta G^0\left(II\right) + RTlnC^0_{Ox}C^0_A = -RTlnK_S\left(OxA_{(S)}\right) + RTlnC^0_{Ox}C^0_A$$

$$\Delta G\left(III\right) = -RTln\alpha_{Red} = -RTln\left(1 + \sum_{i=1}\beta_i\left[L\right]^i\right)$$

$$\Delta G(IV) = -RTln\alpha_A = -RTln\left(1+\sum_{j=1}\beta_j\{H]^j\right)$$

Summing these expressions gives:

$$\Delta G = \Delta G(I)+\Delta G(II)+\Delta G(III)-\Delta G(IV) = \Delta G^0(I)+RTln\frac{\alpha_{Ox}}{\alpha_{Red}}+RTln\frac{C^0_{Red}}{C^0_{Ox}}$$

From here, for the formal electrode potential, the expression is obtained:

$$E = E^0_{Ox/Red}+\frac{RT}{zF}ln\alpha_{Red}\alpha_A-\frac{RT}{zF}lnC^0_{Red}C^0_A$$

Similarly, other schemes of a complex process can be considered. Further, knowing ΔG and E it is not difficult to obtain expressions for the remaining thermodynamic functions.

This approach is applicable in cases where it is necessary to calculate the thermodynamic characteristics of an electrode process involving a thermodynamically stable solid phase. Meanwhile, when changing some parameters, such as pH, the solid phase can become thermodynamically unstable, i.e. dissolve or pass into another solid phase, more stable. Problems of this kind arise, in particular, when constructing potential-pH diagrams or the so-called Pourbaix diagrams (Povar et al., 2019b; Povar & Spinu, 2016b). Under such conditions, the dependence of the change in the Gibbs energy or electrode potential on pH can be calculated in a linear approximation, while determining the boundaries of the stability of solid phases. For this, the reaction of the formation of a solid phase or the transition of one solid phase to another is considered separately.

As an example of the use of the derived expressions, there will be analyzed the calculation of the dependence of the formal electrode potential of the $Co^{3+}-Co^{2+}-NH_3$ system on pH and $!\,^0_{NH_3}$. Authors (Liu et al, 2018) performed a thermodynamic analysis of ammoniacal thiosulphate leaching of gold catalyzed by *Co*(III)/*Co*(II) using Eh-pH and speciation diagrams. The direct effects of NH_3/NH_4^+ concentrations and pH on E can be shown by means of the diagrams $E(pH)$ and $E(!\,^0_{NH_3})$. Thermodynamic data for this system are presented in Table 1. Let's take into account that the electrode potential for this redox pair is equal to

$$Co^{3+}+e=Co^{2+},\ E^0_{Co^{3+}/Co^{2+}}=1.808\text{ V}$$

The dependence of the formal electrode potential on pH and $!\,^0_{NH_3}$ is described by the equations:

$$E_{Co^{3+}/Co^{2+}} = E^0_{Co^{3+}/Co^{2+}}+\frac{RT}{F}ln\frac{\alpha_{Co^{2+}}}{\alpha_{Co^{3+}}}+\frac{RT}{F}ln\frac{C^0_{Co^{3+}}}{C^0_{Co^{2+}}}$$

$$\alpha_{Co^{3+}} = 1 + \sum_{i=1}^{6} \beta_i [NH_3]^i$$

$$\alpha_{Co^{2+}} = 1 + \sum_{j=1}^{6} \beta_j [NH_3]^j$$

$$[NH_3] = \frac{!^{0}_{NH_3}}{1 + \beta_{NH_4^+}[H^+]}$$

The results of calculating the dependence on pH and $!^{0}_{NH_3}$ at $C^{0}_{Co^{3+}} = C^{0}_{Co^{2+}}$ are shown in Fig. 1 and 2. As can be seen, the formation of stable ammonia complexes leads to a strong decrease in the formal electrode potential of this system, both with an increase in pH and with an increase in $!^{0}_{NH_3}$.

Table 1. Equilibrium constants logK(logβ) for the complexation and precipitation of Co(III/II) in ammoniacal solutions at 25±5°C. Data from NIST Critical Stability Constants Database (NIST, 2004) unless stated otherwise.

Metal ion	Equilibrium	logK(logβ)
Co(III)	$Co^{3+}+NH_3=Co(NH_3)^{3+}$	6.7
	$Co^{3+}+2NH_3=Co(NH_3)_2^{3+}$	14.0
	$Co^{3+}+3NH_3=Co(NH_3)_3^{3+}$	20.1
	$Co^{3+}+4NH_3=Co(NH_3)_4^{3+}$	25.7
	$Co^{3+}+5NH_3=Co(NH_3)_5^{3+}$	30.8
	$Co^{3+}+6NH_3=Co(NH_3)_6^{3+}$	35.2
	$Co^{3+}+H_2O=Co(OH)^{2+}+H^+$	-1.31[a]
	$Co(OH)_{3(S)}+3H^+=Co^{3+}+3H_2O$	3.3
Co(II)	$Co^{2+}+NH_3=Co(NH_3)^{2+}$	2.11
	$Co^{2+}+2NH_3=Co(NH_3)_2^{2+}$	3.74
	$Co^{2+}+3NH_3=Co(NH_3)_3^{2+}$	4.79
	$Co^{2+}+4NH_3=Co(NH_3)_4^{2+}$	5.55
	$Co^{2+}+5NH_3=Co(NH_3)_5^{2+}$	5.73
	$Co^{2+}+6NH_3=Co(NH_3)_6^{2+}$	5.11
	$Co^{2+}+H_2O=Co(OH)^{+}+H^+$	-9.2
	$Co^{2+}+2H_2O=Co(OH)_2^{0}+2H^+$	-18.6
	$Co^{2+}+3H_2O=Co(OH)_3^{-}+3H^+$	-31.7
	$Co^{2+}+4H_2O=Co(OH)_4^{2-}+4H^+$	-46.4
	$2Co^{2+}+H_2O=Co_2(OH)^{3+}+H^+$	-9.8
	$4Co^{2+}+4H_2O=Co_4(OH)_4^{4+}+4H^+$	-29.9
	$Co(OH)_{2(S)}+2H^+=Co^{2+}+2H_2O$	-12.2
	$CoO_{(S)}+2H^+=Co^{2+}+H_2O$	-13.78[a]

[a](Liu et al., 2018)

Figure 1. The dependence of the formal electrode potential for the system Co^{3+}– Co^{2+}– NH_3 on pH for different values of ! $^{0}_{NH_3}$, mol/L: 1–1, 2–1·10^{-1}, 3–1·10^{-2}, 4–1·10^{-3}, 5–1·10^{-4}, 6–1·10^{-5}.

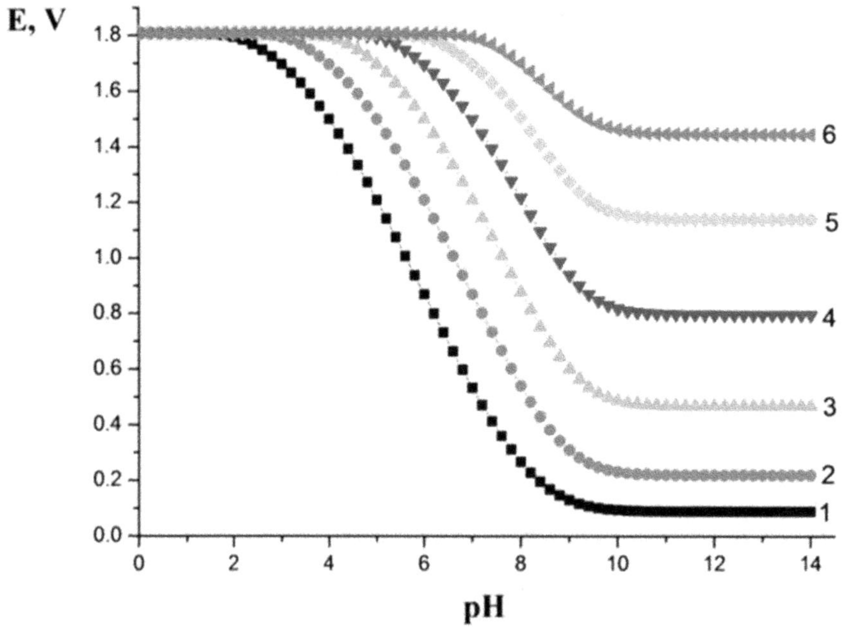

Figure 2. The dependence of the formal electrode potential for the system Co^{3+}– Co^{2+}–NH_3 on ! $^{0}_{NH_3}$ for different values of pH: 1–7, 2–8, 3–9, 4–10.

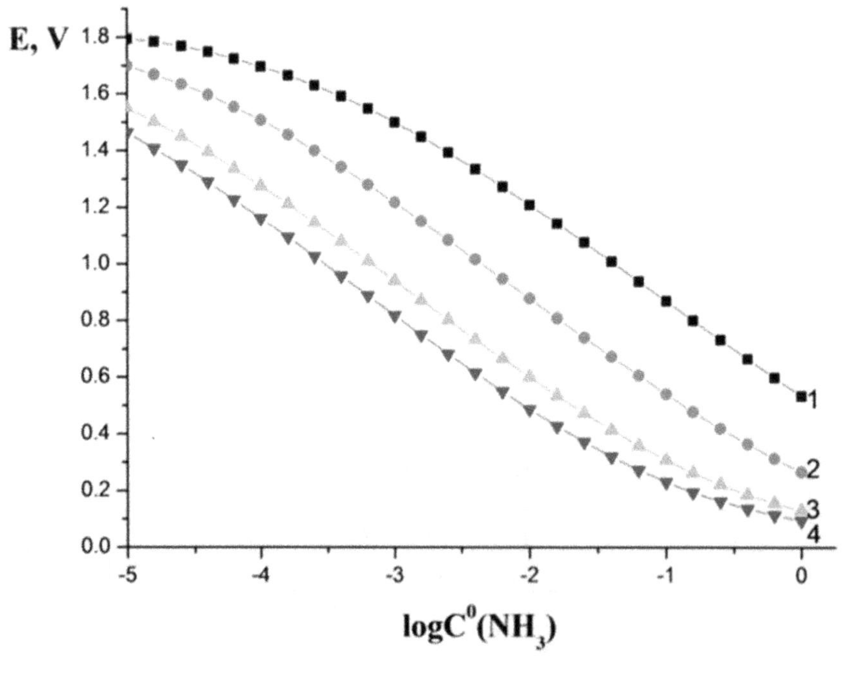

3. SOLUTIONS AND RECOMMENDATIONS

One can conclude that the deduced expressions bring nothing new from a theoretical perspective. However, their advantage is that in cases in which several equilibria occur concurrently, one may follow reasoning quite analogous to that followed in simpler cases in which only a single equilibrium takes place. Their introduction focuses our attention on the reaction we want to study. In brief, using deduced expressions appears to be an elegant way to study the redox complex processes and, moreover, is in harmony with chemical intuition. In derived expressions, the equilibrium concentrations in the system of any degree of complexity may be derived by solving simultaneously a set of nonlinear equations obtained by substitution of the system of extended law of mass-action equations into mass balance conditions. The equilibrium chemical composition can be calculated routinely with a number of software products offered in a variety of computational media.

In summary, this chapter considers the complexation effect on formal redox potentials of ion metal systems. The ligand modified redox potentials could be explored in the design and development of redox systems for novel analytical applications. The adjustment of formal redox potentials of ion metals with selected ligands could be as well a method to prevent oxidative stress, metal ions assimilation by ligands and natural attenuation of the toxic environmental contaminants. Thus, it is possible in many cases to forecast redox stability trends within a natural aqueous system based on tabulated thermodynamic data and soil solution composition relationships. Such predictions can be very helpful in understanding and modeling ground water contamination and remediation. The developed approach provides useful information from which one hopes to increase a stronger understanding of the physicochemical behavior of the environmental systems, in which redox processes happen. The developed approach may also be applied for calculating *potential–pH* (Pourbaix) diagrams, based on the thermodynamic analysis of chemical equilibria in the homogeneous and heterogeneous multicomponent systems.

4. FUTURE RESEARCH DIRECTIONS

These studies may be expanded to more complex multicomponent systems. The derived thermodynamic expressions for redox processes could become valuable tools to soil and environmental scientists for predicting soil–plant nutrient interactions, environmental pollution, and developing more effective nutrient management and remediation technologies. The obtained equations can be successfully used to create new redox buffer systems for redox buffer capacity of ion-Selective electrode solid contacts doped with organometallic complexes (Zhen et al., 2018). In order to facilitate the necessary calculations, the derived expression could be formalized within a rigorous mathematical framework and incorporated in one of existing appropriate computer equilibrium modelling code (Leal et al., 2017; Parkhust & Appelo, 1999).

5. CONCLUSION

A thermodynamic approach for the complex chemical and redox equilibria investigation of mono- and two-phase systems which contain slightly soluble solid phases have been developed. For two-phase systems this approach exploits thermodynamic relations joint with original mass balance restraints, where the solid phases are clearly stated. Based on the derived generalized equation of the overall redox process, where the stoichiometric coefficients are partial molar fractions of the oxidized and reduced forms, the physical meaning of the overall electrode potential, under the conditions other than standard ones, was revealed. Understanding of aqueous redox equilibria will offer new prospects to develop or engineer new approaches for water remediation and is also very important for predicting and protecting environmental well-being. To show the usefulness of the derived expressions, the calculations of formal electrode potentials for the system $Co(\text{III}) - Co(\text{II})\ \text{-}NH_3$ for the different values of pH and ammonium concentrations are accomplished.

ACKNOWLEDGMENT

This work is a part of the Moldovan State Program (2020-2023) "*Study and management of pollution sources to develop recommendations for implementing measures to mitigate the negative impact on environment and human health*", Project number: 20.80009.7007.20.

REFERENCES

Bashir, S., Mustafa, M., Safvi, S. W., Farhad, N. A., & Rizvi, M. A. (2018). Iron Reduces Iron: A Spectroelectrochemical Insight of Ligand Effect on Iron Redox Potential. *Warasan Khana Witthayasat Maha Witthayalai Chiang Mai*, *45*(2), 1087–1098.

Ceresa, A., Bakker, E., Hattendorf, B., Günther, D., & Pretsch, E. (2001). Potentiometric polymeric membrane electrodes for measurement of environmental samples at trace levels: New requirements for selectivities and measuring protocols, and comparison with ICPMS. *Analytical Chemistry*, *73*(2), 343–351. doi:10.1021/ac001034s PMID:11199988

Cuartero, M., & Crespo, G. A. (2018). All-solid-state potentiometric sensors: A new wave for in situ aquatic research. *Current Opinion in Electrochemistry*, *10*, 98–106. doi:10.1016/j.coelec.2018.04.004

Itabashi, H., Umetsu, K., Satoh, K., & Kawashima, T. (1990). Ligand effect on redox reaction of metal ions and its application to potentiometric titration of chromium (VI) and copper (II) with iron (II). *Analytical Sciences*, *6*(5), 721–725. doi:10.2116/analsci.6.721

Itabashi, H., Umetsu, K., Satoh, K., & Kawashima, T. (1991). Indirect spectrophotometric determination of vanadium (IV) by flow injection analysis based on the redox reaction with copper (II) in the presence of neocuproine. *Analytical Letters*, *24*(7), 1219–1230. doi:10.1080/00032719108052965

Itabashi, H., Umetsu, K., Teshima, N., Satoh, K., & Kawashima, T. (1992). Indirect spectrophotometric determination of complexing agents by flow-injection analysis based on redox reaction of copper (II) with iron (II). *Analytica Chimica Acta*, *261*(1-2), 213–218. doi:10.1016/0003-2670(92)80193-B

Katsumata, H., Teshima, N., & Kawashima, T. (1997). Successive potentiometric titration of chromium (VI) and iron (III) with cobalt (II) in the presence of 1, 10-phenanthroline. *Bulletin of the Chemical Society of Japan*, *70*(9), 2151–2154. doi:10.1246/bcsj.70.2151

Leal, A. M., Kulik, D. A., Smith, W. R., & Saar, M. O. (2017). An overview of computational methods for chemical equilibrium and kinetic calculations for geochemical and reactive transport modeling. *Pure and Applied Chemistry*, *89*(5), 597–643. doi:10.1515/pac-2016-1107

Lisak, G. (2021). Reliable environmental trace heavy metal analysis with potentiometric ion sensors-reality or a distant dream. *Environmental Pollution*, *289*, 117882. doi:10.1016/j.envpol.2021.117882 PMID:34364114

Liu, X., Xu, B., Yang, Y., Li, Q., Jiang, T., & He, Y. (2018). Thermodynamic analysis of ammoniacal thiosulphate leaching of gold catalysed by Co (III)/Co (II) using Eh-pH and speciation diagrams. *Hydrometallurgy*, *178*, 240–249. doi:10.1016/j.hydromet.2018.05.014

Maksymiuk, K., Stelmach, E., & Michalska, A. (2020). Unintended changes of ion-selective membranes composition—Origin and effect on analytical performance. *Membranes*, *10*(10), 266. doi:10.3390/membranes10100266 PMID:32998393

McNaught, A. D. (1997). *Compendium of chemical terminology* (Vol. 1669). Blackwell Science.

NIST critically selected stability constants of metal complexes database data. 2004.

Parkhurst, D. L., & Appelo, C. A. J. (1999). User's guide to PHREEQC (Version 2): A computer program for speciation, batch-reaction, one-dimensional transport, and inverse geochemical calculations. Water-resources investigations report, 99(4259), 312.

Povar, I., & Spinu, O. (2016a). Ruthenium redox equilibria 1. Thermodynamic stability of Ru (III) and Ru (IV) hydroxides. *Journal of Electrochemical Science and Engineering*, *6*(1), 123–133. doi:10.5599/jese.226

Povar, I., & Spinu, O. (2016b). Ruthenium redox equilibria 3. Pourbaix diagrams for the systems Ru-H2O and Ru-Cl—H2O. *Journal of Electrochemical Science and Engineering*, *6*(1), 145–153. doi:10.5599/jese.229

Povar, I., & Spinu, O. (2016c). Correlation between global thermodynamic functions and experimental data in multicomponent heterogeneous systems. *Canadian Journal of Chemistry*, *94*(2), 113–119. doi:10.1139/cjc-2015-0411

Povar, I., Spinu, O., Lupascu, T., & Duca, G. (2022). Thermodynamic Stability of Natural Aqueous Systems. In *Research Anthology on Ecosystem Conservation and Preserving Biodiversity* (pp. 531–563). IGI Global., doi:10.4018/978-1-6684-5678-1.ch028

Povar, I., Spinu, O., Zinicovscaia, I., Pintilie, B., & Ubaldini, S. (2019b). Revised Pourbaix diagrams for the vanadium–water system. *Journal of Electrochemical Science and Engineering*, *9*(2), 75–84. doi:10.5599/jese.620

Povar, I., Ubaldini, S., Spinu, O., & Lupascu, T. (2019a). Thermodynamic analysis of the copper (I) homogeneous and heterogeneous speciation in ammonium thiosulfate leaching systems. *Canadian Journal of Chemistry*, *97*(9), 651–658. doi:10.1139/cjc-2018-0446

Rizvi, M. A., Maqsood, R., Jan, R., & Peerzada, G. M. (2011b). Utilizing 1, 10–Phenanthroline Ligand Effect in Estimation of Cobalt (II) and Simultaneous Estimation of Cobalt (II) and Iron (II) with Cerium (IV). *Chemical Educator (Boise, Idaho)*, *16*, 72–75. doi:10.133300897112337a

Rizvi, M. A., Syed, R. M., & Khan, B. (2011a). Complexation effect on redox potential of iron (III)–iron (II) couple: A simple potentiometric experiment. *Journal of Chemical Education*, *88*(2), 220–222. doi:10.1021/ed100339g

Safavi, A., Abdollahi, H., & Nezhad, M. H. (2002). Indirect kinetic spectrophotometric determination of cobalt based on the redox reaction with iron (III) in the presence of 1, 10-phenanthroline. *Spectroscopy Letters*, *35*(5), 681–688. doi:10.1081/SL-120014939

Serjeant, E. P. (1984). Potentiometry and potentiometric titrations. *A series of Monographs on Analytical Chemistry and its Applications, 69.*

Shao, Y., Ying, Y., & Ping, J. (2020). Recent advances in solid-contact ion-selective electrodes: Functional materials, transduction mechanisms, and development trends. *Chemical Society Reviews*, *49*(13), 4405–4465. doi:10.1039/C9CS00587K PMID:32458836

Teshima, N., Itabashi, H., & Kawashima, T. (1992). Simultaneous flow injection determination of vanadium (IV) and (V) based on redox reactions of vanadium (IV) with iron (III) and vanadium (V) with iron (II) in the presence of 1, 10-phenanthroline and diphosphate. *Chemistry Letters*, *21*(11), 2227–2230. doi:10.1246/cl.1992.2227

Teshima, N., Itabashi, H., & Kawashima, T. (1993). Reverse flow injection analysis of complexing agents and its application to estimation of complexing capacity. *Talanta*, *40*(1), 101–106. doi:10.1016/0039-9140(93)80147-J PMID:18965601

Umetsu, K., Itabashi, H., Satoh, K., & Kawashima, T. (1991). Effect of ligands on the redox reaction of metal ions and the use of a ligand buffer for improving the end-point detection in the potentiometric titration of vanadium (V) with iron (II). *Analytical Sciences*, *7*(1), 115–118. doi:10.2116/analsci.7.115

Veerabhadraswamy, M., Devaraj, T. D., & Jayanna, B. K. (2018). Second derivative spectrophotometric determination of iron (II) and ruthenium (III) using 1, 10-phenanthroline. *Analytical Chemistry Letters*, *8*(6), 757–768. doi:10.1080/22297928.2018.1548944

Zhang, Y., & Zhou, M. (2019). A critical review of the application of chelating agents to enable Fenton and Fenton-like reactions at high pH values. *Journal of Hazardous Materials*, *362*, 436–450. doi:10.1016/j.jhazmat.2018.09.035 PMID:30261437

Zhen, X. V., Rousseau, C. R., & Bühlmann, P. (2018). Redox buffer capacity of ion-selective electrode solid contacts doped with organometallic complexes. *Analytical Chemistry*, *90*(18), 11000–11007. doi:10.1021/acs.analchem.8b02595 PMID:30141325

ADDITIONAL READING

Brookins, D. G. (2012). *Eh-pH diagrams for geochemistry*. Springer Science & Business Media.

Burgot, J. L. (2012). *Ionic equilibria in analytical chemistry.* Springer Science & Business Media. doi:10.1007/978-1-4419-8382-4

Butler, J. N. (1998). *Ionic equilibrium: solubility and pH calculations.* John Wiley & Sons.

Comninellis, C., & Chen, G. (Eds.). (2010). *Electrochemistry for the Environment* (Vol. 2015). Springer. doi:10.1007/978-0-387-68318-8

Duca, G., & Vaseashta, A. (Eds.). (2020). *Handbook of Research on Emerging Developments and Environmental Impacts of Ecological Chemistry.* IGI Global., doi:10.4018/978-1-7998-1241-8

Eliaz, N., & Gileadi, E. (2019). *Physical electrochemistry: fundamentals, techniques, and applications.* John Wiley & Sons.

Letcher, T. M. (Ed.). (2007). *Thermodynamics, solubility and environmental issues.* Elsevier.

Pedeferri, P., & Pedeferri, P. (2018). Pourbaix diagrams. *Corrosion Science and Engineering*, 57-72. doi:10.1007/978-3-319-97625-9_4

Rabenhorst, M. C., Hively, W. D., & James, B. R. (2009). Measurements of soil redox potential. *Soil Science Society of America Journal, 73*(2), 668–674. doi:10.2136ssaj2007.0443

Sahrawat, K. L. (2015). Redox potential and pH as major drivers of fertility in submerged rice soils: A conceptual framework for management. *Communications in Soil Science and Plant Analysis, 46*(13), 1597–1606. doi:10.1080/00103624.2015.1043451

Scholz, F., & Kahlert, H. (2019). *Chemical Equilibria in Analytical Chemistry: The Theory of Acid–Base, Complex, Precipitation and Redox Equilibria.* Springer. doi:10.1007/978-3-030-17180-3

Schüring, J., Schulz, H. D., Fischer, W. R., Böttcher, J., & Duijnisveld, W. H. (Eds.). (2000). *Redox: fundamentals, processes, and applications.* Springer Science & Business Media. doi:10.1007/978-3-662-04080-5

Soustelle, M. (2016). *Ionic and Electrochemical Equilibria* (Vol. 6). John Wiley & Sons. doi:10.1002/9781119178606

Stumm, W., & Morgan, J. J. (2012). *Aquatic chemistry: chemical equilibria and rates in natural waters.* John Wiley & Sons.

Tandon, P. K., & Singh, S. B. (2016). Redox processes in water remediation. *Environmental Chemistry Letters, 14*(1), 15–25. doi:10.100710311-015-0540-4

Vaseashta, A., Duca, G., & Travin, S. (Eds.). (2022). *Handbook of Research on Water Sciences and Society* (Vols. 1–2). IGI Global. doi:10.4018/978-1-7998-7356-3

Chapter 9
Dihydroxyfumaric Acid: A Review of Transformations, New Derivatives, Importance, and Applications

Natalia Bolocan
https://orcid.org/0000-0002-1876-9091
Institute of Chemistry, Moldova State University, Moldova

Gheorghe Duca
Institute of Chemistry, Moldova State University, Moldova

ABSTRACT

This chapter presents a review of the research that was conducted on a natural antioxidant - dihydroxyfumaric acid in the frames of the laboratory of physical and quantum chemistry. The structure and chemistry of dihydroxyfumaric acid are presented in view of its importance in biological cycles, as well as in prebiotic chemistry models. Technological applications are briefly reviewed for wine, food, and tobacco industries. Organic syntheses of new derivatives are presented, and their calculated antioxidant/antiradical activities towards the DPPH and ABTS radicals are discussed. Computational chemistry techniques are employed to gain insights into the isomerism and tautomerism of dihydroxyfumaric acid, and indices of global and local reactivity are calculated for the acid and its new derivatives, in order to theoretically explain the experimental results.

INTRODUCTION

The last decades have been characterized by an increased interest in research related to the discovery of disease prevention possibilities, especially related to minimizing the role of free radicals. Studies concerning the biology and medicine of free radicals are expanding rapidly, because these compounds have an important role in human health and, therefore, in the quality of life (Martemucci et al., 2022).

Recent studies have demonstrated that free radicals perform multiple important biological functions, for example, reactive oxygen species (ROS) and reactive nitrogen species (RNS) operate together with reactive halogen species (RHS) as part of the cellular immune response to infection with microorgan-

DOI: 10.4018/978-1-6684-7198-2.ch009

isms (Ferrari et al., 2011; Shields et al., 2021), and even exhibit signaling functions, activating specific biochemical pathways inside the cell (Shields et al., 2021; Stone & Yang, 2006). For good physiological functioning, a balance between free radicals and their deactivation systems is necessary. If free radicals overwhelm the body's ability to regulate them, a condition known as oxidative stress occurs (Shields et al., 2021; Lobo et al., 2010), correlated with an increased incidence of degenerative diseases such as cancer, cardiovascular disease, cataracts, liver disease, inflammation, renal insufficiency etc. (Bhuiyan et al., 2009; Duca & Sychev, 2002). Thus, at high concentrations, the interaction of these species with lipid membranes, nucleic acids, proteins, enzymes, or other small molecules, inevitably leads to cell damage. A proven way to counteract the action of excessive free radical formation is the use of antioxidants and free radical scavengers of natural and synthetic origin, which destroy the damaging species and preserve the cells' healthy condition.

Another direction significantly influenced by the formation of free radicals, equally important and closely related to human health, is the food industry. Food products containing fats and oils, such as meat, cold cuts, vegetable oils, dairy products and bakery products, are susceptible to oxidation due to their chemical composition (Garg et al., 2022). The organoleptic changes occur during long-term storage of food products, especially due to the auto-oxidation process of unsaturated fatty acids, which are more vulnerable to rancidity due to their lower stability compared to saturated fatty acids.

A widely used method to prevent oxidation is to enrich foods with antioxidants during processing and preservation. The mechanisms of action of antioxidants are different, including control of pro-oxidants, inactivation of free radicals and control of oxidation-prone substrates (unsaturated fatty acids and oxygen). However, the industry mostly uses synthetic additives, which are viewed with caution or even negatively by modern consumers, who prefer natural antioxidants and foods without synthetic additives (Poljsak et al., 2021; Pokorny & Smidt, 2001; Bensid et al., 2020).

A large number of organic acids and their salts are already approved for use in the food industry, for example malic (E296), tartaric (E334), ascorbic (E300), lactic (270), succinic (E363) acids (Ben Braiek & Smaoui, 2021). Lactic, citric, and tartaric acids are also widely used in the pharmaceutical and cosmetic industry (Goyal & Jerold, 2021).

Continuous efforts are being made persistently to find new potential compounds with potential antioxidant/antiradical properties with the ultimate goal of enhancing the quality of human life. In this context, the research related to dihydroxyfumaric acid (DHF) and obtaining new derivatives from it, is a promising direction with wide applicability.

DIHYDROXYFUMARIC ACID: WHAT IT IS AND WHY IT IS IMPORTANT

The traditional name of dihydroxyfumaric acid is (2E)-2,3-dihydroxybut-2-enedioic acid according to IUPAC (Favre & Powell, 2013), and it is a naturally occurring antioxidant compound, obtained from tartaric acid by slow dehydrogenation or oxidation processes. The name *dihydroxyfumaric acid* is given to the *trans-* isomer, while the *cis-* isomer is called *dihydroxymaleic acid.* DHF exists in solution in two tautomeric forms in dynamic equilibrium (Fig.1.), and the pH of the solution greatly influences the concentration of the tautomeric forms (Souchay et al., 1967; Secara et al., 2011).

The specialized literature mentions an important disadvantage in the experimental work with the use of DHF, namely the instability of this acid in solution (Fig.1.), due to oxidative transformations leading to the formation of diketosuccinic acid (Wieland & Franke, 1928; Chow & Vennesland, 1958), and decarboxylation with the formation of glycolaldehyde. The reduced solubility of the acid and its Na^+, K^+, and NH_4^+ salts in water are another drawback limiting its applications (Franke & Brathun, 1931; Hay & Harvie, 1965; Secara, 2008).

Figure 1. The main degradation pathways of DHF in solution

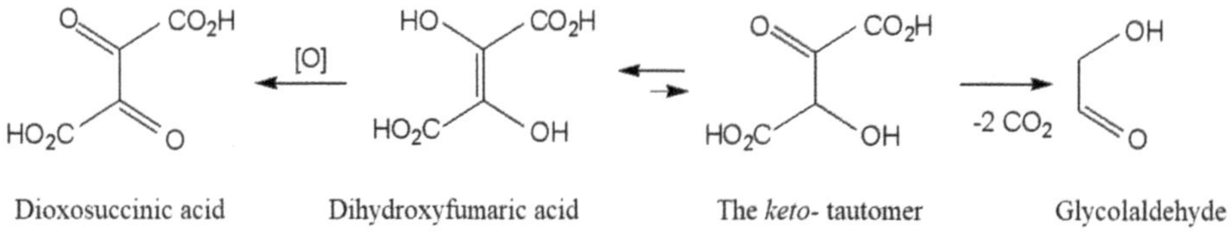

It was shown (Muller et al., 1990; Sagi et al., 2011) that the physical and chemical properties of DHF vary significantly depending on the species, so that DHF possesses very little pronounced nucleophilic character, while the ionization of the carboxylic groups in a basic environment leads to the deprotonation of the acid and the formation of the dicarboxylic ion (DHF^{2-}) with a very pronounced nucleophilic character (Sagi et al., 2011; Eschenmoser, 2007).

Similarly to other hydroxy acids, such as malic, lactic, glycolic, citric, tartaric acids, dihydroxyfumaric acid plays an important role in living nature, being a known intermediate in the cycles of di- and tricarboxylic acids. It is well known that di- and tricarboxylic organic acids play an important role in the metabolism of plants and animals. Being products of carbohydrate transformation, they participate in the biosynthesis of alkaloids, glycosides, amino acids, and other biologically active compounds.

The first evidence of the biological importance of DHF appeared in 1915 when it was observed that DHF was fermented by yeast (Neuberg & Schwenk, 1915). In 1938, an enzyme was extracted from plants, called *dihydroxyfumaric acid oxidase*, the oxidation product of which was later shown to be the diketosuccinic acid (Banga & Szent-Georgyi, 1928; Banga & Philippot, 1939). In 1940 the enzyme peroxidase was discovered in some plants, which oxidized dihydroxyfumaric acid with a strong absorption of oxygen (Swedin & Theorell, 1940).

Research carried out in the 1980s demonstrated that the active centers of *dihydroxyfumarate oxidase* and *peroxidase* are coordinating compounds of copper and iron (Duca, 1979). As previously mentioned, these enzymes catalyze the transformation of DHF into diketosuccinic acid. Taking into account the structural similarities with ascorbic acid, it has been suggested that in biological oxidation DHF can play a role similar to that of ascorbate, namely as an intermediate transporter of hydrogen from substrates to oxygen. Therefore, it has been suggested that in the oxygen + DHF-oxidase system (Fig. 2), the role of DHF is similar to that of ascorbic acid in the ascorbate-oxidase system, leading to the formation of the corresponding α-diketone.

Figure 2. The transformation of DHF into diketosuccinic acid under the action of DHF-oxidase

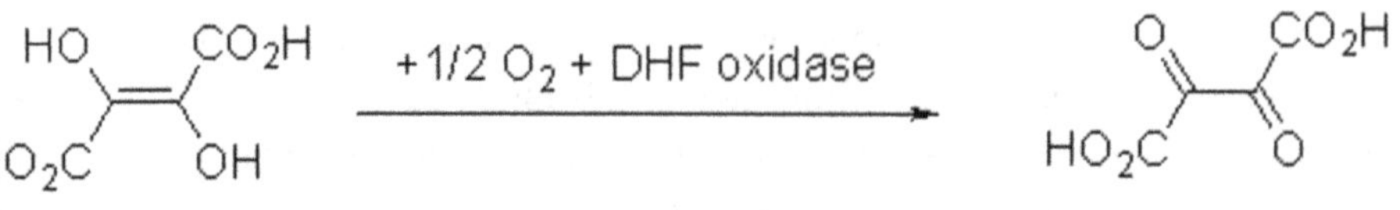

The oxidizing function of peroxidase was later demonstrated against other compounds, such as glutathione, hydro- and naphthoquinone, fluoroglycine and others (Duca, 1979), and a necessary condition for the reaction to proceed was shown to be the presence of cofactors - manganese ions and different phenolic compounds. More information regarding certain enzymatic reactions of dihydroxyfumaric and diketosuccinic acids in plant tissues was obtained in 1954 (Stafford et al., 1954).

The role of DHF in animal metabolism was highlighted for the first time by Stepanow and Kusinin (1934), when the increase in muscle glycogen content after incubation with DHF was noticed. Later, a sequence of enzymatic reactions was discovered, which proved to be a pathway of gluconeogenesis based on the high probability of formation of a pentose (or a pentose phosphate) upon the addition of DHF and glyceraldehydes (or fructose-1,6 -diphosphate and aldolase as a source of glyceraldehyde-3-phosphate), in the rabbit muscle extract. The reaction sequence leading from DHF to 3-ketopentose is shown in Figure 3 (Fukunaga, 1960).

Figure 3. DHF transformations leading to 3-ketopentose formation

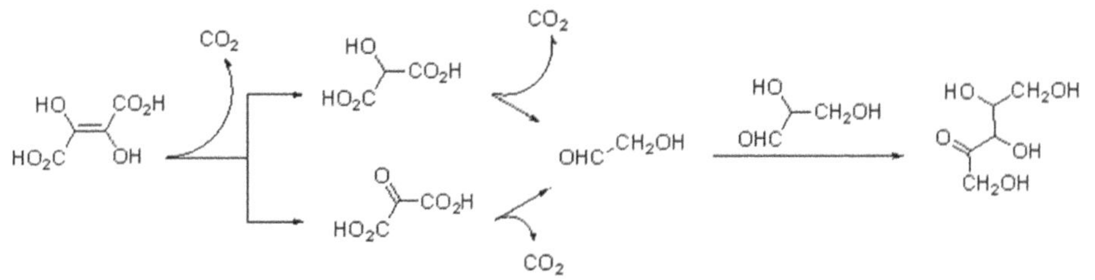

The discovery of this sequence is in line with the research on the origin of life on Earth carried out in the frames of prebiotic chemistry which gained popularity in recent decades. The so-called "glyoxalic scenario" postulated by Eschenmoser suggests that glyoxal and its dimer, DHF, are the central molecules from which the synthesis of biogenic molecules began, giving rise to organic life (Eschenmoser, 2007a,b; Morowitz et al., 2000). According to this hypothesis, carbohydrates, amino acids, nucleotide bases and other compounds that participate in the citric acid cycle could have been formed not only by the classical reaction starting from formic aldehyde (Sagi et al., 2012), but also by the reactions of "aqua-oligomers of carbon monoxide carbon, such as glyoxal and dihydroxyfumaric acid".

Thus, the interest in DHF has increased in recent years as researchers attempted to examine the glyoxalic scenario in more detail. Some mechanistic sequences of the scenario have already been confirmed, for example, reactions in aqueous solutions between dihydroxyfumarate and several short-chain aldehydes showed that the reaction products (dihydroxyacetone, tetrulose and two pentulose) were formed with practically complete quantitative conversion, aldoses being undetectable, which confirmed that reactions between DHF and aldehydes could constitute a pathway for carbohydrate formation (Hay & Harvie, 1965; Rohfling, 1967). The self-condensation of DHF with the formation of pentulosonic acid, a carbohydrate acid that can undergo decarboxylation with the formation of erythrulose, was also examined (Butch et al., 2013). Another process of interest in the context of the glyoxalic scenario is the bis-decarboxylation of DHF in aqueous solutions, with the formation of glycolic aldehyde that can further form higher carbon chain aldoses (Butch et al., 2016).

Although the mentioned above research may seem hypothetical to some degree, it should be mentioned here that DHF is interesting enough from the practical viewpoint, especially for the pharmaceutical and food industry, among others.

DHF is part of the metabolic cycle during grape ripening and is of significant importance in oenology. It is well known that organic acids contribute to the formation of acidity in wines - one of the most important wine characteristics. DHF can occur in amounts of 1 mg/L as a result of the oxidation of tartaric acid, and despite its low content, DHF plays an important role in the complex redox processes that occur in wine (Rodopulo, 1983).

Previous research showed that if air access to wine materials is limited during processing and maturation, and the wine absorbs no more than 3-5 mg/L of oxygen per year, then tartaric acid is oxidized to DHF causing a low redox potential of the wine. The reductone reverses oxidized compounds to their initial state, which contributes to the formation and development of the characteristic taste and aroma of table wines (Rodopulo, 1983; Xu et al., 2021). However, DHF decomposes easily under aerobic conditions, and for this reason the determination of DHF in wines is quite difficult.

During grapes processing, a series of complex redox processes occur, which predetermine the quality of the finished product. To influence these processes, industry uses various preservatives with antioxidant and antiseptic properties (Gonta et al., 1991; Ferrer-Gallego et al., 2018; Jagatic Korenica et al., 2020). In order to increase the stability of wines and improve their hygienic properties, the authors (Gaina et al., 1995) carried out a series of experiments using DHF as a preservative of grape material of the variety "Aligote". The most important polyphenols of this variety are the catechins: (-) catechin, (-) epicatechin, (+) catechin, (-) epigallocatechin, (+)-gallocatechin, epicatechinalate, etc. (Sychev & Duca, 1985). DHF has stronger reducing properties than catechin (Gaina et al., 1995), for this reason DHF in grape drinks is oxidized to diketosuccinic acid following the interaction with oxidized forms of polyphenols (Sychev & Duca, 1983). Further, decarboxylation of diketosuccinic acid occurs (Duca et al., 1987), while natural polyphenols are reduced to their original form. This process reduces the possibility of formation of colloidal particles in the wine, and DHF can be used in certain amounts as a preservative. Wine prepared with the addition of DHF possesses superior taste qualities compared to wine produced using conventional technology, as well as increased stability (Gaina et al., 1995).

Another research (Sychev et al., 1985) was performed on "Onitskansky Belyi" grape material under micro-vinification conditions according to the generally accepted technology, with the use of DHF instead of SO_2 as a preservative. The wine material was stored for 12 months in bottles after their removal from the yeast containing DHF, in the amount of 110-130 mg/L (optimal dose of DHF), and results showed that its organoleptic indicators were comparable to those produced with the use of SO_2, having superior taste qualities and increased stability. According to the obtained results, the content of the components that determine the quality of the product depends on both the amount of DHF and the storage time.

Additionally, the usefulness of DHF was revealed during experiments on using this acid for the inhibition of *N*-nitrosamines formation in model systems, *in vitro* and *in vivo*. Highly carcinogenic compounds, N-nitrosamines are aliphatic or aromatic derivatives of secondary amines, which have a nitroso (–NO) group attached to nitrogen. They are formed by the reaction of nitrite with secondary amines during the fermentation of various foods and cigarette tobacco, as well as by the *in vivo* reaction of nitrites (used as food preservatives) with amines under the acidic conditions of the gastrointestinal tract. It should be noted that nitrates can also contribute to the formation of *N*-nitrosamines through reduction to nitrites by saliva or gastrointestinal tract enzymes. Important sources of N-nitrosamines are prepared meat foods, beer, cheeses, prepared fish, dried fish, powdered milk and cosmetic products.

Because of the toxicity of nitrosamines, manufacturers try to limit their formation in food products by adding antioxidants/antiradical substances. Previously (Gonta, 2009) the antioxidant properties of DHF were studied in the process of inhibiting the nitrosation of secondary aliphatic amines and amides. It was demonstrated that under the action of DHF, the content of nitrosating agents decreased, which led to the reduction of the content of *N*-nitrosamines that can form in the presence of amines. The inhibitory activity of DHF and the acidic sodium salt of DHF was investigated in the process of nitrosation of metabolites formed in gastric juice (model and real system). It was found that during the proteolysis of albumin and casein by proteolytic enzymes, these reagents inhibit the formation of *N*-nitroso compounds without having any influence on the content of the non-hydrolyzed protein substrate, on the concentration of amino acids formed or on the activity of specific enzymes.

Moreover, research showed (Gonta, 2007) that the acidic sodium salt of DHF contributes to reducing the rate of methemoglobin formation in the oxidation reaction of hemoglobin with nitrites, influencing the branching process of the oxidation chain of HbO2, and reducing the rate of the auto-acceleration process in the reaction medium.

Research on the endogenous formation of N-nitrosonornicotine (one of the strongest carcinogenic compounds in unburnt tobacco and cigarette smoke) in rats demonstrated the efficiency of inhibition (86%) of the process by DHF (Porubin et al., 2009, 2007). Likewise, it was found that the use of DHF and tartaric acid in the diet decreases the content of N-nitrosonornicotine, which is formed during the nitrosation of nornicotine with nitrite ions *in vivo*. DHF was successfully used in a model system study as an inhibitor of the nitrosation process of the amine hydrochlorothiazide, which is used as a drug with high diuretic and hypotensive activity.

These results demonstrate the importance and application potential of the research direction related to DHF, especially considering the potential impact in ensuring the health of the population, including through the safety of medicines and food.

NEW DERIVATIVES OF DHF AND THEIR ANTIOXIDANT/ ANTIRADICAL PROPERTIES

It was discussed previously that DHF has low stability in solution due to oxidation, decarboxylation, and self-condensation processes, therefore an important objective concerns the synthesis of new derivatives of DHF, with the aim of obtaining molecular structures that stabilize the enediolic fragment, at the same time maintaining the antioxidant activity.

Due to the presence of carboxyl-, hydroxyl- and olefin fragments in its structure, DHF can be used to create porphyrinic or pyrimidine macrocycles (Bellec et al., 2000; Goldberg et al., 1998). However, despite the obvious progress in this direction, the chemistry of DHF remains a poorly studied field in many aspects. Among the papers related to the modification of DHF, there are limited data on the synthesis of amine derivatives, including those with a heterocyclic fragment (Secara, 2011; Dreher, 2005). The creation of new compounds based on DHF, the search for new synthons for directed synthesis, and the study of the “structure-property” relationship is a current task of great practical interest.

Nitrogen-containing derivatives of DHF can open new possibilities in organic synthesis: the selection of suitable combinations of substituents will allow the adaptation of polarity in broad limits, solvation capacity, and catalytic properties and, therefore, can influence the depth and selectivity of the reaction.

In this order of ideas, Figure 4 shows some of the new derivatives obtained previously by the authors (Secara, 2011), preserving the numbering from the initial publication.

Figure 4. Derivatives of DHF: dimethyl 2,3-dihydroxyfumarate (2), (E)-methyl 2,3-dihydroxy-4-oxo-4-(phenylamino)but-2-enoate (5), 2,3-dihydroxy-N^1,N^4-di(pyridin-2-yl)fumaramide (6), (E)-3-(1H-benzo[d]imidazol-2-yl)-2,3-dihydroxyacrylic acid (12), (E)-1,2-di(1H-benzo[d]imidazol-2-yl)ethene-1,2-diol (13)

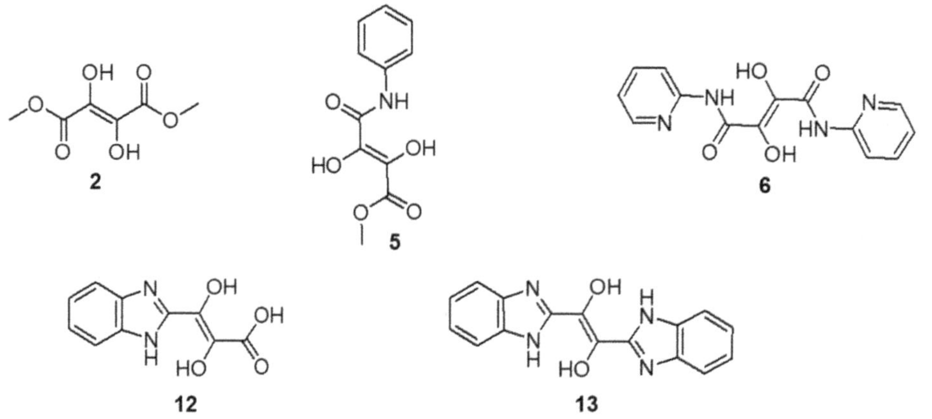

The new derivatives of DHF as presented in Fig. 4., proved good antioxidant activity in the DPPH test. The kinetic curves of DPPH• consumption upon interaction with the studied derivatives, as well as ascorbic acid, as a reference compound, demonstrate the existence of three types of kinetic behavior: fast, mixed (the fast stage is followed by a slower stage, which can be visually observed) and slow kinetic behavior. The kinetic behavior is determined both by the molecular structure of the studied compound and by the stability of the antioxidant radical formed following the reaction with DPPH•.

Ascorbic acid (AAs) was the only compound that showed fast kinetic behavior, and the reaction reached equilibrium within 3 seconds, as demonstrated by stop-flow experiments. Some of the studied compounds demonstrated a mixed kinetic behavior, composed of a fast stage, similar to the behavior of ascorbic acid, in which the concentration of the radical decreases sharply, followed by a slower stage. Such compounds participate in a rapid transfer of hydrogen atoms, and the intermediate formed participates in a slow transfer of hydrogen. Analyzing the obtained kinetic curves, it was fond that DHF, ester 2, MBD acid 12 and BBD diol 13 presented a mixed kinetic behavior. Upon their interaction with DPPH•, first, a rapid decrease in the concentration of the radical was observed (a few seconds for DHF, MBD 12, and up to 1 minute for EMD ester 2, and BBD diol 13), but the equilibrium state was reached in a longer time, compared to AAs (Table 1).

The group of compounds with a slow kinetic behavior included those compounds, which had the property of participating only in a slow hydrogen transfer, and the equilibrium state was reached in a longer time, compared to DHF, depending on the initial concentration of the compound. This group includes anilide 5 and fumaramide 6 (Figure 4).

To determine the antioxidant/antiradical activity of the studied compounds, a series of molar ratios were tested, expressed as moles of antioxidant per mole of DPPH•, in the range from 0.1 to 1. For each molar ratio, the remaining concentration of DPPH• at equilibrium was determined and plotted, allowing subsequent determination from the graph of the molar ratio that reduces half the initial $DPPH^•$ concentration (EC_{50}). For simplicity, the value $1/EC_{50}$, called the antiradical power (PAR), is often used. Thus, the higher the PAR, the more effective the antioxidant.

The stoichiometric constant was defined as the number of molecules of DPPH• reduced by one antioxidant molecule during a complete reaction, and denoted by $\sigma = 1/(2 \times EC_{50})$, according to the reaction:

σ $DPPH^•$ + A-H ® Products (1)

Table 1 presents PAR, EC_{50}, the stoichiometric constant of the tested antioxidants, and the reaction time required to reach equilibrium. Ascorbic and dihydroxyfumaric acids, as well as MBD 12 possess the highest stoichiometric constant, each mole of these compounds cleaves 2 moles of $DPPH^•$ radical. The reaction between the radical and AAs proceeds for several seconds, in the case of the reaction with DHF - up to 2 minutes, and in the case of MBD 12, the reaction reaches equilibrium in up to 10 minutes (Bolocan & Duca, 2022; Secara et al., 2010).

Table 1. Classification of investigated compounds according to the kinetic behavior in the $DPPH^•$ test, and the corresponding values of PAR, EC_{50}, the stoichiometric constant σ, and time to equilibrium (T = 295 K)

	Compound	EC_{50}	PAR	σ	Time to equilibrium
Fast kinetics	AAs	0,25	4	2	< 10-15 s
Mixed kinetics	DHF **1**	0,25	4	2	< 2 min
	Acid MBD **12**	0,25	4	2	< 10 min
	Diol BBD **13**	0,5	2	1	< 10 min
	EMD **2**	0,5	2	1	≈ 20-30 min
Slow kinetics	Anilide **5**	0,7	1.4	0,7	≈ 45-60 min
	Fumaramide **6**	2	0.5	0,25	≈ 45-60 min

Computational Studies of DHF

Isomerism and Tautomerism Studies

Understanding obtained experimental results and predicting chemical reactivity is essential for the synthesis of new derivatives with antioxidant/antiradical properties. Computational methods that allow molecular modeling have extensive applicability in the field of chemical, physical, biological systems, as they are able to predict not only the properties of substances in the reaction process, but are also very useful for the study of drug transporters, as well as in the interactions between these biological substances and molecules.

Despite the recent interest in DHF and its potential for practical utility in various fields, to date there have been no published theoretical works on the conformational isomerism of DHF. Although there is evidence that keto-enolic tautomerism is one of the most studied forms of prototropy (Raczynska et al., 2005), studies of the DHF molecule are very scarce: the keto-enolic tautomerization of DHF acid was studied experimentally by Fleury and Souchay in the 1960s (Souchay et al., 1967), by Sazou et al. (1984), during investigations into the decarboxylation of DHF. In this context, the present paper aims to provide valuable results for further use in both theoretical and practical fields.

In the case of DHF, various conformations are possible for the two tautomeric forms, due to the intramolecular rotations along the single C–C bonds and due to the possible *syn*– and *anti*– periplanar orientation of the hydroxyl hydrogen with respect to the keto oxygen.

The authors showed (Bolocan & Duca, 2021) that there are 22 enediol and 23 keto optimized structures of DHF at the B3LYP/6–311++G(2df,2p) level of theory. The isomers were nominated and arranged from the most stable isomer to the least stable one (according to Gibbs free energy calculations). The enediol isomers were nominated as E1–E22, and the keto forms as K1–K23. The electronic energy (*E*), enthalpy (*H*) and Gibbs energy (*G*) of each conformer were calculated at 298.15 K. Results show that the enediol forms of DHF are more stable than the keto forms, both in gas phase and in solvent (water).

In the gas phase, the enediol structure E1 of DHF is the most stable form and represents the global minimum in the potential energy curve of DHF acid. Its Gibbs free energy is lower than that of the most stable keto structure K1 by 46.7 kJ mol^{-1} at the DFT(B3LYP) level. The stability of E1 may be attributed to the presence of the intramolecular hydrogen bonding effect present in the E1 structure more than in others, and to stabilizing orbital interactions due to the *anti*– periplanar orientation of the hydroxyl hydrogen with respect to the keto oxygen.

In water, the most stable isomer is E2. Its Gibbs free energy is lower than that of the most stable keto structure K1 by 32.7 kJ mol^{-1} at the DFT(B3LYP) level.

The relative abundance (RA) of the isomers was calculated according to the Boltzmann distribution equation (Eq.2), using the ZPE corrected values of internal energy:

$RA = exp\ (-\Delta E/RT)$ (2),

where $\Delta E = \Delta E_{el} + \Delta ZPE$.

The Boltzmann distribution shows that in gas phase, 99.78% of DHF are represented by only 3 enediol structures: the most stable isomer E1 accounts for 87.4%, followed by isomer E2 with a relative abundance of 10.98%, then isomer E3 - 1.4%. In aqueous solution, the three most stable enediol forms account for 97.3% of the acid, with the following relative abundance indices: the most stable isomer E2 accounts for 38.4%, followed by E1 - 31.8% and E3 - 27.1% (Fig.5) (Bolocan & Duca, 2021).

Figure 5. Structures of the three most stable enediol isomers of DHF

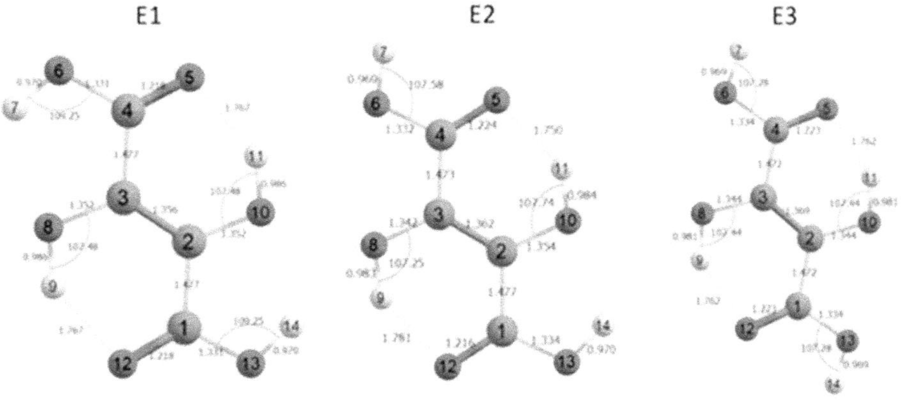

In order to investigate the isomerism and tautomerism of the three most stable enediol structures, a set of thermodynamic and kinetic parameters has been calculated. The computations were performed with the neutral forms of DHF and are therefore applicable to acidic solutions.

The rate constants of the reactions have been calculated using the canonical transition state theory, and Eyring's equation:

$$k = \frac{k_B T}{h} e^{-\Delta G^{\#}/RT} \quad (3),$$

where $\Delta G^{\#}$ is the Gibbs free energy of activation, k_B is the Boltzmann constant, and h is the Planck constant. Thus, the rate constant of the direct reaction (k_1) was calculated using the value $\Delta G_1^{\#} = G_{TS} - G_{reagent}$, and the rate constant of the inverse reaction (k_2) was calculated using the value $\Delta G_2^{\#} = G_{TS} - G_{product.}$

The equilibrium constant K_{eq} was calculated using the equation:

$K_{eq} = exp\ (-\Delta G/RT)$ (4),

where $\Delta G = G^{products} - G^{reagents}$ and the individual thermodynamic parameters are $G = H\text{-}TS$, $H = E_{el} + E_{vib} + E_{rot} + E_{trans} + ZPE + k_B T$ and $S = S_{el} + S_{vib} + S_{rot} + S_{trans}$. All calculations have been performed at 298.15K and 1.0 atm.

Table 2 summarizes the kinetic and thermodynamic data for the transition states of the isomerization and tautomerization reactions of the three most stable enediolic forms, in the gas phase, and in water. In the gas phase, E1 is the most stable species, and the isomerization reactions E1 « E2 and E2 « E3 are characterized by equilibrium constants of 0.13. In water, the most stable isomer is E2, and the isomerization reactions E1 « E2 and E2 « E3 are characterized by equilibrium constants of 1.21 and 0.61, respectively. In water, ΔE, ΔH, ΔG, $\Delta G_1^{\#}$ and $\Delta G_2^{\#}$ are lower by aprox. 2-7 kJ·mol^{-1} for isomerization reactions, and by 10-30 kJ·mol^{-1} for the tautomerization reactions. Thus, in gas phase, the equilibrium constants for the keto-enol tautomeric processes E1 « K5, E2 « K7, E2 « K14 and E3 « K13 are equal to $2.46\cdot10^{-10}$, $10.29\cdot10^{-10}$, $0.61\cdot10^{-10}$ and $5.21\cdot10^{-10}$, respectively. In water, these values are $1.41\cdot10^{-7}$, $2.73\cdot10^{-7}$, $5.61\cdot10^{-7}$,

$3.93 \cdot 10^{-7}$, respectively, demonstrating a significant increase of the direct reaction rates. Of course, we have to bear in mind that current calculations have been performed in an implicit water environment (SMD) and further research is necessary, such as adding explicit water molecules to the calculations, because in water the proton transfers in the tautomerization reactions are most likely to be performed with the aid of water molecules that bridge the proton-donating and the proton-accepting atoms in DHF.

Table 2. Kinetic and thermodynamic parameters for the transition states in the isomerization and tautomerization processes of the three most stable isomers of DHF, in vacuo and in water (SMD model). Energies are presented in kJ mol^{-1} and the rate constants in s^{-1}

	Isomerism				**Keto-enol tautomerism**							
	E1 ↔ E2		**E2 ↔ E3**		**E1 ↔ K5**		**E2 ↔ K7**		**E2 ↔ K14**		**E3 ↔ K13**	
	Gas	**Water**	**Gas**	**Water**	**Gas**	**Water**	**Gas**	**Water**	**Gas**	**Water**	**Gas**	**Water**
ΔE	5.18	-0.59	5.24	-0.01	60.67	35.61	57.66	39.97	65.28	35.84	61.54	39.78
ΔH	5.26	-0.53	5.27	0.58	59.55	40.44	56.50	42.57	63.70	35.38	59.77	42.12
ΔG	5.07	-0.48	4.98	1.24	54.84	39.07	51.30	37.42	58.29	35.64	52.99	36.52
Keq	0.13	1.21	0.13	0.61	$2.46 \cdot 10^{-10}$	$1.41 \cdot 10^{-7}$	$10.29 \cdot 10^{-10}$	$2.73 \cdot 10^{-7}$	$0.61 \cdot 10^{-10}$	$5.61 \cdot 10^{-7}$	$5.21 \cdot 10^{-10}$	$3.93 \cdot 10^{-7}$
$\Delta G_1^{\#}$	47.50	39.90	47.01	40.19	275.13	261.20	272.36	260.72	289.43	270.23	287.50	267.22
$\Delta G_2^{\#}$	42.43	40.38	42.03	38.96	220.29	222.13	220.99	223.29	231.14	234.59	234.51	230.70
k_1	$2.96 \cdot 10^{4}$	$6.29 \cdot 10^{5}$	$3.61 \cdot 10^{4}$	$5.59 \cdot 10^{5}$	$3.91 \cdot 10^{-36}$	$1.01 \cdot 10^{-33}$	$1.19 \cdot 10^{-35}$	$1.23 \cdot 10^{-33}$	$1.22 \cdot 10^{-38}$	$2.66 \cdot 10^{-35}$	$2.66 \cdot 10^{-38}$	$8.96 \cdot 10^{-35}$
k_2	$2.29 \cdot 10^{5}$	$5.18 \cdot 10^{5}$	$2.69 \cdot 10^{5}$	$9.19 \cdot 10^{5}$	$1.59 \cdot 10^{-26}$	$71.85 \cdot 10^{-26}$	$1.19 \cdot 10^{-26}$	$4.49 \cdot 10^{-27}$	$1.99 \cdot 10^{-28}$	$4.70 \cdot 10^{-29}$	$5.11 \cdot 10^{-29}$	$2.26 \cdot 10^{-28}$

Local Reactivity Descriptors

DFT allows the computation of local descriptors, which assign a value to every point in space, providing insight into the reactivity of the molecule at point r in space; they are regioselectivity indicators.

In the frames of DFT, when the chemical reactions are due to mainly covalent interactions, a second order local reactivity descriptor is used, the Fukui function, which allows the identification of the most reactive sites of a molecule, those prone to electrophilic, nucleophilic or radical attacks.

Depending on total electronic densities, the Fukui functions can be defined in two ways, based on the right and left derivatives of $\rho(\vec{r})$ with respect to N (eq. 5 and 6), and the third definition gives the radical attack descriptor (eq. 6), in line with Perdew et al.'s (1996) conclusions:

$$f^{+}(\vec{r})=\left(\frac{\partial\rho(\vec{r})}{\partial N}\right)^{+}_{v(\vec{r})}=\rho_{N+1}(\vec{r})-\rho_{N}(\vec{r}) \tag{5}$$

$$f^{-}(\vec{r})=\left(\frac{\partial\rho(\vec{r})}{\partial N}\right)^{-}_{v(\vec{r})}=\rho_{N}(\vec{r})-\rho_{N-1}(\vec{r}) \tag{6}$$

$$f^{0}(\vec{r})=\frac{1}{2}[\rho_{N+1}(\vec{r})-\rho_{N-1}(\vec{r})] \tag{7}$$

where $\rho(\vec{r})$ is the atomic charge at the rth atomic site, for the neutral (N), anionic ($N+1$) and cationic ($N-1$) chemical species.

The electrophilic descriptor $f^{+}(\vec{r})$ measures the intramolecular reactivity at the site r toward a nucleophilic reagent, and the highest values of $f^{+}(\vec{r})$ will imply the preferred site for a nucleophilic attack (by a nucleophile molecule). The nucleophilic Fukui function $f^{-}(\vec{r})$ measures the intramolecular reactivity at the site r toward an electrophilic reagent, and the highest value indicates the preferred site for an electrophilic attack (Parr & Yang, 1984).

Although the Fukui function is very useful in revealing nucleophilic and electrophilic regions on a molecule, a more accurate local reactivity descriptor has been proposed by Morell et al. (2005), i.e. the dual descriptor, expressed as the derivative of the Fukui function with respect to the electron number:

$$f^{(2)}(\vec{r})=\left(\frac{\partial f(\vec{r})}{\partial N}\right)_{v(\vec{r})}\cong f^{+}(\vec{r})-f^{-}(\vec{r}) \tag{8}$$

Thus, the dual descriptor may be approximated to the difference between the Fukui functions for nucleophilic and electrophilic attacks, and therefore it aids in revealing the true preferred sites for electrophilic and nucleophilic attacks simultaneously. The dual descriptor is especially useful in ambiguous situations, when no exact information can be derived from $f^{+}(\vec{r})$ and $f^{-}(\vec{r})$, because it is a combination of the two. The dual descriptor is positive for locations susceptible to nucleophilic attacks, and negative where an electrophilic attack is more probable.

In terms of densities of frontier molecular orbitals (frontier molecular orbital approximation, FMOA), eq. (8) turns into eq. (9) as follows:

$$f^{(2)}(\vec{r})=\rho_{LUMO}(\vec{r})-\rho_{HOMO}(\vec{r}) \tag{9}$$

In this form, this expression implies that the local reactivity of the molecule depends only on the electronic densities of frontier molecular orbitals. Thus, the electronic densities of the remaining occupied molecular densities are not involved, and the relaxation effects are excluded, making the dual descriptor a more accurate local reactivity descriptor than the $f^{+}(\vec{r})$ and $f^{-}(\vec{r})$ (Morell et al., 2005).

Recently, Luis R. Domingo introduced other local reactivity descriptors, within his concept of Molecular Electron Density Theory (MEDT), i.e. the Parr functions (Domingo et al., 2003) that can be considered as an alternative to the Fukui functions for describing the areas within the molecules where nucleophilic or electrophilic attacks will be favored. The Parr functions can be expressed by the following equations:

$$P^{-}(\vec{r}) = \rho_s^{rc}(\vec{r}) \tag{10}$$

$$P^{+}(\vec{r}) = \rho_s^{ra}(\vec{r}) \tag{11}$$

where $P^{-}(\vec{r})$ and $P^{+}(\vec{r})$ are the nucleophilic and electrophilic Parr functions, respectively, and $\rho_s^{rc}(\vec{r})$ and $\rho_s^{ra}(\vec{r})$ are related to the atomic spin density of the radical cation and radical anion, respectively (Chattaraj et al., 2009, 2002).

It should be noted, however, that Fukui and Parr functions may give poor and even contradictory results for systems that present hyperconjugation, and this effect is stronger in the charged species than in the neutral species.

In order to overcome this possible shortcoming, the reactivity difference descriptor R_k, proposed by Chattaraj et al. (2009) was also employed in this research, allowing to identify the centers with electrophilic (R_k = + n.nn) or nucleophilic (R_k = − n.nn) behavior as well as the ambiphilic (R_k = ± n.nn) behavior, and to eliminate the centers with marginal reactivity [68]:

(a) if ($1 < w_k/N_k < 2$) or ($1 < N_k/w_k < 2$)
 then R_k » $(w_k + N_k)/2$ ® ambiphilic (R_k = ± n.nn)
(b) else R_k » $(w_k - N_k)$
 for $R_k > 0$ ® electrophilic (R_k = + n.nn)
 for $R_k < 0$ ® nucleophilic (R_k = − n.nn)
(c) if $| R_k | < 0.1$, then $R_k = 0$.

The sign (+, −, ±) in the R_k index indicates the electrophilic or/and nucleophilic character of the center k, while the magnitude n.nn provides a measure of the local activation. The local electrophilicity (ωk_j and local nucleophilicity (Nk_j, at the atomic site k are defined in terms of the related condensed Fukui functions, f_k^{+} and f_k^{-}, according to the following:

$$\omega_k = \omega \bullet f_k^{+} \tag{12}$$

and

$$N_k = N \bullet f_k^{-} \tag{13}$$

where N is the nucleophilicity defined relative to tetracyanoethylene (TCE) (Chattaraj et al., 2002). TCE is used as a reference because it presents the lowest HOMO energy in a large series of molecules already investigated, allowing the creation of a normalized scale of positive values:

$$N{=}E_{HOMO(Nu)}(eV) - E_{HOMO(TCE)}(eV) \tag{16}$$

The Fukui functions condensed at the atomic site k, for the nucleophilic attack (f_k^+) and for the electrophilic attack (f_k^-) are given in terms of the respective population q_k (N) of the N− electronic system of k atomic site, as the following:

$$f_k^+ = q_k(N+1) - q_k(N) \tag{14}$$

and

$$f_k^- = q_k(N) - q_k(N-1) \tag{15}$$

Another useful predictor of chemical reactivity, employed in this research for verification purposes, is the electrostatic potential, which shows negative potential regions, expected to be sites of protonation and nucleophilic attack, and positive potential regions, which may indicate electrophilic sites. This is an important predictive tool, since electrostatic attraction/repulsion is long−ranged compared to the electron−transfer effects characterized by the dual descriptor (Cardenas et al., 2009). A classic molecular electrostatic potential (MEP) map indicates the red regions of negative electrostatic potential, the blue sites represent the regions of positive electrostatic potential and the parts of green color represent the regions of zero potential.

The values for f_k^+, f_k^- and f_k^0 of DHF are negative in all investigated media. Basically this means that in the cationic state, the electron density increases and in the anionic state, the electron density decreases on all atoms. Ayres et al., has previously suggested the relation of this behavior with the reduction and oxidation of atomic centers in the molecule (Ayers et al., 2002).

Thus, in the case of DHF no conclusions can be accurately obtained from the electrophilic and nucleophilic Fukui function alone. Instead, the sign of the dual descriptor can describe unambiguously nucleophilic and electrophilic sites within the molecule, as previously mentioned. Specifically, if $f^{(2)}(\vec{r})$ > 0, then the site is favored for a nucleophilic attack, whereas if $f^{(2)}(\vec{r})$ < 0, then the site may be favored for an electrophilic attack (Morell et al., 2005, 2006). The dual descriptor $f^{(2)}(\vec{r})$, and the electrophilic and nucleophilic Parr functions over the atoms of DHF are shown in Table 3. The actual values have been multiplied by 100 for an easier comparison. One of the main criticisms regarding the calculation of Fukui functions is the arbitrariness in the way of choosing the charges (Fuentealba, 2010). In this research, the presented Fukui functions were calculated using Hirshfeld charges (Liu et al., 2014).

For DHF, the data presented in Table 3 shows that in the gas phase, water and methanol, according to the value of the dual descriptor, the preferred sites for the nucleophilic attack are the O8 and O10 atomic sites, while the most susceptible sites for an electrophilic attack are C1 and C4.

The calculated Parr functions indicate that the C1 and C4 sites will be preferred for a nucleophilic attack in all media, while C2 and C3 and O8 and O10 will be the most prone sites for an electrophilic attack for gas; O8 and O10 for water and methanol.

The results for O8 and O10 may seem controversial at first sight, but this may be explained by the fact that dihydroxyfumaric acid is a hyperconjugated system, stabilized by the interactions between the

C=C and C=O double bonds, and the lone pairs of oxygen electrons, and therefore some of the groups may behave ambiphilic.

The MEP map of DFH_4 (Fig. 6) confirms that the region over C1 and C4 is positively charged, so an electrophile would be repelled, favoring these sites for a nucleophilic attack, as suggested by the Parr function, while the electrostatic potential over O8, O10 is very close to neutral, suggesting that these sites could, indeed, be approached by (neutral or positive) electrophiles and (neutral or negative) nucleophiles.

Table 3. The condensed dual descriptor $f_k^{(2)}$, the electrophilic P_k^+ and nucleophilic P_k^- Parr functions, and the local reactivity difference index R_k over the atoms of the dihydroxyfumaric acid molecule, calculated in vacuo and in water. The actual values of $f^{(2)}(\bar{r})$, P_k^+ and P_k^- have been multiplied by 100 for an easier comparison

Atom	$f_k^{(2)}$			P_k^+			P_k^-			R_k		
	Gas	Water	MetOH	Gas	Water	MetOH	Gas	Water	MetOH	Gas	Water	MetOH
1 C	**−7.1**	**−7.1**	**-7.1**	**16.8**	**17.6**	**15.5**	−3.5	0.9	0.8	−0.47	−1.02	-0.93
2 C	3.8	4.2	4.3	15.9	15.0	13.7	**24.3**	21.0	22.1	**±0.37**	−0.66	-0.51
3 C	3.8	3.9	3.9	15.9	15.5	14.0	**24.3**	21.0	20.5	**±0.37**	−0.68	-0.53
4 C	**−7.1**	**−7.4**	**-7.4**	**16.8**	**18.2**	**16.1**	−3.5	0.9	0.8	−0.47	−1.04	-0.95
5 O	−3.4	−4.6	-4.7	12.4	11.4	11.7	7.3	4.2	4.3	−0.41	−0.94	-0.85
6 O	−2.0	−3.2	-3.2	2.7	3.1	4.8	−0.8	-0.1	-0.2	−0.28	−0.57	-0.52
8 O	**8.6**	**10.5**	**10.5**	2.6	2.8	3.9	**23.3**	**23.2**	**23.5**	**±0.34**	**±0.51**	**±0.55**
10 O	**8.6**	**9.7**	**9.8**	2.6	2.7	3.8	**23.3**	**23.2**	**22.2**	**±0.34**	**±0.48**	**±0.52**
12 O	−3.4	−4.3	-4.4	12.4	11.4	11.6	7.3	4.2	4.7	−0.41	−0.94	-0.85
13 O	−2.0	−2.9	-2.9	2.7	3.0	4.4	−0.8	-0.1	-0.3	−0.28	−0.58	-0.53

To verify this statement, the local reactivity difference R_k index was computed (Table 3). The results confirm that O8 and O10 are ambiphilic sites that can act as electrophiles and nucleophiles in the gas phase, water and methanol.

Figure 6. MEP map of DHF (left) and the ambiphilic reactive sites in methanol (right)

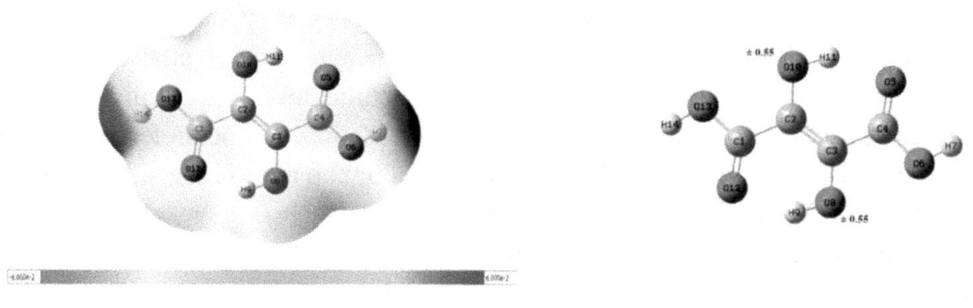

Most Probable Antioxidant Activity Mechanism for DHF Towards DPPH$^{\cdot}$

Intrinsic Thermochemical Parameters

The three mechanisms of antioxidant action by which antioxidants can manifest their protective role have previously been mentioned as follows: the hydrogen atom transfer (HAT) mechanism, the electron transfer followed by proton transfer (SET-PT) mechanism and the proton loss followed by electron transfer (SPLET) mechanism (Szelag et al., 2014; Klein et al., 2009; Baj et al., 2019; Cakmak & Ozbakir, 2020).

In the HAT mechanism, the free radical ($R^{\bullet}$) steals a hydrogen atom from the antioxidant (AH), resulting in the formation of the radical form of the antioxidant. In this case, the reactivity of the antioxidant can be predicted by calculating the O–H bond dissociation enthalpy (BDE).

HAT: DHF ® $DHF^{\bullet}$ + $H^{\bullet}$ (BDE) (16)

The SET-PT mechanism occurs in two steps, in which the electron transfer from the antioxidant compound is followed by proton transfer. The numerical parameter related to this mechanism is the ionization potential (IP) for the first step and the proton dissociation enthalpy (PDE) from the cation radical ($AH^{+\bullet}$) for the second step. Molecules with low PI and PDE values will have high activity (Szelag et al., 2014; Klein et al., 2009; Baj et al., 2019; Cakmak & Ozbakir, 2020). Here, SET is the rate limiting step.

SET: DHF ® $DHF^{\bullet+}$ + e^{-} (IP) (17)

The SPLET mechanism also consists of two steps: first, the dissociation of the antioxidant occurs with the formation of the anionic form and the free proton, and then the ions created in the first reaction react with the free radical. The reaction enthalpy of the first step in this mechanism corresponds to the proton affinity (PA) of the antioxidant anion. The reaction enthalpy of the second step, the electron transfer, corresponds to the electron transfer enthalpy (ETE). Here, PL is the rate limiting step.

PL: DHF ® DHF^{-} + H^{+} (18)

Therefore, in the present research, the thermodynamic parameters BDE, PI, and PA values were calculated in an effort to elucidate the most likely mechanism by which DHF exerts its antioxidant activity. For this purpose, the enthalpy values calculated at 298.15 K at the B3LYP/6–311+G(d,p) level of theory were used, according to the equations below:

$$BDE = H(DHF^{\bullet}) + H(H^{\bullet}) - H(DHF) \quad (19)$$

$$IP = H(DHF^{+\bullet}) + H(e^{-}) - H(DHF) \quad (20)$$

$$PA = H(DHF^{-}) + H(H^{+}) - H(DHF) \quad (21)$$

where the enthalpies of the species H(X) at 298.15K and 1 atm are calculated using the following equation:

$$H(X) = E_0 + ZPE + \Delta H_{trans} + \Delta_{Hrot} + \Delta_{Hvib} + RT \tag{22}$$

where E_0 is the total electronic energy, ZPE - the zero point energy, ΔH_{trans}, ΔH_{rot}, ΔH_{vib} are the translation, rotation and vibration contributions, respectively. The RT term is added to convert energy into enthalpy (Cakmak & Ozbakir, 2020).

To calculate the above indices, the enthalpies of the hydrogen atom, proton and electron in the gas phase and water are needed. The gas-phase enthalpy of the hydrogen atom $H(H^{\bullet})$, proton $H(H^{+})$ and electron $H(e-)$ have been previously calculated: 1306.58, 6.197 and 3.145 kJ mol^{-1}, respectively. These enthalpies in water are: -1310.0, -1083.8 and -101.86 kJ mol^{-1}, respectively (Szelag et al., 2014; Cakmak & Ozbakir, 2020).

In addition, the abundance of different isomers in gas and water must also be taken into account. Thus, *in vacuo* calculations were performed for the E1 isomer (87% relative abundance), and water calculations (SMD solvent model) were performed for all three most abundant isomers, E1, E2, and E3. Another important point is that, due to the geometrical structure of the E2 isomer, the two hydrogen atoms in the hydroxyl groups of the enediol fragment are not equivalent.

Based on the above, the molecular species subjected to research in the case of calculations done in water are presented in Figure 7.

Figure 7. Molecular species involved in the antioxidant activity mechanism of DHF

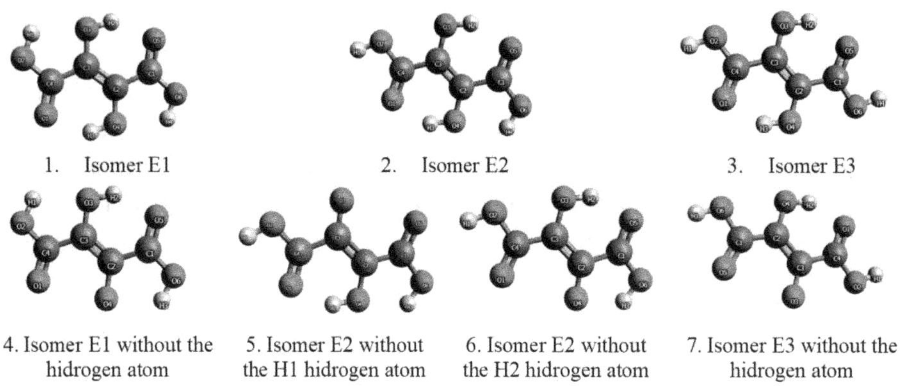

1. Isomer E1 — 2. Isomer E2 — 3. Isomer E3

4. Isomer E1 without the hidrogen atom — 5. Isomer E2 without the H1 hidrogen atom — 6. Isomer E2 without the H2 hidrogen atom — 7. Isomer E3 without the hidrogen atom

The obtained values of BDE, IP, and PA (kJ mol^{-1}) for the DHF isomers (shown in Figure 7) *in vacuo* and in water are shown in Table 4, where the lowest value indicates the preference for a specific mechanism by which antioxidant activity occurs.

The analysis of the obtained data reveals that *in vacuo* the direct hydrogen transfer (HAT) mechanism is the most probable, the BDE value of the E1 isomer (347.13 kJ mol^{-1}) being much lower than the values obtained for the first stages of the SET-PT mechanisms (IP = 864.81 kJ mol^{-1}) and SPLET (PA = 1357.21 kJ mol^{-1}).

Table 4. The calculated values of BDE, IP and PA (kJ·mol^{-1}) for the DHF isomers in vacuo and in water (SMD)

	Species	Mechanism		
		HAT	SET	PL
		BDE	IP	PA
In vacuo	Isomer E1	**347.13**	864.81	1357.21
Water	Isomer E1	329.11	499.99	**81.41**
	Isomer E2^a	327.58	495.91	**84.75**
	Isomer E2^b	332.36	495.91	96.39
	Isomer E3	330.36	485.47	100.66

Note: *The notation E2^a is used to show that upon interaction with the radical, the E2 isomer donates the hydrogen from the OH group bonded to the carbon adjacent to the syn- oriented carboxyl group, and the notation E2^b is used to show that upon interaction with the radical, the isomer E2 donates the hydrogen from the OH group attached to the carbon adjacent to the carboxylic group with anti-orientation.*

In solvent, the HAT mechanism is characterized by BDE values of around 330 100 kJ mol^{-1}, while the PA values in solution are of 80 - 100 kJ mol^{-1}, and the values for IP are around 500 kJ mol^{-1}. Thus, the most probable mechanism by which the antioxidant activity is manifested in a water solution (taking into consideration the SMD model) is the SPLET mechanism, and the most active species is the E1 isomer (PA = 81.41 kJ mol^{-1}), closely followed by the E2 isomer (PA = 84.75 kJ mol^{-1}). It should be noted here that due to the fact that the E2 isomer has two carboxyl groups oriented differently (one *syn-* and one *anti-*), the hydrogens of the hydroxyl groups are not equivalent. Thus, following the interaction with the DPPH$^\bullet$ radical, the E2 isomer can give up one of the two hydrogens from the hydroxyl groups, and the calculations show that the structure shown in Figure 7(5) will be the most likely, with PA = 84.75 kJ mol^{-1}, compared to the structure in Figure 7(6), with PA = 96.39 kJ mol^{-1}. Thus, the hydrogen from the OH group attached to the carbon adjacent to the carboxylic group with *syn-* orientation is most easily transferred to the radical.

DPPH FREE RADICAL SCAVENGING

On the basis of the above information, the half reactions with the DPPH$^\bullet$ radical were written (eq. 23-25) for HAT, SET and PL mechanisms:

HAT: DHF+ DPPH$^\bullet$ ® DHF$^\bullet$ + DPPH−H (23)

SET: DHF+ DPPH$^\bullet$ ® DHF$^{\bullet+}$ + DPPH$^-$ (24)

PL: DHF + DPPH$^\bullet$ ® DHF$^-$ + DPPH-H$^{\bullet+}$ (25)

Standard Gibbs free energies ($\Delta_r G^0$) of the reactions were calculated as follows:

$$\Delta_r G^0\ (HAT) = [G(DHF^{\bullet}) + G(DPPH{-}H)] - [G(DHF) + G(DPPH^{\bullet})] \tag{26}$$

$$\Delta_r G^0\ (SET) = [G(DHF^{\bullet +}) + G(DPPH^{-})] - [G(DPPH^{\bullet}) + G(DHF)] \tag{27}$$

$$\Delta_r G^0\ (PL) = [G(DHF^{-}) + G(DPPH\text{-}H^{\bullet +})] - [G(DHF) + G(DPPH^{\bullet})] \tag{28}$$

As shown in Table 5, in the gas phase, the Gibbs free energy values ($\Delta_r G^0$) are minimal for the HAT mechanism (29.6 kJ·mol^{-1}), indicating on the preference of this reaction in the gas phase. Moreover, the Gibbs free energy values of reactions in water (SMD model) are also minimal for the HAT mechanism, in the range of 17.6 - 20.4 kJ·mol^{-1}. For the SET mechanism, the values of $\Delta_r G^0$ in water are around 150 kJ·mol^{-1} making it the least favored mechanism. For the PL mechanism, the values of $\Delta_r G^0$ are in the range of 58.6-74.4 kJ·mol^{-1}, which is higher than for HAT, but it should be mentioned here that some thermodynamically unfavorable reactions (whose reaction energies do not exceed 50 kJ·mol^{-1}) pass through low-lying transition states, indicating that in such cases reaction kinetics can be the determining factor (Galano et al., 2016; Galano & Alvarez-Idaboy, 2013). Here, the limitations of the SMD model have to be accounted for.

Table 5. The calculated values of $\Delta_r G^0$ for HAT, SET and PL (kJ·mol^{-1}) for the DHF isomers in vacuo and in water (SMD)

		HAT	SET	PL
In vacuo	Isomer E1	29.6	539.2	428.9
Water	Isomer E1	19.9	153.8	58.6
	Isomer E2[a]	20.4	149.3	71.5
	Isomer E2[b]	17.6	149.3	60.4
	Isomer E3	17.9	136.5	74.4

Note: *The notation E2[a] is used to show that upon interaction with the radical, the E2 isomer donates the hydrogen from the OH group bonded to the carbon adjacent to the syn- oriented carboxyl group, and the notation E2[b] is used to show that upon interaction with the radical, the isomer E2 donates the hydrogen from the OH group attached to the carbon adjacent to the carboxylic group with anti-orientation.*

Therefore, considering the results of the intrinsic thermochemical parameters and those obtained for the half-reactions, we can conclude that PL mechanism is the most probable one, where DHF is in the form of isomer E1. Further calculations of transition states could reveal the energy barriers of intermediate species, and adding explicit water molecules could bring more clarity to the issue.

CONCLUSION

This chapter was focused on the review of the research done on dihydroxyfumaric acid in the frames of the Laboratory of Physical and Quantum Chemistry during recent years. The structure and chemistry of dihydroxyfumaric acid have been discussed in view of its importance in biological cycles, as well as in the "glyoxalic scenario" model. Technological applications of DHF have been briefly reviewed, and newly obtained derivatives have been presented. It was shown that computational chemistry techniques can be successfully employed to gain insights into the isomerism and tautomerism of dihydroxyfumaric acid. Local reactivity descriptors and the MEP map of DHF have been computed in order to identify the atomic sites most prone to nucleophilic and electrophilic attacks. Obtained results suggest that the previously obtained antioxidant/antiradical power of investigated compounds towards the DPPH radical may be explained by the highest ambiphilic activation of the oxygen atoms of the hydroxyl groups in the ene-diol moiety. The most probable antioxidant mechanism was determined to describe the interaction of DHF and the DPPH radical; it was shown that the radical is quenched in a two-stage SPLET mechanism. Of course, the limitations of the SMD model have to be accounted for and future work will be focused on modelling the interaction of explicit water molecules. The present research confirms the importance of understanding the chemical interactions for the synthesis of new derivatives with antioxidant/antiradical properties and shows the importance of Density Functional Theory derived concepts in predicting the active reaction sites for the design of useful new derivatives.

ACKNOWLEDGMENT

This work was carried out within the Moldova National Project No. 20.80009.5007.27 "Physicochemical mechanisms of redox processes with electron transfer involved in vital, technological and environmental systems".

REFERENCES

Ayers, P. W., Morrison, R. C., & Roy, R. K. (2002). Variational principles for describing chemical reactions: Condensed reactivity indices. *The Journal of Chemical Physics*, *116*(20), 8731–8744. doi:10.1063/1.1467338

Baj, A., Cedrowski, J., Olchowik-Grabarek, E., Ratkiewicz, A., & Witkowski, S. (2019). Synthesis, DFT calculations, and in vitro antioxidant study on novel carba-analogs of vitamin E. *Antioxidants*, *8*(12), 589. doi:10.3390/antiox8120589 PMID:31779214

Banga, I., & Philippot, E. (1939). Über die dioxymaleinsäureoxydase. *Hoppe-Seyler's Zeitschrift für Physiologische Chemie*, *258*(4), 147–159. doi:10.1515/bchm2.1939.258.4.147

Banga, I., & Szent-Györgyi, A. (1938). Dioxymaleinsäure-oxydase. *Hoppe-Seyler's Zeitschrift für Physiologische Chemie*, *255*(1-3), 57–60. doi:10.1515/bchm2.1938.255.1-3.57

Bellec, N., Garrido Montalban, A., Williams, D. B., Cook, A. S., Anderson, M. E., Feng, X., Barrett, A. G., & Hoffman, B. M. (2000). Porphyrazinediols: Synthesis, characterization, and complexation to group IVB metallocenes. *The Journal of Organic Chemistry*, *65*(6), 1774–1779. doi:10.1021/jo9916840 PMID:10814150

Ben Braïek, O., & Smaoui, S. (2021). Chemistry, safety, and challenges of the use of organic acids and their derivative salts in meat preservation. *Journal of Food Quality*, *2021*, 1–20. doi:10.1155/2021/6653190

Bensid, A., El Abed, N., Houicher, A., Regenstein, J. M., & Özogul, F. (2020). Antioxidant and antimicrobial preservatives: Properties, mechanism of action and applications in food – A Review. *Critical Reviews in Food Science and Nutrition*, *62*(11), 2985–3001. doi:10.1080/10408398.2020.1862046 PMID:33337242

Bhuiyan, M. A., Hoque, M. Z., & Hossain, S. J. (2009). Free Radical Scavenging Activities of Zizyphus mauritiana. *World Journal of Agricultural Sciences*, *5*(3), 318–322.

Bolocan, N., & Duca, G. (2021). DFT study of geometric isomerization and keto–enol tautomerization of dihydroxyfumaric acid. *Studia Universitatis Babes-Bolyai. Chemia*, *66*(4), 341–353. doi:10.24193ubbchem.2021.4.25

Bolocan, N., & Duca, Gh. (2022). *Stopped-flow studies of the interaction of DFH_4 and its derivatives with DPPH•*. National conference with international participation "Life sciences in the dialogue of generations: connections between universities, academia and business community". Chisinau, Moldova.

Butch, C., Cope, E. D., Pollet, P., Gelbaum, L., Krishnamurthy, R., & Liotta, C. L. (2013). Production of tartrates by cyanide-mediated dimerization of glyoxylate: A potential abiotic pathway to the citric acid cycle. *Journal of the American Chemical Society*, *135*(36), 13440–13445. doi:10.1021/ja405103r PMID:23914725

Butch, C. J., Wang, J., Gu, J., Vindas, R., Crowe, J., Pollet, P., Gelbaum, L., Leszczynski, J., Krishnamurthy, R., & Liotta, C. L. (2016). Ph-controlled reaction divergence of decarboxylation versus fragmentation in reactions of dihydroxyfumarate with glyoxylate and formaldehyde: Parallels to biological pathways. *Journal of Physical Organic Chemistry*, *29*(7), 352–360. doi:10.1002/poc.3542

Çakmak, E., & Özbakır Işın, D. (2020). A theoretical evaluation on free radical scavenging activity of 3-styrylchromone derivatives: The DFT study. *Journal of Molecular Modeling*, *26*(5), 98. Advance online publication. doi:10.100700894-020-04368-7 PMID:32279127

Cárdenas, C., Rabi, N., Ayers, P. W., Morell, C., Jaramillo, P., & Fuentealba, P. (2009). Chemical reactivity descriptors for ambiphilic reagents: Dual descriptor, local hypersoftness, and electrostatic potential. *The Journal of Physical Chemistry A*, *113*(30), 8660–8667. doi:10.1021/jp902792n PMID:19580251

Chattaraj, P. K., Chakraborty, A., & Giri, S. (2009). Net electrophilicity. *The Journal of Physical Chemistry A*, *113*(37), 10068–10074. doi:10.1021/jp904674x PMID:19702288

Chattaraj, P. K., Duley, S., & Domingo, L. R. (2012). Understanding local electrophilicity/nucleophilicity activation through a single reactivity difference index. *Organic & Biomolecular Chemistry*, *10*(14), 2855. doi:10.1039/c2ob06943a PMID:22388637

Chow, C. T., & Vennesland, B. (1958). The nonenzymatic decarboxylation of diketosuccinate and oxaloglycolate (dihydroxyfumarate). *The Journal of Biological Chemistry*, *233*(4), 997–1002. doi:10.1016/S0021-9258(18)64693-9 PMID:13587530

Domingo, L. R., Pérez, P., & Sáez, J. A. (2013). Understanding the local reactivity in polar organic reactions through electrophilic and nucleophilic Parr functions. *RSC Advances*, *3*(5), 1486–1494. doi:10.1039/C2RA22886F

Dreher, D. D. (2005). US Patent 0090668. https://www.freepatentsonline.com/y2005/0090668.html

Duca, Gh. (1979). *Mechanisms of ecochimical processes in water medium.* [PhD thesis, Chişinău].

Duca, Gh., Scutaru, I., Sychev, A. (1987). Cinetica interacţiunii iodului cu acidul dihidroxifumaric. *Revista de chimie fizică (In Rus.)*, *61*(8), p. 2266-2268.

Duca, Gh., & Sychev, A. (2002). *Aspecte fundamentale şi aplicative ale catalizei omogene cu compuşii complecşi ai metalelor de tranziţie. Partea I.* Tipografia USM.

Eschenmoser, A. (2007a). On a hypothetical generational relationship between HCN and constituents of the reductive citric acid cycle. *Chemistry & Biodiversity*, *4*(4), 554–573. doi:10.1002/cbdv.200790050 PMID:17443871

Eschenmoser, A. (2007b). The search for the chemistry of life's origin. *Tetrahedron*, *63*(52), 12821–12844. doi:10.1016/j.tet.2007.10.012

Favre, H. A., & Powell, W. H. (2013). *Nomenclature of Organic Chemistry. IUPAC Recommendations and Preferred Name 2013.* The Royal Society of Chemistry. doi:10.1039/9781849733069

Ferrari, C. K., Souto, P. C., França, E. L., & Honorio-França, A. C. (2011). Oxidative and nitrosative stress on phagocytes' function: From effective defense to immunity evasion mechanisms. *Archivum Immunologiae et Therapiae Experimentalis*, *59*(6), 441–448. doi:10.100700005-011-0144-z PMID:21972015

Ferrer-Gallego, R., Puxeu, M., Martín, L., Nart, E., Hidalgo, C., & Andorrà, I. (2018). *Microbiological, physical, and chemical procedures to elaborate high-quality SO_2-free wines.* Grapes and Wines - Advances in Production, Processing, Analysis and Valorization., doi:10.5772/intechopen.71627

Franke, W., & Brathuhn, G. (1931). Über die decarboxylierung Der Dioxymaleinsäure. *Justus Liebigs Annalen der Chemie*, *487*(1), 1–52. doi:10.1002/jlac.19314870102

Fuentealba, P., Florez, E., & Tiznado, W. (2010). Topological analysis of the Fukui function. *Journal of Chemical Theory and Computation*, *6*(5), 1470–1478. doi:10.1021/ct100022w PMID:26615684

Fukunaga, K. (1960). Studies on the metabolism of dihydroxyfumarate, hydroxypyruvate and their related compounds, I. Enzymatic formation of xylulose in rat liver. *Journal of Biochemistry*, *47*(6), 741–744. https://www.jstage.jst.go.jp/article/biochemistry1922/47/6/47_6_741/_pdf. doi:10.1093/oxfordjournals.jbchem.a127120

Gaina, B., Taran, N., Diacon, I., Duca, Gh., Gonţa, M., & Vascan, N. (1995). Studiul proceselor de oxidoreducere în vinuri la şampanizare. Bilanţul activităţii şt. a USM pe anii 1993–1994. Ştiinţe naturale: materialele conf. corpului didactico – şt., 20–27 martie 1995. Chisinau.

Galano, A., & Alvarez-Idaboy, J. R. (2013). A computational methodology for accurate predictions of rate constants in Solution: Application to the assessment of primary antioxidant activity. *Journal of Computational Chemistry*, *34*(28), 2430–2445. doi:10.1002/jcc.23409 PMID:23939817

Galano, A., Mazzone, G., Alvarez-Diduk, R., Marino, T., Alvarez-Idaboy, J. R., & Russo, N. (2016). Food antioxidants: Chemical insights at the molecular level. *Annual Review of Food Science and Technology, 7*(1), 335–352. doi:10.1146/annurev-food-041715-033206 PMID:26772412

Garg, A., Sharma, R., Dey, P., & Kumar, A. (2022). Food auto-oxidation: An overview. *Antioxidants Effects in Health*, 43–68. doi:10.1016/B978-0-12-819096-8.00013-6

Goldberg, D. P., Michel, S. L., White, A. J., Williams, D. J., Barrett, A. G., & Hoffman, B. M. (1998). Molybdocene porphyrazines: A peripheral dithiolene metallacycle fused to a porphyrinic core. *Inorganic Chemistry, 37*(9), 2100–2101. doi:10.1021/ic971498h PMID:11670359

Gonta, M. (2007). The role of the natural antioxidants in the oxihaemoglobin oxidation and the diminution of nitrite oncentration. *Chemistry Journal of Moldova, 1*(2), p. 67-77. https://ibn.idsi.md/ro/vizualizare_articol/6715

Gonta, M. (2009). *Transformarea nitraţilor, nitriţilor, formarea N-nitrozaminelor şi utilizarea inhibitorilor în procese redox* [PhD thesis, Chisinau, the State University of Moldova].

Gonta, M., Sychev, A., Duca, Gh., & Batir, D. (1991). Efectul acidului dihidroxifumaric asupra oxidării (+) catechinei cu reactivul Fenton. [in Russian]. *Chimia Coordinativă, 19*(12), 1654–1657.

Goyal, N., & Jerold, F. (2021). Biocosmetics: Technological Advances and future outlook. *Environmental Science and Pollution Research International, 30*(10), 25148–25169. Advance online publication. doi:10.100711356-021-17567-3 PMID:34825334

Hay, R. W., & Harvie, S. J. (1965). The uncatalysed and nickel(ii) catalysed decarboxylation of dihydroxyfumaric acid. *Australian Journal of Chemistry, 18*(8), 1197. doi:10.1071/CH9651197

Jagatić Korenika, A.-M., Biloš, J., Kozina, B., Tomaz, I., Preiner, D., & Jeromel, A. (2020). Effect of different reducing agents on aromatic compounds, antioxidant and chromatic properties of Sauvignon Blanc wine. *Foods, 9*(8), 996. doi:10.3390/foods9080996 PMID:32722317

Karabinas, P., Sazou, D., & Jannakoudakis, D. (1985). —comparative electrochemical study of L-ascorbic acid and dihydroxyfumaric acid on a mercury electrode in neutral media. *Bioelectrochemistry and Bioenergetics, 14*(4-6), 469–478. doi:10.1016/0302-4598(85)80019-2

Klein, E., Rimarcik, J., & Lukes, V. (2009). DFT/B3LYP study of the O–H bond dissociation enthalpies and proton affinities of para-and metasubstituted phenols in water and benzene. *Acta Chimica Slovenica, 2*, 37–51. http://fez.schk.sk/eserv/changeme:5011/acs_0045.pdf

Liu, S., Rong, C., & Lu, T. (2014). Information conservation principle determines electrophilicity, nucleophilicity, and regioselectivity. *The Journal of Physical Chemistry A, 118*(20), 3698–3704. doi:10.1021/jp5032702 PMID:24784465

Lobo, V., Patil, A., Phatak, A., & Chandra, N. (2010). Free radicals, antioxidants and Functional Foods: Impact on human health. *Pharmacognosy Reviews, 4*(8), 118. doi:10.4103/0973-7847.70902 PMID:22228951

Martemucci, G., Costagliola, C., Mariano, M., D'andrea, L., Napolitano, P., & D'Alessandro, A. G. (2022). Free radical properties, source and targets, antioxidant consumption and health. *Oxygen, 2*(2), 48–78. doi:10.3390/oxygen2020006

Morell, C., Grand, A., & Toro-Labbé, A. (2004). New dual descriptor for chemical reactivity. *The Journal of Physical Chemistry A, 109*(1), 205–212. doi:10.1021/jp046577a PMID:16839107

Morell, C., Grand, A., & Toro-Labbé, A. (2006). Theoretical support for using the Δf(R) descriptor. *Chemical Physics Letters, 425*(4-6), 342–346. doi:10.1016/j.cplett.2006.05.003

Morowitz, H. J., Kostelnik, J. D., Yang, J., & Cody, G. D. (2000). The origin of intermediary metabolism. *Proceedings of the National Academy of Sciences of the United States of America, 97*(14), 7704–7708. doi:10.1073/pnas.110153997 PMID:10859347

Müller, D., Pitsch, S., Kittaka, A., Wagner, E., Wintner, C. E., & Eschenmoser, A. & von a-Aminonitrilen, C. (1990). Aldomerisierung von Glycolaldehyd-phosphat zu racemischen Hexose-2,4,6-triphosphaten und (in Gegenwart von Formaldehyd) racemischen Pentose-2,4-diphosphaten: Rac -Allose-2,4,6-triphosphat und rac-Ribose-2,4-diphosphat sind die Reaktionshauptprodukte. *Helvetica Chimica Acta, 73*, 1410–1468. doi:10.1002/hlca.19900730526

Neuberg, C., & Schwenk, E. (1915). Die Gärung der Dioxymaleinsäure. *The Biochemical Journal, 71*, 104–113.

Parr, R. G., & Yang, W. (1984). Density functional approach to the frontier-electron theory of chemical reactivity. *Journal of the American Chemical Society, 106*(14), 4049–4050. doi:10.1021/ja00326a036

Perdew, J. P., Burke, K., & Ernzerhof, M. (1996). Generalized gradient approximation made simple. *Physical Review Letters, 77*(18), 3865–3868. doi:10.1103/PhysRevLett.77.3865 PMID:10062328

Pokorný, J., & Schmidt, S. (2001). Natural antioxidant functionality during food processing. *Antioxidants in food,* p. 331-354. http://foodclinic.ir/wp-content/uploads/2016/05/Natural-antioxidant-functionality.pdf

Poljsak, B., Kovač, V., & Milisav, I. (2021). Antioxidants, food processing and health. *Antioxidants, 10*(3), 433. doi:10.3390/antiox10030433 PMID:33799844

Porubin, D. (2009). *Utilizarea compuşilor obţinuţi din produse secundare vinicole în inhibiţia proceselor de formare a N-nitrozoaminelor,* [PhD thesis, Chisinau, the State University of Moldova].

Porubin, D., Hecht, S. S., Li, Z., Gonta, M., & Stepanov, I. (2007). Endogenous formation of *n*'-nitrosonornicotine in F344 rats in the presence of some antioxidants and grape seed extract. *Journal of Agricultural and Food Chemistry, 55*(17), 7199–7204. doi:10.1021/jf0712191 PMID:17637060

Raczyńska, E. D., Kosińska, W., Ośmiałowski, B., & Gawinecki, R. (2005). Tautomeric equilibria in relation to PI-electron delocalization. *Chemical Reviews, 105*(10), 3561–3612. doi:10.1021/cr030087h PMID:16218561

Rodopulo, A. K. (1983). *Bases of winemaking biochemistry* (in Russian). Moscow: Light and food industry Publishing.

Rohlfing, D. L. (1967). The catalytic decarboxylation of oxaloacetic acid by thermally prepared poly-α-amino acids. *Archives of Biochemistry and Biophysics*, *118*(2), 468–474. doi:10.1016/0003-9861(67)90375-X PMID:6033722

Sagi, N. V., Karri, P., Hu, F., & Krishnamurthy, R. (2011). Diastereoselective self-condensation of dihydroxyfumaric acid in water: Potential route to sugars. *Angewandte Chemie International Edition*, *50*(35), 8127–8130. doi:10.1002/anie.201102045 PMID:21748829

Sagi, V. N., Punna, V., Hu, F., Meher, G., & Krishnamurthy, R. (2012). Exploratory experiments on the chemistry of the "Glyoxylate Scenario": Formation of ketosugars from Dihydroxyfumarate. *Journal of the American Chemical Society*, *134*(7), 3577–3589. doi:10.1021/ja211383c PMID:22280414

Secara, N. (2008). Dihydroxyfumaric acid transformation, *Chemistry Journal of Moldova, 1*(3), p. 118-119. https://ibn.idsi.md/vizualizare_articol/6787

Secara, N., Duca, Gh., & Macaev, F. (2011). Occurrence and chemistry of dihydroxyfumaric acid. Chemistry Journal of Moldova, 6(1), p. 29-44. https://ibn.idsi.md/ro/vizualizare_articol/12135/datacite

Secara, N., Duca, G., Vlad, L., & Macaev, F. (2010). Observations on the antioxidant activity of novel dihydroxyfumaric acid derivatives. *Chemistry Journal of Moldova*, *5*(2), 59–67. doi:10.19261/cjm.2010.05(2).08

Shields, H. J., Traa, A., & Van Raamsdonk, J. M. (2021). Beneficial and detrimental effects of reactive oxygen species on lifespan: A comprehensive review of comparative and experimental studies. *Frontiers in Cell and Developmental Biology*, *9*, 628157. doi:10.3389/fcell.2021.628157 PMID:33644065

Souchay, P., Fleury, D., & Fleury, M. (1967). Tautomerie enediol-cetol de l'acide dihydroxyfumarique et son intervention dans la decarboxylation de celui-ci. *Comptes Rendus Hebdomadaires des Seances de L'Academie des Sciences Seice C*, *264*(26), 2130.

Stafford, H. A., Magaldi, A., & Vennesland, B. (1954). Enzymatic oxidation of DPNH by diketosuccinate and dihydroxyfumarate. *Science*, *120*(3111), 265–266. doi:10.1126cience.120.3111.265 PMID:13186829

Stepanow, A., & Kusin, A. (1934). Synthesis of a carbohydrate chain by means of enzymes. 6. Resynthesis of glycogen in muscle. *Bernd. deutsch. chem. Gesellsch*, *67*, 723–726.

Stone, J. R., & Yang, S. (2006). Hydrogen peroxide: A signaling messenger. *Antioxidants & Redox Signaling*, *8*(3-4), 243–270. doi:10.1089/ars.2006.8.243 PMID:16677071

Swedin, B., & Theorell, H. (1940). Dioximaleic acid oxidase action of peroxidases. *Nature*, *145*(3663), 71–72. doi:10.1038/145071b0

Sychev, A., Duca, Gh. (1983). Cataliza oxidării acidului dihidroxifumaric. V. Mecanismul oxidării acidului dihidroxifumaric cu oxigenul în prezența ionilor de cupru. *Revista de chimie fizică (în limba rusă), 57*(8), p. 1938-1941.

Sychev, A., Duca, Gh. (1985). Utilizarea acidului dihidroxifumaric pentru îmbunătățirea calității vinurilor seci. *Horticultura, viticultura si vinificatie in Moldova, 12,* p. 38-40.

Sychev, A., Postolatii, T., Duca, Gh., Gonta, M. (1985). *Metodă de stabilizare a materialelor vitivinicole*. Moscow, URSS.

Szeląg, M., Urbaniak, A., & Bluyssen, H. A. R. (2014). A theoretical antioxidant pharmacophore for natural hydroxycinnamic acids. *Open Chemistry*, *13*(1), 000010151520150001. doi:10.1515/chem-2015-0001

Wieland, H., & Franke, W. (1928). Über den mechanismus der oxydationsvorgänge. xiv. Die Aktivierung des sauerstoffs Durch Eisen. *Justus Liebigs Annalen der Chemie*, *464*(1), 101–226. doi:10.1002/jlac.19284640109

Xu, J., Duca, G., Ahmed, S. E., García Márquez, F. P., & Hajiyev, A. (Eds.). (2021). Proceedings of the Fourteenth International Conference on Management Science and Engineering Management. *Advances in Intelligent Systems and Computing*. Springer. 10.1007/978-3-030-49889-4

Chapter 10
Synergistic, Additive, Antagonistic Effects and the Prooxidant Character of Antioxidants:
Interactions in Natural Compounds

Crina Vicol
Institute of Chemistry, Moldova State University, Moldova

Gheorghe Duca
Institute of Chemistry, Moldova State University, Moldova

ABSTRACT

A detailed discussion of the synergistic, additive, and antagonistic antioxidant interactions is presented as the main subject of the chapter. Relevant and recent data is considered in order to emphasize the importance and actuality of the subject. Also, the prooxidant character of antioxidants is shown as a problem that can be eliminated by utilizing low concentrations of antioxidants and by combining them. The reaction mechanisms involved in the antioxidant interactions produced after a combination of various compounds are stated and exemplified. Thus, the regeneration mechanism, the complexes and adducts formation, and complementary effects are responsible for a specific type of antioxidant interaction. On the basis of the beneficial effects of the combination of natural compounds, the general idea of the advantageous consumption of antioxidants from foods is supported.

INTRODUCTION

In the last decades, antioxidants have represented one of the top research subjects worldwide. Basically, this is due to the bioactivity that these compounds manifest, and whose applications offer the possibility to control and operate with different processes from industries; also, this interest in antioxidants is

DOI: 10.4018/978-1-6684-7198-2.ch010

caused by humans` daily intake of antioxidant compounds from foods. The recent popularization of these compounds as inhibitors of oxidative stress resulted in increased consumption of antioxidant supplements that may have many risks. Reaching large concentrations of antioxidants can reveal negative effects caused by their prooxidative action.

By becoming prooxidants, the antioxidant compounds are as harmful as free radicals. This transformation is highly undesirable, therefore, the data reported suggest that by taking low concentrations of antioxidants and by combining them, the prooxidant action can be avoided. These considerations are respected in natural foods like fruits, vegetables, herbs, etc., where antioxidants are present at low concentrations and in different combinations (Cornelli, 2009). In such systems, antioxidant interactions – synergistic, additive, and antagonistic, are present. The synergistic interactions are of most interest because of the enhancement of the total antioxidant activity of the mixture. Studies demonstrated that antioxidant interactions are dependent of many intrinsic and extrinsic factors like the antioxidants` structural features, reducing potentials, interaction mechanisms, concentrations, solubility, solvent, pH, etc. In order to reach the synergistic effect, the regeneration and complex formation mechanisms, that ensure a higher antioxidant activity, need to take place. Equally important for the determination of antioxidant interactions are the methods used for analysis and the calculations performed.

Despite the large number of works that cover these subjects – antioxidant interactions and the prooxidant character, the clear specification of the relationship between them is not shown. Therefore, in this chapter, the problem of the prooxidant effect of antioxidants is discussed and some explanations of the causes of this matter are underlined. The combination of antioxidants, which results in antioxidant interactions, is suggested to be an efficient solution to the elimination of the prooxidant character, and to the maintenance of the antioxidant/prooxidant balance. By this, the idea of the advantageous intake of antioxidants from naturally produced foods is supported and compared to antioxidant supplementation which is still not clear if it is beneficial or harmful (Sotler et al., 2019). In this context, relevant and recent publications on synergistic, additive, and antagonistic interactions are discussed, along with the mechanisms of action proposed by the authors, the methods of analysis, and the particularities of the experiments.

The purpose of the chapter is to broadly discuss the antioxidant interactions between natural compounds and related subjects like the prooxidant character, mechanism of action, etc. Therefore, the objectives of the chapter are: (1) to introduce and discuss the latest data on antioxidant interactions; (2) to identify and exemplify the existent explanations and hypotheses concerning the mechanisms of antioxidant interactions and the influencing reaction factors; (3) to emphasize the link between the prooxidant effect and the antioxidant interactions, and to present evidence that show the antioxidant interactions as solutions to the prooxidant activity; (4) to mention the future research directions and perspectives.

In order to realize this, pertinent literature data from recent articles, book chapters, themed books, and conference proceedings were analyzed and discussed in detail. The research in this area will allow a better understanding of the mechanism of antioxidant interactions and their connection with the prooxidant effect.

RESEARCH BACKGROUND

By definition, antioxidants (AOX) are substances that when presented at low concentrations compared with those of an oxidizable substrate significantly delay or prevent the oxidation of it (Halliwell et al.,

1995). Antioxidant compounds are molecules stable enough to donate an electron to a free radical and neutralize it, thus reducing its capacity to damage (Lobo et al., 2010). Therefore, AOX are compounds that scavenge or neutralize the reactive species like hydroxyl radical (•OH), superoxide anion radical ($O_2^{-\bullet}$), hydrogen peroxide (H_2O_2), alkoxy (RO•) and peroxy (ROO•) radicals, peroxynitrite ($ONOO^-$), etc., that are produced continuously in every natural system.

Antioxidants can be classified in multiple groups according to their chemical structure, polarity, source, function, and mechanism of action. The last one classification is significant for the redox processes since different antioxidants operate with different mechanisms of action. Therefore, a particular antioxidant will be efficient to scavenge a specific type of free radicals according to the reaction conditions, mechanism of action, and other determining features. In this context, antioxidants can scavenge reactive species by several mechanisms: hydrogen-atom transfer (HAT), proton-coupled electron transfer (PCET), radical adduct formation (RAF), single electron transfer (SET), sequential proton-loss electron transfer (SPLET), sequential electron-proton transfer (SEPT), and sequential proton-loss hydrogen-atom transfer (SPLHAT) (Galano et al., 2016).

Natural antioxidants are generally considered beneficial and safe for humans` health compared to the synthetic antioxidants, which are restricted in many countries because of their health risks and toxicity. Fruits, vegetables, plants, nuts and derived products such as wine, tea, chocolate are the main sources of natural antioxidants like vitamins, minerals, phenolic compounds, etc. Due to their ubiquitous distribution, these different classes of compounds are commonly preset together in foods. Therefore, the mutual interactions of food antioxidants are nowadays generally accepted and intensively investigated.

There are three distinctive types of antioxidant interactions: synergistic, additive, and antagonistic. R. Tsao (2015) formulated explicit definitions for the antioxidant interactions as mentioned below:

- synergistic antioxidant interaction – the antioxidant effect of two or more discrete antioxidants when applied together is greater than the sum of the individual antioxidant effects applied separately.
- additive antioxidant interaction – the antioxidant effect of two or more discrete antioxidants when applied together is equal to the sum of the individual antioxidant effects applied separately.
- antagonistic antioxidant interaction – the antioxidant effect of two or more discrete antioxidants when applied together is less than the sum of the individual antioxidant effects applied separately.

The synergistic interactions are advantageous because of the (1) increased efficacy that leads to higher quality and extended shelf life of foods; (2) reduced amount of antioxidants needed, thus a lower production cost; (3) reduced use or total replacement of synthetic antioxidants, which might have adverse health effects; and (4) synergistic combinations may provide more complete distribution of antioxidants in some foods.

The important intrinsic characteristics of antioxidants like the number of electron-donating groups, and reduction potential that directly influences their antioxidant power are not sufficient to explain the mechanistic part of the antioxidant interactions. Despite a large number of publications on this subject, the mechanism of action of synergistic, additive, or antagonistic interactions has not been fully elucidated, mainly due to the reaction`s complexity. However, some observations and subsequent hypotheses have been suggested (Olszowy-Tomczyk, 2020; Chen et al., 2022):

- regeneration phenomenon – regeneration of the strongest antioxidant by the weaker one results in synergistic effect; regeneration of the weaker antioxidant by the strongest one leads to the antagonistic effect. The lack of interaction between the antioxidants compounds results in additive effects.
- formation of antioxidants` intermolecular complexes, dimers or adducts – formation of the intermolecular complexes between antioxidants, of the dimers or adducts with a higher antioxidants property will generate synergistic effects; on the contrary, occurrence of complexes, adducts, polymerization processes that decrease the availability of electron-donating groups will result in antagonistic interactions.
- complementary effects:
    - concentration – depending on the concentration and molar ratios of the compounds in the mixtures, either antioxidant or prooxidant activities prevail.
    - solubility of antioxidants – different solubility of the reducing compounds plays a decisive role to the distribution, localization and orientation of the antioxidants in the system.
    - reaction conditions – solvent and pH influence the compounds` ionization and the intermolecular interaction by binding or deprotonating the antioxidants.
    - combination of compounds with different antioxidant mechanism – various compounds may possess multiple mechanisms of action and can inhibit oxidation at different phases.

Another very important factor for establishing the type of antioxidant interaction is the choice of the adequate assay for analyzing the compounds of interest. This demand comes along with the understanding of the chemistry and intrinsic features of the studied antioxidants, but also with the knowledge of the operational mechanisms of the applied assay (Prior et al., 2005; Abramovič et al., 2017). Therefore, several methods to assess the antioxidant activity are more frequently used: ORAC, TRAP, FRAP, DPPH, ABTS and Folin-Ciocalteu assays.

In short, the ORAC assay (oxygen radical absorption capacity) (Cao et al., 1997), which is based on the HAT mechanism of action, is used to measure the inhibition of peroxyl radical induced by antioxidants. TRAP assay (total peroxyl radical trapping antioxidant parameter) was developed by Wayner et al. (Wayner et al., 1986) and is also based on the HAT reaction mechanism. This method monitors the ability of antioxidant compounds to interfere with the reaction between peroxyl radicals generated by AAPH or ABAP [2,2`-azobis(2-amidinopropane) dihydrochloride] and a target probe (Prior et al., 2005). FRAP assay (ferric reducing antioxidant power) (Benzie & Strain, 1996) is a fast, easy to perform, and accessible method that operates with the SET mechanism. DPPH (2,2-diphenyl-1-picrylhydrazyl) (Brand-Williams et al., 1995), ABTS (2,2'-azino-bis(3-ethylbenzothiazoline-6-sulfonic acid) (Rice-Evans & Miller, 1994) and Folin-Ciocalteu (Singleton & Rossi, 1965) assays are based on both HAT and SET mechanisms (Pior et al., 2005). These methods are widely used for testing the antioxidant activity of single compounds, mixtures, foods, and extracts, due to their accessibility and ease of use.

Starting from the understanding that HAT based methods measure the ability of an antioxidant to quench free radicals by hydrogen donation (Pior et al. 2005) and SET-based methods detect the transfer of an electron from the antioxidant to the free radical, the selection of the suitable assay relies on the initial knowledge of the chemical and structural characteristics of the compounds, reaction conditions like solvent, pH, concentrations, etc. However, analyzing the antioxidant interactions of various combinations of compounds presume the possibility of multiple reaction mechanisms, thus, many antioxidant assays are recommended to be used, and, by comparing the results, some conclusions could be made.

In the literature are described several ways of expressing the synergistic, additive or antagonistic effect – two of them being the most common. Fractional inhibitory concentration index (FIC index) (Meletiadis et al., 2010) is mainly used in biochemistry and pharmacy to determine the interaction between compounds in a mixture. FIC index represents the sum of the FICs of each compound tested. The FIC is determined by dividing the minimum inhibitory concentrations (MIC) of each substance when used in combination by the MIC of each substance when used alone (Eq. 1).

$$\text{FIC index} = \text{FIC}_A + \text{FIC}_B = \left(\frac{C_A}{MIC_A} + \frac{C_B}{MIC_B} \right) \tag{1}$$

where MIC_A and MIC_B are the MICs of substances A and B alone, respectively, and C_A and C_B are the concentration of the substances in combination, respectively, in all of the wells corresponding to am MIC. It is important to mention that a FIC < 1 is characteristic for is a synergistic effect; if FIC = 1, the interaction is additive; and if FIC > 1, there is an antagonistic effect.

This mathematical expression is mostly used to represent the drugs` interactions in the mixtures, however, many researchers successfully used the method to determine the type of antioxidant interaction of single compounds, foods, extracts (Bag & Chattopadhyay, 2015; Enko & Gliszczyńska-Świgło, 2015; Purkait et al., 2020; Vicol et al., 2022).

Another way to numerically express the antioxidant interactions is defined by the ratio of the experimental inhibition of the mixture ($I_{mixture}$) and the theoretical inhibition ($I_{theoretical}$) (Eq. 2 and 3) (Piang-Siong et al., 2017, Quiroga et al., 2019; Vicol & Duca, 2020):

$$\text{SE} = \left(\frac{\%\text{I}_{\text{mixture}}}{\%\text{I}_{\text{theoretical}}} \right) \tag{2}$$

Where

$$\%\text{I}_{\text{theoretical}} = \%\text{I}_A + \%\text{I}_B - \left(\frac{\%\text{I}_A \times \%\text{I}_B}{100} \right) \tag{3}$$

$\%I_A$ and $\%I_B$ represent the inhibition percentages of the antioxidants tested alone. In this case, a synergistic effect (SE) is found when the SE > 1; if SE = 1, then the interaction is additive; and a SE < 1 reveals an antagonistic effect.

Other methods of determining and representing the antioxidant interactions reported in the literature are based on the comparison of the experimental and theoretical antioxidant activity values (expressed as equivalents of standard compounds) (Noguer et al., 2014), or of the experimental and theoretical inhibition percentages (Iacopini et al., 2008; Palafox-Carlos et al., 2012), on the comparison of the half maximal inhibitory concentration (IC_{50}) values of the single compounds and of the mixtures (Choueiri et al., 2012; Mahmoud et al., 2013).

Besides synergistic, additive, and antagonistic interactions, in some conditions, antioxidants can manifest prooxidant character and promote oxidation of the substrate. The problem of the prooxidant effect of antioxidants is a complex one and not completely understood. The shift from antioxidant to prooxidant activity of a compound, and vice versa, is highly dependent on many factors like the intrinsic features of antioxidants, reaction conditions, concentration, presence of transition metals, etc.

THE PROOXIDANT CHARACTER: A MATTER OF CONCENTRATION AND COMBINATION OF ANTIOXIDANTS

The ability of antioxidants to scavenge the radical species and to inhibit oxidation is linked to their ability to interact with free radicals more efficiently than the oxidable substrate, and to produce lower energy antioxidant radicals that do not readily promote further oxidation (Halliwell et al., 1995; Decker et al., 2010). Antioxidants can be effective in various ways: as inhibitors of free radical oxidation reactions (preventive oxidants) by inhibiting the formation of free lipid radicals; by interrupting the propagation of the autoxidation chain reaction (chain-breaking antioxidants); as singlet oxygen quenchers; through synergism with other antioxidants; as reducing agents which convert hydroperoxides into stable compounds; as metal chelators that convert metal prooxidants (iron and copper derivatives) into stable products; and finally as inhibitors of prooxidative enzymes (lipooxigenases) (Carocho & Ferreira, 2013) are present in foods representing the dietary antioxidants – vitamins, phenolic compounds, carotenoids, etc. Despite the fact that a balanced diet can provide a sufficient amount of antioxidants, the intake of antioxidant supplements is highly popularized as a way of maintaining good health by slowing down the biological oxidative processes that contribute to aging and disease risk. Thus, the possibility of reaching high doses of reducing compounds derived from supplements should be taken into consideration.

Large concentrations of antioxidants were found to promote prooxidant effects or to be toxic (Choueiri et al., 2012). By becoming prooxidants, antioxidants act like free radicals, respectively, are highly unstable and active towards chemical reactions with other molecules: proteins, DNA, RNA, lipids, etc. In general, prooxidants are defined as chemicals that induce oxidative stress, usually through the formation of reactive species or by inhibiting antioxidant systems (Carocho & Ferreira, 2013; Puglia & Powell, 1984). They may be classified into several categories: drugs, redox-active metals, pesticides, environmental pollutants, water disinfection products, antioxidants, etc. (Sotler et al., 2019).

The controversy concerning the capacity of antioxidants to act prooxidative depending on their concentration and on the nature of surrounding molecules relate even to the most notorious antioxidants like vitamins C and E, β-carotene, flavonoids, etc. According to Sotler et al. (2019), besides the concentration of the antioxidant, two more factors can influence the function of an antioxidant transforming it to a prooxidant: the presence of metal ions and its redox potential. For example, vitamin C was extensively studied due to its prominent reducing activity, but it can behave as prooxidant at high doses (1000 mg/kg body weight) (Sotler et al., 2019), or in presence of transition metals. Large concentrations of vitamin E also possess prooxidant activity (Rietjens et al., 2002). When reacting with reactive species, it becomes a radical itself, and if there is not enough vitamin C for its regeneration, it remains in the reactive state and promote oxidation. In case of β-carotene, the prooxidant activity depends on the oxygen pressure that produces autoxidation at high oxygen tension, and on the presence of co-antioxidants (Ribeiro et al., 2018).

Phenolic compounds can as well act as prooxidants (Yordi et al., 2012) in systems that contain redox-active metals (Gao et al., 2022). In a study realized by Parker et al. (2010) through the EPR spectroscopy, the authors reported the concentrations of several phenolic compounds that are prooxidant in the reaction with Fenton reagent. Rutin is significantly prooxidative at concentrations of 400 μM and above; ascorbic acid rapidly increases its prooxidant activity at concentrations 0.2 – 400 μM, however, as the concentration of ascorbic acid increased beyond 400 μM, it began to donate hydrogen atoms to the hydroxyl radicals formed from the Fenton reaction to a greater degree than it recycled the Fe3+; gallic acid has a prooxidant effect at concentrations up to approximately 175 μM; catechin is prooxidant at concentrations between 0.75 and 2.3 mM, with greater oxidation at lower concentrations (Parker et al., 2010).

Based on the investigations in human lymphocytes on the antioxidant/prooxidant activity of quercetin, naringenin, hesperidin, and morin, H_2O_2 concentration could not be detected when flavanones naringenin and hesperin were added in the concentration range of 0–200 μM; flavonols quercetin and morin increased H_2O_2 concentration in the concentration range of 25–200 μM and 125–200 μM, respectively (Yen et al., 2003). The generation of superoxide anion radical and products of lipid peroxidation (measured as thiobarbituric acid reacting substances) increased with increasing concentration of each of these flavonoids. According to U. Cornelli (2009), a high concentration of oxidized product – an antioxidant that has given its electron, becomes prooxidant because the tendency to recuperate the lost electron increases in parallel to its concentration as an oxidized product.

On the whole, prooxidant activity is thought to be proportional to the total number of electron-donor groups (like hydroxyl group) in an antioxidant compound (Rahal et al., 2013). Consequently, the structural features that determine the amplitude of the antioxidant activity of a compound equally influence the proportions of its prooxidant character. This duality can be explained if referring to the reduction potentials of antioxidants, precisely – to the change of the redox potential once they undergo oxidative transformations. Antioxidants are characterized by a low reduction potential compared to those free radicals. For example, compounds with low reduction potentials are ascorbic acid (282 mV), quercetin (330 mV), α-tocopherol (500 mV), etc. (Decker et al., 2010). On the other hand, free radicals possess higher energy, and therefore high reduction potentials: H_2O_2 ($O_2^{-\bullet}$, $2H^+$) (940 mV), alkyl peroxyl radical (1000 mV), H_2O_2 ($HOO^\bullet$, H^+) (1060 mV), aliphatic alkoxyl radical (1800 mV).

Antioxidants can inhibit the initiation and propagation of oxidation by scavenging radical species and reducing their energy. As a consequence, the antioxidant radicals formed will have a higher energy than the parent molecules. The resulting antioxidant radicals can act further as reducing agents if the oxidized antioxidants are low-energy radicals and will further donate electrons. High-energy antioxidant radicals will promote oxidation of the substrate like polyunsaturated fatty acids (PUFA) and/or of other surrounding molecules, respectively will act prooxidative (Decker et al., 2010). An efficient antioxidant is considered a compound that results in stable and unreactive antioxidant radicals after oxidation (Prochzkova et al., 2011). This is possible due to different structural and electronic factors, such as the delocalization (throughout of the benzene ring structure, conjugated double bounds) and the spin distribution in the radicalized species (Galano et al., 2016).

Basically, the antioxidant/prooxidant character of compounds depends on the energy of resulting antioxidant radicals. For instance, PUFA with a reducing potential of 600 mV (Decker et al., 2010) can be easily oxidized by high-energy antioxidant radicals, since PUFA itself will act as a reducing compound. In this situation, in order to eliminate the prooxidant effect of antioxidants the presence of other compounds with lower reducing potentials than those of the substrate and/or of the radicalized antioxidants should be considered. Following the same logic line, lower energy compounds will react more rapidly with antioxidant radicals formed, being preferentially oxidized before PUFA. This behavior of mutual regeneration of compounds is reported in the systems formed of many antioxidants and represents the synergistic type of antioxidant interactions. Such a relationship exists between ascorbic acid and α-tocopherol, ascorbic acid, and β-carotene, polyphenols and α-tocopherol, etc. Understanding the thermodynamics of similar redox reactions and intrinsic features of the antioxidants (such as reduction potentials) is important for the prediction of the direction of electron transfer.

Up to this moment, two decisive factors that can eliminate the prooxidant character of antioxidants can be underlined: low concentration of antioxidants and their combination. These considerations are fully respected when antioxidant intake is determined by food intake (Cornelli, 2009). In foods, the usual antioxidants taken are a combination of many different compounds, and the activity derived from them can be either a sum of effect or a synergism (combination effect); the amount of each antioxidant is usually very low, in the range of the recommended dietary allowance, for the most important and known antioxidants in fluid form (concentration effect). These ideas were intensively investigated by scientists in recent years (Sotler et al., 2019; Carocho & Ferreira 2013; Decker et al., 2010; Lo Scalzo, 2008; Freeman et al., 2010; Reber et al., 2011; Palafox-Carlos et al., 2012; Lopez-Martinez et al., 2015; Muhammad et al., 2017). Multi-component systems that originate from food possess multiple mechanisms of action and can inhibit oxidation at many different phases of oxidation, therefore, the resulting antioxidant activity can be more effective than when using a single antioxidant (Decker et al., 2010; Ohigashi & Murakami, 2004). The appropriate examples are combinations of transition metals chelators and free radical scavengers that often result in synergistic interactions and stronger antioxidant activity (Timoshnikov et al., 2020; Lo Scalzo, 2021).

The balance between the production and annihilation of reactive species by antioxidants, or in other words between prooxidant and antioxidant activity, is very delicate, and if this balance tends to the overproduction of reactive species, the cell starts to suffer the consequences of oxidative stress (Carocho & Ferreira, 2013). Choueiri et al. (2012) clearly pointed out that it is still unknown exactly what amounts of antioxidants are needed to have a beneficial antioxidant effect and which dose reflects the safe and appropriate limit for use. However, according to data, people who consume food products rich in antioxidants, which are at low concentrations and in complex combinations, are at low risk of disease incidence (Ohigashi & Murakami, 2004; Seifried et al., 2003).

Antioxidant Interactions of Combined Compounds

As shown in the above subchapter, the combination at certain concentrations of antioxidants represents a solution to the prooxidant effect. At the same time, besides the elimination of the prooxidant character, the combination of various antioxidants or natural compounds produces synergistic, additive, or antagonistic interactions. The mechanisms of these antioxidant interactions are complex and require much research to be completely elucidated. In this subchapter, the most relevant antioxidant interactions reported in the literature are presented and discussed, being taken into consideration their mechanism of action.

Synergistic, Additive, and Antagonistic Interactions of Natural Compounds

Antioxidant interactions, especially synergistic ones, drew significant attention in the past years. The mechanistic investigation of such a phenomenon demonstrated the beauty and complexity of natural processes that involve redox reactions. Nogala-Kalucka et al (2013) revealed the influence of the structure of antioxidants, their concentration, localization in the emulsion, and the factors generating oxidative stress on the synergistic/antagonistic type of interaction between quercetin, rutin, and tocotrienols in the linoleic acid emulsion. The synergistic effect occurred when the effectiveness of individual antioxidants was relatively weak and mainly when the concentration of antioxidants was in the physiologically significant range of 1 μM. Equimolar binary mixtures of resveratrol and other phenolics (gallic acid, caffeic acid, catechin, quercetin) investigated by FRAP, DPPH, and Briggs-Rauscher reaction assays demonstrated the dependence of the antioxidant activity on the mechanism of the applied method (Skroza et al., 2015). The synergistic effect between catechin and resveratrol was confirmed through all three methods. Other phenolic compounds – tyrosol and its derived isomers, exhibited synergistic antioxidant effects when combined with commercial antioxidants like trolox and ascorbic acid (Casadei et al., 2021). Recent investigations showed synergistic effects of *trans*-aconitic acid in the presence of vitamin C, caffeic acid, gallic acid and trolox (Piang-Siong et al., 2017).

The amplitude of the synergistic effect appeared to be dependent on the concentrations of the antioxidants. Two different outcomes from the interaction of α-terpinene or butylated hydroxytoluene (BHT) with acetic, malic, and citric acids were observed by Quiroga et al. (2019). The antioxidant activity of α-terpinene against DPPH radical was enhanced by the presence of organic acids; on the contrary, the BHT's activity was inhibited. Synergism between flavonoids rutin and naringin with trolox or ascorbic acid was demonstrated by using the ORAC method (Tavadyan & Minasyan, 2019). The importance of the O-glucosyl group in the molecular structure was revealed since quercetin and morin that are without glucolsyl group showed antagonism with trolox or ascorbic acid.

By using EPR (Electron Paramagnetic Resonance) spectroscopy, R. Lo Sclazo (2021) studied the antioxidant interactions of ascorbic, chlorogenic acids, and cysteine in combination with unusual antioxidants, such as glucose and citric acid against three free radicals: superoxide anion, hydroxyl radical and *tert*-butyl hydroperoxide radical. Cysteine possesses prooxidant activity in the reaction with *tert*-butyl hydroperoxide radical, but after mixing with other antioxidants, a synergistic effect appeared. The synergistic effects caused by the addition of another organic acid – tartaric acid, to the reaction of ascorbic or dihydroxyfumaric acids with DPPH radical, have been recently reported (Vicol et al., 2021a).

The synergistic interactions that are found to occur between antioxidant compounds from fruit, vegetables and other natural sources are of most interest due to the enhancement of the total antioxidant activity of the mixture. However, the other two types of interactions – additive and antagonistic, should not be underestimated and neglected. Despite the lower antioxidant outcome induced by the regeneration of the weaker antioxidant by the strongest one, or by the formation of the intermolecular complexes between antioxidants, dimers, or adducts with a lower antioxidants power, etc., the experimental data is subtly suggesting that additive and even antagonistic interactions have the potential to eliminate or reduce the prooxidant effect. Although it is not directly stated, it can be deduced that the presence of more than one antioxidant (even in an additive or antagonistic combination) will still reduce the concentration of radicalized free radical scavengers through the above-mentioned mechanisms.

In what follows, recent and eloquent studies on the synergistic, additive, and antagonistic interactions between antioxidants will be detailed. For clarity and distinctness reasons, the data from research works are classified and presented according to the action mechanisms mentioned previously: (a) regeneration mechanism of action; (b) formation of antioxidants` intermolecular complexes, dimers, or adducts; (c) complementary effects.

Regeneration Mechanism

The combination of some antioxidants can follow the regeneration mechanism, which has a positive synergistic effect if the stronger antioxidant is regenerated by the weaker antioxidant (Olszowy-Tomczyk,

2020). This situation was reported by Peyrat-Maillard et al. (2003) when describing the interactions between phenolic acids in the mixture with α-tocopherol. Authors established that caffeic and rosmarinic acids can regenerate α-tocopherol and can be regenerated by quercetin. The evidence of synergistic interactions between α-tocopherol and quercetin, catechin, and epicatechin was demonstrated in different solvents (Pedrielli et al., 2002). Inhibited oxygen intakes measured in the presence of a combination of α-tocopherol and one of the flavonoids confirmed the existence of the antioxidant interactions which appear to be in analogy to the well-known synergism between vitamin E and vitamin C. Amorati et al. (2002), on a basis of the thermochemical data and a large number of kinetic constants proposed a kinetic model for the recycling of α-tocopherol by co-antioxidants in homogeneous solutions (Figure 1). Authors took into consideration three limiting cases: (1) a co-antioxidant irreversibly regenerating α-tocopherol, where the effectiveness of the recycling process depends on the magnitude of the rate constant; (2) a co-antioxidant reversibly recycling α-tocopherol, where regeneration can only be observed if the bond dissociation enthalpy value of the co-antioxidant is lower or at least close to that of the O-H bond of α-tocopherol; (3) the case of a catechol derivative (chosen as a model compound for polyphenolic antioxidants), where recycling of α-tocopherol is feasible even though the BDE value is significantly higher than that of vitamin E. In this case, the driving force for the recycling process is the removal of the semiquinone radical from the catechol derivative by the α-tocopheroxyl radical, which makes the regeneration of α-tocopherol practically irreversible.

Figure 1. The reaction demonstrates the synergistic and antagonistic effects of binary mixtures of antioxidants and co-antioxidants

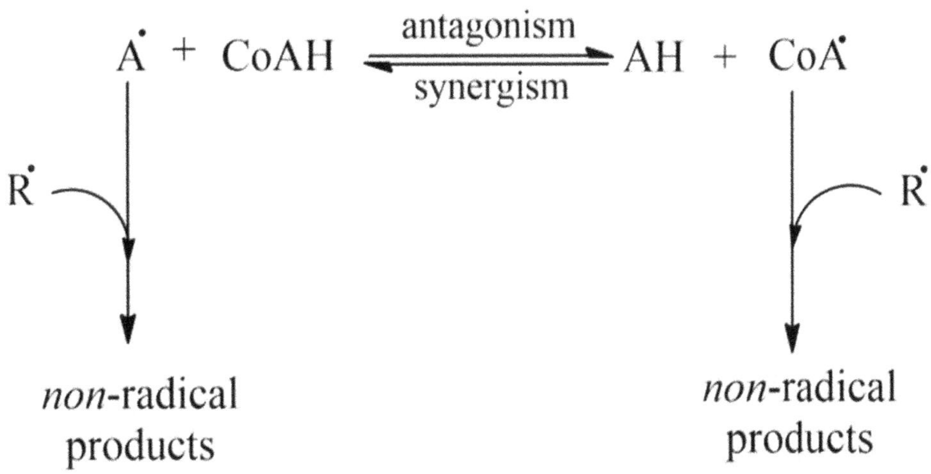

The combination of ascorbic acid with catechin or quercetin results in a synergistic effect, since when ascorbic acid is present in the reaction, it can confer an intermolecular electron delocalization; vitamin C regenerates quercetin and catechin from their oxidized forms, *o*-quinones (Abou Samra et al., 2011). The same regeneration of catechin by ascorbic acid, and other synergistic interactions, was clearly demonstrated by Sawai el al. (2005, 2000, 1998) in their series of works on the determination of the anti-oxidative molecular mechanism of tea antioxidants by using NMR (Nuclear Magnetic Resonance) spectroscopy. This approach offered insight on the intramolecular and intermolecular interactions of catechol, ethyl protocatechuate, quercetin, taxifolin, pyrogallol, myrietin, and ethyl gallate. Binary combinations of myricetin/hesperidin, hesperidin/chlorogenic acid (Freeman et al., 2010), Cat/*p*-coumaric acid, *p*-coumaric acid/quercetin-3-glucoside, and cyaniding/quercetin-3-glucoside (Reber et al., 2011) act synergistically in a regenerating manner.

A significant synergistic interaction was established between Cat and malvidin 3-glucoside and Cat and peonidin 3-glucoside, the first one being regenerated the anthocyanins from their oxidized form (Rossetto et al., 2002). A regeneration mechanism was found between carotenoids (Liang et al., 2009) and carotenoids/tocotrienols (Schroeder et al., 2006); in these situations, the localization and orientation of molecules were demonstrated to be a decisive factor. Palafox – Carlos et al. (2012) showed that binary combinations of gallic, protocatechuic, and chlorogenic acids and various ternary combinations of gallic, protocatechuic, chlorogenic, and vanillic acids manifest synergistic interactions, the results obtained being motivated by the capacity of the phenolic acids to donate hydrogen atoms or electrons not only to the free radicals, by also to the oxidized forms of phenols in order to regenerate them.

On the other hand, the regeneration of the weaker antioxidant by the strongest one results in an antagonistic effect. This was demonstrated to by the case for the combination of α-tocopherol – caffeic acids and α-tocopherol – rosmarinic acid tested by AAPH (2,2'-Azobis(2-amidinopropane) dihydrochloride) – induced oxidation, also for (+)-catechin/caffeic acid and caffeic acid/quercetin (Peyrat-Maillard et al., 2003). Along with the synergistic combinations, Reber et al. (2011) reported the occurrence of the antagonistic interactions of *p*-coumaric acid and pelagronidin. It was suggested that because of the lack of the catechol group in the pelagronidin structure, it cannot recycle the *p*-coumaric, in the same time *p*-coumaric appears to disrupt pelagronidin`s antioxidant activity by drawing away the electrons but not donating them. Ascorbic acid in binary combinations with quercetin, ferulic acid or hesperidin exhibits an antagonistic effect as long as the more efficient compound regenerates the less efficient one (Aoun & Makris, 2012).

Formation of Antioxidants` Intermolecular Complexes, Dimers, or Adducts

The antioxidant interactions can be influenced by the formation of stable intermolecular complexes between the antioxidants (Olszowy-Tomczyk, 2020) which can generate a higher antioxidant outcome. Such combinations of antioxidants are those of quercetin and rosmarinic acid and quercetin and caffeic acid (Peyrat-Maillard et al., 2003). The authors suggest that the result is due to the π-π stacking interactions between the aromatic ring of phenolic acid and the B-ring of flavonol. Finally, this leads to a more efficient regeneration process of rosmarinic and caffeic acids by quercetin. The formation of the antioxidants` dimers with a greater free radical scavenging power was recently reported for the interaction of 1:1 stoichiometric ratio of resveratrol and ascorbic acid studied by NMR spectroscopy (Vicol et al., 2021b). The resveratrol`s oligomerization with the formation of viniferins which possess multiple hydroxyl groups and good antioxidant activity (Mao, 2015) was noticed. The outcome is significantly different when in the system is present ascorbic acid, which may inhibit the oligomerization. A similar effect was observed by Mahmoud et al. (2013) when studying the interaction of various molar ratios of catechin and ascorbic acid. They demonstrated that the reaction could follow different pathways – the regeneration mechanism or the formation of procyanidin structures depending on the concentration of ascorbic acid.

Most reported data suggest that the possible hydrogen-bonding interactions between antioxidant compounds may reduce the total antioxidant activity of the mixture as an effect of decreasing the availability of the hydroxyl groups. This was supposed by Pinelo et al. (2004) in their attempt to rationalize the antagonistic reactions of the mixture of catechin, resveratrol, and quercetin. Antagonistic interactions were also found between various combinations of phenolic compounds in the reaction with DPPH radical (Iacopini et al., 2008). Authors suggested the hydrogen-bonding effect to be a determinant for the occurrence of this kind of antioxidant interaction. Hidalgo et al. (2010) proved that various combinations between five anthocyanins and catechin, epicatechin, myricetin, quercetin or quercetin-3-glucoside have *anti*-synergistic antioxidant interactions because of the intermolecular interactions occurred that caused the decrease of the antioxidant propertied of the tested mixtures. Applying the LC-MS chromatography, the formation of the trimer consisted of two molecules of caffeic acid, and one molecule of quercetin was demonstrated (Choueiri et al., 2012). Because of that, a binary combination of quercetin/caffeic acid possesses antagonistic effect, especially at the 1:2 molar ratios.

Complementary Effects

Antioxidant`s Concentration

Multiple kinds of research initiated on the topic of antioxidant interactions demonstrated the importance of the concentration on the occurrence of synergistic or antagonistic effects. Schroeder et al. (2006) showed that synergistic interactions between carotenoids and tocotrienols are established for carotene–phenol ratios of 1:10 and 1:2 but not for the 1:1 ratio. For the last example, the 1:1 ratio, the regeneration of carotenes from carotene radicals by tocopherols – tocotrienols do not occur. In general, carotenoids behave as antioxidants in the presence of sufficient concentrations of co-antioxidants that act as regenerators; otherwise, the prooxidant activity of carotenoids prevails (Karabulut, 2010).

Various molar ratios of catechin and ascorbic acid demonstrated significantly different outcomes (Mahmoud et al., 2013). The authors concluded that depending on the concentration of ascorbic acid in the system, it supports the regeneration of catechin or the procyanidin structure formation. By increasing the concentration of ascorbic acid in the presence of acetic, malic or citric acids, a better inhibition of the free radical DPPH was noticed, therefore – a greater synergistic effect (Lo Scalzo, 2008). Recent studies revealed a strong dependence of the type of antioxidant interaction on the concentration of each compound from the binary combinations (Vicol, 2022). The author analyzed the influence of the various concentrations of organic acids on the antioxidant activity of some grape bioactive compounds, taken in different concentration, in the reaction with the radical DPPH.

The fact that high concentrations of an antioxidant could be prooxidative, and consequently harmful, is currently taken into consideration by the scientific community. Investigations tend to favor the utilization of small quantities of antioxidants or to mimic the concentrations declared in natural sources like fruits, vegetables, herbs, etc. Such work was performed by an American research team (Freeman et al., 2010; Reder et al., 2011). The authors took as reference the amount of several phenolic compounds found in navel oranges and in strawberries and respected the relative concentrations in their studies. Synergistic and antagonistic interactions, with the prevalence of the synergistic ones, were reported for the combinations of 2, 3 or 4 compounds. The importance of the ratio of cocoa and cinnamon extracts and of their individual compounds was established by Muhammad et al. (2017). The results revealed the fact that by increasing the ratio of the cinnamon extract, the antagonistic effect becomes more pronounced. Also, the type of antioxidant interaction of their constituents is shown to be dependent on the ratio.

Solubility of Antioxidants

The antioxidant activity of natural compounds is directly influenced by their solubility since it determines their location in living organisms or food systems (hydrophilic or lipophilic substrate). In the past decades, the lipid-soluble vitamin E and the water-soluble vitamin C regeneration system was intensively investigated (Niki et al., 1984, Yeum et al., 2009; Kanarovsky & Yaltychenko, 2022). The structural characteristics of α-tocopherol, namely the lipophilic tail with the conjugated double bounds and the hydrophilic head with the heteroatoms, give the molecule the amphiphilic character and determine its localization at the water-lipid interface; ascorbic acid being totally soluble in hydrophilic solvents can annihilate the tocopheroxyl radicals at the interface and, consequently, regenerates them to α-tocopherol.

Similar regeneration processes that take place at the polar/nonpolar solvents` interface are found between α-tocopherol and flavonoids quercetin and (+)-catechin, which can reduce vitamin E from its oxidized form (Pedrielli et al., 2002). It should be mentioned that polyphenols are amphiphilic compounds. This physicochemical property gives polyphenols the ability to interact with both vitamins C and E acting as bridges in the vitamin E regeneration network by vitamin C (Dai & Mumper, 2008; Tsao, 2015). Rosmarinic acid showed to be a good synergist and regenerate α-tocopherol in oil–in–water emulsions (Kittipongpittaya et al., 2016). On the other hand, the experiments in the red palm oil concluded that in the combination with carotenoids, which are lipophilic substances and operate as antioxidants in lipid phases, tocotrienols react as reducing agents and transform carotene radicals in their initial form, by this providing protection against oxidation (Schroeder et al., 2006).

The importance of solubility on the antioxidant activity and on the synergistic effect of a mixture was clearly demonstrated by Liang et al. (2009). Studying the orientation of carotenoids in model lipid membranes, they demonstrated that the combinations astaxanthin – β-carotene and astaxanthin – lycopene possesses significant synergism. Authors suggested the anchoring of astaxanthin in the water/lipid interface to scavenge radicals in the interface and transfer an electron from a non-polar and more reducing carotenoid in the membrane interior.

Reaction Conditions

The solvent and the pH of a reaction are factors that influence considerably the reaction mechanism, the reaction pathway, and, as a final result, the antioxidant activity. The reaction medium is able to influence the competition between the antioxidants in a mixture or between the antioxidants and the oxidable substrate in the reaction of free radicals scavenging. According to Amorati et al. (2012), non-covalent (H-bonding) interactions with surrounding medium involving the reactive –OH moiety of a compound will depress the antioxidant activity if the –OH acts as H-bond donor, while the opposite will generally occur if it acts as H-bond acceptor. Also, remote intermolecular H-bonding, involving a distant –OH group or a ring substituent, may increase or decrease the reactivity of an antioxidant toward free radicals, depending on whether the stabilization produced by the H-bond increases or decreases along the reaction coordinate, on proceeding from reactants to the transition state.

The intermolecular H-bonding stabilization of the oxidized antioxidant is significant for the redox reaction to happen. For example, the aryloxy radicals formed from the oxidation of moieties of phenolic compounds are stabilized in non- or weak hydrogen-bonding solvents by intramolecular H-bonding, bringing about the delocalization of the odd electron over the whole molecule (Apak et al., 2016). As a consequence of H-bonding interactions with the solvent, flavonoids are weak retarders in polar solvents but good chain-breaking antioxidants in apolar media (Pedrielli & Skibsted, 2002). In terms of the mechanism of action, it should be emphasized that a low bound dissociation energy is required for a strong antioxidant essentially acting by H atom donation, whereas a low ionization potential is necessary for a strong antioxidant functioning via electron donation (SET mechanism) (Apak et al., 2016).

Another formulation of the same idea was offered by Prior et al (2005), by mentioning that relative reactivity in SET mechanism is based primarily on deprotonation and ionization potential of the reactive functional group, so SET reactions are pH dependent. On the basis of quantum chemical calculations on quercetin, it was concluded that in nonpolar environments (e.g., lipid bilayer membranes) and at low pH (e.g., as in the stomach), PCET (proton-coupled electron transfer) is the only possible mechanism, whereas in polar solvents and at pH where quercetin is partly deprotonated (e.g., in blood plasma), the faster and therefore more predominant process (SPLET) effectively competes with PCET (Di Meo et al., 2013). The contribution of SPLET mechanism to the overall activity of resveratrol in the solvents that support ionization, such as alcohols and water, was demonstrated in reaction with DPPH radical (Shang et al., 2009). The authors clearly stated that the phenolic compound may be in equilibrium with the corresponding phenolate anion, which is a much stronger electron donor as compared to the parent compound (Shang et al., 2009). When tested in the presence of acetic acid, the antioxidant activity of resveratrol decreases, being concluded that the addition of the acid suppresses the ionization of phenolic compound, and the measured rate constant is decreasing. When the electron donor is the parent molecule and the ET-PT mechanism occurs, the addition of acid will increase or remain the measured rate constant.

The great importance of solvent and pH was demonstrated in a recent work by Konopko and Litwinienko (2021). Resveratrol showed to regenerate synergistically α-tocopherol analogue in micelles and liposomes with the best effectiveness at pH 6 (pH range tested 4 – 10). Compared with the lack of synergism in non-polar solvents, data obtained suggest that the increased persistency of tocopheroxyl radicals in dispersed lipid/water systems and proximal localization of both antioxidants greatly facilitate the possible recovery of α-tocopherol by resveratrol in micellar systems. The authors proposed the mechanism in which resveratrol-derived radicals are coupled to dimers with recovered ability to trap lipid.

Besides resveratrol, rosmarinic acid revealed a pH-dependent synergism with α-tocopherol in oil-in-water emulsions (Kittipongpittaya et al., 2016). Results showed that the combined antioxidants at pH 7 exhibit the strongest synergistic effect due to the formation of caffeic acid as an oxidative degradation product of rosmarinic acid rather than the antioxidant regeneration.

Mukai et al. (2005) studied the influence of pH on the tocopherol-regeneration reaction by catechins in micellar solutions. Reaction rates increased remarkably with increasing pH values and showed one or two peaks around pH 9 – 11. At the higher pH regions, the tocopheroxyl radical reacts with phenolate anions rather than with undissociated phenols, therefore, by increasing the anionic character of catechins, the electron-donating capacity of catechins is also increasing.

The enhancement of the antioxidant activity and synergistic effect at higher pH was reported between lettuce extract with quercetin, green tea extract or grape seed extract against oxidation of peroxidation liposomes (Altunkaya et al., 2016). The deprotonation and stabilization of the phenolate anions in alkaline solutions contribute to their electron-donating ability; their synergistic interactions were suggested to be determined by polymerization reactions. Similar solvent and pH effects are also described for the antioxidant activity of single compounds (Amorati et al., 2006). Recently reported data on the synergistic and *anti*-synergistic interactions of ascorbic and dihydroxyfumaric acids against DPPH radical tested in different media – 96% ethanol and wine matrix, pointed out the importance of the solvent on the antioxidant activity of individual compounds and on the synergistic effect (Vicol et al., 2021a).

Antioxidant Interactions Between Dihydroxyfumaric and Ascorbic Acids in Different Media

Ascorbic acid (AA) is one of the most studied natural antioxidants, while dihydroxyfumaric acid (DHF) (Figure 2) has gained high interest in recent years, both by its plausible role in the "glyoxylate scenario" of primordial metabolism (Sagi et al., 2012) and as one of important participants of dicarbonic acids (Baroud') cycle (Duca, 2012; Cotea et al., 2009). The interaction of AA and DHF becomes naturally possible in fruits, e.g. grapes, where their presence has been reported. Both acids are present in wines mostly by adding them during the wine-making process, as long as their contribution is known to be positive for the quality of the beverage, and respectively, to the total antioxidant capacity of wines (Vacarciuc, 2008; Clark, 2008). As stated previously, the presence of less reactive species like tartaric, citric, malic acids (Vicol & Duca, 2020; Lo Scalzo, 2008), as well as the different media can considerably affect the redox process.

Figure 2. Molecular structure of L – ascorbic acid and dihydroxyfumaric acid

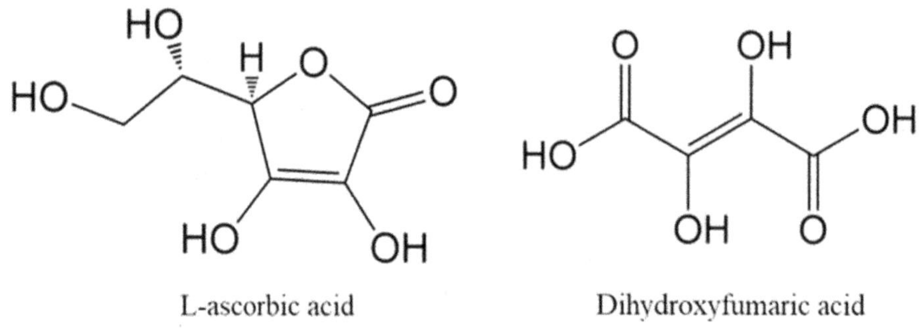

In this regard, the influence of two distinct media – 98% ethanol and wine simulated matrix (5 g/L of tartaric acid in a water–ethanol mixture (12% v/v); the pH set at 3.2 by adding sodium hydroxide), on the antioxidant activity of less studied DHF and of AA was determined within the 2 seconds reaction with the free radical DPPH, by using Stopped-flow spectroscopy (Vicol et al., 2021a). In addition, the synergic or *anti*-synergic effect of various combinations of DHF with AA, both applied at the concentration found in wines, was evaluated in the mentioned two media.

The wine-simulated matrix as a whole has a small antioxidant activity against the DPPH radical, the efficiency being of about 3%. However, it has a significant influence on the antioxidant activity of the tested compounds. As reported in Table 1, the observed rate constants (k_{obs}) of DHF are 10 times higher in the wine matrix compared to 98% ethanol, and the k_{obs} of AA are twice bigger in the wine matrix. It was suggested that such an outcome could be favored by the presence of a high concentration of tartaric acid as a constituent of the wine model solution since the influence of organic acids on the total antioxidant activity of a mixture was reported by others (Lo Scalzo, 2008; Quiroga et al., 2019; Piang-Siong et al., 2017; Vicol & Duca, 2020).

Following the mechanistic explanation of the antioxidant action of some phenolic compounds in the presence of organic acids proposed by Shang et al. (2009), it could be concluded that in this case, the large concentration of tartaric acids` ions from the wine model solution will suppress the ionization of the AA or DHF, and, therefore, will inhibit the SPLET mechanism which, on the contrary, appears to be operative in 98% ethanol. Thus, in wine matrix, the antioxidants scavenge the DPPH free radical by HAT mechanism. On the other hand, as discussed above, the reaction can be considerably affected by the solvent used. In this regard, too many parameters are varying, thus complementary investigations are needed in order to make an objective conclusion.

Table 1. The observed rate constants in ethanol and wine model solution for DHF and AA alone in reaction with DPPH radical

Sample	Concentration, μM	k_{obs}, s^{-1}	
		98% Ethanol	Wine model solution
DHF	60	0.368 ± 0.049	3.656 ± 0.496
	120	0.350 ± 0.030	3.863 ± 0.155
	180	0.383 ± 0.054	3.371 ± 0.091
	400	0.167 ± 0.024	2.277 ± 0,049
	600	0.107 ± 0.076	2.122 ± 0.032
	800	0.040 ± 0.016	2.181 ± 0.057
AA	60	0.328 ± 0.036	0.654 ± 0.021
	120	0.327 ± 0.017	0.653 ± 0.006
	180	0.385 ± 0.020	0.681 ± 0.015

Multiple combinations of the two antioxidants in 98% ethanol and wine matrix were analyzed in order to find out the type of antioxidant interactions. Even though in the wine model solution k_{obs} of various combinations of AA – DHF are higher than in ethanol, the calculations revealed the *anti*-synergistic effect (Vicol et al., 2021a). By comparing the k_{obs} of a single antioxidant and the mixture of antioxidants in the reaction with DPPH• in the wine matrix, it was revealed a decrease in the reaction rate, which caused a reduction of the antioxidant activity. However, the data obtained based on the UV-Vis and EPR spectroscopies on the same combination of compounds: AA – DHF, within 30 min of interaction, showed a strong synergistic effect (Vicol, 2022; Vicol et al., 2022). Based on this, it should be inferred that the reaction is probably not finished within the 2 seconds time of reaction set in the method applied to Stopped – flow technique.

FUTURE RESEARCH DIRECTIONS

The controversy due to antioxidants` capacity to act as prooxidants depending on their concentration and on the nature of surrounding molecules reoriented the scientists toward the investigation of less active compounds, but still important ones. Molecules like organic acids (tartaric, citric, acetic, malic acids) (Lo Scalzo, 2021; Vicol & Duca, 2020; Vicol et al., 2021a), simple and complex sugars (glucose, pectin) (Mercado-Mercado et al., 2020), phospholipids (Doert et al, 2012), enzymes (Liu et al., 2021), etc. possess negligible activities against free radicals, however, in combinations with natural antioxidants, they can exert a strong influence on the mechanism of action and, therefore, can determine the type of antioxidant interactions. It needs to be mentioned that such compounds with low antioxidant activity are physiologically present in natural sources in a higher concentration than that of common free radical scavengers, so their contribution can become significant (Lo Scalzo, 2021).

Another research direction derives from the necessity to develop new techniques for increasing food shelf-life by the addition of preservatives; or as an intention to enhance the antioxidant power of foods. Thus, a large number of data on the synergistic, additive or antagonistic interactions of common antioxidants with different extracts/juices (Graversen et al., 2008; Dai et al., 2008; Colon & Nerin, 2016; Slavova-Kazakova et al., 2021; Stevens-Barrón et al., 2022), along with the results on the antioxidant interactions of various mixtures of foods (Wang et al., 2011; Bag & Chattopadhyay, 2015; Enko & Gliszczyńska-Świgło, 2015; Purkait et al., 2020; Chen et al., 2022; Alves & Tavares, 2019) were reported. According to Çelik & Gökmen (2021), in foods, radical scavengers are present as bound antioxidants, linked to proteins, lipids, and carbohydrates. This fact affects their *in vivo* functional properties, such as bioavailability and bioaccessibility, so bound antioxidants are slowly released from the fiber matrix, in intact form, only in the colon.

In this context, the current investigation of antioxidant interactions between common antioxidants and less active compounds or natural extracts, between various foods or foods extracts can give a better understanding of the influence of the matrix, which naturally occurs the antioxidants, on the mechanism of action and antioxidant power of a mixture; it can better describe the advantage/disadvantage of the combination of certain foods or food extracts; and, finally, this will offer new perspectives on further researches.

CONCLUSION

The study of antioxidants remains current, mainly because of their broad spectrum of utilization, but also due to the lacunae still present at this and derived subjects. In this concern, the antioxidant interactions represent an interesting and complex topic, which demands further investigations and relevant conclusions. Up to this date, the scientific community could hypothesize some effects that occur between antioxidants and determine the type of antioxidant interaction: mutual regeneration, complexes, and adducts formation, and complementary effects which include concentration effect, solubility, reaction conditions, and the combination of antioxidants with different mechanisms of action. Equally important for the determination of the antioxidant activities and antioxidant interactions are the methods used, which implies the utilization of diverse free radicals. The application of more than one method is welcome since different free radicals applied may result in unrelated outcomes due to the distinct mechanisms of action of both antioxidants and radical species. In terms of calculation, several formulas have been developed and used.

The prooxidant character of antioxidants has been widely reported when the antioxidant compounds were present at high concentrations or in the presence of transition metals. On the contrary, at low concentrations and in various combinations, experimental data show the elimination of prooxidant activity. Thus, the link between the prooxidant character of compounds and the antioxidant interactions is revealed, the synergistic, additive and antagonistic effects serving as a solution for the prooxidant activity.

As for future research directions, the literature shows a current interest of scientists in the investigation of antioxidant interactions between common antioxidants and compounds with less or no bioactivity, different natural extracts, and various mixtures of foods.

ACKNOWLEDGMENT

This work has been performed under the Moldova National Research Project Nr. 20.80009.5007.27 "Physical-Chemical Mechanisms of the Redox Processes with Electron Transfer in Vital, Technological and Environmental Systems".

REFERENCES

Abou Samra, M., Chedea, V. S., Economou, A., Calokerinos, A., & Kefalas, P. (2011). Antioxidant/prooxidant properties of model phenolic compounds: Part I. Studies on equimolar mixtures by chemiluminescence and cyclic voltammetry. *Food Chemistry*, *125*(2), 622–629. doi:10.1016/j.foodchem.2010.08.076

Abramovič, H., Grobin, B., Ulrih, N. P., & Cigić, B. (2017). The methodology applied in DPPH, ABTS and Folin-Ciocalteau assays has a large influence on the determined antioxidant potential. *Acta Chimica Slovenica*, *64*(2), 491–499. doi:10.17344/acsi.2017.3408 PMID:28621381

Altunkaya, A., Gökmen, V., & Skibsted, L. H. (2016). pH dependent antioxidant activity of lettuce (L. sativa) and synergism with added phenolic antioxidants. *Food Chemistry*, *190*, 25–32. doi:10.1016/j.foodchem.2015.05.069 PMID:26212937

Alves, A. C., & Tavares, G. M. (2019). Mixing animal and plant proteins: Is this a way to improve protein techno-functionalities? *Food Hydrocolloids*, *97*, 105171. doi:10.1016/j.foodhyd.2019.06.016

Amorati, R., Ferroni, F., Lucarini, M., Pedulli, G. F., & Valgimigli, L. (2002). A quantitative approach to the recycling of α-tocopherol by co-antioxidants. *The Journal of Organic Chemistry*, *67*(26), 9295–9303. doi:10.1021/jo026501f PMID:12492331

Amorati, R., Menichetti, S., Viglianisi, C., & Foti, M. C. (2012). Proton–electron transfer pathways in the reactions of peroxyl and DPPH radicals with hydrogen-bonded phenols. *Chemical Communications*, *48*(97), 11904–11906. doi:10.1039/c2cc36531f PMID:23125978

Amorati, R., Pedulli, G. F., Cabrini, L., Zambonin, L., & Landi, L. (2006). Solvent and pH effects on the antioxidant activity of caffeic and other phenolic acids. *Journal of Agricultural and Food Chemistry*, *54*(8), 2932–2937. doi:10.1021/jf053159+ PMID:16608211

Aoun, M., & Makris, D. P. (2012). Binary mixtures of natural polyphenolic antioxidants with ascorbic acid: Impact of interactions on the antiradical activity. *International Food Research Journal*, *19*(2), 603–606.

Apak, R., Özyürek, M., Güçlü, K., & Çapanoğlu, E. (2016). Antioxidant activity/capacity measurement. 1. Classification, physicochemical principles, mechanisms, and electron transfer (ET)-based assays. *Journal of Agricultural and Food Chemistry*, *64*(5), 997–1027. doi:10.1021/acs.jafc.5b04739 PMID:26728425

Bag, A., & Chattopadhyay, R. R. (2015). Evaluation of synergistic antibacterial and antioxidant efficacy of essential oils of spices and herbs in combination. *PLoS One*, *10*(7), 1–17. doi:10.1371/journal.pone.0131321 PMID:26132146

Benzie, I. F., & Strain, J. J. (1996). The ferric reducing ability of plasma (FRAP) as a measure of "antioxidant power": The FRAP assay. *Analytical Biochemistry*, *239*(1), 70–76. doi:10.1006/abio.1996.0292 PMID:8660627

Brand-Williams, W., Cuvelier, M. E., & Berset, C. L. W. T. (1995). Use of a free radical method to evaluate antioxidant activity. *Lebensmittel-Wissenschaft + Technologie*, *28*(1), 25–30. doi:10.1016/S0023-6438(95)80008-5

Cao, G., Sofic, E., & Prior, R. L. (1997). Antioxidant and prooxidant behavior of flavonoids: Structure-activity relationships. *Free Radical Biology & Medicine*, *22*(5), 749–760. doi:10.1016/S0891-5849(96)00351-6 PMID:9119242

Carocho, M., & Ferreira, I. C. (2013). A review on antioxidants, prooxidants and related controversy: Natural and synthetic compounds, screening and analysis methodologies and future perspectives. *Food and Chemical Toxicology*, *51*, 15–25. doi:10.1016/j.fct.2012.09.021 PMID:23017782

Casadey, R., Challier, C., Altamirano, M., Spesia, M. B., & Criado, S. (2021). Antioxidant and antimicrobial properties of tyrosol and derivative-compounds in the presence of vitamin B2. Assays of synergistic antioxidant effect with commercial food additives. *Food Chemistry*, *335*, 127576. doi:10.1016/j.foodchem.2020.127576 PMID:32739805

Çelik, E. E., & Gökmen, V. (2022). Interactions between free and bound antioxidants under different conditions in food systems. *Critical Reviews in Food Science and Nutrition*, *62*(21), 5766–5782. doi:10.1080/10408398.2021.1892584 PMID:33646856

Chen, X., Li, H., Zhang, B., & Deng, Z. (2022). The synergistic and antagonistic antioxidant interactions of dietary phytochemical combinations. *Critical Reviews in Food Science and Nutrition*, *62*(20), 5658–5677. doi:10.1080/10408398.2021.1888693 PMID:33612011

Choueiri, L., Chedea, V. S., Calokerinos, A., & Kefalas, P. (2012). Antioxidant/pro-oxidant properties of model phenolic compounds. Part II: Studies on mixtures of polyphenols at different molar ratios by chemiluminescence and LC–MS. *Food Chemistry*, *133*(3), 1039–1044. doi:10.1016/j.foodchem.2012.01.057

Clark, A. C. (2008). The production of yellow pigments from (+)-catechin and dihydroxyfumaric acid in a model wine system. *European Food Research and Technology*, *226*(4), 925–931. doi:10.100700217-007-0615-y

Colon, M., & Nerín, C. (2016). Synergistic, antagonistic and additive interactions of green tea polyphenols. *European Food Research and Technology*, *242*(2), 211–220. doi:10.100700217-015-2532-9

Cornelli, U. (2009). Antioxidant uses in nutraceuticals. *Clinics in Dermatology*, *27*(2), 175–194. doi:10.1016/j.clindermatol.2008.01.010 PMID:19167999

Cotea, V. D., Zănoagă, C. V., & Cotea, V. V. (2009). Tratat de oenochimie. [Treatise of oenochemistry], 1, 93-97. Bucharest: Romanian Academy Publisher.

Dai, F., Chen, W. F., & Zhou, B. (2008). Antioxidant synergism of green tea polyphenols with α-tocopherol and l-ascorbic acid in SDS micelles. *Biochimie*, *90*(10), 1499–1505. doi:10.1016/j.biochi.2008.05.007 PMID:18554517

Dai, J., & Mumper, R. J. (2010). Plant phenolics: Extraction, analysis and their antioxidant and anticancer properties. *Molecules (Basel, Switzerland)*, *15*(10), 7313–7352. doi:10.3390/molecules15107313 PMID:20966876

Decker, E. A., Chen, B., Panya, A., & Elias, R. J. (2010). Understanding antioxidant mechanisms in preventing oxidation in foods. In E. A. Decker, R. J. Elias, & D. Julian McClements (Eds.), *Oxidation in foods and beverages and antioxidant applications* (pp. 225–248). Woodhead Publishing. doi:10.1533/9780857090447.2.225

Di Meo, F., Lemaur, V., Cornil, J., Lazzaroni, R., Duroux, J. L., Olivier, Y., & Trouillas, P. (2013). Free radical scavenging by natural polyphenols: Atom versus electron transfer. *The Journal of Physical Chemistry A*, *117*(10), 2082–2092. doi:10.1021/jp3116319 PMID:23418927

Doert, M., Jaworska, K., Moersel, J. T., & Kroh, L. W. (2012). Synergistic effect of lecithins for tocopherols: Lecithin-based regeneration of α-tocopherol. *European Food Research and Technology*, *235*(5), 915–928. doi:10.100700217-012-1815-7

Duca, G. (Ed.). (2012). Homogeneous Catalysis with Metal Complexes. Springer Science & Business Media. doi:10.1007/978-3-642-24629-6

Enko, J., & Gliszczyńska-Świgło, A. (2015). Influence of the interactions between tea (Camellia sinensis) extracts and ascorbic acid on their antioxidant activity: Analysis with interaction indexes and isobolograms. *Food Additives & Contaminants. Part A, Chemistry, Analysis, Control, Exposure & Risk Assessment*, *32*(8), 1234–1242. doi:10.1080/19440049.2015.1049218 PMID:26035225

Freeman, B. L., Eggett, D. L., & Parker, T. L. (2010). Synergistic and antagonistic interactions of phenolic compounds found in navel oranges. *Journal of Food Science*, *75*(6), C570–C576. doi:10.1111/j.1750-3841.2010.01717.x PMID:20722912

Galano, A., Mazzone, G., Alvarez-Diduk, R., Marino, T., Alvarez-Idaboy, J. R., & Russo, N. (2016). Food antioxidants: Chemical insights at the molecular level. *Annual Review of Food Science and Technology*, *7*(1), 335–352. doi:10.1146/annurev-food-041715-033206 PMID:26772412

Gao, Q., Li, Y., Li, Y., Zhang, Z., & Liang, Y. (2022). Antioxidant and prooxidant activities of phenolic acids commonly existed in vegetables and their relationship with structures. *Food Science and Technology (Campinas)*, *42*, 1–9. doi:10.1590/fst.07622

Graversen, H. B., Becker, E. M., Skibsted, L. H., & Andersen, M. L. (2008). Antioxidant synergism between fruit juice and α-tocopherol. A comparison between high phenolic black chokeberry (Aronia melanocarpa) and high ascorbic blackcurrant (Ribes nigrum). *European Food Research and Technology*, *226*(4), 737–743. doi:10.100700217-007-0585-0

Halliwell, B., Aeschbach, R., Löliger, J., & Aruoma, O. I. (1995). The characterization of antioxidants. *Food and Chemical Toxicology*, *33*(7), 601–617. doi:10.1016/0278-6915(95)00024-V PMID:7628797

Hidalgo, M., Sánchez-Moreno, C., & de Pascual-Teresa, S. (2010). Flavonoid–flavonoid interaction and its effect on their antioxidant activity. *Food Chemistry*, *121*(3), 691–696. doi:10.1016/j.foodchem.2009.12.097

Iacopini, P., Baldi, M., Storchi, P., & Sebastiani, L. (2008). Catechin, epicatechin, quercetin, rutin and resveratrol in red grape: Content, in vitro antioxidant activity and interactions. *Journal of Food Composition and Analysis*, *21*(8), 589–598. doi:10.1016/j.jfca.2008.03.011

Kanarovsky, E., & Yaltychenko, O. (2022). Учёт синергии витаминов E и C в кинетической модели перекисного окисления липидов [Determination of the synergy of vitamins E and C in the kinetic model of lipid peroxidation]. *Electronic Materials Processing*, *59*(5), 44–50. PMID:35976784

Karabulut, I. (2010). Effects of α-tocopherol, β-carotene and ascorbyl palmitate on oxidative stability of butter oil triacylglycerols. *Food Chemistry*, *123*(3), 622–627. doi:10.1016/j.foodchem.2010.04.080

Kittipongpittaya, K., Panya, A., Phonsatta, N., & Decker, E. A. (2016). Effects of environmental pH on antioxidant interactions between rosmarinic acid and α-tocopherol in oil-in-water (O/W) emulsions. *Journal of Agricultural and Food Chemistry*, *64*(34), 6575–6583. doi:10.1021/acs.jafc.6b02700 PMID:27494424

Konopko, A., & Litwinienko, G. (2021). Unexpected role of pH and microenvironment on the antioxidant and synergistic activity of resveratrol in model micellar and liposomal systems. *The Journal of Organic Chemistry*, *87*(3), 1698–1709. doi:10.1021/acs.joc.1c01801 PMID:34842421

Liang, J., Tian, Y. X., Yang, F., Zhang, J. P., & Skibsted, L. H. (2009). Antioxidant synergism between carotenoids in membranes. Astaxanthin as a radical transfer bridge. *Food Chemistry*, *115*(4), 1437–1442. doi:10.1016/j.foodchem.2009.01.074

Liu, H., Liu, M., Wang, D., Wang, L., Zhao, Y., Liu, J., Wu, Y., Sun, B., & Zhang, Y. (2021). Competitive binding of synergistic antioxidant chlorogenic acid and (−)-epigallocatechin gallate with lysozyme: Insights from multispectroscopic characterization, molecular docking and activity evaluation. *Journal of Molecular Liquids*, *341*, 117387. doi:10.1016/j.molliq.2021.117387

Lo Scalzo, R. (2008). Organic acids influence on DPPH scavenging by ascorbic acid. *Food Chemistry*, *107*(1), 40–43. doi:10.1016/j.foodchem.2007.07.070

Lo Scalzo, R. (2021). EPR free radical scavenging activity on superoxide, hydroxyl and tert–butyl hydroperoxide radicals by common hydrophilic antioxidants: Effect of mixing and influence of glucose and citric acid. *European Food Research and Technology*, *247*(9), 2253–2265. doi:10.100700217-021-03785-z

Lobo, V., Patil, A., Phatak, A., & Chandra, N. (2010). Free radicals, antioxidants and functional foods: Impact on human health. *Pharmacognosy Reviews*, *4*(8), 118–126. doi:10.4103/0973-7847.70902 PMID:22228951

López-Martínez, L. M., Santacruz-Ortega, H., Navarro, R. E., Sotelo-Mundo, R. R., & González-Aguilar, G. A. (2015). A ^{1}H NMR investigation of the interaction between phenolic acids found in mango (Manguifera indica cv Ataulfo) and papaya (Carica papaya cv Maradol) and 1, 1-diphenyl-2-picrylhydrazyl (DPPH) free radicals. *PLoS One*, *10*(11), 1–11. doi:10.1371/journal.pone.0140242 PMID:26559189

Mahmoud, M. A. A., Chedea, V. S., Detsi, A., & Kefalas, P. (2013). Ascorbic acid modifies the free radical scavenging behaviour of catechin: An insight into the mechanism. *Food Research International*, *51*(2), 907–913. doi:10.1016/j.foodres.2013.02.023

Mao, Q. (2015). *The synthesis and antioxidant capacities of a range of resveratrol and related phenolic glucosides* (090PH M2965) [Doctoral dissertation, University of Adelaide]. digital.library.adelaide.edu.au:2440/9799

Meletiadis, J., Pournaras, S., Roilides, E., & Walsh, T. J. (2010). Defining fractional inhibitory concentration index cutoffs for additive interactions based on self-drug additive combinations, Monte Carlo simulation analysis, and in vitro-in vivo correlation data for antifungal drug combinations against Aspergillus fumigatus. *Antimicrobial Agents and Chemotherapy, 54*(2), 602–609. doi:10.1128/AAC.00999-09 PMID:19995928

Mercado-Mercado, G., Laura, A., & Alvarez-Parrilla, E. (2020). Effect of pectin on the interactions among phenolic compounds determined by antioxidant capacity. *Journal of Molecular Structure, 1199*, 126967. doi:10.1016/j.molstruc.2019.126967

Muhammad, D. R. A., Praseptiangga, D., Van de Walle, D., & Dewettinck, K. (2017). Interaction between natural antioxidants derived from cinnamon and cocoa in binary and complex mixtures. *Food Chemistry, 231*, 356–364. doi:10.1016/j.foodchem.2017.03.128 PMID:28450018

Mukai, K., Mitani, S., Ohara, K., & Nagaoka, S. I. (2005). Structure–activity relationship of the tocopherol-regeneration reaction by catechins. *Free Radical Biology & Medicine, 38*(9), 1243–1256. doi:10.1016/j.freeradbiomed.2005.01.011 PMID:15808422

Niki, E., Saito, T., Kawakami, A., & Kamiya, Y. (1984). Inhibition of oxidation of methyl linoleate in solution by vitamin E and vitamin C. *The Journal of Biological Chemistry, 259*(7), 4177–4182. doi:10.1016/S0021-9258(17)43026-2 PMID:6706998

Nogala-Kałucka, M., Dwiecki, K., Siger, A., Górnaś, P., Polewski, K., & Ciosek, S. (2013). Antioxidant synergism and antagonism between tocotrienols, quercetin and rutin in model system. *Acta Alimentaria, 42*(3), 360–370. doi:10.1556/AAlim.2012.0009

Noguer, M., Cerezo López, A. B., Moyá Morán, M. L., Troncoso González, A. M., & García Parrilla, M. D. C. (2014). Synergism effect between phenolic metabolites and endogenous antioxidants in terms of antioxidant activity. *Advances in Chemical Engineering and Science, 4*(2), 258–265. doi:10.4236/aces.2014.42029

Ohigashi, H., & Murakami, A. (2004). Cancer prevention with food factors: Alone and in combination. *BioFactors (Oxford, England), 22*(1-4), 49–55. doi:10.1002/biof.5520220109 PMID:15630251

Olszowy-Tomczyk, M. (2020). Synergistic, antagonistic and additive antioxidant effects in the binary mixtures. *Phytochemistry Reviews, 19*(1), 63–103. doi:10.100711101-019-09658-4

Palafox-Carlos, H., Gil-Chávez, J., Sotelo-Mundo, R. R., Namiesnik, J., Gorinstein, S., & González-Aguilar, G. A. (2012). Antioxidant interactions between major phenolic compounds found in 'Ataulfo'mango pulp: Chlorogenic, gallic, protocatechuic and vanillic acids. *Molecules (Basel, Switzerland), 17*(11), 12657–12664. doi:10.3390/molecules171112657 PMID:23103532

Parker, T. L., Miller, S. A., Myers, L. E., Miguez, F. E., & Engeseth, N. J. (2010). Evaluation of synergistic antioxidant potential of complex mixtures using oxygen radical absorbance capacity (ORAC) and electron paramagnetic resonance (EPR). *Journal of Agricultural and Food Chemistry*, *58*(1), 209–217. doi:10.1021/jf903080f PMID:19957934

Pedrielli, P., & Skibsted, L. H. (2002). Antioxidant synergy and regeneration effect of quercetin,(−)-epicatechin, and (+)-catechin on α-tocopherol in homogeneous solutions of peroxidating methyl linoleate. *Journal of Agricultural and Food Chemistry*, *50*(24), 7138–7144. doi:10.1021/jf020437l PMID:12428973

Peyrat-Maillard, M. N., Cuvelier, M. E., & Berset, C. (2003). Antioxidant activity of phenolic compounds in 2, 2′-azobis (2-amidinopropane) dihydrochloride (AAPH)-induced oxidation: Synergistic and antagonistic effects. *Journal of the American Oil Chemists' Society*, *80*(10), 1007. doi:10.100711746-003-0812-z

Piang-Siong, W., De Caro, P., Marvilliers, A., Chasseray, X., Payet, B., Sing, A. S. C., & Illien, B. (2017). Contribution of trans-aconitic acid to DPPH scavenging ability in different media. *Food Chemistry*, *214*, 447–452. doi:10.1016/j.foodchem.2016.07.083 PMID:27507497

Pinelo, M., Manzocco, L., Nuñez, M. J., & Nicoli, M. C. (2004). Interaction among phenols in food fortification: Negative synergism on antioxidant capacity. *Journal of Agricultural and Food Chemistry*, *52*(5), 1177–1180. doi:10.1021/jf0350515 PMID:14995117

Prior, R. L., Wu, X., & Schaich, K. (2005). Standardized methods for the determination of antioxidant capacity and phenolics in foods and dietary supplements. *Journal of Agricultural and Food Chemistry*, *53*(10), 4290–4302. doi:10.1021/jf0502698 PMID:15884874

Procházková, D., Boušová, I., & Wilhelmová, N. (2011). Antioxidant and prooxidant properties of flavonoids. *Fitoterapia*, *82*(4), 513–523. doi:10.1016/j.fitote.2011.01.018 PMID:21277359

Puglia, C. D., & Powell, S. R. (1984). Inhibition of cellular antioxidants: A possible mechanism of toxic cell injury. *Environmental Health Perspectives*, *57*, 307–311. doi:10.1289/ehp.8457307 PMID:6094175

Purkait, S., Bhattacharya, A., Bag, A., & Chattopadhyay, R. R. (2020). Synergistic antibacterial, antifungal and antioxidant efficacy of cinnamon and clove essential oils in combination. *Archives of Microbiology*, *202*(6), 1439–1448. doi:10.100700203-020-01858-3 PMID:32185411

Quiroga, P. R., Nepote, V., & Baumgartner, M. T. (2019). Contribution of organic acids to α-terpinene antioxidant activity. *Food Chemistry*, *277*, 267–272. doi:10.1016/j.foodchem.2018.10.100 PMID:30502144

Rahal, A., Kumar, A., Singh, V., Yadav, B., Tiwari, R., Chakraborty, S., & Dhama, K. (2014). Oxidative stress, prooxidants, and antioxidants: The interplay. *BioMed Research International*, *2014*, 1–19. doi:10.1155/2014/761264 PMID:24587990

Reber, J. D., Eggett, D. L., & Parker, T. L. (2011). Antioxidant capacity interactions and a chemical/structural model of phenolic compounds found in strawberries. *International Journal of Food Sciences and Nutrition*, *62*(5), 445–452. doi:10.3109/09637486.2010.549115 PMID:21385007

Ribeiro, D., Freitas, M., Silva, A. M., Carvalho, F., & Fernandes, E. (2018). Antioxidant and pro-oxidant activities of carotenoids and their oxidation products. *Food and Chemical Toxicology*, *120*, 681–699. doi:10.1016/j.fct.2018.07.060 PMID:30077704

Rice-Evans, C., & Miller, N. J. (1994). Total antioxidant status in plasma and body fluids. In Vol. 234, pp. 279–293). Methods in Enzymology. Academic Press. doi:10.1016/0076-6879(94)34095-1

Rietjens, I. M., Boersma, M. G., de Haan, L., Spenkelink, B., Awad, H. M., Cnubben, N. H., Van Zanden, J. J., Van der Woude, H., Alink, G. M., & Koeman, J. H. (2002). The pro-oxidant chemistry of the natural antioxidants vitamin C, vitamin E, carotenoids and flavonoids. *Environmental Toxicology and Pharmacology*, *11*(3-4), 321–333. doi:10.1016/S1382-6689(02)00003-0 PMID:21782615

Rossetto, M., Vanzani, P., Mattivi, F., Lunelli, M., Scarpa, M., & Rigo, A. (2002). Synergistic antioxidant effect of catechin and malvidin 3-glucoside on free radical-initiated peroxidation of linoleic acid in micelles. *Archives of Biochemistry and Biophysics*, *408*(2), 239–245. doi:10.1016/S0003-9861(02)00561-1 PMID:12464277

Sagi, V. N., Punna, V., Hu, F., Meher, G., & Krishnamurthy, R. (2012). Exploratory experiments on the chemistry of the "glyoxylate scenario": Formation of ketosugars from dihydroxyfumarate. *Journal of the American Chemical Society*, *134*(7), 3577–3589. doi:10.1021/ja211383c PMID:22280414

Sawai, Y., & Moon, J. H. (2000). NMR analytical approach to clarify the molecular mechanisms of the antioxidative and radical-scavenging activities of antioxidants in tea using 1, 1-diphenyl-2-picrylhydrazyl. *Journal of Agricultural and Food Chemistry*, *48*(12), 6247–6253. doi:10.1021/jf000500b PMID:11141282

Sawai, Y., Moon, J. H., Sakata, K., & Watanabe, N. (2005). Effects of structure on radical-scavenging abilities and antioxidative activities of tea polyphenols: NMR analytical approach using 1, 1-diphenyl-2-picrylhydrazyl radicals. *Journal of Agricultural and Food Chemistry*, *53*(9), 3598–3604. doi:10.1021/jf040423a PMID:15853407

Sawai, Y., & Sakata, K. (1998). NMR analytical approach to clarify the antioxidative molecular mechanism of catechins using 1, 1-diphenyl-2-picrylhydrazyl. *Journal of Agricultural and Food Chemistry*, *46*(1), 111–114. doi:10.1021/jf970342x PMID:10554204

Schroeder, M. T., Becker, E. M., & Skibsted, L. H. (2006). Molecular mechanism of antioxidant synergism of tocotrienols and carotenoids in palm oil. *Journal of Agricultural and Food Chemistry*, *54*(9), 3445–3453. doi:10.1021/jf053141z PMID:16637706

Seifried, H. E., McDonald, S. S., Anderson, D. E., Greenwald, P., & Milner, J. A. (2003). The antioxidant conundrum in cancer. *Cancer Research*, *63*(15), 4295–4298. PMID:12907593

Shang, Y. J., Qian, Y. P., Liu, X. D., Dai, F., Shang, X. L., Jia, W. Q., Liu, Q., Jian-Guo Fang, J. G., & Zhou, B. (2009). Radical-scavenging activity and mechanism of resveratrol-oriented analogues: Influence of the solvent, radical, and substitution. *The Journal of Organic Chemistry*, *74*(14), 5025–5031. doi:10.1021/jo9007095 PMID:19472994

Singleton, V. L., & Rossi, J. A. Jr. (1965). Colorimetry of total phenolics with phosphomolybdic-phosphotungstic acid reagents. *American Journal of Enology and Viticulture*, *16*(3), 144–158. doi:10.5344/ajev.1965.16.3.144

Skroza, D., Mekinić, I. G., Svilović, S., Šimat, V., & Katalinić, V. (2015). Investigation of the potential synergistic effect of resveratrol with other phenolic compounds: A case of binary phenolic mixtures. *Journal of Food Composition and Analysis*, *38*, 13–18. doi:10.1016/j.jfca.2014.06.013

Slavova-Kazakova, A., Janiak, M. A., Sulewska, K., Kancheva, V. D., & Karamać, M. (2021). Synergistic, additive, and antagonistic antioxidant effects in the mixtures of curcumin with (−)-epicatechin and with a green tea fraction containing (−)-epicatechin. *Food Chemistry, 360*, 129994. doi:10.1016/j.foodchem.2021.129994 PMID:33989877

Sotler, R., Poljšak, B., Dahmane, R., Jukić, T., Pavan Jukić, D., Rotim, C., Polonca Trebše, Starc, A. (2019). Prooxidant activities of antioxidants and their impact on health. *Acta Clinica Croatica, 58*(4.), 726-736.

Stevens-Barrón, J. C., Wall-Medrano, A., Álvarez-Parrilla, E., Olivas-Armendáriz, I., Astiazaran-García, H., Robles-Zepeda, R. E., & De la Rosa, L. A. (2022). Synergistic Interactions between Tocol and Phenolic Extracts from Different Tree Nut Species against Human Cancer Cell Lines. *Molecules (Basel, Switzerland), 27*(10), 3154. doi:10.3390/molecules27103154 PMID:35630629

Tavadyan, L. A., & Minasyan, S. H. (2019). Synergistic and antagonistic co-antioxidant effects of flavonoids with trolox or ascorbic acid in a binary mixture. *Journal of Chemical Sciences, 131*(5), 1–11. doi:10.100712039-019-1618-5

Timoshnikov, V. A., Kobzeva, T. V., Polyakov, N. E., & Kontoghiorghes, G. J. (2020). Redox interactions of vitamin C and iron: Inhibition of the pro-oxidant activity by deferiprone. *International Journal of Molecular Sciences, 21*(11), 3967. doi:10.3390/ijms21113967 PMID:32486511

Tsao, R. (2015). Synergistic interactions between antioxidants used in food preservation. In F. Shahidi (Ed.), *Handbook of antioxidants for food preservation* (pp. 335–347). Woodhead Publishing. doi:10.1016/B978-1-78242-089-7.00013-0

Tsao, R. (2015). Synergistic interactions between antioxidants used in food preservation. In F. Shahidi (Ed.), *Handbook of antioxidants for food preservation* (pp. 335–347). Woodhead Publishing. doi:10.1016/B978-1-78242-089-7.00013-0

Vacarciuc, L. (2008). Potentialul biotehnologic la producerea vinurilor ecologice roze in moldova. *Agricultura – Stiinta si Practica, 57*, 1–8.

Vicol, C. (2022). Study of synergic and anti-synergic interactions of some grape antioxidants with organic acids. In *7th International Conference "Ecological and Environmental Chemistry – 2022"*. 56. Chisinau, CEP MSU.

Vicol, C., Cimpoiu, C., & Duca, G. (2021a). Investigation of synergic/anti-synergic interactions of dihydroxifumaric acid and ascorbic acid with DPPH. *Studia Universitatis Babes-Bolyai. Chemia, 66*(2), 49–58. doi:10.24193ubbchem.2021.02.04

Vicol, C., & Duca, G. (2020). Influenţa acidului tartric asupra procesului de anihilare a radicalilor liberi de către acidul ascorbic [The influence of tartaric acid on the free radical scavenging activity of ascorbic acid]. *Revista de Ştiinţă, Inovare, Cultură şi Artă „Akademos, 56*(1), 39-43.

Vicol, C., Gorincioi, E., & Duca, G. (2021b). 13C NMR investigations of radical scavenging activities of ascorbic acid and resveratrol using 1, 1-diphenyl-2-picrylhydrazyl radical. In *XXIIIrd International Conference New Cryogenic and Isotope Technologies for Energy and Environment.* – EnergEn.

Vicol, C., Iacobescu, I., Fifere, A., & Duca, G. (2022). Study of the synergic interactions of ascorbic and dihydroxyfumaric acids by EPR spectroscopy. In *14th Green Chemistry Postgraduate Summer School (online/in-person.* 90. Venice, Green Sciences for Sustainable Development Foundation. Wang, S., Meckling, K. A., Marcone, M. F., Kakuda, Y., & Tsao, R. (2011). Synergistic, additive, and antagonistic effects of food mixtures on total antioxidant capacities. *Journal of Agricultural and Food Chemistry*, *59*(3), 960–968.

Wayner, D. D. M., Burton, G. W., & Ingold, K. U. (1986). The antioxidant efficiency of vitamin C is concentration-dependent. *Biochimica et Biophysica Acta*, *884*(1), 119–123. doi:10.1016/0304-4165(86)90234-5 PMID:3768406

Yen, G. C., Duh, P. D., Tsai, H. L., & Huang, S. L. (2003). Pro-oxidative properties of flavonoids in human lymphocytes. *Bioscience, Biotechnology, and Biochemistry*, *67*(6), 1215–1222. doi:10.1271/bbb.67.1215 PMID:12843645

Yeum, K. J., Beretta, G., Krinsky, N. I., Russell, R. M., & Aldini, G. (2009). Synergistic interactions of antioxidant nutrients in a biological model system. *Nutrition (Burbank, Los Angeles County, Calif.)*, *25*(7-8), 839–846. doi:10.1016/j.nut.2009.01.011 PMID:19398299

Yordi, E. G., Perez, E. M., Matos, M. J., & Villares, E. U. (2012). Antioxidant and pro-oxidant effects of polyphenolic compounds and structure–activity relationship evidence. In J. Bouayed & T. Bohn (Eds.), *Nutrition, Well-Being and Health* (pp. 23–48). InTech.

Chapter 11
Innovative Materials Produced by the Electrospinning–Calcination Method for Advanced Oxidation:
Kinetic and Empirical Modeling

Petronela Pascariu
Institute of Macromolecular Chemistry Petru Poni, Romania

Corneliu Cojocaru
Institute of Macromolecular Chemistry Petru Poni, Romania

ABSTRACT

Herein, the authors summarize the performance of the innovative semiconducting materials produced by the electrospinning-calcination method and applied in photocatalytic systems for advanced oxidation. Specific details regarding the synthesis, characterization, application, and mechanism of action of these semiconducting catalysts are highlighted. A special emphasis was placed on analyzing the kinetics of the processes related to advanced oxidation. In this regard, the application of the pseudo-first-order kinetic (PFO) model and the Langmuir-Hinshelwood (LH) model was presented and discussed. Likewise, the advantages of adopting the design of experiments methodology are highlighted to build the empirical models which can be used for process optimization.

INTRODUCTION

In recent decades, industrialization, urbanization, population growth, excessive use of non-sustainable resources, and uncontrolled exploitation of natural resources have caused serious damage to the environment which is difficult to repair. A variety of industries and human activities are producing waste that threatens our natural resources. Furthermore, aquatic organisms from lakes, ponds, rivers, seas, and

DOI: 10.4018/978-1-6684-7198-2.ch011

oceans are at major risk. Every day, millions of gallons of wastewater from various industries (such as textiles, paper, pulp, chemicals, fertilizers, pesticides, metal plating, batteries, food processing, refineries, and pharmaceutical) end up in the environment and pollute, sometimes causing ecological disasters (Gusain et al., 2019). Water sources are contaminated by both industrial and domestic wastes. These contaminants induce diverse public health risks to humans by causing serious diseases including, respiratory problems, cancer, diarrhea, neurological disorders, and cardiovascular disease (Ikram et al., 2021).

A wide range of chemical, biological, and physical methods have been engaged in wastewater treatment. Most of them involve filtration, coagulation, reverse osmosis, sedimentation, flotation, oxidation, precipitation, evaporation, and adsorption. Each method has its own set of advantages depending upon its suitability, aiming at a certain category of toxins. One appropriate approach for the removal of industrial effluents, loaded with recalcitrant organic pollutants, is to apply the advanced oxidation processes (AOPs) (Remya & Swain 2019; Klán & Vavrik 2006). Advanced oxidation processes (AOPs) were first developed about four decades ago and were proposed for water treatment. Thus, AOPs were defined as oxidation reactions implying the generation of hydroxyl radicals (•OH) in sufficient amounts to realize water purification (Deng & Zhao 2015). Nowadays, AOPs are employed as modern treatment technologies aimed at decaying and mineralizing the persistent organic pollutant from wastewater *via* reaction with hydroxyl radical (•OH). There are different variants to carry out AOPs, for instance, (1) by homogenous catalysis promoted by iron cations (Fe^{2+}) in the presence of hydrogen peroxide (H_2O_2), giving rise to Fenton systems; (2) by heterogeneous photocatalysis using semiconductor materials as catalysts (TiO_2, ZnO, $BiFeO_3$, etc) under the irradiation of UV/Vis light; (3) by using ozone (O_3) as a strong oxidant, and others (Deng & Zhao 2015). This chapter aims to highlight some recent trends in the production and application of innovative photocatalytic materials devoted to advanced oxidation processes (AOPs) of organic pollutants from contaminated waters.

BACKGROUND

Nowadays, the development of metal oxide-based materials has attracted great attention due to various commercial applications including photocatalysts, batteries, self-cleaning surfaces, fabrics, photovoltaics, cosmetics, antiseptic patches, drug delivery, sunscreen component, ink and paints (Chavali & Nikolova 2019; Velempini et al., 2021). There are a variety of techniques for obtaining these kinds of materials such as the coprecipitation procedure followed by the calcination, hydrothermal, solvothermal route, sol–gel process, electrochemical procedure, electrospinning-calcination method, and so on (Wang et al. 2021a). Nevertheless, one of the most efficient methods of manufacturing these materials at the micro and nanoscale is electrospinning (Ashraf et al., 2019; Wang et al., 2021b). Compared to other nanofibers forms, this is a simple and feasible method to fabricate one-dimensional nanofibers (1D) with different diameters and a large surface area. In general, electrospinning is based on an electrically charged jet of the polymer solution from which are prepared polymer fibers under a voltage of 10–40 kV. The most important parameters to electrospun the intermediate products (ultrathin fibers containing inorganic components) are parameters of the polymeric solution (type of solvent, molecular weight of polymer, viscosity, polymer concentrations, conductivity, and surface tension), working parameters (high voltage source, kV range; tip to collector distance, cm; and feed flow rate, mL/h) and environmental parameters (room temperature and air humidity).

The morphology of the electrospun fibers depends mostly on the ***concentration of the polymer in the doping solution***, and the optimal value is crucial to obtain uniform and continuous fibers, without beads on the surface. Generally, fiber diameters and uniformity get enlarged as the polymer concentration increases in the solution. A minimum conductivity is needed in the electrospinning solution in order to form polymer nanofibers. Electrical conductivity depends on the type of polymer and solvent, polymer concentration, and temperature. Solutions with zero conductivity cannot be electrospun into nanofibers. The second important parameter of the solution is viscosity. When the solution viscosity is too high, it cannot be ejected from the spinneret, while a too-low concentration will produce no fibers. Many studies showed that fiber diameters and uniformity can be adjusted by changing this parameter (Okutan et al., 2014; Cojocaru et al., 2017, Vaseashta, 2007a,b). The applied ***voltage*** plays an extremely important role in the electrospinning process. It is important to note that each polymer has a critical voltage for electrospinning to occur. The researchers report that a higher applied voltage leads to a decrease in the diameter of the nanofibers due to increased electrostatic repulsion forces on the fluid jet. On the contrary, very high voltage can induce nanofibers with large diameters, and sometimes beads can appear on fiber surfaces. In addition, the ***feed flow rate*** of the doping solution (containing solvent, polymer, and additives) influences the jet velocity and transfer rate of the solution (Haider et al., 2018, Bolgen et al. 2022). Therefore, each parameter must be taken into account to obtain materials with good morphology.

According to Figure 1, generally, two-step approaches are used to obtain oxide semiconductor materials. The first step involves the preparation of the intermediate products by the electrospinning technique. The second step requires the calcination of the electrospun intermediates to obtain the targeted inorganic materials. Figure 1 consists of two parts: the electrospinning set-up with the main component parts, followed by the SEM image corresponding to the obtained nanofibers before and after calcination. As an example, the electrospun solution used to obtain the intermediate nanofibers PVP-Zn(II) contained zinc acetate ($ZnC_4H_6O_4$) and polyvinylpyrrolidone (PVP) that were dissolved in a solvent mixture of *N,N*-dimethylformamide (DMF): ethanol (EtOH). This mixture was mechanically stirred for 24 hours, after which it was transferred to the syringe for electrospinning. For each newly synthesized material, electrospinning parameters are established and optimized.

Based on this example, Pascariu et al. (Pascariu et al., 2019a) used the following working parameters: a high voltage source of 25 kV, needle tip–collector distance of 15 cm, and a feed flow rate equal to 0.75mL/h. The resulting electrospun fibers were collected onto stainless steel foil. Next, the membrane was collected from the disk and then calcined for 3 hours at a temperature of 700°C to completely remove the PVP polymer by thermal degradation. This way, ZnO oxide nanostructured materials were prepared *via* the electrospinning-calcination method. We chose polyvinylpyrrolidone (PVP) as a polymer matrix to obtain these oxide nanostructured materials due to the fact that it offers good electrospinnability. In addition, PVP is the most commonly used polymer to obtain these nanofibers since is non-toxic, environmentally benign, low-cost, and can be used in a variety of applications (Turky et al., 2017).

One dimensional (1D) nanoscale semiconductor materials (TiO_2, ZnO) proved to show superior functionality compared to bulk materials due to their enhanced surface area (improved active sites), tunned band gap, shorter electron, and hole diffusion distance improved light absorption and scattering properties, appropriate interface effect, good electronic or ionic charge transport properties, and so on (Prakash et al., 2021). Therefore, this field has been intensively studied with the purpose to develop new 1D oxide semiconductor materials by improving the photocatalytic properties by controlling the crystal structure, morphology, type and percentage of dopants, band gap, and structural defects.

Figure 1. Schematic diagram of the electrospinning-calcination method employed to produce electrospun metal oxides nanostructures

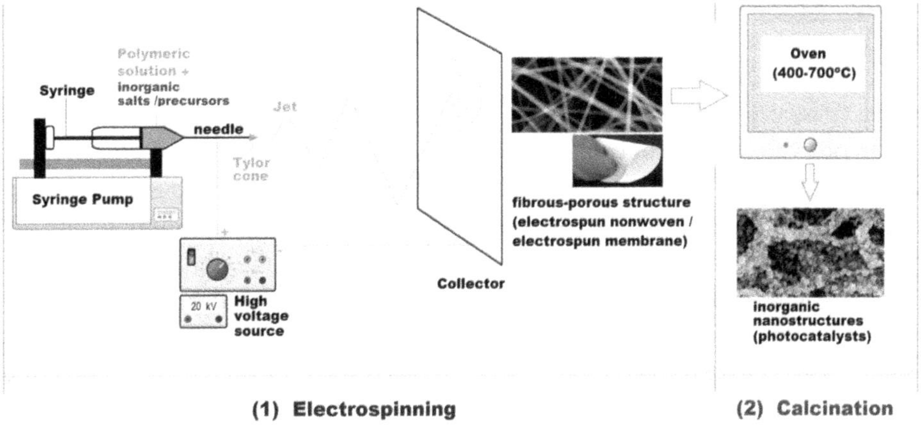

In the scientific literature, there is a variety of semiconductor materials of different shapes and sizes used in photocatalytic applications, which can be classified into *n-type* materials including ZnO, SnO_2, TiO_2, WO_3, MoO_3, In_2O_3, α-Fe_2O_3 and *p-type* materials with the following structures: CuO, NiO, Co_3O_4, Cr_2O_3, Mn_3O_4, Cu_2O (Gusain et al., 2019; Pascariu et al., 2022a). Generally speaking, the majority of carriers of charges in *n*-type materials are preponderantly the electrons (e^-), whereas in *p*-type materials are predominantly the holes (h^+). In this chapter, the focus was placed on the *n*-type semiconductors TiO_2 and ZnO obtained by the electrospinning-calcination technique, owing to their remarkable properties such as non-toxicity, low cost, good chemical stability, and high photocatalytic efficiency.

OXIDE SEMICONDUCTOR MATERIALS

The crystalline structure of oxide semiconductor materials can significantly influence the photocatalytic response. By optimizing the structure, morphology, shape, and size of these nanostructures, the photocatalytic properties are improved, since photocatalytic reactions take place at the active sites located on the surface. In order to increase the number of active sites, the surface area of the photocatalyst as well as the proportion of crystal facets must be increased. For example, 1D structures (nanowires or nanofibers) are more active as photocatalysts than nanocrystals based on the same material due to structural sensitivity or dependence on crystal facets. The morphological characteristics of oxide semiconductor materials obtained by electrospinning can be controlled by changing their obtaining parameters (electrospun solution, working, and environmental conditions), which have a significant impact on the photocatalytic properties of the materials. Moreover, modification of these semiconductors by the inclusion of noble metal, doping with different metal ions, obtaining composite nanostructures including *p-n* types of semiconductors, and developing semiconductor composites and carbon nanomaterials can improve photocatalytic performance. Below we will refer to only two semiconductors studied in this chapter:

- *TiO_2* - is the most important metal oxides semiconductors used in photocatalytic applications. It is found in three structural phases namely anatase (Eg = 3.2 eV), rutile (Eg = 3.0 eV), and brookite

(Eg = 3.2 eV). It is known that the most stable phase is the rutile, but in photocatalytic processes, the predominant anatase phase plays an important role. Besides its special properties, this semiconductor presents limitations regarding the extension of the functional activities beyond the UV region and fast recombination of e^-/h^+ pairs because of the wide band gap.

Pascariu et al. (Pascariu et al., 2020; Pascariu et al., 2021a; Pascariu et al., 2022b,c), showed in previous works that in order to obtain photocatalysts with remarkable performances, it is essential to control the crystal structure through calcination temperature, dopant percentage, band gap fine tunning, defects, and the optimal ratio of anatase/rutile. Furthermore, the authors showed that doping with different transitional metals and lanthanides (Cu, Ag, La, Sm, Er, etc) increases the photocatalytic performance of TiO_2 nanomaterials by extending the optical absorption range of the TiO_2 material.

- *ZnO* - is another oxide semiconductor used successfully with potential applicability for AOPs environmental protection. Its properties are similar to those of TiO_2, with a band gap ranging between 3.00 and 3.37 eV and a high exciton binding energy of 60 meV. The introduction of various metal impurities in the ZnO structure (Pascariu et al., 2020; 2019b,c) can extend the photocatalytic activity in the visible region of the spectrum, as stated for TiO_2. Figure 2 shows, as an example, the morphologies of TiO_2 and ZnO inorganic nanostructures obtained after calcination of the intermediate products PVP-Zn(II) and PVP-Ti(IV), which were previously fabricated by the electrospinning technique.

Figure 2. Micrographs (SEM) of electrospun intermediate products PVP-Zn(II) and PVP-Ti(IV) before calcination, and micrographs (TEM) of resulted inorganic structures ZnO and TiO_2 after calcination

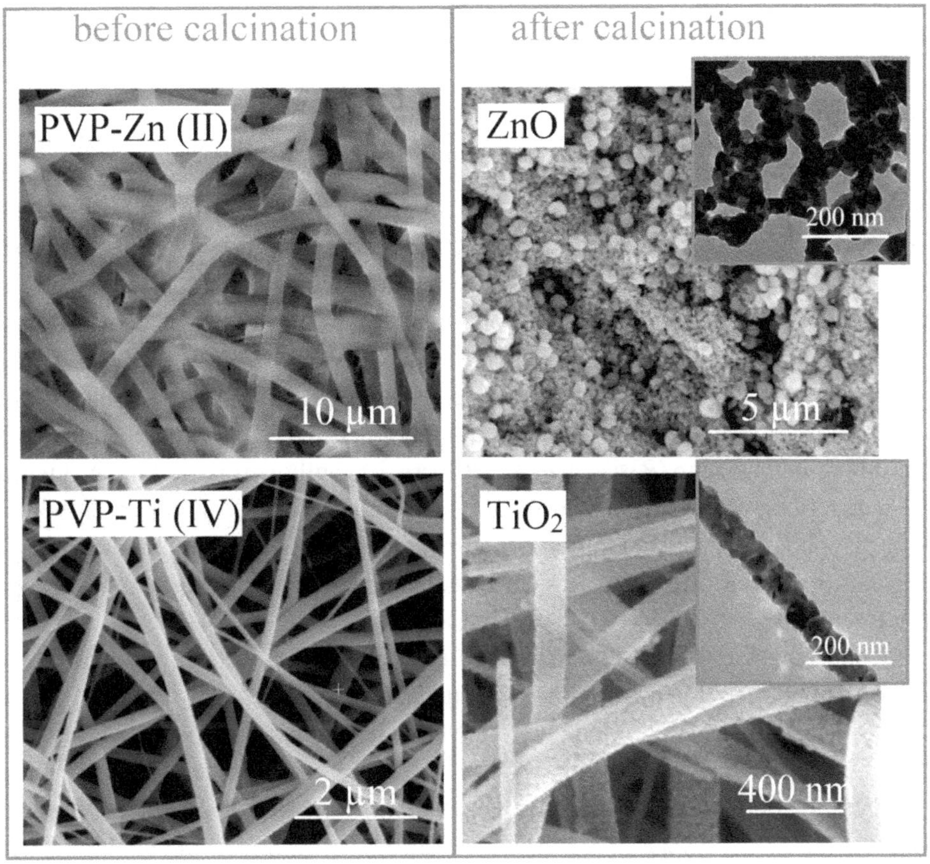

OPTICAL PROPERTIES

Band gap evaluation by optical techniques and studies of photoluminescence (PL) properties of the materials are among the most important investigations for a semiconductor. Recently, Pascariu et al. (2022c) developed original nanostructures based on TiO_2 doped with Cu(0.05-2%), whose band gap values decreased by approximately 10% for the 0.05% Cu doped TiO_2 sample (Eg=2.64 eV), compared with pure TiO_2 material (Eg=3.01 eV). As a result of doping and varying the calcination temperature, the lattice parameters (*a* and *c*) increased and the photocatalytic properties were improved. This was evident from the significant changes in lattice strain. Generally, the increased lattice strain and smaller mean crystallite sizes might be indicative of Cu interstitial insertion into anatase TiO_2 structure during the synthesis of Cu-doped TiO_2 nanostructures. As described in previous works (Pascariu et al., 2022c; Makula et al., 2018), the band gap energy (E_g) can be calculated by measuring UV-Vis reflectance spectra and using the Kubelka-Munk equation (1), and Tauc relation (2).

$$F(R_{\infty}) = \frac{(1-R_{\infty})}{2R_{\infty}} \quad (1)$$

$$(F(R_{\infty})hv)^{1/n} = c(hv - E_g)^{1/n} \quad (2)$$

where $F(R_{\infty})$ represents Kubelka-Munk function and R_{∞} is the absolute reflectance of the samples, $h\nu$ is photon energy and E_g is the band gap energy. The value of the factor n depends on the nature of the electron transition and is equal to 1/2 or 2 for direct and indirect transition band gaps, respectively. From these equations, plotting $(F(R_{\infty})hv)^{1/n}$ against $h\vartheta$ and extrapolating the linear part of the absorption edge to intercept the photon energy axis, the values corresponding to the energy band are obtained. By this approach can be evaluated all semiconducting materials that do not absorb light of sub-band gap energy (or show a slight absorbance). For example, Figure 3 shows the UV-Vis reflectance spectra (a) and Tauc plots of 0.1% Ag-doped TiO2 nanostructured nanofibers, which were carefully characterized by the corresponding research group (Pascariu et al., 2021a). The authors reported the influence of the Ag dopant on TiO2 nanofibers and on the morphological, structural, and optical properties, which were then correlated with the photocatalytic and antibacterial response. Moreover, semiconductor stoichiometry, intrinsic defect concentration, and impurities can affect the band gap values and band edge positions on the photoluminescence spectra.

On the other hand, it is known that ZnO semiconductor shows two main emission bands: the first one is located in the UV region with a maximum of around 380 nm (generated by the free exciton recombination), and the other in the visible region due to the deep level emissions, with the maximum ranging between 410–730 nm. The visible bands are usually the result of intrinsic defects and structural impurities. Impurities or intentional dopants can enhance or diminish the intrinsic defects of nanostructured materials. Doping of ZnO with La ions enhances the emission intensity in the visible region (400–600 nm), where the PL intensity is directly proportional to La ion concentration (Pascariu et al., 2019b). Based on these investigations, it was found that in order to obtain photocatalytic materials with outstanding performance, all the properties described above need to be analyzed and correlated.

Figure 3. UV-Vis reflectance spectra (a) and Tauc plots of 0.1% Ag doped TiO_2 nanostructured nanofibers (b)

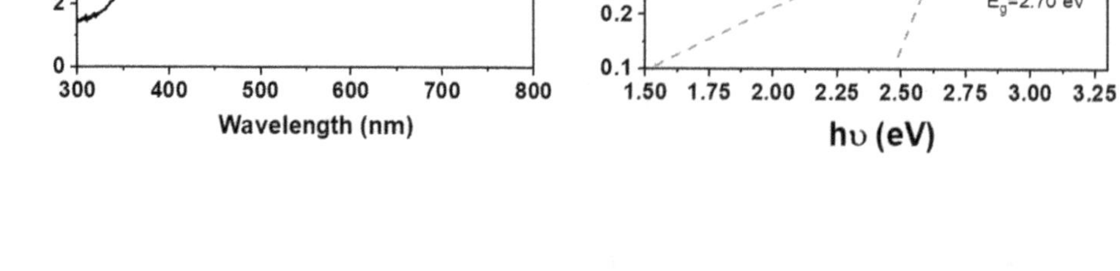

Table 1 (Pascariu et al., 2019a,b; 2020; 2021a; 2022a,b,c; Liu et al., 2018; Wang et al., 2014; Norouzi et al., 2020) are summarized the most relevant studies for pure and doped TiO_2 and ZnO-based nanostructured materials with good performance responses under different degradation conditions and various organic pollutants.

Table 1. A comparison of the photocatalytic performance of ZnO and TiO_2-based materials doped with transition metals or lanthanides

Catalyst	Eg of catalyst (eV)	Pollutant, C_0	Catalyst amount	Time, Light source	Efficiency (%)	Ref.
ZnO-Ag (1%)	3.156	Amaranth, 25 mg/L	0.16 g/L	240 min, UV (150W)	98.40	(Pascariu et al., 2020)
ZnO-La (2%)	2.878	CR, 10.5 mg/L	0.28 g/L	240 min, UV (150W)	97.63	(Pascariu et al., 2019b)
ZnO-Sm (1%)	2.810	CR, 10.7 mg/L	0.24 g/L	240 min, UV (150W)	95.80	(Pascariu et al., 2019a)
ZnO/Fe (1.5%)	3.16	MB, 10 mg/L	0.4 g/l	360 min UV (50W)	~88	(Liu et al., 2018)
TiO_2-La (0.1%)	2.680	MB, 10 mg/L CIP, 10 mg/L	0.2 g/L 0.6 g/L	300 min, visible (400 W)	91 99.5	(Pascariu et al., 2022a)
TiO_2-Sm (0.1%)	2.673	MB, 10 mg/L CIP, 10 mg/L	0.9 g/L	300 min, Visible (400 W)	~99	(Pascariu et al., 2022b)
TiO_2-Er (0.1%)	2.912	MB, 10 mg/L CIP, 10 mg/L	0.9 g/L	300 min, visible (400 W)	~99	
TiO_2-Cu (0.05%)	2.70	Amaranth, 25 mg/L	0.4 g/L	300 min, Visible (400 W)	99.84	(Pascariu et al., 2022c)
TiO_2-Ag (0.1%)	2.70	MB, 10 mg/L	0.4 g/L	300 min, visible (400 W)	97.05	(Pascariu et al., 2021a)
TiO_2-Ag	-	MO, 20 mg/L	0.5 g/L	180 min UV (375 W)	~99	(Wang et al., 2014)
TiO_2-Ag (%)	-	Phenol, 5 mg/L	1.59 g/L	visible (18 W)	78.17	(Norouzi et al., 2020)

MB=Methylene Blue; CR=Congo Red; CIP=Ciprofloxacin; MO=Methyl Orange.

PROPOSED MECHANISM IN PHOTOCATALYSIS

In the previous studies, most of the operating parameters affecting the photocatalytic activity including photocatalyst dose, initial concentration of pollutant, pH solution (Pascariu et al., 2019c), UV or Visible light irradiation, the addition of H_2O_2, direct photolysis, and reaction temperature were investigated (Pascariu et al., 2019a,b; 2020; 2021a,b; 2022a,b,c). Generally, it is known that heterogeneous photocatalysis is one of the cost-effective technologies used for the degradation of recalcitrant organic pollutants. According to specialized literature (Velempini et al., 2021; Pascariu et al., 2020; 2022a), a possible mechanism for the degradation of organic pollutants in the presence of oxide semiconductor nanostructures can be proposed. The process involves the absorption of a certain wavelength of light by *semiconductor oxides* (SO, photocatalysts), inducing the electron excitation from the valence band (VB) to the conduction band (CB) of SO. This electron jump generates positive charge carriers (holes, h^+) in the VB of the SO, as evidenced in reaction (3). The holes in the VB induce the H_2O molecules splitting, from which results hydroxyl-free radicals (OH^*), as given in reaction (4). At the same time, the generated electrons (e^-) in the CB are captured by oxygen molecules to give super oxygen radicals ($O_2^{''-}$), reaction (5).

$$SO + h\nu\left(UV\ or\ visible\ light\right) \rightarrow SO\left\{e^- ... h^+\right\} \rightarrow e_{CB}^- + h_{VB}^+ \quad (3)$$

$$h_{VB}^+ + H_2O \rightarrow H^+ + OH^{''} \quad (4)$$

$$e_{CB}^- + O_2 \rightarrow O_2^{''-} \quad (5)$$

Besides the presented reaction, it can occur the development of various species, which produce more OH^* radicals as shown in reactions (6) – (8).

$$O_2^{''-} + H^+ \rightarrow HO_2^{''} \quad (6)$$

$$2HO_2^{''} \rightarrow H_2O_2 + O_2 \quad (7)$$

$$H_2O_2 \rightarrow 2OH^* \quad (8)$$

The generated OH^* radicals induce the degradation of the organic pollutants initially adsorbed on the surface of the catalyst. At the end of all reactions (photocatalytic process), the main products of mineralization are CO_2 and H_2O according to the reaction (9). However, other by-products also can be formed depending on the pollutant type (Velempini et al., 2021). In some cases, h^+ can directly react with the pollutants to produce intermediates which are in the end degraded to CO_2 and H_2O. A schematic representation of the heterogenous photocatalytic process is highlighted in Figure 4.

$$organic\ pollutants + OH^{''} / O_2^{''-} \rightarrow CO_2 + H_2O + \ldots \quad (9)$$

Figure 4. Possible photocatalytic mechanism for organic pollutants degradation in the presence of metal oxide semiconductor photocatalysts under UV-Vis light irradiation

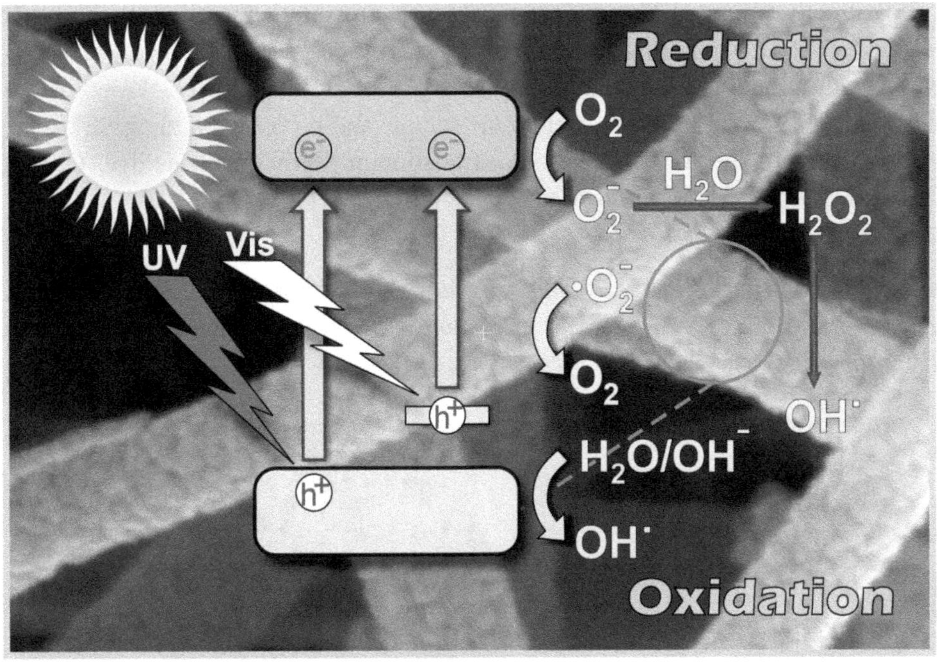

ELECTROSPUN POLYMERIC MEMBRANES EMBEDDING CATALYSTS

The first part of this chapter has focused on the synthesis and characterization of semiconducting inorganic materials with potential applications in photocatalysis. However, it is of practical interest to expand and improve the properties of the materials in order to come as close as possible to applicability in real situations. The difficulties in recovering, recycling, and reusing inorganic nanostructures are one of their drawbacks. Therefore, the development of composite fibrous membranes that can incorporate these inorganic nanostructures is a field of great interest in scientific research. Such a strategy yields to obtaining of composite materials that can be readily manipulated in real-life applications. Also, one of the most important targets is the post-purification separation of these nanostructures from water, a process that is costly and time-consuming. Moreover, the incorporation of semiconducting inorganic nanostructures into polymer matrices contributes to the enhancement of the mechanical characteristics and surface hydrophilicity of the membranes.

Recently, most studies are focused on developing of new polymer membranes, which are generally based on poly(vinylidene fluoride) (PVDF). In these composite membranes are included the inorganic materials (TiO_2/ZnO) for the degradation of organic pollutants. For example, Pascariu et al. (2022d) prepared (TiO_2:Sm)/PVDF/Graphene and (TiO_2:Sm)/PVDF/Graphene oxide composite membranes by electrospinning method, which were evaluated for the photodegradation of methylene blue (MB) under visible light irradiation. The authors found that the hybrid membrane (TiO_2:Sm)(15%)/PVDF/GO(2.5%) tested under optimal conditions (2.8 g/L of catalyst dose, and 0.06 M of H_2O_2) showed remarkable photocatalytic performances (the efficiency reached 100% and the rate constant was of 0.6624 min^{-1}). Therefore, this material can be considered promising for photocatalytic degradation processes due to its stability and reusability. Other fibers-based PVDF membranes embedding ZnO:La inorganic nanostructures with remarkable photocatalytic responses, good recovery and reuse after ten repeating cycles (the photocatalytic activity up to 98%) were obtained by electrospinning method (Pascariu et al. 2021b). Furthermore, Lee et al. (2018) reported promising results for the degradation of methylene blue, bisphenol A and 17α-ethynylestradiol pollutants in the presence of electrospun PVDF/TiO_2 porous membranes. Figure 5 shows a PVDF/ZnO:La fibrous membrane sample and its corresponding SEM image, as well as the UV–Vis absorption spectra for the photodegradation of Methylene Blue dye (C_0 = 10 mg/L) at various irradiation times and color removal efficiency (%) after five reuse cycles.

Figure 5. Morphology and performance of composite fibrous membrane (PVDF/ZnO: La): actual picture, micrograph (SEM image), UV-Vis profiles for methylene blue dye photodegradation, and color removal efficiency (%) after five reuse cycles

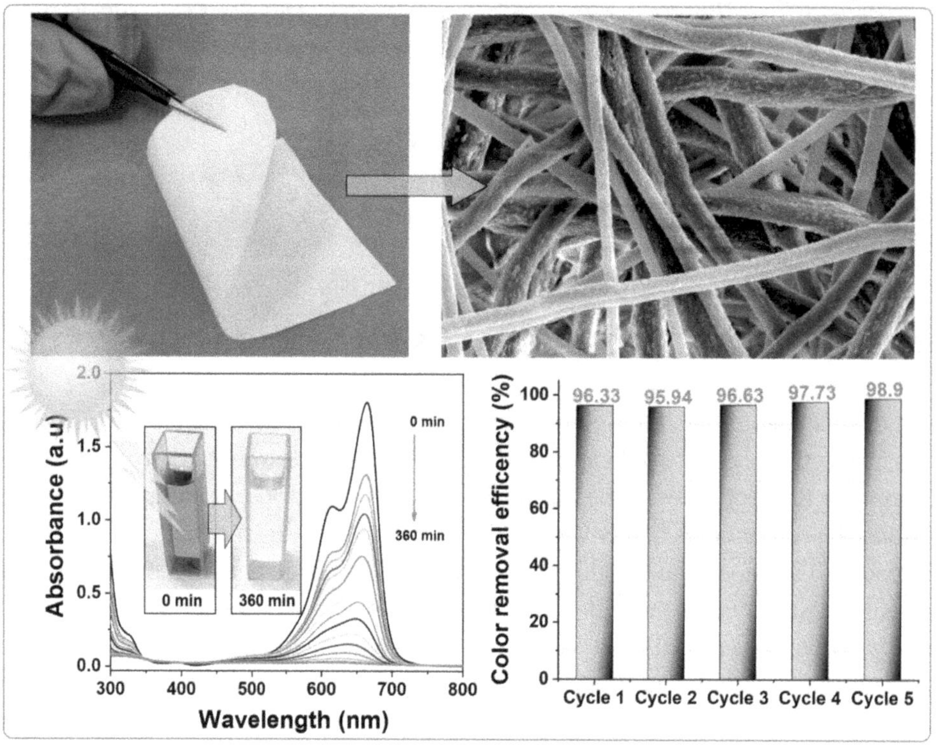

KINETICS MODELING

In the homogeneous and heterogeneous catalysis applied for advanced oxidation of the organic pollutants, the reaction kinetics is frequently modeled by the pseudo-first-order (PFO) kinetic equation. In such applications, the catalyst is used in excess with respect to the concentration of the organic pollutant. Consequently, the concentration of the catalyst can be included in the rate constant, leading to a pseudo-first-order (PFO) rate equation. From another perspective, this mathematical artifact can be seen as a reduction of the second-order rate equation (Pollutant + Catalyst → Product) to a pseudo-first-order rate equation (Pollutant → Product). The differential equation of the PFO model is given in Eq.(10), and the analytical solution derived from integration is given in Eq.(11). Another important kinetic parameter is the *half-life* of the reaction, $\tau_{1/2}$, which represents the time needed for the pollutant concentration to drop by half compared to its initial concentration. In the case of first-order and PFO reactions, the half-life ($\tau_{1/2}$) is estimated as given by Eq.(12).

$$-\frac{dC}{dt} = k \times C \quad (10)$$

$$C(t) = C_0 \times e^{-kt} \quad (11)$$

$$\tau_{1/2} = \frac{\ln(2)}{k} \cong \frac{0.693}{k} \quad (12)$$

where, C denotes the pollutant concentration (mg/L) at any moment of time t (min); C_0 is the initial pollutant concentration (mg/L); k is the pseudo-first-order (PFO) rate constant (min^{-1}); and, ($\tau_{1/2}$) is the half-life of the reaction (min^{-1}). Typically, the PFO rate constant k (min^{-1}) is determined through the regression analysis (nonlinear or linear). Nonlinear regression analysis enables to estimate the rate constant k (min^{-1}) in a more accurate fashion, since the regression procedure is done on the original nonlinear equation (*i.e.*, Eq.11). In case of the linear regression analysis, the Eq.11 must be subjected to linearization (mathematical transformation) that can be expressed as:

$$\ln\left(\frac{C_0}{C}\right) = k \times t \quad (13)$$

From Eq.13 the rate constant (k) can be also determined by representing a linear plot $\ln(C_0/C)$ versus t, from which *slope=k*. Even if linear regression seems to be an easier procedure than nonlinear regression, today's modern software packages include adequate tools for nonlinear regression (e.g., Origin, Minitab, Excel, Scilab, Matlab, Mathcad. SMath Studio, etc.)

Figure 6 shows as an example the kinetics of photodegradation of *ciprofloxacin* (organic pollutant from pharmaceutical effluent), under visible light irradiation in the presence of the doped catalyst TiO_2:La(0.1%) (Pascariu et al., 2022a). The experimental data were fitted to the PFO kinetics model by using regression analysis. In this respect, the nonlinear regression (Figure 6a) was compared with linear regression (Figure 6b) regarding the estimation of the PFO rate constant k (min^{-1}). As one can see from Figure 6, the nonlinear regression method revealed a rate constant of k=1.981×10^{-2} (min^{-1}). In turn, the linear regression technique disclosed a constant somewhat smaller (k=1.814×10^{-2} min^{-1}). This discrepancy represents about 8.4% and might be attributed to the linearization technique, which transforms the original nonlinear equation into a linear equation. Hence, the linearization technique may lead to additional numerical errors of mathematical origin (due to the transformation of equations). Similar conclusions were drawn for adsorption systems when these were modeled by pseudo-first-order and pseudo-second-order kinetics using both nonlinear and linear regression techniques (Kumar, 2006).

Figure 6. Kinetics of ciprofloxacin photodegradation under visible light irradiation and in the presence of TiO_2:La(0.1%) catalyst: (a) nonlinear regression analysis applied for estimation of the PFO rate constant k; (b) linear regression analysis for estimation of the PFO rate constant

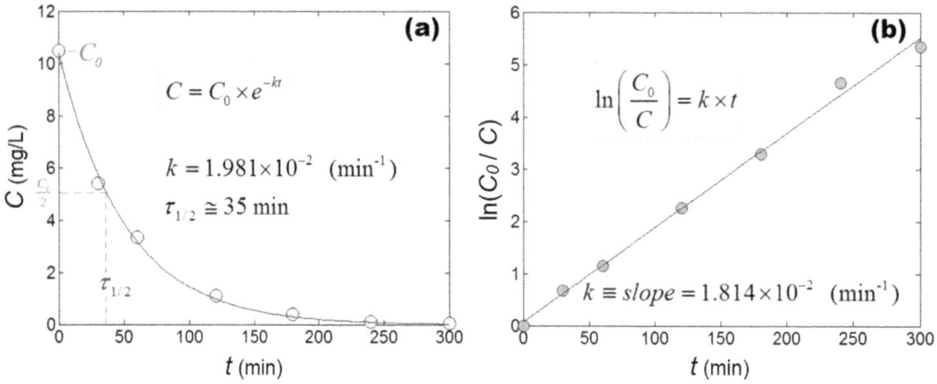

Usually, when applying nonlinear regression the goodness-of-fit (*i.e.*, the agreement between the model and experiment) can be estimated *via* the error function chi-square (χ^2), which can be expressed as:

$$\chi^2 = \sum_{j=1}^{n} \frac{\left(C_j^{calc} - C_j^{obs}\right)^2}{C_j^{calc}} \tag{14}$$

where, C_j^{obs} denotes the observed concentration of pollutant, C_j^{calc} represents the calculated concentration according to the PFO kinetic model, *n* is the number of data points on the kinetic curve, and *j* designates the integer index used for summation. Generally, the lower the χ^2 value, the better the performance of the kinetic model for prediction. For the picture shown in Figure 6a, the the χ^2 value was found to be χ^2=0.101 (Pascariu et al., 2022a). The goodness-of-fit might be expressed by using other residual error functions such as ε - the sum of the absolute errors (EABS), and ε^2 - the sum of the square of the errors (ERRSQ), which can be written as:

$$\varepsilon = \sum_{j=1}^{n} \left|C_j^{calc} - C_j^{obs}\right| \tag{15}$$

$$\varepsilon^2 = \sum_{j=1}^{n} \left(C_j^{calc} - C_j^{obs}\right)^2 \tag{16}$$

For the case of nonlinear regression analysis (Figure 6a), these residual error functions were found to be ε =0.872 and ε^2=0.174. For the data given in Figure 6b (the linear form of the PFO equation), the squared linear coefficient was close to unity (0.996) indicating a good linear correlation.

Table 2 summarizes the values of PFO rate constants (*k*, min^{-1}) and the half-life of the reaction ($\tau_{1/2}$, min) reported in the literature for different photocatalytic systems based mainly on the catalysts obtained through the electrospinning-calcination method. As one may observe from Table 2, the semiconductor catalysts based on doped TiO_2 and ZnO with lanthanides (La, Sm, Er) and transitional metals (Cu, Ag) disclosed reasonable PFO rate constants (k) of order 10^{-3} to 10^{-2} min^{-1} corresponding to a half-life of reaction ($\tau_{1/2}$) ranging from 10 to 160 min. Of course, these kinetic parameters (k and $\tau_{1/2}$) also depend on the applied experimental conditions, *e.g.* catalyst formulation and its dose, type of organic pollutant and its initial concentration as well as the type of light irradiation (visible or UV) (Table 2).

Table 2. Performance of doped catalysts (produced by electrospinning-calcination) for photodegradation of organic pollutants: PFO rate constants and half-life for different conditions

Catalyst (dose %w/v)	Pollutant (initial concentration)	k (min^{-1})	$\tau_{1/2}$, (min)	References
TiO_2:La(0.1%) (0.10%w/v)	MB (10 mg/L) / Vis*)	2.179×10^{-2}	32	(Pascariu et al., 2022a)
TiO_2:La(0.1%) (0.06%w/v)	CIP (10 mg/L) / Vis	1.981×10^{-2}	35	(Pascariu et al., 2022a)
TiO_2:Er(0.7%) (0.02%w/v)	MB (10 mg/L) /UV	5.100×10^{-3}	136	(Lee et al., 2018)
TiO_2:Er(0.1%) (0.02%w/v)	MB (10 mg/L) / Vis	8.990×10^{-3}	77	(Pascariu et al., 2022b)
TiO_2:Sm(0.1%) (0.02%w/v)	MB (10 mg/L) / Vis	1.014×10^{-2}	68	(Pascariu et al., 2022b)
TiO_2:Sm(0.1%) (0.09%w/v) + H_2O_2 (0.11M)	MB (10 mg/L) / Vis	3.496×10^{-1}	2.0	(Pascariu et al., 2022b)
TiO_2:Sm(0.1%) (0.09%w/v)	CIP (10 mg/L) / Vis	2.709×10^{-2}	25	(Pascariu et al., 2022b)
TiO_2:Sm(0.1%) (0.09%w/v) + H_2O_2 (0.11M)	CIP (10 mg/L) / Vis	4.292×10^{-1}	1.6	(Pascariu et al., 2022b)
TiO_2:Cu(0.05%) (0.02%w/v)	Amaranth (25 mg/L) / UV	2.089×10^{-2}	33	(Pascariu et al., 2022c)
TiO_2:Cu(0.05%) (0.02%w/v)	Amaranth (25 mg/L) / Vis	8.650×10^{-3}	80	(Pascariu et al., 2022c)
TiO_2:Cu(0.05%) (0.08%w/v)	Amaranth (25 mg/L) / Vis	2.135×10^{-2}	32	(Pascariu et al., 2022c)
TiO_2:Ag(0.1%) (0.04%w/v)	MB (10 mg/L) / Vis	1.290×10^{-2}	54	(Pascariu et al., 2021a)
TiO_2:Ag(0.1%) (0.04%w/v)	CR (10 mg/L) / Vis	7.280×10^{-2}	10	(Pascariu et al., 2021a)
ZnO:Ag(1%) (0.02% w/v)	Amaranth (25 mg/L) / Vis	9.517×10^{-3}	73	(Pascariu et al., 2020)
ZnO:Ag(1%) (0.02% w/v)	Amaranth (11 mg/L) / Vis	3.229×10^{-2}	21	(Pascariu et al., 2020)
ZnO:La(2%) (0.02% w/v)	CR (10 mg/L) / UV	2.734×10^{-2}	25	(Pascariu et al., 2019b)
ZnO:La(1%) (0.02% w/v)	CR (30 mg/L) / UV	4.320×10^{-3}	160	(Pascariu et al., 2019a)

Continued on following page

Table 2. Continued

Catalyst (dose %w/v)	Pollutant (initial concentration)	k (min^{-1})	$\tau_{1/2}$, (min)	References
ZnO:Er(1%) (0.02% w/v)	CR (30 mg/L) / UV	4.967×10^{-3}	140	(Pascariu et al., 2019a)
ZnO:Sm(1%) (0.02% w/v)	CR (30 mg/L) / UV	6.379×10^{-3}	100	(Pascariu et al., 2019a)
ZnO:Sm(1%) (0.02% w/v)	CR (10 mg/L) / UV	1.337×10^{-2}	48	(Pascariu et al., 2019a)
ZnO:La(2%) /C (0.01% w/v)	RhB (na) / UV	4.270×10^{-2}	16	(Zheng et al., 2013)
PVDF/ZnO:La (0.1% w/v)	MB (10 mg/L) / Vis	6.552×10^{-3}	106	(Pascariu et al., 2021b)
PVDF/ZnO:La (0.1% w/v)	RhB (5 mg/L) / Vis	5.613×10^{-3}	123	(Pascariu et al., 2021b)
PVDF/ZnO:Er (0.1% w/v)	MB (10 mg/L) / Vis	4.437×10^{-3}	156	(Pascariu et al., 2021b)
PVDF/ZnO:Er (0.1% w/v)	RhB (5 mg/L) / Vis	5.222×10^{-3}	133	(Pascariu et al., 2021b)
PVDF/ZnO:Sm (0.1% w/v)	MB (10 mg/L) / Vis	4.985×10^{-3}	139	(Pascariu et al., 2021b)
PVDF/ZnO:Sm (0.1% w/v)	RhB (5 mg/L) / Vis	4.208×10^{-3}	165	(Pascariu et al., 2021b)

Note: *) Type of irradiation of the photocatalytic system: Vis - visible light, and UV - ultraviolet light; MB=Methylene Blue; CR=Congo Red; CIP=Ciprofloxacin; RhB= Rhodamine B.

The encompassing of the inorganic catalyst (ZnO-based) into the polymeric matrix (PVDF) enabled to obtain some composite membranes (readily manipulated) with photo-catalytic properties showing a PFO rate constant of order 10^{-3} min^{-1} and a half-life reaction > 100 min (Table 2). Another interesting aspect is related to process intensification, which has been realized by adding hydrogen peroxide (H_2O_2) into the system followed by the optimization of the ratio between hydrogen peroxide and the catalyst dose. For instance, the addition of hydrogen peroxide to the systems that implied the TiO_2:Sm(0.1%) catalyst had a significant impact on the photodegradation of Methylene blue and Ciprofloxacin. For these systems, the PFO rate constants increased from order 10^{-2} min^{-1} to order 10^{-1} min^{-1}. Consequently, the half-life of the reaction dropped dramatically from tens of minutes (25 and 68 min) to about two minutes (~2 min), which is a remarkable result for the photocatalytic process under visible light irradiation.

Another useful kinetic law that can be applied to interpolate the curves of pollutant decay deals with pseudo-first-order kinetic with a stable component (March et al., 2005). According to this assumption, a portion of the initial pollutant is highly persistent, and this fraction represents a residual concentration of the pollutant noted as C_s (mg/L). Hence, the PFO kinetic model with a stable component may be written as given by (March et al., 2005; Mahu et al., 2020; Pascariu et al 2022b):

$$C(t) = C_s + (C_0 - C_s) \times e^{-kt} \quad (17)$$

For instance, the application of the PFO kinetic model with a stable component was successfully applied for describing the kinetics of photo-degradation of the Congo Red dye and 2,4-D herbicide using the TiO_2 catalyst prepared by ultrasound-assisted sol-gel method (Mahu et al., 2020). For the doped catalyst TiO_2:Sm(0.1%) produced by the electrospinning-calcination technique, the use of the PFO kinetic model with a stable component was found to be appropriate for the description of ciprofloxacin mineralization by photodegradation (Pascariu et al 2022b). In this application, the pollutant concentration was expressed in terms of total organic compounds (TOC) removal (Figure 7).

Figure 7. Kinetics of ciprofloxacin photodegradation in terms of TOC removal: catalyst type TiO_2:Sm(0.1%), catalyst dose=(0.09%w/v)

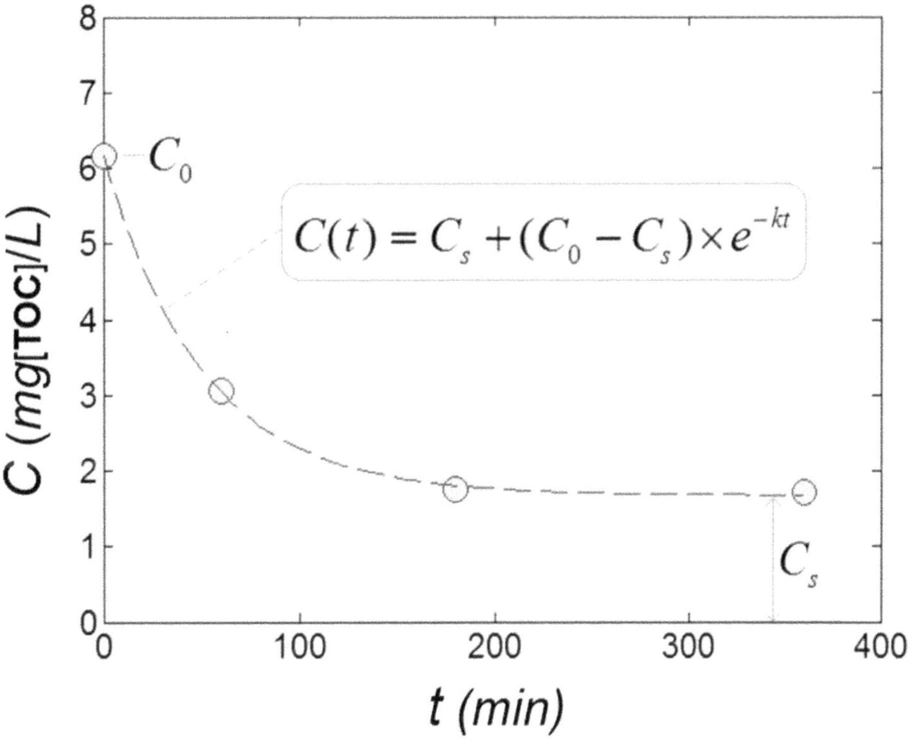

According to Figure 7, the mineralization of ciprofloxacin (CIP) is not complete (even after 6h irradiation), it is about 73%. The residual fraction of 27% represents the stable component that can be expressed in terms of concentration as C_s=1.672 mg(TOC)/L. The PFO kinetic model with a stable component revealed for this system (TiO_2:Sm(0.1%) / CIP) a rate constant of $k=1.976\times10^{-2}$ min^{-1}. In similar conditions, the catalyst TiO_2:Sm(0.1%) was tested for the mineralization of the methylene blue (MB) dye, for which the TOC removal was about 69% and rate constant was found to be $k=1.074\times10^{-2}$ min^{-1} (Pascariu et al 2022b). Thus, the PFO rate constant for the system TiO_2:Sm(0.1%)/MB was found to be about 1.8-fold smaller than for the system TiO_2:Sm(0.1%)/CIP. Even if methylene blue and ciprofloxacin have similar molecular sizes, their chemical structures and electronic configurations make the difference. For example, ciprofloxacin has a core system made of two concatenated rings (aromatic-heterocyclic), which is less conjugated compared with the phenothiazine system in methylene blue (Pascariu et al 2022b). Langmuir-Hinshelwood (LH) model is the most employed mathematical equation to explain the kinetics of heterogeneous catalytic processes (Vasanth Kumar et al. 2008). Using this mathematical expression (LH), it is possible to estimate the contribution of the chemical reaction and the adsorption phenomena to a heterogeneous photocatalytic process. The mathematical equation of the LH model is given by (Vasanth Kumar et al. 2008; Berberidou et al., 2017):

$$r = -\left(\frac{dC}{dt}\right) = \frac{k_r KC}{1+KC} \tag{18}$$

where, r denotes the rate of the reaction that changes within time, C – pollutant concentration that changes within time; k_r and K are parameters of the LH model. Note that the parameter k_r gives the contribution of the chemical reaction to the overall process (Vasanth Kumar et al. 2008). Instead, the parameter K reflects the contribution of the adsorption phenomena during the photocatalytic process. If the condition $KC<<1$ is satisfied, then the LH model might be approximated *via* PFO kinetics. Such an approximation might be assumed when working with a very low concentration of the pollutant and the equilibration experiments ("in the dark") indicated that the adsorption effect is negligible. However, it is not recommended to resort to this technique (reducing LH to PFO). It is better to estimate first both LH parameters (k_r and K) by means of the regression analysis (nonlinear or linear), and based on these to conclude if the adsorption is negligible or not. From a practical viewpoint, it is more facile to represent the LH equation in terms of initial reaction rate (r_0) *versus* initial pollutant concentration (C_0) (Vasanth Kumar et al. 2008; Kumar et al., 2022):

$$r_0 = \frac{k_r KC_0}{1+KC_0} \tag{19}$$

After mathematical re-arrangement, the equation (19) can be re-written in a linear form:

$$\frac{C_0}{r_0} = \frac{1}{k_r K} + \frac{1}{k_r}C_0 \tag{20}$$

Thus, the LH model parameters (k_r and K) can be estimated either by nonlinear regression (using Eq.19) or by linear regression by adopting the equation (20). Of course, nonlinear regression is preferred since it uses the original expression of the LH equation. In a typical procedure for developing the LH model, one needs first to estimate the initial reaction rate (r_0) as a tangent line to the PFO kinetic curve (Figure 8a). Second, this approach must be repeated for several kinetic profiles (Figure 8b). Having these kinetics outcomes, the LH model can be readily developed by the regression analysis.

Figure 8. Development of the Langmuir-Hinshelwood (LH) model: (a) determining the initial reaction rate (r_0) as a tangent line to the PFO kinetic curve; (b) determining several PFO kinetic profiles required for LH model development

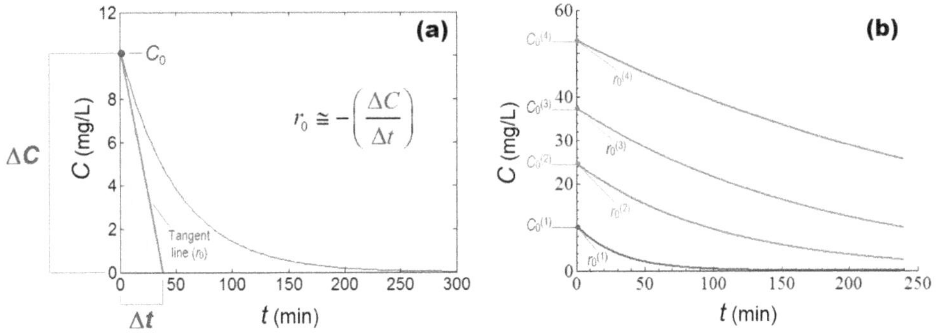

For instance, the kinetics profiles of photodegradation of Amaranth dye under UV-light irradiation and in the presence of ZnO:Ag(1%) catalyst were reported in the reference (Pascariu et al., 2020). Based on these kinetics profiles the LH model was developed both by nonlinear regression (Pascariu et al., 2020) and by linear regression. These outcomes are summarized in Figure 9, illustrating the curves fitting by nonlinear (Figure 9a) and linear (Figure 9b) regression analysis. According to the nonlinear regression analysis, the parameters were found to be k_r=0.324 mg·L^{-1}·min^{-1} and K=0.580 mg^{-1}·L (Pascariu et al., 2020). These results showed that the values of parameters were of the same order of magnitude, suggesting that the adsorption phenomena were significant along with the effect of the chemical reaction. It turns out that in such a scenario the pollutant is firstly adsorbed onto the solid surface of the catalyst and then it is subjected to the photodegradation reaction. Hence, the adsorption phenomena facilitate the photocatalytic decay of the organic pollutants by solid catalysts under UV-Vis light irradiation. A similar conclusion was drawn for the photocatalytic system ZnO:Sm/Congo red (Pascariu et al., 2019a). According to Figure 9, there was a reasonable difference between the LH model parameters determined by nonlinear and linear regression. However, the parameters estimated by nonlinear regression were considered to be more accurate.

Figure 9. Langmuir-Hinshelwood (LH) graphs showing the initial reaction rate (r_0) against initial pollutant concentration (C_0): (a) data interpolation by nonlinear LH equation (k_r=0.324 mg·L^{-1}·min^{-1} and K=0.580 mg^{-1}·L); (b) data linearization and fitting to linear LH equation (k_r=0.326 mg·L^{-1}·min^{-1} and K=0.518 mg^{-1}·L); data related to the photo-catalytic system ZnO:Ag(1%) /Amaranth

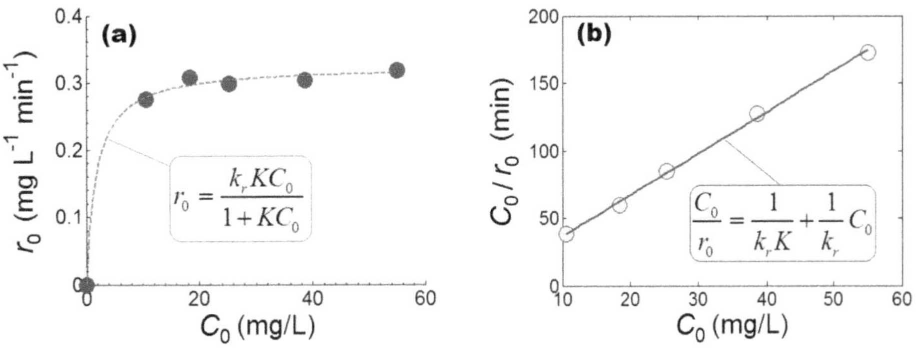

Another practical option to develop the LH model is to express the equation in terms of initial reaction rate (r_0) versus pollutant concentration at equilibrium (C_e), as given by (Caliman et al., 2007; Vasanth Kumar et al., 2008; Kumar et al., 2022; Poulios et al., 2022; Pascariu et al., 2022a):

$$r_0 = \frac{k_r K C_e}{1 + K C_e} \tag{21}$$

$$\frac{C_e}{r_0} = \frac{1}{k_r K} + \frac{1}{k_r} C_e \tag{22}$$

$$\frac{1}{r_0} = \frac{1}{k_r} + \frac{1}{k_r K} \times \frac{1}{C_e} \tag{23}$$

where, equation (21) is the nonlinear expression, while equations (22 and 23) represent the linear forms (two practicable types of linear expressions).

The equation (21) is applicable when the kinetics profiles were investigated starting from a fixed initial concentration (C_0), and the catalyst dose was varied to attain different equilibrium concentrations (C_e). For example, this kind of approach was employed to investigate the effect of the catalyst dose (TiO_2:La(0.1%)) on the performance of the photodegradation of Methylene blue dye (Pascariu et al., 2022a). The LH model developed for this system figured out the relationship between the initial rate of reaction (r_0) and the equilibrium concentration (C_e) of the pollutant. According to LH parameters, the contribution of adsorption phenomena was much more relevant than the effect of the chemical reaction ($K >> k_r$) (Pascariu et al., 2022a).

EMPIRICAL MODELING

Advanced oxidation processes (AOPs) can be optimized by adopting an approach based on the design of experiments (DoE) and empirical modeling (Caliman et al., 2007; Secula et al., 2008; Betianu et al., 2008; Zuorro et al. 2013; Chaker et al., 2021; Mousavi et al., 2021; Murshed et al., 2022). In a larger sense, this complex approach is known as the response surface methodology (RSM). The term "*response*" represents the output variable of a process or system that describes its performance. When a process or system is explored, we actually look for a true response surface of the experiment, which can be approximated by means of an empirical mathematical model. In such an approach, one searches to figure out the functional relationship between controllable input variables (factors) of the process and the response (performance) of the process. The factors (controllable input variables) appearing in AOPs are the catalyst dose (CatDose), the initial concentration of the hydrogen peroxide [H_2O_2] added to boost the process; the pH and temperature of the feed solution, as well as the irradiation time. The typical response subjected to maximization in AOPs is the removal efficiency Y (%) that may be expressed as:

$$Y = \left(1 - \frac{C_t}{C_0}\right) \times 100 \tag{24}$$

where, C_0 is the initial concentration of the organic pollutant in wastewater, and C_t is the residual concentration of the organic pollutant determined after a certain reaction time in the course of the advanced oxidation process.

For modeling by RSM, the actual factors (z_i) are converted to the coded variables (x_i), which are normalized for the same adimensional interval of variation. The scheme of convertion from actual factors to coded variables are given as follows (Bezerra et al., 2008; Mäkelä 2017):

$$z_i^0 = \frac{z_i^{\min} + z_i^{\max}}{2} \tag{25}$$

$$\Delta z = z_i^{\max} - z_i^0 = z_i^0 - z_i^{\min} \tag{26}$$

$$x_i = \frac{z_i - z_i^0}{\Delta z_i} \tag{27}$$

where, z_i^{min} denotes the lower bound (minimum value) of the actual factor; z_i^{min} is the upper bound (maximum value) of the actual factor; z_i^0 – the base (centered) level of the actual factor; Δz_i – the variation range of the factor in actual space; x_i denotes the coded value of the actual factor z_i. According to this relation (Eq.27), the main levels of the coded variable (x_i) imply the lower level denoted by (-1), the base or centered level denoted by (0), and the upper level designated as (+1). Additional axial points ($\pm\alpha$) may extend the lower and upper limits depending on the type of DoE adopted.

In general, the empirical model, developed by RSM, represents a second-order polynomial model with interaction terms, which can be written as (Myers et al., 2004; Bezerra et al., 2008; Khuri and Mukhopadhyay 2010; Mäkelä 2017):

$$\hat{Y} = b_0 + \sum_{i=1}^{n} b_i x_i + \sum_{i=1}^{n} b_{ii} x_i^2 + \sum_{i<j}^{n} b_{ij} x_i x_j \tag{28}$$

where, $\hat{Y}$ designates the estimated response by the model; x_i indicates the coded variables; b_0, b_i, b_{ii}, b_{ij} are the regression coefficients (offset term, first-order, quadratic and interaction coefficients), and n is the total number of factors. Note that the offset coefficient (b_0) suggests the average value of the response. In turn, the first-order coefficients (b_i) describe the main effects of factors on the response. The regression coefficients are calculated by the multiple regression method, as given by (Myers et al., 2004; Bezerra et al., 2008; Khuri and Mukhopadhyay 2010; Mäkelä 2017):

$$\bar{b} = \left(\bar{\bar{X}}^T \bar{\bar{X}}\right)^{-1} \bar{\bar{X}}^T \bar{Y} \tag{29}$$

where, $\bar{b}$ is the vector of regression coefficients; $\bar{\bar{X}}$ is the matrix of the coded values of factors, which was used to design the experiments; $\bar{\bar{X}}^T$ is the transpose of the matrix, *i.e.*, $[\bar{\bar{X}}^T]_{ji}=[\bar{\bar{X}}]_{ij}$; and, $\bar{Y}$ is the column vector of the response (implying the observed values).

For the case of two factors, the equation (28) can be written generically as:

$$\hat{Y} = b_0 + b_1x_1 + b_2x_2 + b_{11}x_1^2 + b_{22}x_2^2 + b_{12}x_1x_2 \tag{30}$$

After substitution of the equation (27) into equation (30), the final empirical model in terms of actual factors can be written as follows:

$$\hat{Y} = b_0 + b_1\left(\frac{z_1 - z_1^0}{\Delta z_1}\right) + b_2\left(\frac{z_2 - z_2^0}{\Delta z_2}\right) + b_{11}\left(\frac{z_1 - z_1^0}{\Delta z_1}\right)^2 + b_{22}\left(\frac{z_2 - z_2^0}{\Delta z_2}\right)^2 + b_{12}\left(\frac{z_1 - z_1^0}{\Delta z_1}\right)\left(\frac{z_2 - z_2^0}{\Delta z_2}\right) \tag{31}$$

Next, after the rearrangement of terms, the final empirical model becomes:

$$\hat{Y} = \beta_0 + \beta_1z_1 + \beta_2z_2 + \beta_{11}z_1^2 + \beta_{22}z_2^2 + \beta_{12}z_1z_2 \tag{32}$$

where, β_0, β_i, β_{ii}, β_{ij} represent the final empirical coefficients obtained after the rearrangement of the regression coefficients (b_0, b_i, b_{ii}, b_{ij}) and of the fixed values z_i^0 and Δz_i. The constructed empirical models are then validated statistically by checking the agreement between the mathematical model and experiment. In this respect, the analysis of variance (ANOVA) and residual analysis are the most employed statistical tests for empirical model validation (Bezerra et al., 2008; Mäkelä 2017).

In the following several applications are mentioned related to data-driven modeling and optimization of AOPs. The first application mentioned herein deals with a doped catalyst TiO_2:Sm(0.1%) sintered at 600°C that was applied for the photodegradation of methylene blue under visible light irradiation (Pascariu et al., 2022b). For empirical modeling and process optimization, a central composite experimental design of rotatable type was employed for experimentation. As a result, the pollutant removal efficiency (at t=15 min irradiation time) was evaluated and modeled as being dependent on two factors (1) the catalyst dose, and (2) the initial concentration of the hydrogen peroxide $[H_2O_2]$ added as a booster for process intensification. The final empirical model developed for this application can be written as (Pascariu et al., 2022b):

$$\hat{Y} = 63.58 + 53.09z_1 + 81.78z_2 - 29.92z_2^2 \tag{33}$$

subjected to: $0.034 \le z_1$=CatDose(TiO_2:Sm) ≤ 1.165 g/L; $0.040 \le z_2=[H_2O_2] \le 0.110$ M

In the second application, the doped catalyst TiO_2:Sm was inserted into the electrospun matrix of the polymer PVDF (polyvinylidene fluoride). Finally, a composite membrane with catalytic properties was produced that contained PVDF polymer, 15% TiO_2:Sm catalyst, and 2.5% graphene oxide (GO). This hybrid membrane TiO_2:Sm/PVDF/GO was tested for the photodegradation of methylene blue, also, under visible light irradiation (Pascariu et al., 2022d). A composite experimental design of a face-centered type was adopted to explore this process. The removal efficiency (recorded at t=60 min irradiation time) was evaluated and empirically modeled as being dependent on the catalyst dose (TiO_2:Sm/PVDF/GO) and initial peroxide concentration. The mathematical expression of the empirical model built for this system is given by (Pascariu et al., 2022d):

$$\hat{Y} = 64.72 + 107.47z_1 + 593.34z_2 - 2672.4z_2^2 - 1059.0z_1z_2 \quad (34)$$

subjected to: $0.1 \leq z_1$=CatDose(TiO_2:Sm/PVDF/GO) ≤ 0.3%w/v; $0 \leq z_2=[H_2O_2] \leq 0.1$ M

Both empirical models (equations 33 and 34) predicting the performance of the (TiO_2:Sm)-based catalysts were employed for the computer-aided simulation and navigation of the factorial space. In this regard, Figure 10 shows (for comparison) the computed contour-lines maps (2D) of the estimated response surfaces. The response surfaces maps highlighted in Figure 10 indicated the best photo-catalytic performance achieved at relevant catalysts doses and at high concentrations of hydrogen peroxide $[H_2O_2]$.

Figure 10. Contour-lines maps (2D) of the estimated response surfaces: (a) 2D map for the photocatalytic system TiO_2:Sm / MB dye; (b) 2D map for the photocatalytic system TiO_2:Sm/PVDF/GO /MB dye

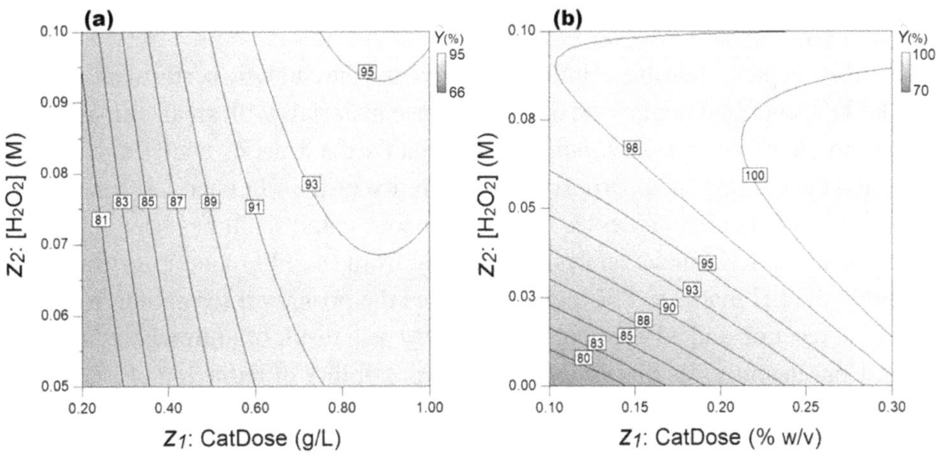

In addition, the constructed empirical models (equations 33 and 34) were employed for process optimization. The model-based optimization was carried out numerically by using the simplex method (the variant of the Nelder-Mead algorithm) (Bezerra et al., 2016). The optimal conditions for the photocatalytic system TiO_2:Sm / MB (Figure 10a) was found to be 0.9g/L catalyst dose and 0.11M $[H_2O_2]$, which ensured a removal efficiency of 96.22% (observed value at t=15 min irradiation time) (Pascariu et al., 2022b). The optimal conditions for the second photocatalytic system (Figure 10b) involving a composite membrane as catalyst (TiO_2:Sm/PVDF/GO) indicated the following values of factors: 0.28% w/v (2.8g/L) catalyst dose and 0.11M $[H_2O_2]$. Under these optimal conditions the maximal observed removal efficiency was 99.90% (for t=60 min irradiation time) (Pascariu et al., 2022d). Generally, a reasonable agreement between model and experiment was observed for both applications (Pascariu et al., 2022b,d).

Likewise, the response surface methodology was successfully applied to optimize other photocatalytic systems. For instance, TiO_2:Ag electrospun nanofibers were tested for phenol photodegradation and process optimization using RSM (Nourozi et al., 2020). For this system, a maximal degradation efficiency of 92.91% was observed for the optimal conditions of 5.62 mg/L initial phenol concentration, 2.06 g/L catalyst dose, and pH 7.87 (Nourozi et al., 2020).

The photocatalytic activity of ZnO:SnO_2 ceramic nanofibers for degradation of the persistent organic dye (Rhodamine B, RhB) was efficiently explored by using the design of experiments followed by process modeling and optimization (Pascariu et al., 2019c). Herein, a functional relationship was reported between the process performance (color removal efficiency) and three factors (irradiation time, temperature, and pH of feed solution). The model-based optimization was performed using the Monte Carlo simulation method. The established optimal conditions were found to be (1) 13.3 h irradiation time; (2) 42 °C temperature of the photodegradation reaction; and (3) pH 9.5 of feed solution. Under these optimal conditions, the process performance (color removal efficiency) was equal to 94.20% (under visible light irradiation) and 99.35% (under UV light irradiation) (Pascariu et al., 2019c). Thus, the reported ceramic material (ZnO:SnO_2) revealed high potential applicability as a photocatalyst for the degradation of persistent organic pollutants under visible- and UV-light irradiation.

SOLUTIONS AND RECOMMENDATIONS

As highlighted in this chapter, a feasible solution for enhancing the catalytic performance of semiconducting materials like TiO_2 and ZnO implies the doping of these materials with small amounts of transitional metals or rare-earth elements (e.g. *lanthanides*). As regards the kinetics analysis, a recommendation has been suggested concerning the appropriate applicability of the PFO and LH kinetic equations for modeling the advanced oxidation processes in heterogeneous catalysis. It has been inferred herein that it is better to perform the LH kinetic analysis separately from the PFO kinetic, without attempting to reduce prematurely the LH model to PFO equation. Since the primary information from the LH model (contribution of the reaction and adsorption) might be lost as a result of untimely reducing LH kinetics to PFO kinetics. Consequently, the discussion about the possibility of reduction of LH equation to PFO equation should be done after the estimation of LH parameters and disclosing that the contribution of adsorption phenomena was negligible.

FUTURE RESEARCH DIRECTIONS

Future research directions in the field should be focused on developing new oxide semiconductor materials as catalyst, expanding the area of applicability by incorporating them into different polymer matrices. Each new obtained material should be carefully characterized, optimized and used in different applications, correlating new results with the existing literature. Regarding the modeling and optimization of the advanced oxidation processes by heterogeneous catalysis, future investigations might be oriented toward employing advanced computational tools based on machine learning (such as *artificial neural networks* and *support vector machines*).

CONCLUSION

This chapter summarized a comprehensive overview of semiconducting materials prepared by the electrospinning-calcination technique, which are used in photocatalytic process based on advanced oxidation. A detailed description of the method for obtaining one-dimensional materials, as well as the parameters that influence the electrospinning process and the morphology of such materials was presented. The improvement of photocatalytic properties by fine-tuning of the morphology, structure, and optical properties was exemplified for semiconductor materials based on TiO_2 and ZnO. Moreover, it was proved that the photocatalytic performances of these materials can be improved by doping with different transition metals and lanthanides. Kinetics modeling of the advanced oxidation processes using semiconducting inorganic catalysts is essential to estimate the rate of the involved chemical reactions. In this respect, the most applied kinetic model in AOPs is the pseudo-first-order (PFO) kinetic equation. Another important kinetic model applied for the heterogeneous catalytic processes is the Langmuir-Hinshelwood (LH) model, which enables estimation of the contribution of the chemical reaction and the adsorption phenomena. The data interpolation to the LH model, according to nonlinear and linear regression techniques, is summarized in this chapter. Moreover, the advanced oxidation processes by heterogeneous catalysis can be empirically modeled and optimized for process intensification, using conventional chemometrics tools such as the design of experiment and response surface methodology.

REFERENCES

Ashraf, R., Sofi, H. S., Malik, A., Beigh, M. A., Hamid, R., & Sheikh, F. A. (2019). Recent trends in the fabrication of starch nanofibers: Electrospinning and non-electrospinning routes and their applications in biotechnology. *Applied Biochemistry and Biotechnology*, *187*(1), 47–74. doi:10.100712010-018-2797-0 PMID:29882194

Berberidou, C., Kitsiou, V., Kazala, E., Lambropoulou, D. A., Kouras, A., Kosma, C. I., Albanis, T. A., & Poulios, I. (2017). Study of the decomposition and detoxification of the herbicide bentazon by heterogeneous photocatalysis: Kinetics, intermediates and transformation pathways. *Applied Catalysis B: Environmental*, *200*, 150–163. doi:10.1016/j.apcatb.2016.06.068

Betianu, C., Caliman, F. A., Gavrilescu, M., Cretescu, I., Cojocaru, C., & Poulios, I. (2008). Response surface methodology applied for Orange II photocatalytic degradation in TiO_2 aqueous suspensions. *Journal of Chemical Technology and Biotechnology*, *83*(11), 1454–1465. doi:10.1002/jctb.1973

Bezerra, M. A., Oliveira dos Santos, Q., Santos, A. G., Novaes, C. G., Costa Ferreira, S. L., & Santos de Souza, V. (2016). Simplex optimization: A tutorial approach and recent applications in analytical chemistry. *Microchemical Journal*, *124*, 45–54. doi:10.1016/j.microc.2015.07.023

Bezerra, M. A., Santelli, R. E., Oliveira, E. P., Villar, L. S., & Escaleira, L. A. (2008). Response surface methodology (RSM) as a tool for optimization in analytical chemistry. *Talanta*, *76*(5), 965–977. doi:10.1016/j.talanta.2008.05.019 PMID:18761143

Bölgen, N., Demir, D., Aşık, M., Sakım, B., & Vaseashta, A. (2022). Introduction and Fundamentals of Electrospinning. In A. Vaseashta & N. Bölgen (Eds.), *Electrospun Nanofibers*. Springer., doi:10.1007/978-3-030-99958-2_1

Caliman, A. F., Cojocaru, C., Antoniadis, A., & Poulios, I. (2007). Optimized photocatalytic degradation of Alcian Blue 8 GX in the presence of TiO_2 suspensions. *Journal of Hazardous Materials, 144*(1-2), 265–273. doi:10.1016/j.jhazmat.2006.10.019 PMID:17112662

Chaker, H., Attar, A. E., Djennas, M., & Fourmentin, S. (2021). A statistical modeling-optimization approach for efficiency photocatalytic degradation of textile azo dye using cerium-doped mesoporous ZnO: A central composite design in response surface methodology. *Chemical Engineering Research & Design, 171*, 198–212. doi:10.1016/j.cherd.2021.05.008

Chavali, M. S., & Nikolova, M. P. (2019). Metal oxide nanoparticles and their applications in nanotechnology. *SN Applied Sciences, 1*(6), 607. doi:10.100742452-019-0592-3

Cojocaru, C., Pascariu Dorneanu, P., Airinei, A., Olaru, N., Samoila, P., & Rotaru, A. (2017). Design and evaluation of electrospun polysulfone fibers and polysulfone/$NiFe_2O_4$ nanostructured composite as sorbents for oil spill cleanup. *Journal of the Taiwan Institute of Chemical Engineers, 70*, 267–281. doi:10.1016/j.jtice.2016.11.005

Deng, Y., & Zhao, R. (2015). Advanced Oxidation Processes (AOPs) in Wastewater Treatment. *Current Pollution Reports, 1*(3), 167–176. doi:10.100740726-015-0015-z

Gusain, R., Gupta, K., Joshi, P., & Khatri, O. P. (2019). Adsorptive removal and photocatalytic degradation of organic pollutants using metal oxides and their composites: A comprehensive review. *Advances in Colloid and Interface Science, 272*, 102009. doi:10.1016/j.cis.2019.102009 PMID:31445351

Haider, A., Haider, S., & Kang, I. K. (2018). A comprehensive review summarizing the effect of electrospinning parameters and potential applications of nanofibers in biomedical and biotechnology. *Arabian Journal of Chemistry, 11*(8), 1165–1188. doi:10.1016/j.arabjc.2015.11.015

Ikram, M., Rashid, M., Haider, A., Naz, S., Haider, J., Raza, A., Ansar, M. T., Uddin, M. K., Ali, N. M., Ahmed, S. S., Imran, M., Dilpazir, S., Khan, Q., & Maqbool, M. (2021). A review of photocatalytic characterization, and environmental cleaning, of metal oxide nanostructured materials. *Sustainable Materials and Technologies, 30*, e00343. doi:10.1016/j.susmat.2021.e00343

Khuri, A. I., & Mukhopadhyay, S. (2010). Response surface methodology. *Wiley Interdisciplinary Reviews: Computational Statistics, 2*(2), 128–149. doi:10.1002/wics.73

Klán, P., & Vavrik, M. (2006). Non-catalytic remediation of aqueous solutions by microwave-assisted photolysis in the presence of H_2O_2. *Journal of Photochemistry and Photobiology A Chemistry, 177*(1), 24–33. doi:10.1016/j.jphotochem.2005.05.008

Kumar, R., George, L., Jun, Z., & Mukherji, S. (2022). Photocatalytic activity of graphene oxide-TiO_2 nanocomposite on dichlorvos and malathion and assessment of toxicity changes due to photodegradation. *Chemosphere, 308*, 136402/1–11. doi:10.1016/j.chemosphere.2022.136402

Lee, C. G., Javed, H., Zhang, D., Kim, J. H., Westerhoff, P., Li, Q., & Alvarez, P. J. J. (2018). Porous electrospun fibers embedding TiO_2 for adsorption and photocatalytic degradation of water pollutants. *Environmental Science & Technology*, *52*(7), 4285–4293. doi:10.1021/acs.est.7b06508 PMID:29553243

Lee, D. Y., Lee, M. H., & Cho, N. I. (2012). Preparation and photocatalytic degradation of erbium doped titanium dioxide nanorods. *Current Applied Physics*, *12*(4), 1229–1233. doi:10.1016/j.cap.2012.03.007

Liu, L., Liu, Z., Yang, Y., Geng, M., Zou, Y., Shahzad, M. B., Dai, Y., & Qi, Y. (2018). Photocatalytic properties of Fe-doped ZnO electrospun nanofibers. *Ceramics International*, *44*(16), 19998–20005. doi:10.1016/j.ceramint.2018.07.268

Mahu, E., Ignat, M., Cojocaru, C., Samoila, P., Coromelci, C., Asaftei, I., & Harabagiu, V. (2020). Development of porous titania structure with improved photocatalytic activity: Response surface modeling and multi-objective optimization. *Nanomaterials*, *10*, 998/1–15.

Mäkelä, M. (2017). Experimental design and response surface methodology in energy applications: A tutorial review. *Energy Conversion and Management*, *151*, 630–640. doi:10.1016/j.enconman.2017.09.021

Makula, P., Pacia, M., & Macyk, W. (2018). How to correctly determine the band gap energy of modified semiconductor photocatalysts based on UV–Vis spectra. *The Journal of Physical Chemistry Letters*, *9*(23), 6814–6817. doi:10.1021/acs.jpclett.8b02892 PMID:30990726

March, J. G., Gual, M., & Ramonell, J. (2005). A kinetic model for chlorine consumption in grey water. *Desalination*, *18*(1-3), 267–273. doi:10.1016/j.desal.2005.04.007

Mousavi, M., Soleimani, M., Hamzehloo, M., Badiei, A., & Ghasemi, J. B. (2021). Photocatalytic degradation of different pollutants by the novel gCN-NS/Black-TiO_2 heterojunction photocatalyst under visible light: Introducing a photodegradation model and optimization by response surface methodology (RSM). *Materials Chemistry and Physics*, *258*, 123912/1–16. doi:10.1016/j.matchemphys.2020.123912

Murshed, M. K., Dursun, A. Y., & Dursun, G. (2022). Application of response surface methodology on photocatalytic degradation of Astrazon Orange G dye by ZnO photocatalyst: Internal mass transfer effects. *Chemical Engineering Research & Design*, *188*, 27–38. doi:10.1016/j.cherd.2022.09.038

Myers, R. H., Montgomery, D. C., Vining, G. G., Borror, C. M., & Kowalski, S. M. (2004). Response surface methodology: A retrospective and literature survey. *Journal of Quality Technology*, *36*(1), 53–77. doi:10.1080/00224065.2004.11980252

Norouzi, M., Fazelia, A., & Tavakoli, O. (2020). Phenol contaminated water treatment by photocatalytic degradation on electrospun Ag/TiO_2 nanofibers: Optimization by the response surface method. *Journal of Water Process Engineering*, *37*, 101489. doi:10.1016/j.jwpe.2020.101489

Okutan, N., Terzi, P., & Altay, F. (2014). Affecting parameters on electrospinning process and characterization of electrospun gelatin nanofibers. *Food Hydrocolloids*, *39*, 19–26. doi:10.1016/j.foodhyd.2013.12.022

Pascariu, P., Cojocaru, C., Airinei, A., Olaru, N., Rosca, I., Koudoumas, E., & Suchea, M.P. (2021a). Innovative Ag-TiO_2 nanofibers offering excellent photocatalytic and antibacterial action. *Catalysts*, 11, 1234/1–20.

Pascariu, P., Cojocaru, C., Homocianu, M., & Samoila, P. (2022b). Tuning of Sm^{3+} and Er^{3+}-doped TiO_2 nanofibers for enhancement of the photocatalytic performance: Optimization of the photodegradation conditions. *Journal of Environmental Management, 316*, 115317. doi:10.1016/j.jenvman.2022.115317 PMID:35658261

Pascariu, P., Cojocaru, C., Homocianu, M., Samoila, P., Dascalu, A., & Suchea, M. (2022a). New La^{3+} doped TiO_2 nanofibers for photocatalytic degradation of organic pollutants: Effects of thermal treatment and doping loadings. *Ceramics International, 48*(4), 4953–4964. doi:10.1016/j.ceramint.2021.11.033

Pascariu, P., Cojocaru, C., Homocianu, M., Samoila, P., Grecu, I., & Bele, A. (2022d). New composite membranes based on PVDF fibers loaded with TiO_2:Sm nanostructures and reinforced with graphene/graphene oxide for photocatalytic applications. *Surfaces and Interfaces, 34*, 102382. doi:10.1016/j.surfin.2022.102382

Pascariu, P., Cojocaru, C., Olaru, N., & Airinei, A. (2019c). Photocatalytical activity of ZnO-SnO_2 ceramic nanofibers for RhB dye degradation: Experimental design, modeling and process optimization. *Physica Status Solidi. B, Basic Research, 1800474*, 1–8.

Pascariu, P., Cojocaru, C., Olaru, N., Samoila, P., Airinei, A., Ignat, M., Sacarescu, L., & Timpu, D. (2019a). Novel rare earth (RE-La, Er, Sm) metal doped ZnO photocatalysts for degradation of Congo-Red dye: Synthesis, characterization and kinetic studies. *Journal of Environmental Management, 239*, 225–234. doi:10.1016/j.jenvman.2019.03.060 PMID:30901700

Pascariu, P., Cojocaru, C., Samoila, P., Airinei, A., Olaru, N., Rotaru, A., Romanitan, C., Tudoran, L. B., & Suchea, M. (2022c). Cu/TiO_2 composite nanofibers with improved photocatalytic performance under UV and UV-visible light irradiation. *Surfaces and Interfaces, 28*, 101644/1–15. doi:10.1016/j.surfin.2021.101644

Pascariu, P., Cojocaru, C., Samoila, P., Airinei, A., Olaru, N., Rusu, D., Rosca, I., & Suchea, M. (2020). Photocatalytic and antimicrobial activity of electrospun ZnO:Ag nanostructures. *Journal of Alloys and Compounds, 834*, 155144. doi:10.1016/j.jallcom.2020.155144

Pascariu, P., Cojocaru, C., Samoila, P., Olaru, N., Bele, A., & Airinei, A. (2021b). Novel electrospun membranes based on PVDF fibers embedding lanthanide doped ZnO for adsorption and photocatalytic degradation of dye organic pollutants. *Materials Research Bulletin, 141*, 111376. doi:10.1016/j.materresbull.2021.111376

Pascariu, P., Homocianu, M., Cojocaru, C., Samoila, P., Airinei, A., & Suchea, M. (2019b). Preparation of La doped ZnO ceramic nanostructures by electrospinning – calcination method: Effect of La^{3+} doping on optical and photocatalytic properties. *Applied Surface Science, 476*, 16–27. doi:10.1016/j.apsusc.2019.01.077

Poulios, I., Micropoulou, E., Panou, R., & Kostopoulou, E. (2003). Photooxidation of eosin Y in the presence of semiconducting oxides. *Applied Catalysis B: Environmental, 41*(4), 345–355. doi:10.1016/S0926-3373(02)00160-1

Prakash, J., Samriti, Kumar, A., Dai, H., Janegitz, B. C., Krishnan, V., Swart, H. C., & Sun, S. (2021). Novel rare earth metaledoped one-dimensional TiO_2 nanostructures: Fundamentals and multifunctional applications. *Materials Today Sustainability*, *13*, 100066. doi:10.1016/j.mtsust.2021.100066

Remya, N., & Swain, A. (2019). Soft drink industry wastewater treatment in microwave photocatalytic system–exploration of removal efficiency and degradation mechanism. *Separation and Purification Technology*, *210*, 600–607. doi:10.1016/j.seppur.2018.08.051

Secula, M. S., Suditu, G. D., Poulios, I., Cojocaru, C., & Cretescu, I. (2008). Response surface optimization of the photocatalytic decolorization of a simulated dyestuff effluent. *Chemical Engineering Journal*, *141*(1-3), 18–26. doi:10.1016/j.cej.2007.10.003

Turky, A. O., Barhoum, A., MohamedRashad, M., & Bechlany, M. (2017). Enhanced the structure and optical properties for ZnO/PVP nanofibers fabricated via electrospinning technique. *Journal of Materials Science Materials in Electronics*, *28*(23), 17526–17532. doi:10.100710854-017-7688-6

Vasanth Kumar, K. (2006). Linear and non-linear regression analysis for the sorption kinetics of methylene blue onto activated carbon. *Journal of Hazardous Materials*, *137*(3), 1538–1544. doi:10.1016/j.jhazmat.2006.04.036 PMID:16730890

Vasanth Kumar, K., Porkodi, K., & Rocha, F. (2008). Langmuir–Hinshelwood kinetics – A theoretical study. *Catalysis Communications*, *9*(1), 82–84. doi:10.1016/j.catcom.2007.05.019

Vaseashta, A. (2007). Controlled formation of multiple Taylor cones in electrospinning process. *Applied Physics Letters*, *90*(9), 093115. doi:10.1063/1.2709958

Vaseashta, A., & Stamatin, I. (2007). Electrospun polymers for controlled release of drugs, vaccine delivery, and system-on-fibers. *Journal of Optoelectronics and Advanced Materials*, *9*(6), 1606–1613.

Velempini, T., Prabakaran, E., & Pillay, K. (2021). Recent developments in the use of metal oxides for photocatalytic degradation of pharmaceutical pollutants in water—A review. *Materials Today. Chemistry*, *19*, 100380. doi:10.1016/j.mtchem.2020.100380

Wang, S., Bai, J., Liang, H., Xu, T., Li, C., Sun, W., & Liu, H. (2014). Synthesis, characterization, and photocatalytic properties of Ag/TiO_2 composite nanofibers prepared by electrospinning. *Journal of Dispersion Science and Technology*, *35*(6), 777–782. doi:10.1080/01932691.2013.813392

Wang, W., Yang, R., Li, T., Komarneni, S., & Liu, B. (2021a). Advances in recyclable and superior photocatalytic fibers: Material, construction, application and future perspective. *Composites. Part B, Engineering*, *205*, 108512. doi:10.1016/j.compositesb.2020.108512

Wang, Y., Yokota, T., & Someya, T. (2021b). Electrospun nanofiber-based soft electronics. *NPG Asia Materials*, *13*(1), 22. doi:10.103841427-020-00267-8

Zheng, W., Miao, Q., Tang, Y., Wei, W., Xu, J., Liu, X., Qian, Q., Xiao, L., Huang, B., & Chen, Q. (2013). La(III)-doped ZnO/C nanofibers with core–shell structure by electrospinning–calcination technology. *Materials Letters*, *98*, 94–97. doi:10.1016/j.matlet.2013.02.004

Zuorro, A., Fidaleo, M., & Lavecchia, R. (2013). Response surface methodology (RSM) analysis of photodegradation of sulfonated diazo dye Reactive Green 19 by UV/H_2O_2 process. *Journal of Environmental Management*, *127*, 28–35. doi:10.1016/j.jenvman.2013.04.023 PMID:23676376

Section 2

Biomedical Aspects of Redox Processes

Chapter 12
Redox Reactions of Coordination Compounds in the Biomedical Environment

Andrew C. Benniston
Newcastle University, UK

Lingli Zeng
Sun Yat-sen University, China

ABSTRACT

Redox balance is important for maintaining physiological activities in organisms. Once the level of redox-active species is abnormal, it may destroy the balance of the internal environment or further induce biological pathological changes. Therefore, it is of great scientific significance to pay attention to the changes of redox-active species in vivo. Coordination compounds are very broad and important molecular systems that can undergo a wide variety of electronic transitions, with the charge transfer from metal to ligands and their associated excited states being particularly significant. These characteristics give coordination compounds the possibility of regulating redox reactions in vivo. After cisplatin, ruthenium complexes have been extensively studied in order to be alternative molecular systems with less toxicity to normal cells. This chapter will mainly introduce in vivo redox systems, redox process detection as well as ruthenium anti-tumor drugs for redox regulation.

INTRODUCTION

Redox balance is fundamental and significant for maintaining physiological activities in organisms. Bioactive small molecules including reactive oxygen species (ROS), reactive nitrogen species (RNS) and reactive sulfur species (RSS) are very important signal molecules involved in redox balance in the human body (Gao et al., 2019). They are vital in human physiology, participating in physiological processes such as cell signal transduction (Zuo et al., 2022), cell protection (Cao et al., 2018), vasodilation (Xu et al., 2022), and immunity and metabolism (Yang et al., 2020) within a normal level. However, abnormal

DOI: 10.4018/978-1-6684-7198-2.ch012

redox species levels can cause disorders in the human body and induce various diseases (Wang et al., 2021). For instance, ROS and RNS are associated with some pathological processes such as Alzheimer's disease (AD) (Bhatt et al., 2021), diabetes (Rendra et al., 2019), cardiovascular diseases (Xu et al., 2019), ageing and cancer (Kim et al., 2019), while RSS play an important role in maintaining redox balance and as antioxidants and a free radical scavenger (Sawa et al., 2022). Therefore, it is meaningful to keep an eye on maintaining the level of redox species in a reasonable range in vivo.

Metal complexes are very broad and important molecular systems. The regulation of redox balance using the structural advantages of metal complexes is of great significance to the diagnosis and treatment of clinical diseases and has aroused extensive attention and interest from researchers. Cis-platin is the most classic example of metal complex antitumor application. However, it is well known that cis-platin will not selectively inhibit tumors. Therefore, its cytotoxicity will cause damage to normal body tissues resulting in detrimental side effects which are often worried about by users (Perse, 2021; Chovanec et al., 2017). To reduce the toxicity of cis-platin, scientists have modified cis-platin with different functional groups in the axial direction, such as a targeted group (Fang et al., 2019) or photoreactive group (Ponte et al., 2019; Gao et al., 2021) to obtain Pt(IV) prodrugs with lower cytotoxicity. Under certain conditions, such as in a highly reducing tumor microenvironment, non-toxic Pt(IV) prodrugs can be reduced to toxic Pt(II) drugs, which further kill tumor cells (Figure 1) (Babu et al., 2020; Karmakar et al., 2019). Compared with cis-platin, ruthenium(II) complexes display more variability in their structural design, ligand exchange ability, good thermodynamic stability, abundant photochemical, photophysical and redox properties, as well as low toxicity and can be easily absorbed and quickly metabolized by the body with a different anti-tumor mechanism than cis-platin. In recent years, ruthenium complexes have attracted wide attention in the field of metal-organic chemistry, making them the most promising antitumor drugs after platinum-based drugs. At the same time, ruthenium (II) complexes are also believed to be able to involve in some redox reactions regulated by intracellular biomolecules such as GSH and NADPH (Liu et al., 2022), which can effectively improve anticancer activity and safety.

Figure 1. Representative examples of platinum prodrugs and drug releasing mechanisms

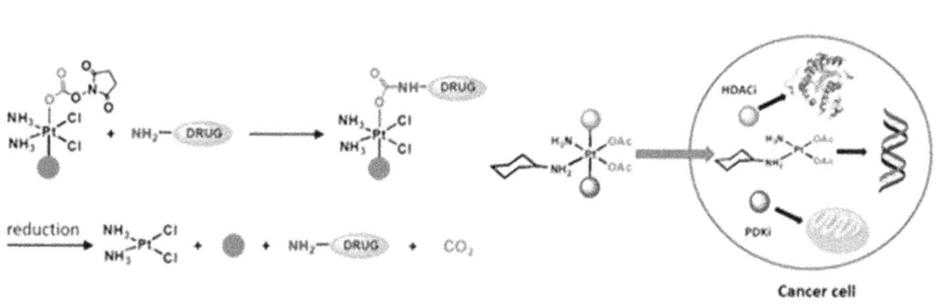

In this chapter, we will give a brief introduction to the main redox species in living organisms and the way of detecting the products generated from the redox reaction. We also report the state-of-the-art on the applications of using ruthenium complexes in regulating the redox process to achieve the ultimate antitumor goals.

REDOX REACTIONS IN THE BIOLOGICAL ENVIRONMENT

Cell redox microenvironment refers to the balanced state of a cell redox reaction, which is the equilibrium of electron transfer among biomolecules from the chemical point of view. The balance of an organism's redox state is mainly maintained by the regulation of enzymes and small molecules. When cells are influenced by the environment or regulated by signals, the redox equilibrium will change when oxidizing active substances or reducing active substances change. Redox homeostasis is essential for normal cell survival. Compared with normal tissues and cells, the redox state in tumors is unbalanced (Jorgenson et al., 2013). The level of ROS in tumor cells increases significantly, which may be related to many factors, such as mitochondrial dysfunction (Murphy, 2013), oncogene activation (Gomez et al., 2019), abnormal oxidative metabolisms as well as hypoxia/reoxygenation cycle (Ghasemi et al., 2020), which makes redox homeostasis easier to be broken, resulting in intracellular accumulation of ROS and eventually leading to apoptosis. The disorder of redox balance also has a close relationship to the occurrence and development of the tumor, including cell proliferation and differentiation, oncogene activation, carcinogenesis, metastasis, reversal and apoptosis of cancer cells (Cheung & Vousden, 2022). Therefore, tumor therapy by regulating cell redox homeostasis has good effectiveness and specificity, which has broad development prospects.

Main Oxidation Substances in Cells

Intracellular active oxidation substances mainly include ROS and RNS, which consist of neutral or anionic active small molecules produced in a variety of cells (Zarkovic, 2020). Specifically, common intracellular oxidation substances include hydrogen peroxide (H_2O_2), hydroxyl radical ($^{\bullet}OH$), superoxide anion ($O_2^{-\bullet}$), singlet oxygen ($^1O_2$), hypochlorous acid ($HOCl/ClO^-$) and $RO^{\bullet}/ROO^-$ (R = hydrogen or alkyl group). Most ROS occur as by-products of mitochondrial oxidative phosphorylation or as intermediates of oxidoreductases and metal-catalyzed oxidation. Theoretically speaking, the O_2 molecule contains two unpaired electrons in different orbitals of the outer shell, therefore it is easy to form free radicals, which lead to the formation of many ROS by adding electrons to continuously reduce O_2. RNS includes NO, $ONOO^-$, NO^+, N_2O_3 and so on (Figure 2) (Ushio-Fukai et al., 2021).

Figure 2. Formation of ROS and RNS by redox processes where SOD = superoxide dismutase and MPO = myeloperoxidase

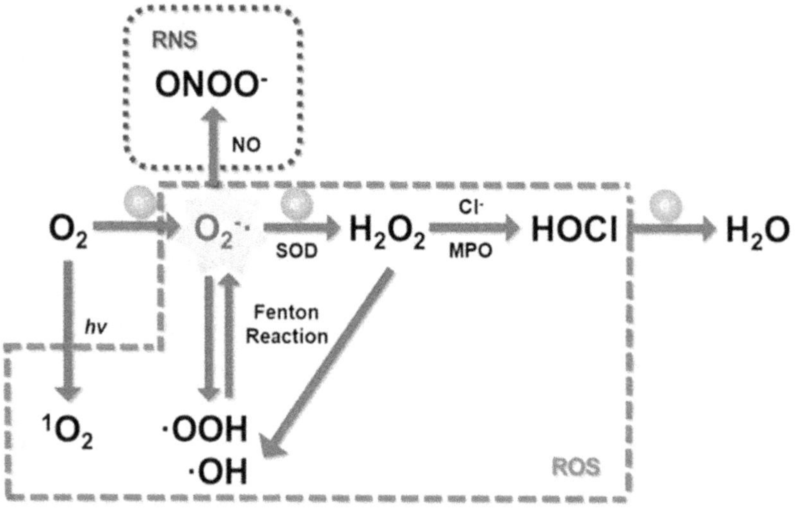

From the organelle level, endoplasmic reticulum (ER) and lysosome are oxidized subcellular compartments in various cells. ER has different oxidation functions in different cell types. For example, in liver cells, the function of ER is to activate O_2 for steroid synthesis. In addition, a more general function of the ER is to introduce structuralized disulfide bonds during protein folding (Malhi & Kaufman, 2011). Meanwhile, a low pH environment in lysosomes prevents the thioredoxin (Trx) or glutathione (GSH) systems from reducing disulfide that normally operates in neutral environments in lysosomes, so lysosomes are also in an oxidized state compared to other suborganelles (Xu & Ren, 2015; Duca et al., 2021; Duca et al, 2022).

At the body level, ROS is produced during the synthesis of ATP from mitochondria, a protein synthesized from the ER, and activation of the NADPH oxidase (NOX) family members. The main source of ROS comes from the substrate end of the respiratory chain in the mitochondrial inner membrane. During the electron transfer process, the mitochondrial electron transport chain (ETC) transfers electrons to O_2. Therefore, part of O_2 is reduced, followed by forming $O_2^{-\bullet}$ and H_2O_2 (Napolitano et al., 2021). Throughout the entire electron transfer process $O_2^{-\bullet}$ is an important intermediate because it is a precursor of most of the ROS, mainly produced by protease complexes I and III in the respiratory chain of the mitochondrial inner membrane. This part, which is a by-product of metabolism having been seen as toxic molecules for a long time, damages biological macromolecules (Circu & Aw, 2010). But in recent years scientists considered that low levels of ROS can be used as a signal of small molecules with physiological functions. Another important source of ROS is NADPH oxidase (NOX), which is usually expressed on the cytoplasmic membrane (Krause, 2018).

ROS plays as a double-edged sword in tumors (Huang et al., 2021). On one hand, ROS activate a variety of redox reactions and signaling pathways, such as NF-κB (Han et al., 2021), MAPK (Lian et al., 2022), P53 (Cao et al., 2020), PI3K/AKT (Yang et al., 2019) and KEAP1-NRF2 (Clerici & Boletta, 2020), which promote tumor genesis, development and metastasis. Under normal physiological conditions, ROS levels in organisms are kept at a reasonably low level, maintaining normal cell growth and apoptosis (Zhang et al., 2016). On the other hand, when ROS is excessive, it can cause cell stress and damage through the oxidative stress mechanism, and finally lead to cell death by activating different cell death pathways, such as apoptosis, necrosis, and autophagy (Cordani et al., 2019; Zhou et al., 2022). Higher levels of ROS may damage biological macromolecules, including DNA, proteins and cell membrane components, resulting in molecular and cellular dysfunction (Tong et al., 2020; Slezak et al., 2021) and even lead to diseases such as Alzheimer's disease (Zhou et al., 2022), cancer (Stieg et al., 2022), cardiovascular disease (Ghosh & Shcherbik, 2020) and ageing (Pichaud et al., 2019). In many types of cancer, the expression of ROS-related sensitive signaling pathway proteins continues to increase, eventually leading to oxidative stress in tumor cells.

Main Reducing Substances in Cells

The main reductive active substances in cells mainly include antioxidants, enzymes and protein systems (Jasek-Gajda et al, 2020) such as small molecule antioxidants like GSH, hydrogen sulfide (H_2S), cysteine (Cys), vitamin C and vitamin E (Olson et al., 2018). GSH is the main reducing compound and is most abundant in tumor cytoplasm (2-10 mM), which is much higher than its extracellular concentration (2-20 μM), while it is at least 4 times higher in a tumor than in normal tissues. At the organelle level the nucleus is normally in a reduced state compared to the cytoplasm.

Studies have shown that the s-glutathionylation of a protein occurs independently in the cytoplasm and the nucleus, respectively. And the GSH/GSSG redox potential may decrease more in the nucleus than in the cytoplasm (Wu et al., 2004). Macromolecular antioxidant enzymes such as catalase (CAT), superoxide dismutase (SOD), glutathione peroxidase (GPx), thioredoxin system (Trx) (Roh et al., 2017) and glutedoxin system are important endogenous antioxidant systems. Disulfide isomerase (PDI) and heat shock protein (HSP) also play important roles in the regulation of redox equilibrium (Kyani et al., 2018). In addition, ferrous ions and lysosomal mercaptan reductase (GILT) are reductive (Figure 3).

Figure 3. Major redox substances found in cells

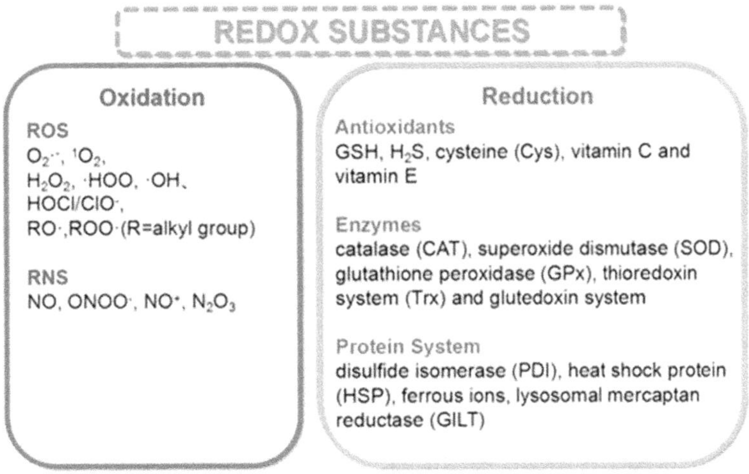

TECHNIQUES FOR DETECTING REDOX PRODUCTS

ROS play a mediating role in cell survival, stress and apoptosis. Different concentrations of ROS play disparate roles in the process, resulting in diverse cell fates. Therefore, reliable measurements for monitoring ROS amount are necessary. The following are several common methods for detecting ROS including the use of an electromagnetic field, UV-Vis spectroscopy, fluorescence spectrophotometry, confocal microscopy and chromatography.

Electron Paramagnetic Resonance (EPR)

EPR spectroscopy, which is also known as electron spin resonance (ESR) spectroscopy, is a method for detecting free radicals and is the "gold standard" for evaluating free radicals generated from oxygen in biological systems (Khramtsov, 2018). EPR signals are usually difficult to detect for highly reactive and short lifetime species (e.g., $O_2^{\bullet-}$, •OH), therefore the combination of EPR and spin traps technology can compensate for this defect (Xie et al., 2018). First, the spin trap reagent reacts with free radicals to produce relatively stable free radical additive products which can be easily detected by EPR, and then followed by the EPR technology determination (Figure 4). It is a powerful and reliable technique for unambiguously measuring $O_2^{\bullet-}$, •OH and NO in biological samples (Suzen et al., 2017; Mikhailov et al., 2019).

Figure 4. Working principles of a typical spin trap reagent 5,5-dimethyl-1-pyrroline-N-oxide (DMPO)

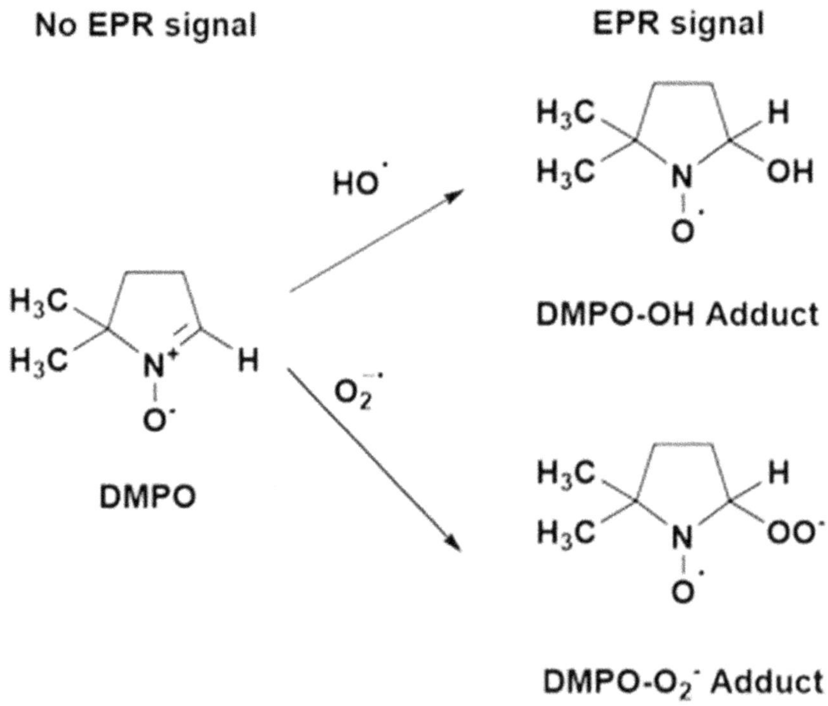

UV-Vis Spectrophotometry

UV-Vis spectrophotometry is a very classic method for detecting ROS. The principle is based on the reaction between free radicals and redox substances. After measuring the absorbance and calculating the difference between substrates and products at different wavelengths after the reaction, it is possible to determine free radicals semi-quantitatively. The most commonly used methods are cytochrome C reduction assay (Li et al., 2022b) and the nitrotetrazolium blue (NBT) test (Figure 5) (Sasahara et al., 2020).

Figure 5. Working principles of a NBT test for ROS detection using redox processes

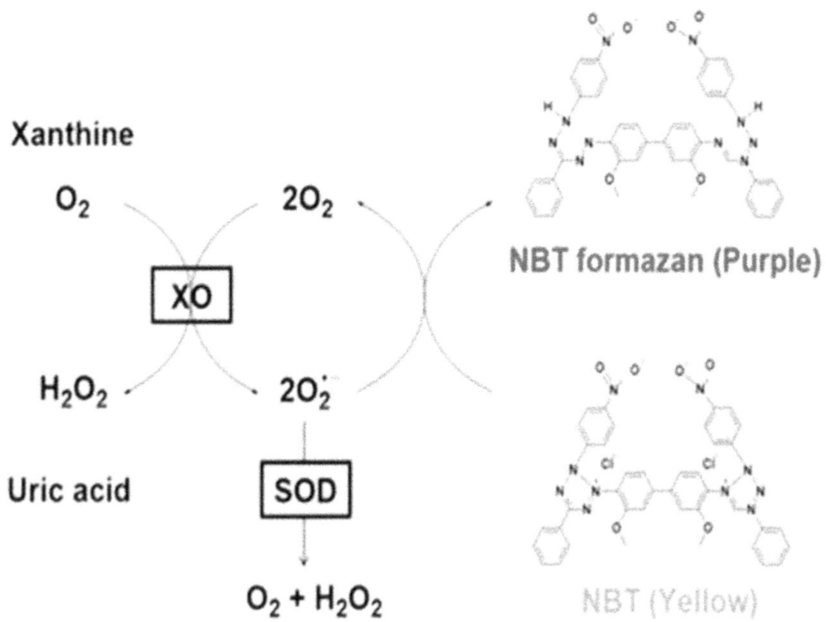

Fluorescent Staining

Fluorescent staining method can also be used to detect ROS. Dihydroethylene (DHE), dichlorodihydrofluorescein diacetate (DCFH-DA) and Amplex Red (cell impermeable) are three types of most commonly used probes (Jiang et al., 2020). The probes are usually sensitive to oxidants and non-fluorescent until oxidized by oxygen. Under certain stimulation, the oxidative state of the cell could be reflected by cell permeable substances, therefore providing useful information on free radicals. The working principle of these probes is usually going through a single electron free radical oxidation process, followed by forming a probe plus free radical intermediate, and eventually obtaining fluorescent final products. For instance, DHE can enter cells and form the specific red fluorescent product 2-hydroxy-ethidium via oxidation by $O_2^{\bullet-}$, or form a non-specific red fluorescent substance ethidium via oxidation by other ROS or RNS such as $^{\bullet}OH$, H_2O, or $OONO^-$. The compound DCFH-DA is another, commonly used to directly measure the whole amount of intracellular ROS. After entering the cells, DCFH-DA can be hydrolyzed by esterase to form the intermediate DCFH, which is membrane impermeable. It can stay and be diffused internally and oxidized by non-specific oxygen radicals, thus generating fluorescent DCF from non-fluorescent intermediate radicals DCFH (Estalayo-Adrián et al., 2020; Stultz et al., 2020). This method is simple but less specific. Amplex Red is used to detect H_2O_2 (Watabe et al., 2011). As measured by fluorescence-based techniques such as fluorescence spectrophotometry or by confocal microscope imaging, the fluorescent intensity reflects the ROS level in the cell directly.

Chemiluminescence (CL)

Chemiluminescence analysis is a technique producing light radiation by a chemical reaction, which can be used for the detection of superoxide (Li et al., 2022a). The essential difference between CL and other luminescence analysis is that the energy source absorbed by the system to produce light irradiation is different. Similar to the fluorescence probe, the CL probe can react with $O_2^{\bullet-}$ to form a photon and be captured by an instrument without using an extra excitation light source. It is a new analytical method with high sensitivity and fast analysis speed. Commonly used CL reagents include luminal (Qiao et al., 2018), lucigenin (Wang et al., 2022) and crustacean luciferin (Syed & Anderson, 2021). The CL reaction involves multiple steps while intermediate plays a key role in the luminescence process, therefore the limitations of the CL probe are similar to those of a fluorescent probe.

Chromatography

Chromatography is mainly used for $^{\bullet}OH$ and the corresponding reaction product separation and identification (Catapano et al., 2019). Compared with EPR, the instrument is more common and can be achieved in the general laboratory. In the system $^{\bullet}OH$ reacts with specific reagents to produce stable compounds, which can be detected by chromatographic analysis, usually using liquid chromatography combined with mass spectrometry (LC-MS). Benzoic acid, salicylic acid and DMSO are three commonly used reagents to stabilize free radicals. Although the chromatography method is easier to determine $^{\bullet}OH$ than EPR, the sensitivity and accuracy are still insufficient. The reaction products should have a certain stability, only then are they conducive to separation and detection.

PROGRESS OF ANTI-TUMOR DRUGS BASED ON ROS REGULATION

Physical Properties of Ruthenium Complexes

Coordination compounds refer to a kind of molecular system formed by metal ions and ligands through coordination bonds. Due to the diversity of electronic structure and bonding properties of transition metal atoms and the abundance of ligands, the number of transition metal complexes can be described in tens of thousands (Imberti et al., 2020). Cis-platin and its derivatives have achieved great success in anti-tumor therapy, but the toxicity and drug resistance of traditional platinum anticancer drugs limit the further development of these compounds in clinical practice. The side effects of platinum drugs have driven researchers to not only look at derivatives of platinum drugs, but also to start developing complexes using metals in the same group as platinum with lower toxicity but efficient anticancer ability. In recent years, non-platinum compounds, such as ruthenium (II), iridium (III) and rhenium (I) complexes with d^6 electronic structure, have attracted more and more attention due to their rich photophysical and photochemical properties, redox properties, diverse geometric configurations and good water solubility (Li & Chen, 2020). These complexes have been developed rapidly into coordination polymers, metal-organic frameworks (MOFs), metal-organic cages (MOCs), coordination supramolecular chemistry and other fields. Applications in catalysis, biological activity and other areas have attracted great attention from scientists (Malinowski et al., 2020). The ruthenium (II) complexes can be easily varied by altering the ligands attached to the metal centre. Also, due to their good thermodynamic stability, beneficial photochemical and redox properties, low toxicity, easy absorption by the body, quick metabolization and different anti-tumor mechanism from the cis-platin, ruthenium (II) complexes have aroused widespread attention in the field of metal organic chemistry, making them the most promising anticancer drugs after platinum (Martinez-Alonso & Gasser, 2021). The most marked one is the complex $[Ru(bpy)_3]^{2+}$ (bpy = bipyridine) and its derivatives. In an aqueous solution, $[Ru(bpy)_3]^{2+}$ has several significant characteristic absorption peaks. The absorption peaks at 452 nm and 425nm are assigned to metal-to-ligand charge transfer (MLCT) from the ruthenium metal center to a bpy ligand. Bands at 352 nm and 320nm belong to a d→d transition at the ruthenium center. At around 287nm, there is the $\pi \rightarrow \pi^*$ transition of the bpy ligand (Ghosh et al., 2009). From the perspective of structural characteristics, studies have shown that ruthenium is suitable for the development of effective antitumor drugs because of the following reasons. First, the ligand exchange rate of ruthenium complex is similar to that of platinum complex, and it has a high kinetic stability (Shahzad et al., 2022). Secondly, the ruthenium complex has a suitable redox potential. Ruthenium usually has three valence states which are Ru(II), Ru(III) and Ru(IV) (Shi et al., 2020). In these three valence states, ruthenium complexes generally adopt six-coordination or eight-coordination configurations. Changing the type and structure of ligands can change their electrochemical properties correspondingly. For instance, Ru(III) and Ru(IV) can be reduced by glutathione (GSH) and ascorbic acid in the hypoxic environment of a biological system (Wiśniewska et al., 2019), and Ru(II) can be oxidized by molecular oxygen and cytochrome oxidase (Sousa et al., 2021). Thirdly, ruthenium can mimic the binding of Fe ions to biological macromolecules, especially serum proteins and transferrin in plasma, which are mediators of Fe transport and can reduce drug toxicity (Simović et al., 2019). The ruthenium complexes with antitumor activity reported to date can be roughly divided into four categories: KP1091 type complexes (Gransbury et al., 2016), NAMI-A type complexes (Yuan et al., 2020), ruthenium (II) arene complexes (Iacopini et al., 2022; Chen et al., 2020), and ruthenium polypyridine complexes (Jiang et al., 2021).

ROS Regulation Mechanisms With Ruthenium Complexes

Based on the needs of cancer treatment, several different ruthenium complexes have been proposed to achieve the regulation of ROS, some being singlet oxygen generators as photodynamic therapy (PDT) photosensitizers, being activated by reductive substances in the body or regulating intracellular metabolism to fight tumors.

Used as PDT Photosensitizers

Since ruthenium (II) complexes have the function of photoinduced charge separation, the regulation of ROS is largely caused by PDT (Wei et al., 2021). The working principle of PDT is that photosensitizer is injected into the tumor site. Under certain wavelength irradiation, electrons are transferred to O_2, followed by generation active the superoxide radical $O_2^{\bullet-}$, damaging DNA and mitochondrial membrane, and eventually inducing tumor cell apoptosis. For instance, Chakravarty *et al.* reported a BODIPY-ruthenium(II) bis-terpyridine complex [Ru(tpy-BODIPY)(tpy-R)]Cl_2 (tpy = 4'-phenyl-2,2':6',2"-terpyridine) for cell imaging and type I/-II PDT. Among the series, complex RU1 which was modified with a pendant biotin moiety as a cancer cell targeting group, demonstrates a remarkable PDT efficacy with a high singlet oxygen quantum yield (Φ_Δ) of 67%, a high phototherapeutic index (PI) which is over 1400, and only required a relatively low light dose activation 400-700 nm, 2.2 J/cm^2. The properties make it a suitable candidate for type I and II dual type PDT (Paul et al., 2021). To design an ideal photosensitizer, a π extension system with low-energy triplet excited states (e.g. ^{3}MLCT, ^{3}IL) are more favourable (Papish & Oladipupo, 2022). For example, Dietzek *et al.* synthesized a series of Ru-ip-nT (n = 1-4) complexes. Complex Ru-ip-3T, which is also called TLD1433 (RU2), is now in phase II human clinical trials for PDT of bladder cancer treatment. Ru-ip-3T demonstrates high cytotoxicity under irradiation due to its long-lived ligand charge transfer triplet state. As the ^{3}IL state can be generated from the initial ^{1}MLCT or ^{1}IL state, and its inherent lifetime exceeding 50 μs, affords the system sufficient time to react with $^3O_2$ (Chettri et al., 2021). The biological tissue optical window (therapeutic window) refers to the wavelength range of the maximum penetration depth of light in biological tissue, generally within the range of near infrared wavelength, therefore the PDT sensitizer must strongly absorb light at relatively long wavelengths. Peng *et al.* reported a red-light-absorbing ruthenium complex [Ru(terpy)(Cl-7-IVQ)I]$^+$ (terpy = 2,2':6',2'-terpyridine, Cl-7-IVQ = 5-chloro7-(2-(1,3,3-trimethyl-3H-indol-1-ium-2-yl)- vinyl)quinolin-8-olate) (RU3) for absorption by cancer cells and selective accumulation in lysosomes. Relatively active wavelength activation occurs at 660 nm (He et al., 2021). Chakravarty *et al.* combined BODIPY and the ruthenium structure. In DMSO, the singlet oxygen quantum yield (Φ_Δ) is around 0.65. Complex RU4 exhibited significant PDT efficacy with an IC_{50} value of around 0.02 μM, and a PI value of more than 5000 within a therapeutic wavelength of 600-720 nm in A549 cells (Paul et al., 2019). Ru-biotin (RU5) synthesized by Chao *et al.* aimed to target cancer cells selectively. A large amount of $^1O_2$ was generated under two-photon excitation at 820 nm causing cell apoptosis (Li et al., 2019). The efficacy of PDT as an agent depends on its ability to bind and release biomolecules such as cell membranes or DNA. To be used in PDT, the ground state complex must be stable, non-toxic, and able to be photochemically transformed into a biologically active form. For instance, Wang *et al.* used an RGD peptide-coupled ruthenium (II) photosensitizer (RU6) to target tumor cells for two-photon photodynamic ablation of a tumor (Zhao et al., 2020).

Figure 6. Representative ruthenium (II) complexes as PDT photosensitizers

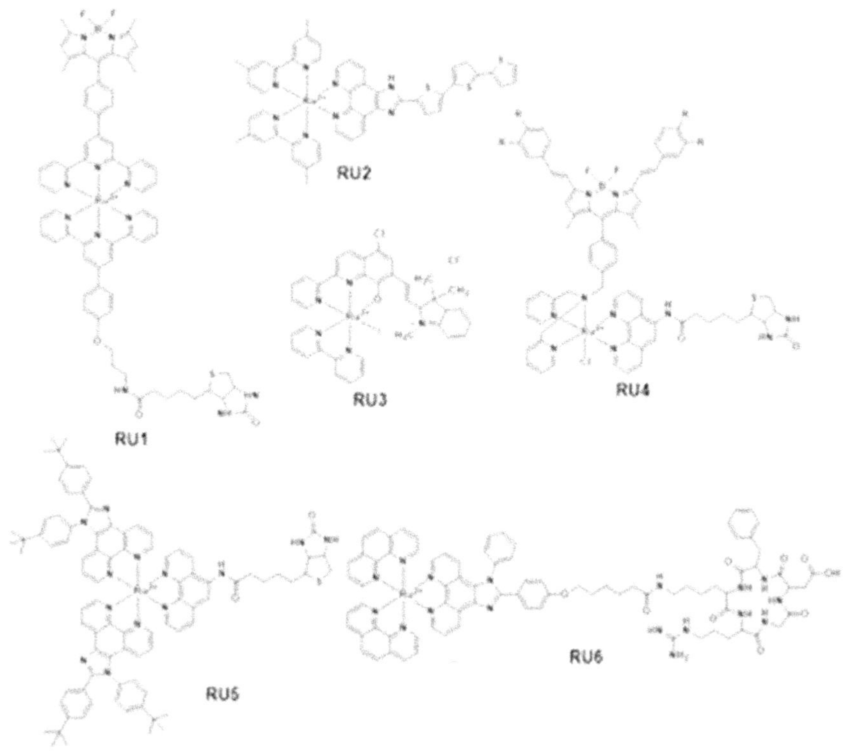

Reduction Activation

It was reported that the first ruthenium complex which entered clinical trials was activated by reduction (Alessio & Messori, 2019). In tumor cells, several typical reductive enzymes such as GSH and NADH are upregulated in cancer cells, providing specific mechanisms for cancer cell death (Liu et al., 2020; Mengele et al., 2021). Both ruthenium (II) and ruthenium (III) have octahedral structures, which is different from platinum (II) with a square planar configuration. Therefore, the mechanism of ruthenium anticancer complexes may be different from that of cis-platin. According to the reduction activation hypothesis, ruthenium (II) complexes can be used as prodrugs and interact with biomolecules after being reductively activated in vivo more rapidly. Tumor cells are in a relatively hypoxic environment due to the rapid consumption of O_2 and nutrients, and the formation speed of new blood vessels usually cannot keep pace with tumor growth (Hompland et al., 2021). To promote rapid proliferation of tumor cells, the cells have to rely more on the glycolysis process, which will produce lactic acid and therefore lower the pH during the process (Thews & Riemann, 2019). Also, the electrochemical potential of tumors, especially in the tumor center, is relatively lower than that of surrounding normal tissues. Ruthenium complexes with a +2 valence state are more likely to be formed inside tumors, because GSH and a large number of reductive proteins can reduce ruthenium (III) complexes in vivo. The ruthenium (II) complexes can be oxidized to ruthenium (III) through O_2, cytochrome oxidase and other oxidants, but this process is relatively rare in tumor cells since it is a hypoxic environment (Figure 7) (Graf & Lippard, 2012).

Figure 7. A schematic diagram of Ru(III) being activated by reduction
Source: Graf and Lippard (2012)

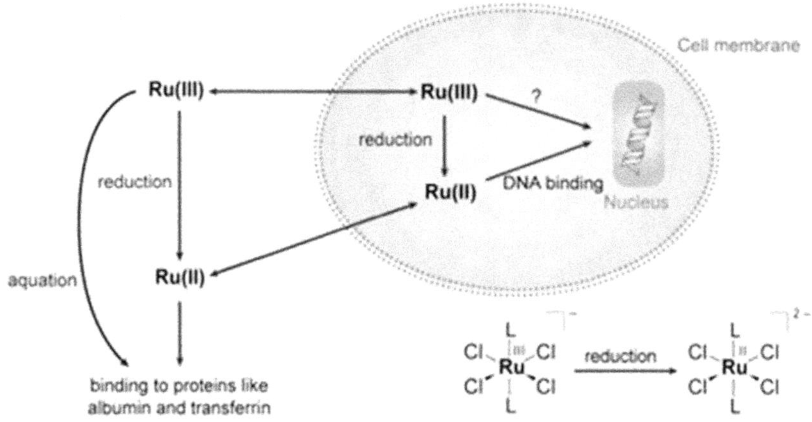

For example, regarding the reductive activation by GSH, Chao *et al.* prepared a ruthenium (II)-azo photosensitizer that can be activated by GSH. The complex displays low toxicity to cells in the dark, while at two photon excitation (810 nm) it exhibits excellent phototoxicity (Figure 8) (Zeng et al., 2017). They also reported another three ruthenium (II) complexes to elevate iron content in triple-negative breast cancer (MDA-MB-231) cells. The iron content contributes to the chemical production of ROS via the Fenton reaction. In addition, ruthenium (II) complexes can reduce GSH in cancer cells, making cells unable to resist oxidative damage. Both of these processes lead to oxidative stress in the tumor cells and ultimately tumor cell death (Xie et al., 2021).

Figure 8. Representation of Ru(II) complex activated by GSH

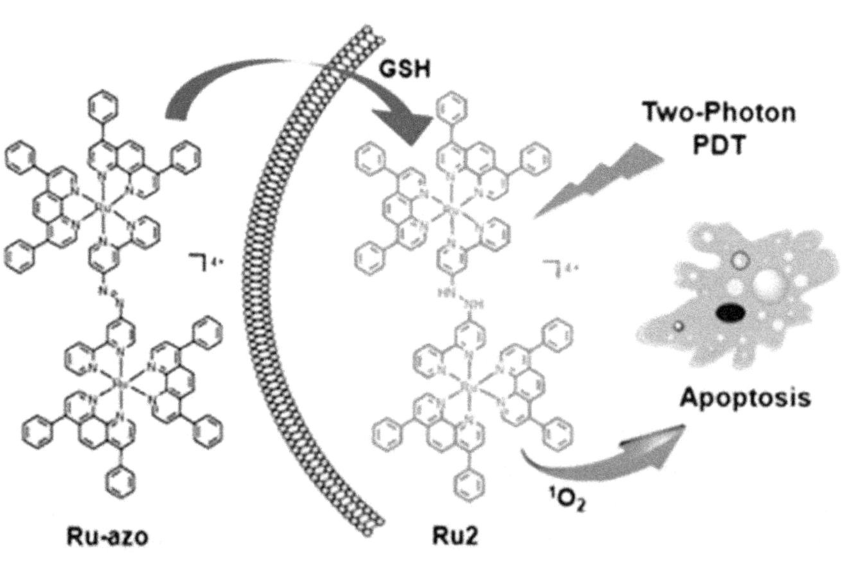

Taking NADH as another example, Zhang and co-workers developed a tris-bipyridine based ruthenium $[Ru(bpy)_3]^{2+}$ photosensitizing complexes with sonodynamic and sonocatalytic therapeutic ability. In this study, the ruthenium tris-bipyridine product was activated by ultrasound and produced energy transfer to O_2 and for $^1O_2$ generation. In the presence of NADH, ultrasonic activation of the complexes can oxidize NADH to produce NAD^+. In vitro and in vivo experiments demonstrated that the $^1O_2$ generated from ultrasonic stimulation and NADH sonoatalytic oxidation, achieved the destruction of breast cancer cells and tumor treatment in mice (Liang et al., 2021).

Figure 9. A representative ruthenium prodrug for regulating cell metabolism
Source: Zhao et al. (2022)

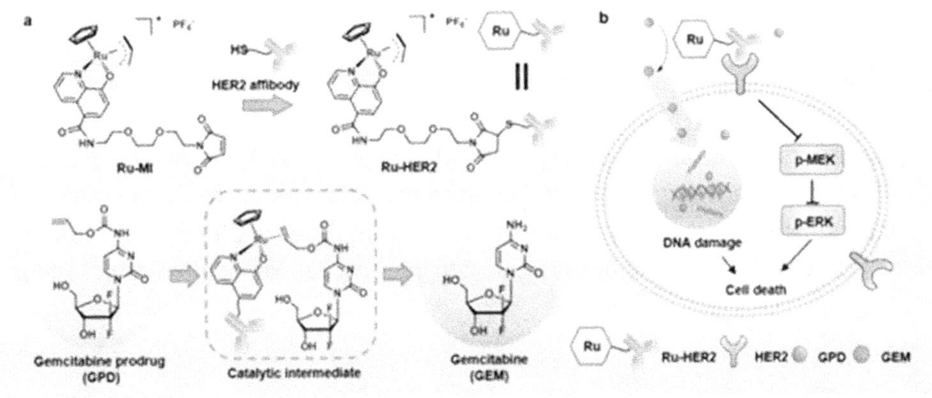

Regulation of Cell Metabolism

Recent studies have shown that metal complexes can inhibit tumor cells through catalytic activation of exogenous prodrugs or oxidation of endogenous metabolic molecules such as NADH. Only a low dose of catalytic dose is required to achieve a good therapeutic effect and effectively reduce the side effects of metal complexes. Prodrug is a promising drug designing strategy. It can release active substances in specific parts of the body through biotransformation to avoid the inactivation of drug molecules before reaching the target, which may help platinum drugs overcome the above shortcomings. However, prodrug systems tend to lack tumor targeting sites and are difficult to accumulate in tumor tissues. To solve this problem, Mao and co-workers proposed a strategy by coupling HER2 affibody fragments with a ruthenium catalyst, which can use antibody fragments to target the metal ruthenium catalyst to the surface of HER2 receptor-positive breast cancer cells, so as to achieve targeted therapy for tumors (Figure 9) (Zhao et al., 2022). Wang *et al.* demonstrated how to inhibit cancer cell growth by downregulation of PI3K/AKT signaling, by a combination of a ruthenium complex and doxorubicin (Lin et al., 2020).

Thioredoxin reductase (TrxR), which is usually overexpressed in human tumor cells, is a major member of the thioredoxin system and important for regulating cellular redox. It could be a target site for anticancer drugs. Therefore, Chen *et al.* reported three ruthenium (II) complexes containing the salicylate ligand. By targeting TrxR, complexes induce ROS-mediated apoptosis. Compared with cis-platin, the complex $[Ru(phen)_2(SA)]$ (phen = 1,10-phenanthroline, SA = salicylate) (**1**) displayed the best antitumor activity in tumor cells and biosafety in BASE-2B normal cell line. To investigate the cell growth inhibition mechanism, in the A549 cell line, complex **1** plays a role in antitumor by inducing tumor cell apoptosis and making the cell arrest in the G0/G1 phase. Also, complex **1** inhibits TrxR selectively, therefore promoting ROS generation and accumulation. ROS then further trigger mitochondrial dysfunction and DNA damage, followed by oxidative stress-sensitive mitogen-activated protein kinase (MAPK) activation and protein kinase B (PKB or AKT) inhibition, eventually leading to the tumor cell death (Chen et al., 2019). These authors synthesized $[Ru(PIP)_3]^{2+}$ (PIP = 2-phenylimidazo[4,5-f][1,10]phenanthroline) which showed higher anti-proliferation activity than cis-platin. TrxR inhibition ability of this complex was more effective than positive control kinofen. This complex is able target mitochondria and induce ROS-mediated apoptosis of cancer cells by targeting TrxR (Luo et al., 2014). Pavan *et al.* reported a novel ruthenium (II) complex which arrests the cell in G2/M phase by downregulating Aurora-B kinase and damaging DNA by ROS-mediated apoptosis in DU-145 cells (De Grandis et al., 2021).

CONCLUSION

Today, the applications of metal complexes in regulating intracellular redox have been expanding. Through this chapter, we hope to give readers a preliminary concept of the intracellular redox system, redox detection, regulation of redox by ruthenium metal complexes and their anti-tumor applications. It will provide some basic information for drug design and mechanism exploration of redox regulation in the future.

ACKNOWLEDGMENT

We thank Newcastle University and Sun Yat-Sen University for financial support.

REFERENCES

Alessio, E., & Messori, L. (2019). NAMI-A and KP1019/1339, two iconic ruthenium anticancer drug candidates face-to-face: A case story in medicinal inorganic chemistry. *Molecules (Basel, Switzerland), 24*(10), 1995–2014. doi:10.3390/molecules24101995 PMID:31137659

Babu, T., Sarkar, A., Karmakar, S., Schmidt, C., & Gibson, D. (2020). Multiaction Pt(IV) carbamate complexes can codeliver Pt(II) drugs and amine containing bioactive molecules. *Inorganic Chemistry, 59*(7), 5182–5193. doi:10.1021/acs.inorgchem.0c00445 PMID:32207294

Bhatt, S., Puli, L., & Patil, C. R. (2021). Role of reactive oxygen species in the progression of Alzheimer's disease. *Drug Discovery Today, 26*(3), 794–803. doi:10.1016/j.drudis.2020.12.004 PMID:33306995

Cao, J., Liu, X., Yang, Y., Wei, B., Li, Q., Mao, G., He, Y., Li, Y., Zheng, L., Zhang, Q., Li, J., Wang, L., & Qi, C. (2020). decylubiquinone suppresses breast cancer growth and metastasis by inhibiting angiogenesis via the ROS/p53/BAI1 signaling pathway. *Angiogenesis, 23*(3), 325–338. doi:10.100710456-020-09707-z PMID:32020421

Cao, X., Nie, X., Xiong, S., Cao, L., Wu, Z., Moore, P. K., & Bian, J.-S. (2018). Renal protective effect of polysulfide in cisplatin-induced nephrotoxicity. *Redox Biology*, *15*, 513–521. doi:10.1016/j.redox.2018.01.012 PMID:29413963

Catapano, M. C., Protti, M., Fontana, T., Mandrioli, R., Mladěnka, P., & Mercolini, L. (2019). An original HPLC method with coulometric detection to monitor hydroxyl radical generation via Fenton chemistry. *Molecules (Basel, Switzerland)*, *24*(17), 3066–3080. doi:10.3390/molecules24173066 PMID:31450723

Chen, C., Xu, C., Li, T., Lu, S., Luo, F., & Wang, H. (2020). Novel NHC-coordinated ruthenium (II) arene complexes achieve synergistic efficacy as safe and effective anticancer therapeutics. *European Journal of Medicinal Chemistry*, *203*, 112605–112615. doi:10.1016/j.ejmech.2020.112605 PMID:32688202

Chen, J.-C., Zhang, Y., Jie, X.-M., She, J., Dongye, G.-Z., Zhong, Y., Deng, Y.-Y., Wang, J., Guo, B.-Y., & Chen, L.-M. (2019). Ruthenium(II) salicylate complexes inducing ROS-mediated apoptosis by targeting thioredoxin reductase. *Journal of Inorganic Biochemistry*, *193*, 112–123. doi:10.1016/j.jinorgbio.2019.01.011 PMID:30711557

Chettri, A., Schneider, K. R. A., Cole, H. D., Roque, J. A. III, Cameron, C. G., McFarland, S. A., & Dietzek, B. (2021). String-attached oligothiophene substituents determine the fate of excited states in ruthenium complexes for photodynamic therapy. *The Journal of Physical Chemistry A*, *125*(32), 6985–6994. doi:10.1021/acs.jpca.1c04900 PMID:34370485

Cheung, E. C., & Vousden, K. H. (2022). The role of ROS in tumour development and progression. *Nature Reviews. Cancer*, *22*(5), 280–297. doi:10.103841568-021-00435-0 PMID:35102280

Chovanec, M., Abu Zaid, M., Hanna, N., El-Kouri, N., Einhorn, L. H., & Albany, C. (2017). Long-term toxicity of cisplatin in germ-cell tumor survivors. *Annals of Oncology : Official Journal of the European Society for Medical Oncology*, *28*(11), 2670–2679. doi:10.1093/annonc/mdx360 PMID:29045502

Circu, M. L., & Aw, T. Y. (2010). Reactive oxygen species, cellular redox systems, and apoptosis. *Free Radical Biology & Medicine*, *48*(6), 749–762. doi:10.1016/j.freeradbiomed.2009.12.022 PMID:20045723

Clerici, S., & Boletta, A. (2020). Role of the KEAP1-NRF2 axis in renal cell carcinoma. *Cancers (Basel)*, *12*(11), 3458–3482. doi:10.3390/cancers12113458 PMID:33233657

Cordani, M., Donadelli, M., Strippoli, R., Bazhin, A. V., & Sanchez-Alvarez, M. (2019). Interplay between ROS and autophagy in cancer and aging: From Molecular Mechanisms to Novel Therapeutic Approaches. *Oxidative Medicine and Cellular Longevity*, *2019*, 1–3. doi:10.1155/2019/8794612 PMID:31467639

De Grandis, R. A., Oliveira, K. M., Guedes, A. P. M., Dos Santos, P. W. S., Aissa, A. F., Batista, A. A., & Pavan, F. R. (2021). a novel ruthenium (II) complex with lapachol induces G2/M phase arrest through Aurora-B kinase down-regulation and ROS-mediated apoptosis in human prostate adenocarcinoma cells. *Frontiers in Oncology*, *11*, 2148–2165. doi:10.3389/fonc.2021.682968 PMID:34249731

Duca, G., & Vaseashta, A. (Eds.). (2020). *Handbook of Research on Emerging Developments and Environmental Impacts of Ecological Chemistry*. IGI Global. doi:10.4018/978-1-7998-1241-8

Duca, G., Vaseashta, A., & Travin, S. (Eds.). (2022). *Handbook of Research on Water Sciences and Society* (Vols. 1–2). IGI Global. doi:10.4018/978-1-7998-7356-3

Estalayo-Adrián, S., Blasco, S., Bright, S. A., McManus, G. J., Orellana, G., Williams, D. C., Kelly, J. M., & Gunnlaugsson, T. (2020). Water-soluble amphiphilic ruthenium(II) polypyridyl complexes as potential light-activated therapeutic agents. *Chemical Communications*, *56*(65), 9332–9335. doi:10.1039/D0CC04397D PMID:32671364

Fang, L., Qin, X., Zhao, J., & Gou, S. (2019). Construction of dual stimuli-responsive platinum(IV) hybrids with NQO1 targeting ability and overcoming cisplatin resistance. *Inorganic Chemistry*, *58*(3), 2191–2200. doi:10.1021/acs.inorgchem.8b03386 PMID:30657321

Gao, P., Pan, W., Li, N., & Tang, B. (2019). Fluorescent probes for organelle-targeted bioactive species imaging. *Chemical Science (Cambridge)*, *10*(24), 6035–6071. doi:10.1039/C9SC01652J PMID:31360411

Gao, Y., Zhang, J., He, Z., Luo, L., & Yan, C. (2021). Platinum-based photosensitizer with near-infrared aggregation-induced emission for synergistic photodynamic-chemo theranostics. *Organic Electronics*, *92*, 106105–106111. doi:10.1016/j.orgel.2021.106105

Ghasemi, H., Amini, M. A., Pegah, A., Azizi, E., Tayebinia, H., Khanverdilou, S., Mousavibahar, S. H., & Alizamir, A. (2020). Overexpression of reactive oxygen species modulator 1 is associated with advanced grades of bladder cancer. *Molecular Biology Reports*, *47*(9), 6497–6505. doi:10.100711033-020-05702-1 PMID:32770525

Ghosh, A., Mandoli, A., Kumar, D. K., Yadav, N. S., Ghosh, T., Jha, B., Thomas, J. A., & Das, A. (2009). DNA binding and cleavage properties of a newly synthesised Ru(II)-polypyridyl complex. *Dalton Transactions (Cambridge, England)*, (42), 9312–9321. doi:10.1039/b906756f PMID:20449211

Ghosh, A., & Shcherbik, N. (2020). Effects of oxidative stress on protein translation: Implications for cardiovascular diseases. *International Journal of Molecular Sciences*, *21*(8), 2661–2685. doi:10.3390/ijms21082661 PMID:32290431

Gomez, M. L., Shah, N., Kenny, T. C., Jenkins, E. C. Jr, & Germain, D. (2019). SOD1 is essential for oncogene-driven mammary tumor formation but dispensable for normal development and proliferation. *Oncogene*, *38*(29), 5751–5765. doi:10.103841388-019-0839-x PMID:31222103

Graf, N., & Lippard, S. J. (2012). Redox activation of metal-based prodrugs as a strategy for drug delivery. *Advanced Drug Delivery Reviews*, *64*(11), 993–1004. doi:10.1016/j.addr.2012.01.007 PMID:22289471

Gransbury, G. K., Kappen, P., Glover, C. J., Hughes, J. N., Levina, A., Lay, P. A., Musgrave, I. F., & Harris, H. H. (2016). Comparison of KP1019 and NAMI-A in tumour-mimetic environments. *Metallomics*, *8*(8), 762–773. doi:10.1039/C6MT00145A PMID:27460862

Han, C., Wang, Z., Chen, S., Li, L., Xu, Y., Kang, W., Wei, C., Ma, H., Wang, M., & Jin, X. (2021). Berbamine suppresses the progression of bladder cancer by modulating the ROS/NF-κB axis. *Oxidative Medicine and Cellular Longevity*, *2021*, 1–6. doi:10.1155/2021/8851763

He, G., Xu, N., Ge, H., Lu, Y., Wang, R., Wang, H., Du, J., Fan, J., Sun, W., & Peng, X. (2021). Red-light-responsive Ru complex photosensitizer for lysosome localization photodynamic therapy. *ACS Applied Materials & Interfaces*, *13*(17), 19572–19580. doi:10.1021/acsami.0c22551 PMID:33900720

Hompland, T., Fjeldbo, C. S., & Lyng, H. (2021). Tumor hypoxia as a barrier in cancer therapy: Why levels matter. *Cancers (Basel)*, *13*(3), 499–520. doi:10.3390/cancers13030499 PMID:33525508

Huang, R., Chen, H., Liang, J., Li, Y., Yang, J., Luo, C., Tang, Y., Ding, Y., Liu, X., Yuan, Q., Yu, H., Ye, Y. C., Xu, W. F., & Xie, X. (2021). Dual role of reactive oxygen species and their application in cancer therapy. *Journal of Cancer*, *12*(18), 5543–5561. doi:10.7150/jca.54699 PMID:34405016

Iacopini, D., Vančo, J., Di Pietro, S., Bordoni, V., Zacchini, S., Marchetti, F., Dvořák, Z., Malina, T., Biancalana, L., Trávníček, Z., & Di Bussoloc, V. (2022). New glycoconjugation strategies for ruthenium (II) arene complexes via phosphane ligands and assessment of their antiproliferative activity. *Bioorganic Chemistry*, *126*, 105901–105916. doi:10.1016/j.bioorg.2022.105901 PMID:35671646

Imberti, C., Zhang, P., Huang, H., & Sadler, P. J. (2020). New designs for phototherapeutic transition metal complexes. *Angewandte Chemie International Edition*, *59*(1), 61–73. doi:10.1002/anie.201905171 PMID:31310436

Jasek-Gajda, E., Jurkowska, H., Jasińska, M., & Lis, G. J. (2020). Targeting the MAPK/ERK and PI3K/AKT signaling pathways affects NRF2, Trx and GSH antioxidant systems in leukemia cells. *Antioxidants*, *9*(7), 633–650. doi:10.3390/antiox9070633 PMID:32709140

Jiang, G.-B., Zhang, W.-Y., He, M., Gu, Y.-Y., Bai, L., Wang, Y.-J., Yi, Q.-Y., & Du, F. (2020). Development of four ruthenium polypyridyl complexes as antitumor agents: Design, biological evaluation and mechanism investigation. *Journal of Inorganic Biochemistry*, *208*, 111104–111118. doi:10.1016/j.jinorgbio.2020.111104 PMID:32485635

Jiang, G.-B., Zhang, W.-Y., He, M., Gu, Y.-Y., Bai, L., Wang, Y.-J., Yi, Q.-Y., & Du, F. (2021). Systematic evaluation of the antitumor activity of three ruthenium polypyridyl complexes. *Journal of Inorganic Biochemistry*, *225*, 111616–111627. doi:10.1016/j.jinorgbio.2021.111616 PMID:34555601

Jorgenson, T. C., Zhong, W., & Oberley, T. D. (2013). Redox imbalance and biochemical changes in cancerredox imbalance in cancer. *Cancer Research*, *73*(20), 6118–6123. doi:10.1158/0008-5472.CAN-13-1117 PMID:23878188

Karmakar, S., Poetsch, I., Kowol, C. R., Heffeter, P., & Gibson, D. (2019). Synthesis and cytotoxicity of water-soluble dual- and triple-action satraplatin derivatives: Replacement of equatorial chlorides of satraplatin by acetates. *Inorganic Chemistry*, *58*(24), 16676–16688. doi:10.1021/acs.inorgchem.9b02796 PMID:31790216

Khramtsov, V. V. (2018). In vivo molecular electron paramagnetic resonance-based spectroscopy and imaging of tumor microenvironment and redox using functional paramagnetic probes. *Antioxidants & Redox Signaling*, *28*(15), 1365–1377. doi:10.1089/ars.2017.7329 PMID:29132215

Kim, E. K., Jang, M., Song, M. J., Kim, D., Kim, Y., & Jang, H. H. (2019). Redox-mediated mechanism of chemoresistance in cancer cells. *Antioxidants*, *8*(10), 471–488. doi:10.3390/antiox8100471 PMID:31658599

Krause, K.-H. (2018). Role of NOX-generated reactive oxygen species in health and disease. *Free Radical Biology & Medicine*, *120*, S7–S7. doi:10.1016/j.freeradbiomed.2018.04.033

Kyani, A., Tamura, S., Yang, S., Shergalis, A., Samanta, S., Kuang, Y., Ljungman, M., & Neamati, N. (2018). Discovery and mechanistic elucidation of a class of protein disulfide isomerase inhibitors for the treatment of glioblastoma. *ChemMedChem*, *13*(2), 164–177. doi:10.1002/cmdc.201700629 PMID:29235250

Li, B., Kim, Y. L., & Lippert, A. R. (2022a). Chemiluminescence measurement of reactive sulfur and nitrogen species. *Antioxidants & Redox Signaling*, *36*(4-6), 337–353. doi:10.1089/ars.2021.0195 PMID:34514838

Li, H., Sun, H., Ding, J., Wu, T., Cai, S., Wang, C., & Yang, R. (2022b). RuO_2/rGO heterostructures as mimic peroxidases for colorimetric detection of glucose. *Mikrochimica Acta*, *189*(7), 1–10. doi:10.100700604-022-05319-0

Li, J., & Chen, T. (2020). Transition metal complexes as photosensitizers for integrated cancer theranostic applications. *Coordination Chemistry Reviews*, *418*, 213355–213374. doi:10.1016/j.ccr.2020.213355

Li, J., Zeng, L., Xiong, K., Rees, T. W., Jin, C., Wu, W., Chen, Y., Ji, L., & Chao, H. (2019). A biotinylated ruthenium (II) photosensitizer for tumor-targeted two-photon photodynamic therapy. *Chemical Communications*, *55*(73), 10972–10975. doi:10.1039/C9CC05826E PMID:31453611

Lian, S., Li, S., Zhu, J., Xia, Y., & Do Jung, Y. (2022). Nicotine stimulates IL-8 expression via ROS/NF-κB and ROS/MAPK/AP-1 axis in human gastric cancer cells. *Toxicology*, *466*, 153062–153075. doi:10.1016/j.tox.2021.153062 PMID:34890707

Liang, C., Xie, J., Luo, S., Huang, C., Zhang, Q., Huang, H., & Zhang, P. (2021). A highly potent ruthenium(II)-sonosensitizer and sonocatalyst for in vivo sonotherapy. *Nature Communications*, *12*(1), 1–9. doi:10.103841467-021-25303-1 PMID:34408151

Lin, K., Rong, Y., Chen, D., Zhao, Z., Bo, H., Qiao, A., Hao, X., & Wang, J. (2020). Combination of ruthenium complex and doxorubicin synergistically inhibits cancer cell growth by down-regulating PI3K/AKT signaling pathway. *Frontiers in Oncology*, *10*, 141–153. doi:10.3389/fonc.2020.00141 PMID:32133289

Liu, C., Liu, J., Zhang, W., Wang, Y. L., Liu, Q., Song, B., Yuan, J., & Zhang, R. (2020). "Two birds with one stone" ruthenium(II) complex probe for biothiols discrimination and detection in vitro and in vivo. *Advancement of Science*, *7*(14), 2000458. doi:10.1002/advs.202000458 PMID:32714756

Liu, C., Zhang, W., Zhang, R., Gao, X., Song, B., & Yuan, J. (2022). Ruthenium(II) complex-based long-lived two-photon luminescence probe for dynamic monitoring of glutathione S-transferases in mouse models of drug-induced liver injury. *Sensors and Actuators. B, Chemical*, *357*, 131440–131449. doi:10.1016/j.snb.2022.131440

Luo, Z., Yu, L., Yang, F., Zhao, Z., Yu, B., Lai, H., Wong, K.-H., Ngai, S.-M., Zheng, W., & Chen, T. (2014). Ruthenium polypyridyl complexes as inducer of ROS-mediated apoptosis in cancer cells by targeting thioredoxin reductase. *Metallomics*, *6*(8), 1480–1490. doi:10.1039/C4MT00044G PMID:24823440

Malhi, H., & Kaufman, R. J. (2011). Endoplasmic reticulum stress in liver disease. *Journal of Hepatology*, *54*(4), 795–809. doi:10.1016/j.jhep.2010.11.005 PMID:21145844

Malinowski, J., Zych, D., Jacewicz, D., Gawdzik, B., & Drzeżdżon, J. (2020). Application of coordination compounds with transition metal ions in the chemical industry - A review. *International Journal of Molecular Sciences*, *21*(15), 5443–5468. doi:10.3390/ijms21155443 PMID:32751682

Martinez-Alonso, M., & Gasser, G. (2021). Ruthenium polypyridyl complex-containing bioconjugates. *Coordination Chemistry Reviews*, *434*, 213736–213776. doi:10.1016/j.ccr.2020.213736

Mengele, A. K., Weixler, D., Chettri, A., Maurer, M., Huber, F. L., Seibold, G. M., Dietzek, B., Eikmanns, B. J., & Rau, S. (2021). Switching the mechanism of NADH photooxidation by supramolecular interactions. *Chemistry (Weinheim an der Bergstrasse, Germany)*, *27*(68), 16840–16845. doi:10.1002/chem.202103029 PMID:34547151

Mikhailov, A. A., Vorobyev, V. A., Nadolinny, V. A., Patrushev, Y. V., Yudina, Y. S., & Kostin, G. A. (2019). Primary and secondary photochemical transformations of biologically active precursor-nitro-nitrosyl ruthenium complex. *Journal of Photochemistry and Photobiology A Chemistry*, *373*, 37–44. doi:10.1016/j.jphotochem.2018.12.037

Murphy, M. P. (2013). Mitochondrial dysfunction indirectly elevates ROS production by the endoplasmic reticulum. *Cell Metabolism*, *18*(2), 145–146. doi:10.1016/j.cmet.2013.07.006 PMID:23931748

Napolitano, G., Fasciolo, G., & Venditti, P. (2021). Mitochondrial Management of reactive oxygen species. *Antioxidants*, *10*(11), 1824–1852. doi:10.3390/antiox10111824 PMID:34829696

Olson, K. R., Gao, Y., Arif, F., Arora, K., Patel, S., DeLeon, E. R., Sutton, T. R., Feelisch, M., Cortese-Krott, M. M., & Straub, K. D. (2018). Metabolism of hydrogen sulfide (H_2S) and production of reactive sulfur species (RSS) by superoxide dismutase. *Redox Biology*, *15*, 74–85. doi:10.1016/j.redox.2017.11.009 PMID:29220697

Papish, E. T., & Oladipupo, O. E. (2022). Factors that influence singlet oxygen formation vs. ligand substitution for light-activated ruthenium anticancer compounds. *Current Opinion in Chemical Biology*, *68*, 102143–102154. doi:10.1016/j.cbpa.2022.102143 PMID:35483128

Paul, S., Kundu, P., Bhattacharyya, U., Garai, A., Maji, R. C., Kondaiah, P., & Chakravarty, A. R. (2019). Ruthenium(II) conjugates of boron-dipyrromethene and biotin for targeted photodynamic therapy in red light. *Inorganic Chemistry*, *59*(1), 913–924. doi:10.1021/acs.inorgchem.9b03178 PMID:31825210

Paul, S., Kundu, P., Kondaiah, P., & Chakravarty, A. R. (2021). BODIPY-ruthenium(II) bis-terpyridine complexes for cellular imaging and type-I/-II photodynamic therapy. *Inorganic Chemistry*, *60*(21), 16178–16193. doi:10.1021/acs.inorgchem.1c01850 PMID:34672556

Perse, M. (2021). Cisplatin Mouse Models: Treatment, toxicity and translatability. *Biomedicines*, *9*(10), 1406–1433. doi:10.3390/biomedicines9101406 PMID:34680523

Pichaud, N., Bérubé, R., Côté, G., Belzile, C., Dufresne, F., Morrow, G., Tanguay, R. M., Rand, D. M., & Blier, P. U. (2019). Age dependent dysfunction of mitochondrial and ROS metabolism induced by mitonuclear mismatch. *Frontiers in Genetics*, *10*, 130–141. doi:10.3389/fgene.2019.00130 PMID:30842791

Ponte, F., Alberto, M. E., De Simone, B. C., Russo, N., & Sicilia, E. (2019). Photophysical exploration of dual-approach Pt-II-BODIPY conjugates: Theoretical insights. *Inorganic Chemistry, 58*(15), 9882–9889. doi:10.1021/acs.inorgchem.9b01002 PMID:31343162

Qiao, Y., Li, Y., Fu, W., Guo, Z., & Zheng, X. (2018). Enhancing the electrochemiluminescence of luminol by chemically modifying the reaction microenvironment. *Analytical Chemistry, 90*(15), 9629–9636. doi:10.1021/acs.analchem.8b02577 PMID:29969897

Rendra, E., Riabov, V., Mossel, D. M., Sevastyanova, T., Harmsen, M. C., & Kzhyshkowska, J. (2019). Reactive oxygen species (ROS) in macrophage activation and function in diabetes. *Immunobiology, 224*(2), 242–253. doi:10.1016/j.imbio.2018.11.010 PMID:30739804

Roh, J.-L., Jang, H., Kim, E. H., & Shin, D. (2017). Targeting of the glutathione, thioredoxin, and Nrf2 antioxidant systems in head and neck cancer. *Antioxidants & Redox Signaling, 27*(2), 106–114. doi:10.1089/ars.2016.6841 PMID:27733046

Sasahara, G. L., Gouveia, F. S., Rodrigues, R. D., Zampieri, D. S., Fonseca, S. G. C., Goncalves, R. D. R., Athaydes, B. R., Kitagawa, R. R., Santos, F. A., Sousa, E. H. S., Nagao-Dias, A. T., & Lopes, L. G. D. (2020). Nitro-imidazole-based ruthenium complexes with antioxidant and anti-inflammatory activities. *Journal of Inorganic Biochemistry, 206*, 111048–111058. doi:10.1016/j.jinorgbio.2020.111048 PMID:32151873

Sawa, T., Takata, T., Matsunaga, T., Ihara, H., Motohashi, H., & Akaike, T. (2022). Chemical biology of reactive sulfur species: Hydrolysis-driven equilibrium of polysulfides as a determinant of physiological functions. *Antioxidants & Redox Signaling, 36*(4-6), 327–336. doi:10.1089/ars.2021.0170 PMID:34409860

Shahzad, K., Asad, M., Asiri, A. M., Irfan, M., & Iqbal, M. A. (2022). In-Vitro Anticancer Profile of Recent ruthenium complexes against liver cancer. *Reviews in Inorganic Chemistry*. Advance online publication. doi:10.1515/revic-2021-0040

Shi, J., Guo, Y. H., Xie, F., Chen, Q. F., & Zhang, M. T. (2020). Redox-active ligand assisted catalytic water oxidation by a RuIV=O intermediate. *Angewandte Chemie International Edition, 59*(10), 4000–4008. doi:10.1002/anie.201910614 PMID:31880387

Simović, A. R., Masnikosa, R., Bratsos, I., & Alessio, E. (2019). Chemistry and reactivity of ruthenium(II) complexes: DNA/protein binding mode and anticancer activity are related to the complex structure. *Coordination Chemistry Reviews, 398*, 113011–113036. doi:10.1016/j.ccr.2019.07.008

Slezak, J., Kura, B., LeBaron, T. W., Singal, P. K., Buday, J., & Barancik, M. (2021). Oxidative stress and pathways of molecular hydrogen effects in medicine. *Current Pharmaceutical Design, 27*(5), 610–625. PMID:32954996

Sousa, S. F., Ertem, M. Z., Faustino, L. A., Machado, A. E. H., Concepcion, J. J., Maia, P. I. S., & Patrocinio, A. O. T. (2021). Mechanistic investigation of the aerobic oxidation of 2-pyridylacetate coordinated to a Ru(II) polypyridyl complex. *Dalton Transactions (Cambridge, England), 50*(42), 15248–15259. doi:10.1039/D1DT02461B PMID:34632989

Stieg, D. C., Wang, Y., Liu, L.-Z., & Jiang, B.-H. (2022). ROS and miRNA dysregulation in ovarian cancer development, angiogenesis and therapeutic resistance. *International Journal of Molecular Sciences*, *23*(12), 6702–6718. doi:10.3390/ijms23126702 PMID:35743145

Stultz, L. K., Hunsucker, A., Middleton, S., Grovenstein, E., O'Leary, J., Blatt, E., Miller, M., Mobley, J., & Hanson, P. K. (2020). Proteomic analysis of the *S. cerevisiae* Response to the anticancer ruthenium complex KP1019. *Metallomics*, *12*(6), 876–890. doi:10.1039/d0mt00008f PMID:32329475

Suzen, S., Gurer-Orhan, H., & Saso, L. (2017). Detection of reactive oxygen and nitrogen species by electron paramagnetic resonance (EPR) technique. *Molecules (Basel, Switzerland)*, *22*(1), 181–189. doi:10.3390/molecules22010181 PMID:28117726

Syed, A. J., & Anderson, J. C. (2021). Applications of bioluminescence in biotechnology and beyond. *Chemical Society Reviews*, *50*(9), 5668–5705. doi:10.1039/D0CS01492C PMID:33735357

Thews, O., & Riemann, A. (2019). Tumor pH and metastasis: A malignant process beyond hypoxia. *Cancer and Metastasis Reviews*, *38*(1-2), 113–129. doi:10.100710555-018-09777-y PMID:30607627

Tong, R., Wu, X., Liu, Y., Liu, Y., Zhou, J., Jiang, X., Zhang, L., He, X., & Ma, L. (2020). Curcumin-induced DNA demethylation in human gastric cancer cells is mediated by the DNA-damage response pathway. *Oxidative Medicine and Cellular Longevity*, *2020*, 1–13. doi:10.1155/2020/2543504 PMID:32617134

Ushio-Fukai, M., Ash, D., Nagarkoti, S., Belin de Chantemele, E. J., Fulton, D. J. R., & Fukai, T. (2021). Interplay between reactive oxygen/reactive nitrogen species and metabolism in vascular biology and disease. *Antioxidants & Redox Signaling*, *34*(16), 1319–1354. doi:10.1089/ars.2020.8161 PMID:33899493

Wang, Y., Du, C., Liu, Z., Pei, K., Zhang, Y., & Qi, W. (2022). Chemiluminescence "Turn-on" detection of tyrosinase activity via in situ generation of dopamine based on a lucigenin and riboflavin system. *New Journal of Chemistry*, *46*(9), 4156–4161. doi:10.1039/D1NJ05628J

Wang, Y., Qi, H., Liu, Y., Duan, C., Liu, X., Xia, T., Chen, D., Piao, H., & Liu, H.-X. (2021). The double-edged roles of ROS in cancer prevention and therapy. *Theranostics*, *11*(10), 4839–4857. doi:10.7150/thno.56747 PMID:33754031

Watabe, S., Sakamoto, Y., Morikawa, M., Okada, R., Miura, T., & Ito, E. (2011). Highly sensitive determination of hydrogen peroxide and glucose by fluorescence correlation spectroscopy. *PLoS One*, *6*(8), e22955–e22959. doi:10.1371/journal.pone.0022955 PMID:21850246

Wei, F., Kuang, S., Rees, T. W., Liao, X., Liu, J., Luo, D., Wang, J., Zhang, X., Ji, L., & Chao, H. (2021). Ruthenium(II) complexes coordinated to graphitic carbon nitride: Oxygen self-sufficient photosensitizers which produce multiple ROS for photodynamic therapy in hypoxia. *Biomaterials*, *276*, 121064. doi:10.1016/j.biomaterials.2021.121064 PMID:34391019

Wiśniewska, J., Fandzloch, M., & Łakomska, I. (2019). The reduction of ruthenium (III) complexes with triazolopyrimidine ligands by ascorbic acid and mechanistic insight into their action in anticancer therapy. *Inorganica Chimica Acta*, *484*, 305–310. doi:10.1016/j.ica.2018.09.051

Wu, G., Fang, Y.-Z., Yang, S., Lupton, J. R., & Turner, N. D. (2004). Glutathione metabolism and its implications for health. *The Journal of Nutrition*, *134*(3), 489–492. doi:10.1093/jn/134.3.489 PMID:14988435

Xie, L., Liu, L., Wang, W., Ma, Z., Xu, L., Zhao, X., & Wang, H. (2018). The structures, spectroscopic properties, and photodynamic reactions of three [RuCl(QN)NO]-Complexes (HQN=8-hydroxyquinoline and its derivatives) as potential no-donating drugs. *Bioinorganic Chemistry and Applications*, *2018*, 1–10. doi:10.1155/2018/7029376 PMID:30627138

Xie, L., Wang, L., Guan, R., Ji, L., & Chao, H. (2021). Anti-metastasis and anti-proliferation effect of mitochondria-accumulating ruthenium(II) complexes via redox homeostasis disturbance and energy depletion. *Journal of Inorganic Biochemistry*, *217*, 111380–111388. doi:10.1016/j.jinorgbio.2021.111380 PMID:33578250

Xu, H., & Ren, D. (2015). Lysosomal physiology. *Annual Review of Physiology*, *77*(1), 57–80. doi:10.1146/annurev-physiol-021014-071649 PMID:25668017

Xu, Q., Chen, G., Chen, G. M., Wu, H. L., Yang, Y. Y., Mai, Z. Y., Sun, R., Luan, P., Guo, C. W., Yu, M., Peng, Z. W., & Yu, Z. Q. (2022). NO-dependent vasodilation and deep tumor penetration for cascade-amplified antitumor performance. *Journal of Controlled Release*, *347*, 389–399. doi:10.1016/j.jconrel.2022.05.022 PMID:35569586

Xu, T., Ding, W., Ji, X. Y., Ao, X., Liu, Y., Yu, W. P., & Wang, J. X. (2019). Oxidative stress in cell death and cardiovascular diseases. *Oxidative Medicine and Cellular Longevity*, *2019*, 1–11. doi:10.1155/2019/9030563 PMID:31781356

Yang, J.-H., Yu, K., Si, X.-K., Li, S., Cao, Y.-J., Li, W., & Zhang, J.-X. (2019). Liensinine inhibited gastric cancer cell growth through ROS generation and the PI3K/AKT pathway. *Journal of Cancer*, *10*(25), 6431–6438. doi:10.7150/jca.32691 PMID:31772676

Yang, Z., Min, Z. J., & Yu, B. (2020). Reactive oxygen species and immune regulation. *International Reviews of Immunology*, *39*(6), 292–298. doi:10.1080/08830185.2020.1768251 PMID:32423322

Yuan, S., Chen, S., Wu, H., Jiang, H., Zheng, S., Zhang, Q., & Liu, Y. (2020). NAMI-A preferentially reacts with the Sp1 protein: Understanding the anti-metastasis effect of the drug. *Chemical Communications*, *56*(9), 1397–1400. doi:10.1039/C9CC08775C PMID:31912815

Zarkovic, N. (2020). Roles and functions of ROS and RNS in cellular physiology and pathology. *Cells*, *9*(3), 767–771. doi:10.3390/cells9030767 PMID:32245147

Zeng, L., Kuang, S., Li, G., Jin, C., Ji, L., & Chao, H. (2017). A GSH-activatable ruthenium(II)-azo photosensitizer for two-photon photodynamic therapy. *Chemical Communications*, *53*(12), 1977–1980. doi:10.1039/C6CC10330H PMID:28119967

Zhang, J., Wang, X., Vikash, V., Ye, Q., Wu, D., Liu, Y., & Dong, W. (2016). ROS and ROS-mediated cellular signaling. *Oxidative Medicine and Cellular Longevity*, *2016*, 1–18. PMID:26998193

Zhao, Z., Qiu, K., Liu, J., Hao, X., & Wang, J. (2020). Two-photon photodynamic ablation of tumour cells using an RGD peptide-conjugated ruthenium(II) photosensitiser. *Chemical Communications*, *56*(83), 12542–12545. doi:10.1039/D0CC04943C PMID:32940288

Zhao, Z., Tao, X., Xie, Y., Lai, Q., Lin, W., Lu, K., Wang, J., Xia, W., & Mao, Z.-W. (2022). In situ prodrug activation by an affibody-ruthenium catalyst hybrid for HER2-targeted chemotherapy. A Angewandte Chemie International Edition, 134(26), e202202855.

Zhou, J., Li, X.-Y., Liu, Y.-J., Feng, J., Wu, Y., Shen, H.-M., & Lu, G.-D. (2022). Full-coverage regulations of autophagy by ROS: From induction to maturation. *Autophagy*, *18*(6), 1240–1255. doi:10.1080/15548627.2021.1984656 PMID:34662529

Zhou, Y., Zhen, Y., Wang, G., & Liu, B. (2022). Deconvoluting the complexity of reactive oxygen species (ROS) in neurodegenerative diseases. *Frontiers in Neuroanatomy*, *16*, 910427–991034. doi:10.3389/fnana.2022.910427 PMID:35756499

Zuo, J., Zhang, Z., Li, M. M., Yang, Y., Zheng, B. H., Wang, P., Huang, C. H., & Zhou, S. T. (2022). The crosstalk between reactive oxygen species and noncoding RNAs: From cancer code to drug role. *Molecular Cancer*, *21*(1), 1–17. doi:10.118612943-021-01488-3 PMID:35081965

KEY TERMS AND DEFINITIONS

Antioxidant: A molecule which will remove free radicals from the body by performing a redox reaction.

Biomedical Environment: A place where cells metabolize directly and the environment where cells live directly.

Cell Metabolism: The series of chemical reactions that occur in living organisms in order to maintain life.

Chemiluminescence: The emission of light from a compound after it undergoes a chemical reaction which produces it in an electronically excited state.

Free Radical: Species which contain unpaired electrons and can be highly reactive towards other molecules by a series of reactions such as H-atom abstraction.

Photodynamic Therapy: A treatment which involves the activation of a medicine (compound) by light to produce a reactive species which eventually causes cell death.

Photosensitizer: A compound which will absorb light and pass on the energy to another near molecule.

Prodrug: A compound covalently linked with a chemical group, fragment, or molecule (carrier) to form a temporary or permanent association that can be converted into a pharmacologically active drug in a specific biological environment before use.

Reactive Oxygen Species: An oxygen containing compound that can undergo facile redox reactions with various other species in solution, often leading to additional side reactions.

Thioredoxin Reductase: An enzyme that will catalyze the reduction of thioredoxin which is a small redox protein found in organisms.

Chapter 13
The Immune Response Modeling to a Viral Load:
A Minimal Model

Olga Yaltychenko
Institute of Applied Physics, Moldova State University, Moldova

Natalya Gorinchoy
https://orcid.org/0000-0003-4529-9061
Institute of Chemistry, Moldova State University, Moldova

Gheorghe Duca
Institute of Chemistry, Moldova State University, Moldova

ABSTRACT

A model of the kinetics of the body's immune response to a viral load is proposed. The model takes into account both cellular and humoral components of the immune response. The interaction of T- and B-type lymphocytes, as well as the effect of delaying the formation of antibodies, were also taken into account. Numerical analysis of the model system with some values of the model parameters showed the possibility of implementing three main modes of regular behavior in the system: 1) a mode of unlimited growth in the amount of antigen and a critical decrease in the number of T-type lymphocytes (unfavorable prognosis of the disease); 2) the mode of dynamic balance in the system "antigen - T-lymphocytes" (satisfactory prognosis of the disease); 3) a mode of rapid reduction in the amount of antigen, adequate growth, and reaching stationary values of the number of immune cells (favorable prognosis of the disease).

DOI: 10.4018/978-1-6684-7198-2.ch013

INTRODUCTION

Mathematical immunology is an independent discipline, which was formed at the intersection of biology, medicine, mathematics, physics and chemistry and began to evolve only in the early 70s of the last century. The founder of modern immunology is considered to be Louis Pasteur, who was the first to formulate the basic principle of protection against the causative agent of any infectious disease and also weakened cultures of pathogens of infectious diseases he called vaccines back in the late 19th century (Aronova, 2006). A large contribution to the development of immunology was made by Elie Metchnikoff, who in 1883 discovered phagocytosis and introduced the concept of "cellular immunity", and in 1898 Paul Ehrlich created the theory of humoral immunity (Perelson & Weisbuch, 1992). These models in immunology can be conceptual, including verbal models, graphs or charts; and experimental, such as a specific cell culture system or mouse strain; or may take the form of quantitative mathematical or computer models (Handel et al., 2020). The most common types of quantitative models are what we term phenomenological models (Lythe & Molina-París, 2018). Phenomenological models are applied to extract patterns or, more broadly, information from data. As such, whenever data are being analyzed in some mathematical manner, this type of model is in play. Computing a correlation coefficient between two quantities of interest is an example of a very simple model that tries to detect a pattern. Regression models, in which a mathematical function is specified and the distance between the data and the function is minimized, also fall into this category (Nickaeen et al., 2019).

In recent years, the increase of available data has led to greater use of more complex phenomenological models, which increasingly go by names such as machine learning or deep learning approaches (Shinde & Kurhekar, 2020). These models do not explicitly describe the mechanisms by which patterns arise, and this is both a strength and a weakness. On the plus side, one can determine correlations, find patterns, deduce potential causation and make predictions without having to understand the underlying mechanisms governing a given system. The drawback is that such models provide, at best, hints regarding potential mechanisms. To study mechanisms and processes, mechanistic models are ideal. Mechanistic simulation models explicitly specify processes describing the mechanisms of interaction between system components (Pélissier et al., 2020). Usually, these models are highly simplified—but if done well, still can be very powerful—for abstractions of the system under study. The advantage of this kind of model is that it can provide mechanistic insights, leading to a better and deeper understanding of the system, to a point where the model might allow for very precise predictions. The main disadvantage of this approach is that model construction requires considerable knowledge (or at least assumptions) about the system and how its components interact.

Both phenomenological and mechanistic models are useful tools with distinct advantages and disadvantages. Deciding which type of model to use depends on the question and study system. It is common to start with phenomenological models, to determine patterns and obtain clues regarding the underlying processes and mechanisms, and then move to a mechanistic model to analyze those processes, their interactions, and the resulting outcomes in more detail. For the purposes of this article, we use the term 'model' meaning a model that describes the dynamics of the components of a system in an explicit and mechanistic way through mathematical equations or computational algorithms. Those models are generally studied by simulating them on a computer. There are different ways that such models can be implemented. The most common types used in immunology are compartmental models, in which each compartment tracks the size of a given biological entity of interest — for example, pathogen load or cytokine concentrations (Mosa, 2020). The most common way to implement a compartmental model is with ordinary differential equations.

In 1964, immunologists F. Burnet and N. Jerne created the clonal selection theory of immunity, which very quickly became one of the main dogmas of immunology (Bell, 1970). In 1970, N. V. Stepanova and colleagues proposed a model of the immune response, in accordance with the theory of Burnet and Jerne (Romanovsky et al., 1984). The model consisted of 4 ordinary differential equations and took into account the stages of changes in B-lymphocytes during the immune response. This direction was continued by G. I. Marchuk with co-authors, who first reduced the number of equations of the system by introducing the effect of delaying the formation of antibodies, and then, as new knowledge about the immune system was accumulated, expanded the model (Bocharov & Marchuk, 2000).

In general, immunology studies the process of immunity, the molecular and cellular mechanisms of the body's response to foreign substances. The very concept of immunity combines a whole set of body responses to the invasion of foreign agents, both individual protein molecules and viruses and bacteria. These foreign agents, called antigens, carry on their surface antigenic determinants that are not characteristic of the given organism. The body's first reaction is nonspecific and develops in the first few days. It consists of the fact that macrophages, with the help of special enzymes, are trying to "digest" a foreign substance. Often this non-specific reaction is not enough. Then the specific immune system reacts. The specific immune response is currently considered to be two-component: cellular, first described by Metchnikoff, and humoral, described by Ehrlich (Aronova, 2006). In the case of a cellular immune response, specific "killer cells" are attached to "enemy cells" and destroy them by lysis. In the case of humoral immunity, special plasma cells produce a large number of protein molecules-antibodies with a spatial configuration complementary to the antigenic determinant. Antibodies are most often polyvalent, and the antigen carries several active determinants, therefore, as a result of the antigen-antibody interaction, large aggregates are formed, which are neutralized by phagocytes and then excreted from the body.

At the moment, there is a wide range of phenomena and experimental and laboratory data that the theory of immunity must consistently describe. Among them are the kinetics of antibody accumulation, immune memory (a faster and more significant response to the reappearance of the same antigen), immunity tolerance in some cases, autoimmune diseases, etc. There are two main populations of lymphocytes, some of which necessarily pass through the thymus (T-lymphocytes), and others, B-lymphocytes, go directly from the bone marrow into the lymph nodes. Antibody-producing plasma cells are descendants of B-lymphocytes, T-cells play an important role in antigen recognition, B-cell instruction, and regulation of the immune response, and are also precursors of "killer cells" (T-killers). A simplified scheme of the immune response is shown below (Figure 1) (Romanovsky et al., 1984).

Figure 1. Simplified scheme of t immune response
Source: IGI (2023)

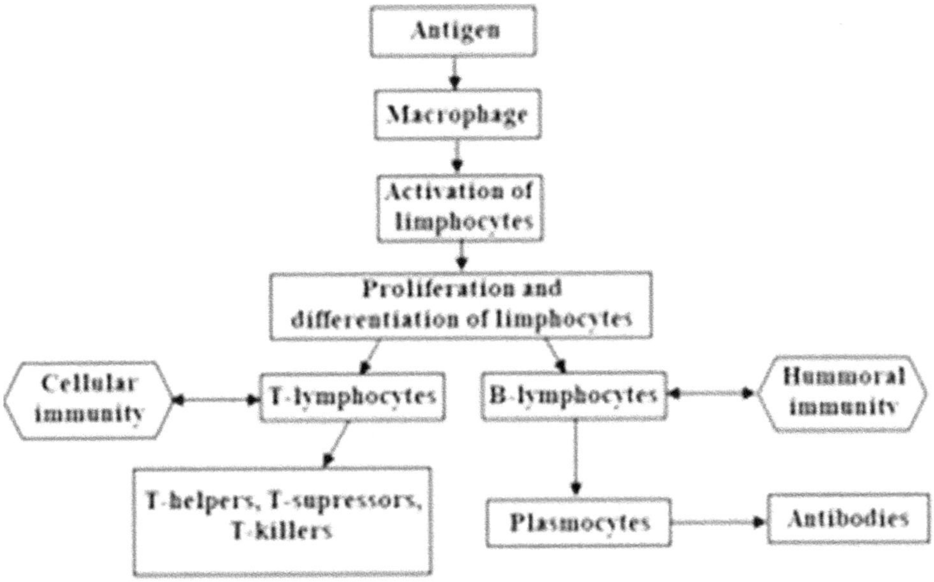

The scheme shows the main stages in the formation of cellular and humoral immune responses. To stimulate B cells, the antigen must be pretreated with T cells (helpers) and macrophages. After antigenic stimulation, B-lymphocytes begin intensive reproduction with the final stage of differentiation into non-proliferating plasma cells capable of producing antibodies. T-cells (suppressors) are engaged in the suppression of the proliferation of T- and B-cells under certain conditions. In addition, the lack of immune response to host antigens is also the merit of T- suppressors. At the last stage of the development of the immune response, memory cells appear that retain the memory of this antigen for a long time and provide a faster immune response to its repeated introduction. Note that in the simplest case, antibodies are considered identical, but in fact their valence and molecular weight change while specificity remains unchanged. First, larger IgM molecules with five or more active sites are produced, and then the cells switch to the production of lighter, bivalent IgG immunoglobulins. It is important to describe the stages that immunocompetent cells (B-lymphocytes) undergo under the action of an antigen. There are three successive states: "X", "Y", and "Z" (Romanovsky et al., 1984), shown in Figure 2.

Figure 2. Scheme of B-cell differentiation under the action of an antigen
Source: IGI (2023)

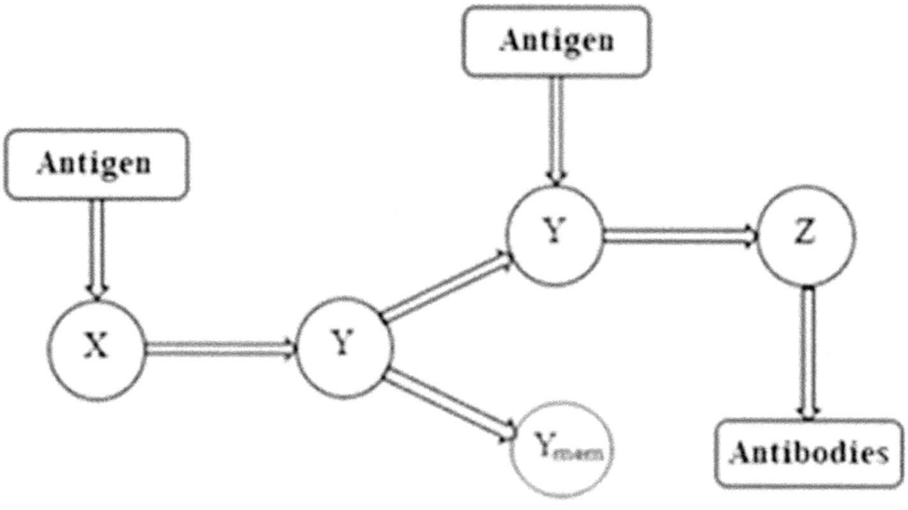

Here "X" are progenitor cells that go into the proliferating state "Y" after interaction with the antigen. These are rapidly dividing cells, which, after repeated interaction with the antigen, pass into the final state "Z" of cells that produce antibodies. The "Y" cells, which have not repeatedly interacted with the Y-cell antigen, go into the state of memory cells, "Y_{mem}", circulating for a long time and providing "immune memory" to a given antigen. It is known that in a state of viral, infectious load, redox processes, including radical oxidation processes, occur especially intensively. This paper describes a model of the body's immune response to a viral load at the level of a cell population, which is accompanied by redox processes at the subcellular level, including biochemical reactions involving radicals.

THEORETICAL APPROACH

The aim of the present work is to study the features of the immune reaction (response) by creating the simplest kinetic model that takes into account both the cellular and humoral components of immunity. With the help of this fairly simple model, through numerical analysis, the main modes in the kinetics of the "virus-immunity" system are identified and qualitatively compared with the available data on the characteristics of the course of the most viral infection. The model takes into account the interaction of T- and B-lymphocytes, as well as the effect of delaying the kinetics of antibodies.

The proposed kinetic model is built in terms of the number of cells involved in the process. For this purpose, the amounts of antigens, T-type lymphocytes, and antibodies are denoted as Ag[t], T[t], Ab[t], respectively. At the first stage of modeling, only the reaction from cellular immunity will be taken into account. In this case, the system of kinetic equations for the amount of antigen and T-type lymphocytes can be written as follows:

$$\frac{dAg[t]}{dt} = \mu_{Ag}Ag[t]\text{-}\gamma_{Ag}Ag[t]T[t],$$
$$\frac{dT[t]}{dt} = \mu_T(Ag[t]\text{-}Ag[t]^2)T[t] - \gamma_T T[t] + \nu_T \tag{1}$$

Here, the first term on the right side of the first equation is responsible for the increase in the amount of antigens, and the second term in the same equation is responsible for the decrease in the amount of antigen due to their destruction by T-type lymphocytes. Accordingly, the first term on the right side of the second equation is responsible for the increase in the number of T-lymphocytes (the second term in brackets takes into account the so-called "crowding effect"), the other two terms are responsible for the natural death of lymphocytes and their constant influx from stem cells. The transition to dimensionless variables and parameters greatly simplifies the form of the system (1):

$$\frac{dAg[t]}{dt} = Ag[t]\text{-}Ag[t]T[t],$$
$$\frac{dT[t]}{dt} = \mu(Ag[t]\text{-}Ag[t]^2)T[t] - \kappa T[t] + \nu \tag{2}$$

It is important to note that the parameters in system (2) are dimensionless and for numerical analysis will be taken from the range 0÷1. This seems appropriate since these parameters are combinations of the parameters of the system (1), which in turn have the meaning of reaction rates and the estimated values from (Rudnev et al., 2007) confirm this range of possible values of the dimensionless parameters. The system (2) of differential equations is non-linear and can be studied using the qualitative theory of Lyapunov differential equations (Nowak & May, 2000; Perelson, 2002; Roberts, 2014). This system has singular points, the position and the type of which depend on the parameters of the system. One singular point on the phase plane T(Ag) lies on the vertical axis T and is defined as (3), while the position of the other two singular points, depending on the parameters, is determined as follows (4).

$$Ag_1 = 0, \quad T_1 = \frac{\nu}{\kappa} \tag{3}$$

$$Ag_{2,3} = 0.5 \pm \sqrt{\frac{1-4(\kappa-\nu)}{4}}, T_{2,3} = 1 \tag{4}$$

The model parameters are combinations of the rates of elementary processes and, at the same time, are dimensionless. Therefore, the parameters used are not tied to the absolute values of the rates of biochemical processes in the immune system but are taken in acceptable ranges of values. The ranges of values are estimated both analytically and consistent with the laboratory values from the work (Rudnev et al., 2007). It is important to note that on the plane of the system parameters κ and ν there are three significant areas in which the behavior of the system differs significantly (Figure 3).

Figure 3. Regions in the space of parameters κ and ν in which the behavior of the system changes significantly
Source: IGI (2023)

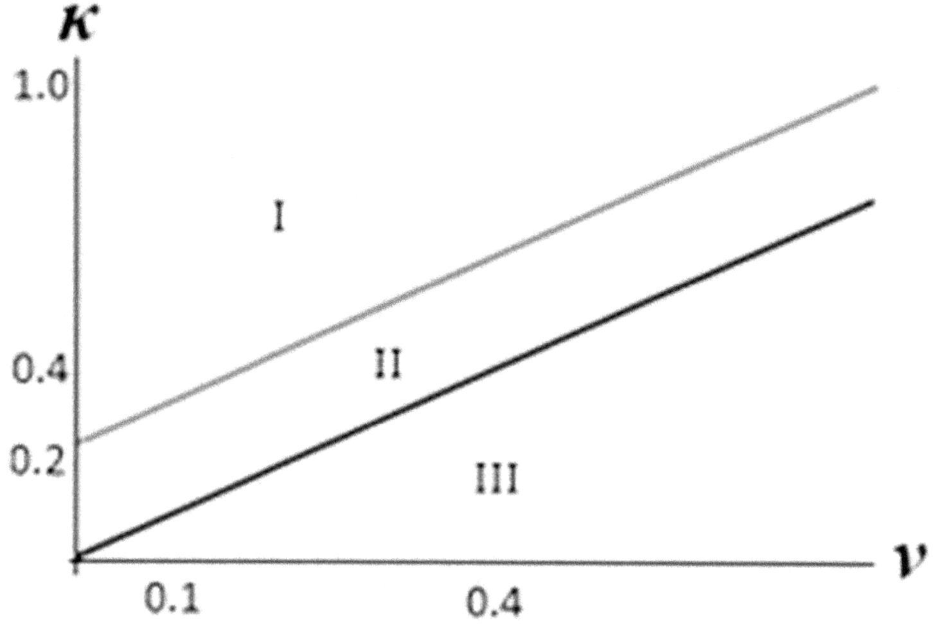

Straight lines κ = ν, and κ = 0.25+ν divide the plane into three regions. So, the region I (here both values that define the coordinate (Ag) for the second and third singular points are not real) corresponds to an absolute lack of immunity. In this range of parameters, the amount of antigens increases indefinitely, while the amount of T-type lymphocytes is rapidly depleted. This region corresponds to cases of a high rate of the lymphocytes death. Region II, Figure 3, corresponds to the case when both values of the coordinate (Ag) at singular points are positive. This is the case when the death of lymphocytes still exceeds their increase, and then two scenarios are possible depending on the initial conditions.

If at the initial moment of time, the concentration of the antigen is high, and there are few lymphocytes, then the outcome of the competition will be in favor of the antigen; if the concentrations of lymphocytes at the initial moment are sufficient, then a kind of dynamic equilibrium will be established, which is much more favorable for the prognosis of recovery than in the region I of system parameters. In region III, Figure 3, only one value of the coordinate (Ag) is positive. The result also strongly depends on the initial conditions. However, the amount of T-type lymphocytes tend to saturation, and the number of antigens tends to zero. This is a case of a full and speedy recovery. Under natural conditions, when there is no targeted stimulation of lymphocytes, case II seems more realistic than case III. Figure 4 shows the phase portraits in the T(Ag) plane corresponding to regions I, II, and III from Figure 3.

Figure 4. Phase portraits on the T(Ag) plane corresponding to regions I, II, and III in Figure 3
Source: IGI (2023)

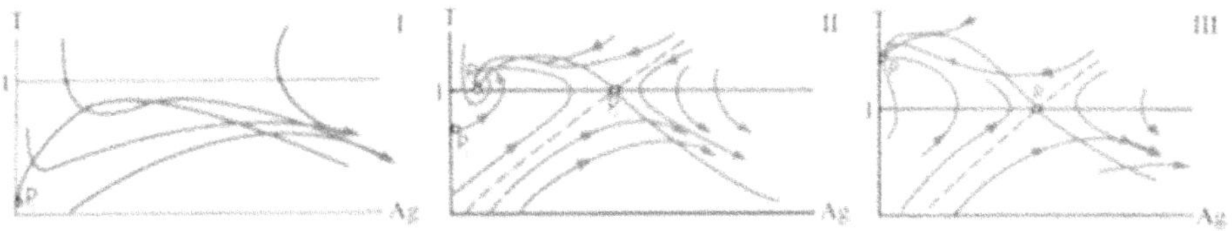

STAGES OF THE NUMERICAL SIMULATION

The results of numerical calculations of the time dependencies and some values of the system parameters, corresponding to the phase portraits in Figure 4, are presented in Figure 5.

Figure 5. Time dependences Ag[t] and T[t] (curves 1 and 2, respectively) at the values of the system parameters: (a) $\kappa = 0.6$, $\nu = 0.2$; (b) $\kappa = 0.3$, $\nu = 0.2$, and (c) $\kappa = 0.2$, $\nu = 0.2$, corresponding to regions I, II, and III in Figure 3. All variables and parameters are in dimensionless units.
Source: IGI (2023)

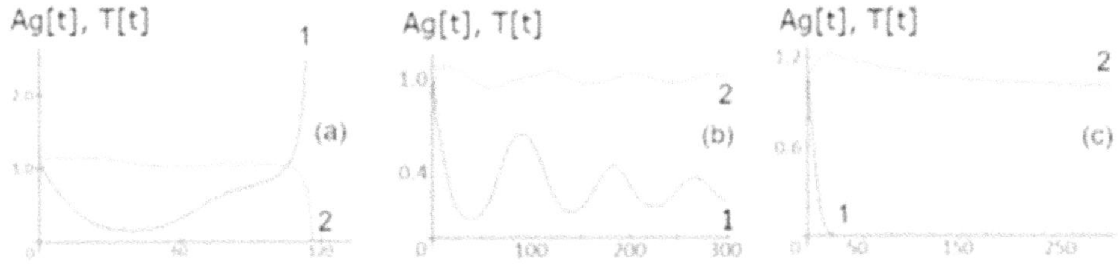

In the first case, Figure 5(a), corresponding to the parameters κ and ν from region I from Figure 3, the immune response is not sufficient to effectively control the disease. It can be seen that starting from a certain point in time, an uncontrolled increase in the amount of antigen and a critical decrease in the amount of T-lymphocytes occur, and as a result, an unfavorable prognosis of the disease is most likely. It can also be seen from the analysis of the figures that the second and third cases correspond to regimes from the regions of parameters II and III. With such ratios of parameters, the immune cellular response is sufficient and leads to the establishment of a dynamic balance between the amounts of antigen and immune cells (case (b), mode II) or to complete recovery, which corresponds to a favorable prognosis of recovery (case (c), mode III).

Further, this minimal model can be extended and improved by taking into account the humoral immune response. For this, the system of equations (2) should be modified. Thus, the term -αAg[t]Ab[t] should be added to the right side of the equation for Ag[t], corresponding to the death of the antigen when it encounters Ab[t] antibodies. It is important to note that antibodies appear with a time delay in the case of primary disease in accordance with the scheme in Figure 2, or are present immediately in a certain amount as a result of a past illness or prior vaccination. When taking into account the influence of antibodies on the prognosis of the disease, it is advisable not to expand the model to a three-dimensional one, remaining within the framework of a two-dimensional system. It is believed that the amount of antibodies is a function of the amount of antigen, which, in turn, is a function of time (5).

$$Ab[t] = Ag[t]/(Ag[t]+\sigma) \quad (5)$$

This choice is physiologically justified since it is known that at low concentrations of the antigen, the amount of antibodies increases linearly, and at high concentrations of the antigen, it does not change significantly. The effect of the interaction of T- and B-lymphocytes, which in some way manifests itself as a synergy of cellular and humoral immunity, should also be taken into account. For this, the term ηAb[t]T[t] is added to the right side of the equation with respect to T[t] of the system of equations (2). Then the modified system of equations (2) will look like this:

$$\begin{aligned} \frac{dAg[t]}{dt} &= Ag[t]-Ag[t]T[t]-\alpha Ag[t]^2 / (Ag[t]+\sigma), \\ \frac{dT[t]}{dt} &= \mu(Ag[t]-Ag[t]^2)T[t]-\kappa T[t]+\nu+\eta Ag[t] / (Ag[t]+\sigma)T[t] \end{aligned} \quad (6)$$

The results of numerical calculations of the system, equations (4) with respect to the time dependences Ag[t], T[t] and Ab[t], with the same values of the system parameters κ and ν, as in Figure 5, are presented in Figure 6.

Figure 6. Time dependencies Ag[t], T[t], and Ab[t] (curves 1, 2, and 3, respectively) of the systems of cellular and humoral immunity for the cases I, II, and III (a, b, c) from Figure 3, without taking into account the effect of "T-B interaction"; α = 1, σ = 0.1. All variables and parameters are in dimensionless units
Source: IGI (2023)

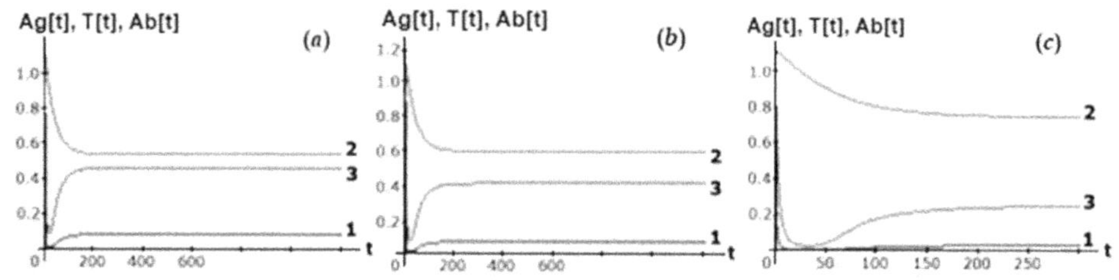

Analysis of the numerical calculation (Figure 6) shows that taking into account the contribution of humoral immunity makes it possible to control the course of the disease even when the cellular immune response is not effective enough. In this case, only 2 modes are implemented (complete recovery, and dynamic equilibrium mode) instead of three scenarios for the development of the disease, as in the case of taking into account the response only from T-cell immunity. At the same time, it is important to note that the load on T-cell immunity is removed, which is also important for further post-infection recovery. Further, it will be shown how taking into account the interaction of T- and B-lymphocytes in this minimal model affects the kinetics of recovery. To do this, the time dependences of Ag[t], T[t] and Ab[t] are calculated for the same values of the system parameters, but taking into account the interaction between the subsystems of T- and B-lymphocytes (Figure 7).

Figure 7. Time dependences Ag[t], T[t], and Ab[t] (curves 1, 2, 3) of the systems of cellular and humoral immunity for cases I and II (a, b) from Figure 3 taking into account the effect of the "T-B interaction"; $\alpha = 1$, $\sigma = 0.1$, $\eta = 0.01$. All variables and parameters are in dimensionless units
Source: IGI (2023)

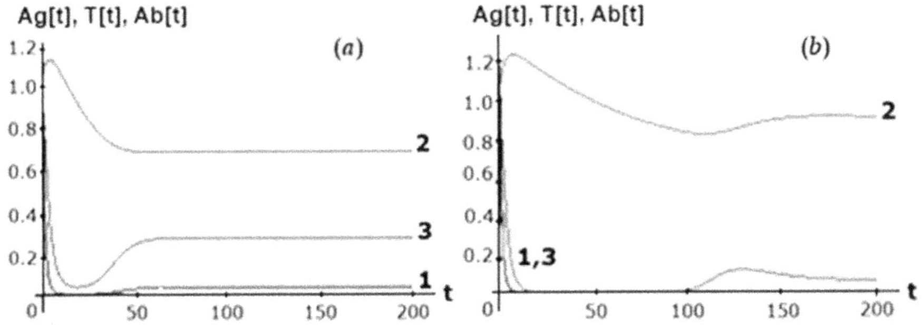

It follows from the analysis of the calculations that taking into account the "T-B interaction" of lymphocytes makes it possible to achieve complete control over the disease with lower amounts of antibodies and in shorter time intervals. This is an important result of this model of cellular humoral immunity. The results described above relate to the situation when at the initial time there is a non-zero amount of antibodies, which refers to cases where an adequate vaccine has been used or the encounter with this antigen is not the first time. The case of no vaccination and disease for the first time, as well as in the absence of innate immunity, must be described separately since there are no antibodies at the initial time, and therefore it is important to take into account the effect of delay in the kinetics of antibodies. This is due to the fact that the process of creating antibodies takes from 3 to 5-6 days in each particular case. The effect of the delay in the formation of antibodies can be taken into account by adding a factor (7) to the terms from the right-hand sides of the system of equations (6), which are responsible for the interaction with antibodies.

$$D[t] = (1 - \exp[-\frac{t}{\tau}]) \tag{7}$$

Here τ is the delay time required for the body to produce antibodies. Then the system equations (6) will look like that in equation (8):

$$\begin{aligned} \frac{\mathrm{dAg[t]}}{\mathrm{dt}} &= \mathrm{Ag[t]\text{-}Ag[t]T[t]}\text{-}\alpha D[t]Ag[t]^2 \,/\, (Ag[t]+\sigma), \\ \frac{\mathrm{dT[t]}}{\mathrm{dt}} &= \mu(\mathrm{Ag[t]\text{-}Ag[t]^2})T[t] - \kappa T[t] + \nu + \eta D[t]Ag[t] \,/\, (Ag[t]+\sigma)T[t] \end{aligned} \tag{8}$$

The corresponding results of numerical calculations are shown in Figure 8.

Figure 8. Time dependences Ag[t], T[t], Ab[t] (curves 1, 2, 3) of cellular and humoral immunity systems for the case II, $\alpha = 1$, $\sigma = 0.1$, $\eta=0.01$, $\tau = 100$. Case (a) - only the T-cell immune response, case (b) - the influence of antibodies is also taken into account, and case (c) - the delay effect is taken into account. All variables and parameters are in dimensionless units.
Source: IGI (2023)

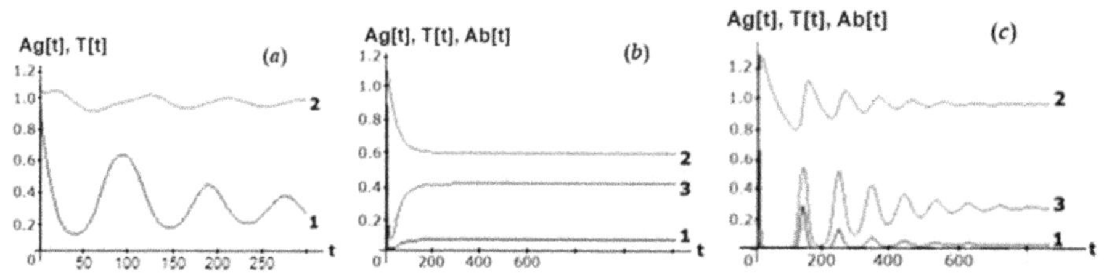

Analyzing the time dependences presented in Figure 8, one can note that case (*b*), where, unlike case (*a*), the influence of antibodies on the kinetics of the process in the system is taken into account, demonstrates quite well the role of humoral immunity in the development of the disease scenario, but it is realistic only if there are antibodies at the moment of the onset of the disease; case (*c*) corresponds to a situation where initially there are no antibodies and there is a delay in their formation.

ASPECTS OF EXPANSION AND APPLICATION OF THE MODEL

The proposed minimal model lends itself well to expansion to take into account new experimental and laboratory data. This model can be used for a qualitative assessment of the main scenarios for the course of a viral disease, including COVID-19, but for this it is necessary to conduct an additional study to identify an unambiguous relationship between the operating parameters of the model and the diagnostic criteria of medical practice. This is the subject of a further separate study.

CONCLUSION

A minimal kinetic model has been proposed for studying the features of the body's immune response to a viral load, taking into account both the cellular and humoral components of the immune response. The model is phenomenological and leads to a system of two differential equations. Numerical analysis revealed the main patterns in the kinetics of the "virus-immunity" system, which can be qualitatively compared with the available data on the features of the course of a viral infection. The model also takes into account the interaction of T- and B-lymphocytes, as well as the effect of delaying the formation of antibodies. The results of numerical calculations were performed only for some values of the parameters of the model system; nevertheless, they show the possibilities of modeling the scenarios of the course of the disease, as well as the possibilities of monitoring and even controlling the process.

ACKNOWLEDGMENT

The authors thank Kanarovskii E. for the discussion of the results. This work was supported within the Moldovan National Projects: Nr. 20.80009.5007.14 and Nr. 20.80009.5007.27.

REFERENCES

Aronova, E. (2006). *Philosophy and experiment: Essays on the history of immunology in the 20th century*. KomKniga.

Bell, G. (1970). Mathematical model of clonal selectional and antibody production. *Journal of Theoretical Biology, 29*(2), 191–232. doi:10.1016/0022-5193(70)90019-6 PMID:5500468

Bocharov, G., & Marchuk, G. (2000). Applied problems of mathematical modeling in immunology. *Журнал вычислительной математики и математической физики, 40*(12), 1905–1920.

Handel, A., La Gruta, N. L., & Thomas, P. G. (2020). Simulation Modelling for Immunologists. *Nature Reviews. Immunology, 20*(3), 186–195. doi:10.103841577-019-0235-3 PMID:31804613

Lythe, G., & Molina-París, C. (2018). Some deterministic and stochastic mathematical models of naïve T-cell homeostasis. *Immunological Reviews, 285*(1), 206–217. doi:10.1111/imr.12696 PMID:30129198

Mosa, A. I. (2020). Antigenic variability. *Frontiers in Immunology, 11*, 2057. doi:10.3389/fimmu.2020.02057 PMID:33013870

Nickaeen, N., Ghaisari, J., Heiner, M., Moein, S., & Gheisari, Y. (2019). Agent-based modeling and bifurcation analysis reveal mechanisms of macrophage polarization and phenotype pattern distribution. *Scientific Reports, 9*(1), 12764. doi:10.103841598-019-48865-z PMID:31484958

Nowak, M. A., & May, R. M. (2000). *Virus dynamics mathematical principles of immunology and virology*. University Press.

Pélissier, A., Akrout, Y., Jahn, K., Kuipers, J., Klein, U., Beerenwinkel, N., & Rodríguez Martínez, M. (2020). Computational model reveals a stochastic mechanism behind germinal center clonal bursts. *Cells, 9*(6), 1448. doi:10.3390/cells9061448 PMID:32532145

Perelson, A. (2002). Modelling viral and immune system dynamics. *Nature Reviews. Immunology, 2*(1), 28–36. doi:10.1038/nri700 PMID:11905835

Perelson, A., & Weisbuch, G. (1992). *Theoretical and experimental insights into immunology*. Springer. doi:10.1007/978-3-642-76977-1

Roberts, E. (2014). Cellular and molecular structure as a unifying framework for whole-cell modeling. *Current Opinion in Structural Biology, 25*, 86–91. doi:10.1016/j.sbi.2014.01.005 PMID:24509245

Romanovsky, Yu. M., Stepanova, N. V., & Chernavsky, D. S. (1984). *Mathematical Biophysics*. Nauka.

Rudnev, S., Romanyukha, A., & Yashin, A. (2007). Modeling of T cell population development and estimation of resource allocation effectiveness. *Mat. Modeling, 19*(11), 25–42.

Shinde, S. B., & Kurhekar, M. P. (2020). Agent-based modeling of the adaptive immune system using netlogo simulation tool. *Soft Comput. Problem Solving*, 463-474.

Chapter 14
Antioxidants in Clinical Treatments:
A Case Study With Vitamin C

Radu Silaghi-Dumitrescu
https://orcid.org/0000-0003-3038-7747
Babes-Bolyai University, Romania

Flavia-Malina Oana (Gadina)
Babes-Bolyai University, Romania

Maria Lehene
Babes-Bolyai University, Romania

ABSTRACT

The ability of antioxidants to scavenge free radicals is well known, but their possible therapeutic action is still unclear because so far most studies are in vitro on cultured cells (where the antioxidant concentrations are typically much larger than those that can be reached in the human body) or in vivo on animals. This chapter is focused on ascorbic acid (vitamin C). Vitamin C exhibits dual reactivity with pro-oxidant activity registered at concentrations above 1 mM while at lower concentrations it has an antioxidant effect. Vitamin C appears to be effective in preventing respiratory diseases, shortening recovery time after surgeries, increasing the efficiency of the immune response to infections, improving glycemic control in type 2 diabetes, decreasing the need for fluids in the treatment of burns, and protecting the skin. Although countless beneficial effects of antioxidants have been proposed, far more studies are required before most of these claims should be taken seriously, possibly to discover new antioxidants that are more effective against the disease.

DOI: 10.4018/978-1-6684-7198-2.ch014

INTRODUCTION

Antioxidants are chemical compounds or molecules that counteract the adverse effects of reactive oxygen or reactive nitrogen species (ROS, RNS). Reactive oxygen/nitrogen species are substances that contain one or more activated/reactive oxygen or nitrogen atoms and can oxidize cell components. ROS/RNS, alongside free radicals (some of which are in fact ROS or RNS, but also possibly carbon-centered radicals or other types), induce and mediate oxidative modification of lipids, proteins, nucleic acids, and carbohydrates. These processes can also be induced by variations in antioxidant compounds and antioxidant enzymes which are nonradical mechanisms (Niki, 2010).

ROS/RNS have various targets, starting from sugars where they induce fragmentation with glycolaldehyde generation and autoxidation, to DNA and RNA where they produce nucleobase modifications, deletions, base-free site production, chain breaks, DNA protein cross-linking, and changes in chromosomal arrangement. On proteins, ROS/RNS induce oxidative modifications, peptide cleavage, and cross-linking, while on lipids peroxidation, membrane damage/destruction, and subsequent carcinogenic products are observed (Niki, 2010).

Antioxidants are at times described as acting in 4 lines in the defense mechanism against free radicals – although this classification can be confusing, with various antioxidants arguably fitting into more than one line. The first line of defense is defined by preventive antioxidants. This is based on the suppression of the formation of reactive species (ROS/RNS), for example by reducing hydrogen peroxide and lipid hydroperoxides (e.g., with peroxidases and catalases) to lipid hydroxides and water, or by sequestering metal ions (Fe and Cu). Enzymes in this category include GPx (glutathione peroxidase), Trx (tyrosine peroxidase), Prx (proline peroxidase), CAT (catalase), LOX (lipoxygenase), or myeloperoxidase; also listed in this class have been SOD (superoxide dismutase), cytochrome P450 and cytochrome *c* – but their placement here is debatable (SOD may better be listed in the second class/line as described below, while the physiological functions of P450 and of cytochrome c do not entail high-affinity/high-efficiency reaction with RNS/ROS) (Niki, 2010).

"Scavenging antioxidants" are defined as the second line of defense. This will rely on the rapid removal of ROS/RNS/free radical species before damage to essential biological molecules can occur. SOD converts superoxide to hydrogen peroxide, while carotenoids physically or chemically capture singlet oxygen. Phenolic compounds and aromatic amines also belong to this category. Many members of this class are thus sacrificial (and thus acting at concentrations closer to those of the RNS/ROS) rather than catalytic antioxidants; the term "scavenging" is often used for this class with the intention of denoting the 1:1 stoichiometry, although one can appreciate that in a very strict sense the "first line" antioxidants also destroy/scavenge ROS/RNS. If complete capture of ROS/RNS is not achieved by the first and second-line antioxidants, then oxidative effects will occur (Niki, 2010). Hydroxyl, alkoxyl, peroxyl, superoxide, nitrous oxide, and radicals containing sulfur or nitrogen groups are free radicals with various properties (biological, chemical, and physical) that may be involved in oxidative stress. The question arises as to whether antioxidants can efficiently eliminate hydroxyl and alkoxyl radicals *in vivo* by antioxidants since these radicals react predominantly with various structures (lipids, proteins, carbohydrates) and with DNA which is much less reactive but in much higher concentration than antioxidants (Niki et al., 2018).

The third line of defense is formed by enzymes that possess various functions such as: repairing oxidative damage to biological molecules, cleaning/removing waste, and reconstituting lost functions. Failure to do so will result in the manifestation of a disease or premature aging (Niki, 2010).

The fourth line is the function of adaptive mechanisms in which antioxidants are generated at the right time and at the right concentration. Antioxidants behave as cellular signaling messengers to regulate the levels of antioxidant compounds and enzymes, which is observed by their increase (Niki, 2010).

Antioxidants can be classified according to various categories; the simplest classification would be enzymatic vs. non-enzymatic. The first category is represented by primary enzymes (superoxide dismutase, catalase, glutathione peroxidase) and secondary enzymes (glutathione reductase, gluco-6-phosphate dehydrogenase). The category of non-enzymatic antioxidants is further divided into other classes such as: flavonoids (flavonols - quercetin, kaempferol; flavanols - catechin, pelargonidin; anthocyanins - cyanidin, pelargonidin, isoflavonoids - genistein, flavanone - hesperidin and flavone-chrysin), phenolic acids (hydroxycinnamic acids - ferulic acid, *p*-coumaric acid and hydroxybenzoic acids - gallic acid, ellagic acid), cofactors (coenzyme Q_{10}), minerals (zinc, selenium), organosulfur compounds (alkyl sulfide, indoles, glutathione), vitamins and derivatives (vitamin A-Retinol, vitamin C - ascorbic acid, vitamin E-tocopherol and tocotrienol, vitamin K, carotenoids (β-carotene, Lycopene, Lutein, Zeaxanthin) and non-protein compounds with nitrogen (uric acid) (Huang et al., 2005).

Another classification of antioxidants is by origin: endogenous (synthesized by the human body - represented by enzymes or small molecules) vs. exogenous (procured from the environment through diet: phenols, flavonoids, phenolic acids, carotenoids, vitamins, and minerals) (Huang et al., 2005). Depending on the mechanism by which they act, one can also classify antioxidants in preventive vs. scavenging. Preventing antioxidants precludes ROS generation or ROS access to important biological sites (cell membrane, UV light filter), inhibits pro-oxidant enzymes (15-lipoxygenase-1), regulates enzymatic antioxidant cofactors (selenium, coenzyme Q_{10}), acts as a catalytic system neutralizing or modifying ROS (SOD, catalase, or glutathione peroxidase), or bind/inactivate metal ions to prevent ROS generation (ferritin, ceruloplasmin, catechin). Free radical scavengers are chain-breaking antioxidants that scavenge or destroy ROS (ascorbic acid, tocopherol, uric acid, glutathione, flavonoids, polyphenols, carotenoids, terpenoids, etc.) (Huang et al., 2005).

Vitamin C: General Considerations

Vitamin C is one of the most important antioxidants and can be synthesized by plants and animals from glucose. Humans and a few other vertebrates (such as the guinea pig or Haplorhini primates) lack the enzyme responsible for vitamin C synthesis, i.e., L-gulono-gamma-lactone oxidase. The active form of vitamin C is L- ascorbic acid. Because vitamin C uptake is limited by an active transport mechanism, only a finite amount is absorbed even if a much higher dose is taken orally (Lykkesfeldt & Tveden-Nyborg, 2019; Liu et al., 2020).

Vitamin C has a 5-hydrocarbonate ring similar to glucose and is a weak acid (cf. Figure 1). In combination with a metal ion, it forms complexes. Of increasing interest is the synthesis of physiologically active and stable ascorbate molecules because L-ascorbic acid is unstable in nature, especially when exposed to light (Procházková et al., 2011).

Figure 1. Structure of ascorbic acid and of its oxidized forms.

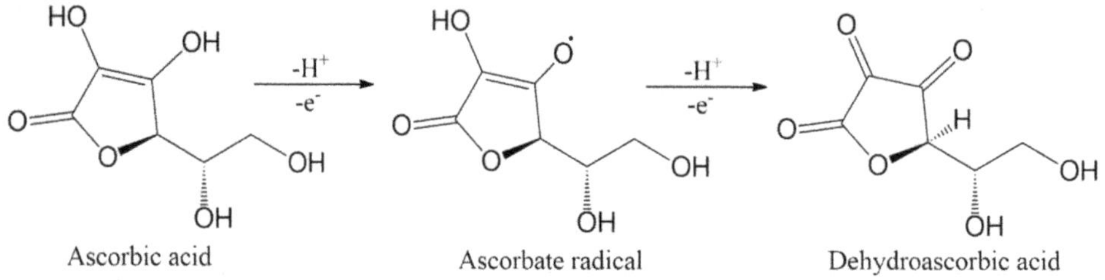

The antioxidant effect of ascorbate is due to its ability to donate electrons (cf. Figure 1), protecting important biomolecules (proteins, lipids, carbohydrates, and nucleic acids) from damage by oxidants generated in cell metabolism as well as from exposure to toxins and pollutants. Vitamin C is also a cofactor for a family of enzymes (monooxygenases and dioxygenases), lysyl and prolyl hydrolases required to stabilize the tertiary structure of collagen, two hydroxylases involved in the biosynthesis of carnitine (a molecule that transports fatty acids into the mitochondria for the generation of metabolic energy), hydrolases involved in the synthesis of catecholamine hormones (norepinephrine), peptide hormones - vasopressin - involved in the cardiovascular response to severe infections (Carr & Maggini, 2017). Ascorbate is thus a complex player in metabolism – so imbalances in its concentration can have far-reaching effects, beyond the simple ROS scavenging activity.

The vitamin C function is influenced by numerous environmental factors and health conditions. For example, the oxidant-antioxidant balance of the body can be affected by exposure to polluting air containing oxidants such as nitrogen dioxide, generating oxidative stress. It can also occur if the antioxidant line of defense is incomplete, as in the case of insufficient vitamin C (Tina Suksmasari, 2015).

Blood tests of participants in one study found that blood serum vitamin C levels were 3 times lower in smokers than in non-smokers, thus recommending an additional vitamin C intake of 35 mg/day, ensuring that there is sufficient ascorbic acid to repair damage caused by oxidants. The chances of developing respiratory disease increase when children and adolescents are exposed to cigarette smoke due to low vitamin C levels (Strauss, 2001).

Severe ascorbate avitaminosis results in serious, sometimes fatal diseases such as scurvy. It manifests itself in weakened collagenous structures, resulting in poor wound healing and impaired immunity. The amount needed to prevent scurvy is 10 mg/day, with the recommended daily dose of vitamin C being a hundred times the amount needed for other vitamins. A diet that provides 100-200 mg/day of vitamin C provides an adequate amount in the plasma and covers what is needed to prevent the risk of chronic disease (Niki, 2010).

On the other hand, an increased concentration of vitamin C in the blood will produce oxidative stress, i.e., it will be in the reactive oxygen species (pro-oxidant) category (Macan et al., 2019). Oxidation and/or degradation of hemoglobin and albumin occurs, and the life span and metabolic activity of leukocytes involved in the antiviral cellular immune response will also be affected. (Burr, 1995) At blood concentrations of 50-150 μM, ascorbate acts as a line of defense, and at concentrations higher than 1 mM (reached only in some cases in intravenous administration), ascorbate shows an acceleration of hemoglobin degradation. The limit at which ascorbate transforms from antioxidant to pro-oxidant is around 1 mM, which is well below the normal physiological concentration in the blood. Other antioxidants that have dual properties depending on concentration will be isoascorbate/erythorbate which are stereoisomers of ascorbate. The reaction between methyl 2-keto-D-gluconate and sodium methoxide will lead to the formation of erythorbic acid which is a stereoisomer of ascorbic acid. Erythorbic acid can also be synthesized from sucrose or by strains of *Penicillium* (Gęgotek & Skrzydlewska, 2022; Lehene et al., 2020).

In the first phase of the inflammatory process, neutrophils will migrate to the affected/injured site to sterilize it until ROS and antimicrobial proteins are released. Neutrophils will undergo apoptosis or be digested by macrophages resulting in an immune response. Vitamin C is thought to influence several important aspects of neutrophil function: migration in response to mediators of inflammation (chemotaxis), phagocytosis, microbial killing, and apoptosis (Carr & Maggini, 2017; Wong et al., 2020).

The proposed clinical applications for vitamin C are very diverse, from engaging and supporting immunity to reducing the body's convalescence. Clinical studies have shown that after surgery, patients require large amounts of vitamin C to restore normal plasma levels. Patients require more than 500 mg/day of vitamin C but taking antioxidant micronutrients can also shorten recovery time (Fukushima & Yamazaki, 2010).

Effect on Leukocyte Function

Intracellular vitamin C values 50-100 times higher than in plasma are found in leukocytes, neutrophils, and monocytes. These cells actively accumulate vitamin C against the concentration gradient, reaching peak concentrations in leukocytes following consumption of ~100 mg/day of vitamin C. Neutrophils contain vitamin C levels of at least 1 mM (Carr & Maggini, 2017).

Vitamin C is a water-soluble antioxidant that scavenges many oxidant reagents and can regenerate important cellular and membrane antioxidants such as glutathione and vitamin E. For example, neutrophils exposed to oxidants may have their motility inhibited. Motility is thought to be impaired due to the oxidation of membrane lipids, thereby affecting cell membrane fluidity (Carr & Maggini, 2017).

Some studies from the 1980s-1990s showed that patients with recurrent infections could have their leukocyte chemotaxis improved with 1 g doses of vitamin C. Also, supplemental administration (400 mg/day) to newborns suspected of septic infections considerably improved neutrophil chemotaxis. Chemotaxis is the migration of neutrophils in response to chemical stimuli, whereas ordinary random migration is called chemokinesis (Carr & Maggini, 2017).

Recurrent infections may also result from genetic alterations in neutrophil function, such as chronic granulomatous disease (CGD), an immunodeficiency disorder that impairs ROS generation by leukocytes, and Chediak-Higashi syndrome (CHS), an autosomal recessive disease that affects vesicular transport. Vitamin C is not expected to influence these genetic diseases. Vitamin C supplementation (200-500 mg/day) was reported to improve neutrophil chemotaxis in two children with CHS (Carr & Maggini, 2017).

Neutrophil Apoptosis

Numerous studies have reported attenuated neutrophil apoptosis in patients with severe infections. Apoptosis is delayed the greater the severity of the disease and is associated with tissue damage in those with sepsis. Immature neutrophils that are released into the blood during severe infections are resistant to apoptosis and have a longer lifespan. Interestingly, high doses of vitamin C can modulate cytokine levels in cancer patients, but this has not been observed in patients with severe infections, which may be another mechanism by which vitamin C modulates neutrophil function (Mikirova et al., 2016). To date, only one study has investigated the effects of vitamin C in neutrophil apoptosis in patients with sepsis (Ferrón-Celma et al., 2009; Doseděl et al., 2021).

Inflammation Mediation

Cytokines are important signaling cells secreted by a variety of immune, young, and adaptive cells in response to infection and inflammation. They comprise a variety of molecules including chemokines, interferons (IFNs), interleukins (ILs), lymphokines, and TNFs, modulating both cellular and humoral immune responses, regulating maturation, growth, and population of specific cells. Supplements of 1 g/day of vitamin C given to healthy volunteers (with and without vitamin E) showed enhanced, increased production of IL-10, IL-1, and TNF-α cells following stimulation with lipopolysaccharide (Carr & Maggini, 2017).

Therefore, the effect of vitamin C in cytokine generation seems to depend on the target cell type or inflammatory stimulus. Recent studies have shown that vitamin C used as a treatment for microglia attenuates cell activation and synthesis of the pro-inflammatory cytokines TNF, IL-6 and IL-1. This is an indicator of the anti-inflammatory phenotype (Portugal et al., 2017).

Diabetes

Diabetes patients are more prone to common infections such as influenza, pneumonia, lower limb infections, which increase their morbidity and mortality. Immune changes are observed in type 2 diabetes. Progression of insulin resistance in this type of diabetes has been associated with persistence of adipose tissue inflammation. Accumulation of pro-inflammatory cytokines (TNF-α) is due to infiltration of pro-inflammatory macrophages and T cells. A decrease in plasma vitamin C has been observed in this type of diabetes, the main cause being hyperglycemia which increases the need for vitamin C due to increased levels of oxidative stress. An inverse correlation has been reported between plasma vitamin C concentrations and diabetes risk, hemoglobin A1C concentration (an index of glucose tolerance), fasting, postprandial glucose levels and oxidative stress. Meta-analysis of studies indicated that vitamin C supplementation may improve glycemic control in type 2 diabetes (Ashor et al., 2017).

Colds and Pneumonia

Shortening the duration of a common cold by 1 to 4 days is achieved by taking vitamin C supplements at daily doses of 200 mg or more. Common colds increase with exposure to physical stress (Heimer et al., 2009).

In the case of pneumonia, beneficial effects in the recovery process have been noted. In patients admitted for pneumonia with very low vitamin C levels, its administration reduced respiratory symptoms in the most severe patients. In other patients with pneumonia, a very low dose of vitamin C (0.25-0.8 g/day) reduced hospitalization by 19% compared with no vitamin C intake. A dose of 0.5-1.6 g/day reduces duration by 36%. Its positive effect was also seen in chest X-ray films, fever and erythrocyte sedimentation rate (Carr & Maggini, 2017; Hemilä et al., 1999, 2017).

Burns

Reactive oxygen species induce endothelial damage to lipids and proteins in the cell membrane. They contribute to increased capillary leakage associated with burns. A reasonable strategy to prevent these phenomena would indeed be to use antioxidants.

Vitamin C is highly water soluble, to the extent that it is unlikely to readily diffuse through (hydrophobic) cell membranes. It is then expected that ascorbate scavenges reactive oxygen species present in the extracellular space and regenerates membrane-bound antioxidants (e.g., vitamin E, membrane-associated cytochromes) to scavenge intracellular ROS. Studies have shown that vitamin C decreases lipid peroxidation - proven by decreased serum levels of malondialdehyde (Rizzo et al., 2016; Dunne et al., 2006).

In addition to scavenging ROS, regenerating vitamin E and attenuating lipid peroxidation, vitamin C in skin burns also appears correlated with decreases in fluid requirements and restoration of edema. Since a reduced volume of isotonic fluids is required by burn patients treated with vitamin C, hypoproteinemic dilution decreases colloid pressure and thus fluids will pass to a small extent into unaffected tissues. A clinical trial included 37 patients with burns over 30% body surface area, with or without high doses of vitamin C (66 mg/kg/h). Fluid requirements were calculated using the Parkland formula, with the addition of 5% albumin at 48 hours after administration and at 72 hours. The group receiving vitamin C required 45% less fluids at 24 hours, demonstrating decreased lipid peroxidation as evidence of decreased malonaldehyde levels and also reduced edema generation. This preliminary study showed no significant difference in mortality rates (Tanaka et al., 2000).

The second study was a retrospective analysis of patients treated over a 3-year period. The same amounts of vitamin C were used as in the previous study. In this case a 25% reduction in fluid required was obtained in the group receiving vitamin C. Although these two studies have not been shown to improve respiratory function or decrease mortality, they have shown that vitamin C can be used to reduce the total per fluid requirement without increasing the risk of kidney failure or other infectious processes (Rizzo et al., 2016; Kahn et al., 2011). A downside of using vitamin C in skin burns is that very high doses will have toxic side-effects: oxalate neuropathy, osmotic diuresis, worsening of excretory system problems prior to vitamin C administration (Rizzo et al., 2016; Kahn et al., 2011).

Dermatology

In connective tissues (as in all other parts of the organism), oxidative stress is generally cited as a key factor in ageing. As discussed above, vitamin C can not only prevent/reduce oxidative stress by quenching ROS, but at the same time it also acts as cofactor for key enzymes in collagen production/maturation by hydroxylation; such hydroxylation reactions, dependent on ascorbate, ensure that enough hydrogen bonds can be formed between collagen chains, which in turns ensures elasticity. (Mumtaz et al., 2021).

The bioavailability of vitamin C in the skin is inadequate when taken orally. When skin is exposed to UV light, free radicals are generated, which are quenched by vitamin C by electron donation. The oxidized form of vitamin C is non-reactive and can return to its original form through the action of dehydroascorbic acid reductase. Exposure to UV light reduces the availability of vitamin C in the skin (Telang, 2013).

Oxidative stress leads to increased production of matrix metalloproteinases that lead to collagen breakdown. It also induces nuclear transcription of factor kappa B (NF-kB), which further produces a number of mediators that contribute to inflammation and aging. It also increases elastin mRNA levels in fibroblasts which accounts for the elastic changes seen in aging skin. UV light destroys elastin, proteoglycans and other dermal cell structures, causes dermal burns, ROS, epidermal mutations and skin cancer. The most common effect is skin ageing and possible melanoma formation. Vitamin C does not absorb UV rays but acts by neutralizing free radicals so protects the skin especially when UV protective lotion is used. It works best in combination with vitamin E which potentiates the action of vitamin C. The hydrophilic and lipophilic compartments are protected by the two vitamins (Telang, 2013; Traikovich, 1999). In the context discussed above of oxidative stress generated by excess vitamin C, the application of vitamin C-based products to the skin can be expected to lead to paradoxical reactions.

Vitamin C is a cofactor in collagen synthesis, specifically the enzymes prolysyl and lysyl hydroxylase, which are responsible for stabilizing and cross-linking collagen molecules. Collagen synthesis is also stimulated by lipid pro-oxidation and the product of the process, malonic dialdehyde, stimulates collagen production. Vitamin C activates the collagen synthesis transcriptome and stabilizes procollagen mRNA, regulating collagen synthesis (Telang, 2013; Traikovich, 1999).

Ascorbic acid interacts with copper ions at the tyrosinase active site, inhibiting tyrosinase activity, decreasing melanin production. It should be noted that it also acts as a perifollicular pigment (Telang, 2013). NF-kB is responsible for activating the cytokines TNF-α, IL-1, IL-6 and IL-8 and vitamin C inhibits this factor. It can be used in cases of acne and post-inflammatory hyperpigmentation. Magnesium ascorbyl phosphate is the most stable and preferred ascorbyl ester. It is a lipophilic molecule easily absorbed into the skin. It has a moisturizing effect and decreases trans-epidermal water loss. It also scavenges free radicals providing protection against UV light and increases collagen production (Telang, 2013; Traikovich, 1999).

Final Considerations

The role of antioxidants in scavenging free radicals generated in the human body is well accepted. What is not yet known are the clear roles they play in treating disease. Most studies so far are conducted *in vitro* on cell cultures or *in vivo* on animals. Very few studies involve human subjects and those are small studies on a small group of people. Indeed, most of the actions known so far are due to the action of antioxidants *in vitro*. What needs to be stressed is that the concentration of antioxidants in cell culture assays is much higher than the concentration that antioxidants can reach in the human body. Under these conditions, many unfounded exaggerated claims may be made about the beneficial effects of ascorbate in medicine. For instance, while foods rich in antioxidants are reasonably expected to increase the ability of the organism to resist oxidative stress, claims have been made (and are well-known under the guise of the "French paradox") that moderate amounts of (preferably red) wine would reduce mortality from coronary artery disease regardless of the subjects' fat intake or blood cholesterol level – all due to the antioxidants from the wine (Burr, 1995). Yet, at the same time, there are clinical studies that show that alcohol consumption increases the risk of developing cancer (Bagnardi et al., 2017). It is in this context important to note that (1) antioxidants can exhibit dual roles (often in dose-dependent manner, switching between beneficial and harmful), and (2) many of the publicly discussed purported benefits of antioxidants/vitamins are supported at most by incomplete data and by consistent and persistent confusion between correlation and causation.

The idea of duality is also observed in the case of vitamin E where it appears to play a role in preventing cardiovascular disease by producing less cholesterol by hepatocytes. On the other hand, there are studies that contradict this and associate the consumption of vitamin E supplements with increased hemorrhagic strokes. It is noted that vitamin E is likely to have anti- cancer properties by inhibiting the growth of cancer cells, protecting the neuronal cell from free radicals, thus preventing the onset of Alzheimer's disease (Rizvi et al., 2014; Wada, 2009).

Vitamin K in addition to its clear role in anticoagulation, also exerts other roles: bone mineralization, insulin resistance, neuronal development, cell proliferation inhibitors with possible anticancer role (Popa et al., 2021; Xv et al., 2018).

The same duality is found in the case of vitamin A, which in addition to its antioxidant effect and beneficial role in the prevention of heart disease also has adverse effects, such as the development of heart disease, so it is not possible to give a clear conclusion about it, thus avoiding the recommendation of dietary supplementation with vitamin A or β-carotene (Palace et al., 1999).

The duality is also noted in the case of vitamin C, which exhibits both antioxidant and pro-oxidant properties depending on the concentration at which it reaches the bloodstream. The limiting concentration at which these two properties will change is 1 mM. Below this concentration vitamin C will be antioxidant and above this concentration vitamin C will be pro-oxidant (Burr, 1995; Lehene et al., 2020).

In smokers, blood serum vitamin C levels are 3 times lower than in non-smokers, thus recommending an additional vitamin C intake of 35 mg/day, ensuring that there is sufficient ascorbic acid to repair oxidant damage. In children and adolescents exposed to cigarette smoke, vitamin C levels are low and the chances of developing respiratory disease increase (Strauss, 2001). Improvement in the severity and duration of a common cold is due to vitamin C supplements of 200 mg or more. In the case of pneumonia, the length of hospitalization has decreased (Carr & Maggini, 2017).

Vitamin C shortens recovery time after surgery, has an effect on neutrophils, influencing several important aspects of neutrophils: migration in response to mediators of inflammation (chemotaxis), phagocytosis, microbial killing and apoptosis. Chemotaxis was enhanced in patients with recurrent infections and in neonates suspected of septic infections (Carr & Maggini, 2017). Improvements in chemotaxis were also seen in diseases where no effect was expected, such as Chediak-Higashi syndrome (CHS).

Vitamin C administration increased the immune response by enhancing and increasing the production of IL-10, IL-1 and TNF-α cells following lipopolysaccharide stimulation. Vitamin C used as a treatment for microglia attenuates cell activation and synthesis of the pro- inflammatory cytokines TNF, IL-6 and IL-1 (Portugal et al., 2017).

It has also been observed that vitamin C supplementation can improve glycemic control in type 2 diabetes (Ashor et al., 2017). As far as burns are concerned, vitamin C, has a beneficial effect by reducing the body's need for fluids, but very high doses that produce this will more or less impair the proper functioning of the body. Adverse effects of taking large amounts of vitamin C have been reported: oxalate neuropathy, osmotic diuresis, aggravation of excretory system problems prior to vitamin C administration (Rizzo et al., 2016; Kahn et al., 2011).

Vitamin C does not absorb UV-light but acts by neutralizing free radicals so it protects the skin especially when UV protective lotion is used. It works best in combination with vitamin E which potentiates the action of vitamin C. It can be used in cases of acne and post-inflammatory hyperpigmentation (Telang, 2013).

In conclusion, ascorbate is one of the best-known vitamins with clear benefits on immune response, body convalescence and in fighting various diseases. We, however, reiterate that the vast majority of publicly-available literature on purported benefits of vitamins and antioxidants (such as vitamin C) relies on limited comparisons rather than on clear clinical trials – and, very often, on confusions between correlation and causation.

REFERENCES

Ashor, A. W., Werner, A. D., Lara, J., Willis, N. D., Mathers, J. C., & Siervo, M. (2017). Effects of Vitamin C supplementation on glycaemic control: A systematic review and meta-analysis of randomised controlled trials. *European Journal of Clinical Nutrition, 71*(12), 1371–1380. doi:10.1038/ejcn.2017.24 PMID:28294172

Bagnardi, V., Blangiardo, M., La Vecchia, C., & Corrao, G. (2017). *Alcohol Consumption and the Risk of Cancer*. Academic Press.

Burr, M. L. (1995). Explaining the French paradox. *The Journal of the Royal Society for the Promotion of Health, 115*(4), 217–219. doi:10.1177/146642409511500404 PMID:7562866

Carr, A. C., & Maggini, S. (2017). Vitamin C and immune function. *Nutrients, 9*(11), 1–25. doi:10.3390/nu9111211 PMID:29099763

Doseděl, M., Jirkovský, E., Macáková, K., Krčmová, L. K., Javorská, L., Pourová, J., Mercolini, L., Remião, F., Nováková, L., & Mladěnka, P. (2021). Vitamin c—sources, physiological role, kinetics, deficiency, use, toxicity, and determination. Nutrients 13(2), 1–36. doi:10.3390/nu13020615

Ferrón-Celma, I., Mansilla, A., Hassan, L., Garcia-Navarro, A., Comino, A. M., Bueno, P., & Ferrón, J. A. (2009). Effect of Vitamin C Administration on Neutrophil Apoptosis in Septic Patients After Abdominal Surgery. *The Journal of Surgical Research, 153*(2), 224–230. doi:10.1016/j.jss.2008.04.024 PMID:18952228

Fukushima, R., & Yamazaki, E. (2010). Vitamin C requirement in surgical patients. *Current Opinion in Clinical Nutrition and Metabolic Care, 13*(6), 669–676. doi:10.1097/MCO.0b013e32833e05bc PMID:20689415

Gęgotek, A., & Skrzydlewska, E. (2022). Antioxidative and Anti-Inflammatory Activity of Ascorbic Acid. Antioxidants, 11(10). doi:10.3390/antiox11101993

Heimer, K. A., Hart, A. M., Martin, L. G., & Rubio-Wallace, S. (2009). Examining the evidence for the use of vitamin C in the prophylaxis and treatment of the common cold. Journal of the American Academy of Nurse Practitioners, 21(5), 295–300. doi:10.1111/j.1745-7599.2009.00409.x

Hemilä, H. (2017). Vitamin C and infections. Nutrients, 9(4). doi:10.3390/nu9040339

Hemilä, H., Douglas, R. M., & Hemilä, H. (1999). Vitamin C and acute respiratory infections. *The International Journal of Tuberculosis and Lung Disease, 3*(9), 756–761. PMID:10488881

Huang, D., Boxin, O. U., & Prior, R. L. (2005). The chemistry behind antioxidant capacity assays. *Journal of Agricultural and Food Chemistry, 53*(6), 1841–1856. doi:10.1021/jf030723c PMID:15769103

Kahn, S. A., Beers, R. J., & Lentz, C. W. (2011). Resuscitation after severe burn injury using high-dose ascorbic acid: A retrospective review. *Journal of Burn Care & Research; Official Publication of the American Burn Association, 32*(1), 110–117. doi:10.1097/BCR.0b013e318204b336 PMID:21131846

Lehene, M., Fische-Fodor, E., Scurtu, F., Hădade, N., Gal, E., Mot, A. C., Matei, A., & Silaghi-Dumitrescu, R. (2020). Excess ascorbate is a chemical stress agent against proteins and cells. *Pharmaceuticals (Basel, Switzerland), 13*(6), 107. doi:10.3390/ph13060107 PMID:32471171

Liu, Y., Liu, C., & Li, J. (2020). Comparison of vitamin c and its derivative antioxidant activity: Evaluated by using density functional theory. *ACS Omega, 5*(39), 25467–25475. doi:10.1021/acsomega.0c04318 PMID:33043226

Lykkesfeldt, J., & Tveden-Nyborg, P. (2019). The pharmacokinetics of vitamin C. *Nutrients, 11*(10), 2412. Advance online publication. doi:10.3390/nu11102412 PMID:31601028

Macan, A. M., Kraljević, T. G., & Raić-malić, S. (2019). Therapeutic perspective of vitamin C and its derivatives. Antioxidants, 8(8). doi:10.3390/antiox8080247

Mikirova, N., Riordan, N., & Casciari, J. (2016). Modulation of cytokines in cancer patients by intravenous ascorbate therapy. *Medical Science Monitor, 22*, 14–25. doi:10.12659/MSM.895368 PMID:26724916

Mumtaz, S., Ali, S., Tahir, H. M., Kazmi, S. A. R., Shakir, H. A., Mughal, T. A., Mumtaz, S., Summer, M., & Farooq, M. A. (2021). Aging and its treatment with vitamin C: a comprehensive mechanistic review. Molecular Biology Reports, 48(12), 8141–8153. doi:10.100711033-021-06781-4

Niki, E. (2010). Assessment of antioxidant capacity in vitro and in vivo. *Free Radical Biology & Medicine, 49*(4), 503–515. doi:10.1016/j.freeradbiomed.2010.04.016 PMID:20416370

Niki, E., The, N., & Tsuchihashi, H. (2018). Interaction among vitamin C, vitamin E and β-carotene. *The American Journal of Clinical Nutrition, 62*(March), 1322–1326. PMID:7495227

Palace, V. P., Khaper, N., Qin, Q., & Singal, P. K. (1999). Antioxidant potentials of vitamin A and carotenoids and their relevance to heart disease. *Free Radical Biology & Medicine, 26*(5–6), 746–761. doi:10.1016/S0891-5849(98)00266-4 PMID:10218665

Popa, D. S., Bigman, G., & Rusu, M. E. (2021). The role of vitamin k in humans: Implication in aging and age-associated diseases. *Antioxidants, 10*(4), 1–31. doi:10.3390/antiox10040566 PMID:33917442

Portugal, C. C., Socodato, R., Canedo, T., Silva, C. M., Martins, T., Coreixas, V. S. M., Loiola, E. C., Gess, B., Röhr, D., Santiago, A. R., Young, P., Minshall, R. D., Paes-De-Carvalho, R., Ambrósio, A. F., & Relvas, J. B. (2017). Caveolin-1-mediated internalization of the Vitamin C transporter SVCT2 in microglia triggers an inflammatory phenotype. *Science Signaling, 10*(472), eaal2005. Advance online publication. doi:10.1126cisignal.aal2005 PMID:28351945

Procházková, D., Boušová, I., & Wilhelmová, N. (2011). Antioxidant and prooxidant properties of flavonoids. *Fitoterapia, 82*(4), 513–523. doi:10.1016/j.fitote.2011.01.018 PMID:21277359

Rizvi, S., Raza, S. T., Ahmed, F., Ahmad, A., Abbas, S., & Mahdi, F. (2014). *The Role of Vitamin E in Human Health and Some Diseases*. Academic Press.

Rizzo, J. A., Rowan, M. P., Driscoll, I. R., Chung, K. K., & Friedman, B. C. (2016). Vitamin C in Burn Resuscitation. *Critical Care Clinics, 32*(4), 539–546. doi:10.1016/j.ccc.2016.06.003 PMID:27600125

Strauss, R. S. (2001). Environmental tobacco smoke and serum vitamin C levels in children. *Pediatrics, 107*(3), 540–542. doi:10.1542/peds.107.3.540 PMID:11230596

Tanaka, H., Matsuda, T., Miyagantani, Y., Yukioka, T., Matsuda, H., & Shimazaki, S. (2000). Reduction of resuscitation fluid volumes in severely burned patients using ascorbic acid administration: A randomized, prospective study. *Archives of Surgery*, *135*(3), 326–331. doi:10.1001/archsurg.135.3.326 PMID:10722036

Telang, P. (2013). Vitamin C in dermatology. *Indian Dermatology Online Journal*, *4*(2), 143. doi:10.4103/2229-5178.110593 PMID:23741676

Tina Suksmasari, B. H. (2015). Multivitamin Supplementation Supports Immune Function and Ameliorates Conditions Triggered by Reduced Air Quality. *Vitamins & Minerals*, *4*(2). Advance online publication. doi:10.4172/2376-1318.1000128

Traikovich, S. S. (1999). Use of topical ascorbic acid and its effects on photodamaged skin topography. *Archives of Otolaryngology—Head & Neck Surgery*, *125*(10), 1091–1098. doi:10.1001/archotol.125.10.1091 PMID:10522500

Wada, S. (2009). Chemoprevention of tocotrienols: The mechanism of antiproliferative effects. *Forum of Nutrition*, *61*, 204–216. doi:10.1159/000212752 PMID:19367124

Wong, S. K., Chin, K. Y., & Ima-Nirwana, S. (2020). Vitamin C: A review on its role in the management of metabolic syndrome. International Journal of Medical Sciences, 17(11), 1625–1638. doi:10.7150/ijms.47103

Xv, F., Chen, J., Duan, L., & Li, S. (2018). Research progress on the anticancer effects of Vitamin K2 [Review]. *Oncology Letters*, *15*(6), 8926–8934. doi:10.3892/ol.2018.8502 PMID:29805627

Chapter 15
Versatility and Specificity of Flavin-Based Oxidoreductases in the Electron Transfer Reactions

Lilia Anghel
https://orcid.org/0000-0002-2388-0958
Institute of Chemistry, Moldova State University, Moldova

Raul Victor Erhan
Horia Hulubei National Institute for R&D in Physics and Nuclear Engineering, Romania

ABSTRACT

Oxidoreductases are a large class of enzymes involved in the catalysis of oxidation/reduction reactions. These oxygen-utilizing biomolecules firmly control the reactions involving molecular oxygen mostly by regulating their cofactors. Flavoenzymes, a large group of oxidoreductases, possess either a flavin mononucleotide or a flavin adenine dinucleotide cofactor non-covalently bound to the protein substrate. Flavin cofactors are versatile due to reactivity and many oxidizing states allowing them to undergo electron transfer reactions. As a result, flavoenzymes can participate in a wide range of biochemical and photochemical processes including electron transport, C-H oxidations, oxidative protein folding, and even repair of cross-linked DNA. Structural studies have shown that the large variety of functions that flavoenzymes possess are limited and dependent on structure and protein folding to ensure the specific electrostatic environment. The present work concerns the structural aspects of specific flavin-dependent oxidoreductases influencing electron transfer reactions.

DOI: 10.4018/978-1-6684-7198-2.ch015

INTRODUCTION

Oxidoreductases are a large class of enzymes involved in the catalysis of a wide range of oxidoreduction reactions, where the oxidation, transfer of electrons from an electron donor molecule which is the reducing agent, takes place with the simultaneous reduction of an electron acceptor molecule that is the oxidizing agent. These versatile catalysts are ubiquitous in nature and participate with high selectivity and specificity in many chemical reactions. They have a central role in catalyzing reactions involving electron transfer, proton/hydrogen release, oxygen insertion and many other important processes (Younus, 2019). Many reactions are energy-yielding processes and the extent of released energy will depend on the redox potential difference between the reduced and the oxidized compound. The redox potential is the tendency of a compound to accept electrons and become reduced; the electrons are transferred to the compound with a higher redox potential (acceptor) from the compound with a lower redox potential (donor). Most of the reactions that oxidoreductases are involved take place due to the enzyme's cofactors that mainly include the nicotinamide adenine dinucleotide (NADP and NADPH) and flavins (flavin mononucleotide - FMN, or flavin adenine dinucleotide - FAD). Other redox centers may include metal ions and complexes such as Cu, Fe, Mo, Fe-S clusters or heme groups (Espina et al., 2021).

Due to their catalytic abilities, the attractiveness of using these enzymes as biocatalysts at an industrial scale has been long time discussed (May & Padgette, 1983). Oxidoreductases could be used in the catalysis of polymer synthesis (Bouldin et al., 2011), enzyme-based diagnostic tests (Soleimany & Bhatia, 2020), biosensors (Sirkar et al., 2000), pollution control (Barber et al., 2020; Bilal et al., 2022; Hazarika et al., 2022) and many other biotechnological applications (May, 1999). However, the application of these enzymes into new manufacturing areas will depend on structure stability, desired selectivity and compatibility with the rigid process conditions and possible reusability. To ensure the industrial and biotechnological applications of oxidoreductases many limitations must be overcome. Special attention has been given to studies on structure characterization and modification, establishing the mechanism of enzymatic activity and optimization of the catalytic properties and conditions of the reaction environment.

Thus, the main objective of this chapter is to put forward the recent developments of studies regarding the structural aspects of specific flavin-dependent oxidoreductases influencing electron transfer reactions.

OXIDOREDUCTASES: AN OVERVIEW

Oxidoreductases are involved in a wide range of biochemical processes performing substrate oxidation or reduction in the reactions of electron transfer, oxygen activation, C-H oxidations and photochemistry, thus providing great support in immune protection, generating movement, transmitting electric impulses, and controlling cell growth and differentiation (Rivas et al., 2020). Based on the type of reactions that are catalyzed, oxidoreductases can be grouped into:

Oxidases: These are the enzymes that catalyze directly the reactions involving the transfer of hydrogen from the substrate towards acceptors (especially dioxygen). Some examples of enzymes from this group include lacasse, aldehyde oxidase, amine oxidase, glucose oxidase, etc.

General reaction: $\text{Substrate}_{(\text{reduced})} + O_2 \rightarrow \text{Substrate}_{(\text{oxidized})} + H_2O_2$

Oxygenases: The group of enzymes that catalyze the reactions involving the incorporation of molecular oxygen into the substrate. This group can be sub-devised into monooxygenases and dioxygenases; examples include the cytochrome P-450 monooxygenase, xylene monooxygenase, cysteine dioxygenase, 2-amino ethanethiol dioxygenase, etc.

General reaction: $\text{Substrate}_{(\text{reduced})} + O_2 \rightarrow \text{Substrate}_{(\text{oxidized})}$

Peroxidases: The enzymes catalyzing the reactions of the substrate's oxidation by hydrogen peroxide; examples enumerate horseradish peroxidase, NADH peroxidase, glutathione peroxidase 4, etc.

General reaction: $\text{Substrate}_{(\text{reduced})} + H_2O_2 \rightarrow \text{Substrate}_{(\text{oxidized})} + H_2O$

Dehydrogenases: The enzymes that mediate the reactions involving the removal of hydride ions from the substrate and their transfer to the acceptor substrate. Several examples from this group of enzymes include alcohol dehydrogenase, succinate dehydrogenase, lactate dehydrogenase, etc.

General reaction: $\text{Substrate1}_{(\text{oxidized})} + \text{Substrate2}_{(\text{reduced})} + H_2O_2 \rightarrow \text{Product1}_{(\text{oxidized})} + \text{Product2}_{(\text{reduced})} + H^+$

Reductases: The group of enzymes catalyzing the reactions of reduction of the substrate; examples include superoxide reductase, periplasmic nitrate reductase, etc.

General reaction: $\text{Substrate} + \text{Acceptor} \rightarrow \text{Product} + \text{Acceptor}_{(\text{reduced})}$

Hydrolases: The group of enzymes that mediate the reactions involving the water as a hydroxyl group donor, used to break a chemical bond. Several representatives of this group include proteases, lipases, esterases, glycosidases, etc.

General reaction: $\text{Substrate} + H_2O \rightarrow \text{Product1-OH} + \text{Product2-H}$

Given the large spectrum of reactions in which these classes of enzymes are involved, they are regarded as potential bio-catalysts for many industrial processes (Espina et al., 2021; May & Padgette, 1983; Younus, 2019). The main issues regarding industrial applications at present have been addressed through combined structural characterization studies, protein engineering, and computational studies to achieve the catalytic activity, selectivity, efficiency, and desired stability to improve their industrial usage.

THE ELECTRON TRANSFER PROCESS AND OXIDOREDUCTASES

Many biochemical reactions, especially those involving cell energy, take place through long-range intra- and interprotein electron transfer, which sometimes can be coupled with proton transfer (Formosinho & Barroso, 2011). These processes are very complex and are influenced by different phenomena including protein structure and dynamics, non-electron transfer processes influencing the rates of transfers, kinetics, etc. (Moser et al., 2010).

Oxidoreductases possess active sites where either oxidation occurs when electrons received from the substrate oxidation reduce the active site, further, the electrons are transferred through the cofactors to an exposed surface site capable of reducing an electron acceptor (EA). In the case when the substrate reduction takes place, the electron donor (ED) transfers the electrons redox site available on the enzyme's surface following the transfer of electrons through an intra-protein pathway to the active site where the substrate reduction may take place (Rivas et al., 2020). Further in the chapter will be exemplified some well-studied oxidoreductases and the mechanism of the main reactions mediated by these enzymes.

STRUCTURAL ASPECTS OF FLAVINS

Flavins such as flavin mononucleotide (FMN) and flavin adenine dinucleotide (FAD) are considered versatile catalysts for many biochemical reactions which couple the electron transfer process to the proton transfer. In comparison to the nicotinamide cofactor, which is known to dissociate from the partner enzyme after each catalytic turnover, the flavin cofactor is bound so tightly that they do not dissociate from the enzyme's structure. This strong association has likely emerged due to the ability of FMN and FAD to undergo rapid and uncontrolled autoxidation outside the enzyme active sites (Valton et al., 2008; Walsh & Wencewicz, 2013). The variety of reactions catalyzed by flavins depends on the activity of the isoalloxazine ring, its availability, i.e. how deeply buried it is in the protein structure, its localization, i.e. the amino acids' active residues in its proximity, and its protonation state. This tricyclic ring can be in the oxidized state (quinone, Fl_{ox}), one-electron-reduced (semiquinone, Fl_{sq}) and the two-electron-reduced state (hydroquinone, Fl_{red}) (Figure 1), thus by modulating the redox state of flavin cofactors; the flavin-based enzymes are able to participate in more thermodynamically challenging reactions (Motteran et al., 2001; Romero et al., 2018; Westphal et al., 2021; Winkler et al., 2007).

Figure 1. The isoalloxazine ring of the flavins in the three protonation states: the oxidized state (quinone, Fl_{ox}), one-electron-reduced (semiquinone, Fl_{sq}) and the two-electron-reduced state (hydroquinone, Fl_{red})
Source: Adapted from Anghel et al. (2023)

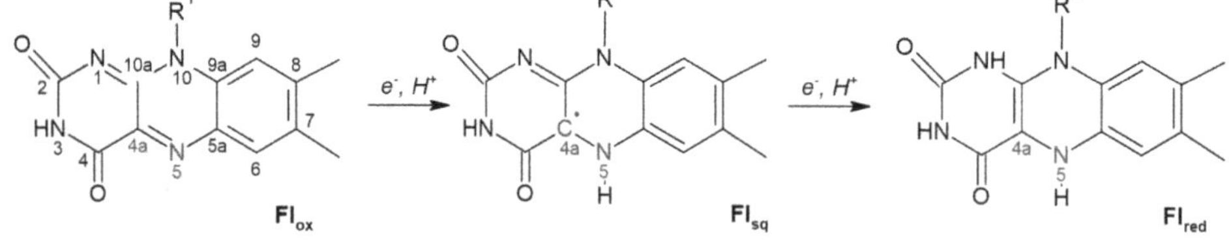

With a couple of exceptions, most flavin-based enzymes use the two-halve reactions mechanisms in which, the first half is the reductive one where the substrate reduces flavin and the second half is the oxidative half-reaction where the flavin is oxidized in return by an electron acceptor. This mechanism allows the reaction to produce that are thermodynamically challenging to overcome; an illustrious example of such reactions is the reaction with molecular oxygen, which is a spin-forbidden reaction according to the law of spin conservation because it involves the molecule of oxygen in the triplet state and the reduced flavin which is in the singlet state. The successive single electron transfers in a step-wise process overcome this restriction and allow the reaction to take place (Massey, 2002; Mattevi, 2006). Certainly, it is difficult to generalize by comparing the reactivity of enzymes with a closely related structure. The large reaction differences could be due to a combination of many factors including the redox potential of the flavin, constraints in movements due to the active sites' environment, polypeptide environment, etc.

MECHANISMS OF THE FLAVIN-BASED ELECTRON BIFURCATION

Since its discovery in 2008, a new term has been introduced - flavin-based electron bifurcation (FBEB) (Herrmann et al., 2008). The term applies to the energy coupling mechanism in which the reduced flavin by a pair of electrons transfers them separately to energetically different acceptors such that one electron lands on a much more reducing electron carrier, compensated by the exergonic transfer of the other electron to the more oxidizing carrier (Kar et al., 2022). In a more simplified manner, in bifurcation, the two electrons, which are obtained as a pair, are separated in coupled transfers in such way that the energy of both is concentrated on one of them (Figure 2). Tight coupling between the two electrons is essential and relies on the energy of the semiquinone state of the isoalloxazine ring.

Figure 2. Schematic representation of the FBEB mechanism

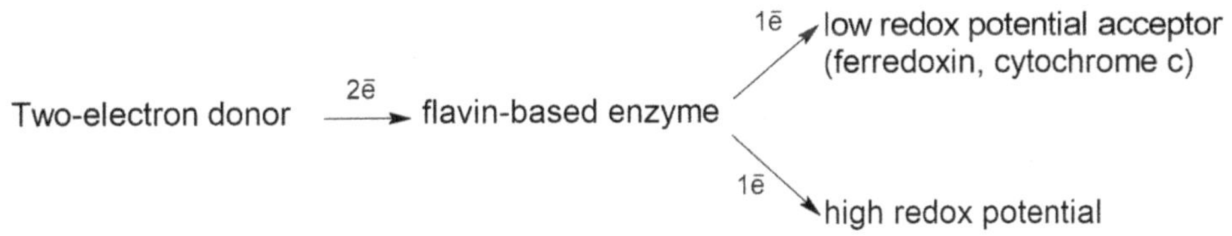

So far, four classes of enzymes have been established: *I.* electron-transfer flavoprotein, *II.* NADPH dependent ferredoxin-NADH reductase, *III.* coenzyme M and B hetero-disulfide reductase, *IV.* NADH dehydrogenase. In FBEB, flavins are included at different sites of the enzyme complexes. Each FBEB enzyme contains at least one flavin acting as the main participant to couple exothermic with endothermic reactions (Kayastha et al., 2021). Flavins are well-fitted for such actions due to their balanced redox potentials $F/(FH\bullet,F\bullet-) \leq (FH\bullet,F\bullet-)/FH-$ because at least one semiquinone is needed for the electron transfer.

VARIETY OF FLAVIN-BASED ENZYMES

Flavin-based enzymes are a large group of oxidoreductases; due to the versatility of the flavin cofactor, these enzymes can catalyze a variety of reactions and therefore facilitate many intracellular processes. Below will be discussed several notorious examples of flavin-based oxidoreductases and the reaction mechanism mediated by these enzymes.

Medium-Chain Acyl-CoA Dehydrogenase

Medium-chain acyl-CoA dehydrogenase (MCAD) is an enzyme with a molecular mass of about 46 kDa, involved in the mitochondrial β-oxidation of long-chain fatty acids, which provides energy demands for heart and skeletal muscles, synthesis or degradation of cellular lipids, and catabolism of aromatic amino acids (Ensenauer et al., 2005; J. J. Kim et al., 1993). This important biochemical process consists of consecutive cycles of oxidation, hydration, oxidation and thiolysis to produce acetyl-CoA and reducing equivalents from the long-chain acyl-CoA esters using the electron transfer flavoprotein as the physiologic electron acceptor (He et al., 2011). The initial step of oxidation is catalyzed by chain-length specific FAD-containing dehydrogenases. The medium-chain acyl-CoA dehydrogenase monomer is folded into three equally sized domains (Figure 3). The N- and C-terminals are mainly folded into α-helices whilst the middle domain contains two orthogonal β-sheets. The isoalloxazine ring of the FAD cofactor is buried between two alpha-helices and a beta-sheet of one subunit. It was determined that the enzymatic activity of the medium-chain acyl-CoA dehydrogenase essentially depends on the dimeric formation. The enzyme can be found also as a tetramer formed as a dimer of the two dimers (J. J. Kim et al., 1993).

Electron Transfer Flavoprotein

Electron transfer flavoprotein (ETF) is another enzyme in the mitochondrial matrix, which participates in the sequential transfer of electrons from a series of dehydrogenation reactions in the mitochondrial β-oxidation of long-chain fatty acids (Henriques et al., 2021). Interestingly, the electron transfer flavoprotein was first described in 1957 as an enzyme having two subunits with two FAD molecules bound to each and a global molecular mass of 57 kDa (Hall & Kamin, 1975). Later it was found that this is a heterodimeric enzyme composed of two subunits alpha and beta (ETF-α and ETF-β) (Figure 3), having a single FAD cofactor and an adenosine monophosphate bound to the heterodimer (Husain & Steenkamp, 1983). The α subunit has a molecular mass of 32 kDa whilst the β subunit has a 27 kDa molecular mass. Studies have shown that the AMP cofactor, has no influence on the enzymatic activity but seems to be significant for the assembly of the holo-structure (Sato et al., 1993).

Electron Transfer Flavoprotein: Ubiquinone Oxidoreductase

Electron transfer flavoprotein: ubiquinone oxidoreductase (ETF:QO) is a component of the mitochondrial matrix also involved in the mitochondrial β-oxidation of long-chain fatty acids. The X-ray crystallographic data on this enzyme show that the monomeric form of this enzyme (Figure 3) has a molecular mass of about 64 kDa containing FAD, Fe and S in a ratio of 1:4:4 (Zhang et al., 2006). The isoalloxazine ring of the FAD cofactor is buried at a distance of 13 Å from the surface of the protein and at 8.5 Å away from the $[4Fe\text{-}4S]^{+1\ +2}$ cluster that is also buried (Watmough & Frerman, 2010). In the process of mitochondrial β-oxidation, the electron transfer flavoprotein participates as the electron acceptor. The iron-sulfur cluster of the electron transfer flavoprotein is considered to be the site of the transfer of electrons from ETF since it is that close to the surface (Page et al., 1999; Usselman et al., 2008).

Figure 3. Graphical representation of the concept of electron transfer process from octanoyl-CoA to ubiquinone (UQ) mediated by the three enzymes: medium-chain acyl-CoA dehydrogenase (MCAD, PDB entry 3MDD), electron transfer flavoprotein (ETF, PDB entry 1EFV) and electron transfer flavoprotein: ubiquinone oxidoreductase (ETF:QO, PDB entry 2GMH). The chains of the enzymes are presented as new cartoon and the FAD cofactor as licorice drawing methods using the VMD program.
Source: Humphrey et al. (1996)

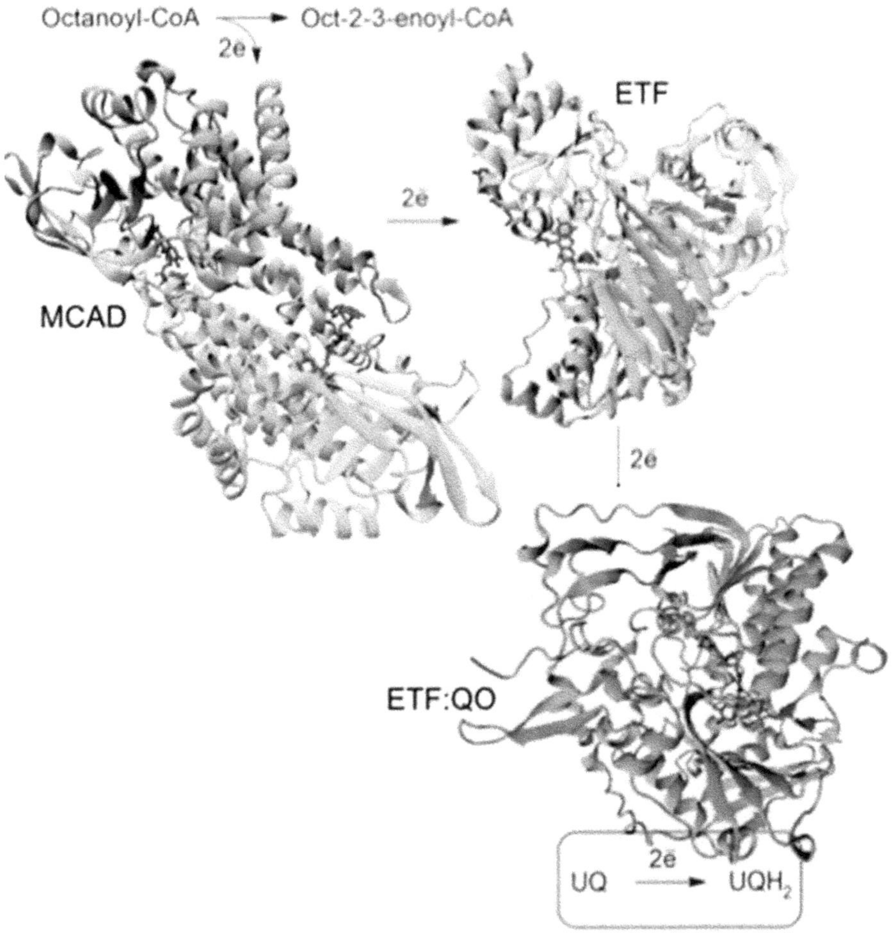

These three types of enzymes, medium-chain acyl-CoA dehydrogenase, electron transfer flavoprotein and electron transfer flavoprotein: ubiquinone oxidoreductase, play an important role in the electron transfer processes in the human body. For example, the electron transfer process from octanoyl-CoA to ubiquinone is mediated by the three enzymes as follows: *1.* first the oxidation of octanoyl-CoA molecule to oct-2-3-enoyl-CoA is catalyzed by medium-chain acyl-CoA dehydrogenase (MCAD), *2.* then, the two-electron reduced flavin cofactor (FADH2) of the medium-chain acyl-CoA dehydrogenase is re-oxidized by two equivalents of electron transfer flavoprotein (ETF) that is a one electron carrier, *3.* in turn, the flavin semiquinone of ETF is re-oxidized by the electron transfer flavoprotein: ubiquinone oxidoreductase (ETF:QO) that catalyzes *4.* the transfer of electrons to the mitochondrial ubiquinone pool.

A similar pathway is also responsible for accepting electrons from sarcosine dehydrogenase (Watmough & Frerman, 2010). The schematic representation of this process is presented in Figure 3.

Deficiencies and mutations of these three types of enzymes can cause various metabolic disorders in human bodies including multiple acyl-CoA dehydrogenase deficiencies (MADD), glutaric acidemia type 2, leukodystrophy, etc., which are mainly associated with the incapacity of the human body to receive energy from proteins and fatty acids (Gautschi et al., 2015; Henriques et al., 2021).

Succinate Dehydrogenase

Succinate dehydrogenase (SDH) is an enzyme participating in the aerobic respiratory chain in eukaryotic and bacterial cells, specifically in the mitochondrial electron transport chain by transferring the electrons from succinate molecules to the ubiquinone pool. Succinate dehydrogenase is known for catalysis of the oxidation of succinate to fumarate (Figure 4) reaction coupled with the reduction of ubiquinone to ubiquinol (H. J. Kim & Winge, 2013).

Figure 4. Graphical representation of the succinate dehydrogenase enzyme (SDH, PDB entry 1YQ3) and schematic representation of the oxidation reaction of succinate to fumarate

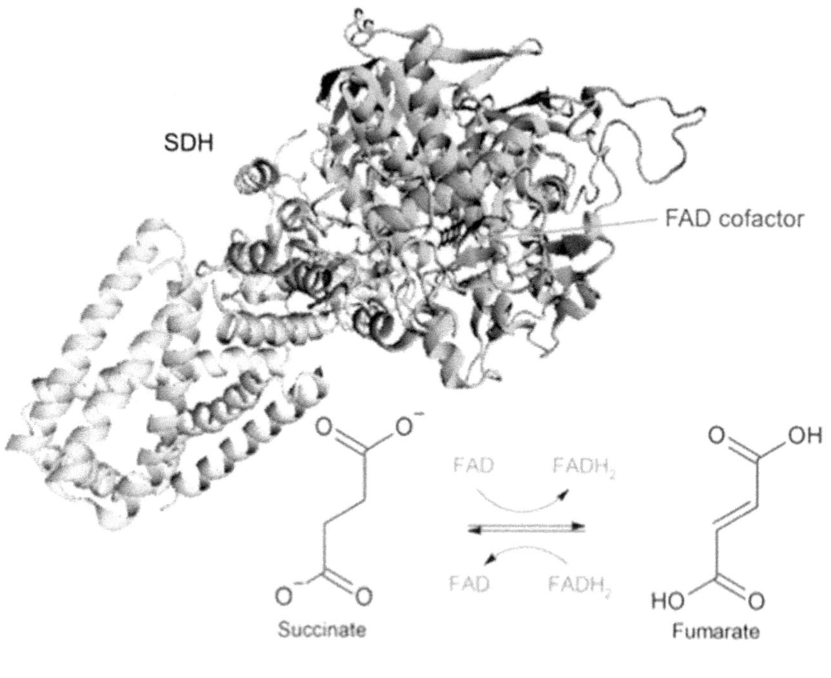

The fumarate product is then utilized in the tricarboxylic acid cycle mediated by fumarate hydratase whilst the ubiquinol product is oxidized by cytochrome *c* reductase in the electron transfer chain. The X-ray crystallographic data show that this is a hetero-enzyme composed of four subunits (Sdh1, Sdh2, Sdh3, Sdh4) (Huang et al., 2006). FAD is covalently bonded to the histidine' active residue from the first subunit (Sdh1) from the enzyme' structure, and it is essential to the enzyme activity in the succinate oxidation. According to the literature data, this covalent bond adds a redox potential of about 60 mV to allow the oxidation of succinate (Lemire & Oyedotun, 2002). The second subunit (Sdh2) contains three different iron-sulfur clusters including: [2Fe–2S], [4Fe–4S], and [3Fe–4S], with the function of transferring sequentially the electrons withdrawn from the succinate dehydrogenation at the active-site FAD (H. J. Kim & Winge, 2013).

Cytochrome C Oxidase

Cytochrome *c* oxidase (Cyt*c*O) is an integral membrane respiratory complex involved in the mechanism of redox-linked proton pumping. Structurally, this is a multi-subunit complex enzyme (composed of a total of 14 subunits) with the largest subunits I, II and III preserved across evolution and species (Scott, 1995). Although this enzyme does not contain a FAD or NAD cofactor, it should be described here due to its important participation in the mitochondrial respiratory chain. Cytochrome *c* oxidase mediates the reaction of transfer of electrons from cytochrome *c* to molecular oxygen to produce water *via* its copper and heme groups, coupled with the reaction of transfer of protons from the inner mitochondrial matrix to the inter-membrane space (Watson & McStay, 2020; Wikström et al., 2018). To do so, the cytochrome *c* oxidase binds at its positive side of a concave surface the cytochrome *c* (Cyt*c*) to produce a larger complex (Cyt*c*-Cyt*c*O, see Figure 5).

Figure 5. Graphical representation cytochrome c oxidase in complex with the cytochrome c (the transparent region of the molecule) (CytcO, PDB entry 5WAU) and the inset portraying the schematic electron transfer via the copper (yellow WdW representation) and heme groups, coupled with the proton transfer across the membrane

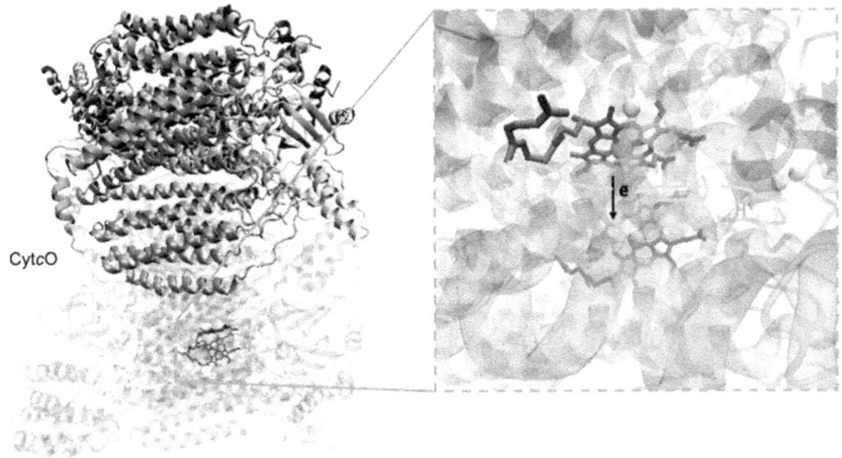

The concave surface of subunit II of Cyt*c*O contains the first site for receiving the electrons from Cyt*c*. According to the literature data (Shimada et al., 2017), the plausible pathway of electron transfer from iron heme to copper is through a series of amino acids active residues available in the proximity, i.e. (Cyt*c*)-Cys14(Cyt*c*)-Lys13(Cyt*c*)-Tyr105(subunit II of Cyt*c*O)-Met207(subunit II of Cyt*c*O)-Cu_A.

Additionally, it was proposed that the Cyt*c*-binding site in the Cyt*c*–Cyt*c*O complex structure is the catalytic site through which electrons are transferred (Shimada et al., 2017).

Flavohemoglobin

Flavohemoglobin (flavoHb) enzymes are mostly present in bacterial systems, although in the 70's when they were firstly reported their function was unclear subsequent studies have shown their implication in the nitric oxide signaling and detoxification (Poole, 2020). Flavohemoglobin plays the role of the free nitric oxide dehydrogenase by mediating the reaction of converting of the free nitric oxide to nitrate in the presence of NADH (nitric oxide dioxygenase activity), thus contributing to the bacterial resistance to the nitrosative stress (Moussaoui et al., 2018). Structurally, flavohemoglobin is related to the hemoglobin and myoglobin of vertebrates, having the heme covalently bonded to the histidine residue of the polypeptide chain and stabilized in the correct orientation through the phenylalanine residue (Bonamore & Boffi, 2007). An interesting structural aspect of this type of enzymes is the C-terminal NAD- and FAD-binding domain, and N-terminal heme-binding domain that is responsible for binding small molecules such as NO, CO and O_2 (Björck et al., 2017). In the course of reaction of nitric oxide transformation to nitrate (Figure 6), NADH of flavohemoglobin reduces FAD, the obtained $FADH_2$ reduces heme iron. Then, the molecular oxygen binds tightly the reduced flavohemoglobin. And finally, the oxygenated enzyme deoxygenates nitric oxide to produce the nitrate (Bonamore & Boffi, 2007).

Figure 6. Graphical representation of the flavohemoglobin (flavoHb, PDB entry 4G1V) and schematic description of the nitric oxide dioxygenase activity taking place on a quinone (Q) substrate

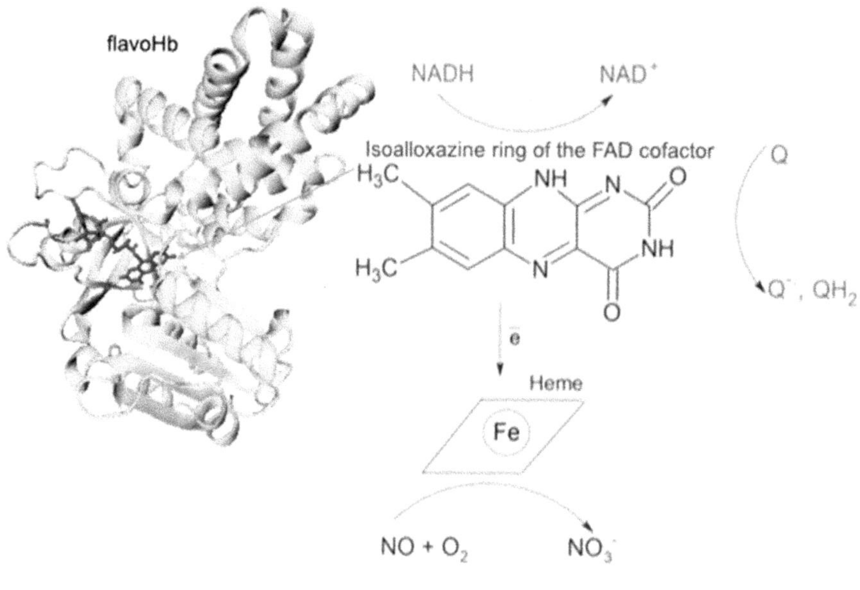

Glucose Oxidase

Glucose oxidase (GOx) is a homodimeric enzyme consisting of two molecules of FAD cofactor, and it is mainly known for the catalysis of the reaction of β-D-glucose oxidation by molecular oxygen to delta-gluconolacctone that subsequently hydrolyses to glycolic acid and hydrogen peroxide (Hecht et al., 1993). Other sugars can be also oxidized, including D-mannose, D-xylose, D-galactose, etc., by using this enzyme as catalyst (Mano, 2019). The structure of this enzyme has been studied extensively since its discovery in 1928. The relative molecular mass of the glucose oxidase dimes ranges between 147 and 180 kDa varying as a function of the extent of glycosylation sites. Each monomer (Figure 7) has a FAD cofactor hurried at distance of 15 Å from the surface of protein. The cofactor is tightly bound however non-covalently and it determines the enzymatic activity. Glucose oxidase is mostly stable at a pH between 5 and 6 (Erhan et al., 2020) and at a temperature of 37 °C, however it can vary, depending on the species.

This enzyme is of industrial and biotechnological importance due to its high specificity, molecular stability and fast mechanism of action. The most well-known biotechnological application of glucose oxidase is its utilisation in management of diabetes as glucose biosensor (Mano, 2019).

Figure 7. Graphical representation glucose oxidase monomer (GOx, PDB entry 13QVR), the inset shows the FAD binding pocket, followed by the scheme of β-D-glucose oxidation reaction mediated by this enzyme

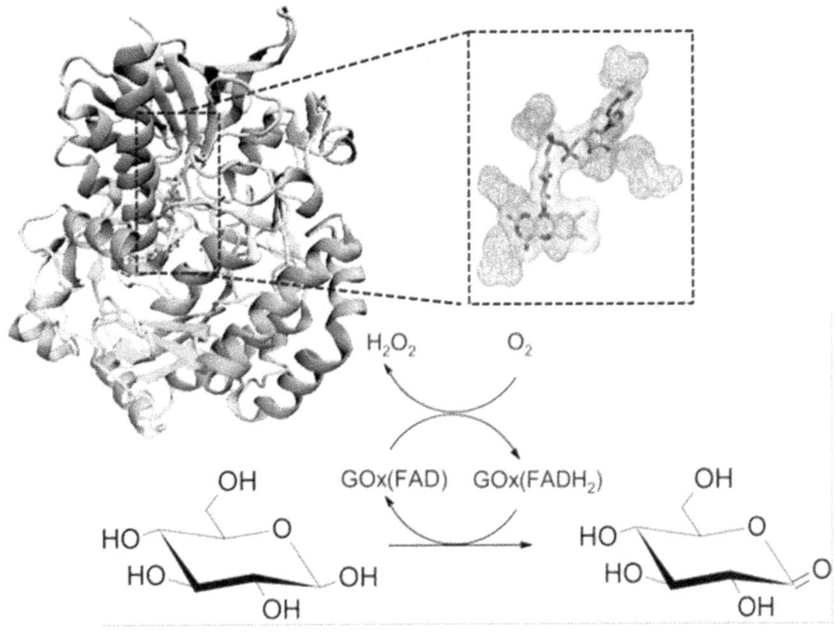

FUTURE PERSPECTIVES

This paper presents a structural survey of oxidoreductases and the schematic mechanism of reactions that these enzymes mediate. The main focus was set on flavin-based oxidoreductases and several most notorious representatives have been exemplified. Even though structurally different, all the exemplified flavoenzymes share the same flavin cofactor which enables them to catalyze diverse sets of reactions of great importance from biological perspectives as well as industrial ones. Constraints such as dependence on the cofactor availability, limited structural stability, and reaction specificity have been addressed by active research that makes progress in overcoming these challenges. Combining experimental and computational tools is the most effective way of identifying structural specificities and can offer information on how to avoid undesired properties of these enzymes. Thus, with the increasing knowledge on the structural specificity of these enzymes and their mechanism of interactions, their properties could be exploited in various industrial and biotechnological applications.

CONCLUSIONS AND OUTLINES

Oxidoreductases are one of the largest classes of enzymes catalyzing redox reactions that take place in biological systems to maintain the structure and function of living cells. These enzymes are involved in important biological processes including the mitochondrial electron transport chain and mitochondrial β-oxidation of long chain fatty acids, which provides energy demands for heart and skeletal muscles, synthesis or degradation of cellular lipids, and catabolism of aromatic amino acids.

- Analysis of the three-dimensional structure shows that the enzyme cofactor and its mode of binding (covalently or non-covalently) will strongly influence the enzymatic activity and respectively the processes that they are involved in.
- Most flavin-based enzymes use the two-halve reactions mechanisms in which, the first half is the reductive one where the substrate reduces flavin and the second half is the oxidative half-reaction where the flavin is oxidized in return by an electron acceptor. This mechanism allows reactions to produce that are thermodynamically challenging to overcome.
- Another, more recently discovered mechanism involving the flavin-based enzymes is the electron bifurcation mechanism, in which, the two electrons, obtained as a pair, are separated in coupled transfers in such way that the energy of both is concentrated on one of them. Tight coupling between the two electrons is essential and relies on the energy of the semiquinone state of the isoalloxazine ring from the flavin structure.
- Certainly, it is difficult to generalize by comparing the reactivity of enzymes with a closely related structure. The large reaction differences could be due to a combination of many factors including the redox potential of the flavin, constraints in movements due to the active sites' environment, polypeptide environment, etc. Complex approaches must be used when studying the structure and properties of oxidoreductases to achieve the catalytic activity, selectivity, efficiency, and desired stability to improve their industrial usage.

ACKNOWLEDGMENT

L.A. gratefully acknowledges the Moldova State Program (2020-2023), project (no.: 20.80009.5007.27) "Physico-chemical mechanisms of redox processes with electron transfer involved in vital, technological and environmental systems". R.V.E. gratefully acknowledges the national research program no: PN 23.21.01.02.

REFERENCES

Anghel, L., Rada, S., & Erhan, R.-V. (2023). Structural factors and electron transfer mechanisms in flavoenzymes. *Analytical Letters*, 1–14. doi:10.1080/00032719.2023.2174131

Barber, E. A., Liu, Z., & Smith, S. R. (2020). Organic contaminant biodegradation by oxidoreductase enzymes in wastewater treatment. *Microorganisms*, *8*(1), 122. doi:10.3390/microorganisms8010122 PMID:31963268

Bilal, M., Bagheri, A. R., Vilar, D. S., Aramesh, N., Eguiluz, K. I. B., Ferreira, L. F. R., Ashraf, S. S., & Iqbal, H. M. N. (2022). Oxidoreductases as a versatile biocatalytic tool to tackle pollutants for clean environment – a review. *Journal of Chemical Technology and Biotechnology*, *97*(2), 420–435. doi:10.1002/jctb.6743

Björck, M. L., Zhou, S., Rydström Lundin, C., Ott, M., Ädelroth, P., & Brzezinski, P. (2017). Reaction of S. cerevisiae mitochondria with ligands: Kinetics of CO and O2 binding to flavohemoglobin and cytochrome c oxidase. *Biochimica et Biophysica Acta (BBA) - Bioenergetics*, *1858*(2), 182–188. doi:10.1016/j.bbabio.2016.11.009 PMID:27871795

Bonamore, A., & Boffi, A. (2007). Flavohemoglobin: Structure and reactivity. *IUBMB Life*, *60*(1), 19–28. doi:10.1002/iub.9 PMID:18379989

Bouldin, R. M., Kyriazidis, L., Fidler, A., Ravichandran, S., Kumar, J., Samuelson, L. A., & Nagarajan, R. (2011). Oxidoreductase catalyzed polymerization of 3-methylpyrrole. *Journal of Macromolecular Science. Journal of Macromolecular Science, Part A: Pure and Applied Chemistry*, *48*(12), 976–982. doi:10.1080/10601325.2011.620413

Ensenauer, R., He, M., Willard, J.-M., Goetzman, E. S., Corydon, T. J., Vandahl, B. B., Mohsen, A.-W., Isaya, G., & Vockley, J. (2005). Human Acyl-CoA dehydrogenase-9 plays a novel role in the mitochondrial β-oxidation of unsaturated fatty acids. *The Journal of Biological Chemistry*, *280*(37), 32309–32316. doi:10.1074/jbc.M504460200 PMID:16020546

Erhan, R. V., Bodnarchuk, V., Radulescu, A., & Anghel, L. (2020). Small angle neutron scattering reveals dimeric glucose oxidase from Aspergillus niger at pH 5.9. *Journal of Surface Investigation. X-ray, Synchrotron and Neutron Techniques*, *14*(S1), S5–S10. Advance online publication. doi:10.1134/S1027451020070125

Espina, G., Atalah, J., & Blamey, J. M. (2021). Extremophilic oxidoreductases for the industry: Five successful examples with promising projections. *Frontiers in Bioengineering and Biotechnology*, *9*, 710035. Advance online publication. doi:10.3389/fbioe.2021.710035 PMID:34458243

Formosinho, S., & Barroso, M. (Eds.). (2011). *Proton-Coupled Electron Transfer*. Royal Society of Chemistry. doi:10.1039/9781849733168

Gautschi, M., Weisstanner, C., Slotboom, J., Nava, E., Zürcher, T., & Nuoffer, J.-M. (2015). Highly efficient ketone body treatment in multiple acyl-CoA dehydrogenase deficiency–related leukodystrophy. *Pediatric Research, 77*(1), 91–98. doi:10.1038/pr.2014.154 PMID:25289702

Hall, C. L., & Kamin, H. (1975). The purification and some properties of electron transfer flavoprotein and general fatty acyl coenzyme A dehydrogenase from pig liver mitochondria. *The Journal of Biological Chemistry, 250*(9), 3476–3486. doi:10.1016/S0021-9258(19)41540-8 PMID:1168197

Hazarika, A., Saikia, S., Devi, B., Yadav, M., & Yadav, H. S. (2022). Oxidoreductase metalloenzymes as green catalyst for phytoremediation of environmental pollutants. In *Phytoremediation Technology for the Removal of Heavy Metals and Other Contaminants from Soil and Water* (pp. 141–172). Elsevier. doi:10.1016/B978-0-323-85763-5.00021-0

He, M., Pei, Z., Mohsen, A.-W., Watkins, P., Murdoch, G., Van Veldhoven, P. P., Ensenauer, R., & Vockley, J. (2011). Identification and characterization of new long chain Acyl-CoA dehydrogenases. *Molecular Genetics and Metabolism, 102*(4), 418–429. doi:10.1016/j.ymgme.2010.12.005 PMID:21237683

Hecht, H. J., Kalisz, H. M., Hendle, J., Schmid, R. D., & Schomburg, D. (1993). crystal structure of glucose oxidase from Aspergillus niger refined at 2·3 Å resolution. *Journal of Molecular Biology, 229*(1), 153–172. doi:10.1006/jmbi.1993.1015 PMID:8421298

Henriques, B. J., Katrine Jentoft Olsen, R., Gomes, C. M., & Bross, P. (2021). Electron transfer flavoprotein and its role in mitochondrial energy metabolism in health and disease. *Gene, 776*, 145407. doi:10.1016/j.gene.2021.145407 PMID:33450351

Herrmann, G., Jayamani, E., Mai, G., & Buckel, W. (2008). energy conservation via electron-transferring flavoprotein in anaerobic bacteria. *Journal of Bacteriology, 190*(3), 784–791. doi:10.1128/JB.01422-07 PMID:18039764

Huang, L., Sun, G., Cobessi, D., Wang, A. C., Shen, J. T., Tung, E. Y., Anderson, V. E., & Berry, E. A. (2006). 3-Nitropropionic acid is a suicide inhibitor of mitochondrial respiration that, upon oxidation by Complex II, forms a covalent adduct with a catalytic base arginine in the active site of the enzyme. *The Journal of Biological Chemistry, 281*(9), 5965–5972. doi:10.1074/jbc.M511270200 PMID:16371358

Humphrey, W., Dalke, A., & Schulten, K. (1996). VMD: Visual molecular dynamics. *Journal of Molecular Graphics, 14*(1), 33–38. doi:10.1016/0263-7855(96)00018-5 PMID:8744570

Husain, M., & Steenkamp, D. J. (1983). Electron transfer flavoprotein from pig liver mitochondria. A simple purification and re-evaluation of some of the molecular properties. *The Biochemical Journal, 209*(2), 541–545. doi:10.1042/bj2090541 PMID:6847633

Kar, R. K., Miller, A., & Mroginski, M. (2022). Understanding flavin electronic structure and spectra. *Wiley Interdisciplinary Reviews. Computational Molecular Science, 12*(2). Advance online publication. doi:10.1002/wcms.1541

Kayastha, K., Vitt, S., Buckel, W., & Ermler, U. (2021). Flavins in the electron bifurcation process. *Archives of Biochemistry and Biophysics, 701*, 108796. doi:10.1016/j.abb.2021.108796 PMID:33609536

Kim, H. J., & Winge, D. R. (2013). Emerging concepts in the flavinylation of succinate dehydrogenase. *Biochimica et Biophysica Acta (BBA) - Bioenergetics, 1827*(5), 627–636. doi:10.1016/j.bbabio.2013.01.012 PMID:23380393

Kim, J. J., Wang, M., & Paschke, R. (1993). Crystal structures of medium-chain acyl-CoA dehydrogenase from pig liver mitochondria with and without substrate. *Proceedings of the National Academy of Sciences of the United States of America, 90*(16), 7523–7527. doi:10.1073/pnas.90.16.7523 PMID:8356049

Lemire, B. D., & Oyedotun, K. S. (2002). The Saccharomyces cerevisiae mitochondrial succinate:ubiquinone oxidoreductase. *Biochimica et Biophysica Acta (BBA) - Bioenergetics, 1553*(1–2), 102–116. doi:10.1016/S0005-2728(01)00229-8 PMID:11803020

Mano, N. (2019). Engineering glucose oxidase for bioelectrochemical applications. *Bioelectrochemistry (Amsterdam, Netherlands), 128*, 218–240. doi:10.1016/j.bioelechem.2019.04.015 PMID:31030174

Massey, V. (2002). The reactivity of oxygen with flavoproteins. *International Congress Series, 1233*, 3–11. doi:10.1016/S0531-5131(02)00519-8

Mattevi, A. (2006). To be or not to be an oxidase: Challenging the oxygen reactivity of flavoenzymes. *Trends in Biochemical Sciences, 31*(5), 276–283. doi:10.1016/j.tibs.2006.03.003 PMID:16600599

May, S. W. (1999). Applications of oxidoreductases. *Current Opinion in Biotechnology, 10*(4), 370–375. doi:10.1016/S0958-1669(99)80067-6 PMID:10449319

May, S. W., & Padgette, S. R. (1983). Oxidoreductase enzymes in biotechnology: Current status and future potential. *Nature Biotechnology, 1*(8), 677–686. doi:10.1038/nbt1083-677

Moser, C. C., Anderson, J. L. R., & Dutton, P. L. (2010). Guidelines for tunneling in enzymes. *Biochimica et Biophysica Acta (BBA) -. Bioenergetics, 1797*(9), 1573–1586. doi:10.1016/j.bbabio.2010.04.441 PMID:20460101

Motteran, L., Pilone, M. S., Molla, G., Ghisla, S., & Pollegioni, L. (2001). Cholesterol oxidase from Brevibacterium sterolicum. *The Journal of Biological Chemistry, 276*(21), 18024–18030. doi:10.1074/jbc.M010953200 PMID:11359791

Moussaoui, M., Misevičienė, L., Anusevičius, Ž., Marozienė, A., Lederer, F., Baciou, L., & Čėnas, N. (2018). Quinones and nitroaromatic compounds as subversive substrates of Staphylococcus aureus flavohemoglobin. *Free Radical Biology & Medicine, 123*, 107–115. doi:10.1016/j.freeradbiomed.2018.05.071 PMID:29793040

Page, C. C., Moser, C. C., Chen, X., & Dutton, P. L. (1999). Natural engineering principles of electron tunnelling in biological oxidation–reduction. *Nature, 402*(6757), 47–52. doi:10.1038/46972 PMID:10573417

Poole, R. K. (2020). Flavohaemoglobin: The pre-eminent nitric oxide–detoxifying machine of microorganisms. *F1000 Research, 9*, 7. doi:10.12688/f1000research.20563.1 PMID:31956400

Rivas, M. G., Gonzalez, P. J., Ferroni, F. M., Rizzi, A. C., & Brondino, C. (2020). Studying electron transfer pathways in oxidoreductases. *Science Reviews - from the End of the World, 1*(2), 6–23. https://doi.org/. doi:10.52712/sciencereviews.v1i2.15

Romero, E., Gómez Castellanos, J. R., Gadda, G., Fraaije, M. W., & Mattevi, A. (2018). Same Substrate, Many Reactions: Oxygen Activation in Flavoenzymes. *Chemical Reviews, 118*(4), 1742–1769. doi:10.1021/acs.chemrev.7b00650 PMID:29323892

Sato, K., Nishina, Y., & Shiga, K. (1993). Electron-transferring flavoprotein has an AMP-binding site in addition to the FAD-binding site1. *Journal of Biochemistry, 114*(2), 215–222. doi:10.1093/oxfordjournals.jbchem.a124157 PMID:8262902

Scott, R. A. (1995). Functional significance of cytochrome c oxidase structure. *Structure (London, England), 3*(10), 981–986. doi:10.1016/S0969-2126(01)00233-7 PMID:8590007

Shimada, S., Shinzawa-Itoh, K., Baba, J., Aoe, S., Shimada, A., Yamashita, E., Kang, J., Tateno, M., Yoshikawa, S., & Tsukihara, T. (2017). Complex structure of cytochrome c –cytochrome c oxidase reveals a novel protein–protein interaction mode. *The EMBO Journal, 36*(3), 291–300. doi:10.15252/embj.201695021 PMID:27979921

Sirkar, K., Revzin, A., & Pishko, M. V. (2000). Glucose and lactate biosensors based on redox polymer/oxidoreductase nanocomposite thin films. *Analytical Chemistry, 72*(13), 2930–2936. doi:10.1021/ac991041k PMID:10905330

Soleimany, A. P., & Bhatia, S. N. (2020). Activity-based diagnostics: An emerging paradigm for disease detection and monitoring. *Trends in Molecular Medicine, 26*(5), 450–468. doi:10.1016/j.molmed.2020.01.013 PMID:32359477

Usselman, R. J., Fielding, A. J., Frerman, F. E., Watmough, N. J., Eaton, G. R., & Eaton, S. S. (2008). Impact of mutations on the midpoint potential of the [4Fe-4S] $^{+1,+2}$ cluster and on catalytic activity in electron transfer flavoprotein-ubiquinone oxidoreductase (ETF-QO). *Biochemistry, 47*(1), 92–100. doi:10.1021/bi701859s PMID:18069858

Valton, J., Mathevon, C., Fontecave, M., Nivière, V., & Ballou, D. P. (2008). Mechanism and regulation of the two-component FMN-dependent monooxygenase ActVA-ActVB from Streptomyces coelicolor. *The Journal of Biological Chemistry, 283*(16), 10287–10296. doi:10.1074/jbc.M709730200 PMID:18245777

Walsh, C. T., & Wencewicz, T. A. (2013). Flavoenzymes: Versatile catalysts in biosynthetic pathways. *Natural Product Reports, 30*(1), 175–200. doi:10.1039/C2NP20069D PMID:23051833

Watmough, N. J., & Frerman, F. E. (2010). The electron transfer flavoprotein: Ubiquinone oxidoreductases. *Biochimica et Biophysica Acta (BBA) -. Bioenergetics, 1797*(12), 1910–1916. doi:10.1016/j.bbabio.2010.10.007 PMID:20937244

Watson, S. A., & McStay, G. P. (2020). Functions of cytochrome c oxidase assembly factors. *International Journal of Molecular Sciences, 21*(19), 7254. doi:10.3390/ijms21197254 PMID:33008142

Westphal, A. H., Tischler, D., & van Berkel, W. J. H. (2021). Natural diversity of FAD-dependent 4-hydroxybenzoate hydroxylases. *Archives of Biochemistry and Biophysics, 702*, 108820. Advance online publication. doi:10.1016/j.abb.2021.108820 PMID:33684360

Wikström, M., Krab, K., & Sharma, V. (2018). Oxygen activation and energy conservation by cytochrome c oxidase. *Chemical Reviews, 118*(5), 2469–2490. doi:10.1021/acs.chemrev.7b00664 PMID:29350917

Winkler, A., Kutchan, T. M., & Macheroux, P. (2007). 6-S-Cysteinylation of bi-covalently attached FAD in berberine bridge enzyme tunes the redox potential for optimal activity. *The Journal of Biological Chemistry*, *282*(33), 24437–24443. doi:10.1074/jbc.M703642200 PMID:17573342

Younus, H. (2019). Oxidoreductases: Overview and practical applications. In Biocatalysis (pp. 39–55). Springer International Publishing. doi:10.1007/978-3-030-25023-2_3

Zhang, J., Frerman, F. E., & Kim, J.-J. P. (2006). Structure of electron transfer flavoprotein-ubiquinone oxidoreductase and electron transfer to the mitochondrial ubiquinone pool. *Proceedings of the National Academy of Sciences of the United States of America*, *103*(44), 16212–16217. doi:10.1073/pnas.0604567103 PMID:17050691

Chapter 16
Electron Transfer:
Mechanisms in Oxidoreductase Enzymes

Anca R. Leonties
Institute of Physical Chemistry Ilie Murgulescu, Romania

Ludmila Aricov
Institute of Physical Chemistry Ilie Murgulescu, Romania

Adina Raducan
https://orcid.org/0000-0002-7420-778X
Faculty of Chemistry, Institute of Physical Chemistry Ilie Murgulescu, Romania

ABSTRACT

Oxidoreductases are a special class of enzymes that use the redox mechanism for the efficient transformation of organic substrates. Most oxidoreductases contain metals in the active site and, for optimal functioning, require the participation of a small co-substrate with the ability to donate electrons. From the multitude of enzymes with economic and applicable potential, the authors focused their attention on three particular classes: catalases, peroxidases, and laccases. Catalases and peroxidases contain heme iron in their active sites and most often require electron donors such as oxygen or hydrogen peroxide while laccase contains copper and demands special co-substrates such as 2,2'-azino-bis(3-ethylbenzothiazoline-6-sulfonic acid) or syringaldehyde. The theoretical aspects regarding the mechanism in which the electron transfer of the three enzymes is involved as well as the practical applications of the selected enzymes in the field of environmental remediation will be the subject of this chapter.

INTRODUCTION

Presently, enzymes are being used as a replacement to traditional catalysts. Specificity, high activity under mild environmental conditions, and catalytic efficiency are some of the potential benefits of biocatalysts. The enzyme market is continually expanding due to advances in production technologies and novel applications (Sellami et al., 2022). Around 75% of all enzymes manufactured are used in the

DOI: 10.4018/978-1-6684-7198-2.ch016

detergent, textile, starch, and food industries (Sellami et al., 2022). The oxidoreductase class of enzymes will be the topic of this chapter. The constituents of this class are oxidases, dehydrogenases, hydroxylases, oxygenases, peroxidases, and reductases. The reactions catalyzed by oxidoreductases occur through the direct transfer of electrons between donors and acceptors, by mediated electron transfer, or in particular cases proton/hydrogen extraction (Sellami et al., 2022). From the multitude of enzymes in the class of oxidoreductases, we will concentrate on catalase, peroxidase and laccase. The electron transfer mechanism that takes place during the catalysis of the substrates with the mentioned enzymes will be detailed.

Catalase is an enzyme with low substrate specificity, able to decompose the hydrogen peroxide into water and molecular oxygen. Therefore, catalase has multiple uses: in medicine is used to prevent and control oxidative processes (Sampayo et al., 2003); in the textile industry for the decomposition of hydrogen peroxide resulting from bleaching treatment of textiles; in the cosmetic industry as an oxygenator for skin rejuvenation [(Sung et al., 2019)], in the food industry as preservative (Patel et al., 2016) etc.

Peroxidases, unlike catalases, oxidize effectively a wide range of reducing substrates, including phenols, aromatic dyes, nicotinamide adenine dinucleotide (oxidized form: NAD^+; reduced form: NADH) and nicotinamide adenine dinucleotide phosphate (oxidized form: $NADP^+$; reduced form: NADPH), lignin, (de Oliveira et al., 2021) etc. The versatility of peroxidase is given by its ability to catalyze peroxidative, oxidative, and hydroxylation reactions involving oxidative and reducing substrates. In typical peroxidase reactions, hydrogen peroxide or hydroperoxides oxidize substrates such phenols or aromatic amines, as well as NADPH and NADH, by transferring a single electron to create coupled radical products. In the absence of an electron source, peroxidases degrade hydrogen peroxide.

Laccases are enzymes with multi-copper active sites that oxidize phenolic compounds and are utilized in the paper, textile, and food industries as well as in bioremediation procedures. The four-electron reduction of molecular oxygen to water is a process that laccases catalyze (Farver & Pecht, 2011). This chapter will focus on the theoretical elements of the mechanism involved in the electron transfer of the three enzymes, as well as the practical applications of the selected enzymes in the field of environmental remediation.

CATALASE

Catalase (EC 1.11.1.6) can work both in the catabolism of hydrogen peroxide water and the peroxide oxidation of small substrates, depending on H_2O_2 concentrations. At high levels of H_2O_2, the decomposition of 2 molecules of hydrogen peroxide to dioxygen is favored, eq (1), while at lower levels of H_2O_2 peroxidation with several hydrogen donors such as small alcohols, phenols, or nitrites occur eq. (2) (Nicholls, 2012):

$$2H_2O_2 \xrightarrow{\text{Catalatic activity}} 2H_2O + O_2 \tag{1}$$

$$2H_2O_2 + SH_2 \xrightarrow{\text{Peroxidatic activity}} 2H_2O + S \tag{2}$$

As early as 1811, when Thénard discovered the existence of hydrogen peroxide in living tissues, there was a question about the existence of a compound in the body that would degrade hydrogen peroxide. Nine years later this compound was isolated and called "catalase" by Loew. Since then, countless studies have been developed on the identification and elucidation of the structure and mechanism of catalases. In 1936 Stern showed that in the enzyme active site, there is a protoporphyrin IX as a prosthetic group, and one year later, the bovine liver catalase was crystallized (Sumner & Dounce, 1937). Chance was the first to differentiate the two mechanisms (catalatic and peroxidatic) according to the initial concentration of hydrogen peroxide (Chance et al., 1952), and identified the main enzyme-substrate complexes: Compound I and Compound II as catalase - H_2O_2 active intermediates (Chance, 1949) and Compound III, as an inactive intermediate (Schonbaum & Chance, 1976), formed in excess of a substrate.

Catalase exists in almost all aerobic organisms since the destruction of reactive oxygen in the cell is vital for the proper development of metabolic processes. There are three classes of catalases, based on their structure. The most studied and also the most widespread in nature are the monofunctional, heme-containing catalases. The second class is composed of bifunctional, heme-containing catalase-peroxidases that are closely related to plant peroxidases. The third class includes the nonheme or Mn-containing catalases (Chelikani et al., 2004; Zámocký & Koller, 1999).

From a structural point of view, all catalases, regardless of their origin, are made up of four almost identical subunits, each one with four principal domains, with different functions. The function of the first domain is to interact with other subunits in a kind of molecular node; histidine residues, located near the C-terminal of this domain, are essential for the proper binding of the enzyme to the substrate. The next domain holds the main channel that leads to the active site and also the most important residues surrounding the heme in the distal part. The third domain, known as the domain of connection and covering, is also well preserved, and its residues surround the active site in the proximal part. The last domain is the helical one, which interacts with the second domain and forms the channel for the binding of NADPH in species that bind to a cofactor. Some catalases also have an additional domain instead of the NADPH, called the "flavodoxin-like" domain (Fita et al., 1986; Kirkman & Gaetani, 1984). The catalase's quaternary structure is crucial to the enzyme's overall activity, and tetramerization may be required to shield the active site from its surroundings. This claim is supported by the finding that the heme stabilizes the binding region and makes it easier for the subunits to engage. Complete tetramers exhibit D2 symmetry, implying that they are made up of two interconnected pairs of polypeptide chains (Alfonso-Prieto et al., 2009).

Each active site of each unit of catalase is a penta-coordinating porphyrin molecule to a central iron atom known as heme. The heme group is deeply buried in the core structure, the distance from the surface being at least 20 Å (Fita & Rossmann, 1985). Besides the heme group, there are several amino acids, from different chains, which are essential for catalytic activity: tyrosine on the proximal side of the heme (Tyr415) which is coordinated to the heme iron, contributing to the oxidation state of an iron atom (3+), and histidine on the distal side (His128) which is essential for the formation of the enzyme-substrate complex. Other amino acids like proximal arginine and distal asparagine and serine are also part of the active site of catalase (Alfonso-Prieto et al., 2012).

Since the heme group is deeply embedded in the catalase structure, there are three channels that link the prosthetic group with the exterior: the main channel, the lateral channel, and the central channel. The main channel is considered to be the primary access route for the hydrogen peroxide substrate, a statement supported also by molecular dynamic studies (Hara et al., 2007). This channel, in its hydrophobic area, shows a lack of water, so it is optimized for hydrogen peroxide binding (Putnam et al., 2000). Also, the favorable orientation of incoming hydrogen peroxide molecules to the active site is facilitated by the electrical potential in the channel. This results in a ''chain-like reaction'', which explains the high rate of catalytic reaction (Díaz et al., 2012). The central channel, which may connect the heme with the cavity at the molecular center, and the lateral one, which approaches the heme laterally, are the other channels. These auxiliary channels most likely contribute to the rapid release of reaction products (H_2O and O_2) (Goc et al., 2021).

The mechanism of hydrogen peroxide decomposition depends, among other variables, on the initial concentration of the substrate. At low concentrations of hydrogen peroxide, the classical mechanism consists of two distinct steps:

$$H_2O_2+Cat \rightarrow H_2O+Cpd\ I \quad (3)$$

$$H_2O_2+Cpd\ I \rightarrow Cat+H_2O+O_2 \quad (4)$$

In the first step, hydrogen peroxide oxidizes the heme to Cpd I, an oxoiron (IV) porphyryl radical; in the second step, this radical is reduced by another hydrogen peroxide to regenerate the native enzyme (Cat), water, and oxygen. Alternatively, the Cpd I of the catalase can oxidize alcohols to aldehydes and water; this is the peroxidatic reaction of the catalase (Sepasi Tehrani & Moosavi-Movahedi, 2018). Cpd I reactivity depends on the shape and nature of the heme distal pocket. Irrespective of the pathway, the reaction does not proceed through a 2-electron step, as commonly assumed, since the two electrons are not transferred simultaneously. Instead, the reaction is better described as two one-electron transfer steps.

At high concentrations of H_2O_2 catalase suffers a process of slow irreversible inactivation (Gray, 1989). In this case, the mechanism consists in two phases called α and β. The enzyme is initially discovered in phase α of activity; nevertheless, as time goes on, the activity reduces to a brand-new stationary phase, called the β phase. During the β phase, it was noticed that the reaction rate gradually decreased as a result of the progressive degradation of the enzyme by hydrogen peroxide (Lardinois et al., 1996). During the β phase, catalase can produce two more intermediates known as Cpd II and Cpd III; however, these compounds lack catalytic activity until they are converted to the ferric version of the enzyme. Similar to Cpd I, Cpd II also contains an iron (IV) ion bonded to oxyporphyrin, but the porphyrin is no longer in its radical state. It can also be created by ferricyanide and radical superoxide being oxidized together with reducing an electron from Cpd I. Compound III can be obtained by reducing native catalase and then adding oxygen, adding hydrogen peroxide to Cpd II, or associating leftover enzyme with superoxide anions (Gan et al., 2022). The nature and shape of the heme distal pocket affect Cpd I reactivity. Regardless of the pathway, since the two electrons are not transported at the same time, the reaction does not go through a 2-electron step as is typically believed. The two one-electron transfer steps are a more accurate description of the reaction (Alfonso-Prieto et al., 2009)

The irreversible inhibition of catalase caused by hydrogen peroxide excess is significantly influenced by compounds II and/or III. and a simple kinetic model was proposed for catalase inactivation at high substrate concentrations (Răducan et al., 2007).

$$H_2O+Cat \rightarrow H_2O+Cpd\ I \quad (5)$$

$$H_2O_2 + Cpd\ I \rightarrow Cat + H_2O + O_2 \quad (6)$$

$$Cpd\ I \rightarrow Cpd\ III \quad (7)$$

The last step includes all possible processes for catalase inactivation; the inactivation constant estimated is around 4 x 10 $^{-3}$ s^{-1}, a value corresponding to a slow inactivation as compared to the main reaction (Oancea et al., 2008).

As well as all enzymes, catalase activity can be inhibited reversible, or irreversible by several compounds. Reversible noncompetitive inhibitors, like as NaF or 4-dimethylaminoantipyrin, bind to the Cpd I directly in the distal heme pocket and displace one or more water molecules that are loosely bound to the conserved H-bonding distal residues (often histidine and asparagine) and the heme iron (Nicholls, 2012). Sodium azide, cyanide, and 3-aminotriazole bind directly to the heme, being competitive irreversible inhibitors of catalase mostly at high concentrations (Lardinois & Rouxhet, 1996). Since catalase regulates the level of oxidative stress in the human body, it is a potential target in many disorders. As a result, numerous research on the impact of drugs on catalase activity have lately been published (Divsalar & Ghobadi, 2021; Dong et al., 2022; Khataee et al., 2022; Shahraki et al., 2020; Wang et al., 2022)

By scavenging excess hydrogen peroxide, which is harmful to cells (Sampayo et al., 2003), catalase is employed in medicine to prevent and reduce oxidative damages linked to some pathophysiological conditions, such as inflammatory disorders, ischemia-reperfusion injury, aging, and cancer (Khorshidi et al., 2022). Also, catalase is used for the development of biosensors for a variety of applications like H_2O_2 sensing or several catalase inhibitors like fluoride and cyanide (Sooch et al., 2014). Additionally, catalase is used in a number of industrial processes, including the food industry in a number of procedures, including milk quality assessment, food product packing, and cheese manufacture (Kaushal et al., 2018).

The catalase has been used in a variety of bioremediation processes. It has been known to act as an indicator of the hydrocarbon degradation of crude oil-polluted soil (Joergensen et al., 1995), where the level of bioremediation occurring in a certain area is determined by comparing the activity of the enzyme before and after the bioremediation method. Catalase also provides the oxygen needed by aerobic bacteria to break down the organic wastes in the aerobic bioremediation process (Achuba & Okoh, 2014).

Catalase is also used to remove hydrogen peroxide from textile industry bleaching effluents instead of hydrosulphite or sodium bisulphite, which could result in an unintentional increase in the salt concentration during the process. Also, using catalase in the textile industry leads to decreased process costs and contamination of residual waters, both of which demand increasing attention in light of the new ecological requirements and from an economic point of view (Tochetto et al., 2022).

Although catalase has one of the highest turnover rates among biocatalysts, the cost of using the enzyme for the breakdown of H_2O_2 and its relatively low stability remain the limiting factors in most bioprocesses (Singh et al., 2023). One strategy is the use of immobilized catalase, using a large variety of carrier materials, including alumina, gelatin, polyacrylamide, hen eggshell, artificial membranes, and carbon materials. The immobilization process leads to better stability at high pH and temperatures (Grigoras, 2017; Liu et al., 2021). Increased thermostability and operational stability of catalase can be achieved at a low cost by adding salts, polyols (polyethylene glycol, glycerol), dextran, bovine serum albumin, polyethyleneimine, polyelectrolytes, organic osmolytes (e.g. betaine), organic solvents (isooctane, cyclohexane, chloroform, benzene) or sugars; the nature of the enzyme influences the selection of the appropriate additive (Butreddy et al., 2021; Cantemir et al., 2013; Xu et al., 2021). Other strategies, more expensive, include the genetic modification of the enzyme (Patel et al., 2016) or identifying alkali thermostable catalase from microorganisms so that the enzyme can be active at high pH values and temperatures (Kumar et al., 2019).

Catalase has been demonstrated to be one of the most abundant and easily accessible enzymes that can be used in a variety of bioremediation processes. One of the main applications of this enzyme in the bioremediation of waste effluents from textile industries has been the removal of hydrogen peroxide from bleaching effluents and allowing the water to be reused for subsequent dyeing steps. Catalase, despite being one of the first enzymes discovered and characterized, will undoubtedly continue to fascinate us with unexpected and intriguing mysteries for generations to come.

PEROXIDASES

Peroxidases are oxidoreductases found in both the animal and plant kingdoms. The most common peroxidases are those that have heme active sites with ferric protoporphyrin IX that functions as a prosthetic group. For most peroxidases, the prosthetic group is composed of four pyrrole rings, together with methane and iron (III) in the center of the molecule (de Oliveira et al., 2021). Other peroxidases contain selenium, vanadium, magnesium, and flavin in the active center (Hofrichter et al., 2010). Peroxidases can catalyze a varied range of reactions such as peroxidations, oxidations, hydroxylation using both oxidizing and reducing substrates (Chen et al., 2014; Fernandes et al., 2020; Flohé et al., 2022).

The peroxidative reaction has several steps and leads to the formation of oxidized intermediates denoted compounds 0, I, II, and III. In the first stage of the reaction, the H_2O_2 interacts with the ferric atom (Fe^{III}) and forms the compound 0 which is an intermediate transient ferric hydroperoxide. Following this stage, the oxygen protonation of compound 0 takes place and leads to the formation of compound I which is represented by an iron oxoderivative coupled with a cationic π radical. Compound II is obtained when compound I is reduced to the oxoderivative of Fe^{IV}. If a phenol or enol is involved in the reaction, compound II may regenerate into the native state. This reaction takes place by electron and proton transfer at the active site concomitant with the reduction of Fe^{IV} to Fe^{III} and the formation of an H_2O molecule. In certain cases, if H_2O_2 is in excess, compound II is not transformed in the native enzyme state but in compound II which is a complex formed by an oxygen molecule bound to the ferrous iron (de Oliveira et al., 2021).

Rich electron substrates such as phenols or aromatic amines, as well as NADPH and NADH are also oxidized in the presence of H_2O_2 or other hydroperoxides via single electron transfer and formation of products with coupled radicals (Chen et al., 2014; Fernandes et al., 2020; Flohé et al., 2022). Another type of peroxidases may use cosubstrates in order to oxidate halide ions. In this type of reaction, the halogenated ion interacts with compound I and produces H_2O_2 (Davies, 2021; Undiano et al., 2022).

One particular class of peroxidases is the Cytochrome P450 which functions as monooxygenases. The ET in Cytochrome P450 takes place differently than in protoporphyrin peroxidases and requires the presence of noncovalently bound heme as a cofactor. Briefly, the ET in Cytochrome P450 happens when one O_2 molecule interacts with the cofactor. As a result, the substrate is oxidized by the insertion of one oxygen atom, the product is formed and one molecule of water exists the system The rate limiting step of this reaction is represented by the two electrons sequential transferred from the redox participant to the cofactor (Mukherjee et al., 2021; Poulos, 2014).

Lignin peroxidase is a metalloenzyme with promising applications able to degrade natural and synthetic polymers (Biko et al., 2020). The enzyme requires the participation of hydrogen peroxide in order to form porphyrin π-radicals and activate the two-electron oxidative transfer. In order to catalyze substrates that exceed its redox potential mediators are needed, and a different type of electron transfer is required. Long-range electron transfer implies that substrates may be oxidized on the surface of the protein by long-ranged electron tunneling from substrates to the activated catalytic (Pérez-Boada et al., 2005). One example of such transfer takes place in the case of the decomposition reaction of propranolol mediated by veratryl alcohol in the presence of lignin peroxidase (Ding et al., 2022). The authors used computational analysis and discovered that long-range electron transfer through the His–Asp to the porphyrin iron is involved in propranolol degradation. So, by using this path the substrate is transformed by single electron transfer or H abstraction while the hydroxyl radicals resulting in the catalysis of the H_2O_2 take part in the hydroxylation of the substrate intermediates. It was confirmed that the veratryl alcohol had a double role being both activators and also protecting the peroxidase to form compound II when the hydrogen peroxide was in excess (Ding et al., 2022).

Due to their versatility, peroxidases may be employed in a variety of reactions in which the electron transfer is involved (Fapyane & Ferapontova, 2016; Okawa et al., 2015). In sulphite oxidase (Elliott et al., 2002) and cytochromes (Mukherjee et al., 2021), the heme is not bound covalently thus allowing the shuttling of electrons to the redox partner of a protein and the catalytic site with internal electron transfer rates higher than the external ones. This property might be utilized to create high-efficiency bioelectronic devices based on assemblies of electrically responsive enzyme molecules immobilized on electrodes. Fapyane and Ferapontova (2016), proposed the reconstitution of horseradish peroxidase onto alkanethiol-conjugated hemin and self-assembled onto the gold nanoparticle nanoparticle-modified electrodes to produce enhanced electron transfer between the gold electrode and the enzyme heme. The researchers also demonstrated that the NP-biohybrid system enables sophisticated programmable design of bioelectronic interfaces and argued that could be applied with good efficiency even for enzymes with the heme not covalently bound (Fapyane & Ferapontova, 2016).

Enzymatic biofuel cells are particular fuel cells that extract energy from biological molecules. These interesting systems may one day be used as an alternative energy source. The Enzymatic biofuel cells could find applications in various domains such as robotics, aerospace, biomedicine, transportation, communication, etc. (Nasar & Perveen, 2019). Biofuel cells may be a multi-enzyme or single enzyme, each having distinct assets and drawbacks (Dutta et al., 2022). Biofuel cell based on horseradish peroxidase may use as immobilization anode copper sulfide, carbon based-nanomaterials, conducting polymers, composites (Šekuljica et al., 2016; Tawalbeh et al., 2022). Improving the rate of electron transfer in those systems is a continuous challenge. Different groups report that the regeneration of cofactors might improve enzymatic biofuel cell efficiency (Stolarczyk et al., 2020; Wang et al., 2017). Also, improvement of electronic transfer for enzymatic biofuel cells could be done via biomimetic enzyme cascade on the electrode surface (Campbell et al., 2012, Herkendell et al., 2019). Nevertheless. adequate immobilization techniques may develop more efficient enzymatic biofuel cells by providing suitable linkers and immobilization supports (Duong et al., 2020; Šekuljica et al., 2016; Pöller et al., 2013).

Depending on the provenience, different applications with peroxidases are found in the literature. One group argues that the peroxidase extracted from Raphanus sativus may be used in the biotransformation of cresols. The biodegradation products formed in the case of o-cresol, m-cresol, and p-cresol were mediated by the H_2O_2 and resulted in harmless degradation products (Saikia et al., 2022). Another application of peroxidases is in the field of green catalysts for bioremediation and biotechnology. A myriad of substrates was efficiently decomposed by peroxidases: pharmaceuticals, phenols, lignin, dyes, polycyclic aromatic hydrocarbons etc. (Besharati Vineh et al., 2018; Calza et al., 2016; Chen & Wan, 2017; Ding et al., 2022; Sellami et al., 2022; Stadlmair et al., 2018; Zhang et al., 2018). Horseradish peroxidase is frequently used in microfluidic analytical devices which are assay platforms with possible applications in medical diagnostics and environmental monitoring (Fedalto et al., 2023). Also, peroxidase may be used in biosensing glucose, as shown in a recent study concerning the electrocatalytic effect of recombinant Mn peroxidase from corn (Izadyar et al., 2022). Glycosylated peroxidase variants find potential applications in the field of cancer therapy (Humer et al., 2021) and in the field of drug activation (Liu et al., 2017).

Peroxidases, which are mostly produced from materials that are cost-effective, renewable, and sustainable, provide operational flexibility in multiple domains from the bioremediation of ecosystems to the monitoring and detection of target molecules by biosensors incorporating. However, it is crucial to recognize the difference between using peroxidases on an industrial scale and in the lab. Practical applications are frequently linked to the necessity of large quantities of enzymes so the industry may benefit from the development of new ways to enhance enzyme expression and production in bacterial and fungal cultures. Recycling can also be used to reduce costs in industrial-scale applications by immobilizing enzymes on mechanically robust supports with high activity retention.

LACCASES

Copper oxidases are also part of the oxidoreductase family. These proteins have molecular weights ranging from 50 to 130 kDa and contain low mass (10 - 20 kDa) copper-like domains characterized by beta-sheets motifs interconnected by turn structures, in which electron transfers predominate (Janusz et al., 2020; Mehra et al., 2018). Depending on the number of cupredoxin-like domains, they can be divided by two-, three- and six-domain enzymes. Laccases are metalloenzymes belonging to the subclass multicopper oxidases with three cupredoxin-like domains. These enzymes are widespread in nature, being found mainly in fungi and bacteria, but also in the plant kingdom in higher plants or in some insects Their primary structure involves a number of approximately 500 – 550 amino acids (I-III) which are consecutively folded in a Greek key β barrel secondary structure stabilized by disulphide bonds between domains I-II and I-III (Janusz et al., 2020; Mehra et al., 2018). The spectroscopic characterization of different laccases revealed that the four copper atoms present per protein unit are likely to be of three categories: paramagnetic “blue” type 1 (T1Cu), paramagnetic ‘‘normal/non-blue” copper type 2 (T2Cu), and binuclear diamagnetic spin-coupled type 3 (T3a and b) (Agrawal et al., 2018). To form the active site, the copper ions are organized into two groups: a mononuclear cluster T1Cu and a trinuclear cluster T2Cu/T3aCu/T3bCu. The main function of T1Cu is to ensure the one-electron oxidation of four equivalents of a given substrate, while the trinuclear site is responsible for the reduction of dioxygen to water molecules (Cardullo et al., 2022). Briefly, laccase-mediated oxidation of a substrate that is able to donate an electron happens at the protein surface, at T1 oxidation occurs with one electron, while O_2 reduction requires 4 electrons to form water at the buried T2/T3 site (Mehra et al., 2018). Another feature of the primary structure of laccases is that 10 to 45% of the total amino acid content can be viewed as potential glycosylation sites (Arregui et al., 2019). This property gives laccases the ability to better retain copper atoms and improves both thermal stability and enzymatic activity (Cardullo et al., 2022). Therefore, the secondary structure of the laccases depends on the source (bacterial, fungal, plant, insect), making them have a high substrate specificity. Thus, laccases participate in both anabolic and catabolic processes, operating on small molecule substrates such as phenols (Akram et al., 2022), but also with large molecules like lignin (Tobimatsu & Schuetz, 2019). An extended list can be found in Table 1.

Table 1. Anabolic and catabolic processes of laccase operating on different substrates

Process Type	Laccase Source	Function	Substrate	Reference
Anabolism	Plants	polymerization	monolignol, phenols	(Jeon et al., 2012; Tobimatsu & Schuetz, 2019)
	Fungi	polymerization, melanogenesis	stilbenoids, melanin	(Cardullo et al., 2022; Jeon et al., 2012)
	Bacterial	oxidation	phenols	(Chauhan et al., 2017)
	Insects	sclerozation	cuticular proteins	(Yang et al., 2022)
Catabolism	Bacterial	polymerization	lignine	(Janusz et al., 2017)
	Fungal	biodelignification	lignocellulose	(Suryadi et al., 2022)

Laccase-catalyzed reactions occur in two ways: by direct electron transfer (DET) (when the substrate has a lower redox potential than laccase, < 400 - 800 mV) or by oxidation involving the presence of a mediator (MET). The DET category includes phenolic substrates, amines, catechols etc. (Jones & Solomon, 2015; Manzano-Nicolas et al., 2020), while the MET category handles various types of pigments, phenols, lignosulphones (Aricov et al., 2020; Mattos et al., 2022; Nugroho Prasetyo et al., 2010). The most used mediators are N-hydroxybenzotriazole, violuric acid, 2,2′-azinobis-(3-ethylbenzylthiozoline 6-sulphate) (ABTS), and 2,2,6,6-tetramethyl-1-piperid-inyloxyl (TEMPO) (Mattos et al., 2022).

Computational methods and kinetics are useful tools that can be successfully used in order to investigate the direct electronic transfer taking place between laccase and substrates. (Manzano-Nicolas et al., 2020) used the chronometric method and obtained the steady state rates for the catalysis of the laccases on catecholamines and related compounds. From the kinetic data, the authors concluded that substrate particularities, mainly the differences between functional groups, strongly affect the catalysis. For example, 1-epinephrine, l-norepinephrine, and isoprenaline are similar in terms of enzyme affinity, but the first two are catalyzed faster, probably due to the greater nucleophilic power owned by the phenolic hydroxyl group. Due to its relation with melanization and sclerotization in the insect cuticle, and because it had the highest catalytic power, dopamine, was further selected for docking studies. Ligand docking showed that the dopamine molecule is stabilized by laccase structure via π-π stacking and hydrogen bonds. The phenyl ring and some of the hydroxyls from dopamine are situated in close vicinity to the amino acids surrounding the active site, therefore the stabilization takes place. Moreover, in case of the oxidative catalysis, the authors showed that the substrate molecules' proximity with histidine from 458 could facilitate the electron transfer towards the blue T1 copper site via the histidine, while the amino tail anchors the dopamine in the active site by forming hydrogen bonds with proline at 163 and phenylalanine from 162. The kinetic-docking correlation underlined that the structural region responsible for binding the substrate is the T1 copper site while the electron transfer is affected by the chemical structure of the substrates which can be considered as ligands on laccase active sites and consequently may play the role of competitive inhibitors for physiological substrates (Manzano-Nicolas et al., 2020).

The high substrate specificity gives laccases countless uses in a variety of fields such as wastewater treatment (Rostami et al., 2022; Villalba-Rodríguez et al., 2022), paper processing (Virk et al., 2012)], biosensing (Shokri et al., 2021), food (Backes et al., 2021) and biofuel processing (Leynaud Kieffer Curran et al., 2022). Although frequently used, the disadvantages derived from the production's high cost and low operational stability limit the use of laccase (Aricov et al., 2020). Strong solvents, pH, salts presence interacts with the laccase structure and interferes in the electron transfer. Furthermore, if not immobilized on a solid support the enzyme recovery and recycling are almost impossible. Therefore, an adequate enzyme immobilization procedure should generate considerable cost reduction by facilitating recycling and also help in overcoming operational limitations (Leontieş et al., 2022).

A parameter that also can influence the DET is the molecular orientation of the enzyme molecules on the immobilization support. Great efforts are made toward obtaining single-oriented enzymes during the immobilization of different materials (Gentil et al., 2020; Guan et al., 2015). A recent study showed that to the laccase molecule was appended a single pyrene group near the T1 copper for optimized control of the enzyme orientation for the immobilization onto nanostructured electrodes (Lalaoui et al., 2016). The electrodes consisting of carbon nanotubes or supramolecular assemblies formed by β-cyclodextrin and carbon nanotube/gold nanoparticle interacted differently with the graphed laccase either by π-stacking or by combined π-stacking and host-guest interaction respectively. The study showed that due to these synergistic interactions and the fact that most of the enzyme coating the first layer of the electrode has a single-oriented population, bio-electrocatalysis of dioxygen conversion is facilitated. Two main differences were found in terms of electron transfer in laccase: carbon nanotubes-electrodes presented heterogeneous DET rates, while the gold nanoparticles acted as electronic bridges, achieving a highly efficient reduction of dioxygen species at low overpotential (Lalaoui et al., 2016).

The fine-tuning of medium conditions such as temperature, pH, buffer type, ionic strength, and additives can positively influence the reaction performance in which immobilized laccase is involved (Dey & Dutta, 2022; El Ichi-Ribault et al., 2017; Lu et al., 2022). An interesting case where laccase DET was mediated by environmental conditions was in the bioremediation for estrogen elimination (Sun, Hong, et al., 2021). Therefore, the *T. versicolor* laccase was employed for estradiol and ethynylestradiol removal in buffered media. The study proved that the low or high pH values strongly affected the velocity constants either by blocking the electron transfer between substrates and the T1- blue copper or by interfering in the intramolecular-electron transfer between T1 - Cu and the trinuclear site. The kinetics and oxidative coupling of estradiol and ethynylestradiol in the presence of *T. versicolor* laccase reached the maximum when the pH was 5, underlining again the optimal pH should be thoroughly investigated when high conversions of substrates are targeted.

A modern way to modulate the electronic transfer is the exposition of enzymes to a magnetic field (Emamdadi et al., 2021). This procedure finds industrial applications in the pharmaceutical industry, materials production, and even the degradation of emerging contaminants (Wang et al., 2022). A particular type of magnetic field is the rotating one. Experimental data suggest that the interaction of such fields with the protein structure may generate changes and implicitly affect the geometry of copper sites in laccase (Wasak et al., 2019). The activity of laccase increased by about 10% when the magnetic rotating field applied was between 10 – 50 Hz suggesting that the DET process was enhanced.

Nowadays, more and more attention is being paid to the environment and sustainable green biofuels, so direct electron transfer can be used for this purpose (Martínez et al., 2017). For example, DET can be applied in the case of immobilized laccase and used as a biosensor in bio-electrocatalysis Mazurenko et al., 2020; Olloqui-Sariego et al., 2021; Yan et al., 2022). Li *et al.* propose the immobilization of the

Ganoderma lucidum laccase in a single-enzyme fuel cell system (Li et al., 2020). The study shows that using intramolecular DET and single orientation species, laccase is able to bio-convert the chemical energy biofuels into electrical energy. Another type of biofuel cell was formed with glucose oxidase as an anode and laccase as a cathode decorated with multi-walled carbon nanotubes with a substrate electrode having gold nanoparticles modified bacterial cellulose (Lv et al., 2018). The inertness, electrochemical stability, and small dimension allow the carbon nanotubes to act as nanowires placed in the enzyme intimacy and so the DET from the active copper towards the electrode is facilitated. The mixed sensor exhibited a broad linear dynamic range and was able to detect micromolar concentrations of glucose in biological media, proving to be a promising flexible electrode in the field of self-powered biosensors. The immobilization of laccase in metal-organic frameworks may improve laccase stability and activity by facilitating the electron transfer between the T1Cu and the trinuclear center as shown by (Jiang et al., 2022). The study shows that the immobilized enzyme has an increased degradation rate for azo dyes and could be a promising green treatment of polluted water from the textile industry.

Expanding the laccase activity over a wide range of substrates may be done by using redox mediators. In this case, the oxidation mechanism takes place by MET. At the base of MET is a complex process in which the oxidation of the mediator by the laccase is followed by the generation of stable and highly reactive radical species. In the next stage, the radicals are released from the active site by diffusion and transformed into electron shuttles between the enzymes and substrates. Therefore, laccase can be viewed as a promising catalyst with broad substrate specificity. Depending on the catalysis mechanism of action and optimal reaction conditions, the mediated transfer happens via hydrogen atom transfer, ionic oxidation or electron transfer, (Zucca et al., 2016) and in this section the discussion will be based on the latter.

When substrates have low-affinity constants and high redox potentials relative to the enzyme, or are too big and suffer from steric hindrances, mediators may be used in enzymatic catalysis (Hickey et al., 2016; Zucca et al., 2016). Ferricyanide, azine dyes, ferrocene compounds, chinones, toluidine blue O, ABTS, p-hydroxybenzyl alcohol, p-hydroxybenzoic acid, vanillin, acetovanillone, syringaldehyde, acetosyringone, syringic acid, p-coumaric, ferulic, TEMPO (and analogues) and sinapic acids are among the used mediators in laccase catalyzed reactions.

Natural substrates such as lignin, humus polymers, naturally occurring phenols, bio-pigments, and so on can be oxidized in systems containing laccase and mediators (Majumdar & Bhowal, 2022; Nugroho Prasetyo et al., 2010; Sun, Li, et al., 2021). Oxidation of the non-phenolic group of lignin, accompanied by bond cleavage or coupling reactions, might occur depending on the mediator utilized and the reaction conditions. Laccases from different sources are reported to polymerize lignosulphones (Euring et al., 2016; Nugroho Prasetyo et al., 2010) or lignin in presence of hydroxybenzotriazole, caffeic acid, vanillyl alcohol, p-coumaric, ferulic and sinapic acid TEMPO or acetosyringone (Sun, Li, et al., 2021). In all cases, the lignin-laccase-mediator-system was viewed as a natural bonding agent which may be of great value for industrial applications involving the treatment of waste lignin under alkaline conditions for obtaining biopolymeric materials. Melanin is a natural class of pigments produced by melanocytes in a process called melanogenesis. Human skin color is owed to its and the term “hyperpigmentation” refers to darker regions of skin caused by an excessive amount of melanin. Different research groups proposed strategies for the depigmentation of melanin using certain enzymes (Baik et al., 2021; Mohorčič et al., 2007). For example, melanin may be decolorized by O_2/laccase/natural mediator systems (Park et al., 2021). The study emphases that the mediated system containing acetosyringone and syringaldehyde had the best decolorization rates and yields as compared with the one lacking the mediator. The enhanced decolorization activity of the systems was owed to the electron-donating groups methoxy and ketone groups of acetosyringone which acted as redox shuttles from the active site of the laccase to the bulk.

Synthetic dyes are a class of compounds that are widely used in many industries. Classic treatment plants are not able to fully degrade them and considerable amounts escape in aqueous media generating hazards for ecosystems (Bhardwaj et al., 2022). For example, malachite green is an aromatic dye that is found undegraded in industrial wastewaters (Zofair et al., 2022), and has a chemical structure that is difficult to breakdown enzymatically, therefore the use of mediators is required (Parra Guardado et al., 2019). In real practice, the ABTS-mediated breakdown of malachite green in the presence of laccase results in the formation of low-mass aromatic molecules, which may further chemically degrade to produce nonhazardous wastewater. (Zofair et al., 2022; Duca et al, 2021' Vaseashta et al, 2022; Vaseashta eta al, 2021). Congo red is another dye that is used in the textile industry and is frequently identified in wastewater run-offs and poses environmental concerns (Oyekanmi et al., 2021). Congo red has an azoic structure and is resistant to physicochemical treatments used in wastewater treatment facilities; nevertheless, decolorization is feasible in the presence of laccase and ABTS, as demonstrated by (Saha & Mukhopadhyay, 2020). The use of co-immobilized laccase and ABTS on geopolymer microspheres was another successful technique for degrading Congo red. According to scientists, the immobilized products' degradation efficiency is generated by the facilitated electron transfer and shorter mass transfer distance between the enzyme and substrate (Shan et al., 2022). Another example of a dye that was successfully oxidized with the aid of mediated electron transfer is the indigo carmine dye (Leontieş et al., 2022). The authors found that the competition for the substrate may develop when ABTS is utilized as an electron transfer mediator in IC discoloration but only after the dye concentration falls below that of the mediator. In the first phase of the reaction, the conversion of ABTS to an intermediate with a high redox potential occurs while in the second substrate (indigo carmine) is oxidized via the electron transfer. The ABTS loses an electron in the first stage of the reaction, forming an intermediate with a high redox potential which is ejected from the active site. As a result, the second substrate, IC, takes electrons and populates the active site. As the IC is oxidized, and the degradation products leave the active site, the mediator recovers and is reduced to its original state, thus becoming the dominant species in the reaction (Cardullo et al., 2022).

Not only dyes are degradable in mediated systems, but also phenols, polycyclic aromatic hydrocarbons, humic substances, etc. (Bhardwaj et al., 2022). One study reports the laccase catalyzed oxidation of 2,6-Dimethoxyphenol in mediated and non-mediated systems (Mattos et al., 2022). The complex electrochemical analysis of laccase-based reactions involving TEMPO (2,2,6,6-tetramethylpiperidin-1-oxyl radical), ABTS, and 2,6-Dimethoxyphenol revealed that the enhanced enzyme activity is not correlated with the high potential differences between laccase and the substrate. Systems containing laccase and TEMPO are employed to selectively oxidize the primary hydroxyl group from oligosaccharides and bioactive glycosides (Pei et al., 2016). Increasing the pH of the medium containing laccase and TEMPO favored electron-mediated transfer and helped the selective oxidation of the chitooligomer by inhibiting amino group protonation and increasing electron donating ability (Pei et al., 2016).

Non-steroidal anti-inflammatory medications are frequently detected as contaminants in aquatic environments, and the present removal methods have a variety of limitations. The biodegradation of pharmaceuticals may be considerably enhanced by laccase-catalyzed treatment, and it may produce degradation products that are less hazardous. Laccase can efficiently degrade diclofenac, naproxen or ketoprofen in the presence of ABTS (Apriceno et al., 2019). Also, the degradation of naproxen and diclofenac can be improved by the presence of HBT or syringaldazine (Lloret et al., 2013).

CONCLUDING REMARKS

The spotlight enzymes of the present chapter, catalase, peroxidase, and laccase, are discussed in terms of how electron transfer occurs in oxidoreductase enzymes. Theoretical features of the mechanism involving the three enzymes' ability to transfer electrons were briefly illustrated, and the practical uses of the chosen enzymes in the field of environmental remediation were discussed. Large efforts were made by researchers to obtain the necessary selectivity, catalytic efficiency, and stability for industrial usage. Research in this area is still needed because it will help us to understand better the processes of O_2 reduction to H_2O, hydrogen peroxide decomposition in the active site, or hydrogen atom transfer are influenced by changes in electron and/or proton transfer rates. A challenge is also the extension of new mediator classes with efficient, biodegradable, and recyclable features, which would increase even more the enzymatic activity against stubborn substrates. If production costs were reduced, the oxidoreductases could be applied in even more industrial processes. One strategy would be to design new immobilization substrates that improve both recycling and the electron transfer of immobilized enzymes.

Conflicts of Interest

The authors declare no conflict of interest.

ACKNOWLEDGMENT

This work was supported by a grant of the Ministry of Research, Innovation and Digitization, CNCS - UEFISCDI, project number TE 68/2022, PN-III-P1-1.1-TE-2021-0418, within PNCDI III. Also, we acknowledge the Romanian Academy within the research program '*Colloids and dispersed systems*' from '*Ilie Murgulescu*' Institute of Physical Chemistry

REFERENCES

Achuba, F. I., & Okoh, P. N. (2014). Effect of petroleum products on soil catalase and dehydrogenase activities. *Open Journal of Social Sciences*, *04*, 399–406.

Agrawal, K., Chaturvedi, V., & Verma, P. (2018). Fungal laccase discovered but yet undiscovered. *Bioresources and Bioprocessing*, *5*(1), 4. doi:10.118640643-018-0190-z

Akram, F., Ashraf, S., Haq, I. ul, Shah, F. I., & Aqeel, A. (2022). eminent industrial and biotechnological applications of laccases from bacterial source: A Current Overview. *Applied Biochemistry and Biotechnology, 194*, 2336-2356.

Alfonso-Prieto, M., Biarnés, X., Vidossich, P., & Rovira, C. (2009). The molecular mechanism of the catalase reaction. *Journal of the American Chemical Society, 131*(33), 11751–11761. doi:10.1021/ja9018572 PMID:19653683

Alfonso-Prieto, M., Vidossich, P., & Rovira, C. (2012). The reaction mechanisms of heme catalases: An atomistic view by ab initio molecular dynamics. *Archives of Biochemistry and Biophysics, 525*(2), 121–130. doi:10.1016/j.abb.2012.04.004 PMID:22516655

Apriceno, A., Astolfi, M. L., Girelli, A. M., & Scuto, F. R. (2019). A new laccase-mediator system facing the biodegradation challenge: Insight into the NSAIDs removal. *Chemosphere, 215*, 535–542. doi:10.1016/j.chemosphere.2018.10.086 PMID:30340161

Aricov, L., Leonties, A. R., Gîfu, I. C., Preda, D., Raducan, A., & Anghel, D. F. (2020). Enhancement of laccase immobilization onto wet chitosan microspheres using an iterative protocol and its potential to remove micropollutants. *Journal of Environmental Management, 276*, 111326. doi:10.1016/j.jenvman.2020.111326 PMID:32891981

Arregui, L., Ayala, M., Gómez-Gil, X., Gutiérrez-Soto, G., Hernández-Luna, C. E., Herrera De Los Santos, M., Levin, L., Rojo-Domínguez, A., Romero-Martínez, D., Saparrat, M. C. N., Trujillo-Roldán, M. A., & Valdez-Cruz, N. A. (2019). Laccases: Structure, function, and potential application in water bioremediation. *Microbial Cell Factories, 18*(1), 200–234. doi:10.118612934-019-1248-0 PMID:31727078

Backes, E., Kato, C. G., Corrêa, R. C. G., Peralta Muniz Moreira, R., Peralta, R. A., Barros, L., Ferreira, I. C. F. R., Zanin, G. M., Bracht, A., & Peralta, R. M. (2021). Laccases in food processing: Current status, bottlenecks and perspectives. *Trends in Food Science & Technology, 115*, 445–460. doi:10.1016/j.tifs.2021.06.052

Baik, J., Purkayastha, A., Park, K. H., & Kang, T. J. (2021). Functional characterization of melanin decolorizing extracellular peroxidase of bjerkandera adusta. *Journal of Fungi (Basel, Switzerland), 7*(9), 762. doi:10.3390/jof7090762 PMID:34575800

Besharati Vineh, M., Saboury, A. A., Poostchi, A. A., Rashidi, A. M., & Parivar, K. (2018). Stability and activity improvement of horseradish peroxidase by covalent immobilization on functionalized reduced graphene oxide and biodegradation of high phenol concentration. *International Journal of Biological Macromolecules, 106*, 1314–1322. doi:10.1016/j.ijbiomac.2017.08.133 PMID:28851646

Bhardwaj, P., Kaur, N., Selvaraj, M., Ghramh, H. A., Al-Shehri, B. M., Singh, G., Arya, S. K., Bhatt, K., Ghotekar, S., Mani, R., Chang, S. W., Ravindran, B., & Awasthi, M. K. (2022). Laccase-assisted degradation of emerging recalcitrant compounds - A review. *Bioresource Technology, 364*, 128031. doi:10.1016/j.biortech.2022.128031 PMID:36167178

Biko, O. D. V., Viljoen-Bloom, M., & van Zyl, W. H. (2020). Microbial lignin peroxidases: Applications, production challenges and future perspectives. *Enzyme and Microbial Technology, 141*, 109669. doi:10.1016/j.enzmictec.2020.109669 PMID:33051019

Butreddy, A., Janga, K. Y., Ajjarapu, S., Sarabu, S., & Dudhipala, N. (2021). Instability of therapeutic proteins — An overview of stresses, stabilization mechanisms and analytical techniques involved in lyophilized proteins. *International Journal of Biological Macromolecules, 167*, 309–325. doi:10.1016/j.ijbiomac.2020.11.188 PMID:33275971

Calza, P., Zacchigna, D., & Laurenti, E. (2016). Degradation of orange dyes and carbamazepine by soybean peroxidase immobilized on silica monoliths and titanium dioxide. *Environmental Science and Pollution Research International, 23*(23), 23742–23749. doi:10.100711356-016-7399-1 PMID:27623850

Campbell, E., Meredith, M., Banta, S., & Minteer, S. D. (2012). Enzymatic biofuel cells utilizing a biomimetic cofactor. *Chemical Communications, 48*(13), 1898–90. doi:10.1039/c2cc16156g PMID:22227738

Cantemir, A. R., Raducan, A., Puiu, M., & Oancea, D. (2013). Kinetics of thermal inactivation of catalase in the presence of additives. *Process Biochemistry (Barking, London, England), 48*(3), 471–477. doi:10.1016/j.procbio.2013.02.013

Cardullo, N., Muccilli, V., & Tringali, C. (2022). Laccase-mediated synthesis of bioactive natural products and their analogues. *RSC Chemical Biology, 3*(6), 614–647. doi:10.1039/D1CB00259G PMID:35755186

Chance, B. (1949). The composition of catalase-peroxide complexes. *The Journal of Biological Chemistry*, 179. PMID:18134593

Chance, B., Greenstein, D. S., & Roughton, F. J. W. (1952). The mechanism of catalase action. I. Steady-state analysis. *Archives of Biochemistry and Biophysics, 37*(2), 301–321. doi:10.1016/0003-9861(52)90194-X PMID:14953443

Chauhan, P. S., Goradia, B., & Saxena, A. (2017). Bacterial laccase: Recent update on production, properties and industrial applications. *Biotech, 7*, 323. PMID:28955620

Chelikani, P., Fita, I., & Loewen, P. C. (2004). Diversity of structures and properties among catalases. *Cellular and Molecular Life Sciences, 61*(2), 192–208. doi:10.100700018-003-3206-5 PMID:14745498

Chen, Z., Li, H., Peng, A., & Gao, Y. (2014). Oxidation of polycyclic aromatic hydrocarbons by horseradish peroxidase in water containing an organic cosolvent. *Environmental Science and Pollution Research International, 21*(18), 10696–10705. doi:10.100711356-014-3005-6 PMID:24894750

Chen, Z., & Wan, C. (2017). Biological valorization strategies for converting lignin into fuels and chemicals. *Renewable & Sustainable Energy Reviews, 73*, 610–621. doi:10.1016/j.rser.2017.01.166

Davies, M. J. (2021). Myeloperoxidase: Mechanisms, reactions and inhibition as a therapeutic strategy in inflammatory diseases. *Pharmacology & Therapeutics, 218*, 107685. doi:10.1016/j.pharmthera.2020.107685 PMID:32961264

de Oliveira, F. K., Santos, L. O., & Buffon, J. G. (2021). Mechanism of action, sources, and application of peroxidases. *Food Research International, 14*, 110266. doi:10.1016/j.foodres.2021.110266 PMID:33992367

Dey, B., & Dutta, T. (2022). Laccases: Thriving the domain of bio-electrocatalysis. *Bioelectrochemistry (Amsterdam, Netherlands), 146*, 108144. doi:10.1016/j.bioelechem.2022.108144 PMID:35550253

Díaz, A., Loewen, P. C., Fita, I., & Carpena, X. (2012). Thirty years of heme catalases structural biology. *Archives of Biochemistry and Biophysics, 252*(2), 102–110. doi:10.1016/j.abb.2011.12.011 PMID:22209752

Ding, Y., Cui, K., Liu, X., Xie, Q., Guo, Z., & Chen, Y. (2022). Lignin peroxidase-catalyzed direct oxidation of trace organic pollutants through a long-range electron transfer mechanism: Using propranolol as an example. *Journal of Hazardous Materials*, *431*, 128544. doi:10.1016/j.jhazmat.2022.128544 PMID:35228075

Divsalar, A., & Ghobadi, R. (2021). The presence of deep eutectic solvents of reline and glyceline on interaction and side effect of anti-cancer drug of 5-fluorouracil: Bovine liver catalase as a target. *Journal of Molecular Liquids*, *323*, 114588. doi:10.1016/j.molliq.2020.114588

Dong, J., Yu, Y., Pei, Y., & Pei, Z. (2022). pH-responsive aminotriazole doped metal organic frameworks nanoplatform enables self-boosting reactive oxygen species generation through regulating the activity of catalase for targeted chemo/chemodynamic combination therapy. *Journal of Colloid and Interface Science*, *607*, 1651–1660. doi:10.1016/j.jcis.2021.09.043 PMID:34592552

Duca, G., & Vaseashta, A. (Eds.). (2020). *Handbook of Research on Emerging Developments and Environmental Impacts of Ecological Chemistry*. IGI Global. doi:10.4018/978-1-7998-1241-8

Duong, N. B., Truong, V. M., Li, Y. S., Wang, C. L., & Yang, H. (2020). improving the immobilization of glucose oxidase on carbon cloth via a hybrid approach of cross-linked chitosan/TPP matrices with Na polymers for high-performance self-pumping enzyme-based biofuel cells. *Energy & Fuels*, *34*(8), 10050–10058. doi:10.1021/acs.energyfuels.0c01033

Dutta, S., Patil, R., & Dey, T. (2022). Electron transfer-driven single and multi-enzyme biofuel cells for self-powering and energy bioscience. *Nano Energy*, *96*, 107074. doi:10.1016/j.nanoen.2022.107074

el Ichi-Ribault, S., Zebda, A., Tingry, S., Petit, M., Suherman, A. L., Boualam, A., Cinquin, P., & Martin, D. K. (2017). Performance and stability of chitosan-MWCNTs-laccase biocathode: Effect of MWCNTs surface charges and ionic strength. *Journal of Electroanalytical Chemistry (Lausanne, Switzerland)*, *799*, 26–33. doi:10.1016/j.jelechem.2017.05.018

Elliott, S. J., McElhaney, A. E., Feng, C., Enemark, J. H., & Armstrong, F. A. (2002). A voltammetric study of interdomain electron transfer within sulfite oxidase. *Journal of the American Chemical Society*, *124*(39), 11612–11613. doi:10.1021/ja027776f PMID:12296723

Emamdadi, N., Gholizadeh, M., & Housaindokht, M. R. (2021). Investigation of static magnetic field effect on horseradish peroxidase enzyme activity and stability in enzymatic oxidation process. *International Journal of Biological Macromolecules*, *170*, 189–195. doi:10.1016/j.ijbiomac.2020.12.034 PMID:33333096

Euring, M., Kirsch, A., Schneider, P., & Kharazipour, A. (2016). Lignin-laccase-mediator-systems (LLMS) for the production of binderless medium density fiberboards (MDF). *Journal of Materials Science Research*, *5*(2), 7. doi:10.5539/jmsr.v5n2p7

Fapyane, D., & Ferapontova, E. E. (2016). Enhanced electron transfer between gold nanoparticles and horseradish peroxidase reconstituted onto alkanethiol-modified hemin. *Electrochemistry Communications*, *70*, 39–42. doi:10.1016/j.elecom.2016.06.016

Farver, O., & Pecht, I. (2011). Electron transfer in blue copper proteins. *Coordination Chemistry Reviews*, *255*(7-8), 757–773. doi:10.1016/j.ccr.2010.08.005

Fedalto, L., de Oliveira, P. R., Agustini, D., Kalinke, C., Banks, C. E., Bergamini, M. F., & Marcolino-Junior, L. H. (2023). Novel and highly stable strategy for the development of microfluidic enzymatic assays based on the immobilization of horseradish peroxidase (HRP) into cotton threads. *Talanta, 252*, 123889. doi:10.1016/j.talanta.2022.123889 PMID:36070669

Fernandes, M., Souza, D. H., Henriques, R. O., Alves, M. V., Skoronski, E., & Junior, A. F. (2020). Obtaining soybean peroxidase from soybean hulls and its application for detoxification of 2,4-dichlorophenol contaminated water. *Journal of Environmental Chemical Engineering*, *8*(3), 103786. doi:10.1016/j.jece.2020.103786

Fita, I., & Rossmann, M. G. (1985). The active center of catalase. *Journal of Molecular Biology*, *185*(1), 21–37. doi:10.1016/0022-2836(85)90180-9 PMID:4046038

Fita, I., Silva, A. M., Murthy, M. R. N., & Rossmann, M. G. (1986). The refined structure of beef liver catalase at 2·5 Å resolution. *Acta Crystallographica. Section B, Structural Science*, *42*(5), 497–515. doi:10.1107/S0108768186097835

Flohé, L., Toppo, S., & Orian, L. (2022). The glutathione peroxidase family: Discoveries and mechanism. *Free Radical Biology & Medicine*, *187*, 113–122. doi:10.1016/j.freeradbiomed.2022.05.003 PMID:35580774

Gan, J. S., Ashraf, S. S., Bilal, M., & Iqbal, H. M. N. (2022). Biodegradation of environmental pollutants using catalase-based biocatalytic systems. *Environmental Research*, *214*, 113914. doi:10.1016/j.envres.2022.113914 PMID:35932834

Gentil, S., Rousselot-Pailley, P., Sancho, F., Robert, V., Mekmouche, Y., Guallar, V., Tron, T., & le Goff, A. (2020). Efficiency of site-specific clicked laccase-carbon nanotubes biocathodes towards O_2 reduction. *Chemistry (Weinheim an der Bergstrasse, Germany)*, *26*(21), 4798–4804. doi:10.1002/chem.201905234 PMID:31999372

Goc, G., Balci, S., Yorke, B. A., Pearson, A. R., & Yuzugullu Karakus, Y. (2021). Probing the role of Val228 on the catalytic activity of Scytalidium catalase. *Biochimica et Biophysica Acta. Proteins and Proteomics*, *1869*(8), 140662. doi:10.1016/j.bbapap.2021.140662 PMID:33887466

Gray, M. R. (1989). Substrate inactivation of enzymes in vitro and in vivo. *Biotechnology Advances*, *7*(4), 527–575. doi:10.1016/0734-9750(89)90722-2 PMID:14542989

Grigoras, A. G. (2017). Catalase immobilization—A review. *Biochemical Engineering Journal*, *117*, 1–20. doi:10.1016/j.bej.2016.10.021

Guan, D., Kurra, Y., Liu, W., & Chen, Z. (2015). A click chemistry approach to site-specific immobilization of a small laccase enables efficient direct electron transfer in a biocathode. *Chemical Communications*, *51*(13), 2522–2525. doi:10.1039/C4CC09179E PMID:25566975

Hara, I., Ichise, N., Kojima, K., Kondo, H., Ohgiya, S., Matsuyama, H., & Yumoto, I. (2007). Relationship between the size of the bottleneck 15 Å from iron in the main channel and the reactivity of catalase corresponding to the molecular size of substrates. *Biochemistry*, *46*(1), 11–22. doi:10.1021/bi061519w PMID:17198371

Herkendell, K., Stemmer, A., & Tel-Vered, R. (2019). Magnetically induced enzymatic cascades-advancing towards multi-fuel direct/mediated bioelectrocatalysis. *Nanoscale Advances*, *1*(5), 1686–1692. doi:10.1039/C8NA00346G PMID:36134209

Hickey, D. P., Milton, R. D., Rasmussen, M., Abdellaoui, S., Nguyen, K., & Minteer, S. D. (2016). Fundamentals and applications of bioelectrocatalysis. *Electrochemistry*, *13*, 97–132. doi:10.1039/9781782620273-00097

Hofrichter, M., Ullrich, R., Pecyna, M. J., Liers, C., & Lundell, T. (2010). New and classic families of secreted fungal heme peroxidases. *Applied Microbiology and Biotechnology, 87.*

Humer, D., Furlanetto, V., Schruef, A. K., Wlodarczyk, A., Kuttke, M., Divne, C., & Spadiut, O. (2021). Potential of unglycosylated horseradish peroxidase variants for enzyme prodrug cancer therapy. *Biomedicine and Pharmacotherapy*, *142*, 112037.

Izadyar, A., Van, M. N., Miranda, M., Weatherford, S., & Hood, E. E. (2022). Electrocatalytic effect of recombinant Mn peroxidase from corn on microbiosensors to detect glucose. *Biocatalysis and Agricultural Biotechnology*, *43*, 102445. doi:10.1016/j.bcab.2022.102445

Janusz, G., Pawlik, A., Sulej, J., Świderska-Burek, U., Jarosz-Wilkolazka, A., & Paszczyński, A. (2017). Lignin degradation: Microorganisms, enzymes involved, genomes analysis and evolution. *FEMS Microbiology Reviews*, *41*(6), 941–962. doi:10.1093/femsre/fux049 PMID:29088355

Janusz, G., Pawlik, A., Świderska-Burek, U., Polak, J., Sulej, J., Jarosz-Wilkołazka, A., & Paszczyński, A. (2020). Laccase properties, physiological functions, and evolution. *International Journal of Molecular Sciences*, *21*(3), 966. doi:10.3390/ijms21030966 PMID:32024019

Jeon, J. R., Baldrian, P., Murugesan, K., & Chang, Y. S. (2012). Laccase-catalysed oxidations of naturally occurring phenols: From in vivo biosynthetic pathways to green synthetic applications. *Microbial Biotechnology*, *5*(3), 318–332. doi:10.1111/j.1751-7915.2011.00273.x PMID:21791030

Jiang, S., Ren, D., Wang, Z., Zhang, S., Zhang, X., & Chen, W. (2022). Improved stability and promoted activity of laccase by One-Pot encapsulation with Cu (PABA) nanoarchitectonics and its application for removal of Azo dyes. *Ecotoxicology and Environmental Safety*, *234*, 113366. doi:10.1016/j.ecoenv.2022.113366 PMID:35272195

Joergensen, R. G., Schmaedeke, F., Windhorst, K., & Meyer, B. (1995). Biomass and activity of microorganisms in a fuel oil contaminated soil. *Soil Biology & Biochemistry*, *27*(9), 1137–1143. doi:10.1016/0038-0717(95)00040-L

Jones, S. M., & Solomon, E. I. (2015). Electron transfer and reaction mechanism of laccases. *Cellular and Molecular Life Sciences*, *72*(5), 869–883. doi:10.100700018-014-1826-6 PMID:25572295

Kaushal, J., Mehandia, S., Singh, G., Raina, A., & Arya, S. K. (2018). Catalase enzyme: Application in bioremediation and food industry. *Biocatalysis and Agricultural Biotechnology*, *16*, 192–199. doi:10.1016/j.bcab.2018.07.035

Khataee, S., Dehghan, G., Yekta, R., Rashtbari, S., Maleki, S., & Khataee, A. (2022). The protective effect of natural phenolic compound on the functional and structural responses of inhibited catalase by a common azo food dye. *Food and Chemical Toxicology*, *160*, 112801. doi:10.1016/j.fct.2021.112801 PMID:34974130

Khorshidi, S., Younesi, S., & Karkhaneh, A. (2022). Peroxide mediated oxygen delivery in cancer therapy. *Colloids and Surfaces. B, Biointerfaces*, *219*, 112832. doi:10.1016/j.colsurfb.2022.112832 PMID:36137337

Kirkman, H. N., & Gaetani, G. F. (1984). Catalase: A tetrameric enzyme with four tightly bound molecules of NADPH. *Proceedings of the National Academy of Sciences of the United States of America*, *81*(14), 4343–4347. doi:10.1073/pnas.81.14.4343 PMID:6589599

Kumar, S., Dangi, A. K., Shukla, P., Baishya, D., & Khare, S. K. (2019). Thermozymes: Adaptive strategies and tools for their biotechnological applications. *Bioresource Technology*, *278*, 372–382. doi:10.1016/j.biortech.2019.01.088 PMID:30709766

Lalaoui, N., Rousselot-Pailley, P., Robert, V., Mekmouche, Y., Villalonga, R., Holzinger, M., Cosnier, S., Tron, T., & le Goff, A. (2016). Direct electron transfer between a site-specific pyrene-modified laccase and carbon nanotube/gold nanoparticle supramolecular assemblies for bioelectrocatalytic dioxygen reduction. *ACS Catalysis*, *6*(3), 1894–1900. doi:10.1021/acscatal.5b02442

Lardinois, O. M., Mestdagh, M. M., & Rouxhet, P. G. (1996). Reversible inhibition and irreversible inactivation of catalase in presence of hydrogen peroxide. *Biochimica et Biophysica Acta. Protein Structure and Molecular Enzymology*, *1295*(2), 222–238. doi:10.1016/0167-4838(96)00043-X PMID:8695649

Lardinois, O. M., & Rouxhet, P. G. (1996). Peroxidatic degradation of azide by catalase and irreversible enzyme inactivation. *Biochimica et Biophysica Acta. Protein Structure and Molecular Enzymology*, *1298*(2), 180–190. doi:10.1016/S0167-4838(96)00130-6 PMID:8980644

Leontieş, A. R., Răducan, A., Cristina Culiţă, D., Alexandrescu, E., Moroşan, A., Eduard Mihaiescu, D., & Aricov, L. (2022). Laccase immobilized on chitosan-polyacrylic acid microspheres as highly efficient biocatalyst for naphthol green B and indigo carmine degradation. *Chemical Engineering Journal*, *439*, 135654. doi:10.1016/j.cej.2022.135654

Leynaud Kieffer Curran, L. M. C., Pham, L. T. M., Sale, K. L., & Simmons, B. A. (2022). Review of advances in the development of laccases for the valorization of lignin to enable the production of lignocellulosic biofuels and bioproducts. *Biotechnology Advances*, *54*, 107809. doi:10.1016/j.biotechadv.2021.107809 PMID:34333091

Li, X., Lv, P., Yao, Y., Feng, Q., Mensah, A., Li, D., & Wei, Q. (2020). A novel single-enzymatic biofuel cell based on highly flexible conductive bacterial cellulose electrode utilizing pollutants as fuel. *Chemical Engineering Journal*, *379*, 122316. doi:10.1016/j.cej.2019.122316

Liu, H., Cao, X., Wang, P., & Ma, X. (2017). pH-sensitive pHluorins as a molecular sensor for in situ monitoring of enzyme-catalyzed prodrug activation. *Biotechnology and Applied Biochemistry*, *64*(4), 482–489. doi:10.1002/bab.1511 PMID:27222443

Liu, S., Bilal, M., Rizwan, K., Gul, I., Rasheed, T., & Iqbal, H. M. N. (2021). Smart chemistry of enzyme immobilization using various support matrices - A review. *International Journal of Biological Macromolecules, 190*, 396–408. doi:10.1016/j.ijbiomac.2021.09.006 PMID:34506857

Lloret, L., Eibes, G., Moreira, M. T., Feijoo, G., & Lema, J. M. (2013). On the use of a high-redox potential laccase as an alternative for the transformation of non-steroidal anti-inflammatory drugs (NSAIDs). *Journal of Molecular Catalysis. B, Enzymatic, 97*, 233–242. doi:10.1016/j.molcatb.2013.08.021

Lu, J., Nie, M., Li, Y., Zhu, H., & Shi, G. (2022). Design of composite nanosupports and applications thereof in enzyme immobilization: A review. *Colloids and Surfaces. B, Biointerfaces, 217*, 112602. doi:10.1016/j.colsurfb.2022.112602 PMID:35660743

Lv, P., Zhou, H., Mensah, A., Feng, Q., Wang, D., Hu, X., Cai, Y., Amerigo Lucia, L., Li, D., & Wei, Q. (2018). A highly flexible self-powered biosensor for glucose detection by epitaxial deposition of gold nanoparticles on conductive bacterial cellulose. *Chemical Engineering Journal, 351*, 177–188. doi:10.1016/j.cej.2018.06.098

Majumdar, S., & Bhowal, J. (2022). Studies on production and evaluation of biopigment and synthetic dye decolorization capacity of laccase produced by A. oryzae cultivated on agro-waste. *Bioprocess and Biosystems Engineering, 45*(1), 45–60. doi:10.100700449-021-02638-z PMID:34591164

Manzano-Nicolas, J., Taboada-Rodriguez, A., Teruel-Puche, J. A., Marin-Iniesta, F., Garcia-Molina, F., Garcia-Canovas, F., Tudela-Serrano, J., & Muñoz-Muñoz, J. L. (2020). Kinetic characterization of the oxidation of catecolamines and related compounds by laccase. *International Journal of Biological Macromolecules, 164*, 1256–1266. doi:10.1016/j.ijbiomac.2020.07.112 PMID:32721460

Martínez, A. T., Ruiz-Dueñas, F. J., Camarero, S., Serrano, A., Linde, D., Lund, H., Vind, J., Tovborg, M., Herold-Majumdar, O. M., Hofrichter, M., Liers, C., Ullrich, R., Scheibner, K., Sannia, G., Piscitelli, A., Pezzella, C., Sener, M. E., Kılıç, S., van Berkel, W. J. H., & Alcalde, M. (2017). Oxidoreductases on their way to industrial biotransformations. *Biotechnology Advances, 35*(6), 815–831. doi:10.1016/j.biotechadv.2017.06.003 PMID:28624475

Mattos, G. J., Schirmann, J. G., Salamanca-Neto, C. A. R., Dekker, R. F. H., Barbosa-Dekker, A. M., & Sartori, E. R. (2022). Electrochemical characterization of the laccase-catalyzed oxidation of 2,6-dimethoxyphenol: An insight into the direct electron transfer by enzyme and enzyme-mediator system. *Applied Biochemistry and Biotechnology, 194*(10), 4348–4361. doi:10.100712010-022-03838-3 PMID:35146637

Mazurenko, I., Hitaishi, V. P., & Lojou, E. (2020). Recent advances in surface chemistry of electrodes to promote direct enzymatic bioelectrocatalysis. *Current Opinion in Electrochemistry, 19*, 113–121. doi:10.1016/j.coelec.2019.11.004

Mehra, R., Muschiol, J., Meyer, A. S., & Kepp, K. P. (2018). A structural-chemical explanation of fungal laccase activity. *Scientific Reports, 8*(1), 17285. doi:10.103841598-018-35633-8 PMID:30470810

Mohorčič, M., Friedrich, J., Renimel, I., André, P., Mandin, D., & Chaumont, J. P. (2007). Production of melanin bleaching enzyme of fungal origin and its application in cosmetics. *Biotechnology and Bioprocess Engineering; BBE, 12*(3), 200–206. doi:10.1007/BF02931093

Moya, R., Saastamoinen, P., Hernández, M., Suurnäkki, A., Arias, E., & Mattinen, M. L. (2011). Reactivity of bacterial and fungal laccases with lignin under alkaline conditions. *Bioresource Technology*, *102*(21), 10006–10012. doi:10.1016/j.biortech.2011.08.046 PMID:21908186

Mukherjee, G., Nandekar, P. P., & Wade, R. C. (2021). An electron transfer competent structural ensemble of membrane-bound cytochrome P450 1A1 and cytochrome P450 oxidoreductase. *Communications Biology*, *4*(1), 55. doi:10.103842003-020-01568-y PMID:33420418

Nasar, A., & Perveen, R. (2019). Applications of enzymatic biofuel cells in bioelectronic devices - A review. *International Journal of Hydrogen Energy*, *44*(29), 15287–15312. doi:10.1016/j.ijhydene.2019.04.182

Nicholls, P. (2012). Classical catalase: Ancient and modern. *Archives of Biochemistry and Biophysics*, *525*(2), 95–101. doi:10.1016/j.abb.2012.01.015 PMID:22326823

Nugroho Prasetyo, E., Kudanga, T., Østergaard, L., Rencoret, J., Gutiérrez, A., del Río, J. C., Ignacio Santos, J., Nieto, L., Jiménez-Barbero, J., Martínez, A. T., Li, J., Gellerstedt, G., Lepifre, S., Silva, C., Kim, S. Y., Cavaco-Paulo, A., Seljebakken Klausen, B., Lutnaes, B. F., Nyanhongo, G. S., & Guebitz, G. M. (2010). Polymerization of lignosulfonates by the laccase-HBT (1-hydroxybenzotriazole) system improves dispersibility. *Bioresource Technology*, *101*(14), 5054–5062. doi:10.1016/j.biortech.2010.01.048 PMID:20176477

Oancea, D., Stuparu, A., Nita, M., Puiu, M., & Raducan, A. (2008). Estimation of the overall kinetic parameters of enzyme inactivation using an isoconversional method. *Biophysical Chemistry*, *138*(1-2), 50–54. doi:10.1016/j.bpc.2008.09.003 PMID:18814949

Okawa, Y., Yokoyama, N., Sakai, Y., & Shiba, F. (2015). Direct electron transfer biosensor for hydrogen peroxide carrying nano-complex composed of horseradish peroxidase and Au-nanoparticle - Characterization and application to bienzyme systems. *Analytical Chemistry Research*, *5*, 1–8. doi:10.1016/j.ancr.2015.05.001

Olloqui-Sariego, J. L., Calvente, J. J., & Andreu, R. (2021). Immobilizing redox enzymes at mesoporous and nanostructured electrodes. *Current Opinion in Electrochemistry, 26*. doi:10.3390u132313264

Oyekanmi, A.A., Ahmad, A., Mohd Setapar, S.H., Alshammari, M.B., Jawaid, M., Hanafiah, M.M., Abdul Khalil, H.P.S., & Vaseashta, A. (2021). Sustainable Durio zibethinus-Derived Biosorbents for Congo Red Removal from Aqueous Solution: Statistical Optimization, Isotherms and Mechanism Studies. *Sustainability (Basel)*, *2021*(13), 13264. doi:10.3390u132313264

Park, S., Jung, D., Do, H., Yun, J., Lee, D., Hwang, S., & Lee, S. H. (2021). Laccase-mediator system using a natural mediator as a whitening agent for the decolorization of melanin. *Polymers*, *13*(21), 100658. doi:10.3390/polym13213671 PMID:34771228

Parra Guardado, A. L., Belleville, M. P., Rostro Alanis, M., Parra Saldivar, R., & Sanchez-Marcano, J. (2019). Effect of redox mediators in pharmaceuticals degradation by laccase: A comparative study. *Process Biochemistry (Barking, London, England)*, *78*, 123–131. doi:10.1016/j.procbio.2018.12.032

Patel, A. K., Singhania, R. R., & Pandey, A. (2016). Novel enzymatic processes applied to the food industry. *Current Opinion in Food Science*, *7*, 64–72. doi:10.1016/j.cofs.2015.12.002

Pei, J., Yin, Y., Shen, Z., Bu, X., & Zhang, F. (2016). Oxidation of primary hydroxyl groups in chitooligomer by a laccase-TEMPO system and physico-chemical characterisation of oxidation products. *Carbohydrate Polymers*, *135*, 234–238. doi:10.1016/j.carbpol.2015.08.099 PMID:26453873

Pérez-Boada, M., Ruiz-Dueñas, F. J., Pogni, R., Basosi, R., Choinowski, T., Martínez, M. J., Piontek, K., & Martínez, A. T. (2005). Versatile peroxidase oxidation of high redox potential aromatic compounds: Site-directed mutagenesis, spectroscopic and crystallographic investigation of three long-range electron transfer pathways. *Journal of Molecular Biology*, *354*(2), 385–402. doi:10.1016/j.jmb.2005.09.047 PMID:16246366

Pöller, S., Koster, D., & Schuhmann, W. (2013). Stabilizing redox polymer films by electrochemically induced crosslinking. *Electrochemistry Communications*, *34*, 327–330. doi:10.1016/j.elecom.2013.07.033

Poulos, T. L. (2014). Heme enzyme structure and function. *Chemical Reviews*, *114*(7), 3919–3962. doi:10.1021/cr400415k PMID:24400737

Putnam, C. D., Arvai, A. S., Bourne, Y., & Tainer, J. A. (2000). Active and inhibited human catalase structures: Ligand and NADPH binding and catalytic mechanism. *Journal of Molecular Biology*, *296*(1), 295–309. doi:10.1006/jmbi.1999.3458 PMID:10656833

Răducan, A., Oancea, D., & Segal, E. (2007). Models for irreversible inactivation in the enzymatic decomposition of hydrogen peroxide and their kinetic treatment by means of graphs. *Revue Roumaine de Chimie*, *52*, 1280–1284.

Rostami, A., Abdelrasoul, A., Shokri, Z., & Shirvandi, Z. (2022). Applications and mechanisms of free and immobilized laccase in detoxification of phenolic compounds - A review. *Korean Journal of Chemical Engineering*, *39*(4), 821–832. doi:10.100711814-021-0984-0

Saha, R., & Mukhopadhyay, M. (2020). Elucidation of the decolorization of Congo Red by Trametes versicolor laccase in presence of ABTS through cyclic voltammetry. *Enzyme and Microbial Technology*, *135*, 109507. doi:10.1016/j.enzmictec.2019.109507 PMID:32146934

Saikia, S., Gogoi, R. D., Yadav, M., & Yadav, H. S. (2022). Isolation, purification and characterization of peroxidase from Raphanus sativus and its applications in biotransformation of cresols. *Biocatalysis and Agricultural Biotechnology*, *46*, 102540. doi:10.1016/j.bcab.2022.102540

Sampayo, J. N., Gill, M. S., & Lithgow, G. J. (2003). Oxidative stress and aging - The use of superoxide dismutase/catalase mimetics to extend lifespan. *Biochemical Society Transactions*, *31*(6), 1305–1307. doi:10.1042/bst0311305 PMID:14641049

Schonbaum, G. R., & Chance, B. (1976). 7 Catalase. *The Enzymes*, *13*, 363–408. doi:10.1016/S1874-6047(08)60245-0

Šekuljica, N., Gvozdenović, M. M., Knežević-Jugović, Z. D., Jugović, B. Z., & Grgur, B. N. (2016). Biofuel cell based on horseradish peroxidase immobilized on copper sulfide as anode for decolorization of anthraquinone AV109 dye. *Journal of Energy Chemistry*, *25*(3), 403–408. doi:10.1016/j.jechem.2016.03.011

Sellami, K., Couvert, A., Nasrallah, N., Maachi, R., Abouseoud, M., & Amrane, A. (2022). Peroxidase enzymes as green catalysts for bioremediation and biotechnological applications: A review. *The Science of the Total Environment*, *806*, 150500. doi:10.1016/j.scitotenv.2021.150500 PMID:34852426

Sepasi Tehrani, H., & Moosavi-Movahedi, A. A. (2018). Catalase and its mysteries. *Progress in Biophysics and Molecular Biology*, *140*, 5–12. doi:10.1016/j.pbiomolbio.2018.03.001 PMID:29530789

Shahraki, S., Saeidifar, M., Delarami, H. S., & Kazemzadeh, H. (2020). Molecular docking and inhibitory effects of a novel cytotoxic agent with bovine liver catalase. *Journal of Molecular Structure*, *1205*, 127590. doi:10.1016/j.molstruc.2019.127590

Shan, H., Wang, X., Ge, Y., & Li, Z. (2022). Homologous amino acids promoted co-immobilization of laccase and mediator onto geopolymer microspheres for enhancing degradation of dyes in water. *Journal of Hazardous Materials*, *423*, 127107. doi:10.1016/j.jhazmat.2021.127107 PMID:34523501

Shokri, Z., Seidi, F., Karami, S., Li, C., Saeb, M. R., & Xiao, H. (2021). Laccase immobilization onto natural polysaccharides for biosensing and biodegradation. *Carbohydrate Polymers*, *262*, 117963. doi:10.1016/j.carbpol.2021.117963 PMID:33838831

Singh, R., Umapathi, A., Patel, G., Patra, C., Malik, U., Bhargava, S. K., & Daima, H. K. (2022). Nanozyme-based pollutant sensing and environmental treatment: Trends, challenges, and perspectives. *The Science of the Total Environment*, *854*, 158771. doi:10.1016/j.scitotenv.2022.158771 PMID:36108853

Sooch, B. S., Kauldhar, B. S., & Puri, M. (2014). Recent insights into microbial catalases: Isolation, production and purification. *Biotechnology Advances*, *32*(8), 1429–1447. doi:10.1016/j.biotechadv.2014.09.003 PMID:25261851

Stadlmair, L. F., Letzel, T., Drewes, J. E., & Grassmann, J. (2018). Enzymes in removal of pharmaceuticals from wastewater: A critical review of challenges, applications and screening methods for their selection. *Chemosphere*, *205*, 649–661. doi:10.1016/j.chemosphere.2018.04.142 PMID:29723723

Stolarczyk, K., Rogalski, J., & Bilewicz, R. (2020). NAD(P)-dependent glucose dehydrogenase: Applications for biosensors, bioelectrodes, and biofuel cells. *Bioelectrochemistry (Amsterdam, Netherlands)*, *135*, 107574. doi:10.1016/j.bioelechem.2020.107574 PMID:32498025

Sumner, J. B., & Dounce, A. L. (1937). Crystalline catalase. *Science*, *85*(2206), 366–367. doi:10.1126cience.85.2206.366 PMID:17776781

Sun, K., Hong, D., Liu, J., Latif, A., Li, S., Chu, G., Qin, W., & Si, Y. (2021). Trametes versicolor laccase-assisted oxidative coupling of estrogens: Conversion kinetics, linking mechanisms, and practical applications in water purification. *The Science of the Total Environment*, *782*, 146917. doi:10.1016/j.scitotenv.2021.146917

Sun, K., Li, S., Si, Y., & Huang, Q. (2021). Advances in laccase-triggered anabolism for biotechnology applications. *Critical Reviews in Biotechnology*, *41*(7), 969–993. doi:10.1080/07388551.2021.1895053 PMID:33818232

Sung, H. J., Khan, M. F., & Kim, Y. H. (2019). Recombinant lignin peroxidase-catalyzed decolorization of melanin using in-situ generated H_2O_2 for application in whitening cosmetics. *International Journal of Biological Macromolecules*, *136*, 20–26. doi:10.1016/j.ijbiomac.2019.06.026 PMID:31175901

Suryadi, H., Judono, J. J., Putri, M. R., Eclessia, A. D., Ulhaq, J. M., Agustina, D. N., & Sumiati, T. (2022). Biodelignification of lignocellulose using ligninolytic enzymes from white-rot fungi. *Heliyon*, *8*(2), e08865. doi:10.1016/j.heliyon.2022.e08865 PMID:35141441

Tawalbeh, M., Javed, R. M. N., Al-Othman, A., & Almomani, F. (2022). The novel advancements of nanomaterials in biofuel cells with a focus on electrodes' applications. *Fuel*, *322*, 124237. doi:10.1016/j.fuel.2022.124237

Tobimatsu, Y., & Schuetz, M. (2019). Lignin polymerization: How do plants manage the chemistry so well? *Current Opinion in Biotechnology*, *56*, 75–81. doi:10.1016/j.copbio.2018.10.001 PMID:30359808

Tochetto, G. A., Aragão, A. M. I., de Oliveira, D., & Immich, A. P. S. (2022). Can enzymatic processes transform textile processes? A critical analysis of the industrial application. *Process Biochemistry (Barking, London, England)*, *123*, 27–35. doi:10.1016/j.procbio.2022.10.030

Undiano, E., Roman, R., Miranda-Molina, A., & Ayala, M. (2022). Halogenation of estrogens catalysed by a fungal chloroperoxidase. *Natural Product Research*, *36*(20), 5353–5357. doi:10.1080/14786419.2021.1925269 PMID:34000929

Vaseashta, A., Duca, G., & Travin, S. (Eds.). (2022). *Handbook of Research on Water Sciences and Society* (Vols. 1–2). IGI Global. doi:10.4018/978-1-7998-7356-3

Vaseashta, A., Gevorgyan, G., Kavaz, D., Ivanov, O., Jawaid, M., & Vasović, D. (2021). Exposome, Biomonitoring, Assessment and Data Analytics to Quantify Universal Water Quality. In A. Vaseashta & C. Maftei (Eds.), *Water Safety, Security and Sustainability. Advanced Sciences and Technologies for Security Applications*. Springer. doi:10.1007/978-3-030-76008-3_4

Villalba-Rodríguez, A. M., Parra-Arroyo, L., González-González, R. B., Parra-Saldívar, R., Bilal, M., & Iqbal, H. M. N. (2022). Laccase-assisted biosensing constructs - Robust modalities to detect and remove environmental contaminants. *Case Studies in Chemical and Environmental Engineering*, *5*, 100180. doi:10.1016/j.cscee.2022.100180

Virk, A. P., Sharma, P., & Capalash, N. (2012). Use of laccase in pulp and paper industry. *Biotechnology Progress*, *28*(1), 21–32. doi:10.1002/btpr.727 PMID:22012940

Wang, F., Liu, Y., Du, C., & Gao, R. (2022). Current Strategies for Real-Time Enzyme Activation. *Biomolecules*, *12*(5), 599. doi:10.3390/biom12050599 PMID:35625527

Wang, M., Chen, B., Fang, Y., & Tan, T. (2017). Cofactor engineering for more efficient production of chemicals and biofuels. *Biotechnology Advances*, *35*(8), 1032–1039. doi:10.1016/j.biotechadv.2017.09.008 PMID:28939499

Wang, X. Y., Lin, C., Chang, W. J., Huang, Y. H., & Mi, F. L. (2022). Thiolated hyaluronic acid and catalase-enhanced CD44-targeting and oxygen self-supplying nanoplatforms with photothermal/photodynamic effects against hypoxic breast cancer cells. *International Journal of Biological Macromolecules*, *221*, 121–134. doi:10.1016/j.ijbiomac.2022.08.164 PMID:36049568

Wasak, A., Drozd, R., Jankowiak, D., & Rakoczy, R. (2019). Rotating magnetic field as tool for enhancing enzymes properties - laccase case study. *Scientific Reports*, *9*(1), 3707. doi:10.103841598-019-39198-y PMID:30842482

Xu, D. M., Fu, R. B., Wang, J. X., Shi, Y. X., & Guo, X. P. (2021). Chemical stabilization remediation for heavy metals in contaminated soils on the latest decade: Available stabilizing materials and associated evaluation methods-A critical review. *Journal of Cleaner Production*, *321*, 128730. doi:10.1016/j.jclepro.2021.128730

Yan, X., Tang, J., Ma, S., Tanner, D., Ludwig, R., Ulstrup, J., & Xiao, X. (2022). Engineering biointerfaces for the direct electron transfer of Myriococcum thermophilum cellobiose dehydrogenase: Towards a mediator-less biosupercapacitor/biofuel cell hybrid. *Biosensors & Bioelectronics*, *210*, 114337. doi:10.1016/j.bios.2022.114337 PMID:35537312

Yang, C. H., Zhang, Q., Zhu, W. Q., Shi, Y., Cao, H. H., Guo, L., Chu, D., Lu, Z., & Liu, T. X. (2022). Involvement of Laccase2 in Cuticle Sclerotization of the Whitefly, Bemisia tabaci Middle East-Asia Minor. *Insects*, *13*(5), 471. doi:10.3390/insects13050471 PMID:35621805

Zámocký, M., & Koller, F. (1999). Understanding the structure and function of catalases: Clues from molecular evolution and in vitro mutagenesis. *Progress in Biophysics and Molecular Biology*, *72*(1), 19–66. doi:10.1016/S0079-6107(98)00058-3 PMID:10446501

Zhang, H., Zhang, J., Zhang, X., & Geng, A. (2018). Purification and characterization of a novel manganese peroxidase from white-rot fungus Cerrena unicolor BBP6 and its application in dye decolorization and denim bleaching. *Process Biochemistry (Barking, London, England)*, *66*, 222–229. doi:10.1016/j.procbio.2017.12.011

Zofair, S. F. F., Ahmad, S., Hashmi, M. A., Khan, S. H., Khan, M. A., & Younus, H. (2022). Catalytic roles, immobilization and management of recalcitrant environmental pollutants by laccases: Significance in sustainable green chemistry. *Journal of Environmental Management*, *309*, 114676. doi:10.1016/j.jenvman.2022.114676 PMID:35151142

Zucca, P., Cocco, G., Sollai, F., & Sanjust, E. (2016). Fungal laccases as tools for biodegradation of industrial dyes. *Biocatalysis*, *1*(1), 82–108. doi:10.1515/boca-2015-0007

Chapter 17
Synthesis of Biologically Active Nitrogen and Sulfur-Containing Terpenoids

Aculina Aricu
Institute of Chemistry, Moldova State University, Moldova

Svetlana Blaja
Institute of Chemistry, Moldova State University, Moldova

Alexandru Ciocarlan
Institute of Chemistry, Moldova State University, Moldova

Elena Secara
Institute of Chemistry, Moldova State University, Moldova

Lidia Lungu
Institute of Chemistry, Moldova State University, Moldova

Nicon Ungur
Institute of Chemistry, Moldova State University, Moldova

Caleria Cucicova
Institute of Chemistry, Moldova State University, Moldova

ABSTRACT

New, original, and ecologically pure methods for the synthesis of compounds with hybrid terpene and heteroatomic functional groups or fragments have been developed, starting from accessible (+)-sclareolide, obtained from (-)-sclareol, well-known labdane diterpenoid diol extracted from the waste of Clary sage (Salvia sclarea L.) remaining after extracting the essential oil. The natural origin of terpene compounds supposes biocompatibility, selective biological activity, and low toxicity. The compounds with combined skeleton were obtained by coupling some terpene derivatives (acids, chloroanhydrides, bromides) with azaheterocyclic compounds or heterocyclization of some intermediates such as hydrazides, hydrazincarbothioamides, or thiosemicarbazones. A series of over 120 newly obtained substances were subjected to biological testing, of which 10 showed pronounced antifungal and antibacterial activity, two amides showed pronounced antioxidant activity, and two derivatives with guanidine fragments showed high antitumor activity.

DOI: 10.4018/978-1-6684-7198-2.ch017

INTRODUCTION

Microbial infections have spread rapidly in recent years, becoming one of the most important concerns in countries around the world (Grahman, 2001; Greenwood et al., 2007). These global trends stimulate the design of new molecular structures with antimicrobial properties, which could lead to new and effective medicinal preparations for their treatment. One of the most important and accessible sources of new biological compounds is natural products due to their origin implies biocompatibility, selective biological activity, and low toxicity. These include terpenoids, a group of compounds of natural and, as well, of synthetic origin, which are widely used in medicine, pharmaceuticals, cosmetics, and agriculture (Fraga, 2013; Jansen & De Groot, 2004). Particular attention is paid to terpenic compounds that possess multiple biological activities, such as anticancer (Allouche et al., 2009), antioxidant, antimicrobial (Komala et al., 2010), antifungal (Zhou et al., 2011), antimalarial (Lhinhatrakool et al., 2011), antidiabetic (Zhao et al., 2012).

Several studies to date confirm that the presence of heteroatoms, especially nitrogen or sulfur, often increases the biological activity of terpenic compounds. Other teams of authors have reported an even greater increase in the activity of terpenes in cases when their molecules contain some heteroatomic functional groups or fragments, or heterocyclic units.

Recently, the synthesis of molecules with a hybrid skeleton has emerged as a powerful tool in the design of drugs and especially preparations with promising biological activity (Viegas-Junior et al., 2007). This approach is based on the combination of several pharmacophores, which produce compounds with a combined molecular structure and have a higher bioactivity than some known drugs.

Our research was focused on the development of original methods for the preparation of new functionalized bicyclic nitrogen and sulfur-containing derivatives based on the available natural diterpenoid - sclareol and the designing of natural chiral molecules of interest for the pharmaceutical industry.

The functionalization of bicyclic sclareol's derivatives, including sesquiterpenoids have been performed both in the side chain and in the C_7 position of cycle B, have led to lactams and other products containing amino groups, or azine, hydrazide, hydrazide and guanidine fragments, as well as heterocyclic units.

BACKGROUND

Synthesis of Norlabdane Compounds With Amino Group

Many drimane sesquiterpenoids, including their prototype drimenol 1, exhibit various biological activities (Kuchkova et al., 2009). Therefore, it seemed interesting to synthesize 11-aminodrim-7-ene 2, an analog of drimenol with an amino group, in order to study its biological activity.

11-Aminodrim-7-ene 2 was synthesized from drimenol in four steps. First, drimenol 1 was oxidized into drimenal 3, and then its oxime 4 was dehydrated by *p*-tosylchloride or acetic anhydride in pyridine to form 9-cyano-11-nordrim-7-ene 5, followed by its reduction with $LiAlH_4$ in the presence of anhydrous $AlCl_3$ produced 11-aminodrim-7-ene. The reaction of 9-cyano-11-nordrim-7-ene 5 with $NaBH_4$, and $CoCl_2 \cdot 6H_2O$ produced a mixture of drimenylamine 2 and 7,8-dihydrodrimenylamine 6 in a 2:1 ratio (Figure 1), that is interesting as compounds with potential biological activity.

Figure 1. Synthesis of 11-aminodrim-7-ene 2 from drimenol 1

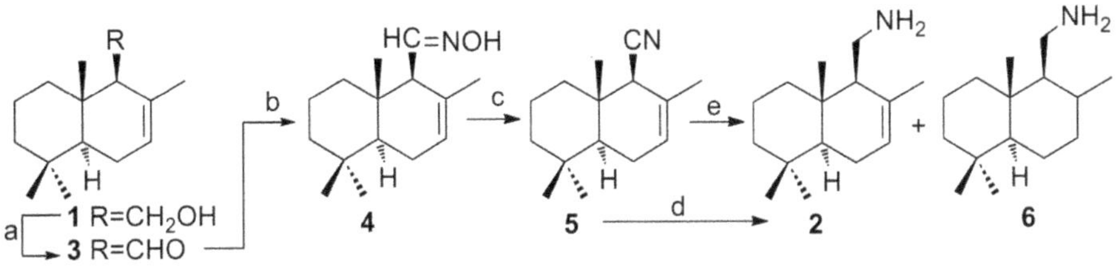

Reagents and reaction conditions: a. P_2O_5, DMSO, 20°C, 95%; b. $NH_2OH·HCl$, EtOH, Py, 80%; c. 1. Ac_2O, Py, 64%, 2. p-TsCl, Py, 90%; d. $LiAlH_4$, $AlCl_3$, Et_2O, 50%; e. $NaBH_4$, $CoCl_2·6H_2O$, MeOH, 20°C, 91%.

In continuation, a series of dihomodrimane sesquiterpenoids containing an NH_2 group were synthesized. Starting from sclareolide 7, 8α-hydroxy-11-dihomodriman-12-one 8 and products of its dehydration 11 and 12 with $\Delta^{8,9}$ and $\Delta^{7,8}$ double bonds, respectively, were obtained (Kuchkova, 2014). Their oximes 13 and 14 were reduced with $LiAlH_4$ to the corresponding amines (Figure 2). According to the ^1H, ^{13}C, and ^{15}N NMR data, oximes 9, 13 and 14 represent mixtures of *Z*- and *E*-isomers which were reduced without separation, since it is known that the *Z*-isomer in a solution easily turns into the *E*-isomer. The ^1H and ^{13}C NMR spectra data confirmed that amines 10, 15, and 16 were obtained as a mixture of two epimers at C_{12}.

Thus, 11-dihomodrimane-12-amine sesquiterpenoids were synthesized starting from the accessible sclareolide 7. The newly obtained amines 10, 15, and 16 are important precursors in the syntheses with the participation of the amine group, which allows for obtaining functionalized compounds with two and more nitrogen atoms.

Figure 2. Synthesis of 11-dihomodriman-12-amine sesquiterpenoids 10, 15, 16 starting from sclareolide 7

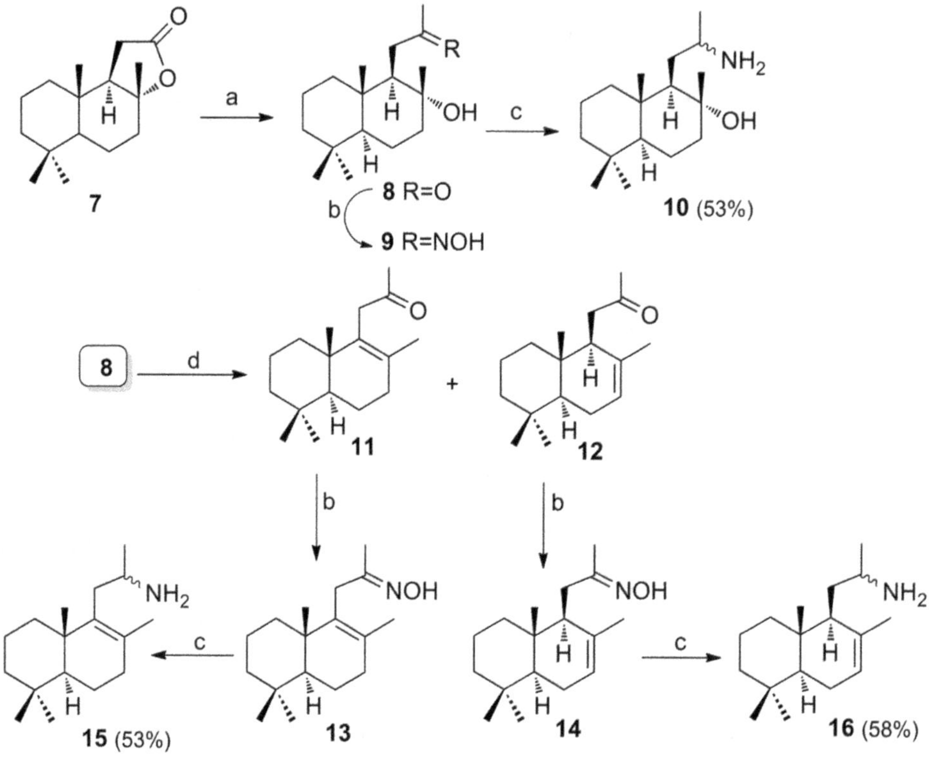

Reagents and reaction conditions: a. CH_3Li, Et_2O, 20°C, 15 min, 65%; b. $NH_2OH{\cdot}HCl$, EtOH-Py (1:1), 20°C, 24 h, 97-99%; c. $LiAlH_4$, THF, 65°C, 10 h.

Four-step synthesis of the 13-amino-14,15-dinorlabd-8(9)-ene 23 from sclareol 17 was performed in a 47% total yield (Figure 3) (Kuchkova et al., 2014). Hydroxy ketone 18 was obtained by oxidation of sclareol 17 using $KMnO_4$ in acetone. Its selective dehydration according to the developed by us procedure upon treatment with $MeSO_3SiMe_3$ in acetonitrile led to 14,15-dinorlabd-8(9)-en-13-one 19. Compared to the procedures described earlier, our method is carried out at room temperature, much faster, and with a high degree of selectivity. It should be noted that the composition of the products significantly depends on the reaction temperature and time. When the reaction is carried out in an ice-water bath for 10 min, sclareol oxide 21 is the major reaction product. The short time heating for 10 min at 60ºC gives, besides unsaturated ketones 19 and 20, sclareol oxide 21 and two unidentified compounds which represent 32% of the total. The reduction of oxime 22, obtained from ketone 19, is a key step in the synthesis of compound 23 from sclareol 17. The mixture of *Z*- and *E*-isomers 22 could be separated by column chromatography, but they can be reduced without purification since *Z*-isomer in solutions easily converts to *E*-isomer. The ^1H and ^{13}C NMR spectra showed that amine 23 was obtained as an inseparable mixture of two diastereomers. The 13-amino-14,15-dinorlabd-8(9)-ene 23, can be of interest as a compound with potential biological activity, as well as an intermediate product, which can be converted to derivatives at the amino group.

Figure 3. Synthesis of 13-amino-14,15-dinorlabd-8(9)-ene 23 from sclareol 17

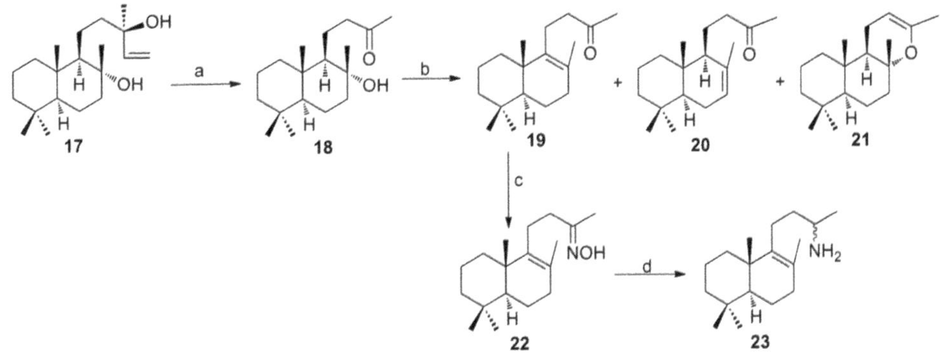

Reagents and reaction conditions: a. $KMnO_4$, CH_3COCH_3, 20°C, 5 h, 80%; b. $CH_3SO_3Si(CH_3)_3$, CH_3CN, 18°C, 10 min, 96%; c. $NH_2OH{\cdot}HCl$, EtOH-Py (1:1), 20°C, 20 h, 99%; d. $LiAlH_4$, THF, Δ, 5 h, 62%.

Next, the results of C_7 nitrogen containing drimanic and homodrimanic compounds synthesis are described (Lungu, 2015). Due to the reactive C_7 carbonyl group, the ketoester 24 and drimenone 25 are important intermediates for the synthesis of the above-mentioned compounds.

Ketones 24 and 25 on treatment with hydroxylamine hydrochloride in a mixture of ethanol: pyridine (ratio 1:1), form drimanic oxime 26 (98%) and homodrimanic oxime 27 (98%), as crystalline compounds (Figure 4). Next, homodrimanic oxime 27 was subjected to the saponification reaction in an alcoholic KOH solution, with the formation of homodrimanic acid oxime 29, obtained with a 95% yield, as a crystalline compound (Lungu, 2015).

Figure 4. Synthesis of C_7 nitrogene containing norlabdane compounds

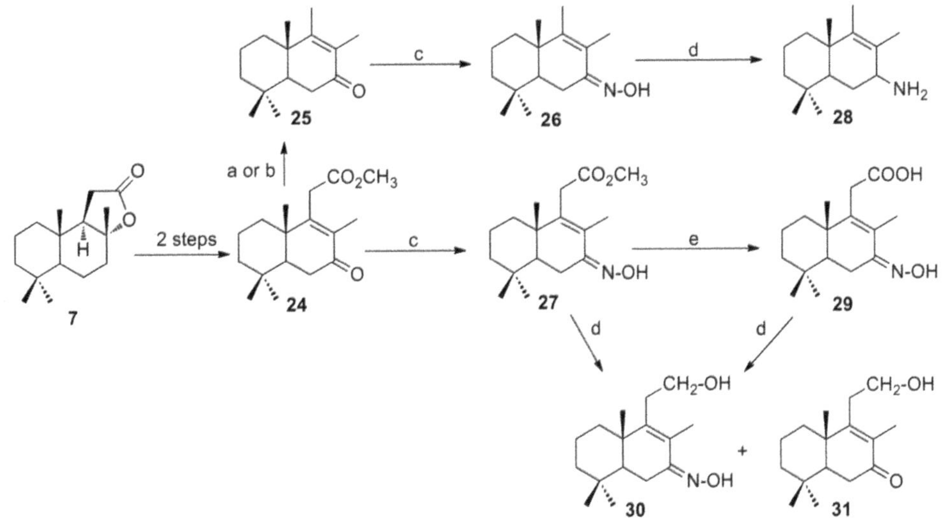

Reagents and reaction conditions: a. KOH, EtOH, reflux, 3h, 98%; b. KOH, EtOH, MW, 1.5h, 98%; c. $NH_2OH{\cdot}HCl$, EtOH, Py, 24h, 96-98%; d. $LiAlH_4$, THF, 5h-24h, 51-60%; e. KOH, MeOH, 95%.

Oximes 26, 27 and 29 were subjected to the reduction reaction with $LiAlH_4$ in anhydrous tetrahydrofuran (THF). In the case of drimanic oxime 26, 7-amino-drim-8(9)-ene 28 was obtained, with a yield of 51%, as a crystalline compound. The reduction of homodrimane oximes 27 and 29 under identical conditions, contrary to expectations did not give homodrimane amines, but mixtures of 7-hydroxyimino-13,14,15,16-tetranorlabd-8(9)-ene-12-ol 30 and 12-hydroxy-13,14,15,16-tetranorlabd-8(9)-en-7-one 31, in 55%: 10% yields for oxime 27 and 60%: 12% yields in the case of oxime 29 (Figure 4).

Other attempts to reduce the oxime group in the molecules of compounds 27 and 29 to the amino group were unsuccessfully, probably because of the steric hindrances that occur in the molecules of the mentioned homodrimane oximes 27 and 29, in contrast to the molecule of the drimane oxime 26.

Synthesis of Norlabdane Guanidine Derivatives

Guanidine-containing compounds can be straightforwardly synthesized starting from amines or from acids. The authors (Duca et al., 2019) described the preparation of guanidine derivatives of 12-amino-11-dihomodrimane-8α-ol 10, 13-amino-14,15-*bis*-norlabd-8(9)-ene 23, and also $\Delta^{8,9}$-bicyclohomofarnesenoic acid 35. Amines 10 were synthesized from sclareolide 7 as described previously by authors (Kuchkova et al., 2014). The synthesis of amine 23 was also reported recently (Kuchkova et al., 2014). The reaction of amines 10 and 23 with sodium hydrogencyanamide in ethanol–water (5:1) solution gave guanidine derivatives 32-34 (Figure 5).

Figure 5. Synthesis of guanidine derivatives of amines

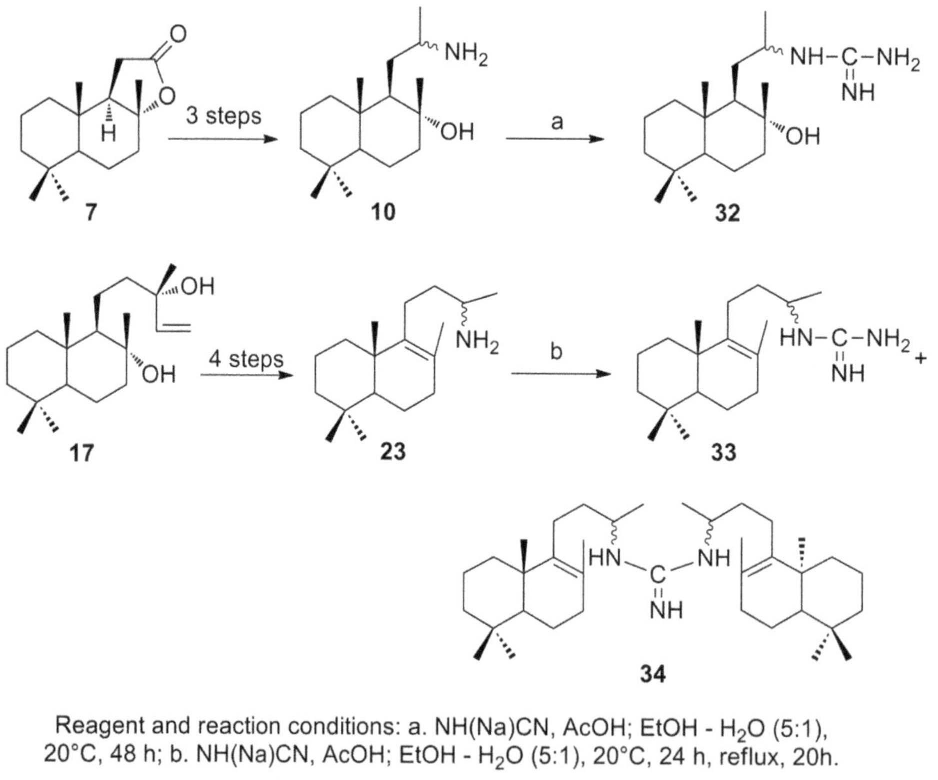

Reagent and reaction conditions: a. NH(Na)CN, AcOH; EtOH - H_2O (5:1), 20°C, 48 h; b. NH(Na)CN, AcOH; EtOH - H_2O (5:1), 20°C, 24 h, reflux, 20h.

The starting material for the guanidines 36 and 37 synthesis was $\Delta^{8,9}$-bicyclohomofarnesenoic acid 35, which was prepared from sclareolide 7 in six steps and 62% overall yield. The acid 35 was activated with CDI in DMF, then treated with a solution of guanidine in DMF at room temperature and generated *N*-($\Delta^{8,9}$-bicyclohomofarnesenoyl)-guanidine 36 and *N*,*N*-di($\Delta^{8,9}$-bicyclohomofarnesenoyl)-guanidine 37 (Figure 6).

Figure 6. Synthesis of $\Delta^{8,9}$-bicyclohomofarnesenic acid guanidine derivatives

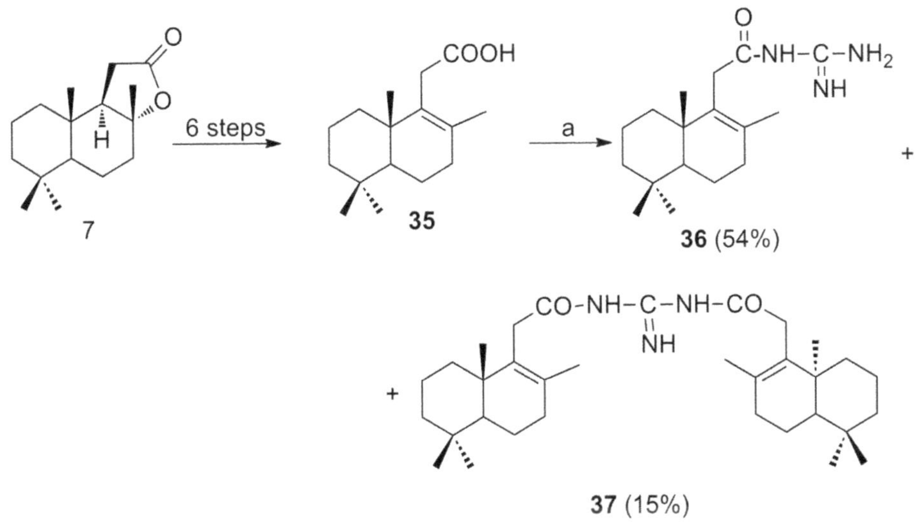

Reagents and reaction conditions: a. CDI, DMF, 20°C, 1h, then addition a guanidine solution in DMF/dioxane (1:1), obtained from guanidine ·HCl and t-BuOK, and the stirring, 20°C, 28h, under Ar.

The assessment of anticancer activity of these compounds showed that two of the tested guanidines 33 and 36 display more potent antiproliferative and cytotoxic activity than cisplatin against Colo 205 and Colo 320 cell lines and maybe act drug lead molecules in cancer chemotherapy (Aricu et al., 2019).

Synthesis of Nitrogen-Containing Compounds by Means of the Beckmann Transposition

The research described in the next article (Kuchkova et al. 2010) refers to the synthesis of nitrogen-containing compounds by means of the Beckmann transposition of oxime 9 in the presence of various reagents. Oxime 9 was obtained from hydroxyketone 8 upon its interaction with hydroxylamine hydrochloride in the mixture of ethanol and pyridine solvents (Figure 7). Then the products of 11-dihomodriman-8α-ol-12-one oxime 9 Beckmann rearrangements with Ac_2O in pyridine, 86% H_3PO_4 and *p*-TsCl in pyridine were investigated. According to the chromatographic and spectral data, the product of the reaction represents the mixture of *Z*- and *E*-isomers. Based on the fact that the *E*-isomer is thermodynamically more stable, and the *Z*-isomer easily transforms into the *E*-isomer, oxime 9 was used in the subsequent reactions as a mixture. It is known that the Beckmann rearrangement proceeds stereospecifically as a result of the anti-migration of the bulkier radical. Thus, the base product of the Beckmann rearrangement of oxime 9 will be an amide. It was established, however, that upon treating oxime 9 with Ac_2O in pyridine at 105-110°C, it does not regroup, but transforms into acetate 38 (90%) and a small amount (5%) of a mixture of the acetates of isomeric oximes 39-41 in the ratio of about 5:4:1. Upon heating the solution of oxime 9 in 86% H_3PO_4, the Beckmann rearrangement also does not occur and as a result of dehydration and cyclization the (1*S*,2*S*,4a*S*,8a*S*)-2,5,5,8a-tetramethyldecahydro-1*H*-naphtho[1,2][5,6]-3-methyl-4,5-dihydro [1,2,6]-oxazine 42 was formed.

Figure 7. Synthesis ofnitrogen-containing driman sesquiterpenoids from 11-dihomodriman-8α-ol-12-one 8

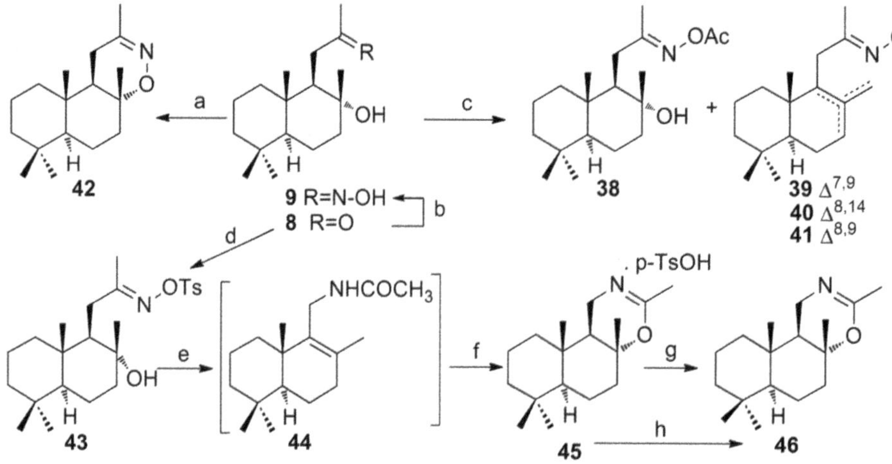

Reagents and reaction conditions: a. 86% H_3PO_4, 60-70°C, 82%; b. $NH_2OH.HCl$, EtOH, Py 94%; c. Ac_2O/Py, reflux; d. p-TsCl/Py, 20°C, 90%; e. CH_3CN, reflux; f. p-TsOH, 95%; g. 10% KOH, MeOH, 95%; h. Al_2O_3, 58%.

Reaction of 9 with *p*-TsCl in Py at room temperature gave the oxime tosylate 43, which underwent Beckmann rearrangement upon brief refluxing in CH_3CN with subsequent loss of water and *p*-toluenesulfonic acid to form amide 44 as an intermediate. Then, this amide, probably due to *p*-TsOH, cyclized and formed the salt of (1*S*,2*S*,4a*S*,8a*S*)2,5,5,8atetramethyldecahydro-1*H*-naphtho[1,2][5,6]-2-methyl-4,5-dihydro[1,3,6]oxazine with *p*-TsOH 45, which after treatment with KOH solution (10%) in MeOH at 20°C or filtration of its salt solution in CH_2Cl_2 through a column with Al_2O_3 produced 1,3,6-oxazine 46. Its structure was confirmed by elemental analysis and spectral data. Oxazine 45 was also characterized by preparing its picrate and hydrochloride.

Thus, as a result of our research, new drimanic nitrogen containing sesquiterpenoids were synthesized, some of which contain 1,2,6- and 1,3,6-oxazine rings. 11-Dihomodriman-8α-ol-12-one 8 served as starting material for the preparation of the newnitrogen-containingg drimanic compounds: 1,2,6-oxazine 42 and *N*-pyroline oxide 47 derivatives (Kuchkova et al., 2011), which are of interest as potentially endowed with promising biological properties compounds. A study of the reaction products of oxime 13 had shown thatit iss heating at 70–80ºC in H_3PO_4 (86%) caused intramolecular cyclization to form (1*S*,2*S*,4a*S*,8a*S*)-2,5,5,8a-tetramethyldecahydro-1*H*-naphtho[1,2-*e*]-3-methyl-4,5-dihydro-[1,2,6]-oxazine

42, which we synthesized previously from 11-dihomodriman-8α-ol-12-one 8, and (1*S*,2*S*,4a*S*,8a*S*)-2,5,5,8atetramethyldecahydro-1*H*-naphtho[1,2-*d*]-2-methylpyrroline-*N*-oxide 47 (Figure 8).

Figure 8. Synthesis of nitrogen-containing homodrimanic sesquiterpenoids from 11-dihomodrim-8(9)-en-12-one 8

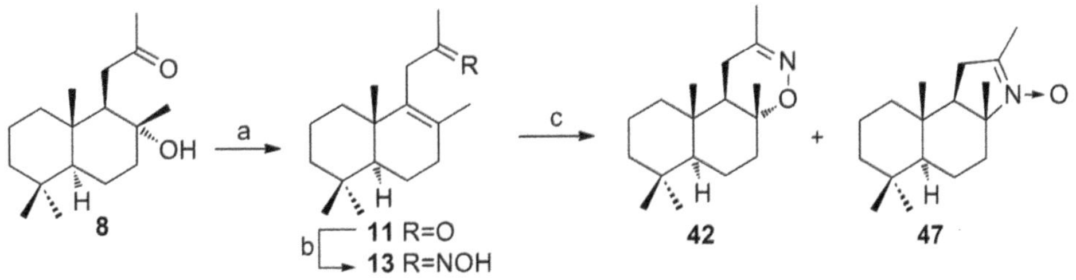

Reagents and reaction conditions: a. $CH_3SO_3Si(CH_3)_3$, 97%; b. $NH_2OH{\cdot}HCl$, EtOH, Py; c. 1. 86% H_3PO_4, 2. CF_3CO_2H.

The synthesis of new drimane and homodrimane lactams 48-52 by the Beckmann rearrangement of the corresponding oximes 26 and 27 is described in the paper (Secara, 2016). Oximes 26 and 27 have been obtained earlier from the commercially available sclareolide 7. The treatment of ketoxime 26 with thionyl chloride in anhydrous dioxane resulted in a mixture of isomeric lactams 48 and 50. The Beckmann rearrangement of oxime 27 under these conditions gave a mixture of lactams 51 and 52 (Figure 9). It is very likely that thiophenolactam 50 is formed as a result of the interaction of the intermediate lactam 49 with thionyl chloride. The presence of the *α,β*–unsaturated carbonyl group in the molecule of lactam 49 activated the C_{11} and C_{12} methyl groups and promoted the reaction of lactam 49 with thionyl chloride.

Figure 9. Beckmann rearrangement of oximes 26 and 27

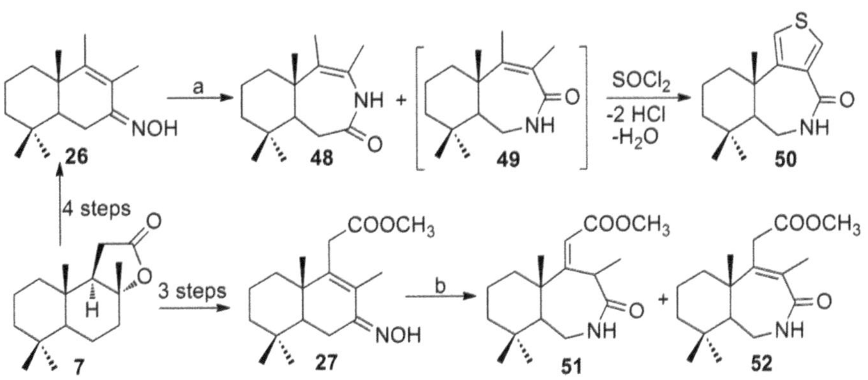

Reagents and reaction conditions: a. $SOCl_2$, dioxan, 50-60°C, 9h; b. $SOCl_2$, dioxan, 60-65°C, 17 h.

Synthesis of Norlabdane Compounds With Azine, Hydrazide, and Dihydrazide Fragments

Several azines and hydrazides are known to possess high and varied biological activity, including anti-tuberculosis and antibacterial. We supposed that adding these fragments to the drimane scaffold could increase its biological potential. The starting material for synthesizing azine 53, dihydrazide 54, and 11-dihomodriman-8α-ol-12-one 8 was prepared from sclareolide 7. A reaction of 8 with hydrazine hydrate in MeOH led to 53. The reflux of 8 with adipic acid dihydrazide in MeOH gave 54, by condensation of two ketones and adipic acid dihydrazide (Aricu et al., 2016) (Figure 10).

Figure 10. Synthesis of norlabdane compounds with azine and dihydrazide fragments

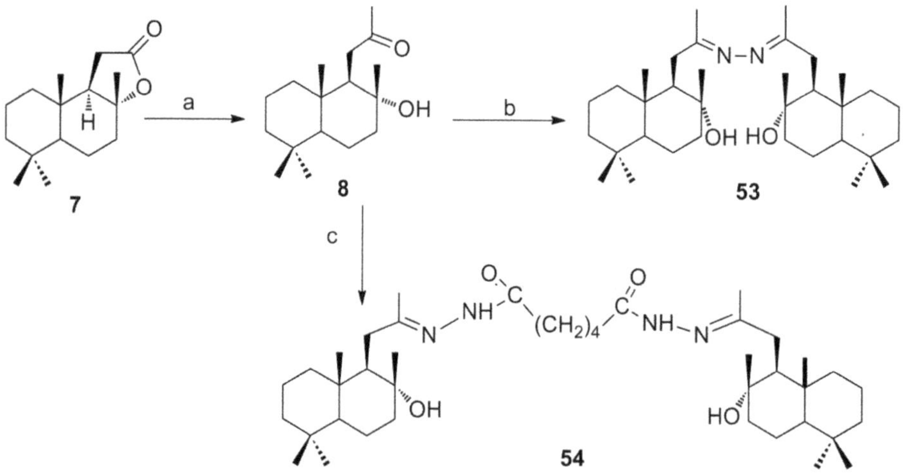

Reagents and reaction conditions: a. CH_3Li, Et_2O, 20°C, 15min, 65%; b. $N_2H_4.H_2O$, CH_3OH, reflux, 10h, 80%; c. $N_2H_4.H_2O$, CH_3OH, 20°C, 48h, reflux, 20h, 31%.

Sclareolide 7 was also used to synthesize $\Delta^{8,13}$-bicyclohomofarnesenoic acid 55 in 60% overall yield. Reaction of its chloroanhydride 56 with hydrazine hydrate in CH_2Cl_2 gave its hydrazide 57 and *N,N-di-*($\Delta^{8,13}$-bicyclohomofarnesenoyl)-hydrazine 58 and saturated hydrazide 59 (Aricu et al., 2016).

Figure 11. Synthesis of norlabdane compounds with hydrazide fragments

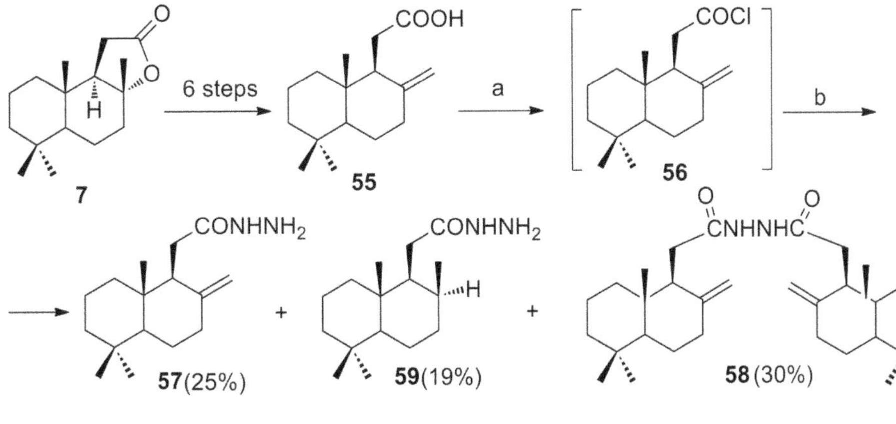

Reagents and reaction conditions: a. $(COCl)_2$, C_6H_6; b. $N_2H_4 \cdot H_2O$, CH_2Cl_2, 20°C, 10h, reflux, 10h; c. $N_2H_4 \cdot H_2O$, CH_2Cl_2.

Compounds 53, 54 and 57-59 were tested *in vitro* for antifungal and antibacterial activity against pure cultures of five fungal species (*Aspergillus flavus*, *Fusarium solani*, *Penicillium chrysogenum*, *P. frequentans*, *Alternaria alternata*) and Gram-negative (*Pseudomonas aeruginosa*) and Gram-positive (*Bacillus* sp.) bacteria. Compound 54 exhibited significant antifungal activity with a minimum inhibiting concentration (MIC) of 0.50 μg/mL, which was comparable with the activity of the known antifungal agent Caspofungin that was used as the reference drug (MIC=0.25 μg/mL), and moderate antibacterial activity (MIC=24 μg/mL) compared with the reference antibiotic Kanamycin (MIC=3 μg/mL) (Aricu et al., 2016).

Synthesis of Norlabdanic Acid Amides Containing Pyrimidine, Pyrazine, 1,2,4-Triazole, and Carbazole Rings

Interest in the synthesis of nitrogen-containing heterocyclic compounds has grown considerably in the last decade due to their use in various fields. Diazines are precisely an important fragment for the construction of new drugs.

$\Delta^{8,13}$-Bicyclohomofarnesenoic acids 55, 13,14,15,16-tetranorlabd-6,8(9)-dien-12-oic 68 drimenoic acid 74 served as initial compounds for the synthesis of terpenic amides containing both norlabdanic and diazinic fragments. The (+)-sclareolide 7 served as a starting material for the synthesis of new optically active acids, e.g. $\Delta^{8,13}$-bicyclohomofarnesenoic acid 55 was obtained in 6 steps with a total yield of 60% (Kuchkova et al., 2014).

Two strategies were approached for the synthesis of bicyclohomofarnesenoic acid amides 61a-c and 62. Using the method I, bicyclohomofarnesenoic acid chloroanhydride 58, obtained *in situ* from acid 57, was treated with the corresponding amines 60a-c (2-aminopyrimidine, 2-aminopyrazine, and 4-aminopyrimidine), which led to the formation of new products with amide group 61a-c and 62 (Figure 12) (Kuchkova et al., 2013, 2014). In the case of 4-aminopyrimidine 60c and aminopyrazine 60b, only monoacyl amides 61b and 61c were obtained, and in the case of 2-aminopyrimidine 60a in line with monoacylamide 61a, *bis*-acylamide 62 was formed. The molecular structure of *bis*-acylamide 62 was established spectrally and its stereochemistry was confirmed based on single crystal X-ray diffration.

Figure 12. Method I. Synthesis of compounds with a combined norlabdane and diazine skeleton 61a-c and 62

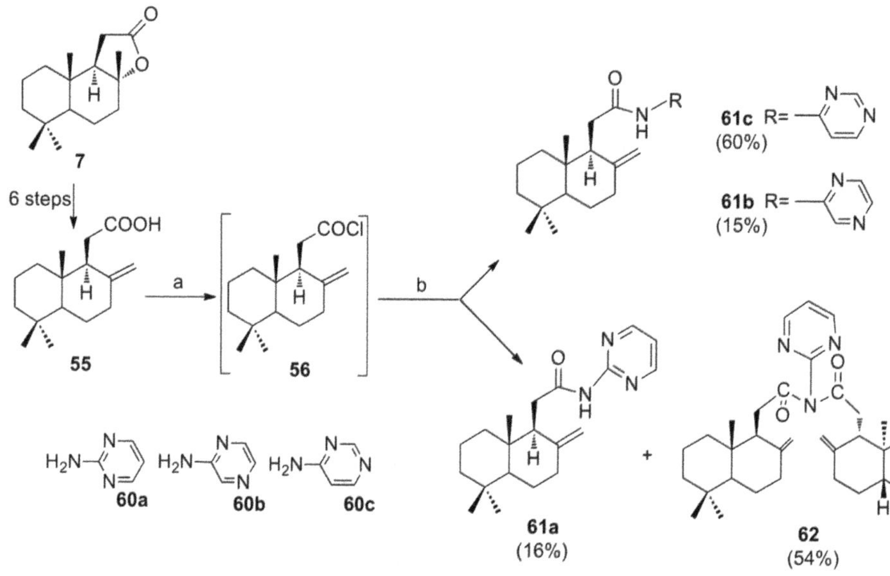

Reagents and reaction conditions: a. $(COCl)_2$, C_6H_6, 20°C, 1 h, reflux 1h (95%); b. $R\text{-}NH_2$ (**60a-c**), CH_2Cl_2, reflux 15h.

In order to establish whether the *bis*-acylation reaction of bicyclohomofarnesenoic acid chloroanhydride 56 with 2-aminopyrimidine 60a could take place under any conditions, it was carried out using another synthetic route (method II, Figure 13).

Acid 55 was directly treated with the corresponding amines 60a-c in the presence of dicyclocarbodiimide (DCC) and 4-dimethylaminopyridine (DMAP). In this case, apart from the expected amides 61a-c, another new compound *N*-$\Delta^{8,13}$-bicyclohomofarnesenoyl-*N,N'*-dicyclohexylurea 62 was obtained (Figure 13).

Figure 13. Method II. Synthesis of compounds with hybrid norlabdanic and diazinic skeletons 61a-c and 62

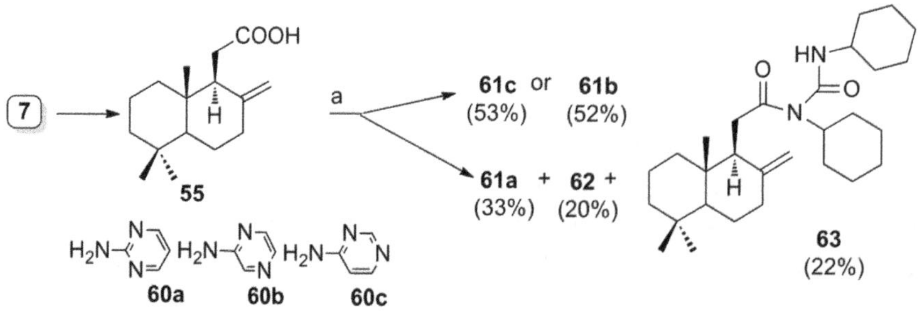

Reagents and reaction conditions: a. $R\text{-}NH_2$, DCC, DMAP, CH_2Cl_2, 20°C, 5 hours, reflux, 15 hours.

Thus, it was established that the reactions carried out by method I give significantly better yields, compared to method II, in which more reaction products are formed, which reduces its effectiveness. Thus, it was established that the *bis*-acylation process occurs only in the case of 2-aminopyrimidine.

The *in vitro* antimicrobial activities of the homodrimane sesquiterpenoids with and without diazine skeleton have been evaluated. The tested compounds have excellent antibacterial activity against Gram-positive strains of *S. aureus* and *B. cereus*. Structure-activity relationship correlation reveals that homodrimane sesquiterpenoids with diazine skeleton possess better antibacterial activity compared with homodrimane sesquiterpenoids without diazine skeleton. Also, in the homodrimane sesquiterpenoid with a diazine skeleton series, the pyrimidine derivatives have a better activity than the pyrazine ones and, the mono-acyl amide has a better activity than the *bis*-acyl amide. None of the compounds tested showed antifungal activity (Kuchkova et al., 2014).

The synthesis of amides 66 and 67 of $\Delta^{8,13}$-bicyclohomofarnesenoic acid 55, which include the 1,2,4-triazole and carbazole rings, was also achieved (Kuchkova et al. 2015). Upon the interaction of acid 55 with $(COCl)_2$, chloroanhydride 56 was obtained *in situ* and its reaction with 3-amino-1,2,4-triazole 64 and *N*-aminocarbazole 65 formed amides 66 and 67 with yields of 58% and 61%, respectively (Figure 14). The structures and configurations of amides 66 and 67 were established based on spectral data, elemental analysis and single crystal X-ray diffraction.

Figure 14. Synthesis of $\Delta^{8,13}$-bicyclohomofarnesenoic acid amides containing the 1,2,4-triazole 66 and carbazole 67 rings

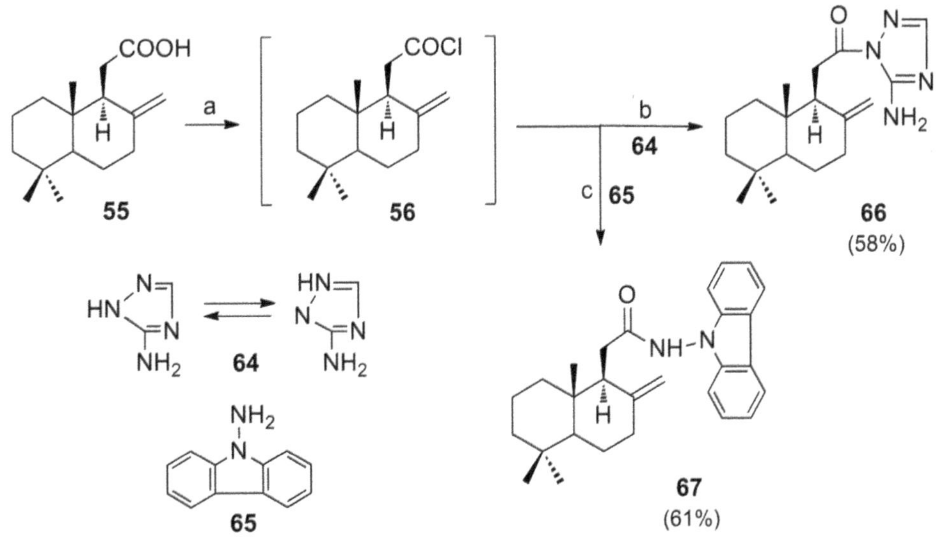

Reagents and reaction conditions: a. $(COCl)_2$, C_6H_6, 20°C, 1h, reflux, 1h; b. amine **64**, CH_2Cl_2, 20°C, 3h; c. amine **65**, CH_2Cl_2, 20°C, 10h, reflux, 5h.

The search for new sources of natural antioxidants that can be used in the prophylaxis and treatment of various human diseases remains a current trend of modern research. The specialized literature describes the recent trend toward the synthesis of substances with antioxidant properties.

The given study describes the original results of the research related to the possibility of changing the antioxidant property of *Nostoc linckia* biomass. It was proved biotechnologically that the newly obtained amides 66, 67 with terpenic and azaheterocyclic hybrid skeleton stimulate the antioxidant activity of the mentioned cyanobacterium. These results were the basis for the development and obtaining of two patents (Cucicova et al., 2015a, 2015b).

These compounds, 66 and 67, in the amount of 50-100 mg/L, increase the antioxidant activity of the 70% ethanolic extracts by 23-42% compared to the control. It is evident that the increase of concentration of ethanolic extract of 70% ethyl alcohol obtained from the *Nostoc linckia* biomass with 1 μg/mL of active substance, increases its antioxidant activity by 2.11-2.73 times compared to the control variant. Therefore, with the increase in the antioxidant activity of the ethanolic extract, the value of *Nostoc linckia* biomass as a producer of antioxidant substances also increases.

In continuation of research on the synthesis of new compounds containing both fragments, terpenic and azaheterocyclic, and with the aim of confirming the formation of the *bis*-acylation product, the synthesis of homodrimanic compounds with pyrimidine, pyrazine, triazole, and carbazole units based on acid 13,14,15,16-tetranorlabd-6,8(9)-dien-12-oic 68 was performed.

(+)-Sclareolide 7 was subjected to several chemical transformations, with the formation of 13,14,15,16-tetranorlabd-6,8(9)-dien-12-oic acid 68, in 94% yield. Treatment of acid 68 *in situ* with oxalyl chloride in anhydrous benzene led to 13,14,15,16-tetranorlabd-6,8(9)-dien-12-oic acid chloroanhydride 69, which further underwent coupling reactions with 2-aminopyrimidine 60a, aminopyrazine 60b and 4-aminopyrimidine 60c in methylene chloride under stirring. In the case of aminopyrazine 60b and 4-aminopyrimidine 60c, only mono-acylamides 70b and 70c were obtained, with yields of 35% and 69%, respectively (Figure 15), while in the case of 2-aminopyrimidine 60a, two reaction products were obtained: mono-acylamide 70a (40%) and *bis*-acylamide 71 (25%) (Duca et al., 2018, Duca et al. 2021).

Figure 15. Synthesis of 13,14,15,16-tetranorlabd-6,8(9)-dien-12-oic acid 68 amides containing the diazine, 1,2,4-triazole and carbazole rings

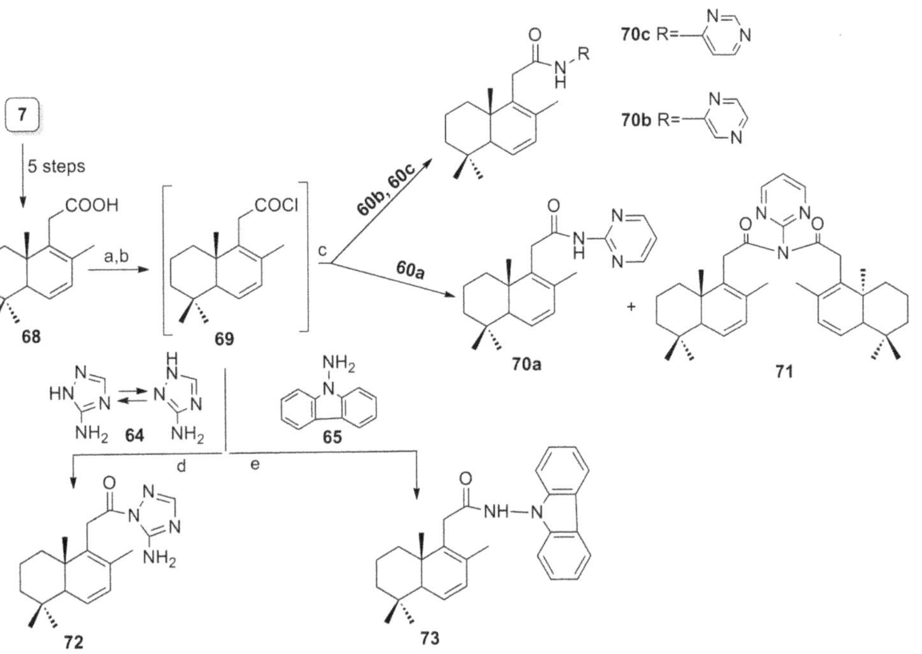

Reagents and reaction conditions: a. C_6H_6 (anhydrous), $(COCl)_2$, reflux, 2h; b. CH_2Cl_2, 40°C, 5-10h, 25-69%; c. R-NH_2, **60a-c**, CH_2Cl_2, 20°C, 5h; d. Amine **64**, CH_2Cl_2, 20°C, 3h; e. Amine **65**, CH_2Cl_2, 20°C, 10h, reflux, 5h.

All *N*-substituted amides, both in the case of $\Delta^{8,13}$-bicyclohomofarnesenoic acid 55 and in the case of 13,14,15,16-tetranorlabd-6,8(9)-dien-12-oic acid 68 result from the condensation of groups of primary amines with acyl chlorides 56 and 69. In turn, all secondary amides can react again with acyl chloride, but according to experimental data only mono-acylamides 61a and 70a undergo *bis*-acylation with the formation of *bis*-acylamides 62 and 71. Presumably, this occurs as a result of the delocalization of the non-bonding electrons of the nitrogen to the adjacent carbonyl group (amide bond resonance) which reduces the reactivity of amides toward amines. In addition, the resonance structures for amides 61a-c and 70a-c also show delocalization towards the aromatic ring such that the aryl substituents determine their reactivity. Presumably, the reaction time is also important for the formation of the *bis*-acylated product.

All newly synthesized azine compounds were tested *in vitro* for antimicrobial activity on pure cultures of fungi and Gram-negative and Gram-positive bacteria.

The *in vitro* antimicrobial activities of norlabdane compounds with and without the diazine backbone were evaluated. The tested compounds have excellent antibacterial activity against Gram-positive strains *S. aureus* and *B. cereus*. The SAR correlation reveals that the norlabdan compounds with the diazine skeleton possess better antibacterial activity compared to the norlabdan compounds without the diazine skeleton. Also, in the series of compounds with combined norlabdane and heterocyclic skeletons, pyrimidine derivatives have better activity than pyrazine ones, and mono-acyl amides have better activity than *bis*-acyl amides. The tested compounds do not possess antifungal activity. Also, the synthesis of pentanorlabdane compounds with pyrimidine and pyrazine fragments, and derivatives containing the 1,2,4-triazole and carbazole ring was carried out.

Drimenoic acid 74 was synthesized from (+)-sclareolide 7 in 6 steps, with a total yield of 55%, according to the method developed by the authors (Aricu et al., 2018).

Figure 16. Synthesis of isodrimenoic and albicanoic acid amides

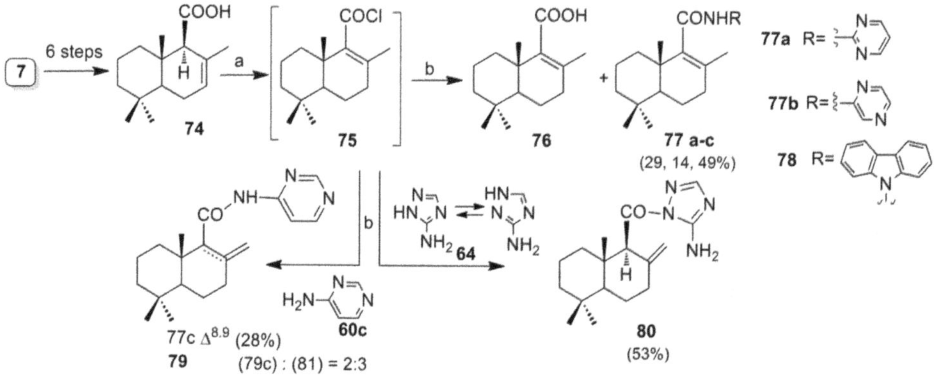

Reagents and reaction conditions: a. $(COCl)_2$, C_6H_6, 20°C, 1h, reflux, 1h (95%); b. $R-NH_2$ (**60a-c**, **64**, **65**), CH_2Cl_2, 20°C, 1-48h, reflux, 10-25h.

Further attempts were made to obtain the amides of drimenoic acid. Contrary to expectations, isodrimenoic acid amides 77a-c, 78 and albicanoic acid amides 79, 80 were obtained (Aricu et al., 2018). It should be noted that during the reaction of drimenoic acid 74 with oxalyl chloride, the isomerization of the double bond $\Delta^{7,8}$ in position $\Delta^{8,9}$ takes place, which leads to a further interaction of chloroanhydride 75 with 2-aminopyrimidine 60a, 2-aminopyrazine 60b, *N*-aminocarbazole 65 to obtain isodrimenoic acid amides 77a,b and 78, albicanoic acid amides 81 and 82. The reaction of intermediate chloroanhydride with 4-aminopyrimidine 60c gave a mixture of amides 77c and 79 in 2:3 ratio. Analysis of the spectral data of the albicanoic acid amide 80 showed that the 3-amino-1,2,4-triazole reacted in the tautomeric form, and the amide 80 contains the amine group and the semicyclic double bond $\Delta^{8,12}$. The structure and configuration of *N*-(isodrimenoylamino)-carbazole 78 were confirmed by single crystal X-ray analysis data.

Synthesis of Norlabdanic Compounds With Pyridazinone Structural Units

In continuation of research with reference to the synthesis of new compounds containing terpene and diazine fragment, a series of norlabdanic compounds with pyridazinone structural units were obtained (Aricu et al., 2016).

Pentanorlabdanic ketones 25 and 81 are important starting compounds for the synthesis of compounds with pentanorlabdanic and heterocyclic hybrid skeletons. These compounds show increased reactivity due to the $\Delta^{8,9}$ double bond that activates the C_{11} and C_{12} carbon atoms.

Next, ketones 25 and 81, obtained from commercially available (+)-sclareolide 7, were treated with *N*-bromosuccinimide (NBS) in carbon tetrachloride to form the bromides: 11-bromo-drim-8(9)-en-7-one 82, 12-bromo-drim-8(9)-en-7-one 83, 11,12-dibromo-drim-8(9)-en-7-one 84 and 11,12-dibromo-drim-5(6),8(9)-dien-7-one 85. Their ratio in the mixture can be changed by varying the amount of *N*-bromosuccinimide and the duration of the reaction (Figure 17).

Figure 17. Synthesis of pentanorlabdanic compounds with p-tolyl-pyridazinonic fragment

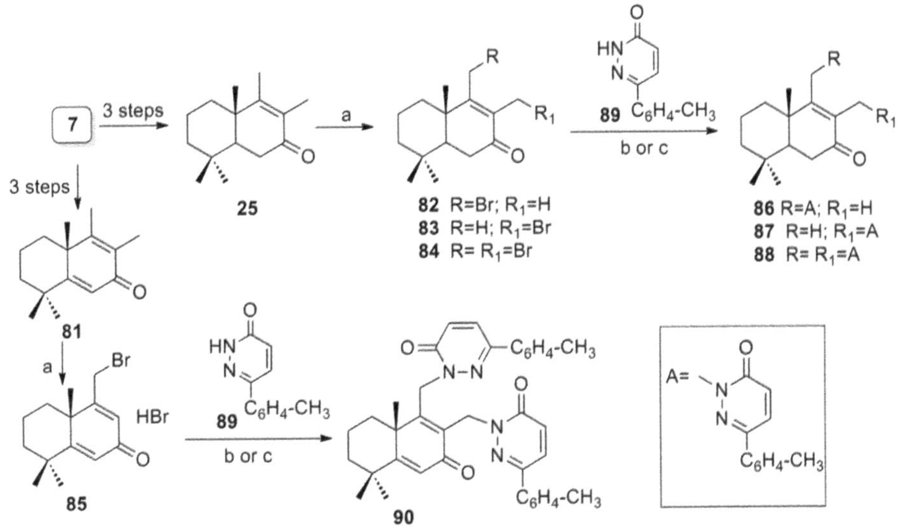

Reagents and reaction conditions: a. NBS, CCl_4, reflux, 2-3h, 70-98%; b).K_2CO_3, DMAA, r.t., 24h, 34-80%; c. K_2CO_3, DMAA, MW, 20 min., 34-84%.

Bromides 82-85 were subjected to the coupling reaction with pyridazinone 89, in basic medium (KOH) in DMAA, both under classical conditions and under microwave irradiation. Chromatographically inseparable bromides 82 and 83 in a 3:2 mixture was coupled with 6-(*p*-tolyl)-3(2H)-pyridazinone 89. The monosubstituted reaction products 11-*p*-tolyl-pyridazonyl-drim-8(9)-en-7-one 86 and 12-*p*-tolyl-pyridazonyl-drim-8(9)-en-7-one 87, unlike the starting bromides 82 and 83, were separated by chromatography on column with silica gel, their summary yield being 85%.

The individual dibromides 84 and 85 on treatment with pyridazinone 89 under the same conditions form 11,12-*p*-tolyl-pyridazonyl-drim-8(9)-en-7-one 88 and 11,12-*p*-tolyl-pyridazonyl-drim-5(6),8(9)-dien-7-one 90. The structures of the new compounds obtained were established based on spectral data (^{1}H and ^{13}C NMR, IR), and the stereochemistry of compound 90 was also confirmed by the X-ray diffraction on a single crystal (Aricu et al. 2016).

Next, in order to obtain the norlabdanic compounds, the methyl ester of 7-oxo-13,14,15,16-tetranorlabd-8-en-12-oic acid 24 served as the starting compound, which can be obtained, in two steps with a yield of 77%, from (+)-sclareolide 7. Ketoester 24 was brominated with NBS (1.5 eq.) to form norlabdanic bromide 92 (Figure 18).

Figure 18. Coupling reaction of ketobromide 24 with p-tolyl-pyridazinones 89 and 91

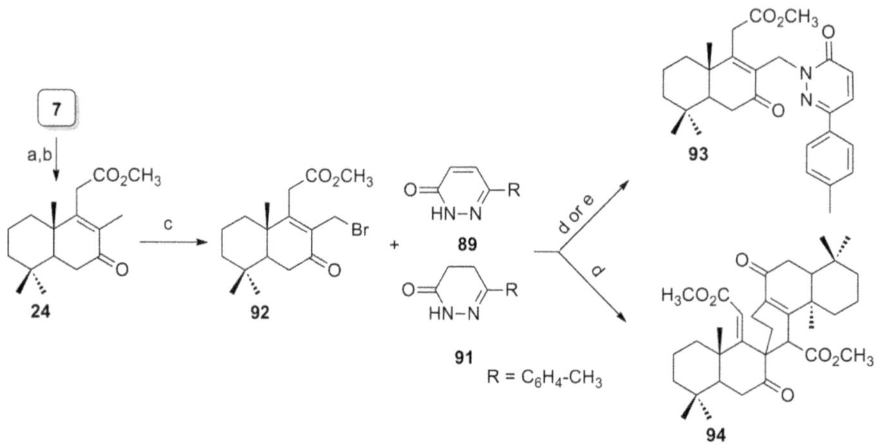

Reagents and reaction conditions: a. H_2SO_4 (conc.), MeOH, reflux, 96h, 90%; b. $LiClO_4$, MeOH, e⁻, 3h, 63%; c. NBS, CCl_4, reflux, 2h, 92%; d. K_2CO_3, DMAA, r.t., 48h, **93** (75%), **94** (70%); e. K_2CO_3, DMAA, MW, 20 min., 79%.

Next the ketobromide 92 was coupled with 6-(*p*-tolyl)-3(2H)-pyridazinone 89 in basic medium (K_2CO_3) in *N,N*-dimethylacetamide (DMAA), with the formation of the norlabdane compound with pyridazinone moiety 93. This reaction is due to the fact that pyridazinone 89 is a conjugated unsaturated substance that increases the reactivity of the N-H bond, easily deprotonable under the basic conditions created (Figure 18). The coupling reaction was performed both under classical conditions and under microwave irradiation.

Contrary to expectations, the interaction of bromide 92 with 6-(*p*-tolyl)-4,5-dihydro-3(2H)-pyridazinone 91 in basic medium does not lead to the expected norlabdanic and pyridazinonic hybrid skeleton compound. This fact is due to a saturated system which substantially decreases the reactivity of the N-H bond. Under these conditions, the dimerization of ketobromide 92 takes place with the formation of compound 94 with a complex structure, with a yield of 70% (Ciocarlan et al., 2017). The structures of these compounds were established based on spectral data (^1H and ^{13}C NMR, IR), and the stereochemistry of compound 94 was also confirmed by single-crystal X-ray diffraction. All the compounds obtained were tested *in vitro* on five strains of fungi and two species of bacteria.

Spiro-compound 94 showed average antifungal activity with MIC values equal to 3.5 μg/mL, which is fourteen times weaker compared to the reference compound Caspofungin (0.25 μg/mL), and antibacterial activity with MIC values equal to 4 μg/mL, compared to the reference compound Kanamycin (3 μg/mL).

Compound 90, which consists of a quinone-analog skeleton and two diazine units, exhibits high antibacterial and antifungal activity. It exhibits antifungal activity at an MIC equal to 5×10^{-3} μg/mL, which is 50 times higher compared to the reference compound Caspofungin (0.25 μg/mL). The MIC value of its antibacterial activity is equal to 3.2×10^{-2} μg/mL, being 90 times higher compared to the reference compound Kanamycin (3 μg/mL) (Aricu et al., 2016).

Figure 19. Antifungal and antibacterial structure-activity correlation in the series of norlabdanic compounds with pyridazinone fragment

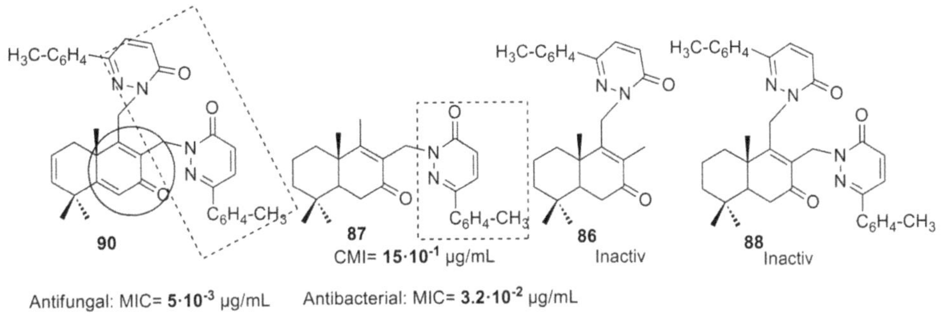

Compound 87, which includes only a diazine ring at C_{12} of the norlabdan moiety and does not contain the analogous quinone moiety, exhibits moderate antifungal activity (MIC=15 × 10^{-1} µg/mL), but six times lower compared to compound reference Caspofungin (0.25 µg/mL). Compounds 86 and 88 do not show inhibitory activity against the mentioned strains of bacteria and fungi (Figure 19).

Synthesis of Norlabdanic Compounds With a Triazole Fragment

In continuation of our research in the field of norlabdanic compounds with biologically active heterocyclic moieties, we set out to synthesize a series of compounds with norlabdanic and 1,2,4-triazole structural units to combine their respective biological potential.

As a starting material, 8α-hydroxy-homodrim-11-hydrazide 95 was used for the synthesis of the mentioned compounds. This compound can be obtained from the commercially available (+)-sclareolide 7 in one step with a yield of 85%, according to Figure 20.

Next, hydrazide 95 was coupled with substituted aryl isothiocyanates in ethanol to form hydrazinecarbothioamides 96a-d. The synthesis of hydrazinecarbothioamides 96a-d by the classical method, which includes heating, presents major drawbacks such as the reaction time (270-300 min) and, consequently, high energy consumption. For these reasons, non-conventional methods were used and the synthesis was carried out by microwave irradiation, at a constant irradiation power and variable temperature.

Figure 20. Synthesis of norlabdanic compounds with 1,2,4-triazole fragment

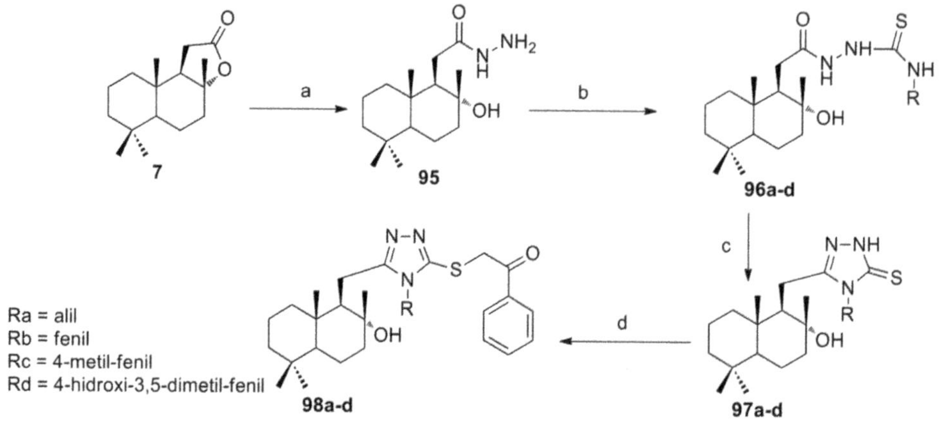

Reagents and reaction conditions: a. NH_2-NH_2·H_2O (95%), EtOH, reflux, 5h, 85%; b. R-NCS, EtOH, r.t. 4-5h, 83-91%, or MW, 5 min., 85-92%; c. NaOH (8%), 70°C, 2-3h, 70-83%; d. Bromoacetophenone, Et_3N, acetone, r.t.,2-3h, 78-80%.

According to the experimental data on microwave irradiation, the reaction time decreases considerably (from several hours to 5 min), the energy consumed also decreases considerably, and the yields increase slightly. Therefore, coupling reactions of hydrazide 95 with aryl isothiocyanates under microwave irradiation are considered environmentally friendly (Lungu et al., 2019).

The hydrazinecarbothioamides 96a-d were treated with aqueous NaOH solution (8%) at 70°C, with the formation of the respective norlabdanic compounds with triazole moiety 97a-d in 70–83% yields. 1,2,4-*S*-Substituted triazoles 98a-d were obtained by the coupling reaction of triazoles 97a-d with bromoacetophenone in acetone in the presence of triethylamine (Et_3N). Theoretically, the N-H functional group in the triazole fragment can be subjected to the condensation reaction with a halogenated aromatic derivative, in our case bromoacetophenone. However, the triazole fragment in compounds 97a-d exists in the form of two resonance structures. One of these is the aromatic zwitterion where N_1 is positively charged and sulfur is negatively charged. The non-bonding electrons make sulfur a nucleophile that readily reacts with bromoacetophenone.

The structures of these compounds were established based on spectral data (^{1}H and ^{13}C NMR, IR), and the stereochemistry of compounds 97c and 97d was definitively confirmed by single crystal X-ray diffraction. The antibacterial and antifungal activity of the synthesized compounds with hybrid norlabdanic and carbothioamide or triazole skeletons 95-97 was tested *in vitro* on pure cultures of bacteria and fungi.

Compounds 96c and 97d show antifungal activity (MIC=0.125 and 0.094 µg/mL) and pronounced antibacterial activity (MIC=0.064 and 0.047 µg/mL). The antibacterial activity of triazole 97d is 63 times higher compared to the reference compound Kanamycin (MIC 3.5 µg/mL). As an antifungal agent, it is three times more active compared to the reference compound Caspofungin (MIC=0.24 µg/mL). Compounds 98a-d, which have the SH thiol group in the 2-position of the triazole moiety substituted with an acetophenone moiety, are inactive (Lungu et al., 2019).

Synthesis of Norlabdanic Compounds With 1,3,4-Oxadiazole and 1,3,4-Thiadiazole Fragments

Oxadiazole and thiadiazole contain a universal heterocyclic nucleus, which has attracted increased attention in medicinal chemistry in search of new therapeutic preparations.

As a starting material, for the synthesis of the mentioned compounds, 8α-hydroxy-homodrim-11-hydrazide 95 was used which, upon interaction with tetramethylthiuram disulfide (TMTD) in DMF, forms a mixture of two compounds 2-thio-5-(11-homodrim-8α-ol)-1,3,4-oxadiazole 99 and 2-mercapto-5-(11-homodrim-8α-ol)-1,3,4-thiadiazole 100 (Figure 21). The ratio of oxadiazole 99 to thiadiazole 100 depends on the amount of TMTD taken into the reaction (Lungu et al. 2019)

When 0.5 mol of TMTD is used in the reaction for 1 mol of hydrazide 95, only 46% of the initial hydrazide reacts, with the formation of oxadiazole 99. When 1 mol of hydrazide 95 is treated with 1 mol of TMTD, a mixture consisting of oxadiazole 99 is formed and thiadiazole 100, in which oxadiazole 99 predominates. On contrary, when 1 mol of hydrazide 95 interacts with 1.5 mol TMTD, the same mixture is obtained, but in this case thiadiazole 100 prevails.

Figure 21. Synthesis of norlabdane compounds with oxadiazole and thiadiazole fragments

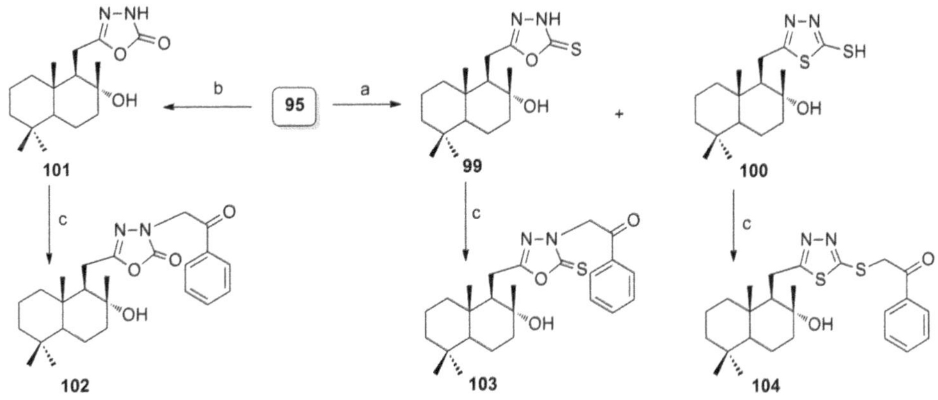

Reagents and reaction conditions: a. TMTD, DMF, 90°C, 1.5h, **99** (20-86%), **100** (5-70%); b. CDI, Et_3N, THF, 0°C, 3h, 74%; c. Br-acetophenone, Et_3N, acetone, 3h, 80-91%.

The structures of the obtained compounds were established on the basis of spectral data (^1H and ^{13}C NMR, IR), the stereochemistry of compound 99 was also confirmed by the single crystal X-ray diffraction method. The synthesis of 5-(11-homodrim-8α-ol)-1,3,4-oxadiazol-2-one 101 consists of treating hydrazide 95 with 1,1'-carbonyldiimidazole (CDI) and Et_3N in tetrahydrofuran (THF), with a yield of 74% (Lungu et al. 2020).

Further oxadiazoles 99, 101 and thiadiazole 100 were subjected to coupling reactions with bromoacetophenone in acetone in the presence of Et_3N to form 3-*N*-acetophenon-5-(11-homodrim-8α-ol)-1,3,4-oxadiazole-2-one 102, with a yield of 80%, 3-*N*-acetophenon-5-(11-homodrim-8α-ol)-1,3,4-oxadiazol-2-thione 103, with a yield of 91% and 2-*S*-acetophenone-5-(11-homodrim-8α-ol)-1,3,4-thiadiazole 104 in 85% yield (Figure 21).

The structures of the obtained compounds were established based on spectral data (^{1}H and ^{13}C NMR, IR), the structure and stereochemistry of compound 104 were also confirmed by single crystal X-ray diffraction. In order to obtain norlabdanic compounds with a thiadiazole fragment, another synthesis method was also used, upon the interaction of norlabdanic hydrazide 97 with isothiocyanate derivatives, without isolating the intermediate compounds. This reaction was carried out in the presence of triethylamine (Et_3N) in water, with the formation of 2-amino-1,3,4-thiadiazoles 105a-c in 70-78% yields (Figure 21) (Lungu et al. 2020).

Another method for the preparation of 2-amino-5-(11-homodrim-8α-ol)-1,3,4-oxadiazole 106 consists in treating hydrazide 95 with cyanogen bromide (CNBr) in aqueous dioxane. As a result, oxadiazole 106 is formed in 80% yield.

Figure 22. Synthesis of norlabdanic compounds with 2-amino-1,3,4-thiadiazole and 2-amino-1,3,4-oxadiazole fragments

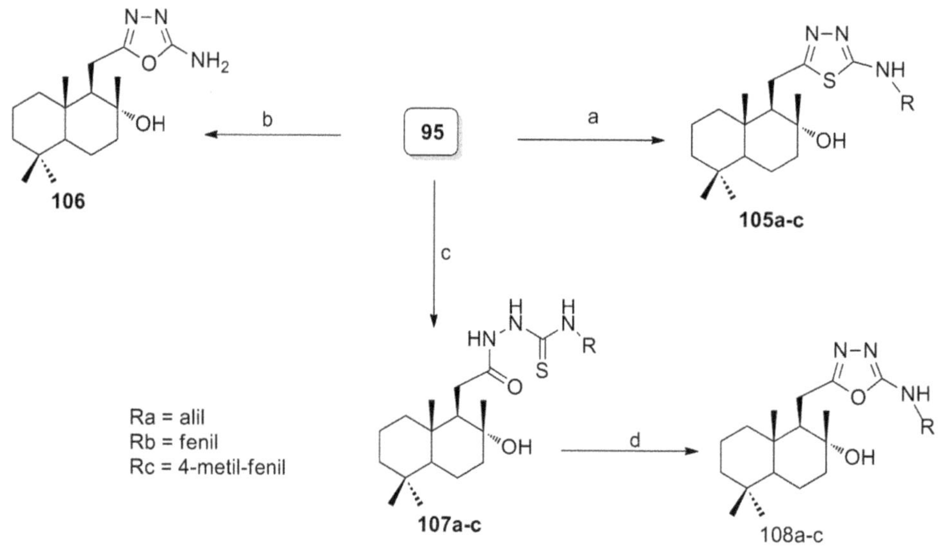

Reagents and reaction conditions: a. R-NCS, Et_3N, H_2O, reflux, 18h, 70-78%; b. CNBr, $NaHCO_3$, aqueous dioxane, 1h, 80%; c. R-NCS, EtOH, 4-5h, 85-92%.

The antibacterial and antifungal activity of the synthesized compounds with hybrid norlabdanic and oxadiazole skeletons 99, 101-103 and 106 or thiadiazole 100, 104 and 105a-c was tested *in vitro* on pure cultures of bacteria and fungi. Thiadiazole 100 exhibits pronounced antifungal activity (MIC=0.032 µg/mL) and antibacterial activity (MIC=0.094 µg/mL) (Aricu et al., 2018). The antibacterial activity of compound 100 is thirty times higher compared to the reference compound, Kanamycin (MIC=3.5 µg/mL). As an antifungal agent, this compound is eight times more active than the reference compound Caspofungin (MIC=0.24 µg/mL). Thiadiazole 105a exhibits antifungal activity (MIC=0.25 µg/mL) and pronounced antibacterial activity (MIC=0.5 µg/mL). The antibacterial activity of compound 105a is six times higher compared to the reference compound, Kanamycin (MIC=3.5 µg/mL).

Oxadiazoles 99 and 102 exhibits significant antifungal activity at a minimum inhibitory concentration (MIC=1.3 and 2 µg/mL), compared to the reference compound Caspofungin (MIC=0.24 µg/mL), while compound 102 also exhibits pronounced antibacterial activity at MIC=0.50 μg/mL which is seven times higher compared to the reference compound, Kanamycin (MIC=3.5 μg/mL).

As a starting material for the synthesis of the new norlabdanic compounds - derivatives of 1,3,4-oxadiazole and 1,3,4-thiadiazole, the hydrazide 109 of the $\Delta^{8,9}$-bicyclohomofarnesenoic acid, obtained from the commercially available sclareolide 7 (seven steps) was used (Aricu et al., 2016), with a total yield of 40% (Figure 23).

In the reaction of hydrazide 109 with cyanogen bromide (CNBr) in aqueous dioxane, 2-amino-5-($\Delta^{8,9}$-bicyclohomofarnesenyl)-1,3,4-oxadiazole 110 was obtained in 91% yield. Reaction of hydrazide 109 with 1,1'-carbonyldiimidazole (CDI) in tetrahydrofuran (THF) in the presence of triethylamine led to 5-(($\Delta^{8,9}$-bicyclohomofarnesenyl)-1,3,4-oxadiazol-2(3*H*)-one 111 with a yield of 92% (Figure 23) (Aricu et al., 2020).

Figure 23. Synthesis of norlabdanic compounds with 1,3,4-thiadiazole and 1,3,4-oxadiazole fragments

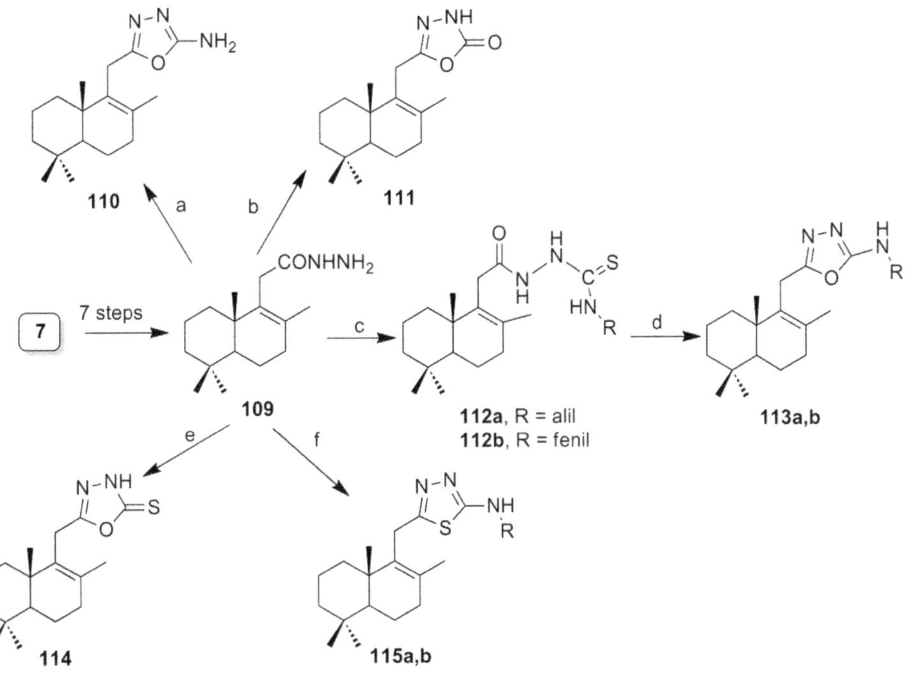

Reagents and reaction conditions: a. BrCN, $NaHCO_3$, dioxane, water, 20°C, 5h, 91%; b. CDI, Et_3N, THF, 0°C, 5h, 92%; c. R-NCS, EtOH, 20°C, 4-5h, 90-92%; d. DCC, acetone, CH_3OH, 5h, 56-70%; e. TMTD, DMF, 90°C, 1.5h, 70%; f. R-NCS, Et_3N, H_2O, reflux 20h, 72-75%.

The reaction of hydrazide 109 with allyl isothiocyanate CH_2=CH-CH_2NCS or phenylisothiocyanate C_6H_5NCS in alcohol led to the formation of hydrazinecarbothioamides 112a,b, which subsequently, upon treatment with *N,N'*-dicyclohexylcarbodiimide (DCC) at reflux in methanol, are transformed into cyclic derivatives of 1,3,4-oxadiazole, containing in position 2 allylamino- and phenylamino- groups 113a,b, respectively (Figure 23). The heating of hydrazide 109 at 90°C in dimethylformamide (DMF) with tetramethylthiuram disulfide $(CH_3)_2$N-(C=S)-SS-(C=S)N$(CH_3)_2$ (TMTD) gave 2-thio-5-($\Delta^{8,9}$-bicyclohomofarnesenyl)-1,3,4-oxadiazole 60 with 70% yield.

A prolonged reflux of hydrazide 109 with allyl isothiocyanate or phenyl isothiocyanate in water in the presence of Et_3N afforded norlabdanic compounds containing the 1,3,4-thiadiazole moiety, 115a,b (Figure 23) (Aricu et al., 2020). To study the structure-biological activity relationship, pentanorlabdane compounds were synthesized containing 1,3,4-oxadiazole and 1,3,4-thiadiazole rings, starting from hydrazide 117 of drimenoic acid 74. The drimenoic acid chloroanhydride 116 was obtained *in situ* by interaction of drimenoic acid 74 with oxalyl chloride $(COCl)_2$, then treated with N_2H_4•H_2O (98%) to form hydrazide 117. Later, starting from hydrazide 117, new norlabdanic derivatives containing 1,3,4-oxadiazole ring 118, 119, 120, 123a,b and 1,3,4-thiadiazole ring 121a,b were obtained (Figure 24) (Aricu et al., 2020). The structures of all newly synthesized substances were established according to elemental analysis and spectral data (IR, ^1H and ^{13}C NMR).

Figure 24. Synthesis of pentanorlabdanic compounds with 1,3,4-thiadiazole and 1,3,4-oxadiazole fragments

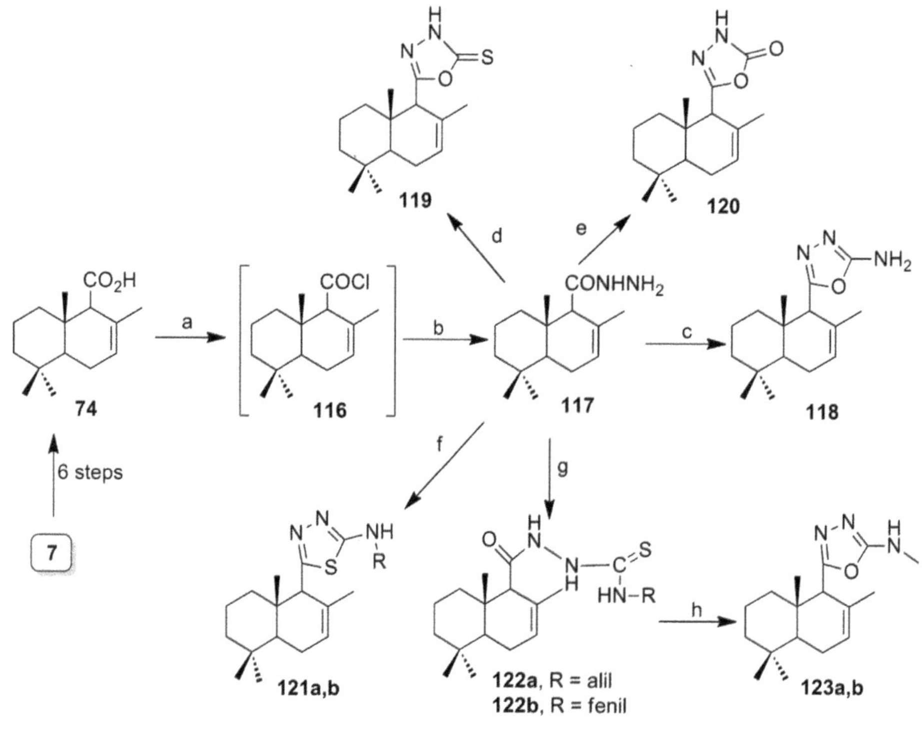

Reagents and reaction conditions: a. $(COCl)_2$, C_6H_6, 20°C, 1h, reflux, 1h; b. N_2H_4?H_2O (98%), CH_2Cl_2, 20°C, 10h, 65%; c. BrCN, $NaHCO_3$, dioxane, water, 20°C, 5h, 89%; d. TMTD, DMF, 90°C, 1.5h, 72%; e. CDI, Et_3N, THF, 0°C, 5h, 91%; f. R-NCS, Et_3N, H_2O, 20h, 70-91%; g. R-NCS, EtOH, 20°C, 4-5h, 97-98%; h. DCC, acetone, CH_3OH, 5, 10h, 83-89%.

Compounds 110, 111, 112a,b, 113a,b, 114, 115a,b, 117-120, 121a,b - 123a,b were tested *in vitro* for antifungal and antibacterial activity against pure cultures of five fungal species (*Aspergillus niger, Fusarium solani, Penicillium chrysogenum, Penicillium frequentans, Alternaria alternata*) and against Gram-negative (*Pseudomonas aeruginosa*) and Gram-positive (*Bacillus polymyxa*) bacteria. Compound 109 possesses significant antifungal activity with minimal inhibitory activity (MIC=0.125 μg/mL), being twice more active as the known antifungal agent Caspofungin (MIC=0.25 μg/mL) and also exhibits significant antibacterial activity (MIC=2.5 μg/mL) compared to the activity of the antibiotic Kanamycin (MIC=4.0 μg/mL). Compound 120 shows moderate antifungal activity (MIC=2.0 μg/mL) and antimicrobial activity (MIC=48.0 μg/mL) compared to Caspofungin and Kanamycin (Aricu et al., 2020).

Synthesis of Norlabdanic Compounds With 1,3-Thiazole Fragment

1,3-Thiazole is a very important fragment for the design of new drugs. Some thiazoles are also known to exhibit anticonvulsant, antimicrobial, anti-inflammatory, antitumor, and other important biological effects. In continuation of research on the synthesis of new compounds, which contain both terpenic and azaheterocyclic fragments, the synthesis of norlabdanic compounds with 1,3-thiazole fragment was carried out.

Figure 25. Synthesis of dinorlabdanic compounds with 1,3-thiazole fragment

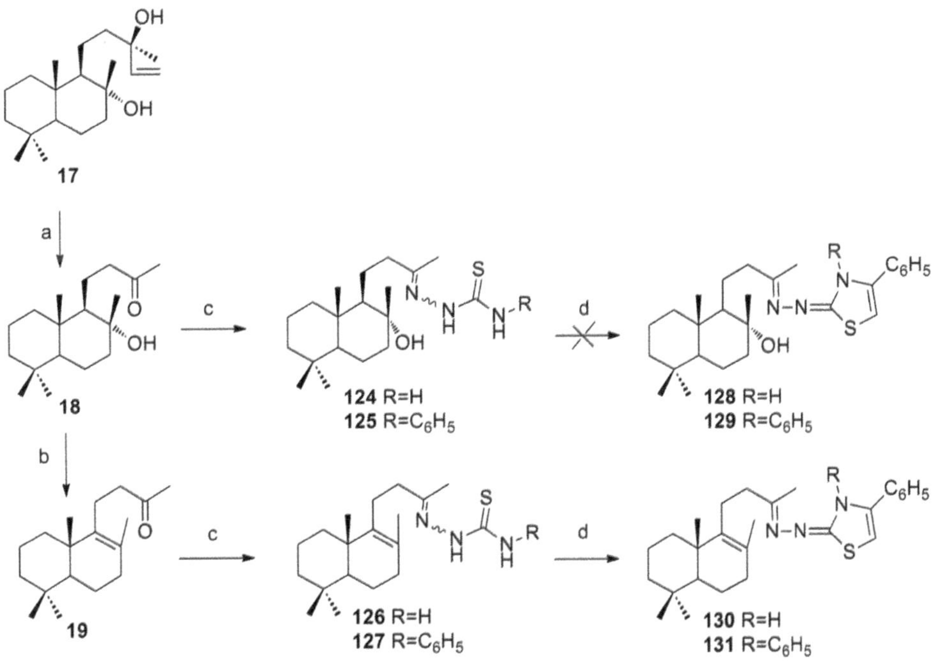

Reagents and reaction conditions: a. $KMnO_4$, acetone, 4h; b. H_2SO_4, benzene, reflux, 3h, c. $NH_2NHCSNH_2$ or $NH_2NHCSNHC_6H_5$, EtOH, 8-24h, 60-80°C, 75-93%; d. $C_6H_5COCH_2Br$, EtOH, 8-14h, 20°C, 58-67%.

(-)-Sclareol 17 was used as starting material for the synthesis of dinorlabdanic compounds; its oxidation with $KMnO_4$ gave 8α-hydroxy-15,16-dinorlab-13-one 18 with 90% yield. Upon interaction of hydroxyketone 18 with trimethylsilylmethanesulfonate ($MeSO_3SiMe_3$) in acetonitrile, 15,16-dinorlabd-8(9)-en-13-one 19 was synthesized in 80% yield (Figure 25). In the reaction of ketones 18 and 19 with thiosemicarbazide or 4-phenylthiosemicarbazide (molar ratio 1:1.1), dinorlabdanic compounds with thiosemicarbazone fragment 124-127 were obtained, each as a mixture of two isomers. The reaction of thiosemicarbazones 126 and 127 with 2-bromoacetophenone in alcohol (1:1 molar ratio) led to the formation of dinorlabdan compounds with 1,3-thiazole moiety 130 and 131 (Blaja et al., 2021). Several attempts to obtain compounds 128 and 129 by heterocyclization of thiosemicarbazones 124 and 125 with 2-bromoacetophenone were unsuccessful.

Figure 26. Synthesis of trinorlabdanic compounds with 1,3-thiazole fragment

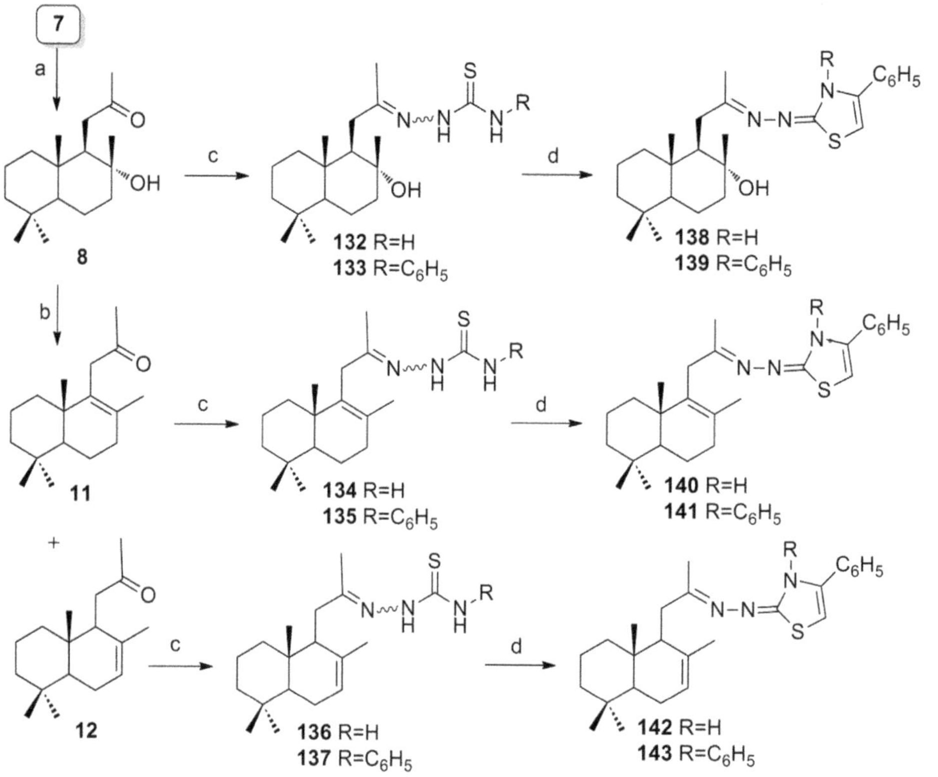

Reagents and reaction conditions: a. CH_3Li, Et_2O, 20°C, 15 min, 65%; b. $MeSO_3SiMe_3$, MeCN, 20°C, 15 min, 80%; c. $NH_2NHCSNH_2$ or $NH_2NHCSNHC_6H_5$, EtOH, 8-24 hours, 60-80°C, 65-83%; d. $C_6H_5COCH_2Br$, EtOH, 8-24h, 20°C, 52-66%.

The commercially available (+)-sclareolide 7 was also used as a starting material for the synthesis of trinorlabdanic compounds with 1,3-thiazole moiety. Interaction of (+)-sclareolide 7 with methyllithium (Me_3Li) gave 8α-hydroxy-14,15,16-trinorlab-12-one 8 in 65% yield according to the methodology. Treatment of hydroxyketone 8 with $MeSO_3SiMe_3$ in acetonitrile gave a mixture of 14,15,16-trinorlabd-8(9)-en-13-one 11 and 14,15,16-trinorlabd-7(8)-en-13-one 12, (ratio 4:1) known in 91% overall yield, which were successfully separated using silica gel column chromatography (Figure 26).

Reaction of ketones 8, 11, 12 with thiosemicarbazide or 4-phenylthiosemicarbazide (1:1.1 molar ratio) afforded trinorlabdane compounds with thiosemicarbazone moiety 132-137. Each of these thiosemicarbazones was obtained in an inseparable mixture of two isomers. Subsequently, in the reaction, the mixture of isomeric thiosemicarbazones was used, since it is known that during the reaction, the *Z*-isomer transforms into the more stable *E*-isomer. The reaction of thiosemicarbazones 132-137 with

2-bromoacetophenone in alcohol (1:1 molar ratio) led to the formation of trinorlabdanic compounds with 1,3-thiazole moiety 138-143.

As a raw material, for the synthesis of tetra- and pentanorlabdanic compounds, the methyl ester of 7-oxo-13,14,15,16-tetranorlabd-8-en-12-oic acid 24 was used, which was obtained from (+)-sclareolide 7, in two steps (with a total yield of 76%). Ketoester 24 after interaction with potassium hydroxide in ethanol for three hours was converted to drim-8(9)-en-7-one 25 in 98% yield (Figure 27).

Figure 27. Synthesis of tetra- and pentanorlabdanic compounds with 1,3-thiazole fragments

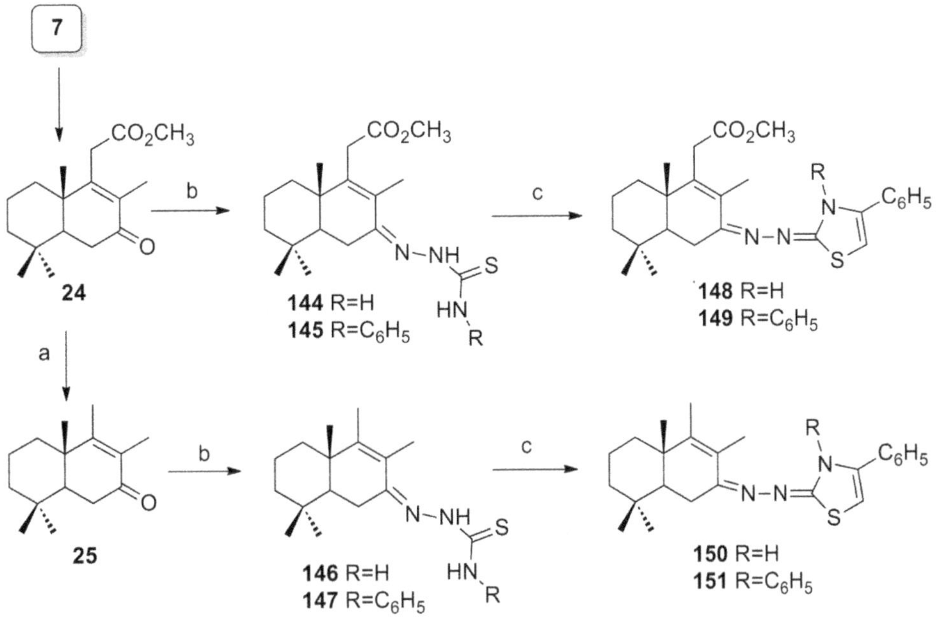

Reagents and reaction conditions: a. KOH, EtOH, reflux, 3h, 98%, b. $NH_2NHCSNH_2$ or $NH_2NHCSNHC_6H_5$, EtOH, 24h, 71-83%, c. $C_6H_5COCH_2Br$, EtOH, 4-6h, 20°C , 81-90%.

Subsequently, in the reaction of ketones 24 and 25 with thiosemicarbazide or 4-phenylthiosemicarbazide in a molar ratio of 1:1.1, tetra- and pentanorlabdanic compounds with thiosemicarbazone fragment 144-147 were obtained. The tetra- and pentanorlabdanic compounds with 1,3-thiazole moiety 148-151 were obtained *via* the heterocyclization reaction of thiosemicarbazones 144-147 with 2-bromoacetophenone. The structures of all newly synthesized compounds were established based on spectral data (IR, ^{1}H, ^{13}C and ^{15}N NMR).

The obtained compounds with thiosemicarbazonic and thiazole fragments were tested *in vitro* for antifungal and antibacterial activity on pure cultures of five species of fungi (*Aspergillus niger*, *Fusarium solani*, *Penicillium chrysogenum*, *Penicillium frequentans*, *Alternaria alternata*) and Gram-negative bacteria (*Pseudomonas aeruginosa*) and Gram-positive (*Bacillus polymyxa*).

Compounds 125 and 132 show antifungal activity at a minimum inhibitory concentration (MIC=0.25 and 0.19 μg/mL, respectively), compared to the activity of the antifungal drug Caspofungin (MIC=0.25 μg/mL) and also show antibacterial activity (MIC=4.0 and 3 μg/mL, respectively), compared to the activity of the antibiotic Kanamycin (MIC=4.0 μg/mL). Compound 144 has moderate antifungal activity with minimum inhibitory concentration (MIC=1.5 μg/mL) compared to the activity of the antifungal agent Caspofungin (MIC=0.2 μg/mL) and also shows significant antibacterial activity (MIC=0.125 μg/mL), that is twenty-four times more active than the well-known antibiotic Kanamycin (MIC=3.0 μg/mL).

CONCLUSION

This paper includes recent data on the syntheses of norlabdanic compounds containing amino groups or azine moieties, hydrazide, dihydrazide, guanidine, and thiosemicarbazones, as well as heterocyclic units such as diazine, 1,2,4-triazole, 1,3,4-oxadiazole, 1,3,4-thiadiazole, 1,3-thiazole which were performed based on the commercially available (+)-sclareolide. The structures of all synthesized compounds were established based on spectral data (^{1}H, ^{13}C and ^{15}N NMR, IR), mass spectroscopy and the stereochemistry of some compounds was also confirmed by single crystal X-ray diffraction. Reported hybrid terpeno-heterocyclic compounds were tested *in vitro* and showed pronounced to good activity against five fungal strains and two bacterial species. Therefore, compounds possessing pronounced antimicrobial activity can be further investigated for their use in the treatment of fungal and bacterial diseases. It was proved biotechnologically that two of the synthesized amides bearing 3-amino-1,2,4-triazole and *N*-aminocarbazole units have a strong stimulating effect on the biosynthesis of antioxidant compounds by the cyanobacteria species *Nostoc linckia*. The assessment of antiproliferative and cytotoxic activity of these compounds showed that two of the tested guanidine derivatives are more potent than cisplatin against Colo 205 and Colo 320 cell lines and may be act drug lead molecules in cancer chemotherapy. A number of the reported compounds were synthesized by combining classical and non-conventional methods such as microwave irradiation, which led to the acceleration of the processes, the increase in yields and the reduction of the amounts of energy and the volumes of solvents used and which have a beneficial impact on the environment.

AKNOWLEDGMENT

This research was funded by the Moldovan National Project "PLANTERAS", Nr. 20.80009.8007.03.

REFERENCES

Allouche, N., Apel, C., Martin, M.-T., Dumontet, V., Guéritte, F., & Litaudon, M. (2009). Cytotoxic sesquiterpenoids from Winteraceae of Caledonian rainforest. *Phytochemistry, 70*(4), 546–553. doi:10.1016/j.phytochem.2009.01.012

Aricu, A., Ciocarlan, A., Lungu, L., Barba, A., Shova, S., Zbancioc, G., Mangalagiu, I. I., D'ambrosio, M., & Vornicu, N. (2016). Synthesis, antibacterial, and antifungal activities of new drimane sesquiterpenoids with azaheterocyclic units. *Medicinal Chemistry Research, 25*(10), 2316–2323. doi:10.100700044-016-1665-0

Aricu, A., Kuchkova, K., Duca, Gh., Secara, E., Barba, A., Dragalin, I., & Ungur, N. (2019). *1-(4-((4aS,8aS)-2,5,5,8a-Tetramethyl-3,4,4a,5,6,7,8,8a-octahydronaphthalen1-yl)butan-2-yl)guanidine compound with antiproliferative and cytotoxic activity.* Patent MD, Nr. 4630, (5), 41.

Aricu, A., Kuchkova, K., Duca, Gh., Secara, E., Barba, A., Dragalin, I., & Ungur, N. (2019). *N-Carbamimidoyl-2-((8aS)-2,5,5,8a-tetramethyl-3,4,4a,5,6,7,8,8a-octahydronaphthalen-1-yl)acetamide compound with antiproliferative and cytotoxic activity.* Patent MD, Nr. 4631, (5), 41-42.

Arîcu, A., Lungu, L., Ciocarlan, A., & Vornicu, N. (2018). *(1R,2R,8aS)-1-((5-Mercapto-1,3,4-thiadiazol-2-yl)methyl)-2,5,5,8a-tetramethyldecahydronaphthalen-2-ol compound with antifungal and antibacterial properties.* Patent MD, Nr. 4580, (7), 37.

Aricu, A., Mangalagiu, I., Ciocarlan, A., Lungu, L., Zbancioc, Gh., & Vornicu, N. (2015). *11,12-Bis-p-tolyl-piridazonil-drim-5(6),8(9)-diene-7-one with antifungal and antibacterial propertie.* Patent MD, Nr. 4370, (9), 24-25.

Aricu, A. N., Kuchkova, K. I., Barba, A. N., Dragalin, I. P., Shova, S. G., Vornicu, N., Gorinchoi, E. K., Sekara, E. S., Lungu, L. V., Niculaua, M., Ungur, N. D., & Vlad, P. F. (2016). Synthesis from norambreinolide, structure, and antimicrobial activity of dihomodrimane sesquiterpenoids with azine, hydrazide, and dihydrazide fragments. *Chemistry of Natural Compounds*, *52*(6), 1029–1036. doi:10.100710600-016-1854-6

Aricu, A. N., Kuchkova, K. I., Secara-Cushnir, E. S., Barba, A. N., Ungur, N. D., & Vornicu, N. (2020). Synthesis and antimicrobial activity of new drimane and homodrimane sesquiterpenoids with oxadiazole and thiadiazole fragments. *Chemistry of Natural Compounds*, *56*(4), 656–662. doi:10.100710600-020-03115-x

Aricu, A. N., Kuchkova, K. I., Sekara, E. S., Barba, A. N., Dragalin, I. P., Ungur, N. D., Melnic, E., & Kravtsov, V. Kh. (2018). Synthesis and structure of drimane sesquiterpenoids containing diazine, 1,2,4-triazole and carbazole rings. *Chemistry of Natural Compounds*, *54*(3), 455–460. doi:10.100710600-018-2378-z

Blaja, S. P., Lungu, L. V., Kuchkova, K. I., Ciocarlan, A. G., Barba, A. N., Vornicu, N., & Aricu, A. N. (2021). Norlabdane compounds containing thiosemicarbazone or 1,3-thiazole fragments: Synthesis and antimicrobial activity. *Chemistry of Natural Compounds*, *57*(1), 101–110. doi:10.100710600-021-03292-3

Ciocarlan, A., Aricu, A., Lungu, L., Edu, C., Barba, A., Shova, S., Mangalagiu, I. I., D'ambrosio, M., Nicolescu, A., Deleanu, C., & Vornicu, N. (2017). Synthesis of novel tetranorlabdane derivatives with unprecedented carbon skeleton. *Synlett*, *28*(5), 565–571. doi:10.1055-0036-1588651

Cucicova, C., Rudic, V., Arîcu, A., Cepoi, L., Rudi, L., Secara, E., Valuţa, A., Barbă, A., Miscu, V., Vlad, P., & Chiriac, T. (2015a). *N-($\Delta^{8,13}$-Bicyclohomofarnesenoylamino)-carbazole and method for cultivating the cyanobacterium Nostoc Linckia, using it.* Patent MD, Nr. 4326 (2), 27.

Cucicova, C., Rudic, V., Arîcu, A., Cepoi, L., Rudi, L., Secara, E., Valuţa, A., Barbă, A., Miscu, V., Vlad, P., & Chiriac, T. (2015b). *N-($\Delta^{8,13}$-Bicyclohomofarnesenoyl)-3-amino-1,2,4-triazole and method for cultivating the cyanobacterium Nostoc Linckia, using it.* Patent MD, Nr. 4327 (2), 27-28.

Duca, G., Aricu, A., Kuchkova, K., Secara, E., Barbă, A., Dragalin, I., Ungur, N., & Spengler, G. (2019). Synthesis, structural elucidation and biological evaluations of new guanidine-containing terpenoids as anticancer agents. *Natural Product Research, 21*(33), 3052–3056. doi:10.1080/14786419.2018.1516658 PMID:30445867

Duca, G., Aricu, A., Lungu, L., Tenu, N., Ciocarlan, A., Gutu, Y., Dragalin, I., & Barba, A. (2018). Synthesis of new homodrimane sesquiterpenoids containing diazine, 1,2,4-triazole and carbazole rings. *Chemistry Journal of Moldova, 13*(1), 69–73. doi:10.19261/cjm.2017.458

Duca, G., & Vaseashta, A. (Eds.). (2020). *Handbook of Research on Emerging Developments and Environmental Impacts of Ecological Chemistry.* IGI Global., doi:10.4018/978-1-7998-1241-8

Fraga, B. M. (2013). Natural sesquiterpenoids. *Natural Product Reports, 30*(9), 1226-1264. doi:10.1039/c3np70047j

Grahman, P. L. (2001). *An Introduction to Medicinal Chemistry* (2nd ed.). Oxford University Press.

Greenwood, D., Finch, R., Davey, P., & Wilcox, M. (2007). Antimicrobial Chemotherapy (5th ed.). Oxford University Press.

Jansen, B. J. M., & De Groot, A. (2004). Occurrence, biological activity and synthesis of drimane sesquiterpenoids. *Natural Product Reports, 21*(4), 449-447. doi:10.1039/b311170a

Komala, I., Ito, T., Nagashima, F., Yagi, Y., & Asakawa, Y. (2010). Cytotoxic, radical scavenging and antimicrobial activities of sesquiterpenoids from the Tahitian liverwort *Mastigophora diclados* (Brid.) Nees (Mastigophoraceae). *Journal of Natural Medicines, 64*(4), 417-422. doi:10.1007/s11418-010-0423-8

Kuchkova, K., Aricu, A., Barba, A., Vlad, P., Shova, S., Secara, E., Ungur, N., Zbancioc, G., & Mangalagiu, I. I. (2013). An efficient and straightforward method to new organic compounds: Homodrimane sesquiterpenoids with diazine units. *Synlett, 24*(6), 697–700. doi:10.1055-0032-1318253

Kuchkova, K., Aricu, A., Secara, E., Barba, A., Vlad, P., Macaev, F., Melnic, E., & Kravtov, V. (2015). Synthesis and structure of homodrimanic sesquiterpenoids containing 1,2,4-triazole and carbazole rings. *Chemistry of Natural Compounds, 51*(14), 684–688. doi:10.100710600-015-1384-7

Kuchkova, K., Aricu, A., Secara, E., Barba, A., Vlad, P., Ungur, N., & Mangalagiu, I. I. (2014). Design, synthesis, and antimicrobial activity of some novel homodrimane sesquiterpenoids with diazine skeleton. *Medicinal Chemistry Research, 23*(3), 1559–1568. doi:10.100700044-013-0720-3

Kuchkova, K. I., & Aricu, A. N. (2009). Synthesis of 11-aminodrim-7-ene from drimenol. *Chemistry of Natural Compounds, 45*(3), 367–370. doi:10.100710600-009-9341-y

Kuchkova, K. I., Aricu, A. N., Barba, A. N., Sekara, E. S., Vlad, P. F., & Ungur, N. D. (2014). Synthesis of 12-amino-11-dihomodrimane sesquiterpenoids from norambreinolide. *Chemistry of Natural Compounds, 50*(3), 458–461. doi:10.100710600-014-0986-9

Kuchkova, K. I., Aricu, A. N., Barba, A. N., Vlad, P. F., Lipkovskii, J., Simonov, Yu. A., & Kravtov, V. Kh. (2011). Synthesis of nitrogen-containing drimane sesquiterpenoids from 11-dihomodrim-8(9)-en-12-one. *Chemistry of Natural Compounds, 47*(2), 223–228. doi:10.100710600-011-9888-2

Kuchkova, K. I., Aricu, A. N., Vlad, P. F., Deleanu, C., & Nicolescu, A. (2010). Synthesis of *N*-containing drimane sesquiterpenoids from 11-dihomodriman-8α-ol-12-one. *Chemistry of Natural Compounds, 46*(4), 539–544. doi:10.100710600-010-9671-9

Kuchkova, K. I., Aryku, A. N., Sekara, E. S., Barba, A. N., Dragalin, I. P., Vlad, P. F., & Ungur, N. D. (2014). Synthesis of 13-amino-14,15-dinorlabd-8(9)-ene from sclareol. Russian Chemical Bulletin, International Edition, 63(9), 1-3. doi:10.100711172-014-0703-7

Lhinhatrakool, T., Prabpai, S., Kongsaeree, P., & Sutthivaiyakit, S. (2011). Antiplasmodial sesquiterpene alkaloids from the roots of *Maytenus mekongensis*. *Journal of Natural Products, 74*(6), 1386–1391. doi:10.1021/np200014k

Lungu, L. (2015). Synthesis of new nitrogen-containing drimane and homodrimane sesquiterpenoids from sclareolide. *Chemistry Journal of Moldova, 10*(2), 58–61. doi:10.19261/cjm.2015.10(2).07

Lungu, L., Ciocarlan, A., Barba, A., Shova, S., Pogrebnoi, S., Mangalagiu, I., Moldoveanu, C., Vornicu, N., Ambrosio, M., Babak, M. V., Arion, V., & Aricu, A. (2019). Synthesis and evaluation of biological activity of homodrimane sesquiterpenoids bearing hydrazinecarbothioamide or 1,2,4-triazole unit. *Chemistry of Heterocyclic Compounds, 55*(8), 716–724. doi:10.100710593-019-02526-1

Lungu, L., Ciocarlan, A., Smigon, C., Ozer, I., Shova, S., Gutu, I., Vornicu, N., Mangalagiu, I., D'ambrosio, M., & Aricu, A. (2020). Synthesis and evaluation of biological activity of homodrimane sesquiterpenoids bearing 1,3,4-oxadiazole or 1,3,4-thiadiazole units. *Chemistry of Heterocyclic Compounds, 56*(5), 578–585. doi:10.100710593-020-02703-7

Secara, E. (2016). Synthesis of new drimane and homodrimane lactams by Beckmann rearrangement of some ketoximes. *Chemistry Journal of Moldova, 11*(1), 50–54. doi:10.19261/cjm.2016.11(1).06

Viegas-Junior, C., Danuello, A., Da Silva Bolzani, V., Barreiro, E. J., & Fraga, C. A. M. (2007). Molecular hybridization: A useful tool in the design of new drug prototypes. *Current Medicinal Chemistry, 14*(17), 1829–1852. doi:10.2174/092986707781058805

Zhao, G., Li, X., Chen, W., Xi, Z., & Sun, L. (2012). Three new sesquiterpenes from *Tithonia diversifolia* and their anti-hyperglycemic activity. *Fitoterapia, 83*(8), 1590–1597. doi:10.1016/j.fitote.2012.09.007

Zhou, H., Zhu, T., Cai, S., Gu, Q., & Li, D. (2011). Drimane sesquiterpenoids from the mangrove-derived Fungus *Aspergillus ustus*. *Chemical & Pharmaceutical Bulletin, 59*(6), 762–766. doi:10.1248/cpb.59.762

Chapter 18
Redox Technologies in Wastewater Treatment for Removal of Pharmaceutical (Cephalexin) Contaminants

Maria Vasile Gonta
Moldova State University, Moldova

Larisa Mocanu
Moldova State University, Moldova

Gheorghe Duca
Moldova State University, Moldova

ABSTRACT

This chapter presents the study of kinetics and mechanisms of transformation of EC antibiotics by advanced oxidation processes. The aim of this chapter was to investigate the degradation of cephalexin in wastewater by redox technologies. The removal kinetics of the antibiotic CEX could be described by a pseudo first-order reaction with a reaction rate of (1.42 · 10-6 M-1s-1), a rate constant of 0.020 s-1, and a half-life of 35 s. At heterogeneous oxidation, the degradation/mineralization kinetics decreases, so the reaction rate is (0.93 · 10-6 M-1s-1), the rate constant is 0.008 s-1, and the half-life time is 87 s. Thus, the CEX concentration has reduced by 80% and the COD values by 88% for 30 min. The data obtained reveal that the homogeneous oxidation Fe2+/H2O2/UV-C is a promising treatment for the degradation of cephalexin with an initial concentration range of (50-100) mg/L comparative with heterogeneous oxidation TiO2/H2O2/UV-A.

DOI: 10.4018/978-1-6684-7198-2.ch018

INTRODUCTION

β-lactam antibiotics, including cephalosporin, are among the most widely used antibiotics in human and veterinary medicine. The most quantity of β-lactam antibiotics are excreted unchanged in wastewater and, since they are not biodegradable, can enter the aquatic environment. The presence of these compounds in the environment raises concerns about their toxicity to aquatic organisms and the emergence of antibiotic-resistant bacteria (Benarab et al., 2020).

Contamination of natural water by pharmaceutical wastewater plays an essential role in the anthropogenesis impacts. Under the action of different external factors, there comes to pass degradation of toxicants and the formation of new substances including the more toxic ones. Removal of pharmaceutical compounds from the aqueous environment is one of the most significant priorities in the field of wastewater treatment. Having a complex composition, these pollutants are difficult to remove by conventional or biological methods, and the application of photochemical methods is an effective alternative. From this point of view, various advanced oxidation processes (AOPs) have shown the potential to remove antibiotics from an aqueous solution. Due to high non-degradability, non-toxicity, universal viability, and acceptability AOPs have become a promising option for efficiently removing persistent pharmaceutical compounds. According to the studied references, it was concluded that catalytic and photocatalytic processes may be potential methods for the removal cephalexin (CEX) in optimal conditions at low concentrations of antibiotics. However, this process must be adapted to real substrate concentrations of CEX.

In this context, the main purpose of research is to study the degradation of CEX from aqueous solution by applying homogeneous and heterogeneous oxidation. Further studies are, however, required to optimize the operating conditions for maximum degradation of multiple pharmaceuticals in wastewater under realistic conditions and on an industrial scale. Frequently investigated operational parameters include pH of the solution, reaction time, catalyst and oxidant concentration, catalyst loading, initial concentration of the pharmaceutical under investigation, wavelength, and light intensity. After establishing the optimal conditions, it has released degradation/mineralization of the CEX antibiotic over the initial concentrations range of 50÷200 mg/L which represents the real concentrations in the residual effluents.

BACKGROUND

As maintained by WHO and UNICEF, 780 million people do not have access to safe drinking water, of which 185 million use surface water to meet their daily needs. The risk associated with pollution with compounds derived from the pharmaceutical industry is highly topical. In the last decades, pollution by emerging contaminants has been paid more attention by the academic community (Gonta et al., 2022). On the one hand, the discharged pharmaceutical pollutants are toxic and are found in a considerable diversity, which makes the treatment of wastewater difficult (Verma et al., 2020; Mocanu et al., 2021). Concern about emerging contaminants, such as pharmaceutical compounds (PCs) has increased due to their negative impact on the ecosystem, as effluents from the pharmaceutical industries are discharged into the sewers of domestic wastewater, natural water flows, soil, etc. (Tran et al., 2018; Gonta et al., 2022; Vaseashta et al, 2021; Vaseashta et al, 2022). Also, it has been shown that pharmaceutical compounds in wastewater have negative environmental impacts. It was found that some bacteria and genes acquired resistance to antibiotics (Hansen et al., 2020,). Lethal effects on aquatic organisms, interference with the natural decomposition of organic matter, and a decrease in the diversity of microbial communities in various environments have been observed (Felis et al., 2019).

Among pharmaceutical compounds, antibiotics are receiving increased attention because they are responsible for the development of antibiotic-resistant bacteria and genes. Their presence in water bodies affects the aquatic ecosystem, and also poses a threat to human health (Pieri et al., 2020, Gonta et al., 2022). On the report of WHO (2015), ≈10% of the drugs used by patients in the Republic of Moldova are antibiotics. At the same time, WHO standards provide for the consumption of any type of antibiotics by six percent per country. As a result, since April 2017, these drugs can only be purchased with a prescription. This law came after alarms sounded about the uncontrolled use of antibiotics, starting in 2013.

A 2014 study showed that Moldavian people most often used the antibiotics ampicillin, amoxicillin, cephalothin, cefamandole etc. (Mocanu et al., 2021). According to point prevalence survey of healthcare-associated infections and antimicrobial use in hospitals of the Republic of Moldova, conducted in 2018 under the auspices of the Ministry of Health, Labour and Social Protection, 43% of patients receive antibiotic treatment. The study was conducted on a sample of 67 hospitals, 546 wards, including 10594 patients. Thus, it has been calculated that the highest prevalence of infections in intensive care units is 20%. Most often, antibiotics are prescribed to patients for treatment (74%): including the respiratory system - 47%, urinary tract infections - 12%, and eye/ear/nose/throat infections - 8.8% (Mocanu et al., 2020).

In nature, CFs are found in very low concentrations (from micrograms/liter to nanograms/liter), but their influence on the environment and human health is little studied (Cuerda-Correa et al., 2019; Sodhi et al., 2021). Antibiotics are used in both human and veterinary medicine, therefore CFs are identified in groundwater, surface water, soil etc. (Wang et al., 2016). Due to their bioaccumulation potential and their persistent nature, these compounds have a longer life span and activity in the environment. Pharmaceutical compounds (PCs) are found in effluents from wastewater treatment plants, and due to non-biodegradability enter natural waters in concentrations ranging from ng to µg/L (Hansen et al., 2020). In general, drugs are not fully disposed of at wastewater treatment plants. For example, it was observed a 7% removal for carbamazepine, and 99% for acetylsalicylic acid. Consequently, studies have shown that juvenile fish of the Rutilus species which have been gradually exposed to drugs from residual effluents for 150 days have experienced changes in the reproductive system (Gonta et al., 2022).

Despite the multitude of antibiotics, cephalexin is a preparation belonging to the class of first-generation cephalosporins, having a heteroatomic ring in its chemical structure, consisting of three carbon atoms and one nitrogen atom, characterizing cephalexin as β-lactam (Ribeiro et al., 2017). It is one of the most widely used antibiotics. This drug is commonly prescribed to treat several respiratory and urinary tract infections because it has a broad spectrum of antimicrobial action, thus being one of the most widely consumed antibiotics in the world (Sodhi et al., 2021). Studies show that only 10% of the consumed cephalexin is metabolized by the body. In fact of its therapeutic effect, the accumulation of this contaminant (C) in the environment leads to the appearance of multidrug-resistant pathogens, causing a major impact on human health and aquatic ecosystems.

Therefore, the study of CEX removal processes is important. Studies have shown that conventional processes have different removal rates ranging from 20 to 90%. In addition, large amounts of waste and by-products are formed that may be more toxic than the original antibiotic (Pulicharla et al., 2015). To roll back the inconveniences related to the application of conventional methods, the application of AOPs is an effective alternative (Figure 1). It is known that the viability of these methods depends on the source of hydroxyl (OH·) radical formation (Mocanu et al., 2021; Masmoudi et al., 2022).

$$C_{16}H_{17}N_3O_4S + HO\cdot \rightarrow CO_2 + H_2O + \text{intermidiate products} \quad (1)$$

Figure 1. AOPs for degradation/mineralization of cephalexin

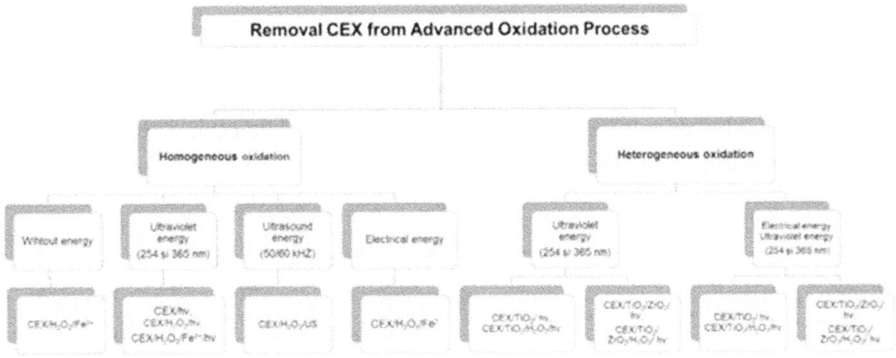

AOPs are of scientific interest because they have great potential to transform most organic pharmaceutical compounds into low molecular mass compounds (Masmoudi et al., 2022; Cordeiro et al., 2021; Gonzaga et al., 2021; Almasi et al., 2020; Ding et al., 2020; Antonin et al., 2019; Tavasol et al., 2020; Serna-Galvis et al., 2020). Indirect photolysis involves reactions with photo-induced reactive species, such as hydroxyl radicals, singlet oxygen, the triplet excited state of chromophoric dissolved organic matter formed in natural waters after light absorption by different species (nitrite, nitrate, dissolved organic matter, iron complexes), among others (Cordeiro et al., 2021; Gonta et al., 2022). Reactive oxygen species are generated, mainly hydroxyl (OH·) radicals, which are extremely oxidizing agents (E^0=2.8 V) and very reactive (half-life ~1 ns) (Masmoudi et al., 2022). These radicals react with organic compounds by electrophilic substitution, often resulting in contaminants that are less harmful than the original compound (Serna-Galvis et al., 2020). Moreover, complete mineralization of CEX (conversion up to CO_2, H_2O, and inorganic ions) can also be achieved without phase transfer of pollutants. The main aim of this research is to study the degradation of CEX in an aqueous solution by applying homogeneous and heterogeneous oxidation.

EXPERIMENTAL SECTION

The research was carried out within the Research Laboratory on Ecological Chemistry and Modern Chemical Technologies, Moldova State University.

Materials and Solutions

Deionized water has been used to provide solutions. All chemicals used in this study were chemically pure from Merck (Germany). The powder of cephalexin (99.8% purity) was supplied from S.A. Balkan Pharmaceuticals (Republic of Moldova) and used without any processing. Hydrogen peroxide solution

(30%) in stable form was used as a source of hydroxyl radicals. Titanium dioxide (mixture of rutile and anatase) and ammonium iron (II) sulphate hexahydrate sulfate (Fe^{2+}) were used as photocatalysts. The pH value of solutions was adjusted by using H_2SO_4 or NaOH 1.0 M until pH 2.0 and 3.8 (final condition).

Preparation of Cephalexin Stock Solution

The precise amount of $5.76 \cdot 10^{-4}$ M cephalexin (with technical specification in Table 1) was weighed with a digital weighing scale (Kern) and quantitatively transferred into a 100 mL volumetric flask to give a cephalexin stock solution with a concentration of 2.0 g/L. The main solution was provided daily because of the possible volatility of the drug.

Table 1. Chemical/molecular structure, technical specification of cephalexin, and pKa's values

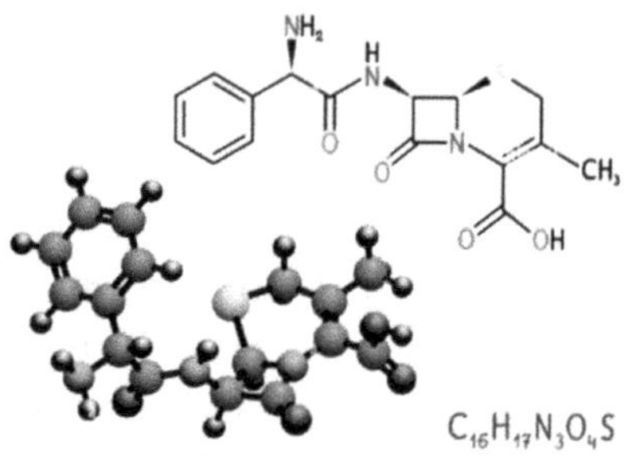

Molecular structure

Chemical structure	$C_{16}H_{17}N_3O_4S$
Molecular weight	347.4 g/mol
pKa' s	2.68, 7.23, and 11.91
Form	Solid
Solubility	10 g/L (la 298 ^{0}K)
Antibiotic activity spectrum	Gram-negative bacteria Gram-positive bacteria

Source: Ribeiro et al. (2017)

Experimental Set-Up and Procedure for Photo-Catalytic Oxidation of Cephalexin

The photochemical experiments were carried out using a lab-scale stainless steel photoreactor (Vilbert Lourmant VL 215 LC (50/60 Hz). The UV irradiation source was a low-pressure mercury vapor lamp (15 W, Philips Company). The lamp was placed horizontally above the cylindrical quartz reactor. All experiments were performed in a batch thermostatic reactor with a capacity of 250 mL (made of quartz glass, inner diameter: 15 cm, height: 10 cm). Uniform mixing was provided using a magnetic stirrer (MSH-20D). The water temperature in the reactor was also kept constant during the reaction by a thermostatic bath (Julabo). The working volume of antibiotic solution in the reactor was set to 100 mL to adsorb UV energy when it was placed in the device. The working concentration (50 mg/L) was chosen based on real values found in pharmaceutical industrial wastewater. The pH value was measured by means of a Consort C3030 pH meter and was adjusted with H_2SO_4 or NaOH solution.

For *homogenous oxidation of cephalexin,* the required amount of iron (II) was added to the reactor which contains 50 mg/L CEX in an aqueous solution and the reaction began with the addition of a certain quantity of H_2O_2 solution. Aliquots were withdrawn from the reactor at selected intervals (1, 5, 10, 30, 60, and 120 min) for COD_t and C_t. *For heterogeneous oxidation of cephalexin, t*he required amount of catalyst (titanium dioxide) was added to the reactor which contains 50 mg/L CEX in an aqueous solution, and was stirred for 2 min (500 rpm). The reaction began with the addition of a certain quantity of H_2O_2 solution. Aliquots were withdrawn from the reactor at selected intervals (1, 5, 10, 30, 60, and 120 min) for COD_t and C_t. Prior to analysis, the samples taken from photocatalytic experiments were centrifuged for 2 minutes to remove TiO_2.

Chemical Analysis

The spectrophotometric determinations were performed using the UV/Vis T80+ spectrophotometer equipped with spectrum recording and data processing software. Spectrophotometry was applied for the qualitative and quantitative determinations of cephalexin present in synthetic effluents: antibiotic concentration (C_t) and COD_t was determined according to the methods presented (Bansal, 2018; Surpateanu et al., 2014).

The substrate concentration (CEX) was determined from the calibration curve, according to the regression equation: A = 0.0298 x + 0.00458, where the correlation coefficient is 0.9994. Beers law is obeyed over the concentration range of 1.0÷100.0 mg/L. The degradation rate of CEX (%) was determined by Eq. 2.

$$\boldsymbol{CEX(\%)} = \frac{(C_0 - C_t)}{C_0} \cdot 100\% \qquad (2)$$

where C_0, and C_t represent the initial and final concentration (mg/L) after different times.

Kinetic Analysis

Chemical kinetics is applicable for investigating the rate of chemical compound reactions - the isolation method. The rate of a reaction could be expressed by decreasing the concentration of a reactant in reaction time or increasing the concentration of their product per reaction time (Seid-Mohammadi et al., 2019). In this study, the pseudo-first-order kinetics models was applicable for CEX removal and was calculated using Eq. 3. Thus, the kinetics of the CEX antibiotic removal was studied in optimum conditions by using the above-mentioned models,

$$\boldsymbol{ln\frac{c_0}{c_t} = -kt}\text{ , s}^{-1} \qquad (3)$$

where C_0 and C_t are the initial and final concentrations of CEX antibiotic (mg/L), respectively, and k is a removal constant. Finally, k value is equal to the slope of the plot of ln (C_0/C_t) versus time t.

Mineralization of CEX

The chemical oxygen demand (COD, mgO/L) was investigated to determine the efficiency of removal organic compounds. COD values were determined from the calibration curve, according to the regression equation: A = 0.0016 x + 0.003, where the correlation coefficient is 0.9998. The oxidation/mineralization rate was determined by Eq. 4,

$$Oxid/min\ (\%) = \frac{(COD_0 - COD_t)}{COD_0} \cdot 100\% \tag{4}$$

where COD_0 and COD_t represent the initial and final concentrations of organic compounds after different times.

MAIN FOCUS OF THE CHAPTER

To optimize the oxidation process of CEX with the Photo-Fenton reagent and titanium dioxide, the working concentration of the antibiotic in aqueous solution was 50 mg/L. This initial concentration was chosen based on actual values determined in pharmaceutical industrial wastewater (Cuerda-Correa et al., 2019).

Effect of the Temperature

The performance of removal cephalexin by AOPs has insignificantly influenced by increasing temperature from 10 up to 40°C. If the temperature has exceeded 40°C, the oxidation/mineralization has decreased. It has been found that the optimum Photo-catalytic reaction temperature has been optimum at 25÷30°C (Gonta et al., 2022).

Effect of the pH Value

The photo-oxidation performance of the system containing CEX depends significantly on the reaction medium. The effects of varying experimental conditions such as pH values, initial concentration of oxidant, catalyst, and substrate, reaction time on the removal of CEX has also examined.

Recent studies show the high efficiency removal of antibiotics at pH 3.3 (Hirte et al., 2016; Mocanu et al., 2021; Gonta et al., 2022) and recommend performing AOPs of antibiotics in the pH range of 2÷4. Since at $pH < 2$, the stability of H_2O_2 decreases due to the formation of oxonium ions ($H_3O_2^+$), which leads to a decrease in its reactivity with catalysts.

Figure 2. Effect of pH on the degradation efficiency of cephalexin at using Photo-Fenton reagent (left) and titanium dioxide (right); $[CEX]_0 = 50$ mg/L, $[Fe^{2+}]_0 = 0.56$ mg/L, $[TiO_2]_0 = 25$ mg/L, $[H_2O_2]_0 = 3.4$ mg/L, UV = 254 and 365 nm

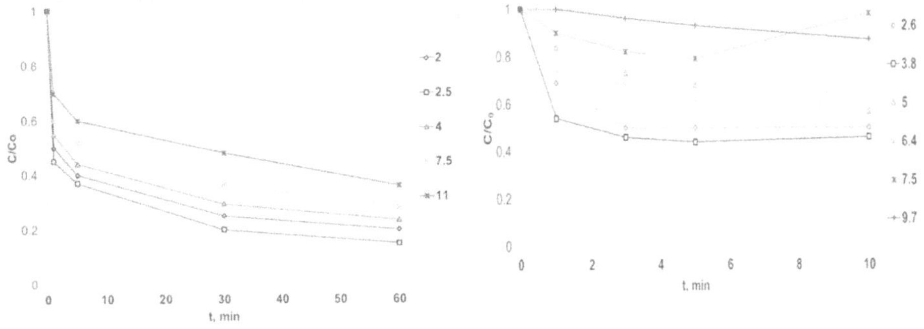

According to Figure 2, the removal of CEX in an aqueous solution depends on the pH value, which directly influences the structure of the antibiotic, the rate constant (k), as well as the production of hydroxyl (HO•) radicals. Different species of iron ions are formed in the reaction system, which are the main components responsible for the decomposition of hydrogen peroxide and the generation of hydroxyl radicals. The effect of initial pH on the CEX removal rate in the Photo-Fenton system was examined under different pH values ranging from 2 to 11. This range was made in accordance with the pKa's values of the CEX (Table 1).

Thus, in the first minute, 50% of CEX had degraded by irradiation of UV-C. After which there is a decrease of another 20% for 30 min. For this study, CEX removal has obtained 88% in the acidic conditions (at pH 2.5) and, respectively, decreases non-essential with increasing pH values (83%). The decreased CEX concentration is due, on the one hand, to the hydrolysis and dimerization reaction (by nucleophilic attack of the amino side chain on the β-lactam carbonyl) with the formation of intermediates, and on the other hand it depends on the number of hydroxyl radicals.

As stated above, the optimum pH has been 2.5. This fact is also confirmed by the magnitude of the rate constant k_2 which is 0.0165 min^{-1} (pH 2.5) compared to the constants k_1 (pH 2) and k_3 (pH 4) which are 0.0109 and 0.0161 min^{-1} respectively. Decreasing the rate constant with increasing pH values is explained by the molecular structure of CEX that tends to absorb the positive ion. Therefore, when the pH is strongly acidic, increases the positive charge in the reaction medium. By absorbing the positive charge changes its structure, because the structure of CEX contains carboxyl groups, i.e. It has two proton binding sites, which have a low acidity constant with a β-lactam group of the carboxylic acid (pKa's 2.7 and 3.6) (Ajoudanian et al., 2015). For this reason, the CEX molecules are more available in the medium with positive charges, which led to the removal of the antibiotic up to 88%.

The effect of pH on the photocatalytic reaction by titanium dioxide has been extensively studied due to the fact that photocatalytic water treatment is highly dependent on the pH as it affects the charge on the catalyst particles, size of aggregates, the position of conductance and valance bands. Furthermore, the surface of the dioxide titanium can be protonated or deprotonated under acidic or alkaline conditions. Therefore, the surface of the TiO_2 is positively charged under acidic conditions and negatively charged under alkaline conditions (Hirte et al., 2016; Gonta et al., 2022).

In order to study the effect of pH on heterogeneous degradation of CEX, a series of experimental research have performed in the pH range from 2.5 to 9.7. As a result, it was found that CEX removal in the presence of hydrogen peroxide, catalyzed by titanium dioxide upon irradiation with UV-A rays in the pH range of (3.5÷4). Degradation has reached up to 50% for 10 min. More than that, with increasing reaction time or pH value, the CEX removal has decreased (30%), which has indicated that the catalyst is inactivated under these working conditions.

According to the experimental results, the efficiency of photocatalytic process depends on several variables. It may be concluded that the highest CEX performance of oxidation was achieved at pH 2.5 for homogeneous oxidation and pH 3.8 for heterogeneous oxidation.

Effect of UV Irradiation

Photo-oxidation treatment can degrade the antibiotics using only irradiation UV. Direct photo-oxidation occurs through the interaction of the UV radiation emitted by the lamp with the antibiotic molecule, which absorbs photons and excites electrons to their maximum electron energy, breaking the molecule and degrading it through the generation of free radicals (Paniagua et al., 2019; Cordeiro et al., 2021). Indirect photolysis, on the other hand, involves reactions with photo-induced reactive species, such as hydroxyl radicals, singlet oxygen, and the triplet excited state of chromophore dissolved organic matter formed in natural waters after light absorption by different species (nitrite, nitrate, dissolved organic matter, iron complexes) (Lastre-Acosta et al., 2019).

To observe the effect of indirect photolysis of cephalexin due to UV irradiation, the experiment was conducted at 50 mg/L initial concentration of substrate of various systems (Figure 3). After 60 s UV-C irradiation, the degradation rate of CEX by applying homogenous oxidation was 60, 72, and 78%, respectively. The degradation was due to antibiotic hydrolysis. The hydrolysis reaction would proceed through the attack of the nucleophile H_2O to the β-lactam ring followed by ring opening (Cordeiro et al., 2021). Upon cephalexin oxidation, fragmentation of the molecule takes place, yielding kinetically unstable by-products that are more toxic than the precursor (Mocanu et al., 2021).

Degradation of cephalexin by heterogeneous oxidation was achieved at pH 3.8 (which was experimentally established as the optimum pH), CEX concentration of 50 mg/L, and H_2O_2 concentration of 3.4 mg/L by UV irradiation. According to the experimental research (Figure 3 left), it has been established that heterogeneous processes have high efficiency when irradiated with monochromatic rays of the UV-A because the titanium dioxide semiconductor can only be activated by light with a wavelength ≤ 390 nm (Tavasol et al., 2020; Al-Furaiji et al., 2020). Based on the results shown in Figure 3, it can be observed that the highest reaction rate ($1.42 \cdot 10^{-6}$, $M^{-1}s^{-1}$), half-life (35 s) and degradation (78%) has reached at the homogeneous oxidation of CEX ($CEX/H_2O_2/Fe^{2+}/UV\text{-}C$) compared to the heterogeneous oxidation ($CEX/H_2O_2/TiO_2/UV\text{-}A$). For the system, the kinetic parameters have been lower: reaction rate - $0.93 \cdot 10^{-6}$, $M^{-1}s^{-1}$, half-life - 87 s and degradation - 38%. In conclusion, homogeneous oxidation of CEX is much more efficient than heterogeneous oxidation.

Figure 3. Effect of UV radiation on the degradation efficiency of cephalexin for homogenous oxidation (left) and heterogeneous oxidation (right); $[CEX]_0 = 50$ mg/L, $[Fe^{2+}]_0 = 0.56$ mg/L, $[TiO_2]_0 = 25$ mg/L, $[H_2O_2]_0 = 3.4$ mg/L, t = 60 s

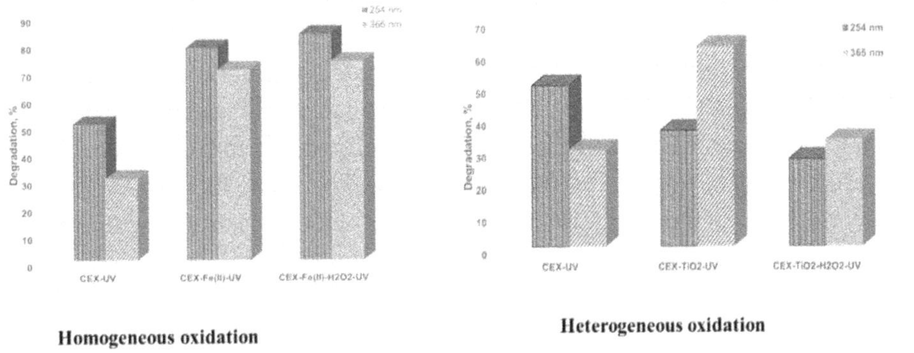

Effect of Oxidant Addition

The concentration of hydrogen peroxide is another parameter that influences the photo-catalytic process (Figure 4). The optimum concentration of hydrogen peroxide depends on the concentration of the substrate. Theoretically, 4.38 mg/L of hydrogen peroxide is required for the complete mineralization of one mg/L of CEX. Despite of, an excess of hydrogen peroxide for the complete removal of CEX from wastewater is required.

Thus, the removal efficiency of 50 mg/L CEX in aqueous solution in the four systems was, on average, 60, 78, 85, and 81% respectively. The degradation was achieved with a reaction rate of 1.3, 1.1, 1.4, and $1.5 \cdot 10^{-6}$ M s^{-1} respectively and the COD removal was 85, 78, 83, and 80%.

It can be concluded that high concentrations of peroxide hydrogen act as an inhibitor of hydroxyl radical formation and, as a result, reduce the efficiency of the oxidation process of CEX.

Figure 4. Photo-degradation of CEX by UV 254 nm (left) and kinetic adjustment for the pseudo first order model for CEX degradation (right) at different H_2O_2 concentration using Photo-Fenton reactor; $[CEX]_0 = 50$ mg/L, $[Fe^{2+}]_0 = 0.56$ mg/L, pH = 2.5

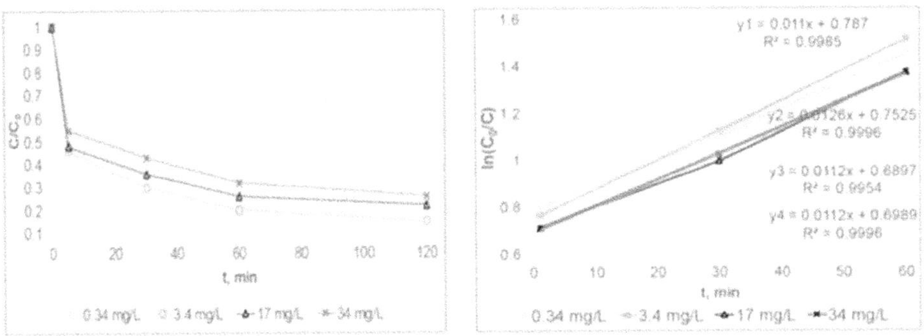

Effect of Catalyst Addition

To determine the optimum catalyst concentration, the concentration of hydrogen peroxide (3.4 mg/L) was kept constant and the concentration of iron(II) ions was varied in the range from 0.56 to 28 mg/L, taking into account that the maximum allowable concentration (MAC) of iron(II) ions) is 10 mg/L.

Based on the obtained results (Figure 5 left), it is observed that as the concentration of iron(II) ions increases at 0.56 to 28 mg/L, the photo-degradation performance of CEX decreases, on average from 75 to 45% after 60 min of reaction. This inefficiency is also confirmed by the rate constants $k_1 > k_2 > k_4 > k_3$, which have values as follows: 0.0121, 0.01, 0.0064 and 0.0085 min^{-1} for 0.56, 2.8, 5.6, and 28 mg/L iron(II) ions. Therefore, the photo-degradation performance of CEX increased with increasing iron(II) ions concentration until an optimum level 0÷0.56 mg/L and then started to decrease 2.8÷28 mg/L.

It has established that the degradation of the CEX in aqueous solution is proportional to the concentration of OH radicals. As the catalyst concentration increases, the excess iron(II) ions react with the hydroxyl radicals and reduce their concentration, thus reducing the efficiency of the process (Cordeiro et al., 2021).

Figure 5. Photo-degradation of CEX by UV 254 nm (left) and Kinetic adjustment for the pseudo first order model for CEX degradation (right) at different iron(II) ions concentration using Photo-Fenton reactor; $[CLX]_0 = 50$ mg/L, $[H_2O_2]_0 = 3.4$ mg/L, pH = 2.5

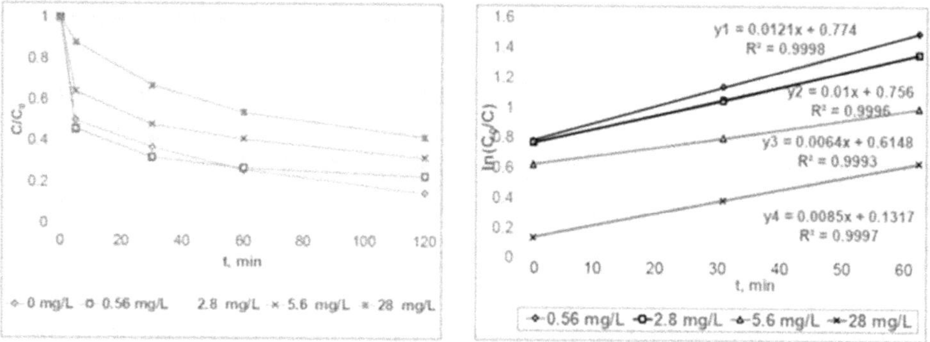

Effect of Substrate Concentration

The rate of photocatalytic degradation of certain pollutants depends on its nature, concentration and other existing compounds in the water matrix. In order to optimize the initial CEX concentration, the removal effect and the degree of mineralization oxidation in the concentration range of (50÷200) mg/L has investigated. According to Figure 6a, the efficiency of CEX degradation decreased as the concentration substrate increased from 50 to 200 mg/L. Moreover, a high degradation was achieved at 120 min of reaction at a concentration of 50 mg/L (90%). In addition, Figure 6b displays that the rate constants, likewise, decrease with increasing substrate concentration, as follows: $k_1 > k_2 > k_3$. Hydroxyl radicals can easily remove a high concentration of contaminants; but, as the concentration of CEX increases, the amount of hydroxyl ions produced is not sufficient for the complete removal of the antibiotic (Ajoudanian et al., 2015). These results are in line with the results of other groups of researchers (Cordeiro et al., 2021; Serna-Galvis et al., 2020; Abdolmotaleb et al., 2019).

The same law is observed for chemical oxygen demand (COD): COD_t values increase, on average, from 14 to 30 mgO/L with increasing of CEX concentration. Although, the COD removal increased with the increase of CEX concentration from 82, 88, and 91% for the concentration of 50, 100, and 200 mg/L CEX respectively (Figure 6c), the COD_t values do not fall within the maximum available concentration, MAC = 6÷8 mg O/L.

The results of this study, as well as data from the specialized literature, indicate a high efficiency of photocatalytic methods in the oxidative degradation of organic pollutants in aquatic environments. The experimental results reveal that the CEX removal process could be described by pseudo first order reaction.

According to the obtained data, as a result of the degradation, the antibacterial activity is reduced by the action of hydroxyl radicals at the nucleus of penicillin, the toxicity decreases and two stereoisomers of penicillinoic acid are formed.

Figure 6. Photo-degradation/oxidation of CEX by UV-C irradiation at different cephalexin concentrations using Photo-Fenton reactor; $[Fe^{2+}]_0 = 0.56$ mg/L, $[H_2O_2]0 = 3.4$ mg/L, pH = 2.5

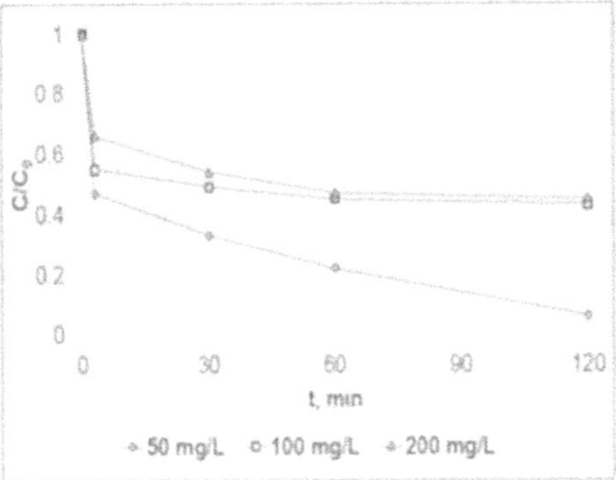

a – CEX removal

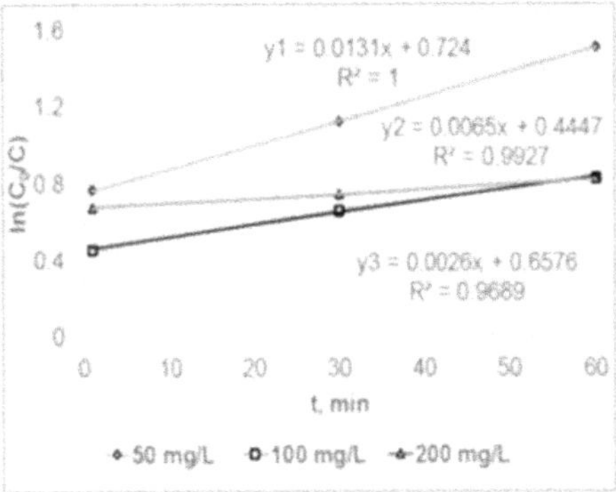

b – kinetic adjustment for the pseudo first order model for CEX

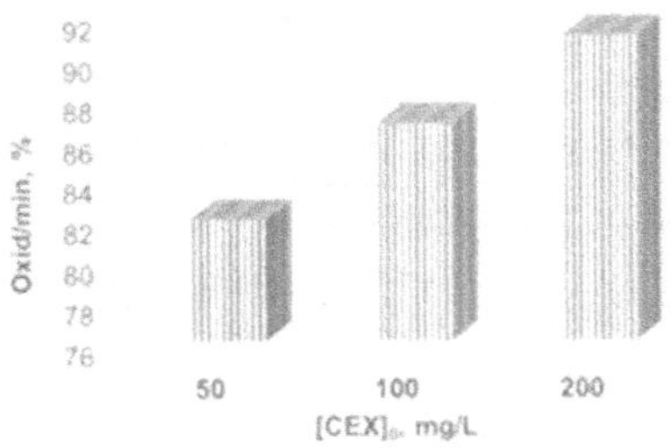

c – COD removal

SOLUTIONS AND RECOMMENDATIONS

The data obtained reveal that the homogenous and heterogeneous studied process are a promising treatment for the degradation of cephalexin in an aqueous solution with an initial concentration of substrate ranging of 50÷200 mg/L. CEX removal has reached, on average, 80% and COD removal 88% respectively.

According to the results of this research, it can have concluded that homogenous and heterogeneous processes can be a potential method for the degradation/mineralization CEX in optimal conditions at real concentrations. The research results can be used to develop university courses in the field, as well as serve as starting points for the submission of national and international projects within the strategic direction of Innovative materials, technologies, and products, Ecological chemistry of water, etc. This important to study the mechanisms of degradation and mineralization of various antibiotics and compare their performance in the same conditions.

FUTURE RESEARCH DIRECTIONS

AOPs are promising and emerging techniques for the degradation and mineralization of antibiotics in the aquatic environment. Furthermore, Photo-Fenton reagents are easily available, safe to use, easy to handle, and can be performed easily at ambient temperature and pressure. In addition, homogenous oxidation is also one of the emerging methods for the treatment of pharmaceutical wastewater. The efficiency of the homogenous and heterogeneous process depends on several variables, such as temperature, pH value, hydrogen peroxide, catalyst loading, reaction time, and substrate concentration. To achieve high performances of treatment, experimental conditions must be optimized. At present, the removal CEX by homogenous and heterogonous processes has made some progress and future research should be focused on the following aspects:

- It needs to improve the adaptability and practicability of the process.
- Using solar light for photocatalytic oxidation.
- Optional, it should use pyrite, titanium dioxide films, or fibers as catalysts.
- Combination of the Photo-Fenton process with coagulation or electroflotation could be an effective and environmentally friendly option that can be realized.
- Pay more attention to the generation mechanism of free radicals and the degradation mechanism.
- To reuse waste and treated water from some processing industries.
- Due to the complex composition of pharmaceutical wastewater, it needs to pay more attention to the practical wastewater and finally fulfill the industrial application.

CONCLUSION

1. Based on the experimental results, the efficiency of photocatalytic process depends on several variables. It may be concluding that the effect of each independent variable depends on the value

of the other due to the existence of simultaneous interactions. Thus, any of the variables could positively or negatively affect both the degradation and mineralization rate of cephalexin.
2. The highest CEX removal rate was achieved at pH 2.5 (homogeneous oxidation) and pH 3.8 (heterogeneous oxidation).
3. The removal process of CEX could be described by the pseudo-first-order kinetic.
4. Optimal reactant concentrations have been established: 50 mg/L of cephalexin is required to oxidize 3.4 mg/L of hydrogen peroxide and 5.6 mg/L of iron(II) ions.
5. It was concluded that the photocatalytic studied process can be a potential method for the removal CEX in optimal conditions at real concentrations.

ACKNOWLEDGMENT

This research was supported by the Moldovan National Research Project Nr.20.80009.5007.27.

REFERENCES

Abdolmotaleb, S. M., Mina, B., Sana, O., & Fateme, A. (2019). Removal of cephalexin from aqueous solutions using magnesium oxide/granular activated carbon hybrid photocatalytic process. *Avicenna Journal of Environmental Health Engineering*, *6*(1), 24–32. doi:10.34172/ajehe.2019.04

Ajoudanian, N., & Nezamzadeh-Ejhieh, A. (2015). Enhanced photocatalytic activity of nickel oxide supported on clinoptilolite nanoparticles for the photodegradation of aqueous cephalexin. *Materials Science in Semiconductor Processing*, *36*, 62–69. doi:10.1016/j.mssp.2015.03.042

Al-Furaiji, M., & Kalash, K. (2020). Advanced oxidation of antibiotics polluted water using titanium dioxide in solar photo catalysis reactor. *Journal of Engineering*, *26*(2), 1–13. doi:10.31026/j.eng.2020.02.01

Almasi, A., Esmaeilpoor, R., Hoseini, H., Abtin, V., & Mohammadi, M. (2020). Photocatalytic degradation of cephalexin by UV activated persulfate and Fenton in synthetic wastewater: Optimization, kinetic study, reaction pathway and intermediate products. *Journal of Environmental Health Science & Engineering*, *18*(15), 1359–1373. doi:10.100740201-020-00553-1 PMID:33312648

Antonin, V. S., Aquino, J. M., Silva, B. F., Silva, A. J., & Rocha-Filho, R. C. (2019). Comparative study on the degradation of cephalexin by four electrochemical advanced oxidation processes: Evolution of oxidation intermediates and antimicrobial activity. *Chemical Engineering Journal*, *372*, 1104–1112. doi:10.1016/j.cej.2019.04.185

Bansal, P., Verma, A., Mehta, C., Singla, J., & Toor, A. P. (2018). Assessment of integrated binary process by coupling photocatalysis and Photo-Fenton for the removal of cephalexin from aqueous solution. *Journal of Materials Science*, *53*(10), 7326–7343. doi:10.100710853-018-2094-x

Benarab, N., & Fangninou, F. F. (2020). The issues of Antibiotics: Cephalexin Antibiotic as emerging environment contaminant. *Int. J. Sci. Res. Publ*, *10*(2), 306–318. doi:10.29322/IJSRP.10.02.2020.p9843

Cordeiro, S. G., Ziem, R., Schweizer, Y. A., Costa, B., Kuhn, D., Haas, P., Weber, A. C., Heidrich, D., Ethur, E. M., Steffens, C., & Hoehne, L. (2021). Degradation of micropollutant cephalexin by ultraviolet (UV) and assessment of residual antimicrobial activity of transformation products. *Water Science and Technology*, *84*(2), 374–383. doi:10.2166/wst.2021.170 PMID:34312344

Cuerda-Correa, E. M., Alexandre-Franco, M. F., & Fernández-González, C. (2019). Advanced oxidation processes for the removal of antibiotics from water. An Overview. *Water (Basel)*, *12*(1), 102. doi:10.3390/w12010102

Ding, Y., Jiang, W., Liang, B., Han, J., Cheng, H., Haider, M. R., Wang, H., Liu, W., Liu, S., & Wang, A. (2020). UV photolysis as an efficient pretreatment method for antibiotics decomposition and their antibacterial activity elimination. *Journal of Hazardous Materials*, *392*, 122321. doi:10.1016/j.jhazmat.2020.122321 PMID:32092653

Felis, E., Kalka, J., Sochacki, A., Kowalska, K., & ... . (2019). Antimicrobial pharmaceuticals in the aquatic environment - occurrence and environmental implications. *European Journal of Pharmacology*, 172813. PMID:31751574

Gonta, M., Mocanu, L., & Porubin-Schimbator, V. (2022). *Advanced oxidation processes for mineralization of amoxicillin in aqueous solutions. In Handbook of Research on Water Sciences and Society*. IGI GLOBAL.

Gonzaga, I. M. D., Moratalla, A., Eguiluz, K. I. B., Salazar-Banda, G. R., Cañizares, P., Rodrigo, M. A., & Saez, C. (2021). Novel Ti/RuO_2IrO_2 anode to reduce the dangerousness of antibiotic polluted urines by Fenton-based processes. *Chemosphere*, *270*, 129344. doi:10.1016/j.chemosphere.2020.129344 PMID:33395582

Hansen, E., Karslake, J., Woods, R. J., Read, A. F., & Wood, K. B. (2020). Antibiotics can be used to contain drug-resistant bacteria by maintaining sufficiently large sensitive populations. *PLoS Biology*, *18*(5), e3000713. doi:10.1371/journal.pbio.3000713 PMID:32413038

Hirte, K., Seiwert, B., Schüürmann, G., & Reemtsma, T. (2016). New hydrolysis products of the beta-lactam antibiotic amoxicillin, their pH-dependent formation and search in municipal wastewater. *Water Research*, *88*, 880–888. doi:10.1016/j.watres.2015.11.028 PMID:26613181

Lastre-Acosta, A. M., Barberato, B., Parizi, M. P. S., & Teixeira, A. C. S. C. (2019). Direct and indirect photolysis of the antibiotic enoxacin: Kinetics of oxidation by reactive photo-induced species and simulations. *Environmental Science and Pollution Research International*, *26*(5), 4337–4347. doi:10.100711356-018-2555-4 PMID:29931641

Masmoudi, R., Khettaf, S., Soltani, A., & Dibi, A. (2022). Cephalexin degradation initiated by OH radicals: theoretical prediction of the mechanisms and the toxicity of byproducts. *J Mol Model.*, *28*(6), 141.

Mocanu, L., Gonta, M., & Matveevici, V. (2020). Optimizarea procesului de oxidare catalitică a amoxicilinei în soluții apoase. *Studia Universitatis*, *1*(131), 38–47.

Mocanu, L., Gonta, M., Matveevici, V., Duca, G., & Porubin-Schimbator, V. (2021). Îndepărtarea cefalexinei din soluții apoase utilizând procesul fotocatalitic Fenton. *Revista de Ştiinţă, Inovare, Cultură şi Artă "Akademos"*, *2*(61), 53-60.

Paniagua, C. E. S., Ricardo, I. A., Marson, E. O., Gonçalves, B. R., & Trovó, A. G. (2019). Simultaneous degradation of the pharmaceuticals gemfibrozil, hydrochlorothiazide and naproxen and toxicity changes during UV-C and UV-C/H_2O_2 processes in different aqueous matrices. *Journal of Environmental Chemical Engineering*, *7*(3), 103164. doi:10.1016/j.jece.2019.103164

Pieri, A., Aschbacher, R., Fasani, G., Mariella, J., Brusetti, L., Pagani, E., Sartelli, M., & Pagani, L. (2020). Country income is only one of the tiles: The global journey of antimicrobial resistance among humans, animals, and environment. *Antibiotics (Basel, Switzerland)*, *9*(8), 473. doi:10.3390/antibiotics9080473 PMID:32752276

Pulicharla, R., Brar, S. K., Drogui, P., Verma, M., & Surampalli, R. Y. (2015). Removal processes of antibiotics in waters and wastewaters: Crucial link to physical-chemical properties and degradation. *Journal of Hazardous, Toxic and Radioactive Waste*, *19*(4), 04015008. doi:10.1061/(ASCE)HZ.2153-5515.0000285

Ribeiro, A. R., & Schmidt, T. C. (2017). Determination of acid dissociation constants (pKa) of cephalosporin antibiotics: Computational and experimental approaches. *Chemosphere*, *169*, 524–533. doi:10.1016/j.chemosphere.2016.11.097 PMID:27898325

Seid-Mohammadi, A., Asgarai, G., Ghorbanian, Z., & Dargahi, A. (2019). The removal of cephalexin antibiotic in aqueous solutions by ultrasonic waves/hydrogen peroxide/nickel oxide nanoparticles (US/H_2O_2/NiO) hybrid process. *Separation Science and Technology*, 1–11.

Serna-Galvis, E. A., Cáceres-Peña, A. C., & Torres-Palma, R. A. (2020). Elimination of representative fluoroquinolones, penicillins, and cephalosporins by solar Photo-Fenton: Degradation routes, primary transformations, degradation improvement by citric acid addition, and antimicrobial activity evolution. *Environmental Science and Pollution Research International*, *33*(33), 4138–41393. doi:10.100711356-020-10069-8 PMID:32683623

Sodhi, K. K., Kumar, M., Balan, B., Dhaulaniya, A. S., Shree, P., Sharma, N., & Singh, D. K. (2021). Perspectives on antibiotic contamination, resistance, metabolomics, and systemic remediation. *SN Applied Sciences*, *269*(3), 269. doi:10.100742452-020-04003-3

Surpateanu, M., Zaharia, C. (2013-2014). Determinarea substantelor oxidabile din apă. *Chimia Mediului: teste de control în laborator și probleme,* (1), 45-46.

Tavasol, F., Tabatabaie, T., Ramavandi, B., & Amiri, F. (2020). Design a new photocatalyst of sea sediment/titanate to remove cephalexin antibiotic from aqueous media in the presence of sonication/ultraviolet/hydrogen peroxide: Pathway and mechanism for degradation. *Ultrasonics Sonochemistry*, *65*, 105062. doi:10.1016/j.ultsonch.2020.105062 PMID:32172148

Tran, N., Reinhard, M., & Gin, K. (2018). Occurrence and fate of emerging contaminants in municipal wastewater treatment plants from different geographical regions - A review. *Water Resources*, *133*, 182–207. PMID:29407700

Vaseashta, A., Duca, G., & Travin, S. (Eds.). (2022). *Handbook of Research on Water Sciences and Society* (Vols. 1–2). IGI Global. doi:10.4018/978-1-7998-7356-3

Vaseashta, A., Gevorgyan, G., Kavaz, D., Ivanov, O., Jawaid, M., & Vasović, D. (2021). Exposome, Biomonitoring, Assessment and Data Analytics to Quantify Universal Water Quality. In A. Vaseashta & C. Maftei (Eds.), *Water Safety, Security and Sustainability. Advanced Sciences and Technologies for Security Applications*. Springer. doi:10.1007/978-3-030-76008-3_4

Verma, M., & Haritash, A. K. (2020). Review of advanced oxidation processes (AOPs) for treatment of pharmaceutical wastewater. *Advances in Environmental Research*, *9*(1), 1–17.

Wang, J., & Wang, S. (2016). Removal of pharmaceuticals and personal care products (PPCPs) from wastewater: A review. *Journal of Environmental Management*, *182*, 620–640. doi:10.1016/j.jenvman.2016.07.049 PMID:27552641

WHO. (2015). *Global Action Plan on Antimicrobial Resistance*. Geneva: World Health Organization. http://www.wpro.who.int/entity/drug_resistance/ resources/global_action_plan_eng.pdf

Compilation of References

Abdollahi, M., & Hosseini, A. (2014). Hydrogen peroxide. Encyclopedia of toxicology. doi:10.1016/B978-0-12-386454-3.00736-3

Abdolmotaleb, S. M., Mina, B., Sana, O., & Fateme, A. (2019). Removal of cephalexin from aqueous solutions using magnesium oxide/granular activated carbon hybrid photocatalytic process. *Avicenna Journal of Environmental Health Engineering*, *6*(1), 24–32. doi:10.34172/ajehe.2019.04

Abdur Rauf, S. A. (2015). The Effect of n-Methylation of Amino Acids (Ac-X-OMe) on Solubility and Conformation: A DFT Study. *Organic & Biomolecular Chemistry*, *13*, 9993–10006. doi:10.1039/C5OB01565K PMID:26289381

Abe, M. (2013). Diradicals. *Chemical Reviews*, *113*(9), 7011–7088. doi:10.1021/cr400056a PMID:23883325

Abou Samra, M., Chedea, V. S., Economou, A., Calokerinos, A., & Kefalas, P. (2011). Antioxidant/prooxidant properties of model phenolic compounds: Part I. Studies on equimolar mixtures by chemiluminescence and cyclic voltammetry. *Food Chemistry*, *125*(2), 622–629. doi:10.1016/j.foodchem.2010.08.076

Abragam, A., & Bleaney, B. (1970). *Electron Paramagnetic Resonance*. Clarendon Press.

Abrahams, S. C., Collin, R. L., & Lipscomb, W. N. (1951). The cristal structure of hydrogen peroxide. *Acta Crystallographica*, *4*(1), 15–20. doi:10.1107/S0365110X51000039

Abramovič, H., Grobin, B., Ulrih, N. P., & Cigić, B. (2017). The methodology applied in DPPH, ABTS and Folin-Ciocalteau assays has a large influence on the determined antioxidant potential. *Acta Chimica Slovenica*, *64*(2), 491–499. doi:10.17344/acsi.2017.3408 PMID:28621381

Achuba, F. I., & Okoh, P. N. (2014). Effect of petroleum products on soil catalase and dehydrogenase activities. *Open Journal of Social Sciences*, *04*, 399–406.

Agrawal, K., Chaturvedi, V., & Verma, P. (2018). Fungal laccase discovered but yet undiscovered. *Bioresources and Bioprocessing*, *5*(1), 4. doi:10.118640643-018-0190-z

Ajoudanian, N., & Nezamzadeh-Ejhieh, A. (2015). Enhanced photocatalytic activity of nickel oxide supported on clinoptilolite nanoparticles for the photodegradation of aqueous cephalexin. *Materials Science in Semiconductor Processing*, *36*, 62–69. doi:10.1016/j.mssp.2015.03.042

Akram, F., Ashraf, S., Haq, I. ul, Shah, F. I., & Aqeel, A. (2022). eminent industrial and biotechnological applications of laccases from bacterial source: A Current Overview. *Applied Biochemistry and Biotechnology*, *194*, 2336-2356.

Alessio, E., & Messori, L. (2019). NAMI-A and KP1019/1339, two iconic ruthenium anticancer drug candidates face-to-face: A case story in medicinal inorganic chemistry. *Molecules (Basel, Switzerland)*, *24*(10), 1995–2014. doi:10.3390/molecules24101995 PMID:31137659

Alfonso-Prieto, M., Biarnés, X., Vidossich, P., & Rovira, C. (2009). The molecular mechanism of the catalase reaction. *Journal of the American Chemical Society*, *131*(33), 11751–11761. doi:10.1021/ja9018572 PMID:19653683

Alfonso-Prieto, M., Vidossich, P., & Rovira, C. (2012). The reaction mechanisms of heme catalases: An atomistic view by ab initio molecular dynamics. *Archives of Biochemistry and Biophysics*, *525*(2), 121–130. doi:10.1016/j.abb.2012.04.004 PMID:22516655

Al-Furaiji, M., & Kalash, K. (2020). Advanced oxidation of antibiotics polluted water using titanium dioxide in solar photo catalysis reactor. *Journal of Engineering*, *26*(2), 1–13. doi:10.31026/j.eng.2020.02.01

Allen, M. P., & Tildesley, D. J. (2017). *Computer Simulation of Liquids* (2nd ed.). Oxford University Press. doi:10.1093/oso/9780198803195.001.0001

Allinger, N. L., & Burkert, U. (1982). *Molecular Mechanics*. American Chemical Society Publication.

Allouche, N., Apel, C., Martin, M.-T., Dumontet, V., Guéritte, F., & Litaudon, M. (2009). Cytotoxic sesquiterpenoids from Winteraceae of Caledonian rainforest. *Phytochemistry, 70*(4), 546–553. doi:10.1016/j.phytochem.2009.01.012

Almasi, A., Esmaeilpoor, R., Hoseini, H., Abtin, V., & Mohammadi, M. (2020). Photocatalytic degradation of cephalexin by UV activated persulfate and Fenton in synthetic wastewater: Optimization, kinetic study, reaction pathway and intermediate products. *Journal of Environmental Health Science & Engineering*, *18*(15), 1359–1373. doi:10.100740201-020-00553-1 PMID:33312648

Altunkaya, A., Gökmen, V., & Skibsted, L. H. (2016). pH dependent antioxidant activity of lettuce (L. sativa) and synergism with added phenolic antioxidants. *Food Chemistry*, *190*, 25–32. doi:10.1016/j.foodchem.2015.05.069 PMID:26212937

Alves, A. C., & Tavares, G. M. (2019). Mixing animal and plant proteins: Is this a way to improve protein techno-functionalities? *Food Hydrocolloids*, *97*, 105171. doi:10.1016/j.foodhyd.2019.06.016

Amman, M. (2018). Principles of Redox Reactions: Basic concepts with Resolved Questions and Problems. *Create Space Publishing Platform,* 26.

Amorati, R., Ferroni, F., Lucarini, M., Pedulli, G. F., & Valgimigli, L. (2002). A quantitative approach to the recycling of α-tocopherol by co-antioxidants. *The Journal of Organic Chemistry*, *67*(26), 9295–9303. doi:10.1021/jo026501f PMID:12492331

Amorati, R., Menichetti, S., Viglianisi, C., & Foti, M. C. (2012). Proton–electron transfer pathways in the reactions of peroxyl and DPPH radicals with hydrogen-bonded phenols. *Chemical Communications*, *48*(97), 11904–11906. doi:10.1039/c2cc36531f PMID:23125978

Amorati, R., Pedulli, G. F., Cabrini, L., Zambonin, L., & Landi, L. (2006). Solvent and pH effects on the antioxidant activity of caffeic and other phenolic acids. *Journal of Agricultural and Food Chemistry*, *54*(8), 2932–2937. doi:10.1021/jf053159+ PMID:16608211

Anderson, P. W. (1959). New Approach to the Theory of Superexchange Interactions. *Physical Review*, *115*(1), 2–13. doi:10.1103/PhysRev.115.2

Andrade, M. F. C., Ko, H. Y., Zhang, L., Car, R., & Selloni, A. (2020). Free energy of proton transfer at the water–TiO 2 interface from ab initio deep potential molecular dynamics. *Chemical Science (Cambridge)*, *11*(9), 2335–2341. doi:10.1039/C9SC05116C PMID:34084393

Andrés, C. M. C., Pérez de la Lastra, J. M., Juan, C. A., Plou, F. J., & Pérez-Lebeña, E. (2022). Chemistry of hydrogen peroxide formation and elimination in mammalian cells, and its role in various pathologies. *Stresses*, *2*(3), 256–274. doi:10.3390tresses2030019

Anghel, L., Rada, S., & Erhan, R.-V. (2023). Structural factors and electron transfer mechanisms in flavoenzymes. *Analytical Letters*, 1–14. doi:10.1080/00032719.2023.2174131

Antonin, V. S., Aquino, J. M., Silva, B. F., Silva, A. J., & Rocha-Filho, R. C. (2019). Comparative study on the degradation of cephalexin by four electrochemical advanced oxidation processes: Evolution of oxidation intermediates and antimicrobial activity. *Chemical Engineering Journal*, *372*, 1104–1112. doi:10.1016/j.cej.2019.04.185

Aoun, M., & Makris, D. P. (2012). Binary mixtures of natural polyphenolic antioxidants with ascorbic acid: Impact of interactions on the antiradical activity. *International Food Research Journal*, *19*(2), 603–606.

Apak, R., Özyürek, M., Güçlü, K., & Çapanoğlu, E. (2016). Antioxidant activity/capacity measurement. 1. Classification, physicochemical principles, mechanisms, and electron transfer (ET)-based assays. *Journal of Agricultural and Food Chemistry*, *64*(5), 997–1027. doi:10.1021/acs.jafc.5b04739 PMID:26728425

Apriceno, A., Astolfi, M. L., Girelli, A. M., & Scuto, F. R. (2019). A new laccase-mediator system facing the biodegradation challenge: Insight into the NSAIDs removal. *Chemosphere*, *215*, 535–542. doi:10.1016/j.chemosphere.2018.10.086 PMID:30340161

Ardo, S., Achey, D., Morris, A. J., Abrahamsson, M., & Meyer, G. J. (2011). Non-Nernstian two-electron transfer photocatalysis at metalloporphyrin–TiO2 interfaces. *Journal of the American Chemical Society*, *133*(41), 16572–16580. doi:10.1021/ja206139n PMID:21888402

Aricov, L., Leonties, A. R., Gîfu, I. C., Preda, D., Raducan, A., & Anghel, D. F. (2020). Enhancement of laccase immobilization onto wet chitosan microspheres using an iterative protocol and its potential to remove micropollutants. *Journal of Environmental Management*, *276*, 111326. doi:10.1016/j.jenvman.2020.111326 PMID:32891981

Aricu, A., Kuchkova, K., Duca, Gh., Secara, E., Barba, A., Dragalin, I., & Ungur, N. (2019). *1-(4-((4aS,8aS)-2,5,5,8a-Tetramethyl-3,4,4a,5,6,7,8,8a-octahydronaphthalen1-yl)butan-2-yl)guanidine compound with antiproliferative and cytotoxic activity*. Patent MD, Nr. 4630, (5), 41.

Aricu, A., Kuchkova, K., Duca, Gh., Secara, E., Barba, A., Dragalin, I., & Ungur, N. (2019). *N-Carbamimidoyl-2-((8aS)-2,5,5,8a-tetramethyl-3,4,4a,5,6,7,8,8a-octahydronaphthalen-1-yl)acetamide compound with antiproliferative and cytotoxic activity*. Patent MD, Nr. 4631, (5), 41-42.

Arîcu, A., Lungu, L., Ciocarlan, A., & Vornicu, N. (2018). *(1R,2R,8aS)-1-((5-Mercapto-1,3,4-thiadiazol-2-yl)methyl)-2,5,5,8a-tetramethyldecahydronaphthalen-2-ol compound with antifungal and antibacterial properties*. Patent MD, Nr. 4580, (7), 37.

Aricu, A., Mangalagiu, I., Ciocarlan, A., Lungu, L., Zbancioc, Gh., & Vornicu, N. (2015). *11,12-Bis-p-tolyl-piridazonil-drim-5(6),8(9)-diene-7-one with antifungal and antibacterial propertie*. Patent MD, Nr. 4370, (9), 24-25.

Aricu, A. N., Kuchkova, K. I., Barba, A. N., Dragalin, I. P., Shova, S. G., Vornicu, N., Gorinchoi, E. K., Sekara, E. S., Lungu, L. V., Niculaua, M., Ungur, N. D., & Vlad, P. F. (2016). Synthesis from norambreinolide, structure, and antimicrobial activity of dihomodrimane sesquiterpenoids with azine, hydrazide, and dihydrazide fragments. *Chemistry of Natural Compounds*, *52*(6), 1029–1036. doi:10.100710600-016-1854-6

Aricu, A. N., Kuchkova, K. I., Secara-Cushnir, E. S., Barba, A. N., Ungur, N. D., & Vornicu, N. (2020). Synthesis and antimicrobial activity of new drimane and homodrimane sesquiterpenoids with oxadiazole and thiadiazole fragments. *Chemistry of Natural Compounds*, *56*(4), 656–662. doi:10.100710600-020-03115-x

Aricu, A. N., Kuchkova, K. I., Sekara, E. S., Barba, A. N., Dragalin, I. P., Ungur, N. D., Melnic, E., & Kravtsov, V. Kh. (2018). Synthesis and structure of drimane sesquiterpenoids containing diazine, 1,2,4-triazole and carbazole rings. *Chemistry of Natural Compounds*, *54*(3), 455–460. doi:10.100710600-018-2378-z

Aricu, A., Ciocarlan, A., Lungu, L., Barba, A., Shova, S., Zbancioc, G., Mangalagiu, I. I., D'ambrosio, M., & Vornicu, N. (2016). Synthesis, antibacterial, and antifungal activities of new drimane sesquiterpenoids with azaheterocyclic units. *Medicinal Chemistry Research*, *25*(10), 2316–2323. doi:10.100700044-016-1665-0

Aronova, E. (2006). *Philosophy and experiment: Essays on the history of immunology in the 20th century*. KomKniga.

Arregui, L., Ayala, M., Gómez-Gil, X., Gutiérrez-Soto, G., Hernández-Luna, C. E., Herrera De Los Santos, M., Levin, L., Rojo-Domínguez, A., Romero-Martínez, D., Saparrat, M. C. N., Trujillo-Roldán, M. A., & Valdez-Cruz, N. A. (2019). Laccases: Structure, function, and potential application in water bioremediation. *Microbial Cell Factories*, *18*(1), 200–234. doi:10.118612934-019-1248-0 PMID:31727078

Arsene, I. (2021). DFT Evidence of Ionic Character of Interaction of Dihydroxyfumaric Acid with the Free Radical DPPH*. *Proceedings of the conference Research-based education for a prosperous society,* (*vol. 2*, pp. 9-13). IDSI.

Arsene, I. (2008). The theoretical study of some reactions with the participation of OH and HO_2 radicals. *Chemistry Journal of Moldova*, *3*(2), 109–113. doi:10.19261/cjm.2008.03(2).18

Arsene, I., & Gorinchoy, N. (2019). DFT study of the entire reaction cycle of H_2O_2 decomposition and O_2 generation catalyzed by Fenton reagent. *Chemistry Journal of Moldova*, *14*(1), 88–97. doi:10.19261/cjm.2018.543

Ashor, A. W., Werner, A. D., Lara, J., Willis, N. D., Mathers, J. C., & Siervo, M. (2017). Effects of Vitamin C supplementation on glycaemic control: A systematic review and meta-analysis of randomised controlled trials. *European Journal of Clinical Nutrition*, *71*(12), 1371–1380. doi:10.1038/ejcn.2017.24 PMID:28294172

Ashraf, R., Sofi, H. S., Malik, A., Beigh, M. A., Hamid, R., & Sheikh, F. A. (2019). Recent trends in the fabrication of starch nanofibers: Electrospinning and non-electrospinning routes and their applications in biotechnology. *Applied Biochemistry and Biotechnology*, *187*(1), 47–74. doi:10.100712010-018-2797-0 PMID:29882194

Atkinson, R. (1997). Gas-Phase Tropospheric Chemistry of Volatile Organic Compounds: 1. Alkanes and Alkenes. *Journal of Physical and Chemical Reference Data*, *26*(2), 215–290. doi:10.1063/1.556012

Auber, B. (2001). Observation of CP Violation in the B^0 Meson System. *Physical Review Letters*, *87*(9), 091801. doi:10.1103/PhysRevLett.87.091801 PMID:11531560

Ayers, P. W., Morrison, R. C., & Roy, R. K. (2002). Variational principles for describing chemical reactions: Condensed reactivity indices. *The Journal of Chemical Physics*, *116*(20), 8731–8744. doi:10.1063/1.1467338

Baalen, C. V., & Marler, J. E. (1966). Occurrence of hydrogen peroxide in seawater. *Nature*, *211*(5052), 951–951. doi:10.1038/211951a0

Babu, T., Sarkar, A., Karmakar, S., Schmidt, C., & Gibson, D. (2020). Multiaction Pt(IV) carbamate complexes can codeliver Pt(II) drugs and amine containing bioactive molecules. *Inorganic Chemistry*, *59*(7), 5182–5193. doi:10.1021/acs.inorgchem.0c00445 PMID:32207294

Backes, E., Kato, C. G., Corrêa, R. C. G., Peralta Muniz Moreira, R., Peralta, R. A., Barros, L., Ferreira, I. C. F. R., Zanin, G. M., Bracht, A., & Peralta, R. M. (2021). Laccases in food processing: Current status, bottlenecks and perspectives. *Trends in Food Science & Technology*, *115*, 445–460. doi:10.1016/j.tifs.2021.06.052

Bag, A., & Chattopadhyay, R. R. (2015). Evaluation of synergistic antibacterial and antioxidant efficacy of essential oils of spices and herbs in combination. *PLoS One*, *10*(7), 1–17. doi:10.1371/journal.pone.0131321 PMID:26132146

Bagnardi, V., Blangiardo, M., La Vecchia, C., & Corrao, G. (2017). *Alcohol Consumption and the Risk of Cancer*. Academic Press.

Bahadir, A. M., & Duca, Gh. (2009). *The role of ecological chemistry in pollution research and sustainable development.* Springer. https://link.springer.com/content/pdf/10.1007/978-90-481-2903-4.pdf doi:10.1007/978-90-481-2903-4

Baik, J., Purkayastha, A., Park, K. H., & Kang, T. J. (2021). Functional characterization of melanin decolorizing extracellular peroxidase of bjerkandera adusta. *Journal of Fungi (Basel, Switzerland), 7*(9), 762. doi:10.3390/jof7090762 PMID:34575800

Baj, A., Cedrowski, J., Olchowik-Grabarek, E., Ratkiewicz, A., & Witkowski, S. (2019). Synthesis, DFT calculations, and in vitro antioxidant study on novel carba-analogs of vitamin E. *Antioxidants, 8*(12), 589. doi:10.3390/antiox8120589 PMID:31779214

Balan, I., Gorbachev, M., Gorinchoy, N., & Arsene, I. (2022). *Pseudo Jahn-Teller origin of the proton-transfer energy barrier in the proton-bound dimers.* In Proceedings of the International Scientific-Practical Conference Training by research for a prosperous society. Chişinău, Moldova.

Balan, I., & Gorinchoy, N. (2011). Ab initio study of chemical activation and hydrogenation of white phosphorus in reaction with rhodium trihydride complex. *Chemistry Journal of Moldova, 6*(2), 84–90. doi:10.19261/cjm.2011.06(2).18

Banga, I., & Philippot, E. (1939). Über die dioxymaleinsäureoxydase. *Hoppe-Seyler's Zeitschrift für Physiologische Chemie, 258*(4), 147–159. doi:10.1515/bchm2.1939.258.4.147

Banga, I., & Szent-Györgyi, A. (1938). Dioxymaleinsäure-oxydase. *Hoppe-Seyler's Zeitschrift für Physiologische Chemie, 255*(1-3), 57–60. doi:10.1515/bchm2.1938.255.1-3.57

Bansal, P., Verma, A., Mehta, C., Singla, J., & Toor, A. P. (2018). Assessment of integrated binary process by coupling photocatalysis and Photo-Fenton for the removal of cephalexin from aqueous solution. *Journal of Materials Science, 53*(10), 7326–7343. doi:10.100710853-018-2094-x

Barbaro, P., Caporali, M., Ienco, A., Mealli, C., Peruzzini, M., & Vizza, F. (2008). Dynamic behaviour of the [(Triphos) $Rh(\eta^1:\eta^2-P_4RR¢)]^{n+}$ complexes [Triphos = $MeC(CH_2PPh_2)_3$; R = H, Alkyl, Aryl; R¢ = Lone Pair, H, Me; *n* = 0, 1]: NMR and computational studies. *European Journal of Inorganic Chemistry, 2008*(9), 1392–1399. doi:10.1002/ejic.200701070

Barbaro, P., Ienco, A., Mealli, C., Peruzzini, M., Scherer, O. J., Schmitt, G., Vizza, F., & Wolmershäuser, G. (2003). Activation and functionalization of white phosphorus at rhodium: experimental and computational analysis of the [(triphos) Rh ($\eta^1:\eta^2-P_4RR¢$)]Y complexes (triphos = $MeC(CH_2PPh_2)_3$; R = H, Alkyl, Aryl; R¢= 2 Electrons, H, Me). *Chemistry (Weinheim an der Bergstrasse, Germany), 9*(21), 5195–5210. doi:10.1002/chem.200305091 PMID:14613129

Barbaro, P., Peruzzini, M., Ramirez, J. A., & Vizza, F. (1999). Rhodium-mediated functionalization of white phosphorus: A novel formation of C–P bonds. *Organometallics, 18*(21), 4237–4240. doi:10.1021/om990550+

Barber, E. A., Liu, Z., & Smith, S. R. (2020). Organic contaminant biodegradation by oxidoreductase enzymes in wastewater treatment. *Microorganisms, 8*(1), 122. doi:10.3390/microorganisms8010122 PMID:31963268

Bashir, S., Mustafa, M., Safvi, S. W., Farhad, N. A., & Rizvi, M. A. (2018). Iron Reduces Iron: A Spectroelectrochemical Insight of Ligand Effect on Iron Redox Potential. *Warasan Khana Witthayasat Maha Witthayalai Chiang Mai, 45*(2), 1087–1098.

Becke, A. D. (1992). Density-functional thermochemistry. I. The effect of the exchange-only gradient correction. *The Journal of Chemical Physics, 96*(3), 2155–2160. doi:10.1063/1.462066

Bellec, N., Garrido Montalban, A., Williams, D. B., Cook, A. S., Anderson, M. E., Feng, X., Barrett, A. G., & Hoffman, B. M. (2000). Porphyrazinediols: Synthesis, characterization, and complexation to group IVB metallocenes. *The Journal of Organic Chemistry*, *65*(6), 1774–1779. doi:10.1021/jo9916840 PMID:10814150

Bell, G. (1970). Mathematical model of clonal selectional and antibody production. *Journal of Theoretical Biology*, *29*(2), 191–232. doi:10.1016/0022-5193(70)90019-6 PMID:5500468

Ben Braïek, O., & Smaoui, S. (2021). Chemistry, safety, and challenges of the use of organic acids and their derivative salts in meat preservation. *Journal of Food Quality*, *2021*, 1–20. doi:10.1155/2021/6653190

Benarab, N., & Fangninou, F. F. (2020). The issues of Antibiotics: Cephalexin Antibiotic as emerging environment contaminant. *Int. J. Sci. Res. Publ*, *10*(2), 306–318. doi:10.29322/IJSRP.10.02.2020.p9843

Bencini, A., & Gatteschi, D. (1990). *EPR of Exchange Coupled Systems*. Springer-Verlag.

Bender, C. L., Schröder, D., Pinedo, R., Adelhelm, P., & Janek, J. (2016). One-or Two-Electron Transfer? The Ambiguous Nature of the Discharge Products in Sodium–Oxygen Batteries. *Angewandte Chemie International Edition*, *55*(15), 4640–4649. doi:10.1002/anie.201510856 PMID:26970321

Benelli, C., & Gatteschi, D. (2002). Magnetism of lanthanides in molecular materials with transition-metal ions and organic radicals. *Chemical Reviews*, *102*(6), 2369–2388. doi:10.1021/cr010303r PMID:12059272

Bensid, A., El Abed, N., Houicher, A., Regenstein, J. M., & Özogul, F. (2020). Antioxidant and antimicrobial preservatives: Properties, mechanism of action and applications in food – A Review. *Critical Reviews in Food Science and Nutrition*, *62*(11), 2985–3001. doi:10.1080/10408398.2020.1862046 PMID:33337242

Benzie, I. F., & Strain, J. J. (1996). The ferric reducing ability of plasma (FRAP) as a measure of "antioxidant power": The FRAP assay. *Analytical Biochemistry*, *239*(1), 70–76. doi:10.1006/abio.1996.0292 PMID:8660627

Berberidou, C., Kitsiou, V., Kazala, E., Lambropoulou, D. A., Kouras, A., Kosma, C. I., Albanis, T. A., & Poulios, I. (2017). Study of the decomposition and detoxification of the herbicide bentazon by heterogeneous photocatalysis: Kinetics, intermediates and transformation pathways. *Applied Catalysis B: Environmental*, *200*, 150–163. doi:10.1016/j.apcatb.2016.06.068

Bergman, P., Parise, B., Liseau, R., Larsson, B., Olofsson, H., Menten, K. M., & Güsten, R. (2011). Detection of interstellar hydrogen peroxide. *Astronomy & Astrophysics*, *531*, L8. doi:10.1051/0004-6361/201117170

Bersuker, I. B. (1980). Are activated complexes of chemical reactions experimentally observable ones? *New Journal of Chemistry*, *4*(3), 139–145.

Bersuker, I. B. (1993). The concept of vibronic interactions. *New Journal of Chemistry*, *17*(1-2), 3–6.

Bersuker, I. B. (2006). *The Jahn-Teller Effect*. Cambridge University Press. doi:10.1017/CBO9780511524769

Bersuker, I. B. (2013). Pseudo Jahn–Teller Effect – A two-state paradigm in formation, deformation, and transformation of molecular systems and solids. *Chemical Reviews*, *113*(3), 1351–1390. doi:10.1021/cr300279n PMID:23301718

Bersuker, I. B. (2016). Spontaneous symmetry breaking in matter induced by degeneracies and pseudodegeneracies. *Advances in Chemical Physics*, *160*, 159–208. doi:10.1002/9781119165156.ch3

Bersuker, I. B. (2021). Jahn–Teller and Pseudo-Jahn–Teller effects: From particular features to general tools in exploring molecular and solid state properties. *Chemical Reviews*, *121*(3), 1463–1512. doi:10.1021/acs.chemrev.0c00718 PMID:33353296

Bersuker, I. B., Gorinchoi, N. N., & Polinger, V. Z. (1984). On the origin of dynamic instability of molecular systems. *Theoretica Chimica Acta*, *66*(3), 161–172. doi:10.1007/BF00549666

Bersuker, I. B., Gorinchoi, N. N., & Polinger, V. Z. (1987). Vibronic origin of basic state instability and stable excited states of transition complex of the H + H_2 ® H_2 + H chemical reaction. *Khimicheskaya Fizika*, *6*(6), 733–736.

Bersuker, I. B., Gorinchoi, N. N., & Polinger, V. Z. (1992). Pseudo Jahn-Teller origin of square-planar configuration instability of main-group-element hydrides. *Journal of Molecular Structure*, *270*, 369–380. doi:10.1016/0022-2860(92)85040-N

Bersuker, I. B., & Polinger, V. Z. (1989). *Vibronic Interactions in Molecules and Crystals*. Springer-Verlag. doi:10.1007/978-3-642-83479-0

Bersuker, I. B., & Stavrov, S. S. (1988). Structure and properties of metalloporphyrins and hemoproteins: The vibronic approach. *Coordination Chemistry Reviews*, *88*, 1–68. doi:10.1016/0010-8545(88)80001-8

Bersuker, I. B., Stavrov, S. S., & Vekhter, B. G. (1979). Pseudo Jahn-Teller origin of the trigger mechanism of conformational changes in the process of hemoglobin oxygenation. *Biofizica*, *24*, 413–418.

Bertin, D., Gigmes, D., Marque, S. R., & Tordo, P. (2011). Kinetic subtleties of nitroxide mediated polymerization. *Chemical Society Reviews*, *40*(5), 2189–2198. doi:10.1039/c0cs00110d PMID:21234490

Besharati Vineh, M., Saboury, A. A., Poostchi, A. A., Rashidi, A. M., & Parivar, K. (2018). Stability and activity improvement of horseradish peroxidase by covalent immobilization on functionalized reduced graphene oxide and biodegradation of high phenol concentration. *International Journal of Biological Macromolecules*, *106*, 1314–1322. doi:10.1016/j.ijbiomac.2017.08.133 PMID:28851646

Betianu, C., Caliman, F. A., Gavrilescu, M., Cretescu, I., Cojocaru, C., & Poulios, I. (2008). Response surface methodology applied for Orange II photocatalytic degradation in TiO_2 aqueous suspensions. *Journal of Chemical Technology and Biotechnology*, *83*(11), 1454–1465. doi:10.1002/jctb.1973

Bezerra, M. A., Oliveira dos Santos, Q., Santos, A. G., Novaes, C. G., Costa Ferreira, S. L., & Santos de Souza, V. (2016). Simplex optimization: A tutorial approach and recent applications in analytical chemistry. *Microchemical Journal*, *124*, 45–54. doi:10.1016/j.microc.2015.07.023

Bezerra, M. A., Santelli, R. E., Oliveira, E. P., Villar, L. S., & Escaleira, L. A. (2008). Response surface methodology (RSM) as a tool for optimization in analytical chemistry. *Talanta*, *76*(5), 965–977. doi:10.1016/j.talanta.2008.05.019 PMID:18761143

Bhardwaj, P., Kaur, N., Selvaraj, M., Ghramh, H. A., Al-Shehri, B. M., Singh, G., Arya, S. K., Bhatt, K., Ghotekar, S., Mani, R., Chang, S. W., Ravindran, B., & Awasthi, M. K. (2022). Laccase-assisted degradation of emerging recalcitrant compounds - A review. *Bioresource Technology*, *364*, 128031. doi:10.1016/j.biortech.2022.128031 PMID:36167178

Bhatt, S., Puli, L., & Patil, C. R. (2021). Role of reactive oxygen species in the progression of Alzheimer's disease. *Drug Discovery Today*, *26*(3), 794–803. doi:10.1016/j.drudis.2020.12.004 PMID:33306995

Bhuiyan, M. A., Hoque, M. Z., & Hossain, S. J. (2009). Free Radical Scavenging Activities of Zizyphus mauritiana. *World Journal of Agricultural Sciences*, *5*(3), 318–322.

Bhunia, S., Rana, A., Ghosh, S., Ivancich, A., & Dey, A. (2020). A designed second-sphere hydrogen-bond interaction that critically influences the O–O bond activation for heterolytic cleavage in ferric iron–porphyrin complexes. *Chemical Science, The Royal Society of Chemistry, 11*(10), 2681-2695. https://hal.archives-ouvertes.fr/hal-02887080/document

Biasin, E., Fox, Z. W., Andersen, A., Ledbetter, K., Kjær, K. S., Alonso-Mori, R., & Khalil, M. (2021). Direct observation of coherent femtosecond solvent reorganization coupled to intramolecular electron transfer. *Nature Chemistry, 13*(4), 343–349. doi:10.103841557-020-00629-3 PMID:33589787

Biko, O. D. V., Viljoen-Bloom, M., & van Zyl, W. H. (2020). Microbial lignin peroxidases: Applications, production challenges and future perspectives. *Enzyme and Microbial Technology, 141*, 109669. doi:10.1016/j.enzmictec.2020.109669 PMID:33051019

Bilal, M., Bagheri, A. R., Vilar, D. S., Aramesh, N., Eguiluz, K. I. B., Ferreira, L. F. R., Ashraf, S. S., & Iqbal, H. M. N. (2022). Oxidoreductases as a versatile biocatalytic tool to tackle pollutants for clean environment – a review. *Journal of Chemical Technology and Biotechnology, 97*(2), 420–435. doi:10.1002/jctb.6743

Björck, M. L., Zhou, S., Rydström Lundin, C., Ott, M., Ädelroth, P., & Brzezinski, P. (2017). Reaction of S. cerevisiae mitochondria with ligands: Kinetics of CO and O2 binding to flavohemoglobin and cytochrome c oxidase. *Biochimica et Biophysica Acta (BBA) - Bioenergetics, 1858*(2), 182–188. doi:10.1016/j.bbabio.2016.11.009 PMID:27871795

Blaja, S. P., Lungu, L. V., Kuchkova, K. I., Ciocarlan, A. G., Barba, A. N., Vornicu, N., & Aricu, A. N. (2021). Norlabdane compounds containing thiosemicarbazone or 1,3-thiazole fragments: Synthesis and antimicrobial activity. *Chemistry of Natural Compounds, 57*(1), 101–110. doi:10.100710600-021-03292-3

Block, S. S. (1991). *Peroxide hydrogen compounds in disinfection, sterilization and preservation* (S. S. Block, Ed.). Vol. 167). Lea & Febinger.

Bocharov, G., & Marchuk, G. (2000). Applied problems of mathematical modeling in immunology. *Журнал вычислительной математики и математической физики, 40*(12), 1905–1920.

Bogani, L., & Wernsdorfer, W. (2010). Molecular spintronics using single-molecule magnets. *Nanoscience and technology: a collection of reviews from nature journals*, 194-201.

Bölgen, N., Demir, D., Aşık, M., Sakım, B., & Vaseashta, A. (2022). Introduction and Fundamentals of Electrospinning. In A. Vaseashta & N. Bölgen (Eds.), *Electrospun Nanofibers*. Springer., doi:10.1007/978-3-030-99958-2_1

Bolocan, N., & Duca, Gh. (2022). *Stopped-flow studies of the interaction of DFH_4 and its derivatives with DPPH•.* National conference with international participation "Life sciences in the dialogue of generations: connections between universities, academia and business community". Chisinau, Moldova.

Bolocan, N., & Duca, G. (2021). DFT study of geometric isomerization and keto–enol tautomerization of dihydroxyfumaric acid. *Studia Universitatis Babes-Bolyai. Chemia, 66*(4), 341–353. doi:10.24193ubbchem.2021.4.25

Bomfleur, B., Sironi, M., Raimondi, M., & Voitländer, J. (2000). Ab initio calculation of the zero-field splitting parameter D of benzene and naphthalene. *The Journal of Chemical Physics, 112*(3), 1066–1069. doi:10.1063/1.480660

Bonamore, A., & Boffi, A. (2007). Flavohemoglobin: Structure and reactivity. *IUBMB Life, 60*(1), 19–28. doi:10.1002/iub.9 PMID:18379989

Bouldin, R. M., Kyriazidis, L., Fidler, A., Ravichandran, S., Kumar, J., Samuelson, L. A., & Nagarajan, R. (2011). Oxidoreductase catalyzed polymerization of 3-methylpyrrole. *Journal of Macromolecular Science. Journal of Macromolecular Science, Part A: Pure and Applied Chemistry, 48*(12), 976–982. doi:10.1080/10601325.2011.620413

Brand-Williams, W., Cuvelier, M. E., & Berset, C. L. W. T. (1995). Use of a free radical method to evaluate antioxidant activity. *Lebensmittel-Wissenschaft + Technologie, 28*(1), 25–30. doi:10.1016/S0023-6438(95)80008-5

Buchachenko, A. L., & Berdinsky, V. L. (2002). Electron Spin Catalysis. *Chemical Reviews, 102*(3), 603–612. doi:10.1021/cr010370l PMID:11890751

Buhro, W. E., Patton, A. T., Strouse, C. E., Gladysz, J. A., McCormick, F. B., & Etter, M. C. (1983). Syntheses, chemical properties, and x-ray crystal structures of rhenium formaldehyde and thioformaldehyde complexes. *Journal of the American Chemical Society*, *105*(4), 1056–1058. doi:10.1021/ja00342a070

Bunduchi, E., Duca, Gh., & Gladchi, V. (2022). New kinetic parameters for natural water quality assessment. In *Handbook of research on water sciences and society* (Vol. 14). IGI Global. doi:10.4018/978-1-7998-7356-3.ch011

Burr, M. L. (1995). Explaining the French paradox. *The Journal of the Royal Society for the Promotion of Health*, *115*(4), 217–219. doi:10.1177/146642409511500404 PMID:7562866

Buta, M. C., Frecus, B., Enache, M., Humelnicu, I., Toader, A. M., & Cimpoesu, F. (2021). Intra- and Inter-Molecular Spin Coupling in Phenalenyl Dimeric Systems. *The Journal of Physical Chemistry A*, *125*(32), 6893–6901. doi:10.1021/acs.jpca.1c02705 PMID:34353026

Buta, M. C., Toader, A. M., Frecus, B., Oprea, C. I., Cimpoesu, F., & Ionita, G. (2019). Molecular and Supramolecular Interactions in Systems with Nitroxide-based Radicals. *International Journal of Molecular Sciences*, *20*(19), 4733. doi:10.3390/ijms20194733 PMID:31554219

Butch, C. J., Wang, J., Gu, J., Vindas, R., Crowe, J., Pollet, P., Gelbaum, L., Leszczynski, J., Krishnamurthy, R., & Liotta, C. L. (2016). Ph-controlled reaction divergence of decarboxylation versus fragmentation in reactions of dihydroxyfumarate with glyoxylate and formaldehyde: Parallels to biological pathways. *Journal of Physical Organic Chemistry*, *29*(7), 352–360. doi:10.1002/poc.3542

Butch, C., Cope, E. D., Pollet, P., Gelbaum, L., Krishnamurthy, R., & Liotta, C. L. (2013). Production of tartrates by cyanide-mediated dimerization of glyoxylate: A potential abiotic pathway to the citric acid cycle. *Journal of the American Chemical Society*, *135*(36), 13440–13445. doi:10.1021/ja405103r PMID:23914725

Butreddy, A., Janga, K. Y., Ajjarapu, S., Sarabu, S., & Dudhipala, N. (2021). Instability of therapeutic proteins — An overview of stresses, stabilization mechanisms and analytical techniques involved in lyophilized proteins. *International Journal of Biological Macromolecules*, *167*, 309–325. doi:10.1016/j.ijbiomac.2020.11.188 PMID:33275971

Çakmak, E., & Özbakır Işın, D. (2020). A theoretical evaluation on free radical scavenging activity of 3-styrylchromone derivatives: The DFT study. *Journal of Molecular Modeling*, *26*(5), 98. Advance online publication. doi:10.100700894-020-04368-7 PMID:32279127

Caliman, A. F., Cojocaru, C., Antoniadis, A., & Poulios, I. (2007). Optimized photocatalytic degradation of Alcian Blue 8 GX in the presence of TiO_2 suspensions. *Journal of Hazardous Materials*, *144*(1-2), 265–273. doi:10.1016/j.jhazmat.2006.10.019 PMID:17112662

Calzado, C. J., Cabrero, J., Malrieu, J. P., & Caballol, R. (2002). Analysis of the magnetic coupling in binuclear complexes. I. Physics of the coupling. *The Journal of Chemical Physics*, *116*(7), 2728–2747. doi:10.1063/1.1430740

Calza, P., Zacchigna, D., & Laurenti, E. (2016). Degradation of orange dyes and carbamazepine by soybean peroxidase immobilized on silica monoliths and titanium dioxide. *Environmental Science and Pollution Research International*, *23*(23), 23742–23749. doi:10.100711356-016-7399-1 PMID:27623850

Campbell, D. W., II. (1997). Hydrogen peroxide: medical miracle. Ed.: Hampton Roads Publishing Company.

Campbell, E., Meredith, M., Banta, S., & Minteer, S. D. (2012). Enzymatic biofuel cells utilizing a biomimetic cofactor. *Chemical Communications*, *48*(13), 1898–90. doi:10.1039/c2cc16156g PMID:22227738

Caneschi, A., Ferraro, F., Gatteschi, D., le Lirzin, A., Novak, M. A., Rentschler, E., & Sessoli, R. (1995). Ferromagnetic order in the sulfur-containing nitronyl nitroxide radical, 2-(4-thiomethyl) phenyl-4, 4, 5, 5-tetramethylimidazoline-l-oxyl-3-oxide, NIT(SMe)Ph. *Advanced Materials*, *7*(5), 476–478. doi:10.1002/adma.19950070515

Caneschi, A., Gatteschi, D., Lalioti, N., Sangregorio, C., Sessoli, R., Venturi, G., Vindigni, A., Rettori, A., Pini, M. G., & Novak, M. A. (2001). Cobalt (II)-nitronyl nitroxide chains as molecular magnetic nanowires. *Angewandte Chemie International Edition*, *40*(9), 1760–1763. doi:10.1002/1521-3773(20010504)40:9<1760::AID-ANIE17600>3.0.CO;2-U PMID:11353503

Cantemir, A. R., Raducan, A., Puiu, M., & Oancea, D. (2013). Kinetics of thermal inactivation of catalase in the presence of additives. *Process Biochemistry (Barking, London, England)*, *48*(3), 471–477. doi:10.1016/j.procbio.2013.02.013

Cao, G., Sofic, E., & Prior, R. L. (1997). Antioxidant and prooxidant behavior of flavonoids: Structure-activity relationships. *Free Radical Biology & Medicine*, *22*(5), 749–760. doi:10.1016/S0891-5849(96)00351-6 PMID:9119242

Cao, J., Liu, X., Yang, Y., Wei, B., Li, Q., Mao, G., He, Y., Li, Y., Zheng, L., Zhang, Q., Li, J., Wang, L., & Qi, C. (2020). decylubiquinone suppresses breast cancer growth and metastasis by inhibiting angiogenesis via the ROS/p53/BAI1 signaling pathway. *Angiogenesis*, *23*(3), 325–338. doi:10.100710456-020-09707-z PMID:32020421

Cao, X., Nie, X., Xiong, S., Cao, L., Wu, Z., Moore, P. K., & Bian, J.-S. (2018). Renal protective effect of polysulfide in cisplatin-induced nephrotoxicity. *Redox Biology*, *15*, 513–521. doi:10.1016/j.redox.2018.01.012 PMID:29413963

Caproiu, M. T., Ionita, G., Draghici, C., & Ionita, P. (2008). Synthesis and characterisation of several di-, tri-, and tetra-radicals linked by flexible or rigid linkers. *ARKIVOC*, *14*(14), 158–165. doi:10.3998/ark.5550190.0009.e16

Cárdenas, C., Rabi, N., Ayers, P. W., Morell, C., Jaramillo, P., & Fuentealba, P. (2009). Chemical reactivity descriptors for ambiphilic reagents: Dual descriptor, local hypersoftness, and electrostatic potential. *The Journal of Physical Chemistry A*, *113*(30), 8660–8667. doi:10.1021/jp902792n PMID:19580251

Cardullo, N., Muccilli, V., & Tringali, C. (2022). Laccase-mediated synthesis of bioactive natural products and their analogues. *RSC Chemical Biology*, *3*(6), 614–647. doi:10.1039/D1CB00259G PMID:35755186

Carocho, M., & Ferreira, I. C. (2013). A review on antioxidants, prooxidants and related controversy: Natural and synthetic compounds, screening and analysis methodologies and future perspectives. *Food and Chemical Toxicology*, *51*, 15–25. doi:10.1016/j.fct.2012.09.021 PMID:23017782

Carr, A. C., & Maggini, S. (2017). Vitamin C and immune function. *Nutrients*, *9*(11), 1–25. doi:10.3390/nu9111211 PMID:29099763

Casadey, R., Challier, C., Altamirano, M., Spesia, M. B., & Criado, S. (2021). Antioxidant and antimicrobial properties of tyrosol and derivative-compounds in the presence of vitamin B2. Assays of synergistic antioxidant effect with commercial food additives. *Food Chemistry*, *335*, 127576. doi:10.1016/j.foodchem.2020.127576 PMID:32739805

Catapano, M. C., Protti, M., Fontana, T., Mandrioli, R., Mladěnka, P., & Mercolini, L. (2019). An original HPLC method with coulometric detection to monitor hydroxyl radical generation via Fenton chemistry. *Molecules (Basel, Switzerland)*, *24*(17), 3066–3080. doi:10.3390/molecules24173066 PMID:31450723

Catling, D. C., Glein, C. R., Zahnle, K. J., & McKay, C. P. (2005). Why O_2 is required by complex life on habitable planets and the concept of planetary" oxygenation time. *Astrobiology*, *5*(3), 415–438. doi:10.1089/ast.2005.5.415 PMID:15941384

Çelik, E. E., & Gökmen, V. (2022). Interactions between free and bound antioxidants under different conditions in food systems. *Critical Reviews in Food Science and Nutrition*, *62*(21), 5766–5782. doi:10.1080/10408398.2021.1892584 PMID:33646856

Ceresa, A., Bakker, E., Hattendorf, B., Günther, D., & Pretsch, E. (2001). Potentiometric polymeric membrane electrodes for measurement of environmental samples at trace levels: New requirements for selectivities and measuring protocols, and comparison with ICPMS. *Analytical Chemistry*, *73*(2), 343–351. doi:10.1021/ac001034s PMID:11199988

Chaker, H., Attar, A. E., Djennas, M., & Fourmentin, S. (2021). A statistical modeling-optimization approach for efficiency photocatalytic degradation of textile azo dye using cerium-doped mesoporous ZnO: A central composite design in response surface methodology. *Chemical Engineering Research & Design*, *171*, 198–212. doi:10.1016/j.cherd.2021.05.008

Chance, B. (1949). The composition of catalase-peroxide complexes. *The Journal of Biological Chemistry*, •••, 179. PMID:18134593

Chance, B., Greenstein, D. S., & Roughton, F. J. W. (1952). The mechanism of catalase action. I. Steady-state analysis. *Archives of Biochemistry and Biophysics*, *37*(2), 301–321. doi:10.1016/0003-9861(52)90194-X PMID:14953443

Chan, W. L., Tritsch, J. R., & Zhu, X. Y. (2012). Harvesting singlet fission for solar energy conversion: One-versus two-electron transfer from the quantum mechanical superposition. *Journal of the American Chemical Society*, *134*(44), 18295–18302. doi:10.1021/ja306271y PMID:23066740

Chapyshev, S. V., Walton, R., Sanborn, J. A., & Lahti, P. M. (2000). Quintet and Septet State Systems Based on Pyridylnitrenes: Effects of Substitution on Open-Shell High-Spin States. *Journal of the American Chemical Society*, *122*(8), 1580–1588. doi:10.1021/ja993131c

Chattaraj, P. K., Chakraborty, A., & Giri, S. (2009). Net electrophilicity. *The Journal of Physical Chemistry A*, *113*(37), 10068–10074. doi:10.1021/jp904674x PMID:19702288

Chattaraj, P. K., Duley, S., & Domingo, L. R. (2012). Understanding local electrophilicity/nucleophilicity activation through a single reactivity difference index. *Organic & Biomolecular Chemistry*, *10*(14), 2855. doi:10.1039/c2ob06943a PMID:22388637

Chatterjee, A., & König, B. (2019). Birch-Type Photoreduction of Arenes and Heteroarenes by Sensitized Electron Transfer. *Angewandte Chemie*, *131*(40), 14427–14432. doi:10.1002/ange.201905485 PMID:31379035

Chauhan, P. S., Goradia, B., & Saxena, A. (2017). Bacterial laccase: Recent update on production, properties and industrial applications. *Biotech*, *7*, 323. PMID:28955620

Chavali, M. S., & Nikolova, M. P. (2019). Metal oxide nanoparticles and their applications in nanotechnology. *SN Applied Sciences*, *1*(6), 607. doi:10.100742452-019-0592-3

Chelikani, P., Fita, I., & Loewen, P. C. (2004). Diversity of structures and properties among catalases. *Cellular and Molecular Life Sciences*, *61*(2), 192–208. doi:10.100700018-003-3206-5 PMID:14745498

Chen, C., Shi, T., Chang, W., & Zhao, J. (2015). Essential roles of proton transfer in photocatalytic redox reactions. *ChemCatChem*, *7*(5), 724–731. doi:10.1002/cctc.201402880

Chen, C., Xu, C., Li, T., Lu, S., Luo, F., & Wang, H. (2020). Novel NHC-coordinated ruthenium (II) arene complexes achieve synergistic efficacy as safe and effective anticancer therapeutics. *European Journal of Medicinal Chemistry*, *203*, 112605–112615. doi:10.1016/j.ejmech.2020.112605 PMID:32688202

Chen, G. F., Yuan, Y., Jiang, H., Ren, S. Y., Ding, L. X., Ma, L., & Wang, H. (2020). Electrochemical reduction of nitrate to ammonia via direct eight-electron transfer using a copper–molecular solid catalyst. *Nature Energy*, *5*(8), 605–613. doi:10.103841560-020-0654-1

Chen, J.-C., Zhang, Y., Jie, X.-M., She, J., Dongye, G.-Z., Zhong, Y., Deng, Y.-Y., Wang, J., Guo, B.-Y., & Chen, L.-M. (2019). Ruthenium(II) salicylate complexes inducing ROS-mediated apoptosis by targeting thioredoxin reductase. *Journal of Inorganic Biochemistry, 193*, 112–123. doi:10.1016/j.jinorgbio.2019.01.011 PMID:30711557

Chen, X., Li, H., Zhang, B., & Deng, Z. (2022). The synergistic and antagonistic antioxidant interactions of dietary phytochemical combinations. *Critical Reviews in Food Science and Nutrition, 62*(20), 5658–5677. doi:10.1080/10408398.2021.1888693 PMID:33612011

Chen, Z., Li, H., Peng, A., & Gao, Y. (2014). Oxidation of polycyclic aromatic hydrocarbons by horseradish peroxidase in water containing an organic cosolvent. *Environmental Science and Pollution Research International, 21*(18), 10696–10705. doi:10.100711356-014-3005-6 PMID:24894750

Chen, Z., & Wan, C. (2017). Biological valorization strategies for converting lignin into fuels and chemicals. *Renewable & Sustainable Energy Reviews, 73*, 610–621. doi:10.1016/j.rser.2017.01.166

Chettri, A., Schneider, K. R. A., Cole, H. D., Roque, J. A. III, Cameron, C. G., McFarland, S. A., & Dietzek, B. (2021). String-attached oligothiophene substituents determine the fate of excited states in ruthenium complexes for photodynamic therapy. *The Journal of Physical Chemistry A, 125*(32), 6985–6994. doi:10.1021/acs.jpca.1c04900 PMID:34370485

Cheung, E. C., & Vousden, K. H. (2022). The role of ROS in tumour development and progression. *Nature Reviews. Cancer, 22*(5), 280–297. doi:10.103841568-021-00435-0 PMID:35102280

Choueiri, L., Chedea, V. S., Calokerinos, A., & Kefalas, P. (2012). Antioxidant/pro-oxidant properties of model phenolic compounds. Part II: Studies on mixtures of polyphenols at different molar ratios by chemiluminescence and LC–MS. *Food Chemistry, 133*(3), 1039–1044. doi:10.1016/j.foodchem.2012.01.057

Chovanec, M., Abu Zaid, M., Hanna, N., El-Kouri, N., Einhorn, L. H., & Albany, C. (2017). Long-term toxicity of cisplatin in germ-cell tumor survivors. *Annals of Oncology : Official Journal of the European Society for Medical Oncology, 28*(11), 2670–2679. doi:10.1093/annonc/mdx360 PMID:29045502

Chow, C. T., & Vennesland, B. (1958). The nonenzymatic decarboxylation of diketosuccinate and oxaloglycolate (dihydroxyfumarate). *The Journal of Biological Chemistry, 233*(4), 997–1002. doi:10.1016/S0021-9258(18)64693-9 PMID:13587530

Christou, G., Gatteschi, D., Hendrickson, D. N., & Sessoli, R. (2000). Single-molecule magnets. *MRS Bulletin, 25*(11), 66–71. doi:10.1557/mrs2000.226

Ciocarlan, A., Aricu, A., Lungu, L., Edu, C., Barba, A., Shova, S., Mangalagiu, I. I., D'ambrosio, M., Nicolescu, A., Deleanu, C., & Vornicu, N. (2017). Synthesis of novel tetranorlabdane derivatives with unprecedented carbon skeleton. *Synlett, 28*(5), 565–571. doi:10.1055-0036-1588651

Circu, M. L., & Aw, T. Y. (2010). Reactive oxygen species, cellular redox systems, and apoptosis. *Free Radical Biology & Medicine, 48*(6), 749–762. doi:10.1016/j.freeradbiomed.2009.12.022 PMID:20045723

Clark, A. C. (2008). The production of yellow pigments from (+)-catechin and dihydroxyfumaric acid in a model wine system. *European Food Research and Technology, 226*(4), 925–931. doi:10.100700217-007-0615-y

Clerici, S., & Boletta, A. (2020). Role of the KEAP1-NRF2 axis in renal cell carcinoma. *Cancers (Basel), 12*(11), 3458–3482. doi:10.3390/cancers12113458 PMID:33233657

Closs, G. L., Calcaterra, L. T., Green, N. J., Penfield, K. W., & Miller, J. R. (1986). Distance, stereoelectronic effects, and the Marcus inverted region in intramolecular electron transfer in organic radical anions. *Journal of Physical Chemistry, 90*(16), 3673–3683. doi:10.1021/j100407a039

Cojocaru, C., Pascariu Dorneanu, P., Airinei, A., Olaru, N., Samoila, P., & Rotaru, A. (2017). Design and evaluation of electrospun polysulfone fibers and polysulfone/$NiFe_2O_4$ nanostructured composite as sorbents for oil spill cleanup. *Journal of the Taiwan Institute of Chemical Engineers*, *70*, 267–281. doi:10.1016/j.jtice.2016.11.005

Colon, M., & Nerín, C. (2016). Synergistic, antagonistic and additive interactions of green tea polyphenols. *European Food Research and Technology*, *242*(2), 211–220. doi:10.100700217-015-2532-9

Commission Implementing Regulation (EU) (2017). Approving the Basic Substance Hydrogen Peroxide in Accordance With Regulation (EC) No 1107/2009 of the European Parliament and of the Council concerning the placing of plant protection products on the market and amending the Annex to Commission Implementing Regulation (EU) No 540/2011. *Official Journal of EU.09.03.2017*, L63/95. https://eur-lex.europa.eu/legal-content/EN/TXT/PDF/?uri=CELEX:32017R0409&from=LT

Constantin, M. M., Corbu, C. G., Tanase, C., Codrici, E., Mihai, S., Popescu, I. D., Enciu, A. M., Mocanu, S., Matei, I., & Ionita, G. (2019). Spin probe method of electron paramagnetic resonance spectroscopy–a qualitative test for measuring the evolution of dry eye syndrome under treatment. *Analytical Methods*, *11*(7), 965–972. doi:10.1039/C8AY02783H

Cooper, D. L. (Ed.). (2001). *Valence Bond Theory*. Elsevier Science.

Cooper, D. L., Gerratt, J., & Raimondi, M. (1991). Applications of spin-coupled valence bond theory. *Chemical Reviews*, *91*(5), 929–964. doi:10.1021/cr00005a014

Cooper, D. L., Thorsteinsson, T., & Gerratt, J. (1999). Modern VB representations of CASSCF wave functions and the fully-variational optimization of modern VB wave functions using the CASVB strategy. *Advances in Quantum Chemistry*, *32*, 51–67. doi:10.1016/S0065-3276(08)60406-2

Cooper, W. J., Zika, R. G., Petasne, R. G., & Plane, J. M. C. (1988). Photochemical formation of hydrogen peroxide in natural waters exposed to sunlight. *Environmental Science & Technology*, *22*(10), 1156–1160. doi:10.1021/es00175a004 PMID:22148607

Cordani, M., Donadelli, M., Strippoli, R., Bazhin, A. V., & Sanchez-Alvarez, M. (2019). Interplay between ROS and autophagy in cancer and aging: From Molecular Mechanisms to Novel Therapeutic Approaches. *Oxidative Medicine and Cellular Longevity*, *2019*, 1–3. doi:10.1155/2019/8794612 PMID:31467639

Cordeiro, S. G., Ziem, R., Schweizer, Y. A., Costa, B., Kuhn, D., Haas, P., Weber, A. C., Heidrich, D., Ethur, E. M., Steffens, C., & Hoehne, L. (2021). Degradation of micropollutant cephalexin by ultraviolet (UV) and assessment of residual antimicrobial activity of transformation products. *Water Science and Technology*, *84*(2), 374–383. doi:10.2166/wst.2021.170 PMID:34312344

Cornaton, Y., & Marquardt, R. (2016). A global analytical representation of the potential energy surface of the FHF^- anion. *The Journal of Physical Chemistry A*, *120*(30), 5959–5968. doi:10.1021/acs.jpca.6b05325 PMID:27400137

Cornelli, U. (2009). Antioxidant uses in nutraceuticals. *Clinics in Dermatology*, *27*(2), 175–194. doi:10.1016/j.clindermatol.2008.01.010 PMID:19167999

Coronado, E., Delhaès, P., Gatteschi, D., & Miller, J. S. (1996). *Molecular Magnetism: From Molecular Assemblies to the Devices*. Springer. doi:10.1007/978-94-017-2319-0

Costa, J. R., Ramos, R., Costa, K., Brasil, D., Silva, C., Ferreira, E., Borges, R., Campos, J., Macêdo, W., & Santos, C. (2018). An in silico study of the antioxidant ability for two caffeine analogs using molecular docking and quantum chemical methods. *Molecules (Basel, Switzerland)*, *23*(11), 1–17. doi:10.3390/molecules23112801 PMID:30380600

Cotea, V. D., Zănoagă, C. V., & Cotea, V. V. (2009). Tratat de oenochimie. [Treatise of oenochemistry], 1, 93-97. Bucharest: Romanian Academy Publisher.

Cotton, F. A., Wilkinson, G., & Gaus, P. L. (1994). *Basic inorganic chemistry* (3rd ed.). John Wiley and Sons, Inc.

Cotton, F. A., Wilkinson, G., & Murillo, C. A. (1999). *Manfred Bochmann Advanced Inorganic Chemistry* (6th ed.). Wiley-Interscience.

Crawford, V. H., Richardson, H. W., Wasson, J. R., Hodgson, D. J., & Hatfield, W. E. (1976). Relation between the singlet-triplet splitting and the copper-oxygen-copper bridge angle in hydroxo-bridged copper dimers. *Inorganic Chemistry, 15*(9), 2107–2110. doi:10.1021/ic50163a019

Creutz, C., & Taube, H. (1969). Direct approach to measuring the Franck-Condon barrier to electron transfer between metal ions. *Journal of the American Chemical Society, 91*(14), 3988–3989. doi:10.1021/ja01042a072

Cuartero, M., & Crespo, G. A. (2018). All-solid-state potentiometric sensors: A new wave for in situ aquatic research. *Current Opinion in Electrochemistry, 10*, 98–106. doi:10.1016/j.coelec.2018.04.004

Cucicova, C., Rudic, V., Arîcu, A., Cepoi, L., Rudi, L., Secara, E., Valuţa, A., Barbă, A., Miscu, V., Vlad, P., & Chiriac, T. (2015a). *N-($\Delta^{8,13}$-Bicyclohomofarnesenoylamino)-carbazole and method for cultivating the cyanobacterium Nostoc Linckia, using it.* Patent MD, Nr. 4326 (2), 27.

Cucicova, C., Rudic, V., Arîcu, A., Cepoi, L., Rudi, L., Secara, E., Valuţa, A., Barbă, A., Miscu, V., Vlad, P., & Chiriac, T. (2015b). *N-($\Delta^{8,13}$-Bicyclohomofarnesenoyl)-3-amino-1,2,4-triazole and method for cultivating the cyanobacterium Nostoc Linckia, using it.* Patent MD, Nr. 4327 (2), 27-28.

Cuerda-Correa, E. M., Alexandre-Franco, M. F., & Fernández-González, C. (2019). Advanced oxidation processes for the removal of antibiotics from water. An Overview. *Water (Basel), 12*(1), 102. doi:10.3390/w12010102

Dai, F., Chen, W. F., & Zhou, B. (2008). Antioxidant synergism of green tea polyphenols with α-tocopherol and l-ascorbic acid in SDS micelles. *Biochimie, 90*(10), 1499–1505. doi:10.1016/j.biochi.2008.05.007 PMID:18554517

Dai, J., & Mumper, R. J. (2010). Plant phenolics: Extraction, analysis and their antioxidant and anticancer properties. *Molecules (Basel, Switzerland), 15*(10), 7313–7352. doi:10.3390/molecules15107313 PMID:20966876

Darcy, J. W., Koronkiewicz, B., Parada, G. A., & Mayer, J. M. (2018). A continuum of proton-coupled electron transfer reactivity. *Accounts of Chemical Research, 51*(10), 2391–2399. doi:10.1021/acs.accounts.8b00319 PMID:30234963

Davies, M. J. (2021). Myeloperoxidase: Mechanisms, reactions and inhibition as a therapeutic strategy in inflammatory diseases. *Pharmacology & Therapeutics, 218*, 107685. doi:10.1016/j.pharmthera.2020.107685 PMID:32961264

De Grandis, R. A., Oliveira, K. M., Guedes, A. P. M., Dos Santos, P. W. S., Aissa, A. F., Batista, A. A., & Pavan, F. R. (2021). a novel ruthenium (II) complex with lapachol induces G2/M phase arrest through Aurora-B kinase down-regulation and ROS-mediated apoptosis in human prostate adenocarcinoma cells. *Frontiers in Oncology, 11*, 2148–2165. doi:10.3389/fonc.2021.682968 PMID:34249731

de Nooy, A. E., Besemer, A. C., & van Bekkum, H. (1996). On the use of stable organic nitroxyl radicals for the oxidation of primary and secondary alcohols. *Synthesis, 1996*(10), 1153–1176. doi:10.1055-1996-4369

de Oliveira, F. K., Santos, L. O., & Buffon, J. G. (2021). Mechanism of action, sources, and application of peroxidases. *Food Research International, 14*, 110266. doi:10.1016/j.foodres.2021.110266 PMID:33992367

Decker, E. A., Chen, B., Panya, A., & Elias, R. J. (2010). Understanding antioxidant mechanisms in preventing oxidation in foods. In E. A. Decker, R. J. Elias, & D. Julian McClements (Eds.), *Oxidation in foods and beverages and antioxidant applications* (pp. 225–248). Woodhead Publishing. doi:10.1533/9780857090447.2.225

Deng, W., Zhang, P., Seger, B., & Gong, J. (2022). Unraveling the rate-limiting step of two-electron transfer electrochemical reduction of carbon dioxide. *Nature Communications*, *13*(1), 1–9. doi:10.103841467-022-28436-z PMID:35145084

Deng, Y., & Zhao, R. (2015). Advanced Oxidation Processes (AOPs) in Wastewater Treatment. *Current Pollution Reports*, *1*(3), 167–176. doi:10.100740726-015-0015-z

Desplanches, C., Ruiz, E., Rodriguez-Fortea, A., & Alvarez, S. (2002). Exchange Coupling of Transition-Metal Ions through Hydrogen Bonding: A Theoretical Investigation. *Journal of the American Chemical Society*, *124*(18), 5197–5205. doi:10.1021/ja0178160 PMID:11982384

Dey, B., & Dutta, T. (2022). Laccases: Thriving the domain of bio-electrocatalysis. *Bioelectrochemistry (Amsterdam, Netherlands)*, *146*, 108144. doi:10.1016/j.bioelechem.2022.108144 PMID:35550253

Di Meo, F., Lemaur, V., Cornil, J., Lazzaroni, R., Duroux, J. L., Olivier, Y., & Trouillas, P. (2013). Free radical scavenging by natural polyphenols: Atom versus electron transfer. *The Journal of Physical Chemistry A*, *117*(10), 2082–2092. doi:10.1021/jp3116319 PMID:23418927

Díaz, A., Loewen, P. C., Fita, I., & Carpena, X. (2012). Thirty years of heme catalases structural biology. *Archives of Biochemistry and Biophysics*, *252*(2), 102–110. doi:10.1016/j.abb.2011.12.011 PMID:22209752

Diken, E. G., Headrick, J. M., Roscioli, J. R., Bopp, J. C., Johnson, M. A., & McCoy, A. B. (2005). Fundamental excitations of the shared proton in the H_3O_2-and $H_5O_2^+$ complexes. *The Journal of Physical Chemistry A*, *109*(8), 1487–1490. doi:10.1021/jp044155v PMID:16833469

Ding, Y., Cui, K., Liu, X., Xie, Q., Guo, Z., & Chen, Y. (2022). Lignin peroxidase-catalyzed direct oxidation of trace organic pollutants through a long-range electron transfer mechanism: Using propranolol as an example. *Journal of Hazardous Materials*, *431*, 128544. doi:10.1016/j.jhazmat.2022.128544 PMID:35228075

Ding, Y., Jiang, W., Liang, B., Han, J., Cheng, H., Haider, M. R., Wang, H., Liu, W., Liu, S., & Wang, A. (2020). UV photolysis as an efficient pretreatment method for antibiotics decomposition and their antibacterial activity elimination. *Journal of Hazardous Materials*, *392*, 122321. doi:10.1016/j.jhazmat.2020.122321 PMID:32092653

Ditchfield, R., Hehre, W. J., & Pople, J. A. (1971). Self-Consistent Molecular Orbital Methods. 9. Extended Gaussian-type basis for molecular-orbital studies of organic molecules. *The Journal of Chemical Physics*, *54*(2), 724–728. doi:10.1063/1.1674902

Divsalar, A., & Ghobadi, R. (2021). The presence of deep eutectic solvents of reline and glyceline on interaction and side effect of anti-cancer drug of 5-fluorouracil: Bovine liver catalase as a target. *Journal of Molecular Liquids*, *323*, 114588. doi:10.1016/j.molliq.2020.114588

Djeradi, H. R., Rahmouni, A., & Cheriti, A. (2014). Antioxidant activity of flavonoids: A QSAR modeling using Fukui indices descriptors. *Journal of Molecular Modeling*, *20*(10), 1–9. doi:10.100700894-014-2476-1 PMID:25311723

Doert, M., Jaworska, K., Moersel, J. T., & Kroh, L. W. (2012). Synergistic effect of lecithins for tocopherols: Lecithin-based regeneration of α-tocopherol. *European Food Research and Technology*, *235*(5), 915–928. doi:10.100700217-012-1815-7

Dogonadze, R. R. (Ed.). (1985). *The Chemical Physics of Solvation*. Elsevier.

Dogonadze, R. R., Kuznetsov, A. M., & Chernenko, A. A. (1965). Theory of homogeneous and heterogeneous electronic processes in liquids. *Russian Chemical Reviews*, *34*(10), 759–775. doi:10.1070/RC1965v034n10ABEH001561

Domingo, L. R., Pérez, P., & Sáez, J. A. (2013). Understanding the local reactivity in polar organic reactions through electrophilic and nucleophilic Parr functions. *RSC Advances*, *3*(5), 1486–1494. doi:10.1039/C2RA22886F

Dong, J., Yu, Y., Pei, Y., & Pei, Z. (2022). pH-responsive aminotriazole doped metal organic frameworks nanoplatform enables self-boosting reactive oxygen species generation through regulating the activity of catalase for targeted chemo/chemodynamic combination therapy. *Journal of Colloid and Interface Science*, *607*, 1651–1660. doi:10.1016/j.jcis.2021.09.043 PMID:34592552

Doseděl, M., Jirkovský, E., Macáková, K., Krčmová, L. K., Javorská, L., Pourová, J., Mercolini, L., Remião, F., Nováková, L., & Mladěnka, P. (2021). Vitamin c—sources, physiological role, kinetics, deficiency, use, toxicity, and determination. Nutrients 13(2), 1–36. doi:10.3390/nu13020615

Dougherty, D. A., & Anslyn, E. V. (2005). Modern physical organic chemistry. *Universitas Scientiarum*, 122.

Dreher, D. D. (2005). US Patent 0090668. https://www.freepatentsonline.com/y2005/0090668.html

Duca, G. (2012). Homogeneous Catalysis with Metal Complexes: Fundamentals and Applications. Berlin; Heidelberg: Springer.

Duca, G., & Travin, S. (2023). Hydrogen Peroxide, A Substance That Conquered the World. In Duca and Vaseashta (eds.) Handbook of Research on Redox Processes Within Environmental, Technological, and Biomedical Contexts. IGI Publishers.

Duca, Gh. (1979). *Mechanisms of ecochimical processes in water medium.* [PhD thesis, Chişinău].

Duca, Gh., & Romanciuc, L. (2017). An overview of ecological chemistry conferences within 1985-2016. In: *Abstract Book of the 6th International Conference on Ecological & Environmental Chemistry*, 21-28. Chisinau.

Duca, Gh., Romanciuc, L., Goreacea, N., & Borodaev, R. (1997). *Model of self-purification processing in natural water.* In: Metal compounds in environment and life, VII-th Int. Hans Wolfgang Nurnberg Memorial Symp., Modena (Italy), 51.

Duca, Gh., Scutaru, I., Sychev, A. (1987). Cinetica interacţiunii iodului cu acidul dihidroxifumaric. *Revista de chimie fizică (In Rus.), 61*(8), p. 2266-2268.

Duca, Gh., Skurlatov, Yu., & Miziti, A. (1994). Introduction in ecological chemistry (Rus.). Moscow: Vysshaia shkola, 400p.

Duca, Gh., Skurlatov, Yu., Sychev, A., & Romanciuc, L. (1988). *Method of plants growth stimulation.* Patent Nr.1544334 (SU).

Duca, M., Port, A., Romanciuc, L., & Duca, Gh. (1995). The redox state of aquatic medium and plants germination process regulation. Self-purification processes in natural waters, 245-256. Chisinau.

Duca, G. (2007). Ecological chemistry in new independent states of the former Soviet Union. *Clean (Weinheim)*, *35*(1), 17–25. doi:10.1002/clen.200600007

Duca, G. (2017). Ecological and environmental chemistry. *Chemistry Journal of Moldova*, *12*(1), 9–19. doi:10.19261/cjm.2017.427

Duca, G. (2021). Hydrogen peroxide in ecological and environmental chemistry. *Chemistry Journal of Moldova*, *16*(2), 28–45. doi:10.19261/cjm.2021.918

Duca, G., Aricu, A., Kuchkova, K., Secara, E., Barbă, A., Dragalin, I., Ungur, N., & Spengler, G. (2019). Synthesis, structural elucidation and biological evaluations of new guanidine-containing terpenoids as anticancer agents. *Natural Product Research*, *21*(33), 3052–3056. doi:10.1080/14786419.2018.1516658 PMID:30445867

Duca, G., Aricu, A., Lungu, L., Tenu, N., Ciocarlan, A., Gutu, Y., Dragalin, I., & Barba, A. (2018). Synthesis of new homodrimane sesquiterpenoids containing diazine, 1,2,4-triazole and carbazole rings. *Chemistry Journal of Moldova, 13*(1), 69–73. doi:10.19261/cjm.2017.458

Duca, G., Blonschi, V., Gladchi, V., & Travin, S. (2020). Dynamics of different sulfur forms in natural waters and their influence on the redox state. *American Journal of Physical Chemistry, 9*(3), 52. doi:10.11648/j.ajpc.20200903.12

Duca, G., Mereuta, A., Velisco, N., Tanaselia, C., & Mitina, T. (2019). Determination of quality indicators of prut river water. *Chemistry Journal of Moldova, 14*(1), 61–67. doi:10.19261/cjm.2018.541

Duca, G., Secara, N., & Duca, D. (2008). Physical methods of fast reactions investigation. *Chemistry Journal of Moldova, 3*(1), 22–30. doi:10.19261/cjm.2008.03(1).15

Duca, G., & Travin, S. (2020). Reactions' mechanisms and applications of hydrogen peroxide. *American Journal of Physical Chemistry, 9*(2), 36–44. doi:10.11648/j.ajpc.20200902.13

Duca, Gh. (2012). *Homogeneous catalysis with metal complexes: fundamentals and applications*. Springer. doi:10.1007/978-3-642-24629-6

Duca, Gh., Skurlatov, Yu., & Sychev, A. (2002). *Redox catalysis and ecological chemistry*. CEP USM.

Duca, Gh., & Sychev, A. (2002). *Aspecte fundamentale şi aplicative ale catalizei omogene cu compuşii complecşi ai metalelor de tranziţie. Partea I*. Tipografia USM.

Duca, Gh., & Vaseashta, A. (2020). *Handbook of research on emerging developments and environmental impacts of ecological chemistry*. IGI Global., doi:10.4018/978-1-7998-1241-8

Duong, N. B., Truong, V. M., Li, Y. S., Wang, C. L., & Yang, H. (2020). improving the immobilization of glucose oxidase on carbon cloth via a hybrid approach of cross-linked chitosan/TPP matrices with Na polymers for high-performance self-pumping enzyme-based biofuel cells. *Energy & Fuels, 34*(8), 10050–10058. doi:10.1021/acs.energyfuels.0c01033

Dutta, S., Patil, R., & Dey, T. (2022). Electron transfer-driven single and multi-enzyme biofuel cells for self-powering and energy bioscience. *Nano Energy, 96*, 107074. doi:10.1016/j.nanoen.2022.107074

Eberson, L. (1982). Electron-transfer reactions in organic chemistry. Advances in Physical Organic Chemistry. Academic Press.

Eiko Toyota, E., & Matsudera, N. (1980). Paramagnetic Co(III) Complexes with Organic Ligands. V. The Isolation of Properties of Salts of the Paramagnetis Bis(2,2-bipyridine)(5nitro-salycilato)cobalt(III) and Paramagnetic "1 and 2"-Isomers of the 5-Nitrosalycitato(triethilenetetramine)-cobalt(III) and Their Derivatives. *Bulletin of the Chemical Society of Japan, 53*, 3517–3521. doi:10.1246/bcsj.53.3517

el Ichi-Ribault, S., Zebda, A., Tingry, S., Petit, M., Suherman, A. L., Boualam, A., Cinquin, P., & Martin, D. K. (2017). Performance and stability of chitosan-MWCNTs-laccase biocathode: Effect of MWCNTs surface charges and ionic strength. *Journal of Electroanalytical Chemistry (Lausanne, Switzerland), 799*, 26–33. doi:10.1016/j.jelechem.2017.05.018

Elliott, J. A., & Paddison, S. J. (2007). Modelling of morphology and proton transport in PFSA membranes. *Physical Chemistry Chemical Physics, 9*(21), 2602–2618. doi:10.1039/b701234a PMID:17627306

Elliott, S. J., McElhaney, A. E., Feng, C., Enemark, J. H., & Armstrong, F. A. (2002). A voltammetric study of inter-domain electron transfer within sulfite oxidase. *Journal of the American Chemical Society, 124*(39), 11612–11613. doi:10.1021/ja027776f PMID:12296723

Emamdadi, N., Gholizadeh, M., & Housaindokht, M. R. (2021). Investigation of static magnetic field effect on horseradish peroxidase enzyme activity and stability in enzymatic oxidation process. *International Journal of Biological Macromolecules*, *170*, 189–195. doi:10.1016/j.ijbiomac.2020.12.034 PMID:33333096

Enami, S., Sakamoto, Y., & Colussi, A. J. (2013). Fenton chemistry at aqueous interfaces. *Proceedings of the National Academy of Sciences of the United States of America*, *111*(2), 623–628. doi:10.1073/pnas.1314885111 PMID:24379389

Enko, J., & Gliszczyńska-Świgło, A. (2015). Influence of the interactions between tea (Camellia sinensis) extracts and ascorbic acid on their antioxidant activity: Analysis with interaction indexes and isobolograms. *Food Additives & Contaminants. Part A, Chemistry, Analysis, Control, Exposure & Risk Assessment*, *32*(8), 1234–1242. doi:10.1080/19440049.2015.1049218 PMID:26035225

Ensenauer, R., He, M., Willard, J.-M., Goetzman, E. S., Corydon, T. J., Vandahl, B. B., Mohsen, A.-W., Isaya, G., & Vockley, J. (2005). Human Acyl-CoA dehydrogenase-9 plays a novel role in the mitochondrial β-oxidation of unsaturated fatty acids. *The Journal of Biological Chemistry*, *280*(37), 32309–32316. doi:10.1074/jbc.M504460200 PMID:16020546

Erhan, R. V., Bodnarchuk, V., Radulescu, A., & Anghel, L. (2020). Small angle neutron scattering reveals dimeric glucose oxidase from Aspergillus niger at pH 5.9. *Journal of Surface Investigation. X-ray, Synchrotron and Neutron Techniques*, *14*(S1), S5–S10. Advance online publication. doi:10.1134/S1027451020070125

Eschenmoser, A. (2007a). On a hypothetical generational relationship between HCN and constituents of the reductive citric acid cycle. *Chemistry & Biodiversity*, *4*(4), 554–573. doi:10.1002/cbdv.200790050 PMID:17443871

Eschenmoser, A. (2007b). The search for the chemistry of life's origin. *Tetrahedron*, *63*(52), 12821–12844. doi:10.1016/j.tet.2007.10.012

Espina, G., Atalah, J., & Blamey, J. M. (2021). Extremophilic oxidoreductases for the industry: Five successful examples with promising projections. *Frontiers in Bioengineering and Biotechnology*, *9*, 710035. Advance online publication. doi:10.3389/fbioe.2021.710035 PMID:34458243

Estalayo-Adrián, S., Blasco, S., Bright, S. A., McManus, G. J., Orellana, G., Williams, D. C., Kelly, J. M., & Gunnlaugsson, T. (2020). Water-soluble amphiphilic ruthenium(II) polypyridyl complexes as potential light-activated therapeutic agents. *Chemical Communications*, *56*(65), 9332–9335. doi:10.1039/D0CC04397D PMID:32671364

Euring, M., Kirsch, A., Schneider, P., & Kharazipour, A. (2016). Lignin-laccase-mediator-systems (LLMS) for the production of binderless medium density fiberboards (MDF). *Journal of Materials Science Research*, *5*(2), 7. doi:10.5539/jmsr.v5n2p7

Evans, D. H. (2008). One-electron and two-electron transfers in electrochemistry and homogeneous solution reactions. *Chemical Reviews*, *108*(7), 2113–2144. doi:10.1021/cr068066l PMID:18620364

Eyring, H. (1935). The Activated Complex in Chemical Reactions. *The Journal of Chemical Physics*, *3*(2), 107–115. doi:10.1063/1.1749604

Fang, L., Qin, X., Zhao, J., & Gou, S. (2019). Construction of dual stimuli-responsive platinum(IV) hybrids with NQO1 targeting ability and overcoming cisplatin resistance. *Inorganic Chemistry*, *58*(3), 2191–2200. doi:10.1021/acs.inorgchem.8b03386 PMID:30657321

Fapyane, D., & Ferapontova, E. E. (2016). Enhanced electron transfer between gold nanoparticles and horseradish peroxidase reconstituted onto alkanethiol-modified hemin. *Electrochemistry Communications*, *70*, 39–42. doi:10.1016/j.elecom.2016.06.016

Farver, O., & Pecht, I. (2011). Electron transfer in blue copper proteins. *Coordination Chemistry Reviews*, *255*(7-8), 757–773. doi:10.1016/j.ccr.2010.08.005

Favre, H. A., & Powell, W. H. (2013). *Nomenclature of Organic Chemistry. IUPAC Recommendations and Preferred Name 2013*. The Royal Society of Chemistry. doi:10.1039/9781849733069

Fedalto, L., de Oliveira, P. R., Agustini, D., Kalinke, C., Banks, C. E., Bergamini, M. F., & Marcolino-Junior, L. H. (2023). Novel and highly stable strategy for the development of microfluidic enzymatic assays based on the immobilization of horseradish peroxidase (HRP) into cotton threads. *Talanta*, *252*, 123889. doi:10.1016/j.talanta.2022.123889 PMID:36070669

Felis, E., Kalka, J., Sochacki, A., Kowalska, K., & … . (2019). Antimicrobial pharmaceuticals in the aquatic environment - occurrence and environmental implications. *European Journal of Pharmacology*, 172813. PMID:31751574

Fenton, H. J. H. (1894). Oxidation of tartaric acid in presence of iron. *Journal of the Chemical Society Transactions*, *65*(65), 899–911. doi:10.1039/CT8946500899

Ferbinteanu, M., & Cimpoesu, F. (2014). Quantum Nanosystems. Structure, Properties, and Interactions. In *Magnetic Anisotropy in Case Studies*. Apple Academics.

Ferbinteanu, M., Miyasaka, H., Wernsdorfer, W., Nakata, K., Sugiura, K. I., Yamashita, M., Coulon, C., & Clérac, R. (2005). Single-chain magnet (NEt4)[Mn2 (5-MeOsalen) 2Fe (CN) 6] made of MnIII– FeIII– MnIII trinuclear single-molecule magnet with an ST= 9/2 spin ground state. *Journal of the American Chemical Society*, *127*(9), 3090–3099. doi:10.1021/ja0468123 PMID:15740148

Fernandes, M., Souza, D. H., Henriques, R. O., Alves, M. V., Skoronski, E., & Junior, A. F. (2020). Obtaining soybean peroxidase from soybean hulls and its application for detoxification of 2,4-dichlorophenol contaminated water. *Journal of Environmental Chemical Engineering*, *8*(3), 103786. doi:10.1016/j.jece.2020.103786

Ferrando-Soria, J., Vallejo, J., Castellano, M., Martinez-Lillo, J., Pardo, E., Cano, J., Castro, I., Lloret, F., Ruiz-Garcia, R., & Julve, M. (2017). Molecular magnetism, quo vadis? A historical perspective from a coordination chemist viewpoint. *Coordination Chemistry Reviews*, *339*, 17–103. doi:10.1016/j.ccr.2017.03.004

Ferrari, C. K., Souto, P. C., França, E. L., & Honorio-França, A. C. (2011). Oxidative and nitrosative stress on phagocytes' function: From effective defense to immunity evasion mechanisms. *Archivum Immunologiae et Therapiae Experimentalis*, *59*(6), 441–448. doi:10.100700005-011-0144-z PMID:21972015

Ferrer-Gallego, R., Puxeu, M., Martín, L., Nart, E., Hidalgo, C., & Andorrà, I. (2018). *Microbiological, physical, and chemical procedures to elaborate high-quality SO_2-free wines*. Grapes and Wines - Advances in Production, Processing, Analysis and Valorization., doi:10.5772/intechopen.71627

Ferrón-Celma, I., Mansilla, A., Hassan, L., Garcia-Navarro, A., Comino, A. M., Bueno, P., & Ferrón, J. A. (2009). Effect of Vitamin C Administration on Neutrophil Apoptosis in Septic Patients After Abdominal Surgery. *The Journal of Surgical Research*, *153*(2), 224–230. doi:10.1016/j.jss.2008.04.024 PMID:18952228

Filipović, M. M.-M., Marković, Z., Đorović, J., Marković, J. D., Lučić, B., & Amić, D. (2015). QSAR of the free radical scavenging potency of selected hydroxybenzoic acids and simple phenolics. *Comptes Rendus. Chimie*, *18*(5), 492–498. doi:10.1016/j.crci.2014.09.001

Fischer, H. (2001). The persistent radical effect: A principle for selective radical reactions and living radical polymerizations. *Chemical Reviews*, *101*(12), 3581–3610. doi:10.1021/cr990124y PMID:11740916

Fita, I., & Rossmann, M. G. (1985). The active center of catalase. *Journal of Molecular Biology, 185*(1), 21–37. doi:10.1016/0022-2836(85)90180-9 PMID:4046038

Fita, I., Silva, A. M., Murthy, M. R. N., & Rossmann, M. G. (1986). The refined structure of beef liver catalase at 2·5 Å resolution. *Acta Crystallographica. Section B, Structural Science, 42*(5), 497–515. doi:10.1107/S0108768186097835

Flohé, L., Toppo, S., & Orian, L. (2022). The glutathione peroxidase family: Discoveries and mechanism. *Free Radical Biology & Medicine, 187*, 113–122. doi:10.1016/j.freeradbiomed.2022.05.003 PMID:35580774

Formosinho, S., & Barroso, M. (Eds.). (2011). *Proton-Coupled Electron Transfer.* Royal Society of Chemistry. doi:10.1039/9781849733168

Fourmond, V., Wiedner, E. S., Shaw, W. J., & Léger, C. (2019). Understanding and design of bidirectional and reversible catalysts of multielectron, multistep reactions. *Journal of the American Chemical Society, 141*(28), 11269–11285. doi:10.1021/jacs.9b04854 PMID:31283209

Fraga, B. M. (2013). Natural sesquiterpenoids. *Natural Product Reports, 30*(9), 1226-1264. doi:10.1039/c3np70047j

Franke, W., & Brathuhn, G. (1931). Über die decarboxylierung Der Dioxymaleinsäure. *Justus Liebigs Annalen der Chemie, 487*(1), 1–52. doi:10.1002/jlac.19314870102

Freeman, B. L., Eggett, D. L., & Parker, T. L. (2010). Synergistic and antagonistic interactions of phenolic compounds found in navel oranges. *Journal of Food Science, 75*(6), C570–C576. doi:10.1111/j.1750-3841.2010.01717.x PMID:20722912

Frisch, M. J., Trucks, G. W., Schlegel, H. B., Scuseria, G. E., Robb, M. A., Cheeseman, J. R., Scalmani, G., Barone, V., Mennucci, B., Petersson, G. A., Nakatsuji, H., Caricato, M., Li, X., Hratchian, H. P., Izmaylov, A. F., Bloino, J., Zheng, G., Sonnenberg, J. L., [...] & Fox, D. J. (2009). *Gaussian 09 Revision A.2.* Scientific Research Publishing.

Frog, B. N., Skurlatov, Yu. I., Shtamm, E. V., & Vichutinskaya, E. V. (2012). The influence of water-soluble sulfur compounds on the toxic properties of natural and waste waters (Rus.). *Vestnik MGSU, 6*, 105–113. doi:10.22227/1997-0935.2012.6.105-113

Fuentealba, P., Florez, E., & Tiznado, W. (2010). Topological analysis of the Fukui function. *Journal of Chemical Theory and Computation, 6*(5), 1470–1478. doi:10.1021/ct100022w PMID:26615684

Fujii, M., & Otani, E. (2017). Photochemical generation and decay kinetics of superoxide and hydrogen peroxide in the presence of standard humic and fulvic acids. *Water Research, 123*, 642–654. doi:10.1016/j.watres.2017.07.015 PMID:28710981

Fukunaga, K. (1960). Studies on the metabolism of dihydroxyfumarate, hydroxypyruvate and their related compounds, I. Enzymatic formation of xylulose in rat liver. *Journal of Biochemistry, 47*(6), 741–744. https://www.jstage.jst.go.jp/article/biochemistry1922/47/6/47_6_741/_pdf. doi:10.1093/oxfordjournals.jbchem.a127120

Fukushima, R., & Yamazaki, E. (2010). Vitamin C requirement in surgical patients. *Current Opinion in Clinical Nutrition and Metabolic Care, 13*(6), 669–676. doi:10.1097/MCO.0b013e32833e05bc PMID:20689415

Gaina, B., Taran, N., Diacon, I., Duca, Gh., Gonţa, M., & Vascan, N. (1995). Studiul proceselor de oxidoreducere în vinuri la şampanizare. Bilanţul activităţii şt. a USM pe anii 1993–1994. Ştiinţe naturale: materialele conf. corpului didactico – şt., 20–27 martie 1995. Chisinau.

Galano, A., & Alvarez-Idaboy, J. R. (2013). A computational methodology for accurate predictions of rate constants in Solution: Application to the assessment of primary antioxidant activity. *Journal of Computational Chemistry, 34*(28), 2430–2445. doi:10.1002/jcc.23409 PMID:23939817

Galano, A., Mazzone, G., Alvarez-Diduk, R., Marino, T., Alvarez-Idaboy, J. R., & Russo, N. (2016). Food antioxidants: Chemical insights at the molecular level. *Annual Review of Food Science and Technology*, *7*(1), 335–352. doi:10.1146/annurev-food-041715-033206 PMID:26772412

Galy, J., Jaud, J., Kahn, O., & Tola, P. (1979). Crystal structure and magnetic properties of Cu2(fsa)2en, CH3OH, with H4(fsa)2en = N,N¢-bis(2-hydroxy, 3-carboxybenzilidene)-1,2-diaminoethane. *Inorganica Chimica Acta*, *36*, 229–236. doi:10.1016/S0020-1693(00)89398-6

Gambardella, F., Ganzeveld, I., Jos, W., & Heeres, H. J. (2005). Kinetics of the reaction of FeII (EDTA) with oxygen in aqueous solutions. *Industrial & Engineering Chemistry Research*, *44*(22), 8190–8198. doi:10.1021/ie050448e

Gan, J. S., Ashraf, S. S., Bilal, M., & Iqbal, H. M. N. (2022). Biodegradation of environmental pollutants using catalase-based biocatalytic systems. *Environmental Research*, *214*, 113914. doi:10.1016/j.envres.2022.113914 PMID:35932834

Gao, P., Pan, W., Li, N., & Tang, B. (2019). Fluorescent probes for organelle-targeted bioactive species imaging. *Chemical Science (Cambridge)*, *10*(24), 6035–6071. doi:10.1039/C9SC01652J PMID:31360411

Gao, Q., Li, Y., Li, Y., Zhang, Z., & Liang, Y. (2022). Antioxidant and prooxidant activities of phenolic acids commonly existed in vegetables and their relationship with structures. *Food Science and Technology (Campinas)*, *42*, 1–9. doi:10.1590/fst.07622

Gao, Y., Zhang, J., He, Z., Luo, L., & Yan, C. (2021). Platinum-based photosensitizer with near-infrared aggregation-induced emission for synergistic photodynamic-chemo theranostics. *Organic Electronics*, *92*, 106105–106111. doi:10.1016/j.orgel.2021.106105

García-Fernández, P., García-Canales, L., García-Lastra, J. M., Junquera, J., Moreno, M., & Aramburu, J. A. (2008). Pseudo-Jahn-Teller origin of the low barrier hydrogen bond in $N_2H_7^+$. *The Journal of Chemical Physics*, *129*(12), 124313. doi:10.1063/1.2980053 PMID:19045029

Garg, A., Sharma, R., Dey, P., & Kumar, A. (2022). Food auto-oxidation: An overview. *Antioxidants Effects in Health*, 43–68. doi:10.1016/B978-0-12-819096-8.00013-6

Gateschi, D., & Sessoli, R. (2003). Quantum Tunneling of Magnetization and Related Phenomena in Molecular Materials. *Angewandte Chemie International Edition*, *42*(3), 268–297. doi:10.1002/anie.200390099 PMID:12548682

Gautschi, M., Weisstanner, C., Slotboom, J., Nava, E., Zürcher, T., & Nuoffer, J.-M. (2015). Highly efficient ketone body treatment in multiple acyl-CoA dehydrogenase deficiency–related leukodystrophy. *Pediatric Research*, *77*(1), 91–98. doi:10.1038/pr.2014.154 PMID:25289702

Gęgotek, A., & Skrzydlewska, E. (2022). Antioxidative and Anti-Inflammatory Activity of Ascorbic Acid. Antioxidants, 11(10). doi:10.3390/antiox11101993

Gentil, S., Rousselot-Pailley, P., Sancho, F., Robert, V., Mekmouche, Y., Guallar, V., Tron, T., & le Goff, A. (2020). Efficiency of site-specific clicked laccase-carbon nanotubes biocathodes towards O_2 reduction. *Chemistry (Weinheim an der Bergstrasse, Germany)*, *26*(21), 4798–4804. doi:10.1002/chem.201905234 PMID:31999372

Geru, I. I. (2018). Time-Reversal Symmetry: Seven Time-Reversal Operators for Spin Containing Systems. Springer Nature. doi:10.1007/978-3-030-01210-6

Geru, I., Gorinchoy, N., & Balan, I. (2012). Pseudo Jahn–Teller origin of the proton tunneling in Zundel cation containing water clusters. *Ukrainian Journal of Physics*, *57*(11), 1149–1149.

Ghasemi, H., Amini, M. A., Pegah, A., Azizi, E., Tayebinia, H., Khanverdilou, S., Mousavibahar, S. H., & Alizamir, A. (2020). Overexpression of reactive oxygen species modulator 1 is associated with advanced grades of bladder cancer. *Molecular Biology Reports*, *47*(9), 6497–6505. doi:10.100711033-020-05702-1 PMID:32770525

Gherman, O., & Saliu, L. (1976). *Fizică statistică*. Editura Tehnică.

Ghosh, A., Mandoli, A., Kumar, D. K., Yadav, N. S., Ghosh, T., Jha, B., Thomas, J. A., & Das, A. (2009). DNA binding and cleavage properties of a newly synthesised Ru(II)-polypyridyl complex. *Dalton Transactions (Cambridge, England)*, (42), 9312–9321. doi:10.1039/b906756f PMID:20449211

Ghosh, A., & Shcherbik, N. (2020). Effects of oxidative stress on protein translation: Implications for cardiovascular diseases. *International Journal of Molecular Sciences*, *21*(8), 2661–2685. doi:10.3390/ijms21082661 PMID:32290431

Giguere, P. A. (1950). The infra-red spectrum of hydrogen peroxide. *The Journal of Chemical Physics*, *18*(1), 88–92. doi:10.1063/1.1747464

Gladchi, V., Duca, G., Goreaceva, N., Bunduchi, E., & Lis, A. (2013). Chemical composition from the Dniester River tributaries. *Chemistry Journal of Moldova*, *8*(1), 23–32. doi:10.19261/cjm.2013.08(1).03

Gladchi, V., Goreaceva, N., Duca, G., Bunduchi, E., Romanciuc, L., Mardari, I., & Borodaev, R. (2008). The study of redox conditions in the Dniester River. *Chemistry Journal of Moldova*, *3*(1), 70–76. doi:10.19261/cjm.2008.03(1).14

Glaser, T., Theil, H., Liratzis, I., Weyhermüller, T., & Bill, E. (2006). Ferromagnetic Coupling by Orthogonal Magnetic Orbitals in a Heterodinuclear CuIIVIV=O Complex and in a Homodinuclear CuIICuII Complex. *Inorganic Chemistry*, *45*(13), 4889–4891. doi:10.1021/ic0606328 PMID:16780307

Goc, G., Balci, S., Yorke, B. A., Pearson, A. R., & Yuzugullu Karakus, Y. (2021). Probing the role of Val228 on the catalytic activity of Scytalidium catalase. *Biochimica et Biophysica Acta. Proteins and Proteomics*, *1869*(8), 140662. doi:10.1016/j.bbapap.2021.140662 PMID:33887466

Goldberg, D. P., Michel, S. L., White, A. J., Williams, D. J., Barrett, A. G., & Hoffman, B. M. (1998). Molybdocene porphyrazines: A peripheral dithiolene metallacycle fused to a porphyrinic core. *Inorganic Chemistry*, *37*(9), 2100–2101. doi:10.1021/ic971498h PMID:11670359

Goldsmith, Z. K., Soudackov, A. V., & Hammes-Schiffer, S. (2019). Theoretical analysis of the inverted region in photoinduced proton-coupled electron transfer. *Faraday Discussions*, *216*, 363–378. doi:10.1039/C8FD00240A PMID:31017599

Gomez, M. L., Shah, N., Kenny, T. C., Jenkins, E. C. Jr, & Germain, D. (2019). SOD1 is essential for oncogene-driven mammary tumor formation but dispensable for normal development and proliferation. *Oncogene*, *38*(29), 5751–5765. doi:10.103841388-019-0839-x PMID:31222103

Gonta, M. (2007). The role of the natural antioxidants in the oxihaemoglobin oxidation and the diminution of nitrite oncentration. *Chemistry Journal of Moldova, 1*(2), p. 67-77. https://ibn.idsi.md/ro/vizualizare_articol/6715

Gonta, M. (2009). *Transformarea nitraţilor, nitriţilor, formarea N-nitrozaminelor şi utilizarea inhibitorilor în procese redox* [PhD thesis, Chisinau, the State University of Moldova].

Gonta, M., Mocanu, L., & Porubin-Schimbator, V. (2022). *Advanced oxidation processes for mineralization of amoxicillin in aqueous solutions. In Handbook of Research on Water Sciences and Society*. IGI GLOBAL.

Gonta, M., Sychev, A., Duca, Gh., & Batir, D. (1991). Efectul acidului dihidroxifumaric asupra oxidării (+) catechinei cu reactivul Fenton. [in Russian]. *Chimia Coordinativă*, *19*(12), 1654–1657.

Gonzaga, I. M. D., Moratalla, A., Eguiluz, K. I. B., Salazar-Banda, G. R., Cañizares, P., Rodrigo, M. A., & Saez, C. (2021). Novel Ti/RuO_2IrO_2 anode to reduce the dangerousness of antibiotic polluted urines by Fenton-based processes. *Chemosphere*, *270*, 129344. doi:10.1016/j.chemosphere.2020.129344 PMID:33395582

Gorbachev, M. G. (2015). Antioxidant Properties of Dihydroxyfumaric Acid and Its Dimethyl Ether: a Comparative DFT Study of Their Reactions with the Stable Radical DPPH*. *Chemistry Journal of Moldova, 10*(1).

Gorbachev, M. G. (2022). Molecular Orbital Nature of the Atmospheric Reactions between the NO3• Radical and C3 – C10 n-Alkanes: DFT Study. *The 7th International Conference: Ecological and Environmental Chemistry-2022*. IDSI.

Gorbachev, M. B., Budei, O., Arsene, I., & Gorinchoy, N. N. (2014). Surface Tension of Liquids with π-Conjugated Planar Molecules: DFT Calculation. *Journal of Molecular Liquids*, *190*, 50–53. doi:10.1016/j.molliq.2013.10.017

Gorbachev, M. G. (2022). Molecular orbital nature of solubility of shot chain n-alkanes in water and their reaction rate constants with nitronium cations: A DFT study. *International Journal of Organic Chemistry*, *12*, 92–101. doi:10.4236/ijoc.2022.122008

Gorbachev, M. G., Gorinchoy, N. N., & Arsene, I. (2017). Surface Tension of Near-Boiling Liquids with Intramolecular πσ*-Hyperconjugation: DFT Calculations. *Journal of Molecular Liquids*, *237*, 108–112. doi:10.1016/j.molliq.2017.04.069

Gorbachev, M. G., Gorinchoy, N., & Balan, I. (2022). Some particularities of the reaction between antioxidant phenolic acids and the free radical ABTS•+: A comparativeDFT study for the gas phase and ethanol. *Chemistry Journal of Moldova*, *17*(1), 24–30. doi:10.19261/cjm.2021.919

Gorbachev, M., Gorinchoy, N. N., & Arsene, I. (2021). Key Role of Some Specific Occupied Molecular Orbitals of Short Chain n-Alkanes in Their Surface Tension and Reaction Rate Constants with Hydroxyl Radicals: DFT Study. *International Journal of Organic Chemistry*, *11*(1), 1–13. doi:10.4236/ijoc.2021.111001

Gorinchoy, N. (2014). Jahn-Teller and pseudo Jahn-Teller origin of structural distortions of coordinated molecules. *Revue Roumaine de Chimie*, *59*(3-4), 273–280.

Gorinchoy, N. N., Balan, I. I., & Bersuker, I. B. (2011). Jahn–Teller, pseudo Jahn–Teller, and Renner–Teller effects in systems with fractional charges. *Computational & Theoretical Chemistry*, *976*(1-3), 113–119. doi:10.1016/j.comptc.2011.08.013

Gorinchoy, N., Balan, I., Polinger, V., & Bersuker, I. (2021). Pseudo Jahn-Teller origin of the proton-transfer energy barrier in the hydrogen-bonded [FHF]-system. *Chemistry Journal of Moldova*, *16*(1), 115–120. doi:10.19261/cjm.2021.834

Gould, I. R., Ege, D., Mattes, S. L., & Farid, S. (1987). Return electron transfer within geminate radical ion pairs. Observation of the Marcus inverted region. *Journal of the American Chemical Society*, *109*(12), 3794–3796. doi:10.1021/ja00246a055

Goyal, N., & Jerold, F. (2021). Biocosmetics: Technological Advances and future outlook. *Environmental Science and Pollution Research International*, *30*(10), 25148–25169. Advance online publication. doi:10.100711356-021-17567-3 PMID:34825334

Grabowski, S. J. (2015). $[FHF]^-$ – The strongest hydrogen bond under the influence of external interactions. *Crystals*, *6*(1), 3. doi:10.3390/cryst6010003

Graf, N., & Lippard, S. J. (2012). Redox activation of metal-based prodrugs as a strategy for drug delivery. *Advanced Drug Delivery Reviews*, *64*(11), 993–1004. doi:10.1016/j.addr.2012.01.007 PMID:22289471

Grahman, P. L. (2001). *An Introduction to Medicinal Chemistry* (2nd ed.). Oxford University Press.

Grampp, G. (1993). The Marcus inverted region from theory to experiment. *Angewandte Chemie International Edition in English*, *32*(5), 691–693. doi:10.1002/anie.199306911

Gransbury, G. K., Kappen, P., Glover, C. J., Hughes, J. N., Levina, A., Lay, P. A., Musgrave, I. F., & Harris, H. H. (2016). Comparison of KP1019 and NAMI-A in tumour-mimetic environments. *Metallomics*, *8*(8), 762–773. doi:10.1039/C6MT00145A PMID:27460862

Graversen, H. B., Becker, E. M., Skibsted, L. H., & Andersen, M. L. (2008). Antioxidant synergism between fruit juice and α-tocopherol. A comparison between high phenolic black chokeberry (Aronia melanocarpa) and high ascorbic blackcurrant (Ribes nigrum). *European Food Research and Technology*, *226*(4), 737–743. doi:10.100700217-007-0585-0

Gray, M. R. (1989). Substrate inactivation of enzymes in vitro and in vivo. *Biotechnology Advances*, *7*(4), 527–575. doi:10.1016/0734-9750(89)90722-2 PMID:14542989

Greenwood, D., Finch, R., Davey, P., & Wilcox, M. (2007). Antimicrobial Chemotherapy (5th ed.). Oxford University Press.

Grigoras, A. G. (2017). Catalase immobilization—A review. *Biochemical Engineering Journal*, *117*, 1–20. doi:10.1016/j.bej.2016.10.021

Grinstead, R. R. (1960). The oxidation of ascorbic acid by hydrogen peroxide. Catalysis by ethylenediaminetetraacetato-Iron(III). *Journal of the American Chemical Society*, *82*(13), 3464–3471. https://pubs.acs.org/doi/10.1021/ja01498a057. doi:10.1021/ja01498a057

Guan, D., Kurra, Y., Liu, W., & Chen, Z. (2015). A click chemistry approach to site-specific immobilization of a small laccase enables efficient direct electron transfer in a biocathode. *Chemical Communications*, *51*(13), 2522–2525. doi:10.1039/C4CC09179E PMID:25566975

Gusain, R., Gupta, K., Joshi, P., & Khatri, O. P. (2019). Adsorptive removal and photocatalytic degradation of organic pollutants using metal oxides and their composites: A comprehensive review. *Advances in Colloid and Interface Science*, *272*, 102009. doi:10.1016/j.cis.2019.102009 PMID:31445351

Haider, A., Haider, S., & Kang, I. K. (2018). A comprehensive review summarizing the effect of electrospinning parameters and potential applications of nanofibers in biomedical and biotechnology. *Arabian Journal of Chemistry*, *11*(8), 1165–1188. doi:10.1016/j.arabjc.2015.11.015

Hall, C. L., & Kamin, H. (1975). The purification and some properties of electron transfer flavoprotein and general fatty acyl coenzyme A dehydrogenase from pig liver mitochondria. *The Journal of Biological Chemistry*, *250*(9), 3476–3486. doi:10.1016/S0021-9258(19)41540-8 PMID:1168197

Halliwell, B., Aeschbach, R., Löliger, J., & Aruoma, O. I. (1995). The characterization of antioxidants. *Food and Chemical Toxicology*, *33*(7), 601–617. doi:10.1016/0278-6915(95)00024-V PMID:7628797

Halliwell, B., & Gutteridge, J. M. C. (2015). *Antioxidant defends synthesized in vivo'. Free Radicals in Biology and Medicine* (5th ed.). Academic. doi:10.1093/acprof:oso/9780198717478.003.0003

Han, C., Wang, Z., Chen, S., Li, L., Xu, Y., Kang, W., Wei, C., Ma, H., Wang, M., & Jin, X. (2021). Berbamine suppresses the progression of bladder cancer by modulating the ROS/NF-κB axis. *Oxidative Medicine and Cellular Longevity*, *2021*, 1–6. doi:10.1155/2021/8851763

Handel, A., La Gruta, N. L., & Thomas, P. G. (2020). Simulation Modelling for Immunologists. *Nature Reviews. Immunology*, *20*(3), 186–195. doi:10.103841577-019-0235-3 PMID:31804613

Hansen, E., Karslake, J., Woods, R. J., Read, A. F., & Wood, K. B. (2020). Antibiotics can be used to contain drug-resistant bacteria by maintaining sufficiently large sensitive populations. *PLoS Biology*, *18*(5), e3000713. doi:10.1371/journal.pbio.3000713 PMID:32413038

Hara, I., Ichise, N., Kojima, K., Kondo, H., Ohgiya, S., Matsuyama, H., & Yumoto, I. (2007). Relationship between the size of the bottleneck 15 Å from iron in the main channel and the reactivity of catalase corresponding to the molecular size of substrates. *Biochemistry*, *46*(1), 11–22. doi:10.1021/bi061519w PMID:17198371

Harper, W. S., & Daigler, D. W. (1968). *Production of hydrogen peroxide from aqueous acidic solution obtained by hydrolysis.* Patent US 1968-01-23. https://patents.google.com/patent/US3694154A/en

Harrison, J. F. (1971). An Ab Initio Study of the Zero Field Splitting Parameters of 3 B 1 Methylene. *The Journal of Chemical Physics*, *54*(12), 5413–5417. doi:10.1063/1.1674842

Hay, P. J., Thibeault, J. C., & Hoffmann, R. (1975). Orbital Interactions in Metal Dimer Complexes. *Journal of the American Chemical Society*, *97*(17), 4884–4899. doi:10.1021/ja00850a018

Hay, R. W., & Harvie, S. J. (1965). The uncatalysed and nickel(ii) catalysed decarboxylation of dihydroxyfumaric acid. *Australian Journal of Chemistry*, *18*(8), 1197. doi:10.1071/CH9651197

Hayyan, M., Hashim, M. A., & Al Nashef, I. M. (2016). Superoxide ion: Generation and chemical implications. *Chemical Reviews*, *116*(5), 3029–3085. doi:10.1021/acs.chemrev.5b00407 PMID:26875845

Hazarika, A., Saikia, S., Devi, B., Yadav, M., & Yadav, H. S. (2022). Oxidoreductase metalloenzymes as green catalyst for phytoremediation of environmental pollutants. In *Phytoremediation Technology for the Removal of Heavy Metals and Other Contaminants from Soil and Water* (pp. 141–172). Elsevier. doi:10.1016/B978-0-323-85763-5.00021-0

Hecht, H. J., Kalisz, H. M., Hendle, J., Schmid, R. D., & Schomburg, D. (1993). crystal structure of glucose oxidase from Aspergillus niger refined at 2·3 Å resolution. *Journal of Molecular Biology*, *229*(1), 153–172. doi:10.1006/jmbi.1993.1015 PMID:8421298

He, G., Xu, N., Ge, H., Lu, Y., Wang, R., Wang, H., Du, J., Fan, J., Sun, W., & Peng, X. (2021). Red-light-responsive Ru complex photosensitizer for lysosome localization photodynamic therapy. *ACS Applied Materials & Interfaces*, *13*(17), 19572–19580. doi:10.1021/acsami.0c22551 PMID:33900720

Heimer, K. A., Hart, A. M., Martin, L. G., & Rubio-Wallace, S. (2009). Examining the evidence for the use of vitamin C in the prophylaxis and treatment of the common cold. Journal of the American Academy of Nurse Practitioners, 21(5), 295–300. doi:10.1111/j.1745-7599.2009.00409.x

Heisenberg, W. (1928). On the theory of ferromagnetism. *The European Physical Journal A*, *49*(9-10), 619–636. doi:10.1007/BF01328601

Heitler, W., & London, F. (1927). Wechselwirkung neutraler Atome und homöopolare Bindung nach der Quantenmechanik. *The European Physical Journal A*, *44*(6-7), 455–472. doi:10.1007/BF01397394

He, M., Pei, Z., Mohsen, A.-W., Watkins, P., Murdoch, G., Van Veldhoven, P. P., Ensenauer, R., & Vockley, J. (2011). Identification and characterization of new long chain Acyl-CoA dehydrogenases. *Molecular Genetics and Metabolism*, *102*(4), 418–429. doi:10.1016/j.ymgme.2010.12.005 PMID:21237683

Hemilä, H. (2017). Vitamin C and infections. Nutrients, 9(4). doi:10.3390/nu9040339

Hemilä, H., Douglas, R. M., & Hemilä, H. (1999). Vitamin C and acute respiratory infections. *The International Journal of Tuberculosis and Lung Disease*, *3*(9), 756–761. PMID:10488881

Henriques, B. J., Katrine Jentoft Olsen, R., Gomes, C. M., & Bross, P. (2021). Electron transfer flavoprotein and its role in mitochondrial energy metabolism in health and disease. *Gene*, *776*, 145407. doi:10.1016/j.gene.2021.145407 PMID:33450351

Her, J.-H., Stephens, P. W., Ribas-Ariño, J., Novoa, J. J., Shum, W. W., & Miller, J. S. (2009). Structure and Magnetic Interactions in the Organic-Based Ferromagnet Decamethylferrocenium Tetracyanoethenide, [FeCp*2]•+[TCNE]•−. *Inorganic Chemistry*, *48*(8), 3296–3307. doi:10.1021/ic801679m PMID:19361235

Herkendell, K., Stemmer, A., & Tel-Vered, R. (2019). Magnetically induced enzymatic cascades-advancing towards multi-fuel direct/mediated bioelectrocatalysis. *Nanoscale Advances*, *1*(5), 1686–1692. doi:10.1039/C8NA00346G PMID:36134209

Hermoso, W., Ilkhani, A. R., & Bersuker, I. B. (2014). Pseudo Jahn–Teller origin of instability of planar configurations of hexa-heterocycles $C_4N_2H_4X_2$ (X = H, F, Cl, Br). *Computational & Theoretical Chemistry*, *1049*, 109–114. doi:10.1016/j.comptc.2014.10.007

Herrmann, G., Jayamani, E., Mai, G., & Buckel, W. (2008). energy conservation via electron-transferring flavoprotein in anaerobic bacteria. *Journal of Bacteriology*, *190*(3), 784–791. doi:10.1128/JB.01422-07 PMID:18039764

Hickey, D. P., Milton, R. D., Rasmussen, M., Abdellaoui, S., Nguyen, K., & Minteer, S. D. (2016). Fundamentals and applications of bioelectrocatalysis. *Electrochemistry*, *13*, 97–132. doi:10.1039/9781782620273-00097

Hicks, R. G. (2007). What's new in stable radical chemistry? *Organic & Biomolecular Chemistry*, *5*(9), 1321–1338. doi:10.1039/b617142g PMID:17464399

Hicks, R. G. (Ed.). (2010). *Stable Radicals*. Wiley. doi:10.1002/9780470666975

Hidalgo, M., Sánchez-Moreno, C., & de Pascual-Teresa, S. (2010). Flavonoid–flavonoid interaction and its effect on their antioxidant activity. *Food Chemistry*, *121*(3), 691–696. doi:10.1016/j.foodchem.2009.12.097

Hinzmann, A., Stricker, M., Busch, J., Glinski, S., Oike, K., & Gröger, H. (2020). Selective TEMPO-Oxidation of Alcohols to Aldehydes in Alternative Organic Solvents. *European Journal of Organic Chemistry*, *2020*(16), 2399–2408. doi:10.1002/ejoc.201901365

Hirte, K., Seiwert, B., Schüürmann, G., & Reemtsma, T. (2016). New hydrolysis products of the beta-lactam antibiotic amoxicillin, their pH-dependent formation and search in municipal wastewater. *Water Research*, *88*, 880–888. doi:10.1016/j.watres.2015.11.028 PMID:26613181

Hofrichter, M., Ullrich, R., Pecyna, M. J., Liers, C., & Lundell, T. (2010). New and classic families of secreted fungal heme peroxidases. *Applied Microbiology and Biotechnology, 87.*

Holland, H. D. (2006). The oxygenation of the atmosphere and oceans. *Philosophical Transactions of the Royal Society of London. Series B, Biological Sciences*, *361*(1470), 903–915. doi:10.1098/rstb.2006.1838 PMID:16754606

Hompland, T., Fjeldbo, C. S., & Lyng, H. (2021). Tumor hypoxia as a barrier in cancer therapy: Why levels matter. *Cancers (Basel)*, *13*(3), 499–520. doi:10.3390/cancers13030499 PMID:33525508

Hongshin, L., Hyejin, L., Jiwon, S., Hyungeun, K., Yun, S., Jaehong, K., & Changha, L. (2016). Activation of oxygen and hydrogen peroxide by copper (II) coupled with hydroxylamine for oxidation of organic contaminants. *Environmental Science & Technology*, *50*(15), 8231–8238. doi:10.1021/acs.est.6b02067 PMID:27387011

Huang, D., Boxin, O. U., & Prior, R. L. (2005). The chemistry behind antioxidant capacity assays. *Journal of Agricultural and Food Chemistry*, *53*(6), 1841–1856. doi:10.1021/jf030723c PMID:15769103

Huang, L., Sun, G., Cobessi, D., Wang, A. C., Shen, J. T., Tung, E. Y., Anderson, V. E., & Berry, E. A. (2006). 3-Nitropropionic acid is a suicide inhibitor of mitochondrial respiration that, upon oxidation by Complex II, forms a covalent adduct with a catalytic base arginine in the active site of the enzyme. *The Journal of Biological Chemistry*, *281*(9), 5965–5972. doi:10.1074/jbc.M511270200 PMID:16371358

Huang, R., Chen, H., Liang, J., Li, Y., Yang, J., Luo, C., Tang, Y., Ding, Y., Liu, X., Yuan, Q., Yu, H., Ye, Y. C., Xu, W. F., & Xie, X. (2021). Dual role of reactive oxygen species and their application in cancer therapy. *Journal of Cancer*, *12*(18), 5543–5561. doi:10.7150/jca.54699 PMID:34405016

Huang, X., Braams, B. J., Carter, S., & Bowman, J. M. (2004). Quantum calculations of vibrational energies of $H_3O_2^-$ on an ab initio potential. *Journal of the American Chemical Society*, *126*(16), 5042–5043. doi:10.1021/ja049801i PMID:15099067

Humer, D., Furlanetto, V., Schruef, A. K., Wlodarczyk, A., Kuttke, M., Divne, C., & Spadiut, O. (2021). Potential of unglycosylated horseradish peroxidase variants for enzyme prodrug cancer therapy. *Biomedicine and Pharmacotherapy*, *142*, 112037.

Humphrey, W., Dalke, A., & Schulten, K. (1996). VMD: Visual molecular dynamics. *Journal of Molecular Graphics*, *14*(1), 33–38. doi:10.1016/0263-7855(96)00018-5 PMID:8744570

Hunt, R. H., Leacock, R. A., Peters, C. W., & Hecht, K. T. (1965). Internal-rotation in hydrogen peroxide: The far-infrared spectrum and the determination of the hindering potential. *The Journal of Chemical Physics*, *42*(6), 1931–1946. doi:10.1063/1.1696228

Husain, M., & Steenkamp, D. J. (1983). Electron transfer flavoprotein from pig liver mitochondria. A simple purification and re-evaluation of some of the molecular properties. *The Biochemical Journal*, *209*(2), 541–545. doi:10.1042/bj2090541 PMID:6847633

Hush, N. S. (1985). Distance dependence of electron transfer rates. *Coordination Chemistry Reviews*, *64*, 135–157. doi:10.1016/0010-8545(85)80047-3

Hydrogen Peroxide. Safety and Handling . (2006). Technical data sheet HH_2323, Revised 0305 Solvay Chemicals, Inc. www.solvaychemicals.us

Iacopini, D., Vančo, J., Di Pietro, S., Bordoni, V., Zacchini, S., Marchetti, F., Dvořák, Z., Malina, T., Biancalana, L., Trávníček, Z., & Di Bussoloc, V. (2022). New glycoconjugation strategies for ruthenium (II) arene complexes via phosphane ligands and assessment of their antiproliferative activity. *Bioorganic Chemistry*, *126*, 105901–105916. doi:10.1016/j.bioorg.2022.105901 PMID:35671646

Iacopini, P., Baldi, M., Storchi, P., & Sebastiani, L. (2008). Catechin, epicatechin, quercetin, rutin and resveratrol in red grape: Content, in vitro antioxidant activity and interactions. *Journal of Food Composition and Analysis*, *21*(8), 589–598. doi:10.1016/j.jfca.2008.03.011

Ikram, M., Rashid, M., Haider, A., Naz, S., Haider, J., Raza, A., Ansar, M. T., Uddin, M. K., Ali, N. M., Ahmed, S. S., Imran, M., Dilpazir, S., Khan, Q., & Maqbool, M. (2021). A review of photocatalytic characterization, and environmental cleaning, of metal oxide nanostructured materials. *Sustainable Materials and Technologies*, *30*, e00343. doi:10.1016/j.susmat.2021.e00343

Ilan Y. A., Czapski G., Meisel D. (1976) The one-electron transfer redox potentials of free radicals. *I. The oxygen/superoxide system //Biochimica et Biophysica Acta (BBA)-Bioenergetics, 430*(2), 209-224.

Ilkhani, A. R., Hermoso, W., & Bersuker, I. B. (2015). Pseudo Jahn-Teller origin of instability of planar configurations of hexa-heterocycles. Application to compounds with 1,2- and 1,4-C_4X_2 skeletons (X= O, S, Se, Te). *Chemical Physics*, *460*, 75–82. doi:10.1016/j.chemphys.2015.02.017

Imberti, C., Zhang, P., Huang, H., & Sadler, P. J. (2020). New designs for phototherapeutic transition metal complexes. *Angewandte Chemie International Edition*, *59*(1), 61–73. doi:10.1002/anie.201905171 PMID:31310436

Ionita, G., Ariciu, A. M., Smith, D. K., & Chechik, V. (2015). Ion exchange in alginate gels–dynamic behaviour revealed by electron paramagnetic resonance. *Soft Matter*, *11*(46), 8968–8974. doi:10.1039/C5SM02062J PMID:26399427

Ionita, G., Ariciu, A. M., Turcu, I. M., & Chechik, V. (2014). Properties of polyethylene glycol/cyclodextrin hydrogels revealed by spin probes and spin labelling methods. *Soft Matter*, *10*(11), 1778–1783. doi:10.1039/c3sm52004h PMID:24651650

Ionita, G., & Chechik, V. (2010). Exploring polyethylene glycol/cyclodextrin hydrogels with spin probes and EPR spectroscopy. *Chemical Communications*, *46*(43), 8255–8257. doi:10.1039/c0cc02759f PMID:20890495

Itabashi, H., Umetsu, K., Satoh, K., & Kawashima, T. (1990). Ligand effect on redox reaction of metal ions and its application to potentiometric titration of chromium (VI) and copper (II) with iron (II). *Analytical Sciences*, *6*(5), 721–725. doi:10.2116/analsci.6.721

Itabashi, H., Umetsu, K., Satoh, K., & Kawashima, T. (1991). Indirect spectrophotometric determination of vanadium (IV) by flow injection analysis based on the redox reaction with copper (II) in the presence of neocuproine. *Analytical Letters*, *24*(7), 1219–1230. doi:10.1080/00032719108052965

Itabashi, H., Umetsu, K., Teshima, N., Satoh, K., & Kawashima, T. (1992). Indirect spectrophotometric determination of complexing agents by flow-injection analysis based on redox reaction of copper (II) with iron (II). *Analytica Chimica Acta*, *261*(1-2), 213–218. doi:10.1016/0003-2670(92)80193-B

Izadyar, A., Van, M. N., Miranda, M., Weatherford, S., & Hood, E. E. (2022). Electrocatalytic effect of recombinant Mn peroxidase from corn on microbiosensors to detect glucose. *Biocatalysis and Agricultural Biotechnology*, *43*, 102445. doi:10.1016/j.bcab.2022.102445

Jacquemin, D., Zuniga, J., Requena, A., & Céron-Carrasco, J. P. (2014). Assessing the importance of proton transfer reactions in DNA. *Accounts of Chemical Research*, *47*(8), 2467–2474. doi:10.1021/ar500148c PMID:24849375

Jagatić Korenika, A.-M., Biloš, J., Kozina, B., Tomaz, I., Preiner, D., & Jeromel, A. (2020). Effect of different reducing agents on aromatic compounds, antioxidant and chromatic properties of Sauvignon Blanc wine. *Foods*, *9*(8), 996. doi:10.3390/foods9080996 PMID:32722317

James, T., & Wales, D. J. (2005). Protonated water clusters described by an empirical valence bond potential. *The Journal of Chemical Physics*, *122*(13), 024516. doi:10.1063/1.1869987 PMID:15847464

Jansen, B. J. M., & De Groot, A. (2004). Occurrence, biological activity and synthesis of drimane sesquiterpenoids. *Natural Product Reports*, *21*(4), 449-447. doi:10.1039/b311170a

Janusz, G., Pawlik, A., Sulej, J., Świderska-Burek, U., Jarosz-Wilkolazka, A., & Paszczyński, A. (2017). Lignin degradation: Microorganisms, enzymes involved, genomes analysis and evolution. *FEMS Microbiology Reviews*, *41*(6), 941–962. doi:10.1093/femsre/fux049 PMID:29088355

Janusz, G., Pawlik, A., Świderska-Burek, U., Polak, J., Sulej, J., Jarosz-Wilkołazka, A., & Paszczyński, A. (2020). Laccase properties, physiological functions, and evolution. *International Journal of Molecular Sciences*, *21*(3), 966. doi:10.3390/ijms21030966 PMID:32024019

Jasek-Gajda, E., Jurkowska, H., Jasińska, M., & Lis, G. J. (2020). Targeting the MAPK/ERK and PI3K/AKT signaling pathways affects NRF2, Trx and GSH antioxidant systems in leukemia cells. *Antioxidants*, *9*(7), 633–650. doi:10.3390/antiox9070633 PMID:32709140

Jeon, J. R., Baldrian, P., Murugesan, K., & Chang, Y. S. (2012). Laccase-catalysed oxidations of naturally occurring phenols: From in vivo biosynthetic pathways to green synthetic applications. *Microbial Biotechnology*, *5*(3), 318–332. doi:10.1111/j.1751-7915.2011.00273.x PMID:21791030

Jiang, G.-B., Zhang, W.-Y., He, M., Gu, Y.-Y., Bai, L., Wang, Y.-J., Yi, Q.-Y., & Du, F. (2020). Development of four ruthenium polypyridyl complexes as antitumor agents: Design, biological evaluation and mechanism investigation. *Journal of Inorganic Biochemistry*, *208*, 111104–111118. doi:10.1016/j.jinorgbio.2020.111104 PMID:32485635

Jiang, G.-B., Zhang, W.-Y., He, M., Gu, Y.-Y., Bai, L., Wang, Y.-J., Yi, Q.-Y., & Du, F. (2021). Systematic evaluation of the antitumor activity of three ruthenium polypyridyl complexes. *Journal of Inorganic Biochemistry*, *225*, 111616–111627. doi:10.1016/j.jinorgbio.2021.111616 PMID:34555601

Jiang, S., Ren, D., Wang, Z., Zhang, S., Zhang, X., & Chen, W. (2022). Improved stability and promoted activity of laccase by One-Pot encapsulation with Cu (PABA) nanoarchitectonics and its application for removal of Azo dyes. *Ecotoxicology and Environmental Safety*, *234*, 113366. doi:10.1016/j.ecoenv.2022.113366 PMID:35272195

Jin, Y., Zhang, Q., Zhang, Y., & Duan, C. (2020). Electron transfer in the confined environments of metal–organic coordination supramolecular systems. *Chemical Society Reviews*, *49*(15), 5561–5600. doi:10.1039/C9CS00917E PMID:32643720

Joergensen, R. G., Schmaedeke, F., Windhorst, K., & Meyer, B. (1995). Biomass and activity of microorganisms in a fuel oil contaminated soil. *Soil Biology & Biochemistry*, *27*(9), 1137–1143. doi:10.1016/0038-0717(95)00040-L

Jones, S. M., & Solomon, E. I. (2015). Electron transfer and reaction mechanism of laccases. *Cellular and Molecular Life Sciences*, *72*(5), 869–883. doi:10.100700018-014-1826-6 PMID:25572295

Jorgensen, W. L., Maxwell, D. S., & Tirado-Rives, J. (1996). Development and testing of the OPLS all-atom force field on conformational energetics and properties of organic liquids. *Journal of the American Chemical Society*, *118*(45), 11225–11236. doi:10.1021/ja9621760

Jorgenson, T. C., Zhong, W., & Oberley, T. D. (2013). Redox imbalance and biochemical changes in cancerredox imbalance in cancer. *Cancer Research*, *73*(20), 6118–6123. doi:10.1158/0008-5472.CAN-13-1117 PMID:23878188

Kahn, O. (1993). *Molecular Magnetism*. Wiley-VCH.

Kahn, O., & Charlot, M. F. (1980). Overlap density in binuclear complexes: A topological approach of the exchange interaction. *Nuov. J. Chim.*, *4*, 567–576. doi:10.1007/978-94-010-9716-1_11

Kahn, O., Pei, Y., & Journaux, Y. (1989). Orthogonality of the Magnetic Orbitals and Ferromagnetic Interaction. *Molecular Crystals and Liquid Crystals Incorporating Nonlinear Optics*, *176*(1), 429–441. doi:10.1080/00268948908037500

Kahn, S. A., Beers, R. J., & Lentz, C. W. (2011). Resuscitation after severe burn injury using high-dose ascorbic acid: A retrospective review. *Journal of Burn Care & Research; Official Publication of the American Burn Association*, *32*(1), 110–117. doi:10.1097/BCR.0b013e318204b336 PMID:21131846

Kanarovsky, E., & Yaltychenko, O. (2022). Учёт синергии витаминов Е и С в кинетической модели перекисного окисления липидов [Determination of the synergy of vitamins E and C in the kinetic model of lipid peroxidation]. *Electronic Materials Processing*, *59*(5), 44–50. PMID:35976784

Karabinas, P., Sazou, D., & Jannakoudakis, D. (1985). —comparative electrochemical study of L-ascorbic acid and dihydroxyfumaric acid on a mercury electrode in neutral media. *Bioelectrochemistry and Bioenergetics*, *14*(4-6), 469–478. doi:10.1016/0302-4598(85)80019-2

Karabulut, I. (2010). Effects of α-tocopherol, β-carotene and ascorbyl palmitate on oxidative stability of butter oil triacylglycerols. *Food Chemistry*, *123*(3), 622–627. doi:10.1016/j.foodchem.2010.04.080

Karamoschos, N., & Tasis, D. (2022). Photocatalytic evolution of hydrogen peroxide: A minireview. *Energies, 15*(17), 6202. doi:10.3390/en15176202

Karmakar, S., Poetsch, I., Kowol, C. R., Heffeter, P., & Gibson, D. (2019). Synthesis and cytotoxicity of water-soluble dual- and triple-action satraplatin derivatives: Replacement of equatorial chlorides of satraplatin by acetates. *Inorganic Chemistry, 58*(24), 16676–16688. doi:10.1021/acs.inorgchem.9b02796 PMID:31790216

Kar, R. K., Miller, A., & Mroginski, M. (2022). Understanding flavin electronic structure and spectra. *Wiley Interdisciplinary Reviews. Computational Molecular Science, 12*(2). Advance online publication. doi:10.1002/wcms.1541

Katsumata, H., Teshima, N., & Kawashima, T. (1997). Successive potentiometric titration of chromium (VI) and iron (III) with cobalt (II) in the presence of 1, 10-phenanthroline. *Bulletin of the Chemical Society of Japan, 70*(9), 2151–2154. doi:10.1246/bcsj.70.2151

Kaushal, J., Mehandia, S., Singh, G., Raina, A., & Arya, S. K. (2018). Catalase enzyme: Application in bioremediation and food industry. *Biocatalysis and Agricultural Biotechnology, 16*, 192–199. doi:10.1016/j.bcab.2018.07.035

Kayastha, K., Vitt, S., Buckel, W., & Ermler, U. (2021). Flavins in the electron bifurcation process. *Archives of Biochemistry and Biophysics, 701*, 108796. doi:10.1016/j.abb.2021.108796 PMID:33609536

Kendall, R. A., Dunning, T. H. Jr, & Harrison, R. J. (1992). Electron affinities of the first-row atoms revisited. Systematic basis sets and wave functions. *The Journal of Chemical Physics, 96*(9), 6796–6806. doi:10.1063/1.462569

Khataee, S., Dehghan, G., Yekta, R., Rashtbari, S., Maleki, S., & Khataee, A. (2022). The protective effect of natural phenolic compound on the functional and structural responses of inhibited catalase by a common azo food dye. *Food and Chemical Toxicology, 160*, 112801. doi:10.1016/j.fct.2021.112801 PMID:34974130

Khorshidi, S., Younesi, S., & Karkhaneh, A. (2022). Peroxide mediated oxygen delivery in cancer therapy. *Colloids and Surfaces. B, Biointerfaces, 219*, 112832. doi:10.1016/j.colsurfb.2022.112832 PMID:36137337

Khramtsov, V. V. (2018). In vivo molecular electron paramagnetic resonance-based spectroscopy and imaging of tumor microenvironment and redox using functional paramagnetic probes. *Antioxidants & Redox Signaling, 28*(15), 1365–1377. doi:10.1089/ars.2017.7329 PMID:29132215

Khuri, A. I., & Mukhopadhyay, S. (2010). Response surface methodology. *Wiley Interdisciplinary Reviews: Computational Statistics, 2*(2), 128–149. doi:10.1002/wics.73

Kim, E. K., Jang, M., Song, M. J., Kim, D., Kim, Y., & Jang, H. H. (2019). Redox-mediated mechanism of chemoresistance in cancer cells. *Antioxidants, 8*(10), 471–488. doi:10.3390/antiox8100471 PMID:31658599

Kim, H. J., & Winge, D. R. (2013). Emerging concepts in the flavinylation of succinate dehydrogenase. *Biochimica et Biophysica Acta (BBA) - Bioenergetics, 1827*(5), 627–636. doi:10.1016/j.bbabio.2013.01.012 PMID:23380393

Kim, H., Kim, J. S., Heo, J. M., Pei, M., Park, I. H., Liu, Z., & Lee, T. W. (2020). Proton-transfer-induced 3D/2D hybrid perovskites suppress ion migration and reduce luminance overshoot. *Nature Communications, 11*(1), 1–13. doi:10.103841467-020-17072-0 PMID:32632144

Kim, J. J., Wang, M., & Paschke, R. (1993). Crystal structures of medium-chain acyl-CoA dehydrogenase from pig liver mitochondria with and without substrate. *Proceedings of the National Academy of Sciences of the United States of America, 90*(16), 7523–7527. doi:10.1073/pnas.90.16.7523 PMID:8356049

Kirkman, H. N., & Gaetani, G. F. (1984). Catalase: A tetrameric enzyme with four tightly bound molecules of NADPH. *Proceedings of the National Academy of Sciences of the United States of America, 81*(14), 4343–4347. doi:10.1073/pnas.81.14.4343 PMID:6589599

Kittipongpittaya, K., Panya, A., Phonsatta, N., & Decker, E. A. (2016). Effects of environmental pH on antioxidant interactions between rosmarinic acid and α-tocopherol in oil-in-water (O/W) emulsions. *Journal of Agricultural and Food Chemistry*, *64*(34), 6575–6583. doi:10.1021/acs.jafc.6b02700 PMID:27494424

Klán, P., & Vavrik, M. (2006). Non-catalytic remediation of aqueous solutions by microwave-assisted photolysis in the presence of H_2O_2. *Journal of Photochemistry and Photobiology A Chemistry*, *177*(1), 24–33. doi:10.1016/j.jphotochem.2005.05.008

Klein, E., Rimarcik, J., & Lukes, V. (2009). DFT/B3LYP study of the O–H bond dissociation enthalpies and proton affinities of para-and metasubstituted phenols in water and benzene. *Acta Chimica Slovenica*, *2*, 37–51. http://fez.schk.sk/eserv/changeme:5011/acs_0045.pdf

Klopman, G. (1968). Chemical Reactivity and the Concept of Charge- and Frontier-Controlled Reactions. *Journal of the American Chemical Society*, *90*(2), 223–234. doi:10.1021/ja01004a002

Koch, W., & Holthausen, M. C. (2001). *A Chemist's Guide to Density Functional Theory*. VCH. doi:10.1002/3527600043

Kollmar, C., & Kahn, O. (1993). Ferromagnetic spin alignment in molecular systems: An orbital approach. *Accounts of Chemical Research*, *26*(5), 259–265. doi:10.1021/ar00029a006

Komala, I., Ito, T., Nagashima, F., Yagi, Y., & Asakawa, Y. (2010). Cytotoxic, radical scavenging and antimicrobial activities of sesquiterpenoids from the Tahitian liverwort *Mastigophora diclados* (Brid.) Nees (Mastigophoraceae). *Journal of Natural Medicines, 64*(4), 417-422. doi:10.1007/s11418-010-0423-8

Konopko, A., & Litwinienko, G. (2021). Unexpected role of pH and microenvironment on the antioxidant and synergistic activity of resveratrol in model micellar and liposomal systems. *The Journal of Organic Chemistry*, *87*(3), 1698–1709. doi:10.1021/acs.joc.1c01801 PMID:34842421

Koper, M. T. M. (2011). Thermodynamic theory of multi-electron transfer reactions: Implications for electrocatalysis. *Journal of Electroanalytical Chemistry (Lausanne, Switzerland)*, *660*(2), 254–260. doi:10.1016/j.jelechem.2010.10.004

Kotelnikov, A. I., Medvedev, E. S., & Goreachev, N. S. (2011). Structural and Dtnamic Aspects of Electron Transfer in Proteins – Highly Organized Natural Nanostructures (in Rus.). *Proc.AS. Ser.Chim*, *7*, 1295.

Krause, K.-H. (2018). Role of NOX-generated reactive oxygen species in health and disease. *Free Radical Biology & Medicine*, *120*, S7–S7. doi:10.1016/j.freeradbiomed.2018.04.033

Kresge, A. J. (1975). What makes proton transfer fast. *Accounts of Chemical Research*, *8*(10), 354–360. doi:10.1021/ar50094a006

Kresin, V. Z., & William, A. (1987). On the mechanism of singlet-triplet transitions. *Chemical Physics Letters*, *138*(1), 59–64. doi:10.1016/0009-2614(87)80342-1

Krumova, K., & Cosa, G. (2016). Chapter 1: Overview of reactive oxygen species, In: Singlet oxygen: Applications in biosciences and nanosciences, 1, 1-2. Royal Society of Chemistry. doi:10.1039/9781782622208-00001

Kuchkova, K. I., Aryku, A. N., Sekara, E. S., Barba, A. N., Dragalin, I. P., Vlad, P. F., & Ungur, N. D. (2014). Synthesis of 13-amino-14,15-dinorlabd-8(9)-ene from sclareol. Russian Chemical Bulletin, International Edition, 63(9), 1-3. doi:10.100711172-014-0703-7

Kuchkova, K. I., & Aricu, A. N. (2009). Synthesis of 11-aminodrim-7-ene from drimenol. *Chemistry of Natural Compounds*, *45*(3), 367–370. doi:10.100710600-009-9341-y

Kuchkova, K. I., Aricu, A. N., Barba, A. N., Sekara, E. S., Vlad, P. F., & Ungur, N. D. (2014). Synthesis of 12-amino-11-dihomodrimane sesquiterpenoids from norambreinolide. *Chemistry of Natural Compounds*, *50*(3), 458–461. doi:10.100710600-014-0986-9

Kuchkova, K. I., Aricu, A. N., Barba, A. N., Vlad, P. F., Lipkovskii, J., Simonov, Yu. A., & Kravtov, V. Kh. (2011). Synthesis of nitrogen-containing drimane sesquiterpenoids from 11-dihomodrim-8(9)-en-12-one. *Chemistry of Natural Compounds*, *47*(2), 223–228. doi:10.100710600-011-9888-2

Kuchkova, K. I., Aricu, A. N., Vlad, P. F., Deleanu, C., & Nicolescu, A. (2010). Synthesis of *N*-containing drimane sesquiterpenoids from 11-dihomodriman-8α-ol-12-one. *Chemistry of Natural Compounds*, *46*(4), 539–544. doi:10.100710600-010-9671-9

Kuchkova, K., Aricu, A., Barba, A., Vlad, P., Shova, S., Secara, E., Ungur, N., Zbancioc, G., & Mangalagiu, I. I. (2013). An efficient and straightforward method to new organic compounds: Homodrimane sesquiterpenoids with diazine units. *Synlett*, *24*(6), 697–700. doi:10.1055-0032-1318253

Kuchkova, K., Aricu, A., Secara, E., Barba, A., Vlad, P., Macaev, F., Melnic, E., & Kravtov, V. (2015). Synthesis and structure of homodrimanic sesquiterpenoids containing 1,2,4-triazole and carbazole rings. *Chemistry of Natural Compounds*, *51*(14), 684–688. doi:10.100710600-015-1384-7

Kuchkova, K., Aricu, A., Secara, E., Barba, A., Vlad, P., Ungur, N., & Mangalagiu, I. I. (2014). Design, synthesis, and antimicrobial activity of some novel homodrimane sesquiterpenoids with diazine skeleton. *Medicinal Chemistry Research*, *23*(3), 1559–1568. doi:10.100700044-013-0720-3

Kulkarni, A. D., Pathak, R. K., & Bartolotti, L. J. (2006). Effect of additional hydrogen peroxide to H_2O_2...$(H_2O)n$, n=1 and 2 complexes: Quantum chemical study. *The Journal of Chemical Physics*, *124*(21), 214309. doi:10.1063/1.2202098 PMID:16774409

Kumar, R., George, L., Jun, Z., & Mukherji, S. (2022). Photocatalytic activity of graphene oxide-TiO_2 nanocomposite on dichlorvos and malathion and assessment of toxicity changes due to photodegradation. *Chemosphere*, *308*, 136402/1–11. doi:10.1016/j.chemosphere.2022.136402

Kumar, S., Dangi, A. K., Shukla, P., Baishya, D., & Khare, S. K. (2019). Thermozymes: Adaptive strategies and tools for their biotechnological applications. *Bioresource Technology*, *278*, 372–382. doi:10.1016/j.biortech.2019.01.088 PMID:30709766

Kyani, A., Tamura, S., Yang, S., Shergalis, A., Samanta, S., Kuang, Y., Ljungman, M., & Neamati, N. (2018). Discovery and mechanistic elucidation of a class of protein disulfide isomerase inhibitors for the treatment of glioblastoma. *ChemMedChem*, *13*(2), 164–177. doi:10.1002/cmdc.201700629 PMID:29235250

Lafridi, H. O. (2020). QSAR modeling of antiradical properties of phenolic compounds using DFT calculations. *Moroccan. J. Chem.*, *8*(4), 841–852.

Lalaoui, N., Rousselot-Pailley, P., Robert, V., Mekmouche, Y., Villalonga, R., Holzinger, M., Cosnier, S., Tron, T., & le Goff, A. (2016). Direct electron transfer between a site-specific pyrene-modified laccase and carbon nanotube/gold nanoparticle supramolecular assemblies for bioelectrocatalytic dioxygen reduction. *ACS Catalysis*, *6*(3), 1894–1900. doi:10.1021/acscatal.5b02442

Lardinois, O. M., Mestdagh, M. M., & Rouxhet, P. G. (1996). Reversible inhibition and irreversible inactivation of catalase in presence of hydrogen peroxide. *Biochimica et Biophysica Acta. Protein Structure and Molecular Enzymology*, *1295*(2), 222–238. doi:10.1016/0167-4838(96)00043-X PMID:8695649

Lardinois, O. M., & Rouxhet, P. G. (1996). Peroxidatic degradation of azide by catalase and irreversible enzyme inactivation. *Biochimica et Biophysica Acta. Protein Structure and Molecular Enzymology*, *1298*(2), 180–190. doi:10.1016/S0167-4838(96)00130-6 PMID:8980644

Lastre-Acosta, A. M., Barberato, B., Parizi, M. P. S., & Teixeira, A. C. S. C. (2019). Direct and indirect photolysis of the antibiotic enoxacin: Kinetics of oxidation by reactive photo-induced species and simulations. *Environmental Science and Pollution Research International*, *26*(5), 4337–4347. doi:10.100711356-018-2555-4 PMID:29931641

Launay, J. P. (2001). Long-distance intervalence electron transfer. *Chemical Society Reviews*, *30*(6), 386–397. doi:10.1039/b101377g

Lazauskas, R., Song, Y. H., & Gudkov, V. (2014). Parity and time-reversal violation in A = 2 – 4 nuclei. *EPJ Web of Conferences*, *66*, 05014. doi:10.1051/epjconf/20146605014

Leal, A. M., Kulik, D. A., Smith, W. R., & Saar, M. O. (2017). An overview of computational methods for chemical equilibrium and kinetic calculations for geochemical and reactive transport modeling. *Pure and Applied Chemistry*, *89*(5), 597–643. doi:10.1515/pac-2016-1107

Lee, C. G., Javed, H., Zhang, D., Kim, J. H., Westerhoff, P., Li, Q., & Alvarez, P. J. J. (2018). Porous electrospun fibers embedding TiO_2 for adsorption and photocatalytic degradation of water pollutants. *Environmental Science & Technology*, *52*(7), 4285–4293. doi:10.1021/acs.est.7b06508 PMID:29553243

Lee, C., Yang, W., & Parr, R. G. (1988). Development of the Colle-Salvetti correlation-energy formula into a functional of the electron density. *Physical Review. B*, *37*(2), 785–789. doi:10.1103/PhysRevB.37.785 PMID:9944570

Lee, D. Y., Lee, M. H., & Cho, N. I. (2012). Preparation and photocatalytic degradation of erbium doped titanium dioxide nanorods. *Current Applied Physics*, *12*(4), 1229–1233. doi:10.1016/j.cap.2012.03.007

Lehene, M., Fische-Fodor, E., Scurtu, F., Hădade, N., Gal, E., Mot, A. C., Matei, A., & Silaghi-Dumitrescu, R. (2020). Excess ascorbate is a chemical stress agent against proteins and cells. *Pharmaceuticals (Basel, Switzerland)*, *13*(6), 107. doi:10.3390/ph13060107 PMID:32471171

Lelieveld, J., Dentener, F. J., Peters, W., & Krol, M. C. (2004). On the role of hydroxyl radicals in the self-cleansing capacity of the troposphere. *Atmospheric Chemistry and Physics*, *4*(9/10), 2337–2344. www.atmos-chem-phys.org/acp/4/2337/. doi:10.5194/acp-4-2337-2004

Lemire, B. D., & Oyedotun, K. S. (2002). The Saccharomyces cerevisiae mitochondrial succinate:ubiquinone oxidoreductase. *Biochimica et Biophysica Acta (BBA) - Bioenergetics*, *1553*(1–2), 102–116. doi:10.1016/S0005-2728(01)00229-8 PMID:11803020

Leontieş, A. R., Răducan, A., Cristina Culiţă, D., Alexandrescu, E., Moroşan, A., Eduard Mihaiescu, D., & Aricov, L. (2022). Laccase immobilized on chitosan-polyacrylic acid microspheres as highly efficient biocatalyst for naphthol green B and indigo carmine degradation. *Chemical Engineering Journal*, *439*, 135654. doi:10.1016/j.cej.2022.135654

Levich, V. G., Dogonadze, R. R., German, E. D., Kuznetsov, A. M., & Kharkats, Y. I. (1970). Theory of homogeneous reactions involving proton transfer. *Electrochimica Acta*, *15*(2), 353–367. doi:10.1016/0013-4686(70)80027-5

Leynaud Kieffer Curran, L. M. C., Pham, L. T. M., Sale, K. L., & Simmons, B. A. (2022). Review of advances in the development of laccases for the valorization of lignin to enable the production of lignocellulosic biofuels and bioproducts. *Biotechnology Advances*, *54*, 107809. doi:10.1016/j.biotechadv.2021.107809 PMID:34333091

Lhinhatrakool, T., Prabpai, S., Kongsaeree, P., & Sutthivaiyakit, S. (2011). Antiplasmodial sesquiterpene alkaloids from the roots of *Maytenus mekongensis. Journal of Natural Products*, *74*(6), 1386–1391. doi:10.1021/np200014k

Liang, C., Xie, J., Luo, S., Huang, C., Zhang, Q., Huang, H., & Zhang, P. (2021). A highly potent ruthenium(II)-sonosensitizer and sonocatalyst for in vivo sonotherapy. *Nature Communications*, *12*(1), 1–9. doi:10.103841467-021-25303-1 PMID:34408151

Liang, J., Tian, Y. X., Yang, F., Zhang, J. P., & Skibsted, L. H. (2009). Antioxidant synergism between carotenoids in membranes. Astaxanthin as a radical transfer bridge. *Food Chemistry*, *115*(4), 1437–1442. doi:10.1016/j.foodchem.2009.01.074

Liang, N., Miller, J. R., & Closs, G. L. (1990). Temperature-independent long-range electron transfer reactions in the Marcus inverted region. *Journal of the American Chemical Society*, *112*(13), 5353–5354. doi:10.1021/ja00169a054

Lian, S., Li, S., Zhu, J., Xia, Y., & Do Jung, Y. (2022). Nicotine stimulates IL-8 expression via ROS/NF-κB and ROS/MAPK/AP-1 axis in human gastric cancer cells. *Toxicology*, *466*, 153062–153075. doi:10.1016/j.tox.2021.153062 PMID:34890707

Li, B., Kim, Y. L., & Lippert, A. R. (2022a). Chemiluminescence measurement of reactive sulfur and nitrogen species. *Antioxidants & Redox Signaling*, *36*(4-6), 337–353. doi:10.1089/ars.2021.0195 PMID:34514838

Libby, W. F. (1952). Theory of Electron Exchange Reactions in Aqueous Solution. *Journal of Physical Chemistry*, *56*(7), 863–868. doi:10.1021/j150499a010

Liew, K. B., Daud, W. R. W., Ghasemi, M., Leong, J. X., Lim, S. S., & Ismail, M. (2014). Non-Pt catalyst as oxygen reduction reaction in microbial fuel cells: A review. *International Journal of Hydrogen Energy*, *39*(10), 4870–4883. doi:10.1016/j.ijhydene.2014.01.062

Li, H., Sun, H., Ding, J., Wu, T., Cai, S., Wang, C., & Yang, R. (2022b). RuO_2/rGO heterostructures as mimic peroxidases for colorimetric detection of glucose. *Mikrochimica Acta*, *189*(7), 1–10. doi:10.100700604-022-05319-0

Li, J., & Chen, T. (2020). Transition metal complexes as photosensitizers for integrated cancer theranostic applications. *Coordination Chemistry Reviews*, *418*, 213355–213374. doi:10.1016/j.ccr.2020.213355

Li, J., Duke, B., & McWeeny, R. (2007). *VB2000 Version 2.0.* SciNet Technologies.

Li, J., & McWeeny, R. (2002). VB2000: Pushing valence bond theory to new limits. *International Journal of Quantum Chemistry*, *89*(4), 208–216. doi:10.1002/qua.10293

Li, J., Zeng, L., Xiong, K., Rees, T. W., Jin, C., Wu, W., Chen, Y., Ji, L., & Chao, H. (2019). A biotinylated ruthenium (II) photosensitizer for tumor-targeted two-photon photodynamic therapy. *Chemical Communications*, *55*(73), 10972–10975. doi:10.1039/C9CC05826E PMID:31453611

Li, L., Turnbull, M. M., Landee, C. P., Jornet, J., Deumal, M., Novoa, J. J., & Wikaira, J. L. (2007). Synthesis, structure, and magnetic behavior of bis (2-amino-5-fluoropyridinium) tetrachlorocuprate (II). *Inorganic Chemistry*, *46*(26), 11254–11265. doi:10.1021/ic701645y PMID:18020329

Lim, D. W., Sadakiyo, M., & Kitagawa, H. (2019). Proton transfer in hydrogen-bonded degenerate systems of water and ammonia in metal–organic frameworks. *Chemical Science (Cambridge)*, *10*(1), 16–33. doi:10.1039/C8SC04475A PMID:30746070

Lin, K., Rong, Y., Chen, D., Zhao, Z., Bo, H., Qiao, A., Hao, X., & Wang, J. (2020). Combination of ruthenium complex and doxorubicin synergistically inhibits cancer cell growth by down-regulating PI3K/AKT signaling pathway. *Frontiers in Oncology*, *10*, 141–153. doi:10.3389/fonc.2020.00141 PMID:32133289

Lisak, G. (2021). Reliable environmental trace heavy metal analysis with potentiometric ion sensors-reality or a distant dream. *Environmental Pollution*, *289*, 117882. doi:10.1016/j.envpol.2021.117882 PMID:34364114

Lismont, C., Revenco, Iu., & Fransen, M. (2019). Peroxisomal hydrogen peroxide metabolism and signaling in health and disease. *International Journal of Molecular Sciences*, *20*(15), 3673. doi:10.3390/ijms20153673 PMID:31357514

Liu, C., Liu, J., Zhang, W., Wang, Y. L., Liu, Q., Song, B., Yuan, J., & Zhang, R. (2020). "Two birds with one stone" ruthenium(II) complex probe for biothiols discrimination and detection in vitro and in vivo. *Advancement of Science*, *7*(14), 2000458. doi:10.1002/advs.202000458 PMID:32714756

Liu, C., Ren, L., Yan, B., Luo, L., Zhang, J., & Awasthi, M. K. (2021). Electron transfer and mechanism of energy production among syntrophic bacteria during acidogenic fermentation: A review. *Bioresource Technology*, *323*, 124637. doi:10.1016/j.biortech.2020.124637 PMID:33421831

Liu, C., Zhang, W., Zhang, R., Gao, X., Song, B., & Yuan, J. (2022). Ruthenium(II) complex-based long-lived two-photon luminescence probe for dynamic monitoring of glutathione S-transferases in mouse models of drug-induced liver injury. *Sensors and Actuators. B, Chemical*, *357*, 131440–131449. doi:10.1016/j.snb.2022.131440

Liu, H., Cao, X., Wang, P., & Ma, X. (2017). pH-sensitive pHluorins as a molecular sensor for in situ monitoring of enzyme-catalyzed prodrug activation. *Biotechnology and Applied Biochemistry*, *64*(4), 482–489. doi:10.1002/bab.1511 PMID:27222443

Liu, H., Liu, M., Wang, D., Wang, L., Zhao, Y., Liu, J., Wu, Y., Sun, B., & Zhang, Y. (2021). Competitive binding of synergistic antioxidant chlorogenic acid and (−)-epigallocatechin gallate with lysozyme: Insights from multispectroscopic characterization, molecular docking and activity evaluation. *Journal of Molecular Liquids*, *341*, 117387. doi:10.1016/j.molliq.2021.117387

Liu, L., Liu, Z., Yang, Y., Geng, M., Zou, Y., Shahzad, M. B., Dai, Y., & Qi, Y. (2018). Photocatalytic properties of Fe-doped ZnO electrospun nanofibers. *Ceramics International*, *44*(16), 19998–20005. doi:10.1016/j.ceramint.2018.07.268

Liu, R., Liang, Z., Gong, Z., & Yang, Y. (2019). Research Progress in Multielectron Reactions in Polyanionic Materials for Sodium-Ion Batteries. *Small Methods*, *3*(4), 1800221. doi:10.1002mtd.201800221

Liu, S., Bilal, M., Rizwan, K., Gul, I., Rasheed, T., & Iqbal, H. M. N. (2021). Smart chemistry of enzyme immobilization using various support matrices - A review. *International Journal of Biological Macromolecules*, *190*, 396–408. doi:10.1016/j.ijbiomac.2021.09.006 PMID:34506857

Liu, S., Rong, C., & Lu, T. (2014). Information conservation principle determines electrophilicity, nucleophilicity, and regioselectivity. *The Journal of Physical Chemistry A*, *118*(20), 3698–3704. doi:10.1021/jp5032702 PMID:24784465

Liu, X., Xu, B., Yang, Y., Li, Q., Jiang, T., & He, Y. (2018). Thermodynamic analysis of ammoniacal thiosulphate leaching of gold catalysed by Co (III)/Co (II) using Eh-pH and speciation diagrams. *Hydrometallurgy*, *178*, 240–249. doi:10.1016/j.hydromet.2018.05.014

Liu, Y., Bersuker, I. B., Zou, W., & Boggs, J. E. (2010). Pseudo Jahn–Teller versus Renner–Teller effects in the instability of linear molecules. *Chemical Physics*, *376*(1-3), 30–35. doi:10.1016/j.chemphys.2010.07.029

Liu, Y., Liu, C., & Li, J. (2020). Comparison of vitamin c and its derivative antioxidant activity: Evaluated by using density functional theory. *ACS Omega*, *5*(39), 25467–25475. doi:10.1021/acsomega.0c04318 PMID:33043226

Li, X. B., Tung, C. H., & Wu, L. Z. (2019). Quantum dot assembly for light-driven multielectron redox reactions, such as hydrogen evolution and CO_2 reduction. *Angewandte Chemie*, *131*(32), 10918–10925. doi:10.1002/ange.201901267 PMID:30821022

Li, X., Lv, P., Yao, Y., Feng, Q., Mensah, A., Li, D., & Wei, Q. (2020). A novel single-enzymatic biofuel cell based on highly flexible conductive bacterial cellulose electrode utilizing pollutants as fuel. *Chemical Engineering Journal, 379*, 122316. doi:10.1016/j.cej.2019.122316

Li, Z., Vinayan, B. P., Jankowski, P., Njel, C., Roy, A., Vegge, T., & Zhao-Karger, Z. (2020). Multi-Electron Reactions Enabled by Anion-Based Redox Chemistry for High-Energy Multivalent Rechargeable Batteries. *Angewandte Chemie, 132*(28), 11580–11587. doi:10.1002/ange.202002560 PMID:32220137

Lloret, L., Eibes, G., Moreira, M. T., Feijoo, G., & Lema, J. M. (2013). On the use of a high-redox potential laccase as an alternative for the transformation of non-steroidal anti-inflammatory drugs (NSAIDs). *Journal of Molecular Catalysis. B, Enzymatic, 97*, 233–242. doi:10.1016/j.molcatb.2013.08.021

Lo Scalzo, R. (2008). Organic acids influence on DPPH scavenging by ascorbic acid. *Food Chemistry, 107*(1), 40–43. doi:10.1016/j.foodchem.2007.07.070

Lo Scalzo, R. (2021). EPR free radical scavenging activity on superoxide, hydroxyl and tert–butyl hydroperoxide radicals by common hydrophilic antioxidants: Effect of mixing and influence of glucose and citric acid. *European Food Research and Technology, 247*(9), 2253–2265. doi:10.100700217-021-03785-z

Lobo, V., Patil, A., Phatak, A., & Chandra, N. (2010). Free radicals, antioxidants and functional foods: Impact on human health. *Pharmacognosy Reviews, 4*(8), 118–126. doi:10.4103/0973-7847.70902 PMID:22228951

López-Martínez, L. M., Santacruz-Ortega, H., Navarro, R. E., Sotelo-Mundo, R. R., & González-Aguilar, G. A. (2015). A ^{1}H NMR investigation of the interaction between phenolic acids found in mango (Manguifera indica cv Ataulfo) and papaya (Carica papaya cv Maradol) and 1, 1-diphenyl-2-picrylhydrazyl (DPPH) free radicals. *PLoS One, 10*(11), 1–11. doi:10.1371/journal.pone.0140242 PMID:26559189

Lu, J., Nie, M., Li, Y., Zhu, H., & Shi, G. (2022). Design of composite nanosupports and applications thereof in enzyme immobilization: A review. *Colloids and Surfaces. B, Biointerfaces, 217*, 112602. doi:10.1016/j.colsurfb.2022.112602 PMID:35660743

Lungu, L. (2015). Synthesis of new nitrogen-containing drimane and homodrimane sesquiterpenoids from sclareolide. *Chemistry Journal of Moldova, 10*(2), 58–61. doi:10.19261/cjm.2015.10(2).07

Lungu, L., Ciocarlan, A., Barba, A., Shova, S., Pogrebnoi, S., Mangalagiu, I., Moldoveanu, C., Vornicu, N., Ambrosio, M., Babak, M. V., Arion, V., & Aricu, A. (2019). Synthesis and evaluation of biological activity of homodrimane sesquiterpenoids bearing hydrazinecarbothioamide or 1,2,4-triazole unit. *Chemistry of Heterocyclic Compounds, 55*(8), 716–724. doi:10.100710593-019-02526-1

Lungu, L., Ciocarlan, A., Smigon, C., Ozer, I., Shova, S., Gutu, I., Vornicu, N., Mangalagiu, I., D'ambrosio, M., & Aricu, A. (2020). Synthesis and evaluation of biological activity of homodrimane sesquiterpenoids bearing 1,3,4-oxadiazole or 1,3,4-thiadiazole units. *Chemistry of Heterocyclic Compounds, 56*(5), 578–585. doi:10.100710593-020-02703-7

Luo, Z., Yu, L., Yang, F., Zhao, Z., Yu, B., Lai, H., Wong, K.-H., Ngai, S.-M., Zheng, W., & Chen, T. (2014). Ruthenium polypyridyl complexes as inducer of ROS-mediated apoptosis in cancer cells by targeting thioredoxin reductase. *Metallomics, 6*(8), 1480–1490. doi:10.1039/C4MT00044G PMID:24823440

Luther, G. W. III. (2005). Manganese (II) oxidation and Mn (IV) reduction in the environment—Two one-electron transfer steps versus a single two-electron step. *Geomicrobiology Journal, 22*(3-4), 195–203. doi:10.1080/01490450590946022

Luther, G. W. III. (2010). The role of one-and two-electron transfer reactions in forming thermodynamically unstable intermediates as barriers in multi-electron redox reactions. *Aquatic Geochemistry, 16*(3), 395–420. doi:10.100710498-009-9082-3

Lv, P., Zhou, H., Mensah, A., Feng, Q., Wang, D., Hu, X., Cai, Y., Amerigo Lucia, L., Li, D., & Wei, Q. (2018). A highly flexible self-powered biosensor for glucose detection by epitaxial deposition of gold nanoparticles on conductive bacterial cellulose. *Chemical Engineering Journal, 351*, 177–188. doi:10.1016/j.cej.2018.06.098

Lykkesfeldt, J., & Tveden-Nyborg, P. (2019). The pharmacokinetics of vitamin C. *Nutrients, 11*(10), 2412. Advance online publication. doi:10.3390/nu11102412 PMID:31601028

Lyons, T. W., Reinhard, C. T., & Planavsky, N. J. (2014). The rise of oxygen in Earth's early ocean and atmosphere. *Nature, 506*(7488), 307–315. www.nature.com/doifinder/10.1038/nature13068. doi:10.1038/nature13068 PMID:24553238

Lythe, G., & Molina-París, C. (2018). Some deterministic and stochastic mathematical models of naïve T-cell homeostasis. *Immunological Reviews, 285*(1), 206–217. doi:10.1111/imr.12696 PMID:30129198

Macan, A. M., Kraljević, T. G., & Raić-malić, S. (2019). Therapeutic perspective of vitamin C and its derivatives. Antioxidants, 8(8). doi:10.3390/antiox8080247

Mackerell, A. D. (2004). Empirical force fields for biological macromolecules: Overview and issues. *Journal of Computational Chemistry, 25*(13), 1584–1604. doi:10.1002/jcc.20082 PMID:15264253

Mahmoud, M. A. A., Chedea, V. S., Detsi, A., & Kefalas, P. (2013). Ascorbic acid modifies the free radical scavenging behaviour of catechin: An insight into the mechanism. *Food Research International, 51*(2), 907–913. doi:10.1016/j.foodres.2013.02.023

Mahu, E., Ignat, M., Cojocaru, C., Samoila, P., Coromelci, C., Asaftei, I., & Harabagiu, V. (2020). Development of porous titania structure with improved photocatalytic activity: Response surface modeling and multi-objective optimization. *Nanomaterials, 10*, 998/1–15.

Majumdar, S., & Bhowal, J. (2022). Studies on production and evaluation of biopigment and synthetic dye decolorization capacity of laccase produced by A. oryzae cultivated on agro-waste. *Bioprocess and Biosystems Engineering, 45*(1), 45–60. doi:10.100700449-021-02638-z PMID:34591164

Mäkelä, M. (2017). Experimental design and response surface methodology in energy applications: A tutorial review. *Energy Conversion and Management, 151*, 630–640. doi:10.1016/j.enconman.2017.09.021

Makita, H., & Hastings, G. (2017). Inverted-region electron transfer as a mechanism for enhancing photosynthetic solar energy conversion efficiency. *Proceedings of the National Academy of Sciences of the United States of America, 114*(35), 9267–9272. doi:10.1073/pnas.1704855114 PMID:28814630

Maksymiuk, K., Stelmach, E., & Michalska, A. (2020). Unintended changes of ion-selective membranes composition—Origin and effect on analytical performance. *Membranes, 10*(10), 266. doi:10.3390/membranes10100266 PMID:32998393

Makula, P., Pacia, M., & Macyk, W. (2018). How to correctly determine the band gap energy of modified semiconductor photocatalysts based on UV–Vis spectra. *The Journal of Physical Chemistry Letters, 9*(23), 6814–6817. doi:10.1021/acs.jpclett.8b02892 PMID:30990726

Malhi, H., & Kaufman, R. J. (2011). Endoplasmic reticulum stress in liver disease. *Journal of Hepatology, 54*(4), 795–809. doi:10.1016/j.jhep.2010.11.005 PMID:21145844

Malinowski, J., Zych, D., Jacewicz, D., Gawdzik, B., & Drzeżdżon, J. (2020). Application of coordination compounds with transition metal ions in the chemical industry - A review. *International Journal of Molecular Sciences, 21*(15), 5443–5468. doi:10.3390/ijms21155443 PMID:32751682

Malrieu, J. P., Caballol, R., Calzado, C. J., De Graaf, C., & Guihery, N. (2014). Magnetic interactions in molecules and highly correlated materials: Physical content, analytical derivation, and rigorous extraction of magnetic Hamiltonians. *Chemical Reviews*, *114*(1), 429–492. doi:10.1021/cr300500z PMID:24102410

Mandal, M., Kawashima, K., Saito, K., & Ishikita, H. (2020). Redox potential of the oxygen-evolving complex in the electron transfer cascade of photosystem II. *The Journal of Physical Chemistry Letters*, *11*(1), 249–255. doi:10.1021/acs.jpclett.9b02831 PMID:31729876

Mano, N. (2019). Engineering glucose oxidase for bioelectrochemical applications. *Bioelectrochemistry (Amsterdam, Netherlands)*, *128*, 218–240. doi:10.1016/j.bioelechem.2019.04.015 PMID:31030174

Mansikkamäki, A., & Tuononen, H. M. (2018). The role of orbital symmetries in enforcing ferromagnetic ground state in mixed radical dimers. *The Journal of Physical Chemistry Letters*, *9*(13), 3624–3630. doi:10.1021/acs.jpclett.8b01507 PMID:29897759

Manzano-Nicolas, J., Taboada-Rodriguez, A., Teruel-Puche, J. A., Marin-Iniesta, F., Garcia-Molina, F., Garcia-Canovas, F., Tudela-Serrano, J., & Muñoz-Muñoz, J. L. (2020). Kinetic characterization of the oxidation of catecolamines and related compounds by laccase. *International Journal of Biological Macromolecules*, *164*, 1256–1266. doi:10.1016/j.ijbiomac.2020.07.112 PMID:32721460

Mao, Q. (2015). *The synthesis and antioxidant capacities of a range of resveratrol and related phenolic glucosides* (090PH M2965) [Doctoral dissertation, University of Adelaide]. digital.library.adelaide.edu.au:2440/9799

March, J. G., Gual, M., & Ramonell, J. (2005). A kinetic model for chlorine consumption in grey water. *Desalination*, *18*(1-3), 267–273. doi:10.1016/j.desal.2005.04.007

Marcotte, S., Villers, D., Guillet, N., Roue, L., & Dodelet, J. P. (2004). Electroreduction of oxygen on Co-based catalysts: Determination of the parameters affecting the two-electron transfer reaction in an acid medium. *Electrochimica Acta*, *50*(1), 179–188. doi:10.1016/j.electacta.2004.07.029

Marcus, R. A. (1964). Chemical and electrochemical electron-transfer theory. *Annual Review of Physical Chemistry*, *15*(1), 155–196. doi:10.1146/annurev.pc.15.100164.001103

Marcus, R. A. (1993). Electron Transfer Reactions in Chemistry: Theory and Experiment (Nobel Lecture). *Angewandte Chemie International Edition in English*, *32*(8), 1111–1121. doi:10.1002/anie.199311113

Maretti, L., Ferbinteanu, M., Cimpoesu, F., Islam, S. S., Ohba, Y., Kajiwara, T., Yamashita, M., & Yamauchi, S. (2007). Spin Coupling in the Supramolecular Structure of a New Tetra (Quinoline– TEMPO) Yttrium (III) Complex. *Inorganic Chemistry*, *46*(3), 660–669. doi:10.1021/ic0610841 PMID:17257008

Martemucci, G., Costagliola, C., Mariano, M., D'andrea, L., Napolitano, P., & D'Alessandro, A. G. (2022). Free radical properties, source and targets, antioxidant consumption and health. *Oxygen*, *2*(2), 48–78. doi:10.3390/oxygen2020006

Martínez, A. T., Ruiz-Dueñas, F. J., Camarero, S., Serrano, A., Linde, D., Lund, H., Vind, J., Tovborg, M., Herold-Majumdar, O. M., Hofrichter, M., Liers, C., Ullrich, R., Scheibner, K., Sannia, G., Piscitelli, A., Pezzella, C., Sener, M. E., Kılıç, S., van Berkel, W. J. H., & Alcalde, M. (2017). Oxidoreductases on their way to industrial biotransformations. *Biotechnology Advances*, *35*(6), 815–831. doi:10.1016/j.biotechadv.2017.06.003 PMID:28624475

Martinez-Alonso, M., & Gasser, G. (2021). Ruthenium polypyridyl complex-containing bioconjugates. *Coordination Chemistry Reviews*, *434*, 213736–213776. doi:10.1016/j.ccr.2020.213736

Masmoudi, R., Khettaf, S., Soltani, A., & Dibi, A. (2022). Cephalexin degradation initiated by OH radicals: theoretical prediction of the mechanisms and the toxicity of byproducts. *J Mol Model.*, *28*(6), 141.

Massey, V. (2002). The reactivity of oxygen with flavoproteins. *International Congress Series*, *1233*, 3–11. doi:10.1016/S0531-5131(02)00519-8

Matsushita, M. M., Izuoka, A., Sugawara, T., Kobayashi, T., Wada, N., Takeda, N., & Ishikawa, M. (1997). Hydrogen-bonded organic ferromagnet. *Journal of the American Chemical Society*, *119*(19), 4369–4379. doi:10.1021/ja964083o

Mattevi, A. (2006). To be or not to be an oxidase: Challenging the oxygen reactivity of flavoenzymes. *Trends in Biochemical Sciences*, *31*(5), 276–283. doi:10.1016/j.tibs.2006.03.003 PMID:16600599

Mattos, G. J., Schirmann, J. G., Salamanca-Neto, C. A. R., Dekker, R. F. H., Barbosa-Dekker, A. M., & Sartori, E. R. (2022). Electrochemical characterization of the laccase-catalyzed oxidation of 2,6-dimethoxyphenol: An insight into the direct electron transfer by enzyme and enzyme-mediator system. *Applied Biochemistry and Biotechnology*, *194*(10), 4348–4361. doi:10.100712010-022-03838-3 PMID:35146637

May, S. W. (1999). Applications of oxidoreductases. *Current Opinion in Biotechnology*, *10*(4), 370–375. doi:10.1016/S0958-1669(99)80067-6 PMID:10449319

May, S. W., & Padgette, S. R. (1983). Oxidoreductase enzymes in biotechnology: Current status and future potential. *Nature Biotechnology*, *1*(8), 677–686. doi:10.1038/nbt1083-677

Mazurenko, I., Hitaishi, V. P., & Lojou, E. (2020). Recent advances in surface chemistry of electrodes to promote direct enzymatic bioelectrocatalysis. *Current Opinion in Electrochemistry*, *19*, 113–121. doi:10.1016/j.coelec.2019.11.004

McGrady, G. E., & Stranger, R. (1997). Redox-Induced Changes in the Geometry and Electronic Structure of Di-. *Journal of the American Chemical Society*, *119*(36), 8512–8522. doi:10.1021/ja964360r

McLean, A. D., & Chandler, G. S. (1980). Contracted Gaussian-basis sets for molecular calculations. 1. 2nd row atoms, Z=11-18. *The Journal of Chemical Physics*, *72*(10), 5639–5648. doi:10.1063/1.438980

McNaught, A. D. (1997). *Compendium of chemical terminology* (Vol. 1669). Blackwell Science.

Mehra, R., Muschiol, J., Meyer, A. S., & Kepp, K. P. (2018). A structural-chemical explanation of fungal laccase activity. *Scientific Reports*, *8*(1), 17285. doi:10.103841598-018-35633-8 PMID:30470810

Meletiadis, J., Pournaras, S., Roilides, E., & Walsh, T. J. (2010). Defining fractional inhibitory concentration index cutoffs for additive interactions based on self-drug additive combinations, Monte Carlo simulation analysis, and in vitro-in vivo correlation data for antifungal drug combinations against Aspergillus fumigatus. *Antimicrobial Agents and Chemotherapy*, *54*(2), 602–609. doi:10.1128/AAC.00999-09 PMID:19995928

Mengele, A. K., Weixler, D., Chettri, A., Maurer, M., Huber, F. L., Seibold, G. M., Dietzek, B., Eikmanns, B. J., & Rau, S. (2021). Switching the mechanism of NADH photooxidation by supramolecular interactions. *Chemistry (Weinheim an der Bergstrasse, Germany)*, *27*(68), 16840–16845. doi:10.1002/chem.202103029 PMID:34547151

Mercado-Mercado, G., Laura, A., & Alvarez-Parrilla, E. (2020). Effect of pectin on the interactions among phenolic compounds determined by antioxidant capacity. *Journal of Molecular Structure*, *1199*, 126967. doi:10.1016/j.molstruc.2019.126967

Mikhailov, A. A., Vorobyev, V. A., Nadolinny, V. A., Patrushev, Y. V., Yudina, Y. S., & Kostin, G. A. (2019). Primary and secondary photochemical transformations of biologically active precursor-nitro-nitrosyl ruthenium complex. *Journal of Photochemistry and Photobiology A Chemistry*, *373*, 37–44. doi:10.1016/j.jphotochem.2018.12.037

Mikirova, N., Riordan, N., & Casciari, J. (2016). Modulation of cytokines in cancer patients by intravenous ascorbate therapy. *Medical Science Monitor*, *22*, 14–25. doi:10.12659/MSM.895368 PMID:26724916

Mikulski, D. E., Eder, K., & Molski, M. (2014). Quantum-chemical study on relationship between structure and antioxidant properties of hepatoprotective compounds occurring in Cynara scolymus and Silybum marianum. *Journal of Theoretical and Computational Chemistry*, *13*(1), 1–24. doi:10.1142/S0219633614500047

Miller, G. W., Morgan, C. A., Kieber, D. J., King, D. W., Snow, J. A., Heikes, B. G., Mopper, K., & Kiddle, J. J. (2005). Hydrogen peroxide method intercomparision study in seawater. *Marine Chemistry*, *97*(1–2), 4–13. doi:10.1016/j.marchem.2005.07.001

Miller, J. R., Calcaterra, L. T., & Closs, G. L. (1984). Intramolecular long-distance electron transfer in radical anions. The effects of free energy and solvent on the reaction rates. *Journal of the American Chemical Society*, *106*(10), 3047–3049. doi:10.1021/ja00322a058

Miller, J. S., & Drillon, M. (Eds.). (2001). *Magnetism: Molecules to Materials*. Wiley-VCH.

Minaev, B. F. (2007). Electronic mechanisms of molecular oxygen activation. *Russian Chemical Reviews*, *76*(11), 989–1011. doi:10.1070/RC2007v076n11ABEH003720

Minaev, B., & Agren, H. (1996). Spin-catalysis phenomena. *International Journal of Quantum Chemistry*, *57*(3), 519–532. doi:10.1002/(SICI)1097-461X(1996)57:3<519::AID-QUA25>3.0.CO;2-X

Misochko, E. Y., Korchagin, D. V., Bozhenko, K. V., Chapyshev, S. V., & Aldoshin, S. M. (2010). A density functional theory study of the zero-field splitting in high-spin nitrenes. *The Journal of Chemical Physics*, *133*(6), 064101. doi:10.1063/1.3474574 PMID:20707555

Mitani, M., Mori, H., Takano, Y., Yamaki, D., Yoshioka, Y., & Yamaguchi, K. (2000). Density functional study of intramolecular ferromagnetic interaction through m-phenylene coupling unit (I): UBLYP, UB3LYP and UHF calculation. *The Journal of Chemical Physics*, *113*(10), 4035–4050. doi:10.1063/1.1286418

Mocanu, L., Gonta, M., Matveevici, V., Duca, G., & Porubin-Schimbator, V. (2021). Îndepărtarea cefalexinei din soluții apoase utilizând procesul fotocatalitic Fenton. *Revista de Ştiinţă, Inovare, Cultură şi Artă "Akademos"*, *2*(61), 53-60.

Mocanu, L., Gonta, M., & Matveevici, V. (2020). Optimizarea procesului de oxidare catalitică a amoxicilinei în soluții apoase. *Studia Universitatis*, *1*(131), 38–47.

Mohorčič, M., Friedrich, J., Renimel, I., André, P., Mandin, D., & Chaumont, J. P. (2007). Production of melanin bleaching enzyme of fungal origin and its application in cosmetics. *Biotechnology and Bioprocess Engineering; BBE*, *12*(3), 200–206. doi:10.1007/BF02931093

Morales, R., Charon, M. H., Kachalova, G., Serre, L., Medina, M., Gómez-Moreno, C., & Frey, M. (2000). A redox-dependent interaction between two electron-transfer partners involved in photosynthesis. *EMBO Reports*, *1*(3), 271–276. doi:10.1093/embo-reports/kvd057 PMID:11256611

Morell, C., Grand, A., & Toro-Labbé, A. (2004). New dual descriptor for chemical reactivity. *The Journal of Physical Chemistry A*, *109*(1), 205–212. doi:10.1021/jp046577a PMID:16839107

Morell, C., Grand, A., & Toro-Labbé, A. (2006). Theoretical support for using the Δf(R) descriptor. *Chemical Physics Letters*, *425*(4-6), 342–346. doi:10.1016/j.cplett.2006.05.003

Morowitz, H. J., Kostelnik, J. D., Yang, J., & Cody, G. D. (2000). The origin of intermediary metabolism. *Proceedings of the National Academy of Sciences of the United States of America*, *97*(14), 7704–7708. doi:10.1073/pnas.110153997 PMID:10859347

Mosa, A. I. (2020). Antigenic variability. *Frontiers in Immunology*, *11*, 2057. doi:10.3389/fimmu.2020.02057 PMID:33013870

Moser, C. C., Anderson, J. L. R., & Dutton, P. L. (2010). Guidelines for tunneling in enzymes. *Biochimica et Biophysica Acta (BBA) -. Bioenergetics*, *1797*(9), 1573–1586. doi:10.1016/j.bbabio.2010.04.441 PMID:20460101

Moser, C. C., Keske, J. M., Warncke, K., Farid, R. S., & Dutton, P. L. (1992). Nature of biological electron transfer. *Nature*, *355*(6363), 796–802. doi:10.1038/355796a0 PMID:1311417

Motteran, L., Pilone, M. S., Molla, G., Ghisla, S., & Pollegioni, L. (2001). Cholesterol oxidase from Brevibacterium sterolicum. *The Journal of Biological Chemistry*, *276*(21), 18024–18030. doi:10.1074/jbc.M010953200 PMID:11359791

Mousavi, M., Soleimani, M., Hamzehloo, M., Badiei, A., & Ghasemi, J. B. (2021). Photocatalytic degradation of different pollutants by the novel gCN-NS/Black-TiO_2 heterojunction photocatalyst under visible light: Introducing a photodegradation model and optimization by response surface methodology (RSM). *Materials Chemistry and Physics*, *258*, 123912/1–16. doi:10.1016/j.matchemphys.2020.123912

Moussaoui, M., Misevičienė, L., Anusevičius, Ž., Marozienė, A., Lederer, F., Baciou, L., & Čėnas, N. (2018). Quinones and nitroaromatic compounds as subversive substrates of Staphylococcus aureus flavohemoglobin. *Free Radical Biology & Medicine*, *123*, 107–115. doi:10.1016/j.freeradbiomed.2018.05.071 PMID:29793040

Moya, R., Saastamoinen, P., Hernández, M., Suurnäkki, A., Arias, E., & Mattinen, M. L. (2011). Reactivity of bacterial and fungal laccases with lignin under alkaline conditions. *Bioresource Technology*, *102*(21), 10006–10012. doi:10.1016/j.biortech.2011.08.046 PMID:21908186

Muhammad, D. R. A., Praseptiangga, D., Van de Walle, D., & Dewettinck, K. (2017). Interaction between natural antioxidants derived from cinnamon and cocoa in binary and complex mixtures. *Food Chemistry*, *231*, 356–364. doi:10.1016/j.foodchem.2017.03.128 PMID:28450018

Mukai, K., Mitani, S., Ohara, K., & Nagaoka, S. I. (2005). Structure–activity relationship of the tocopherol-regeneration reaction by catechins. *Free Radical Biology & Medicine*, *38*(9), 1243–1256. doi:10.1016/j.freeradbiomed.2005.01.011 PMID:15808422

Mukherjee, G., Nandekar, P. P., & Wade, R. C. (2021). An electron transfer competent structural ensemble of membrane-bound cytochrome P450 1A1 and cytochrome P450 oxidoreductase. *Communications Biology*, *4*(1), 55. doi:10.103842003-020-01568-y PMID:33420418

Muliadi, M. B., Basimin, M. Q., & Jayali, A. M. (2021). Density Functional Theory for QSAR Antioxidant Compound Myristicin Derivatives. *Indo.J.Chem.Res.*, *9*(1), 49–56. doi:10.30598//ijcr.2021.9-mul

Müller, D., Pitsch, S., Kittaka, A., Wagner, E., Wintner, C. E., & Eschenmoser, A. & von a-Aminonitrilen, C. (1990). Aldomerisierung von Glycolaldehyd-phosphat zu racemischen Hexose-2,4,6-triphosphaten und (in Gegenwart von Formaldehyd) racemischen Pentose-2,4-diphosphaten: Rac -Allose-2,4,6-triphosphat und rac-Ribose-2,4-diphosphat sind die Reaktionshauptprodukte. *Helvetica Chimica Acta*, *73*, 1410–1468. doi:10.1002/hlca.19900730526

Mumtaz, S., Ali, S., Tahir, H. M., Kazmi, S. A. R., Shakir, H. A., Mughal, T. A., Mumtaz, S., Summer, M., & Farooq, M. A. (2021). Aging and its treatment with vitamin C: a comprehensive mechanistic review. Molecular Biology Reports, 48(12), 8141–8153. doi:10.100711033-021-06781-4

Murphy, M. P. (2013). Mitochondrial dysfunction indirectly elevates ROS production by the endoplasmic reticulum. *Cell Metabolism*, *18*(2), 145–146. doi:10.1016/j.cmet.2013.07.006 PMID:23931748

Murshed, M. K., Dursun, A. Y., & Dursun, G. (2022). Application of response surface methodology on photocatalytic degradation of Astrazon Orange G dye by ZnO photocatalyst: Internal mass transfer effects. *Chemical Engineering Research & Design*, *188*, 27–38. doi:10.1016/j.cherd.2022.09.038

Myers, R. H., Montgomery, D. C., Vining, G. G., Borror, C. M., & Kowalski, S. M. (2004). Response surface methodology: A retrospective and literature survey. *Journal of Quality Technology*, *36*(1), 53–77. doi:10.1080/00224065.2004.11980252

Nadezhdin, A. D., Kozlov, Y. N., & Purmal, A. P. (1975). Reactions of one-and two-electron donors with TNM. *Zh. Fiz. Khim*, *49*(9), 2263–2269.

Napolitano, G., Fasciolo, G., & Venditti, P. (2021). Mitochondrial Management of reactive oxygen species. *Antioxidants*, *10*(11), 1824–1852. doi:10.3390/antiox10111824 PMID:34829696

Nasar, A., & Perveen, R. (2019). Applications of enzymatic biofuel cells in bioelectronic devices - A review. *International Journal of Hydrogen Energy*, *44*(29), 15287–15312. doi:10.1016/j.ijhydene.2019.04.182

Neuberg, C., & Schwenk, E. (1915). Die Gärung der Dioxymaleinsäure. *The Biochemical Journal*, *71*, 104–113.

Nicholls, P. (2012). Classical catalase: Ancient and modern. *Archives of Biochemistry and Biophysics*, *525*(2), 95–101. doi:10.1016/j.abb.2012.01.015 PMID:22326823

Nickaeen, N., Ghaisari, J., Heiner, M., Moein, S., & Gheisari, Y. (2019). Agent-based modeling and bifurcation analysis reveal mechanisms of macrophage polarization and phenotype pattern distribution. *Scientific Reports*, *9*(1), 12764. doi:10.103841598-019-48865-z PMID:31484958

Niki, E. (2010). Assessment of antioxidant capacity in vitro and in vivo. *Free Radical Biology & Medicine*, *49*(4), 503–515. doi:10.1016/j.freeradbiomed.2010.04.016 PMID:20416370

Niki, E., Saito, T., Kawakami, A., & Kamiya, Y. (1984). Inhibition of oxidation of methyl linoleate in solution by vitamin E and vitamin C. *The Journal of Biological Chemistry*, *259*(7), 4177–4182. doi:10.1016/S0021-9258(17)43026-2 PMID:6706998

Niki, E., The, N., & Tsuchihashi, H. (2018). Interaction among vitamin C, vitamin E and β-carotene. *The American Journal of Clinical Nutrition*, *62*(March), 1322–1326. PMID:7495227

Nimura, S., & Yabe, A. (1999). In P. M. Lahti (Ed.), *Molecular Magnetism of Organic-Based Materials* (p. 127). Dekker.

Nishimi, T., Kamachi, T., Kato, K., Kato, T., & Yoshizawa, K. (2011). Mechanistic Study on the Production of Hydrogen Peroxide in the Anthraquinone Process. *European Journal of Organic Chemistry*, *22*(22), 4113–4120. doi:10.1002/ejoc.201100300

NIST critically selected stability constants of metal complexes database data. 2004.

Nogala-Kałucka, M., Dwiecki, K., Siger, A., Górnaś, P., Polewski, K., & Ciosek, S. (2013). Antioxidant synergism and antagonism between tocotrienols, quercetin and rutin in model system. *Acta Alimentaria*, *42*(3), 360–370. doi:10.1556/AAlim.2012.0009

Noguer, M., Cerezo López, A. B., Moyá Morán, M. L., Troncoso González, A. M., & García Parrilla, M. D. C. (2014). Synergism effect between phenolic metabolites and endogenous antioxidants in terms of antioxidant activity. *Advances in Chemical Engineering and Science*, *4*(2), 258–265. doi:10.4236/aces.2014.42029

Noodleman, L. (1981). Valence bond description of antiferromagnetic coupling in transition metal dimers. *The Journal of Chemical Physics*, *74*(10), 5737–5743. doi:10.1063/1.440939

Noodleman, L., & Davidson, E. R. (1986). Ligand spin polarization and antiferromagnetic coupling in transition metal dimers. *Chemical Physics*, *109*(1), 131–143. doi:10.1016/0301-0104(86)80192-6

Norouzi, M., Fazelia, A., & Tavakoli, O. (2020). Phenol contaminated water treatment by photocatalytic degradation on electrospun Ag/TiO_2 nanofibers: Optimization by the response surface method. *Journal of Water Process Engineering*, *37*, 101489. doi:10.1016/j.jwpe.2020.101489

Novoa, J. J., Deumal, M., & Jornet-Somoza, J. (2011). Calculation of microscopic exchange interactions and modeling of macroscopic magnetic properties in molecule-based magnets. *Chemical Society Reviews*, *40*(6), 3182–3212. doi:10.1039/c0cs00112k PMID:21321725

Nowak, M. A., & May, R. M. (2000). *Virus dynamics mathematical principles of immunology and virology*. University Press.

Nugroho Prasetyo, E., Kudanga, T., Østergaard, L., Rencoret, J., Gutiérrez, A., del Río, J. C., Ignacio Santos, J., Nieto, L., Jiménez-Barbero, J., Martínez, A. T., Li, J., Gellerstedt, G., Lepifre, S., Silva, C., Kim, S. Y., Cavaco-Paulo, A., Seljebakken Klausen, B., Lutnaes, B. F., Nyanhongo, G. S., & Guebitz, G. M. (2010). Polymerization of lignosulfonates by the laccase-HBT (1-hydroxybenzotriazole) system improves dispersibility. *Bioresource Technology*, *101*(14), 5054–5062. doi:10.1016/j.biortech.2010.01.048 PMID:20176477

Oancea, D., Stuparu, A., Nita, M., Puiu, M., & Raducan, A. (2008). Estimation of the overall kinetic parameters of enzyme inactivation using an isoconversional method. *Biophysical Chemistry*, *138*(1-2), 50–54. doi:10.1016/j.bpc.2008.09.003 PMID:18814949

Ogurtsov I., Gorinchoy N., Balan I., Munteanu G. (2003) Comparative ab initio study of the pseudo Jahn-Teller instability of planar D_{3h} configuration of $H_3O^{\bullet}$ radical and H_3O^+ cation. *Scientific annals of the State University of Moldova. Chemical-Biological sciences series,* 366-371. University of Moldova.

Ogurtsov, I., Gorinchoy, N., & Balan, I. (2007). Vibronic origin of the $H_3O^{\bullet}$ metastability. *Journal of Molecular Structure*, *838*(1-3), 107–111. doi:10.1016/j.molstruc.2007.01.065

Ohigashi, H., & Murakami, A. (2004). Cancer prevention with food factors: Alone and in combination. *BioFactors (Oxford, England)*, *22*(1-4), 49–55. doi:10.1002/biof.5520220109 PMID:15630251

Okawa, Y., Yokoyama, N., Sakai, Y., & Shiba, F. (2015). Direct electron transfer biosensor for hydrogen peroxide carrying nano-complex composed of horseradish peroxidase and Au-nanoparticle - Characterization and application to bienzyme systems. *Analytical Chemistry Research*, *5*, 1–8. doi:10.1016/j.ancr.2015.05.001

Okutan, N., Terzi, P., & Altay, F. (2014). Affecting parameters on electrospinning process and characterization of electrospun gelatin nanofibers. *Food Hydrocolloids*, *39*, 19–26. doi:10.1016/j.foodhyd.2013.12.022

Olloqui-Sariego, J. L., Calvente, J. J., & Andreu, R. (2021). Immobilizing redox enzymes at mesoporous and nanostructured electrodes. *Current Opinion in Electrochemistry, 26*. doi:10.3390u132313264

Olson, M. V. (2022). Oxidation-reduction reaction. *Encyclopedia Britannica*. https://www.britannica.com/science/oxidation-reduction-reaction

Olson, K. R., Gao, Y., Arif, F., Arora, K., Patel, S., DeLeon, E. R., Sutton, T. R., Feelisch, M., Cortese-Krott, M. M., & Straub, K. D. (2018). Metabolism of hydrogen sulfide (H_2S) and production of reactive sulfur species (RSS) by superoxide dismutase. *Redox Biology*, *15*, 74–85. doi:10.1016/j.redox.2017.11.009 PMID:29220697

Olsson, L., Levit, G. S., & Hosfeld, U. (2017). *The ''Biogenetic Law' in zoology: from Ernst Haeckel's formulation to current approaches. Theory Biosci*. Springer-Verlag Berlin Heidelberg. doi:10.100712064-017-0243-4

Olszowy-Tomczyk, M. (2020). Synergistic, antagonistic and additive antioxidant effects in the binary mixtures. *Phytochemistry Reviews*, *19*(1), 63–103. doi:10.100711101-019-09658-4

Olszyna, K. J., Meagher, J. F., & Bailey, E. M. (1988). Gas-phase, cloud and rain-water measurements of hydrogen peroxide at a high-elevation site. *Atmospheric Environment*, *8*(8), 1699–1706. doi:10.1016/0004-6981(88)90398-8

Onishi, T., Takano, Y., Kitagawa, Y., Kawakami, T., Yoshioka, Y., & Yamaguchi, K. (2001). Theoretical study of the magnetic interaction for M–O–M type metal oxides. Comparison of broken-symmetry approaches. *Polyhedron*, *20*(11-14), 1177–1184. doi:10.1016/S0277-5387(01)00591-5

Oparin, A. I. (1968). Life, its nature, origin and development (2nd). Ed.: M.: Nauka.

Orabi, E. (2021). Molecular dynamics investigation of the structural flexibility of H_2O_2 and H_2S_2 in response to medium polarity. *Journal of Molecular Liquids*, *329*, 115469. doi:10.1016/j.molliq.2021.115469

Ovchenkova, A. E., Safonova, E. A., Martynov, A. G., & Senchykhin, I. N., Gorbunova, G., Tsyvadze, A. (2021). New Crown-Substituted Phtalo- Oxanthrocyanate of Ittrium: Synthesis and Study of Supramolecular Aggregation (in Rus.). *Conf.Proc.*

Page, C. C., Moser, C. C., Chen, X., & Dutton, P. L. (1999). Natural engineering principles of electron tunnelling in biological oxidation–reduction. *Nature*, *402*(6757), 47–52. doi:10.1038/46972 PMID:10573417

Palace, V. P., Khaper, N., Qin, Q., & Singal, P. K. (1999). Antioxidant potentials of vitamin A and carotenoids and their relevance to heart disease. *Free Radical Biology & Medicine*, *26*(5–6), 746–761. doi:10.1016/S0891-5849(98)00266-4 PMID:10218665

Palafox-Carlos, H., Gil-Chávez, J., Sotelo-Mundo, R. R., Namiesnik, J., Gorinstein, S., & González-Aguilar, G. A. (2012). Antioxidant interactions between major phenolic compounds found in 'Ataulfo'mango pulp: Chlorogenic, gallic, protocatechuic and vanillic acids. *Molecules (Basel, Switzerland)*, *17*(11), 12657–12664. doi:10.3390/molecules171112657 PMID:23103532

Paniagua, C. E. S., Ricardo, I. A., Marson, E. O., Gonçalves, B. R., & Trovó, A. G. (2019). Simultaneous degradation of the pharmaceuticals gemfibrozil, hydrochlorothiazide and naproxen and toxicity changes during UV-C and UV-C/H_2O_2 processes in different aqueous matrices. *Journal of Environmental Chemical Engineering*, *7*(3), 103164. doi:10.1016/j.jece.2019.103164

Panich, A. M. (1995). NMR study of the F–H···F hydrogen bond. Relation between hydrogen atom position and F–H···F bond length. *Chemical Physics*, *196*(3), 511–519. doi:10.1016/0301-0104(95)00128-B

Papish, E. T., & Oladipupo, O. E. (2022). Factors that influence singlet oxygen formation vs. ligand substitution for light-activated ruthenium anticancer compounds. *Current Opinion in Chemical Biology*, *68*, 102143–102154. doi:10.1016/j.cbpa.2022.102143 PMID:35483128

Parada, G. A., Goldsmith, Z. K., Kolmar, S., Pettersson Rimgard, B., Mercado, B. Q., Hammarström, L., & Mayer, J. M. (2019). Concerted proton-electron transfer reactions in the Marcus inverted region. *Science*, *364*(6439), 471–475. doi:10.1126cience.aaw4675 PMID:30975771

Parida, S. K., Mandal, T., Das, S., Hota, S. K., De Sarkar, S., & Murarka, S. (2021). Single electron transfer-induced redox processes involving N-(acyloxy) phthalimides. *ACS Catalysis*, *11*(3), 1640–1683. doi:10.1021/acscatal.0c04756

Parker, T. L., Miller, S. A., Myers, L. E., Miguez, F. E., & Engeseth, N. J. (2010). Evaluation of synergistic antioxidant potential of complex mixtures using oxygen radical absorbance capacity (ORAC) and electron paramagnetic resonance (EPR). *Journal of Agricultural and Food Chemistry*, *58*(1), 209–217. doi:10.1021/jf903080f PMID:19957934

Parkhurst, D. L., & Appelo, C. A. J. (1999). User's guide to PHREEQC (Version 2): A computer program for speciation, batch-reaction, one-dimensional transport, and inverse geochemical calculations. Water-resources investigations report, 99(4259), 312.

Park, S., Jung, D., Do, H., Yun, J., Lee, D., Hwang, S., & Lee, S. H. (2021). Laccase-mediator system using a natural mediator as a whitening agent for the decolorization of melanin. *Polymers, 13*(21), 100658. doi:10.3390/polym13213671 PMID:34771228

Parra Guardado, A. L., Belleville, M. P., Rostro Alanis, M., Parra Saldivar, R., & Sanchez-Marcano, J. (2019). Effect of redox mediators in pharmaceuticals degradation by laccase: A comparative study. *Process Biochemistry (Barking, London, England), 78*, 123–131. doi:10.1016/j.procbio.2018.12.032

Parr, R. G., & Yang, W. (1984). Density functional approach to the frontier-electron theory of chemical reactivity. *Journal of the American Chemical Society, 106*(14), 4049–4050. doi:10.1021/ja00326a036

Parr, R. G., & Yang, W. (1989). *Density functional theory of atoms and molecules.* Oxford University, Press.

Pascariu, P., Cojocaru, C., Airinei, A., Olaru, N., Rosca, I., Koudoumas, E., & Suchea, M.P. (2021a). Innovative Ag-TiO_2 nanofibers offering excellent photocatalytic and antibacterial action. *Catalysts*, 11, 1234/1–20.

Pascariu, P., Cojocaru, C., Homocianu, M., & Samoila, P. (2022b). Tuning of Sm^{3+} and Er^{3+}-doped TiO_2 nanofibers for enhancement of the photocatalytic performance: Optimization of the photodegradation conditions. *Journal of Environmental Management, 316*, 115317. doi:10.1016/j.jenvman.2022.115317 PMID:35658261

Pascariu, P., Cojocaru, C., Homocianu, M., Samoila, P., Dascalu, A., & Suchea, M. (2022a). New La^{3+} doped TiO_2 nanofibers for photocatalytic degradation of organic pollutants: Effects of thermal treatment and doping loadings. *Ceramics International, 48*(4), 4953–4964. doi:10.1016/j.ceramint.2021.11.033

Pascariu, P., Cojocaru, C., Homocianu, M., Samoila, P., Grecu, I., & Bele, A. (2022d). New composite membranes based on PVDF fibers loaded with TiO_2:Sm nanostructures and reinforced with graphene/graphene oxide for photocatalytic applications. *Surfaces and Interfaces, 34*, 102382. doi:10.1016/j.surfin.2022.102382

Pascariu, P., Cojocaru, C., Olaru, N., & Airinei, A. (2019c). Photocatalytical activity of ZnO-SnO_2 ceramic nanofibers for RhB dye degradation: Experimental design, modeling and process optimization. *Physica Status Solidi. B, Basic Research, 1800474*, 1–8.

Pascariu, P., Cojocaru, C., Olaru, N., Samoila, P., Airinei, A., Ignat, M., Sacarescu, L., & Timpu, D. (2019a). Novel rare earth (RE-La, Er, Sm) metal doped ZnO photocatalysts for degradation of Congo-Red dye: Synthesis, characterization and kinetic studies. *Journal of Environmental Management, 239*, 225–234. doi:10.1016/j.jenvman.2019.03.060 PMID:30901700

Pascariu, P., Cojocaru, C., Samoila, P., Airinei, A., Olaru, N., Rotaru, A., Romanitan, C., Tudoran, L. B., & Suchea, M. (2022c). Cu/TiO_2 composite nanofibers with improved photocatalytic performance under UV and UV-visible light irradiation. *Surfaces and Interfaces, 28*, 101644/1–15. doi:10.1016/j.surfin.2021.101644

Pascariu, P., Cojocaru, C., Samoila, P., Airinei, A., Olaru, N., Rusu, D., Rosca, I., & Suchea, M. (2020). Photocatalytic and antimicrobial activity of electrospun ZnO:Ag nanostructures. *Journal of Alloys and Compounds, 834*, 155144. doi:10.1016/j.jallcom.2020.155144

Pascariu, P., Cojocaru, C., Samoila, P., Olaru, N., Bele, A., & Airinei, A. (2021b). Novel electrospun membranes based on PVDF fibers embedding lanthanide doped ZnO for adsorption and photocatalytic degradation of dye organic pollutants. *Materials Research Bulletin, 141*, 111376. doi:10.1016/j.materresbull.2021.111376

Pascariu, P., Homocianu, M., Cojocaru, C., Samoila, P., Airinei, A., & Suchea, M. (2019b). Preparation of La doped ZnO ceramic nanostructures by electrospinning – calcination method: Effect of La^{3+} doping on optical and photocatalytic properties. *Applied Surface Science*, *476*, 16–27. doi:10.1016/j.apsusc.2019.01.077

Patel, A. K., Singhania, R. R., & Pandey, A. (2016). Novel enzymatic processes applied to the food industry. *Current Opinion in Food Science*, *7*, 64–72. doi:10.1016/j.cofs.2015.12.002

Paul, S., Kundu, P., Bhattacharyya, U., Garai, A., Maji, R. C., Kondaiah, P., & Chakravarty, A. R. (2019). Ruthenium(II) conjugates of boron-dipyrromethene and biotin for targeted photodynamic therapy in red light. *Inorganic Chemistry*, *59*(1), 913–924. doi:10.1021/acs.inorgchem.9b03178 PMID:31825210

Paul, S., Kundu, P., Kondaiah, P., & Chakravarty, A. R. (2021). BODIPY-ruthenium(II) bis-terpyridine complexes for cellular imaging and type-I/-II photodynamic therapy. *Inorganic Chemistry*, *60*(21), 16178–16193. doi:10.1021/acs.inorgchem.1c01850 PMID:34672556

Pedrielli, P., & Skibsted, L. H. (2002). Antioxidant synergy and regeneration effect of quercetin,(−)-epicatechin, and (+)-catechin on α-tocopherol in homogeneous solutions of peroxidating methyl linoleate. *Journal of Agricultural and Food Chemistry*, *50*(24), 7138–7144. doi:10.1021/jf0204371 PMID:12428973

Pei, J., Yin, Y., Shen, Z., Bu, X., & Zhang, F. (2016). Oxidation of primary hydroxyl groups in chitooligomer by a laccase-TEMPO system and physico-chemical characterisation of oxidation products. *Carbohydrate Polymers*, *135*, 234–238. doi:10.1016/j.carbpol.2015.08.099 PMID:26453873

Pélissier, A., Akrout, Y., Jahn, K., Kuipers, J., Klein, U., Beerenwinkel, N., & Rodríguez Martínez, M. (2020). Computational model reveals a stochastic mechanism behind germinal center clonal bursts. *Cells*, *9*(6), 1448. doi:10.3390/cells9061448 PMID:32532145

Penney, W. G., & Sutherland, G. B. (1934). A note on the structure of H_2O_2 and H_4N_2 with particular reference to electric moments and free rotation. *Transactions of the Faraday Society*, *30*, 898–902. doi:10.1039/tf934300898b

Perdew, J. P., Burke, K., & Ernzerhof, M. (1996). Generalized gradient approximation made simple. *Physical Review Letters*, *77*(18), 3865–3868. doi:10.1103/PhysRevLett.77.3865 PMID:10062328

Perelson, A. (2002). Modelling viral and immune system dynamics. *Nature Reviews. Immunology*, *2*(1), 28–36. doi:10.1038/nri700 PMID:11905835

Perelson, A., & Weisbuch, G. (1992). *Theoretical and experimental insights into immunology*. Springer. doi:10.1007/978-3-642-76977-1

Pérez-Boada, M., Ruiz-Dueñas, F. J., Pogni, R., Basosi, R., Choinowski, T., Martínez, M. J., Piontek, K., & Martínez, A. T. (2005). Versatile peroxidase oxidation of high redox potential aromatic compounds: Site-directed mutagenesis, spectroscopic and crystallographic investigation of three long-range electron transfer pathways. *Journal of Molecular Biology*, *354*(2), 385–402. doi:10.1016/j.jmb.2005.09.047 PMID:16246366

Perse, M. (2021). Cisplatin Mouse Models: Treatment, toxicity and translatability. *Biomedicines*, *9*(10), 1406–1433. doi:10.3390/biomedicines9101406 PMID:34680523

Peruzzini, M., Ramirez, J. A., & Vizza, F. (1998). Hydrogenation of white phosphorus to phosphane with rhodium and iridium trihydrides. *Angewandte Chemie International Edition*, *37*(16), 2255–2257. doi:10.1002/(SICI)1521-3773(19980904)37:16<2255::AID-ANIE2255>3.0.CO;2-H PMID:29711443

Peyrat-Maillard, M. N., Cuvelier, M. E., & Berset, C. (2003). Antioxidant activity of phenolic compounds in 2, 2′-azobis (2-amidinopropane) dihydrochloride (AAPH)-induced oxidation: Synergistic and antagonistic effects. *Journal of the American Oil Chemists' Society*, *80*(10), 1007. doi:10.100711746-003-0812-z

Piang-Siong, W., De Caro, P., Marvilliers, A., Chasseray, X., Payet, B., Sing, A. S. C., & Illien, B. (2017). Contribution of trans-aconitic acid to DPPH scavenging ability in different media. *Food Chemistry*, *214*, 447–452. doi:10.1016/j.foodchem.2016.07.083 PMID:27507497

Pichaud, N., Bérubé, R., Côté, G., Belzile, C., Dufresne, F., Morrow, G., Tanguay, R. M., Rand, D. M., & Blier, P. U. (2019). Age dependent dysfunction of mitochondrial and ROS metabolism induced by mitonuclear mismatch. *Frontiers in Genetics*, *10*, 130–141. doi:10.3389/fgene.2019.00130 PMID:30842791

Pieri, A., Aschbacher, R., Fasani, G., Mariella, J., Brusetti, L., Pagani, E., Sartelli, M., & Pagani, L. (2020). Country income is only one of the tiles: The global journey of antimicrobial resistance among humans, animals, and environment. *Antibiotics (Basel, Switzerland)*, *9*(8), 473. doi:10.3390/antibiotics9080473 PMID:32752276

Pinelo, M., Manzocco, L., Nuñez, M. J., & Nicoli, M. C. (2004). Interaction among phenols in food fortification: Negative synergism on antioxidant capacity. *Journal of Agricultural and Food Chemistry*, *52*(5), 1177–1180. doi:10.1021/jf0350515 PMID:14995117

Piwowarska, D., Gnulek, P., & Rudowicz, C. (2019). Origin of the Ground Kramers Doublets for $C0^{2+}d^7$. *Applied Magnetic Resonance*, *50*(6), 797–808. doi:10.100700723-018-1080-4

Pizzino, G., Irrera, N., Cucinotta, M., Pallio, G., Mannino, F., Arcoraci, V., Squadrito, F., Altavilla, D., & Bitto, A. (2017). Oxidative stress: Harms and benefits for human health. *Oxidative Medicine and Cellular Longevity*, *8416763*, 1–13. doi:10.1155/2017/8416763 PMID:28819546

Plauck, A., Stangland, E. E., Dumesic, J. A., & Mavrikakis, M. (2016). Active sites and mechanisms for H_2O_2 decomposition over Pd catalysts. *Proceedings of the National Academy of Sciences of the United States of America*, *113*(14), E1973–E1982. doi:10.1073/pnas.1602172113 PMID:27006504

Pokorný, J., & Schmidt, S. (2001). Natural antioxidant functionality during food processing. *Antioxidants in food,* p. 331-354. http://foodclinic.ir/wp-content/uploads/2016/05/Natural-antioxidant-functionality.pdf

Polinger, V. Z., Gorinchoi, N. N., & Bersuker, I. B. (1992). MO LCAO analysis of the vibronic instability of the $CuCl_5^{3-}$ trigonal bipyramidal configuration. Critical view on the angular overlap model in vibronic problems. *Chemical Physics*, *159*(1), 75–87. doi:10.1016/0301-0104(92)80061-Y

Poljsak, B., Kovač, V., & Milisav, I. (2021). Antioxidants, food processing and health. *Antioxidants*, *10*(3), 433. doi:10.3390/antiox10030433 PMID:33799844

Pöller, S., Koster, D., & Schuhmann, W. (2013). Stabilizing redox polymer films by electrochemically induced crosslinking. *Electrochemistry Communications*, *34*, 327–330. doi:10.1016/j.elecom.2013.07.033

Ponte, F., Alberto, M. E., De Simone, B. C., Russo, N., & Sicilia, E. (2019). Photophysical exploration of dual-approach Pt-II-BODIPY conjugates: Theoretical insights. *Inorganic Chemistry*, *58*(15), 9882–9889. doi:10.1021/acs.inorgchem.9b01002 PMID:31343162

Poole, R. K. (2020). Flavohaemoglobin: The pre-eminent nitric oxide–detoxifying machine of microorganisms. *F1000 Research*, *9*, 7. doi:10.12688/f1000research.20563.1 PMID:31956400

Popa, D. S., Bigman, G., & Rusu, M. E. (2021). The role of vitamin k in humans: Implication in aging and age-associated diseases. *Antioxidants*, *10*(4), 1–31. doi:10.3390/antiox10040566 PMID:33917442

Portugal, C. C., Socodato, R., Canedo, T., Silva, C. M., Martins, T., Coreixas, V. S. M., Loiola, E. C., Gess, B., Röhr, D., Santiago, A. R., Young, P., Minshall, R. D., Paes-De-Carvalho, R., Ambrósio, A. F., & Relvas, J. B. (2017). Caveolin-1-mediated internalization of the Vitamin C transporter SVCT2 in microglia triggers an inflammatory phenotype. *Science Signaling*, *10*(472), eaal2005. Advance online publication. doi:10.1126cisignal.aal2005 PMID:28351945

Porubin, D. (2009). *Utilizarea compuşilor obţinuţi din produse secundare vinicole în inhibiţia proceselor de formare a N-nitrozoaminelor,* [PhD thesis, Chisinau, the State University of Moldova].

Porubin, D., Hecht, S. S., Li, Z., Gonta, M., & Stepanov, I. (2007). Endogenous formation of *n*'-nitrosonornicotine in F344 rats in the presence of some antioxidants and grape seed extract. *Journal of Agricultural and Food Chemistry*, *55*(17), 7199–7204. doi:10.1021/jf0712191 PMID:17637060

Poulios, I., Micropoulou, E., Panou, R., & Kostopoulou, E. (2003). Photooxidation of eosin Y in the presence of semiconducting oxides. *Applied Catalysis B: Environmental*, *41*(4), 345–355. doi:10.1016/S0926-3373(02)00160-1

Poulos, T. L. (2014). Heme enzyme structure and function. *Chemical Reviews*, *114*(7), 3919–3962. doi:10.1021/cr400415k PMID:24400737

Povar, I., & Spinu, O. (2016a). Ruthenium redox equilibria 1. Thermodynamic stability of Ru (III) and Ru (IV) hydroxides. *Journal of Electrochemical Science and Engineering*, *6*(1), 123–133. doi:10.5599/jese.226

Povar, I., & Spinu, O. (2016b). Ruthenium redox equilibria 3. Pourbaix diagrams for the systems Ru-H2O and Ru-Cl—H2O. *Journal of Electrochemical Science and Engineering*, *6*(1), 145–153. doi:10.5599/jese.229

Povar, I., & Spinu, O. (2016c). Correlation between global thermodynamic functions and experimental data in multicomponent heterogeneous systems. *Canadian Journal of Chemistry*, *94*(2), 113–119. doi:10.1139/cjc-2015-0411

Povar, I., Spinu, O., Lupascu, T., & Duca, G. (2022). Thermodynamic Stability of Natural Aqueous Systems. In *Research Anthology on Ecosystem Conservation and Preserving Biodiversity* (pp. 531–563). IGI Global., doi:10.4018/978-1-6684-5678-1.ch028

Povar, I., Spinu, O., Zinicovscaia, I., Pintilie, B., & Ubaldini, S. (2019b). Revised Pourbaix diagrams for the vanadium–water system. *Journal of Electrochemical Science and Engineering*, *9*(2), 75–84. doi:10.5599/jese.620

Povar, I., Ubaldini, S., Spinu, O., & Lupascu, T. (2019a). Thermodynamic analysis of the copper (I) homogeneous and heterogeneous speciation in ammonium thiosulfate leaching systems. *Canadian Journal of Chemistry*, *97*(9), 651–658. doi:10.1139/cjc-2018-0446

Prakash, J., Samriti, Kumar, A., Dai, H., Janegitz, B. C., Krishnan, V., Swart, H. C., & Sun, S. (2021). Novel rare earth metaledoped one-dimensional TiO_2 nanostructures: Fundamentals and multifunctional applications. *Materials Today Sustainability*, *13*, 100066. doi:10.1016/j.mtsust.2021.100066

Pratim, K. (2009). *Chemical Reactivity Theory. A Density Functional View*. CRC Press.

Price, D., Worsfold, P. J., Fauzi, R., & Mantoura, C. (1994). Determination of hydrogen peroxide in sea water by flow-injection analysis with chemiluminescence detection. *Analytica Chimica Acta*, *298*(1), 121–128. doi:10.1016/0003-2670(94)90050-7

Prigozhin, I., & Stengers, I. (1986). *Order from Chaos. A new human dialogue with nature*. Progress.

Prior, R. L., Wu, X., & Schaich, K. (2005). Standardized methods for the determination of antioxidant capacity and phenolics in foods and dietary supplements. *Journal of Agricultural and Food Chemistry, 53*(10), 4290–4302. doi:10.1021/jf0502698 PMID:15884874

Procházková, D., Boušová, I., & Wilhelmová, N. (2011). Antioxidant and prooxidant properties of flavonoids. *Fitoterapia, 82*(4), 513–523. doi:10.1016/j.fitote.2011.01.018 PMID:21277359

Puglia, C. D., & Powell, S. R. (1984). Inhibition of cellular antioxidants: A possible mechanism of toxic cell injury. *Environmental Health Perspectives, 57*, 307–311. doi:10.1289/ehp.8457307 PMID:6094175

Pulicharla, R., Brar, S. K., Drogui, P., Verma, M., & Surampalli, R. Y. (2015). Removal processes of antibiotics in waters and wastewaters: Crucial link to physical-chemical properties and degradation. *Journal of Hazardous, Toxic and Radioactive Waste, 19*(4), 04015008. doi:10.1061/(ASCE)HZ.2153-5515.0000285

Purkait, S., Bhattacharya, A., Bag, A., & Chattopadhyay, R. R. (2020). Synergistic antibacterial, antifungal and antioxidant efficacy of cinnamon and clove essential oils in combination. *Archives of Microbiology, 202*(6), 1439–1448. doi:10.100700203-020-01858-3 PMID:32185411

Putnam, C. D., Arvai, A. S., Bourne, Y., & Tainer, J. A. (2000). Active and inhibited human catalase structures: Ligand and NADPH binding and catalytic mechanism. *Journal of Molecular Biology, 296*(1), 295–309. doi:10.1006/jmbi.1999.3458 PMID:10656833

Putz, M. V., Cimpoesu, F., & Ferbinteanu, M. (2018). *Structural Chemistry, Principles, Methods, and Case Studies.* Springer.

Pylaeva, S. A., Elgabarty, H., Sebastiani, D., & Tolstoy, P. M. (2017). Symmetry and dynamics of FHF− anion in vacuum, in CD_2Cl_2 and in CCl_4. Ab initio MD study of fluctuating solvent–solute hydrogen and halogen bonds. *Physical Chemistry Chemical Physics, 19*(38), 26107–26120. doi:10.1039/C7CP04493C PMID:28929144

Qiao, Y., Li, Y., Fu, W., Guo, Z., & Zheng, X. (2018). Enhancing the electrochemiluminescence of luminol by chemically modifying the reaction microenvironment. *Analytical Chemistry, 90*(15), 9629–9636. doi:10.1021/acs.analchem.8b02577 PMID:29969897

Quiroga, P. R., Nepote, V., & Baumgartner, M. T. (2019). Contribution of organic acids to α-terpinene antioxidant activity. *Food Chemistry, 277*, 267–272. doi:10.1016/j.foodchem.2018.10.100 PMID:30502144

Rackers, J. A., Laury, M. L., Lu, C., Wang, Z., Lagardère, L., Piquemal, J.-P., Ren, P., & Ponder, J. W. (2018). TINKER 8: A Modular Software Package for Molecular Design and Simulation. *Journal of Chemical Theory and Computation, 14*, 5273–5289. doi:10.1021/acs.jctc.8b00529 PMID:30176213

Raczyńska, E. D., Kosińska, W., Ośmiałowski, B., & Gawinecki, R. (2005). Tautomeric equilibria in relation to PI-electron delocalization. *Chemical Reviews, 105*(10), 3561–3612. doi:10.1021/cr030087h PMID:16218561

Raducan, A., Leonties, A. R., Puiu, M., & Oancea, D. (2012). Kinetics of hydrogen peroxide decomposition by catalase: Hydroxylic solvent effects. *Bioprocess and Biosystems Engineering, 35*(9), 1523–1530. doi:10.100700449-012-0742-0 PMID:22565543

Răducan, A., Oancea, D., & Segal, E. (2007). Models for irreversible inactivation in the enzymatic decomposition of hydrogen peroxide and their kinetic treatment by means of graphs. *Revue Roumaine de Chimie, 52*, 1280–1284.

Rahal, A., Kumar, A., Singh, V., Yadav, B., Tiwari, R., Chakraborty, S., & Dhama, K. (2014). Oxidative stress, prooxidants, and antioxidants: The interplay. *BioMed Research International, 2014*, 1–19. doi:10.1155/2014/761264 PMID:24587990

Raimondi, M., & Cooper, D. L. (1999). Ab initio modern valence bond theory. *Topics in Current Chemistry*, *203*, 105–120. doi:10.1007/3-540-48972-X_6

Ramaswamy, N., & Mukerjee, S. (2011). Influence of inner- and outer-sphere electron transfer mechanisms during electrocatalysis of oxygen reduction in alkaline media I. *The Journal of Physical Chemistry C*, *115*(36), 18015–18026. doi:10.1021/jp204680p

Rapaport, D. C. (2004). *The Art of Molecular Dynamics Simulation* (2nd ed.). Cambridge University Press. doi:10.1017/CBO9780511816581

Reber, J. D., Eggett, D. L., & Parker, T. L. (2011). Antioxidant capacity interactions and a chemical/structural model of phenolic compounds found in strawberries. *International Journal of Food Sciences and Nutrition*, *62*(5), 445–452. doi:10.3109/09637486.2010.549115 PMID:21385007

Rehm, D., & Weller, A. (1970). Kinetics of Fluorescence Quenching by Electron and H-Atom Transfer. *Israel Journal of Chemistry*, *8*(2), 259–271. doi:10.1002/ijch.197000029

Remya, N., & Swain, A. (2019). Soft drink industry wastewater treatment in microwave photocatalytic system–exploration of removal efficiency and degradation mechanism. *Separation and Purification Technology*, *210*, 600–607. doi:10.1016/j.seppur.2018.08.051

Rendra, E., Riabov, V., Mossel, D. M., Sevastyanova, T., Harmsen, M. C., & Kzhyshkowska, J. (2019). Reactive oxygen species (ROS) in macrophage activation and function in diabetes. *Immunobiology*, *224*(2), 242–253. doi:10.1016/j.imbio.2018.11.010 PMID:30739804

Ren, W., Xiong, L., Nie, G., Zhang, H., Duan, X., & Wang, S. (2019). Insights into the electron-transfer regime of peroxydisulfate activation on carbon nanotubes: The role of oxygen functional groups. *Environmental Science & Technology*, *54*(2), 1267–1275. doi:10.1021/acs.est.9b06208 PMID:31846314

Ribeiro, A. R., & Schmidt, T. C. (2017). Determination of acid dissociation constants (pKa) of cephalosporin antibiotics: Computational and experimental approaches. *Chemosphere*, *169*, 524–533. doi:10.1016/j.chemosphere.2016.11.097 PMID:27898325

Ribeiro, D., Freitas, M., Silva, A. M., Carvalho, F., & Fernandes, E. (2018). Antioxidant and pro-oxidant activities of carotenoids and their oxidation products. *Food and Chemical Toxicology*, *120*, 681–699. doi:10.1016/j.fct.2018.07.060 PMID:30077704

Rice-Evans, C., & Miller, N. J. (1994). Total antioxidant status in plasma and body fluids. In Vol. 234, pp. 279–293). Methods in Enzymology. Academic Press. doi:10.1016/0076-6879(94)34095-1

Richardson, D. E., & Taube, H. (1984). Mixed-valence molecules: Electronic delocalization and stabilization. *Coordination Chemistry Reviews*, *60*, 107–129. doi:10.1016/0010-8545(84)85063-8

Rietjens, I. M., Boersma, M. G., de Haan, L., Spenkelink, B., Awad, H. M., Cnubben, N. H., Van Zanden, J. J., Van der Woude, H., Alink, G. M., & Koeman, J. H. (2002). The pro-oxidant chemistry of the natural antioxidants vitamin C, vitamin E, carotenoids and flavonoids. *Environmental Toxicology and Pharmacology*, *11*(3-4), 321–333. doi:10.1016/S1382-6689(02)00003-0 PMID:21782615

Rigamonti, P., Zardi, S., Carlino, S., Demartin, F., Castellano, C., Pigani, L., Ponti, A., Ferretti, A. M., & Pasini, A. (2020). Selective formation, reactivity, redox and magnetic properties of. *International Journal of Molecular Sciences*, *21*(21), 7882. doi:10.3390/ijms21217882 PMID:33114262

Riplinger, C., & Neese, F. (2011). The Reaction Mechanism of Cytochrome P450 NO Reductase: A Detailed Quantum Mechanics/Molecular Mechanics Study. *ChemPhysChem*, *12*(17), 3192–3203. doi:10.1002/cphc.201100523 PMID:22095732

Rivas, M. G., Gonzalez, P. J., Ferroni, F. M., Rizzi, A. C., & Brondino, C. (2020). Studying electron transfer pathways in oxidoreductases. *Science Reviews - from the End of the World*, *1*(2), 6–23. https://doi.org/. doi:10.52712/sciencereviews.v1i2.15

Rizvi, S., Raza, S. T., Ahmed, F., Ahmad, A., Abbas, S., & Mahdi, F. (2014). *The Role of Vitamin E in Human Health and Some Diseases*. Academic Press.

Rizvi, M. A., Maqsood, R., Jan, R., & Peerzada, G. M. (2011b). Utilizing 1, 10–Phenanthroline Ligand Effect in Estimation of Cobalt (II) and Simultaneous Estimation of Cobalt (II) and Iron (II) with Cerium (IV). *Chemical Educator (Boise, Idaho)*, *16*, 72–75. doi:10.133300897112337a

Rizvi, M. A., Syed, R. M., & Khan, B. (2011a). Complexation effect on redox potential of iron (III)– iron (II) couple: A simple potentiometric experiment. *Journal of Chemical Education*, *88*(2), 220–222. doi:10.1021/ed100339g

Rizzo, J. A., Rowan, M. P., Driscoll, I. R., Chung, K. K., & Friedman, B. C. (2016). Vitamin C in Burn Resuscitation. *Critical Care Clinics*, *32*(4), 539–546. doi:10.1016/j.ccc.2016.06.003 PMID:27600125

Roberts, B. M., Dzuba, V. A., & Flambaum, V. V. (2015). Parity and Time-Reversal Violation in Atomic Systems. *Annual Review of Nuclear and Particle Science*, *65*(1), 63–86. doi:10.1146/annurev-nucl-102014-022331

Roberts, E. (2014). Cellular and molecular structure as a unifying framework for whole-cell modeling. *Current Opinion in Structural Biology*, *25*, 86–91. doi:10.1016/j.sbi.2014.01.005 PMID:24509245

Rodopulo, A. K. (1983). *Bases of winemaking biochemistry* (in Russian). Moscow: Light and food industry Publishing.

Roh, J.-L., Jang, H., Kim, E. H., & Shin, D. (2017). Targeting of the glutathione, thioredoxin, and Nrf2 antioxidant systems in head and neck cancer. *Antioxidants & Redox Signaling*, *27*(2), 106–114. doi:10.1089/ars.2016.6841 PMID:27733046

Rohlfing, D. L. (1967). The catalytic decarboxylation of oxaloacetic acid by thermally prepared poly-α-amino acids. *Archives of Biochemistry and Biophysics*, *118*(2), 468–474. doi:10.1016/0003-9861(67)90375-X PMID:6033722

Romanciuc, L., Gladchi, V., Goreaceva, N., Bunduchi, E., & Duca, Gh. (2009). Water quality monitoring and management in Moldova. In: Bahadir, A.M., Duca, G. (eds) The role of ecological chemistry in pollution research and sustainable development. NATO Science for Peace and Security Series C: Environmental Security. Springer, Dordrecht. doi:10.1007/978-90-481-2903-4_6

Romanovsky, Yu. M., Stepanova, N. V., & Chernavsky, D. S. (1984). *Mathematical Biophysics*. Nauka.

Romero, E., Gómez Castellanos, J. R., Gadda, G., Fraaije, M. W., & Mattevi, A. (2018). Same Substrate, Many Reactions: Oxygen Activation in Flavoenzymes. *Chemical Reviews*, *118*(4), 1742–1769. doi:10.1021/acs.chemrev.7b00650 PMID:29323892

Rosi, M., Sgamellotti, A., Tarantelli, F., & Floriani, C. (1988). Interaction between iron (0) and formaldehyde, thioformaldehyde, and acetone: Ab initio calculations on the model compounds $Fe(CO)_2(Ph_3)_2(\eta^2\text{-}CH_2O)$, $Fe(CO)_2(Ph_3)_2(\eta^2\text{-}CH_2S)$, $Fe(Ph_3)_4(\eta^2\text{-}CH_2O)$, and $Fe(Ph_3)_4(\eta^2\text{-}CMe_2O)$. *Inorganic Chemistry*, *27*(1), 69–73. doi:10.1021/ic00274a016

Rossetto, M., Vanzani, P., Mattivi, F., Lunelli, M., Scarpa, M., & Rigo, A. (2002). Synergistic antioxidant effect of catechin and malvidin 3-glucoside on free radical-initiated peroxidation of linoleic acid in micelles. *Archives of Biochemistry and Biophysics*, *408*(2), 239–245. doi:10.1016/S0003-9861(02)00561-1 PMID:12464277

Rostami, A., Abdelrasoul, A., Shokri, Z., & Shirvandi, Z. (2022). Applications and mechanisms of free and immobilized laccase in detoxification of phenolic compounds - A review. *Korean Journal of Chemical Engineering*, *39*(4), 821–832. doi:10.100711814-021-0984-0

Rost, D., Welker, A., Welker, J., Millonig, G., Berger, I., Autschbach, F., Schuppan, D., & Mueller, S. (2007). Liver-homing of purified glucose oxidase: A novel in vivo model of physiological hepatic oxidative stress (H_2O_2). *Journal of Hepatology*, *46*(3), 482–491. doi:10.1016/j.jhep.2006.09.025 PMID:17188390

Rudakov, E. V., Volkova, L. K., & Lutsyk, A. I. (2011). Reactivity of C3-C8 alkanes with nitronium ions in sulfuric acid. *Theoretical and Experimental Chemistry*, *47*(4), 244–250. doi:10.100711237-011-9211-0

Rudnev, S., Romanyukha, A., & Yashin, A. (2007). Modeling of T cell population development and estimation of resource allocation effectiveness. *Mat. Modeling*, *19*(11), 25–42.

Ruiz, E., Alemany, P., Alvarez, S., & Cano, J. (1997). Structural Modeling and Magneto-Structural Correlations for Hydroxo-Bridged Copper(II) Binuclear Complexes. *Inorganic Chemistry*, *36*(17), 3683–3688. doi:10.1021/ic970310r PMID:11670060

Ruiz, E., Cano, J., Alvarez, S., & Alemany, P. (1999). Broken symmetry approach to calculation of exchange coupling constants for homobinuclear and heterobinuclear transition metal complexes. *Journal of Computational Chemistry*, *20*(13), 1391–1400. doi:10.1002/(SICI)1096-987X(199910)20:13<1391::AID-JCC6>3.0.CO;2-J

Rutledge, H. L., & Tezcan, F. A. (2020). Electron transfer in nitrogenase. *Chemical Reviews*, *120*(12), 5158–5193. doi:10.1021/acs.chemrev.9b00663 PMID:31999100

Safavi, A., Abdollahi, H., & Nezhad, M. H. (2002). Indirect kinetic spectrophotometric determination of cobalt based on the redox reaction with iron (III) in the presence of 1, 10-phenanthroline. *Spectroscopy Letters*, *35*(5), 681–688. doi:10.1081/SL-120014939

Sagi, N. V., Karri, P., Hu, F., & Krishnamurthy, R. (2011). Diastereoselective self-condensation of dihydroxyfumaric acid in water: Potential route to sugars. *Angewandte Chemie International Edition*, *50*(35), 8127–8130. doi:10.1002/anie.201102045 PMID:21748829

Sagi, V. N., Punna, V., Hu, F., Meher, G., & Krishnamurthy, R. (2012). Exploratory experiments on the chemistry of the "Glyoxylate Scenario": Formation of ketosugars from Dihydroxyfumarate. *Journal of the American Chemical Society*, *134*(7), 3577–3589. doi:10.1021/ja211383c PMID:22280414

Saha, R., & Mukhopadhyay, M. (2020). Elucidation of the decolorization of Congo Red by Trametes versicolor laccase in presence of ABTS through cyclic voltammetry. *Enzyme and Microbial Technology*, *135*, 109507. doi:10.1016/j.enzmictec.2019.109507 PMID:32146934

Saikia, S., Gogoi, R. D., Yadav, M., & Yadav, H. S. (2022). Isolation, purification and characterization of peroxidase from Raphanus sativus and its applications in biotransformation of cresols. *Biocatalysis and Agricultural Biotechnology*, *46*, 102540. doi:10.1016/j.bcab.2022.102540

Sampayo, J. N., Gill, M. S., & Lithgow, G. J. (2003). Oxidative stress and aging - The use of superoxide dismutase/catalase mimetics to extend lifespan. *Biochemical Society Transactions*, *31*(6), 1305–1307. doi:10.1042/bst0311305 PMID:14641049

Sariciftci, N. S., Smilowitz, L., Heeger, A. J., & Wudl, F. (1992). Photoinduced electron transfer from a conducting polymer to buckminsterfullerene. *Science*, *258*(5087), 1474–1476. doi:10.1126cience.258.5087.1474 PMID:17755110

Sarkar, A. M., Middya, T. R., & Jana, A. D. (2012). A QSAR study of radical scavenging antioxidant activity of a series of flavonoids using DFT based quantum chemical descriptors – the importance of group frontier electron density. *Journal of Molecular Modeling*, *18*(6), 2621–2631. doi:10.100700894-011-1274-2 PMID:22080306

Sasahara, G. L., Gouveia, F. S., Rodrigues, R. D., Zampieri, D. S., Fonseca, S. G. C., Goncalves, R. D. R., Athaydes, B. R., Kitagawa, R. R., Santos, F. A., Sousa, E. H. S., Nagao-Dias, A. T., & Lopes, L. G. D. (2020). Nitro-imidazole-based ruthenium complexes with antioxidant and anti-inflammatory activities. *Journal of Inorganic Biochemistry*, *206*, 111048–111058. doi:10.1016/j.jinorgbio.2020.111048 PMID:32151873

Sato, K., Nishina, Y., & Shiga, K. (1993). Electron-transferring flavoprotein has an AMP-binding site in addition to the FAD-binding site1. *Journal of Biochemistry*, *114*(2), 215–222. doi:10.1093/oxfordjournals.jbchem.a124157 PMID:8262902

Sawai, Y., & Moon, J. H. (2000). NMR analytical approach to clarify the molecular mechanisms of the antioxidative and radical-scavenging activities of antioxidants in tea using 1, 1-diphenyl-2-picrylhydrazyl. *Journal of Agricultural and Food Chemistry*, *48*(12), 6247–6253. doi:10.1021/jf000500b PMID:11141282

Sawai, Y., Moon, J. H., Sakata, K., & Watanabe, N. (2005). Effects of structure on radical-scavenging abilities and anti-oxidative activities of tea polyphenols: NMR analytical approach using 1, 1-diphenyl-2-picrylhydrazyl radicals. *Journal of Agricultural and Food Chemistry*, *53*(9), 3598–3604. doi:10.1021/jf040423a PMID:15853407

Sawai, Y., & Sakata, K. (1998). NMR analytical approach to clarify the antioxidative molecular mechanism of catechins using 1, 1-diphenyl-2-picrylhydrazyl. *Journal of Agricultural and Food Chemistry*, *46*(1), 111–114. doi:10.1021/jf970342x PMID:10554204

Sawa, T., Takata, T., Matsunaga, T., Ihara, H., Motohashi, H., & Akaike, T. (2022). Chemical biology of reactive sulfur species: Hydrolysis-driven equilibrium of polysulfides as a determinant of physiological functions. *Antioxidants & Redox Signaling*, *36*(4-6), 327–336. doi:10.1089/ars.2021.0170 PMID:34409860

Scholes, G., & Weiss, J. (1959). *Oxygen effects and formation of peroxides in aqueous solutions. Radiation research supplement*. Proceedings of the International Congress of Radiation Research, Burlington, Vermont. https://www.jstor.org/stable/3583638

Schonbaum, G. R., & Chance, B. (1976). 7 Catalase. *The Enzymes*, *13*, 363–408. doi:10.1016/S1874-6047(08)60245-0

Schroeder, M. T., Becker, E. M., & Skibsted, L. H. (2006). Molecular mechanism of antioxidant synergism of tocotrienols and carotenoids in palm oil. *Journal of Agricultural and Food Chemistry*, *54*(9), 3445–3453. doi:10.1021/jf053141z PMID:16637706

Scott, R. A. (1995). Functional significance of cytochrome c oxidase structure. *Structure (London, England)*, *3*(10), 981–986. doi:10.1016/S0969-2126(01)00233-7 PMID:8590007

Secara, N. (2008). Dihydroxyfumaric acid transformation, *Chemistry Journal of Moldova, 1*(3), p. 118-119. https://ibn.idsi.md/vizualizare_articol/6787

Secara, N., Duca, Gh., & Macaev, F. (2011). Occurrence and chemistry of dihydroxyfumaric acid. Chemistry Journal of Moldova, 6(1), p. 29-44. https://ibn.idsi.md/ro/vizualizare_articol/12135/datacite

Secara, E. (2016). Synthesis of new drimane and homodrimane lactams by Beckmann rearrangement of some ketoximes. *Chemistry Journal of Moldova*, *11*(1), 50–54. doi:10.19261/cjm.2016.11(1).06

Secara, N. (2010). Stopped-Flow Spectrophotometric Studies of the Kinetics of Interaction of Dihydroxyfumaric Acid with the DPPH Free Radical. *Chemistry Journal of Moldova*, *5*(2), 83–87. doi:10.19261/cjm.2010.05(2).11

Secara, N., Duca, G., Vlad, L., & Macaev, F. (2010). Observations on the antioxidant activity of novel dihydroxyfumaric acid derivatives. *Chemistry Journal of Moldova*, *5*(2), 59–67. doi:10.19261/cjm.2010.05(2).08

Secula, M. S., Suditu, G. D., Poulios, I., Cojocaru, C., & Cretescu, I. (2008). Response surface optimization of the photocatalytic decolorization of a simulated dyestuff effluent. *Chemical Engineering Journal*, *141*(1-3), 18–26. doi:10.1016/j.cej.2007.10.003

Sedgwick, A. C., Wu, L., Han, H. H., Bull, S. D., He, X. P., James, T. D., & Yoon, J. (2018). Excited-state intramolecular proton-transfer (ESIPT) based fluorescence sensors and imaging agents. *Chemical Society Reviews*, *47*(23), 8842–8880. doi:10.1039/C8CS00185E PMID:30361725

Seid-Mohammadi, A., Asgarai, G., Ghorbanian, Z., & Dargahi, A. (2019). The removal of cephalexin antibiotic in aqueous solutions by ultrasonic waves/hydrogen peroxide/nickel oxide nanoparticles (US/H_2O_2/NiO) hybrid process. *Separation Science and Technology*, 1–11.

Seifried, H. E., McDonald, S. S., Anderson, D. E., Greenwald, P., & Milner, J. A. (2003). The antioxidant conundrum in cancer. *Cancer Research*, *63*(15), 4295–4298. PMID:12907593

Šekuljica, N., Gvozdenović, M. M., Knežević-Jugović, Z. D., Jugović, B. Z., & Grgur, B. N. (2016). Biofuel cell based on horseradish peroxidase immobilized on copper sulfide as anode for decolorization of anthraquinone AV109 dye. *Journal of Energy Chemistry*, *25*(3), 403–408. doi:10.1016/j.jechem.2016.03.011

Sellami, K., Couvert, A., Nasrallah, N., Maachi, R., Abouseoud, M., & Amrane, A. (2022). Peroxidase enzymes as green catalysts for bioremediation and biotechnological applications: A review. *The Science of the Total Environment*, *806*, 150500. doi:10.1016/j.scitotenv.2021.150500 PMID:34852426

Sepasi Tehrani, H., & Moosavi-Movahedi, A. A. (2018). Catalase and its mysteries. *Progress in Biophysics and Molecular Biology*, *140*, 5–12. doi:10.1016/j.pbiomolbio.2018.03.001 PMID:29530789

Serjeant, E. P. (1984). Potentiometry and potentiometric titrations. *A series of Monographs on Analytical Chemistry and its Applications, 69*.

Serna-Galvis, E. A., Cáceres-Peña, A. C., & Torres-Palma, R. A. (2020). Elimination of representative fluoroquinolones, penicillins, and cephalosporins by solar Photo-Fenton: Degradation routes, primary transformations, degradation improvement by citric acid addition, and antimicrobial activity evolution. *Environmental Science and Pollution Research International*, *33*(33), 4138–41393. doi:10.100711356-020-10069-8 PMID:32683623

Serra-Maia, R., Bellier, M., Chastka, S., Tranhuu, K., Subowo, A., Rimstidt, J. D., Usov, P. M., Morris, A. J., & Michel, F. M. (2018). Mechanism and kinetics of hydrogen peroxide decomposition on platinum nanocatalysts. *ACS Applied Materials & Interfaces*, *10*(25), 21224–21234. doi:10.1021/acsami.8b02345 PMID:29851338

Shahraki, S., Saeidifar, M., Delarami, H. S., & Kazemzadeh, H. (2020). Molecular docking and inhibitory effects of a novel cytotoxic agent with bovine liver catalase. *Journal of Molecular Structure*, *1205*, 127590. doi:10.1016/j.molstruc.2019.127590

Shahzad, K., Asad, M., Asiri, A. M., Irfan, M., & Iqbal, M. A. (2022). In-Vitro Anticancer Profile of Recent ruthenium complexes against liver cancer. *Reviews in Inorganic Chemistry*. Advance online publication. doi:10.1515/revic-2021-0040

Shaik, S., Cohen, S., Wang, Y., Chen, H., Kumar, D., & Thiel, W. (2010). P450 enzymes: Their structure, reactivity, and selectivity-modeled by QM/MM calculations. *Chemical Reviews*, *110*(2), 949–1017. doi:10.1021/cr900121s PMID:19813749

Shaik, S., & Hiberty, P. C. (2008). *A Chemist's Guide to Valence Bond Theory*. John Wiley & Sons Inc.

Shang, Y. J., Qian, Y. P., Liu, X. D., Dai, F., Shang, X. L., Jia, W. Q., Liu, Q., Jian-Guo Fang, J. G., & Zhou, B. (2009). Radical-scavenging activity and mechanism of resveratrol-oriented analogues: Influence of the solvent, radical, and substitution. *The Journal of Organic Chemistry*, *74*(14), 5025–5031. doi:10.1021/jo9007095 PMID:19472994

Shan, H., Wang, X., Ge, Y., & Li, Z. (2022). Homologous amino acids promoted co-immobilization of laccase and mediator onto geopolymer microspheres for enhancing degradation of dyes in water. *Journal of Hazardous Materials*, *423*, 127107. doi:10.1016/j.jhazmat.2021.127107 PMID:34523501

Shao, Y., Ying, Y., & Ping, J. (2020). Recent advances in solid-contact ion-selective electrodes: Functional materials, transduction mechanisms, and development trends. *Chemical Society Reviews*, *49*(13), 4405–4465. doi:10.1039/C9CS00587K PMID:32458836

Sharma, V. K. (2010). Oxidation of inorganic compounds by ferrate (VI) and ferrate (V): One-electron and two-electron transfer steps. *Environmental Science & Technology*, *44*(13), 5148–5152. doi:10.1021/es1005187 PMID:20527775

Shen, X., Liu, W., Gao, X., Lu, Z., Wu, X., & Gao, X. (2015). Mechanisms of oxidase and superoxide dismutation-like activities of gold, silver, platinum, and palladium, and their alloys: A general way to the activation of molecular oxygen. *Journal of the American Chemical Society*, *137*(50), 15882–15891. doi:10.1021/jacs.5b10346 PMID:26642084

Shields, H. J., Traa, A., & Van Raamsdonk, J. M. (2021). Beneficial and detrimental effects of reactive oxygen species on lifespan: A comprehensive review of comparative and experimental studies. *Frontiers in Cell and Developmental Biology*, *9*, 628157. doi:10.3389/fcell.2021.628157 PMID:33644065

Shi, J., Guo, Y. H., Xie, F., Chen, Q. F., & Zhang, M. T. (2020). Redox-active ligand assisted catalytic water oxidation by a RuIV=O intermediate. *Angewandte Chemie International Edition*, *59*(10), 4000–4008. doi:10.1002/anie.201910614 PMID:31880387

Shimada, S., Shinzawa-Itoh, K., Baba, J., Aoe, S., Shimada, A., Yamashita, E., Kang, J., Tateno, M., Yoshikawa, S., & Tsukihara, T. (2017). Complex structure of cytochrome c –cytochrome c oxidase reveals a novel protein–protein interaction mode. *The EMBO Journal*, *36*(3), 291–300. doi:10.15252/embj.201695021 PMID:27979921

Shinde, S. B., & Kurhekar, M. P. (2020). Agent-based modeling of the adaptive immune system using netlogo simulation tool. *Soft Comput. Problem Solving*, 463-474.

Shokri, Z., Seidi, F., Karami, S., Li, C., Saeb, M. R., & Xiao, H. (2021). Laccase immobilization onto natural polysaccharides for biosensing and biodegradation. *Carbohydrate Polymers*, *262*, 117963. doi:10.1016/j.carbpol.2021.117963 PMID:33838831

Shtamm, E., Purmal, A., & Skurlatov, Yu. (1991). The role of hydrogen peroxide in natural water environment (Rus.). *Successes of Chemistry*, *60*(11), 2373–2406.

Shvydkiy, V., Dolgov, S., Dubovik, A., Kozlov, M., Povkh, A., Shishkina, L., & Duca, G. (2022). New aspects for the estimation of the state of natural water. *Chemistry Journal of Moldova*, *17*(2), 35–42. doi:10.19261/cjm.2022.973

Sigel, A., Sigel, H., & Sigel, R. K. O. (2007). *The Ubiquitous Roles of Cytochrome P450 Proteins*. Wiley. doi:10.1002/9780470028155

Sigg, A., & Neftel, A. (1988). Seasonal variations in hydrogen peroxide in polar ice cores. *Annals of Glaciology*, *10*, 157–162. doi:10.3189/S0260305500004353

Simović, A. R., Masnikosa, R., Bratsos, I., & Alessio, E. (2019). Chemistry and reactivity of ruthenium(II) complexes: DNA/protein binding mode and anticancer activity are related to the complex structure. *Coordination Chemistry Reviews*, *398*, 113011–113036. doi:10.1016/j.ccr.2019.07.008

Sinel'nikov, V. E. (1968). Luminescent analysis of natural and polluted waters (Rus.). Obninsk: Gidrometcentr SSSR.

Sinel'nikov, V. E. (1971). Hydrogen peroxide level in river water, and methods for detecting it (Rus.). *Hydrobiological Journal*, *7*(1), 115–119.

Singh, D. V. (2018). Investigation on antioxidant potential of phenolic acids and flavonoids: The common phytochemical ingredients in plants. *Journal of Plant Biochemistry & Physiology*, *6*(3), 1–5. doi:10.4172/2329-9029.1000219

Singh, R., Umapathi, A., Patel, G., Patra, C., Malik, U., Bhargava, S. K., & Daima, H. K. (2022). Nanozyme-based pollutant sensing and environmental treatment: Trends, challenges, and perspectives. *The Science of the Total Environment*, *854*, 158771. doi:10.1016/j.scitotenv.2022.158771 PMID:36108853

Singleton, V. L., & Rossi, J. A. Jr. (1965). Colorimetry of total phenolics with phosphomolybdic-phosphotungstic acid reagents. *American Journal of Enology and Viticulture*, *16*(3), 144–158. doi:10.5344/ajev.1965.16.3.144

Sinha, S., Sahad, M. E., & Mondal, R. (2022). A Singlet-Diradical Co(III)-Dimer as a Nonvolatile Resistive Switching Device: Synthesis, Redox-Induced Interconversion, and Current–Voltage Characteristics. *Journal of the American Chemical Society*, *144*(44), 20442–20451. doi:10.1021/jacs.2c08941 PMID:36282681

Sirkar, K., Revzin, A., & Pishko, M. V. (2000). Glucose and lactate biosensors based on redox polymer/oxidoreductase nanocomposite thin films. *Analytical Chemistry*, *72*(13), 2930–2936. doi:10.1021/ac991041k PMID:10905330

Skinner, J. R., & Stinle, S. E. (1953). *Method for the production of hydrogen peroxide*. Patent US, 1953-12-23. https://patents.google.com/patent/US2871103A/en

Skroza, D., Mekinić, I. G., Svilović, S., Šimat, V., & Katalinić, V. (2015). Investigation of the potential synergistic effect of resveratrol with other phenolic compounds: A case of binary phenolic mixtures. *Journal of Food Composition and Analysis*, *38*, 13–18. doi:10.1016/j.jfca.2014.06.013

Skulachev, V. P. (1996). Role of uncoupled and non-coupled oxidations in the maintenance of safely low levels of oxygen and its one-electron reductants. *Quarterly Reviews of Biophysics*, *29*(2), 169–202. doi:10.1017/S0033583500005795 PMID:8870073

Skurlatov, Yu., Duca, Gh., Batir, D., & Travin, S. (1989). Partial charge transfer coordination compounds in the redox processes (Rus.). *Coordination Chemistry*, *15*(3), 291–307.

Skurlatov, Yu., Ernestova, L., Shtamm, E., Shpotova, T., & Kalinin, V. (1984). Redox state and seasonal toxicity of natural water. *Proc. of Acad. Sci. USSR*, *276*(4), 1014–1016.

Slavova-Kazakova, A., Janiak, M. A., Sulewska, K., Kancheva, V. D., & Karamać, M. (2021). Synergistic, additive, and antagonistic antioxidant effects in the mixtures of curcumin with (−)-epicatechin and with a green tea fraction containing (−)-epicatechin. *Food Chemistry*, *360*, 129994. doi:10.1016/j.foodchem.2021.129994 PMID:33989877

Slesak, I., Slesak, H., & Kruk, J. (2012). Oxygen and hydrogen peroxide in the early evolution of life on earth: In silico comparative analysis of biochemical pathways. *Astrobiology*, *12*(8), 775–784. doi:10.1089/ast.2011.0704 PMID:22970865

Slezak, J., Kura, B., LeBaron, T. W., Singal, P. K., Buday, J., & Barancik, M. (2021). Oxidative stress and pathways of molecular hydrogen effects in medicine. *Current Pharmaceutical Design*, *27*(5), 610–625. PMID:32954996

Sodhi, K. K., Kumar, M., Balan, B., Dhaulaniya, A. S., Shree, P., Sharma, N., & Singh, D. K. (2021). Perspectives on antibiotic contamination, resistance, metabolomics, and systemic remediation. *SN Applied Sciences*, *269*(3), 269. doi:10.100742452-020-04003-3

Soleimany, A. P., & Bhatia, S. N. (2020). Activity-based diagnostics: An emerging paradigm for disease detection and monitoring. *Trends in Molecular Medicine*, *26*(5), 450–468. doi:10.1016/j.molmed.2020.01.013 PMID:32359477

Song, L., Chen, Z., Ying, F., Song, J., Chen, X., Su, P., Mo, Y., Zhang, Q., & Wu, W. (2012). XMVB 2.0: An ab initio Non-Orthogonal Valence Bond Program. Xiamen University, China.

Song, L., Mo, Y., Zhang, Q., & Wu, W. (2005). XMVB: A program for ab initio nonorthogonal valence bond computations. *Journal of Computational Chemistry*, *26*(5), 514–521. doi:10.1002/jcc.20187 PMID:15704237

Sooch, B. S., Kauldhar, B. S., & Puri, M. (2014). Recent insights into microbial catalases: Isolation, production and purification. *Biotechnology Advances*, *32*(8), 1429–1447. doi:10.1016/j.biotechadv.2014.09.003 PMID:25261851

Sotler, R., Poljšak, B., Dahmane, R., Jukić, T., Pavan Jukić, D., Rotim, C., Polonca Trebše, Starc, A. (2019). Prooxidant activities of antioxidants and their impact on health. *Acta Clinica Croatica, 58*(4.), 726-736.

Souchay, P., Fleury, D., & Fleury, M. (1967). Tautomerie enediol-cetol de l'acide dihydroxyfumarique et son intervention dans la decarboxylation de celui-ci. *Comptes Rendus Hebdomadaires des Seances de L'Academie des Sciences Seice C, 264*(26), 2130.

Sousa, S. F., Ertem, M. Z., Faustino, L. A., Machado, A. E. H., Concepcion, J. J., Maia, P. I. S., & Patrocinio, A. O. T. (2021). Mechanistic investigation of the aerobic oxidation of 2-pyridylacetate coordinated to a Ru(II) polypyridyl complex. *Dalton Transactions (Cambridge, England)*, *50*(42), 15248–15259. doi:10.1039/D1DT02461B PMID:34632989

Spiegel, M. (2022). Current Trends in Computational Quantum Chemistry Studies on Antioxidant Radical Scavenging Activity. *Journal of Chemical Information and Modeling*, *62*(11), 2639–2658. doi:10.1021/acs.jcim.2c00104 PMID:35436117

Sridharan, U. C., Reimann, B., & Kaufman, F. (1980). Kinetics of the reaction OH + $H_2O_2 \rightarrow HO_2$ + H_2O. *The Journal of Chemical Physics*, *73*(3), 1286–1293. doi:10.1063/1.440240

Stadlmair, L. F., Letzel, T., Drewes, J. E., & Grassmann, J. (2018). Enzymes in removal of pharmaceuticals from wastewater: A critical review of challenges, applications and screening methods for their selection. *Chemosphere*, *205*, 649–661. doi:10.1016/j.chemosphere.2018.04.142 PMID:29723723

Stafford, H. A., Magaldi, A., & Vennesland, B. (1954). Enzymatic oxidation of DPNH by diketosuccinate and dihydroxyfumarate. *Science*, *120*(3111), 265–266. doi:10.1126cience.120.3111.265 PMID:13186829

Stepanow, A., & Kusin, A. (1934). Synthesis of a carbohydrate chain by means of enzymes. 6. Resynthesis of glycogen in muscle. *Bernd. deutsch. chem. Gesellsch*, *67*, 723–726.

Stephens, P. D., Devlin, F. J., Chabalowski, C. F., & Frisch, M. J. (1994). Ab Initio calculation of vibrational absorption and circular dichroism spectra using density functional force fields. *Journal of Physical Chemistry*, *98*(45), 11623–11627. doi:10.1021/j100096a001

Stevens-Barrón, J. C., Wall-Medrano, A., Álvarez-Parrilla, E., Olivas-Armendáriz, I., Astiazaran-García, H., Robles-Zepeda, R. E., & De la Rosa, L. A. (2022). Synergistic Interactions between Tocol and Phenolic Extracts from Different Tree Nut Species against Human Cancer Cell Lines. *Molecules (Basel, Switzerland)*, *27*(10), 3154. doi:10.3390/molecules27103154 PMID:35630629

Stieg, D. C., Wang, Y., Liu, L.-Z., & Jiang, B.-H. (2022). ROS and miRNA dysregulation in ovarian cancer development, angiogenesis and therapeutic resistance. *International Journal of Molecular Sciences*, *23*(12), 6702–6718. doi:10.3390/ijms23126702 PMID:35743145

Stolarczyk, K., Rogalski, J., & Bilewicz, R. (2020). NAD(P)-dependent glucose dehydrogenase: Applications for biosensors, bioelectrodes, and biofuel cells. *Bioelectrochemistry (Amsterdam, Netherlands), 135*, 107574. doi:10.1016/j.bioelechem.2020.107574 PMID:32498025

Stone, J. R., & Yang, S. (2006). Hydrogen peroxide: A signaling messenger. *Antioxidants & Redox Signaling, 8*(3-4), 243–270. doi:10.1089/ars.2006.8.243 PMID:16677071

Stone, J., Edgar, J. O., Gould, J. A., & Telling, J. (2022). Tectonically-driven oxidant production in the hot biosphere. *Nature Communications, 13*(1), 1–10. doi:10.103841467-022-32129-y PMID:35941147

Strauss, R. S. (2001). Environmental tobacco smoke and serum vitamin C levels in children. *Pediatrics, 107*(3), 540–542. doi:10.1542/peds.107.3.540 PMID:11230596

Stultz, L. K., Hunsucker, A., Middleton, S., Grovenstein, E., O'Leary, J., Blatt, E., Miller, M., Mobley, J., & Hanson, P. K. (2020). Proteomic analysis of the *S. cerevisiae* Response to the anticancer ruthenium complex KP1019. *Metallomics, 12*(6), 876–890. doi:10.1039/d0mt00008f PMID:32329475

Sugawara, T., Matsushita, M. M., Izuoka, A., Wada, N., Takeda, N., & Ishikawa, M. (1994). An organic ferromagnet: A-phase crystal of 2-(2′, 5′-dihydroxyphenyl)-4, 4, 5, 5-tetramethyl-4, 5-dihydro-1 H-imidazolyl-1-oxy-3-oxide (α-HQNN). *Journal of the Chemical Society. Chemical Communications*, (14), 1723–1724. doi:10.1039/C39940001723

Sumner, J. B., & Dounce, A. L. (1937). Crystalline catalase. *Science, 85*(2206), 366–367. doi:10.1126cience.85.2206.366 PMID:17776781

Sung, H. J., Khan, M. F., & Kim, Y. H. (2019). Recombinant lignin peroxidase-catalyzed decolorization of melanin using in-situ generated H_2O_2 for application in whitening cosmetics. *International Journal of Biological Macromolecules, 136*, 20–26. doi:10.1016/j.ijbiomac.2019.06.026 PMID:31175901

Sun, K., Hong, D., Liu, J., Latif, A., Li, S., Chu, G., Qin, W., & Si, Y. (2021). Trametes versicolor laccase-assisted oxidative coupling of estrogens: Conversion kinetics, linking mechanisms, and practical applications in water purification. *The Science of the Total Environment, 782*, 146917. doi:10.1016/j.scitotenv.2021.146917

Sun, K., Li, S., Si, Y., & Huang, Q. (2021). Advances in laccase-triggered anabolism for biotechnology applications. *Critical Reviews in Biotechnology, 41*(7), 969–993. doi:10.1080/07388551.2021.1895053 PMID:33818232

Sun, K., Xu, W., Lin, X., Tian, S., Lin, W. F., Zhou, D., & Sun, X. (2021). Electrochemical oxygen reduction to hydrogen peroxide via a two-electron transfer pathway on carbon-based single-atom catalysts. *Advanced Materials Interfaces, 8*(8), 2001360. doi:10.1002/admi.202001360

Suppan P. (1992). The Marcus inverted region. *Photoinduced Electron Transfer IV*, 95-130.

Surpateanu, M., Zaharia, C. (2013-2014). Determinarea substantelor oxidabile din apă. *Chimia Mediului: teste de control în laborator și probleme,* (1), 45-46.

Suryadi, H., Judono, J. J., Putri, M. R., Eclessia, A. D., Ulhaq, J. M., Agustina, D. N., & Sumiati, T. (2022). Biodelignification of lignocellulose using ligninolytic enzymes from white-rot fungi. *Heliyon, 8*(2), e08865. doi:10.1016/j.heliyon.2022.e08865 PMID:35141441

Sutin, N. (1986). Marcus' contributions to electron-transfer theory. *Journal of Physical Chemistry, 90*(16), 3465–3466. doi:10.1021/j100407a002

Suzen, S., Gurer-Orhan, H., & Saso, L. (2017). Detection of reactive oxygen and nitrogen species by electron paramagnetic resonance (EPR) technique. *Molecules (Basel, Switzerland), 22*(1), 181–189. doi:10.3390/molecules22010181 PMID:28117726

Swedin, B., & Theorell, H. (1940). Dioximaleic acid oxidase action of peroxidases. *Nature, 145*(3663), 71–72. doi:10.1038/145071b0

Sychev, A., & Travin, S. Duca Gh, & Skurlatov Yu. (1983). Catalytic reactions and the protection of the environment. Chisinau: Stiinta Publishing.

Sychev, A., Duca, Gh. (1983). Cataliza oxidării acidului dihidroxifumaric. V. Mecanismul oxidării acidului dihidroxifumaric cu oxigenul în prezența ionilor de cupru. *Revista de chimie fizică (în limba rusă), 57*(8), p. 1938-1941.

Sychev, A., Duca, Gh. (1985). Utilizarea acidului dihidroxifumaric pentru îmbunătățirea calității vinurilor seci. *Horticultura, viticultura si vinificatie in Moldova, 12,* p. 38-40.

Sychev, A., Postolatii, T., Duca, Gh., Gonta, M. (1985). *Metodă de stabilizare a materialelor vitivinicole*. Moscow, URSS.

Sychev, A., Travin, C., Duca, Gh., & Skurlatov, Iu. (1983). *Catalytic reactions and environmental protection*. Stiinta.

Syed, A. J., & Anderson, J. C. (2021). Applications of bioluminescence in biotechnology and beyond. *Chemical Society Reviews, 50*(9), 5668–5705. doi:10.1039/D0CS01492C PMID:33735357

Szeląg, M., Urbaniak, A., & Bluyssen, H. A. R. (2014). A theoretical antioxidant pharmacophore for natural hydroxycinnamic acids. *Open Chemistry, 13*(1), 000010151520150001. doi:10.1515/chem-2015-0001

Tachikawa, H. (2021). Formation of hydrogen peroxide from $O^{-}(H_2O)_n$ clusters. *The Journal of Physical Chemistry A, 125*(21), 4598–4605. doi:10.1021/acs.jpca.1c02883 PMID:34033487

Tamura, M., Nakazawa, Y., Shiomi, D., Nozawa, K., Hosokoshi, Y., Ishikawa, M., Takahashi, M., & Kinoshita, M. (1991). Bulk ferromagnetism in the β-phase crystal of the p-nitrophenyl nitronyl nitroxide radical. *Chemical Physics Letters, 186*(4-5), 401–404. doi:10.1016/0009-2614(91)90198-I

Tanaka, H., Matsuda, T., Miyagantani, Y., Yukioka, T., Matsuda, H., & Shimazaki, S. (2000). Reduction of resuscitation fluid volumes in severely burned patients using ascorbic acid administration: A randomized, prospective study. *Archives of Surgery, 135*(3), 326–331. doi:10.1001/archsurg.135.3.326 PMID:10722036

Taube, H. (1984). Electron Transfer between Metal Complexes— A Retrospective View(Nobel Lecture). *Angewandte Chemie International Edition in English, 23*(5), 329–339. doi:10.1002/anie.198403293

Tavadyan, L. A., & Minasyan, S. H. (2019). Synergistic and antagonistic co-antioxidant effects of flavonoids with trolox or ascorbic acid in a binary mixture. *Journal of Chemical Sciences, 131*(5), 1–11. doi:10.100712039-019-1618-5

Tavasol, F., Tabatabaie, T., Ramavandi, B., & Amiri, F. (2020). Design a new photocatalyst of sea sediment/titanate to remove cephalexin antibiotic from aqueous media in the presence of sonication/ultraviolet/hydrogen peroxide: Pathway and mechanism for degradation. *Ultrasonics Sonochemistry, 65*, 105062. doi:10.1016/j.ultsonch.2020.105062 PMID:32172148

Tawalbeh, M., Javed, R. M. N., Al-Othman, A., & Almomani, F. (2022). The novel advancements of nanomaterials in biofuel cells with a focus on electrodes' applications. *Fuel, 322*, 124237. doi:10.1016/j.fuel.2022.124237

Telang, P. (2013). Vitamin C in dermatology. *Indian Dermatology Online Journal, 4*(2), 143. doi:10.4103/2229-5178.110593 PMID:23741676

Teshima, N., Itabashi, H., & Kawashima, T. (1992). Simultaneous flow injection determination of vanadium (IV) and (V) based on redox reactions of vanadium (IV) with iron (III) and vanadium (V) with iron (II) in the presence of 1, 10-phenanthroline and diphosphate. *Chemistry Letters, 21*(11), 2227–2230. doi:10.1246/cl.1992.2227

Teshima, N., Itabashi, H., & Kawashima, T. (1993). Reverse flow injection analysis of complexing agents and its application to estimation of complexing capacity. *Talanta, 40*(1), 101–106. doi:10.1016/0039-9140(93)80147-J PMID:18965601

Theil, S., Yerande, R., Chikate, R., Dahan, F., Bousseksou, A., Padhye, S., & Tuchagues, J.-P. (1997). Synthesis, structure, and magnetic and redox properties of linear bis-dinuclear complexes afforded by schiff base ligants containing catecholate and pyridine or imidazole groups. *Inorganic Chemistry, 36*(27), 6279–6286. doi:10.1021/ic9703538

Thews, O., & Riemann, A. (2019). Tumor pH and metastasis: A malignant process beyond hypoxia. *Cancer and Metastasis Reviews, 38*(1-2), 113–129. doi:10.100710555-018-09777-y PMID:30607627

Thorsteinsson, T., & Cooper, D. L. (1998). Modern valence bond descriptions of molecular excited states: An application of CASVB. *International Journal of Quantum Chemistry, 70*(4-5), 637–650. doi:10.1002/(SICI)1097-461X(1998)70:4/5<637::AID-QUA10>3.0.CO;2-#

Tian, X., Luo, T., Nie, Y., Shi, J., Tian, Y., Dionysiou, D. D., & Wang, Y. (2022). New insight into a Fenton-like reaction mechanism over sulfidated β-FeOOH: Key role of sulfidation in efficient iron (III) reduction and sulfate radical generation. *Environmental Science & Technology, 56*(9), 5542–5551. doi:10.1021/acs.est.2c00132 PMID:35412804

Timoshnikov, V. A., Kobzeva, T. V., Polyakov, N. E., & Kontoghiorghes, G. J. (2020). Redox interactions of vitamin C and iron: Inhibition of the pro-oxidant activity by deferiprone. *International Journal of Molecular Sciences, 21*(11), 3967. doi:10.3390/ijms21113967 PMID:32486511

Tina Suksmasari, B. H. (2015). Multivitamin Supplementation Supports Immune Function and Ameliorates Conditions Triggered by Reduced Air Quality. *Vitamins & Minerals, 4*(2). Advance online publication. doi:10.4172/2376-1318.1000128

Titov, V., Mosyagin, N. S., Petrov, A. N., Isaev, T. A., & DeMille, D. (2006). P,T-Parity Violation Effects in Polar Heavy-Atom Molecules. In J.-P. Julien (Eds.), *Recent Advances in the Theory of Chemical and Physical Systems. Progress in Theoretical Chemistry and Physics* (Vol. 15, pp. 253–283). Springer.

Toader, A. M., Buta, C. M., Frecus, B., Mischie, A., & Cimpoesu, F. (2019). Valence Bond Account of Triangular Polyaromatic Hydrocarbons with Spin: Combining Ab Initio and Phenomenological Approaches. *The Journal of Physical Chemistry C, 123*(11), 6869–6880. doi:10.1021/acs.jpcc.8b12250

Tobimatsu, Y., & Schuetz, M. (2019). Lignin polymerization: How do plants manage the chemistry so well? *Current Opinion in Biotechnology, 56*, 75–81. doi:10.1016/j.copbio.2018.10.001 PMID:30359808

Tochetto, G. A., Aragão, A. M. I., de Oliveira, D., & Immich, A. P. S. (2022). Can enzymatic processes transform textile processes? A critical analysis of the industrial application. *Process Biochemistry (Barking, London, England), 123*, 27–35. doi:10.1016/j.procbio.2022.10.030

Tong, R., Wu, X., Liu, Y., Liu, Y., Zhou, J., Jiang, X., Zhang, L., He, X., & Ma, L. (2020). Curcumin-induced DNA demethylation in human gastric cancer cells is mediated by the DNA-damage response pathway. *Oxidative Medicine and Cellular Longevity, 2020*, 1–13. doi:10.1155/2020/2543504 PMID:32617134

Traikovich, S. S. (1999). Use of topical ascorbic acid and its effects on photodamaged skin topography. *Archives of Otolaryngology—Head & Neck Surgery, 125*(10), 1091–1098. doi:10.1001/archotol.125.10.1091 PMID:10522500

Tran, N., Reinhard, M., & Gin, K. (2018). Occurrence and fate of emerging contaminants in municipal wastewater treatment plants from different geographical regions - A review. *Water Resources, 133*, 182–207. PMID:29407700

Travin S. O. (2018) Modern methods of mathematical modeling of chemical and biological processes in the environment. *Chemical safety, 2*(1), 22-49.

Travin, S., Duca, G., & Gladchi, V. (2019). Self-purification of aquatic media from hexachlorocyclohexane in a radical process. *Chemistry Journal of Moldova, 14*(1), 47–53. doi:10.19261/cjm.2018.537

Triki, S., Gomez-Garcia, C. J., Ruiz, E., & Sala-Pala, J. (2005). Asymmetric azido-copper(II) bridges: Ferro- or antiferromagnetic? Experimental and theoretical magneto-structural studies. *Inorganic Chemistry*, *44*(15), 5501–5508. doi:10.1021/ic0504543 PMID:16022548

Tritsch, J. R., Chan, W. L., Wu, X., Monahan, N. R., & Zhu, X. Y. (2013). Harvesting singlet fission for solar energy conversion via triplet energy transfer. *Nature Communications*, *4*(1), 1–7. doi:10.1038/ncomms3679 PMID:24145737

Troiani, F., & Affronte, M. (2011). Molecular spins for quantum information technologies. *Chemical Society Reviews*, *40*(6), 3119–3129. doi:10.1039/c0cs00158a PMID:21336365

Tsao, R. (2015). Synergistic interactions between antioxidants used in food preservation. In F. Shahidi (Ed.), *Handbook of antioxidants for food preservation* (pp. 335–347). Woodhead Publishing. doi:10.1016/B978-1-78242-089-7.00013-0

Tsuneda, T., & Taketsugu, T. (2018). Theoretical investigations on hydrogen peroxide decomposition in aquo. *Physical Chemistry Chemical Physics*, *20*(38), 24992–24999. doi:10.1039/C8CP04299C PMID:30239545

Turky, A. O., Barhoum, A., MohamedRashad, M., & Bechlany, M. (2017). Enhanced the structure and optical properties for ZnO/PVP nanofibers fabricated via electrospinning technique. *Journal of Materials Science Materials in Electronics*, *28*(23), 17526–17532. doi:10.100710854-017-7688-6

Umetsu, K., Itabashi, H., Satoh, K., & Kawashima, T. (1991). Effect of ligands on the redox reaction of metal ions and the use of a ligand buffer for improving the end-point detection in the potentiometric titration of vanadium (V) with iron (II). *Analytical Sciences*, *7*(1), 115–118. doi:10.2116/analsci.7.115

Undiano, E., Roman, R., Miranda-Molina, A., & Ayala, M. (2022). Halogenation of estrogens catalysed by a fungal chloroperoxidase. *Natural Product Research*, *36*(20), 5353–5357. doi:10.1080/14786419.2021.1925269 PMID:34000929

Ushio-Fukai, M., Ash, D., Nagarkoti, S., Belin de Chantemele, E. J., Fulton, D. J. R., & Fukai, T. (2021). Interplay between reactive oxygen/reactive nitrogen species and metabolism in vascular biology and disease. *Antioxidants & Redox Signaling*, *34*(16), 1319–1354. doi:10.1089/ars.2020.8161 PMID:33899493

Usselman, R. J., Fielding, A. J., Frerman, F. E., Watmough, N. J., Eaton, G. R., & Eaton, S. S. (2008). Impact of mutations on the midpoint potential of the $[4Fe\text{-}4S]^{+1,+2}$ cluster and on catalytic activity in electron transfer flavoprotein-ubiquinone oxidoreductase (ETF-QO). *Biochemistry*, *47*(1), 92–100. doi:10.1021/bi701859s PMID:18069858

Vacarciuc, L. (2008). Potentialul biotehnologic la producerea vinurilor ecologice roze in moldova. *Agricultura – Stiinta si Practica, 57*, 1–8.

Valadbeigi, Y., Farrokhpour, H., & Tabrizchi, M. (2014). Effect of hydration on the kinetics of proton-bound dimer formation: Experimental and theoretical study. *The Journal of Physical Chemistry A*, *118*(36), 7663–7671. doi:10.1021/jp506140m PMID:25137125

Valadbeigi, Y., Ilbeigi, V., Michalczuk, B., Sabo, M., & Matejcik, S. (2019). Effect of basicity and structure on the hydration of protonated molecules, proton-bound dimer and cluster formation: An ion mobility-time of flight mass spectrometry and theoretical study. *Journal of the American Society for Mass Spectrometry*, *30*(7), 1242–1253. doi:10.100713361-019-02180-z PMID:31049871

Valton, J., Mathevon, C., Fontecave, M., Nivière, V., & Ballou, D. P. (2008). Mechanism and regulation of the two-component FMN-dependent monooxygenase ActVA-ActVB from Streptomyces coelicolor. *The Journal of Biological Chemistry*, *283*(16), 10287–10296. doi:10.1074/jbc.M709730200 PMID:18245777

Van Niel, C. B. (1932). On the morphology and physiology of the purple and green sulphur bacteria. *Archives of Microbiology*, *3*(1), 1–112. doi:10.1007/BF00454965 van Vleck, J. H., & Sherman, A. (1935). The Quantum Theory of Valence. *Reviews of Modern Physics*, *7*(3), 167–228. doi:10.1103/RevModPhys.7.167

Vasanth Kumar, K. (2006). Linear and non-linear regression analysis for the sorption kinetics of methylene blue onto activated carbon. *Journal of Hazardous Materials*, *137*(3), 1538–1544. doi:10.1016/j.jhazmat.2006.04.036 PMID:16730890

Vasanth Kumar, K., Porkodi, K., & Rocha, F. (2008). Langmuir–Hinshelwood kinetics – A theoretical study. *Catalysis Communications*, *9*(1), 82–84. doi:10.1016/j.catcom.2007.05.019

Vaseashta, A. (2007). Controlled formation of multiple Taylor cones in electrospinning process. *Applied Physics Letters*, *90*(9), 093115. doi:10.1063/1.2709958

Vaseashta, A., Duca, G., & Travin, S. (Eds.). (2022). *Handbook of Research on Water Sciences and Society* (Vols. 1–2). IGI Global. doi:10.4018/978-1-7998-7356-3

Vaseashta, A., Gevorgyan, G., Kavaz, D., Ivanov, O., Jawaid, M., & Vasović, D. (2021). Exposome, Biomonitoring, Assessment and Data Analytics to Quantify Universal Water Quality. In A. Vaseashta & C. Maftei (Eds.), *Water Safety, Security and Sustainability. Advanced Sciences and Technologies for Security Applications*. Springer. doi:10.1007/978-3-030-76008-3_4

Vaseashta, A., & Stamatin, I. (2007). Electrospun polymers for controlled release of drugs, vaccine delivery, and system-on-fibers. *Journal of Optoelectronics and Advanced Materials*, *9*(6), 1606–1613.

Veerabhadraswamy, M., Devaraj, T. D., & Jayanna, B. K. (2018). Second derivative spectrophotometric determination of iron (II) and ruthenium (III) using 1, 10-phenanthroline. *Analytical Chemistry Letters*, *8*(6), 757–768. doi:10.1080/22297928.2018.1548944

Velempini, T., Prabakaran, E., & Pillay, K. (2021). Recent developments in the use of metal oxides for photocatalytic degradation of pharmaceutical pollutants in water—A review. *Materials Today. Chemistry*, *19*, 100380. doi:10.1016/j.mtchem.2020.100380

Veprek-Siska, J. (1979). The role of metal ions in oxygen activation. *Acta Biologica et Medica Germanica*, *38*(2-3), 357–361. PMID:517007

Verdaguer, M., Bleuzen, A., Marvaud, V., Vaissermann, J., Seuleiman, M., Desplanches, C., Scuiller, A., Train, C., Garde, R., Gelly, G., Lomenech, C., Rosenman, I., Veillet, P., Cartier, C., & Villain, F. (1999). Molecules to build solids: High TC molecule-based magnets by design and recent revival of cyano complexes chemistry. *Coordination Chemistry Reviews*, *190–192*, 1023–1047. doi:10.1016/S0010-8545(99)00156-3

Verma, M., & Haritash, A. K. (2020). Review of advanced oxidation processes (AOPs) for treatment of pharmaceutical wastewater. *Advances in Environmental Research*, *9*(1), 1–17.

Vesselli, E. (2021). Stabilization and activation of molecular oxygen at biomimetic tetrapyrroles on surfaces: From UHV to near-ambient pressure. *Nanoscale Advances*, *3*(5), 1319–1330. doi:10.1039/D0NA00827C PMID:36132852

Vicol, C. (2022). Study of synergic and anti-synergic interactions of some grape antioxidants with organic acids. In *7th International Conference "Ecological and Environmental Chemistry – 2022"*. 56. Chisinau, CEP MSU.

Vicol, C., & Duca, G. (2020). Influența acidului tartric asupra procesului de anihilare a radicalilor liberi de către acidul ascorbic [The influence of tartaric acid on the free radical scavenging activity of ascorbic acid]. *Revista de Ştiinţă, Inovare, Cultură şi Artă „Akademos*, *56*(1), 39-43.

Vicol, C., Cimpoiu, C., & Duca, G. (2021a). Investigation of synergic/anti-synergic interactions of dihydroxifumaric acid and ascorbic acid with DPPH. *Studia Universitatis Babes-Bolyai. Chemia*, *66*(2), 49–58. doi:10.24193ubbchem.2021.02.04

Vicol, C., Gorincioi, E., & Duca, G. (2021b). 13C NMR investigations of radical scavenging activities of ascorbic acid and resveratrol using 1, 1-diphenyl-2-picrylhydrazyl radical. In *XXIIIrd International Conference New Cryogenic and Isotope Technologies for Energy and Environment.* – EnergEn.

Vicol, C., Iacobescu, I., Fifere, A., & Duca, G. (2022). Study of the synergic interactions of ascorbic and dihydroxyfumaric acids by EPR spectroscopy. In *14th Green Chemistry Postgraduate Summer School (online/in-person.* 90. Venice, Green Sciences for Sustainable Development Foundation. Wang, S., Meckling, K. A., Marcone, M. F., Kakuda, Y., & Tsao, R. (2011). Synergistic, additive, and antagonistic effects of food mixtures on total antioxidant capacities. *Journal of Agricultural and Food Chemistry, 59*(3), 960–968.

Viegas-Junior, C., Danuello, A., Da Silva Bolzani, V., Barreiro, E. J., & Fraga, C. A. M. (2007). Molecular hybridization: A useful tool in the design of new drug prototypes. *Current Medicinal Chemistry, 14*(17), 1829–1852. doi:10.2174/092986707781058805

Villalba-Rodríguez, A. M., Parra-Arroyo, L., González-González, R. B., Parra-Saldívar, R., Bilal, M., & Iqbal, H. M. N. (2022). Laccase-assisted biosensing constructs - Robust modalities to detect and remove environmental contaminants. *Case Studies in Chemical and Environmental Engineering, 5*, 100180. doi:10.1016/j.cscee.2022.100180

Villaverde, J. J., Sevilla-Morán, B., López-Goti, C., Alonso-Prados, J. L., & Sandín-España, P. (2020). QSAR/QSPR models based on quantum chemistry for risk assessment of pesticides according to current European legislation. *SAR and QSAR in Environmental Research, 31*(1), 49–72. doi:10.1080/1062936X.2019.1692368 PMID:31766890

Virk, A. P., Sharma, P., & Capalash, N. (2012). Use of laccase in pulp and paper industry. *Biotechnology Progress, 28*(1), 21–32. doi:10.1002/btpr.727 PMID:22012940

Vitulli, G., Bernini, M., Bertozzi, S., Pitzalis, E., Salvadori, P., Coluccia, S., & Martra, G. (2002). Nanoscale copper particles derived from solvated Cu atoms in the activation of molecular oxygen. *Chemistry of Materials, 14*(3), 1183–1186. doi:10.1021/cm011199x

Volkov, A. G., Bookal, A., Hairston, J. S., Roberts, J., Taengwa, G., & Patel, D. (2021). Mechanisms of multielectron reactions at the plasma/water interface: Interfacial catalysis, RONS, nitrogen fixation, and plasma-activated water. *Electrochimica Acta, 385*, 138441. doi:10.1016/j.electacta.2021.138441

Wada, S. (2009). Chemoprevention of tocotrienols: The mechanism of antiproliferative effects. *Forum of Nutrition, 61*, 204–216. doi:10.1159/000212752 PMID:19367124

Walsh, C. T., & Wencewicz, T. A. (2013). Flavoenzymes: Versatile catalysts in biosynthetic pathways. *Natural Product Reports, 30*(1), 175–200. doi:10.1039/C2NP20069D PMID:23051833

Wang, F., Liu, Y., Du, C., & Gao, R. (2022). Current Strategies for Real-Time Enzyme Activation. *Biomolecules, 12*(5), 599. doi:10.3390/biom12050599 PMID:35625527

Wang, J., Ding, T., Gao, K., Wang, L., Zhou, P., & Wu, K. (2021). Marcus inverted the region of charge transfer from low-dimensional semiconductor materials. *Nature Communications, 12*(1), 1–9. doi:10.103841467-021-26705-x PMID:34732730

Wang, J., Ding, T., & Wu, K. (2020). Coulomb Barrier for Sequential Two-Electron Transfer in a Nanoengineered Photocatalyst. *Journal of the American Chemical Society, 142*(32), 13934–13940. doi:10.1021/jacs.0c06256 PMID:32672949

Wang, J., & Wang, S. (2016). Removal of pharmaceuticals and personal care products (PPCPs) from wastewater: A review. *Journal of Environmental Management, 182*, 620–640. doi:10.1016/j.jenvman.2016.07.049 PMID:27552641

Wang, M., Chen, B., Fang, Y., & Tan, T. (2017). Cofactor engineering for more efficient production of chemicals and biofuels. *Biotechnology Advances*, *35*(8), 1032–1039. doi:10.1016/j.biotechadv.2017.09.008 PMID:28939499

Wang, S., Bai, J., Liang, H., Xu, T., Li, C., Sun, W., & Liu, H. (2014). Synthesis, characterization, and photocatalytic properties of Ag/TiO_2 composite nanofibers prepared by electrospinning. *Journal of Dispersion Science and Technology*, *35*(6), 777–782. doi:10.1080/01932691.2013.813392

Wang, W., & Lee, D. J. (2021). Direct interspecies electron transfer mechanism in enhanced methanogenesis: A mini review. *Bioresource Technology*, *330*, 124980. doi:10.1016/j.biortech.2021.124980 PMID:33743275

Wang, W., Yang, R., Li, T., Komarneni, S., & Liu, B. (2021a). Advances in recyclable and superior photocatalytic fibers: Material, construction, application and future perspective. *Composites. Part B, Engineering*, *205*, 108512. doi:10.1016/j.compositesb.2020.108512

Wang, X. Y., Lin, C., Chang, W. J., Huang, Y. H., & Mi, F. L. (2022). Thiolated hyaluronic acid and catalase-enhanced CD44-targeting and oxygen self-supplying nanoplatforms with photothermal/photodynamic effects against hypoxic breast cancer cells. *International Journal of Biological Macromolecules*, *221*, 121–134. doi:10.1016/j.ijbiomac.2022.08.164 PMID:36049568

Wang, Y., Du, C., Liu, Z., Pei, K., Zhang, Y., & Qi, W. (2022). Chemiluminescence "Turn-on" detection of tyrosinase activity via in situ generation of dopamine based on a lucigenin and riboflavin system. *New Journal of Chemistry*, *46*(9), 4156–4161. doi:10.1039/D1NJ05628J

Wang, Y., Liu, Y., & Bersuker, I. B. (2019). Sudden polarization and zwitterion formation as a pseudo-Jahn–Teller effect: A new insight into the photochemistry of alkenes. *Physical Chemistry Chemical Physics*, *21*(20), 10677–10692. doi:10.1039/C9CP01023H PMID:31086864

Wang, Y., Qi, H., Liu, Y., Duan, C., Liu, X., Xia, T., Chen, D., Piao, H., & Liu, H.-X. (2021). The double-edged roles of ROS in cancer prevention and therapy. *Theranostics*, *11*(10), 4839–4857. doi:10.7150/thno.56747 PMID:33754031

Wang, Y., Yokota, T., & Someya, T. (2021b). Electrospun nanofiber-based soft electronics. *NPG Asia Materials*, *13*(1), 22. doi:10.103841427-020-00267-8

Wasak, A., Drozd, R., Jankowiak, D., & Rakoczy, R. (2019). Rotating magnetic field as tool for enhancing enzymes properties - laccase case study. *Scientific Reports*, *9*(1), 3707. doi:10.103841598-019-39198-y PMID:30842482

Watabe, S., Sakamoto, Y., Morikawa, M., Okada, R., Miura, T., & Ito, E. (2011). Highly sensitive determination of hydrogen peroxide and glucose by fluorescence correlation spectroscopy. *PLoS One*, *6*(8), e22955–e22959. doi:10.1371/journal.pone.0022955 PMID:21850246

Watmough, N. J., & Frerman, F. E. (2010). The electron transfer flavoprotein: Ubiquinone oxidoreductases. *Biochimica et Biophysica Acta (BBA) -. Bioenergetics*, *1797*(12), 1910–1916. doi:10.1016/j.bbabio.2010.10.007 PMID:20937244

Watson, S. A., & McStay, G. P. (2020). Functions of cytochrome c oxidase assembly factors. *International Journal of Molecular Sciences*, *21*(19), 7254. doi:10.3390/ijms21197254 PMID:33008142

Wayner, D. D. M., Burton, G. W., & Ingold, K. U. (1986). The antioxidant efficiency of vitamin C is concentration-dependent. *Biochimica et Biophysica Acta*, *884*(1), 119–123. doi:10.1016/0304-4165(86)90234-5 PMID:3768406

Wei, F., Kuang, S., Rees, T. W., Liao, X., Liu, J., Luo, D., Wang, J., Zhang, X., Ji, L., & Chao, H. (2021). Ruthenium(II) complexes coordinated to graphitic carbon nitride: Oxygen self-sufficient photosensitizers which produce multiple ROS for photodynamic therapy in hypoxia. *Biomaterials*, *276*, 121064. doi:10.1016/j.biomaterials.2021.121064 PMID:34391019

Weinberg, D. R., Gagliardi, C. J., Hull, J. F., Murphy, C. F., Kent, C. A., Westlake, B. C., Paul, A., Ess, D. H., McCafferty, D. G., & Meyer, T. J. (2012). Proton-coupled electron transfer. *Chemical Reviews, 112*(7), 4016–4093. doi:10.1021/cr200177j PMID:22702235

Werner, H., Steinmetz, M., Peters, K., & von Schnering, H. G. (1998). A novel route to formaldehyde and thioformaldehyde rhodium complexes. *European Journal of Inorganic Chemistry, 1998*(11), 1605–1617. doi:10.1002/(SICI)1099-0682(199811)1998:11<1605::AID-EJIC1605>3.0.CO;2-H

Westphal, A. H., Tischler, D., & van Berkel, W. J. H. (2021). Natural diversity of FAD-dependent 4-hydroxybenzoate hydroxylases. *Archives of Biochemistry and Biophysics, 702*, 108820. Advance online publication. doi:10.1016/j.abb.2021.108820 PMID:33684360

WHO. (2015). *Global Action Plan on Antimicrobial Resistance.* Geneva: World Health Organization. http://www.wpro.who.int/entity/drug_resistance/ resources/global_action_plan_eng.pdf

Wieland, H., & Franke, W. (1928). Über den mechanismus der oxydationsvorgänge. xiv. Die Aktivierung des sauerstoffs Durch Eisen. *Justus Liebigs Annalen der Chemie, 464*(1), 101–226. doi:10.1002/jlac.19284640109

Wigner, E. (1959). Theory Groups and Its Applications to the Quantum Mechanics of Atomic Spectra. Cambridge University Press.

Wigner, E. P. (1970). *Symmetries and Reflections.* Indiana University Press.

Wikström, M., Krab, K., & Sharma, V. (2018). Oxygen activation and energy conservation by cytochrome c oxidase. *Chemical Reviews, 118*(5), 2469–2490. doi:10.1021/acs.chemrev.7b00664 PMID:29350917

Willett, R. D., Gatteschi, D., & Kahn, O. (1985). *Magneto-structural correlations in exchange coupled systems.* Reidel Publishing Co.

Wilson, E. B., Decius, J. C., & Cross, P. C. (1980). *Molecular vibrations: the theory of infrared and Raman vibrational spectra.* Dover Publications.

Winkler, A., Kutchan, T. M., & Macheroux, P. (2007). 6-S-Cysteinylation of bi-covalently attached FAD in berberine bridge enzyme tunes the redox potential for optimal activity. *The Journal of Biological Chemistry, 282*(33), 24437–24443. doi:10.1074/jbc.M703642200 PMID:17573342

Wiśniewska, J., Fandzloch, M., & Łakomska, I. (2019). The reduction of ruthenium (III) complexes with triazolopyrimidine ligands by ascorbic acid and mechanistic insight into their action in anticancer therapy. *Inorganica Chimica Acta, 484*, 305–310. doi:10.1016/j.ica.2018.09.051

Wong, S. K., Chin, K. Y., & Ima-Nirwana, S. (2020). Vitamin C: A review on its role in the management of metabolic syndrome. International Journal of Medical Sciences, 17(11), 1625–1638. doi:10.7150/ijms.47103

Woodham, A. P., Meijer, G., & Fielicke, A. (2012). Activation of molecular oxygen by anionic gold clusters. *Angewandte Chemie International Edition, 51*(18), 4444–4447. doi:10.1002/anie.201108958 PMID:22461317

Woodham, A. P., Meijer, G., & Fielicke, A. (2013). Charge separation promoted activation of molecular oxygen by neutral gold clusters. *Journal of the American Chemical Society, 135*(5), 1727–1730. doi:10.1021/ja312223t PMID:23327337

Wu, G., Fang, Y.-Z., Yang, S., Lupton, J. R., & Turner, N. D. (2004). Glutathione metabolism and its implications for health. *The Journal of Nutrition, 134*(3), 489–492. doi:10.1093/jn/134.3.489 PMID:14988435

Xie, L., Liu, L., Wang, W., Ma, Z., Xu, L., Zhao, X., & Wang, H. (2018). The structures, spectroscopic properties, and photodynamic reactions of three [RuCl(QN)NO]-Complexes (HQN=8-hydroxyquinoline and its derivatives) as potential no-donating drugs. *Bioinorganic Chemistry and Applications*, *2018*, 1–10. doi:10.1155/2018/7029376 PMID:30627138

Xie, L., Wang, L., Guan, R., Ji, L., & Chao, H. (2021). Anti-metastasis and anti-proliferation effect of mitochondria-accumulating ruthenium(II) complexes via redox homeostasis disturbance and energy depletion. *Journal of Inorganic Biochemistry*, *217*, 111380–111388. doi:10.1016/j.jinorgbio.2021.111380 PMID:33578250

Xie, Y., Mo, W., Xu, D., Shen, Z., Sun, N., Hu, B., & Hu, X. (2007). Efficient NO equivalent for activation of molecular oxygen and its applications in transition-metal-free catalytic aerobic alcohol oxidation. *The Journal of Organic Chemistry*, *72*(11), 4288–4291. doi:10.1021/jo0705824 PMID:17447820

Xu, J., Duca, G., Ahmed, S. E., García Márquez, F. P., & Hajiyev, A. (Eds.). (2021). Proceedings of the Fourteenth International Conference on Management Science and Engineering Management. *Advances in Intelligent Systems and Computing*. Springer. 10.1007/978-3-030-49889-4

Xu, C., Zi, Y., Wang, A. C., Zou, H., Dai, Y., He, X., Wang, Z. L., Wang, Y.-C., Feng, P., Li, D., & Wang, Z. L. (2018). On the electron-transfer mechanism in the contact-electrification effect. *Advanced Materials*, *30*(15), 1706790. doi:10.1002/adma.201706790 PMID:29508454

Xu, D. M., Fu, R. B., Wang, J. X., Shi, Y. X., & Guo, X. P. (2021). Chemical stabilization remediation for heavy metals in contaminated soils on the latest decade: Available stabilizing materials and associated evaluation methods-A critical review. *Journal of Cleaner Production*, *321*, 128730. doi:10.1016/j.jclepro.2021.128730

Xu, H., & Ren, D. (2015). Lysosomal physiology. *Annual Review of Physiology*, *77*(1), 57–80. doi:10.1146/annurev-physiol-021014-071649 PMID:25668017

Xu, Q., Chen, G., Chen, G. M., Wu, H. L., Yang, Y. Y., Mai, Z. Y., Sun, R., Luan, P., Guo, C. W., Yu, M., Peng, Z. W., & Yu, Z. Q. (2022). NO-dependent vasodilation and deep tumor penetration for cascade-amplified antitumor performance. *Journal of Controlled Release*, *347*, 389–399. doi:10.1016/j.jconrel.2022.05.022 PMID:35569586

Xu, T., Ding, W., Ji, X. Y., Ao, X., Liu, Y., Yu, W. P., & Wang, J. X. (2019). Oxidative stress in cell death and cardiovascular diseases. *Oxidative Medicine and Cellular Longevity*, *2019*, 1–11. doi:10.1155/2019/9030563 PMID:31781356

Xv, F., Chen, J., Duan, L., & Li, S. (2018). Research progress on the anticancer effects of Vitamin K2 [Review]. *Oncology Letters*, *15*(6), 8926–8934. doi:10.3892/ol.2018.8502 PMID:29805627

Yang, C. H., Zhang, Q., Zhu, W. Q., Shi, Y., Cao, H. H., Guo, L., Chu, D., Lu, Z., & Liu, T. X. (2022). Involvement of Laccase2 in Cuticle Sclerotization of the Whitefly, Bemisia tabaci Middle East-Asia Minor. *Insects*, *13*(5), 471. doi:10.3390/insects13050471 PMID:35621805

Yang, J., Feng, L., Pi, S., Cui, D., Ma, F., Zhao, H. P., & Li, A. (2020). A critical review of aerobic denitrification: Insights into the intracellular electron transfer. *The Science of the Total Environment*, *731*, 139080. doi:10.1016/j.scitotenv.2020.139080 PMID:32417477

Yang, J.-H., Yu, K., Si, X.-K., Li, S., Cao, Y.-J., Li, W., & Zhang, J.-X. (2019). Liensinine inhibited gastric cancer cell growth through ROS generation and the PI3K/AKT pathway. *Journal of Cancer*, *10*(25), 6431–6438. doi:10.7150/jca.32691 PMID:31772676

Yang, Z., Min, Z. J., & Yu, B. (2020). Reactive oxygen species and immune regulation. *International Reviews of Immunology*, *39*(6), 292–298. doi:10.1080/08830185.2020.1768251 PMID:32423322

Yan, X., Tang, J., Ma, S., Tanner, D., Ludwig, R., Ulstrup, J., & Xiao, X. (2022). Engineering bio-interfaces for the direct electron transfer of Myriococcum thermophilum cellobiose dehydrogenase: Towards a mediator-less biosupercapacitor/ biofuel cell hybrid. *Biosensors & Bioelectronics*, *210*, 114337. doi:10.1016/j.bios.2022.114337 PMID:35537312

Yee, G. M., & Tolman, W. B. (2015). Transition Metal Complexes and the Activation of Dioxygen. In P. Kroneck & M. Sosa Torres (Eds.), *Sustaining Life on Planet Earth: Metalloenzymes Mastering Dioxygen and Other Chewy Gases. Metal Ions in Life Sciences* (Vol. 15). Springer. doi:10.1007/978-3-319-12415-5_5

Yen, G. C., Duh, P. D., Tsai, H. L., & Huang, S. L. (2003). Pro-oxidative properties of flavonoids in human lymphocytes. *Bioscience, Biotechnology, and Biochemistry*, *67*(6), 1215–1222. doi:10.1271/bbb.67.1215 PMID:12843645

Yeum, K. J., Beretta, G., Krinsky, N. I., Russell, R. M., & Aldini, G. (2009). Synergistic interactions of antioxidant nutrients in a biological model system. *Nutrition (Burbank, Los Angeles County, Calif.)*, *25*(7-8), 839–846. doi:10.1016/j.nut.2009.01.011 PMID:19398299

Yoneda, Y., Mora, S. J., Shee, J., Wadsworth, B. L., Arsenault, E. A., Hait, D., & Fleming, G. R. (2021). Electron–nuclear dynamics accompanying proton-coupled electron transfer. *Journal of the American Chemical Society*, *143*(8), 3104–3112. doi:10.1021/jacs.0c10626 PMID:33601880

Yordanov, N. (1996). Is Our Knowledge about the Chemical and Physical Properties of DPPH Enough to Consider It as a Primary Standard for Quantitative EPR Spectrometry. *Applied Magnetic Resonance*, *10*(1), 339–350. doi:10.1007/BF03163117

Yordi, E. G., Perez, E. M., Matos, M. J., & Villares, E. U. (2012). Antioxidant and pro-oxidant effects of polyphenolic compounds and structure–activity relationship evidence. In J. Bouayed & T. Bohn (Eds.), *Nutrition, Well-Being and Health* (pp. 23–48). InTech.

Younus, H. (2019). Oxidoreductases: Overview and practical applications. In Biocatalysis (pp. 39–55). Springer International Publishing. doi:10.1007/978-3-030-25023-2_3

Yuan, S., Chen, S., Wu, H., Jiang, H., Zheng, S., Zhang, Q., & Liu, Y. (2020). NAMI-A preferentially reacts with the Sp1 protein: Understanding the anti-metastasis effect of the drug. *Chemical Communications*, *56*(9), 1397–1400. doi:10.1039/C9CC08775C PMID:31912815

Yu, C. Y., & Yang, Z. (2011). A systemic investigation of hydrogen peroxide clusters $(H_2O_2)_n$ (n = 1–6) and liquid-state hydrogen peroxide: Based on atom-bond electronegativity. *The Journal of Physical Chemistry A*, *115*(12), 2615–2626. doi:10.1021/jp111284t PMID:21388113

Yu, J., Tang, L., Pang, Y., Zeng, G., Feng, H., Zou, J., & Tan, J. (2020). Hierarchical porous biochar from shrimp shell for persulfate activation: A two-electron transfer path and key impact factors. *Applied Catalysis B: Environmental*, *260*, 118160. doi:10.1016/j.apcatb.2019.118160

Zámocký, M., & Koller, F. (1999). Understanding the structure and function of catalases: Clues from molecular evolution and in vitro mutagenesis. *Progress in Biophysics and Molecular Biology*, *72*(1), 19–66. doi:10.1016/S0079-6107(98)00058-3 PMID:10446501

Zarkovic, N. (2020). Roles and functions of ROS and RNS in cellular physiology and pathology. *Cells*, *9*(3), 767–771. doi:10.3390/cells9030767 PMID:32245147

Zeng, L., Kuang, S., Li, G., Jin, C., Ji, L., & Chao, H. (2017). A GSH-activatable ruthenium(II)-azo photosensitizer for two-photon photodynamic therapy. *Chemical Communications*, *53*(12), 1977–1980. doi:10.1039/C6CC10330H PMID:28119967

Zhang, H., Zhang, J., Zhang, X., & Geng, A. (2018). Purification and characterization of a novel manganese peroxidase from white-rot fungus Cerrena unicolor BBP6 and its application in dye decolorization and denim bleaching. *Process Biochemistry (Barking, London, England)*, *66*, 222–229. doi:10.1016/j.procbio.2017.12.011

Zhang, J., Frerman, F. E., & Kim, J.-J. P. (2006). Structure of electron transfer flavoprotein-ubiquinone oxidoreductase and electron transfer to the mitochondrial ubiquinone pool. *Proceedings of the National Academy of Sciences of the United States of America*, *103*(44), 16212–16217. doi:10.1073/pnas.0604567103 PMID:17050691

Zhang, J., Wang, X., Vikash, V., Ye, Q., Wu, D., Liu, Y., & Dong, W. (2016). ROS and ROS-mediated cellular signaling. *Oxidative Medicine and Cellular Longevity*, *2016*, 1–18. PMID:26998193

Zhang, Y., Fan, J., Yang, B., Huang, W., & Ma, L. (2017). Copper–catalyzed activation of molecular oxygen for oxidative destruction of acetaminophen: The mechanism and superoxide-mediated cycling of copper species. *Chemosphere*, *166*, 89–95. doi:10.1016/j.chemosphere.2016.09.066 PMID:27689888

Zhang, Y., & Zhou, M. (2019). A critical review of the application of chelating agents to enable Fenton and Fenton-like reactions at high pH values. *Journal of Hazardous Materials*, *362*, 436–450. doi:10.1016/j.jhazmat.2018.09.035 PMID:30261437

Zhao, G., Li, X., Chen, W., Xi, Z., & Sun, L. (2012). Three new sesquiterpenes from *Tithonia diversifolia* and their anti-hyperglycemic activity. *Fitoterapia*, *83*(8), 1590–1597. doi:10.1016/j.fitote.2012.09.007

Zhao, Z., Tao, X., Xie, Y., Lai, Q., Lin, W., Lu, K., Wang, J., Xia, W., & Mao, Z.-W. (2022). In situ prodrug activation by an affibody-ruthenium catalyst hybrid for HER2-targeted chemotherapy. A Angewandte Chemie International Edition, 134(26), e202202855.

Zhao, F., Harnisch, F., Schröder, U., Scholz, F., Bogdanoff, P., & Herrmann, I. (2006). Challenges and constraints of using oxygen cathodes in microbial fuel cells. *Environmental Science & Technology*, *40*(17), 5193–5199. doi:10.1021/es060332p PMID:16999088

Zhao, Z., Qiu, K., Liu, J., Hao, X., & Wang, J. (2020). Two-photon photodynamic ablation of tumour cells using an RGD peptide-conjugated ruthenium(II) photosensitiser. *Chemical Communications*, *56*(83), 12542–12545. doi:10.1039/D0CC04943C PMID:32940288

Zheng, W., Miao, Q., Tang, Y., Wei, W., Xu, J., Liu, X., Qian, Q., Xiao, L., Huang, B., & Chen, Q. (2013). La(III)-doped ZnO/C nanofibers with core–shell structure by electrospinning–calcination technology. *Materials Letters*, *98*, 94–97. doi:10.1016/j.matlet.2013.02.004

Zhen, X. V., Rousseau, C. R., & Bühlmann, P. (2018). Redox buffer capacity of ion-selective electrode solid contacts doped with organometallic complexes. *Analytical Chemistry*, *90*(18), 11000–11007. doi:10.1021/acs.analchem.8b02595 PMID:30141325

Zhou, H., Zhu, T., Cai, S., Gu, Q., & Li, D. (2011). Drimane sesquiterpenoids from the mangrove-derived Fungus *Aspergillus ustus. Chemical & Pharmaceutical Bulletin*, *59*(6), 762–766. doi:10.1248/cpb.59.762

Zhou, J., Li, X.-Y., Liu, Y.-J., Feng, J., Wu, Y., Shen, H.-M., & Lu, G.-D. (2022). Full-coverage regulations of autophagy by ROS: From induction to maturation. *Autophagy*, *18*(6), 1240–1255. doi:10.1080/15548627.2021.1984656 PMID:34662529

Zhou, Y., Zhen, Y., Wang, G., & Liu, B. (2022). Deconvoluting the complexity of reactive oxygen species (ROS) in neurodegenerative diseases. *Frontiers in Neuroanatomy*, *16*, 910427–991034. doi:10.3389/fnana.2022.910427 PMID:35756499

Zhu, J. X., & Ting, C. S. (1998). Tunneling spectrum into a *d*-wave superconductor with a time-reversal-symmetry-broken surface state. *Physical Review B: Condensed Matter*, *57*(5), 3038–3044. doi:10.1103/PhysRevB.57.3038

Zhu, M., Zhao, C., Liu, X., Wang, X., Zhou, F., Wang, J., Hu, Y., Zhao, Y., Yao, T., Yang, L.-M., & Wu, Y. (2021). Single Atomic Cerium Sites with a High Coordination Number for Efficient Oxygen Reduction in Proton-Exchange Membrane Fuel Cells. *ACS Catalysis*, *11*(7), 3923–3929. doi:10.1021/acscatal.0c05503

Zhu, Q., Lian, Y., Thyagarajan, S., Rokita, S. E., Karlin, K. D., & Blough, N. V. (2008). Hydrogen peroxide and dioxygen activation by dinuclear copper complexes in aqueous solution: Hydroxyl radical production initiated by internal electron transfer. *Journal of the American Chemical Society*, *130*(20), 6304–6305. doi:10.1021/ja800080z PMID:18433125

Zhu, S. L., Xu, X., Ou, S., Zhao, M., He, W. L., & Wu, C. D. (2016). Assembly of a metalloporphyrin–polyoxometalate hybrid material for highly efficient activation of molecular oxygen. *Inorganic Chemistry*, *55*(15), 7295–7300. doi:10.1021/acs.inorgchem.6b00971 PMID:27408952

Zofair, S. F. F., Ahmad, S., Hashmi, M. A., Khan, S. H., Khan, M. A., & Younus, H. (2022). Catalytic roles, immobilization and management of recalcitrant environmental pollutants by laccases: Significance in sustainable green chemistry. *Journal of Environmental Management*, *309*, 114676. doi:10.1016/j.jenvman.2022.114676 PMID:35151142

Zucca, P., Cocco, G., Sollai, F., & Sanjust, E. (2016). Fungal laccases as tools for biodegradation of industrial dyes. *Biocatalysis*, *1*(1), 82–108. doi:10.1515/boca-2015-0007

Zuo, J., Zhang, Z., Li, M. M., Yang, Y., Zheng, B. H., Wang, P., Huang, C. H., & Zhou, S. T. (2022). The crosstalk between reactive oxygen species and noncoding RNAs: From cancer code to drug role. *Molecular Cancer*, *21*(1), 1–17. doi:10.118612943-021-01488-3 PMID:35081965

Zuorro, A., Fidaleo, M., & Lavecchia, R. (2013). Response surface methodology (RSM) analysis of photodegradation of sulfonated diazo dye Reactive Green 19 by UV/H_2O_2 process. *Journal of Environmental Management*, *127*, 28–35. doi:10.1016/j.jenvman.2013.04.023 PMID:23676376

About the Contributors

Ashok Vaseashta is CEO/CTO and Executive Director for research with the International Clean Water Institute in Virginia, USA. He also serves in several honorary positions, such as a Professor at the Transylvania University of Brasov; Chaired Professor of Nanotechnology at the Ghitu Institute of Electrical Engineering and Nanotechnologies, Academy of Sciences of Moldova; Academician at Euro-Mediterranean Academy of Arts and Sciences, member of CIRET, France, and Senior Strategic Research Advisor for several organizations. Inspired by nature and guided by societal necessities, he strives for technological innovations to address the global challenges of the 21st century. His research interests include critical infrastructure safety, security, and environmental sustainability - all using nexus of advanced technological solution platforms. He is a scholar, visionary, strategist, and dedicated futurist providing strategic leadership to promote and advance research initiatives and priorities using data-driven decisions. He received PhD from the Virginia Polytechnic Institute and State University, Blacksburg, VA in 1990 followed by Kobe post-doctoral fellowship. Following his Ph.D., he served as a professor and researcher at Virginia Tech and Marshall University. He also served as the Director of Research at the Institute for Advanced Sciences Convergence and International Clean Water Institute for Norwich University Applied Research Institutes, Vice-Provost (Rector) for Research in South Carolina, and Executive Director and Chair of the Institutional Review Board at a State University in New Jersey. His earlier honorary positions included a Visiting Professor at the Riga Technical University, Latvia, 3 Nano-SAE Research Centre, University of Bucharest, Romania, and Visiting Scientist at the Weizmann Institute of Science, Israel. He served the U.S. Department of State in two rotations, as a strategic S&T advisor in the Bureau of International Security and Nonproliferation, Office of Weapons of Mass Destruction and Terrorism, and U.S. diplomat. His research interests span foresight, nanotechnology, environmental/ecological science, and critical infrastructure safety and security. He is the author/editor of 14 books and has published over 300 articles in scientific journals, book chapters, and conferences. He serves on the editorial boards of several international journals and is an active member of various professional organizations.

* * *

Aricu Aculina is a scientific researcher in organic and bioorganic chemistry; chemistry of natural and physiologically active compounds; elaboration of chemo-, regio- and stereoselective methods of the synthesis of optically active norlabdanic and drimanic compounds, including nitrogen and halogen-containing derivatives, as well as terpeno-heterocyclic hybrid compounds on the basis of accessible natural labdane diterpenoid sclareol; the elucidation of reaction mechanisms and optimal conditions of obtaining the mentioned compounds; the targeted synthesis of new compounds with high the potential of biological activity, with the application of computational methods for forecasting activity, with further possibilities for industrial applications.

Lilia Anghel is a qualified researcher, particularly interested in identifying the protein modifications induced by various extrinsic factors and understanding the relationship between the protein structure, dynamics and properties. Experienced in planning SANS and SAXS experiments, writing proposals and preparing suitable samples for small-angle scattering experiments to study the protein structure in solution, protein conformation stability in solution, the interaction of proteins in various systems, protein-protein interactions, protein-polyelectrolyte (polysaccharides) interactions.

Ludmila Aricov received her Ph.D. in chemistry from the School of Advanced Studies of the Romanian Academy in 2017. Currently, she works as a scientific researcher at Colloid Chemistry Laboratory of the "Ilie Murgulescu" Institute of Physical Chemistry. Her research interests are related to the physicochemical properties and application of advanced materials. Her current research topics involve enzyme stabilization using polymeric matrices and biodegradation of recalcitrant textile dyes.

Ion Arsene is an Assoc. Prof. and senior scientific researcher. Research areas: Chemical physics; Study of electronic structure and stability of molecular systems; Study of potential energy surfaces, dynamics, energy profile and mechanism of chemical reactions.

Iolanta Balan is the head of the Laboratory of Physical and Quantum Chemistry. Research areas include Computational chemistry, molecular modeling, quantum chemistry, quantum-chemical calculations of electronic structure, Jahn-Teller effect and pseudoJahn-Teller effect.

Andy Benniston has spent most of his academic career at Newcastle University, after a spell at Glasgow University where he started as a lecturer. He is a professor in chemistry and has held several positions including Head of Chemistry, Deputy Head of School of Natural and Environmental Sciences and now the Associate Dean of Strategic Projects. His work has covered many different aspects including coordination chemistry of macrocycles, the photophysical properties of organic and inorganic compounds, artificial photosynthesis and the study of redoxactive complexes. He has published over 160 papers and co-written several review articles.

Natalia Bolocan, born in Moldova to a family of doctors, discovered her passion about science in early childhood. Driven by the desire to tackle the mystery of how the matter is organized at the most intimate level, and how living cells use chemical compounds as building blocks, messengers, and energy carriers, Natalia chose chemistry for her professional life, and successfully defended her PhD in Physical Chemistry in 2023. For the last 15 years, she has been involved in the synthesis, characterization, and study of novel antioxidant compounds with potential applications in the pharmaceutical and food industries. Her activity implies experimental, as well as computational research, and the key interests are related to the computational design and study of novel organic compounds with useful properties.

Elena Bunduchi graduated from Moldova State University, Faculty of Chemistry, in June 1996. She holds a PhD in "Redox-Catalytic transformations of hydroquinone and glyoxalic acid in aquatic environments" in 2010 at Moldova State University. From 1997 she began her teaching activity at the Department of Industrial and Ecological Chemistry at Moldova State University, becoming associate professor in 2014. Currently she is Chief of Department of Industrial and Ecological Chemistry "Acad. Gh. Duca" at Faculty of Chemistry and Chemical Technology at Moldova State University. She's research focuses on the natural waters self-purification process assured through oxygen active forms (H2O2, OH radicals) and their role in natural waters as a living environment for hydrobiontites.

Maria Cristina Buta is scientific researcher at the Institute of Physical Chemistry "I. Murgulescu" of Romanian Academy, Theoretical and Structural Chemistry Group. Know-how in quantum chemistry calculations using methods such as: Density Functional Theory (DFT), Complete Active Space Self-Consistent Field (CASSCF), Valence Bond (VB), Ligand Field theory (LF); programming in Excel/VBA, Matlab and Mathematica codes; experimental handling and modeling of Electron Paramagnetic Resonance (EPR) spectra.

Fanica Cimpoesu obtained PhD on a thesis devoted to orbital models of vibronic coupling, supervised by prof. I. B. Bersuker (a coryphaeus in the Jahn-Teller and related effects). Then, self-didactically, approached other several topics such as organometallic stereochemistry and molecular magnetism, trying continuously to enlarge his paradigmatic area. The trademark of FC's works is finding new methodological clues and heuristic viewpoints, at the borderline between theory, computation and real chemistry. Research stages at universities of Leuven (Prof. A. Ceulemans), Goettingen (Prof. H. Roesky), Tokyo (Prof. K. Hirao) and Fribourg (Prof. C. Daul) are acknowledged as emulative events in the curriculum vitae.

Corneliu Cojocaru is an experienced scientific researcher in the field of modeling and optimization of physicochemical processes (adsorption, capillarity, adsorption-ultrafiltration, and (photo)catalysis). He studied chemistry and chemical technology at the State University of Moldova, where he graduated in 1999. Then, he joined the group of Prof. Dr. Eng. Matei Macoveanu (at "Gh. Asachi" Technical University of Iasi, Romania) for his doctoral research (1999-2005) in modeling and optimization of processes in environmental chemical engineering. He received his Ph.D. in 2005. During 2006-2011, he was involved in multidisciplinary research activities as a postdoctoral and visiting scientist at different research groups from Iasi, Warsaw, and Madrid. In 2012, he was employed as an experienced researcher at the "Petru Poni" Institute of Macromolecular Chemistry from Iasi, where he is still active. His research focuses on the following areas of interest: polymer materials and polymer-based composites for applications in pollution control and environmental protection; kinetic models; empirical models; models based on artificial neural networks; chemometric optimization methods; non-linear regression; molecular modeling and computational chemistry. He is a co-author of 122 articles published in ISI-rated international journals.

Raul Victor Erhan is an experienced research scientist with a demonstrated history of working in the research industry. Skilled in Mathematical Modeling, Experimental Physics, Materials Science, Spectroscopy, and Nanotechnology. Strong research professional with a Doctor of Philosophy (Ph.D.) focused in Physics from University of Bucharest.

Ion Geru was born on December 8, 1937 in the village of Cotiujeni, Soroca County, Romania. After graduating from middle school No. 1 in Cotiujeni, which was already part of the Cotiujeni district of the Moldovan SSR, he studied at the Physics Department of the Faculty of Physics and Mathematics (1954-1959) of the Chisinau State University. In 1959 he started working as a laboratory assistant, then as a senior laboratory assistant, at the Institute of Chemistry of the Moldovan branch in Chisinau of the USSR Academy of Sciences. In 1962-1965 he was a PhD student at the Institute of Semiconductors of the Academy of Sciences of the Ukrainian SSR (Kyiv). From 1965 to 1969 he was a senior engineer, then a senior scientific researcher, in the Moldovan Basic Laboratory of the Union Institute for Research on Current Sources of the USSR (Chisinau). He defended his doctoral dissertation in physics and mathematics in 1967, and from 1969 to 1977 he worked as a senior researcher in the Department of Semiconductor Theory and Quantum Electronics at the Institute of Applied Physics of the MSSR Academy of Sciences. From 1977 to 1984 he worked as an assistant professor at the Faculty of Physics of the State University of Moldova and from 1984 to 2004 as a university professor at the same faculty. From January 1996 to July 1997 he was head of the General Physics Department. He was the founder and head of the "Superconductivity and Magnetism" laboratory (1994-2004) of the State University of Moldova. From 2005 to 2008 he was director of the Center for Automation and Metrology of the Academy of Sciences of Moldova, transformed on February 14, 2006 into a Center for Metrology and Analytical Research Methods (CMARM). In January 2006 he founded the "Magnetic Resonance and Laser Spectroscopy" laboratory. On August 1, 2008 he was employed at the Institute of Chemistry of the Academy of Sciences of Moldova as head of the Laboratory of Magnetic Resonance and Laser Spectroscopy, which was transmitted from the CMARM to the Institute of Chemistry. On March 1, 2013, he was transferred to the position of Principal Investigator of the same institute. Scientific interests: ESR, NMR, diluted magnetic semiconductors, coordination compounds, high-Tc superconductors, fullerenes, carbon nanotubes (including mesoporous carbon nanotubes), carbon quanrum dots, A2B6 quantum dots, the microwave spectroscopy of excitons in semiconductors, the theory of groups (including groups of magnetic symmetry), the total and partial time-renersal symmetry, the structural distortions caused by time-reversal symmetry, the experimental realization of the virtual reversal of time, the restoration of the reduced time-reversal symmetry of spin systems by involving metaparticles, the cosmological aspects of metaparticles.

Viorica Gladchi, doctor in chemical sciences, university lecturer, dean of the Faculty of Chemistry and Chemical Technology of the State University of Moldova. Specialist in the field of ecological chemistry and environmental protection, holder of 10 university subjects, author of over 100 scientific works, 5 monographs, 6 methodical-didactic works for students. He participated in the realization of more than 10 international and national scientific projects. Field of scientific interests: the study of chemical self-purification processes of natural waters; catalytic redox and photochemical transformations of pollutants in natural waters; dynamics of the redox state of waters; the quality of surface and underground waters; ecological information and training.

Mikhail Gorbachev is a senior scientific researcher. Research areas: physical chemistry: quantum chemistry, physical chemistry of liquids, investigation of quantitative structure – activity relationships (QSAR).

Maria Lehene is a PhD student under the direction of Prof. Radu-Silaghi-Dumitrescu at the Babes-Bolyai University in Romania, studying redox activation of small molecules by bioinorganic centers.

Anca Ruxandra Leontieş received her Ph.D. in chemistry from the University of Bucharest in 2013 and now is a researcher at the Colloid department at the "Ilie Murgulescu" Institute of Physical Chemistry. Her research interests are related to the physicochemical properties of proteins and polymers. The current research topics are centered on strategies for the preparation of improved biocatalysts by enzyme immobilization, stabilization, and reaction design.

Flavia-Malina Oana received her MSc degrees from the Babes-Bolyai University in Romania (2022, clinical chemistry and forensic chemistry) under the direction of Prof. Radu Silaghi-Dumitrescu working on clinical aspects of vitamin C, and is currently pursuing a career as a chemistry teacher.

Petronela Pascariu is a scientific researcher working in the area of semiconducting (nano)materials produced by electrospinning-calcination techniques and electrochemical methods. She received a Ph.D. in Physics at "Al. I. Cuza" University from Iasi (Romania) in the field of functional multilayered nanostructures for spintronic applications (September 2011). Currently, she is an experienced researcher (CSIII) at the "Petru Poni" Institute of Macromolecular Chemistry from Iasi, Romania. Her main scientific interests involve the design, synthesis, and investigation of nanoparticles, thin films, and multilayers nanowires, metal oxide ceramic nanofibers based on ZnO and TiO2 doped with transitional metals and lanthanides, composites, polymer/inorganic nanoparticles nanostructures, with specific optical, electrical, photocatalytic and magnetic properties for use in various modern applications (photocatalysis, sensors, electronics and optoelectronics, electrochemical supercapacitors, etc.). She is a co-author of 67 scientific articles published in ISI-rated international journals. Also, she is a member of the editorial board of the international journal Surfaces and Interfaces (Elsevier).

Victor Polinger graduated from the Department of Physics and Mathematics of the Kishinev State University (now the State University of Moldova). In 1974 V. Polinger earned his doctorate (Russian: Kandidat of science) and in 1987 his habilitation degree (Russian: Doctor of science) in physical and mathematical sciences from the Acad. Sci. USSR. Since 1973 he was a research fellow, then a leading researcher at the Institute of Chemistry, Acad. Sci. Mo. SSR. In 1992 he moved to the United States where he was employed as a physics instructor, adjunct professor, and affiliate professor at different colleges and universities. At present, V. Polinger is an Emeritus instructor at Bellevue College (Bellevue, WA) and an Affiliate Professor at the Department of Chemistry of the University of Washington in Seattle, WA.

Igor Povar, head of the department "Physicochemical Methods of Research and Analysis" at the Institute of Chemistry, Republic of Moldova, is a well-known specialist in the area of modeling and theoretical research, mainly, involving chemical reaction equilibrium in natural systems. During the 2001-2009 years he worked at the University of Western Ontario, London, Canada, where he was actively involved in research in chemical kinetics and thermodynamics of reactions in the organic chemistry using various ab-initio theoretical techniques, modeling and theoretical research.

Adina Răducan received her doctorate in Ph.D. at the University of Bucharest in 1999, and currently, she is the Head of the Department of Analytical Chemistry and Physical Chemistry, Faculty of Chemistry,

University of Bucharest. Her research interests are in the chemical kinetics of complex enzymatic and non-enzymatic reactions. She is focused on modeling the kinetics of catalytic processes, discrimination between rival models, and simulating time-evolving chemical systems by using numerical analysis of experimental data and statistical analysis.

Lidia Romanciuc is doctor in chemical sciences, senior scientific researcher at the Moldova State University/Institute of Chemistry, Chisinau, Republic of Moldova. She has a scientific background in physical chemistry (kinetics and catalysis) and ecological chemistry. The main field of scientific interest is focused on the research of water self-purification mechanisms, chemical and photochemical transformation of natural water pollutants, redox processes in the aquatic environment, water quality, hydrochemistry and monitoring of small rivers. She is a graduate of the Moldova State University. She paticipated in several scientific scholarships in Europe, Japan, USA. She has about 100 scientific publications and invention patents. She contributed to the organization of a series of international scientific conferences of ecological and environmental chemistry, with the participation of experienced and young scientists from the CIS, Europe, USA, etc. She contributed to the development of bilateral, regional and international scientific cooperation relations.

Radu Silaghi-Dumitrescu received his PhD degrees from the University of Georgia in Athens, GA, U.S.A. (2004, experimental chemistry) and from the Babeş-Bolyai University in Romania (2005, computational chemistry). After a postdoctoral stage at the University of Essex (Colchester, England), he returned to the "Babeş-Bolyai" University in 2007, where he teaches biochemistry, computational chemistry and related subjects, while also serving as president of the Scientific Council (2012-) and director of the Center for University Strategies (2020-). His research explores the mechanisms of redox stress in biology and medicine (especially free radicals and metals) in contexts such as blood substitutes, anticancer drugs, or plant extracts – but also interdisciplinary applications of exact science tools in scientometrics or in cultural studies.

Oxana Spinu is a scientific researcher at the Institute of Chemistry, Republic of Moldova. Areas of interests: 1. Development and use of the theory of thermodynamic stability in homogeneous and heterogeneous systems to assess the state of pollutants in the short and long term in contaminated waters. 2. Determining the equilibrium composition of multiphase and multicomponent systems of various initial composition. 3. Determining the optimal conditions for separating the solid phases from the studied heterogeneous mixtures. 4. Thermodynamic analysis of pollutant species and determination of optimal conditions for their removal from wastewater.

Sergey Travin is graduated at Moscow Institute of Physics and Technology (MIPT), faculty of molecular and chemical physics in 1976, awarding diploma with honors. Ph.D. in Chemistry was obtained in 1979. He is Doctor of Sciences in Economics, Professor since 2001. Prof. Travin is full member of the Russian Academy of Natural Sciences in the section "Education and sustainable development" since 2002. Current position: senior scientist in N.N. Semenov Federal Research Center for Chemical Physics Russian Academy of Sciences. Area of interest: Mathematical modeling in Chemical kinetics, Chemistry and Physics, Computer science, Information systems.

Lingli Zeng is currently a postdoctoral researcher at the Sun Yat-sen Memorial Hospital, Sun Yat-sen University. She obtained her PhD degree from Newcastle University in May 2020 under the supervision of Prof. Andrew Benniston. Her research interests include synthesis and applications of light-activated molecular systems, multimodal bioimaging, tumor diagnosis, drug delivery with nanosystems and the related disciplines. She has several publications in the fields of organic molecules, coordination compounds and nanomaterials for detection and tumor therapy.

Index

Q

R

S

T

V

W

www.ingramcontent.com/pod-product-compliance
Lightning Source LLC
La Vergne TN
LVHW062103300926
849161LV00010B/2367
9781668471982